Dolphin 7/83

Center for American Archeology

DAVID L. ASCH, EDITOR

At Northwestern University
1911 Ridge Avenue
P.O. Box 1499
Evanston, Illinois 60204

Research Series

Judith Droessler. Craniometry and Biological Distance: Biocultural Continuity and Change at the Late-Woodland–Mississippian Interface

Christopher Carr. Handbook on Soil Resistivity Surveying: Interpretation of Data from Earthen Archeological Sites

Kampsville Seminars in Archeology

Robert Whallon and *James A. Brown.* (Editors). Essays on Archaeological Typology

Distributor of Research Volumes for Northwestern University Archeological Program

David L. Asch. The Middle Woodland Population of the Lower Illinois Valley: A Study in Paleodemographic Methods

Jane E. Buikstra. Hopewell in the Lower Illinois Valley: A Regional Approach to the Study of Human Biological Variability and Prehistoric Behavior

Jane E. Buikstra (Editor). Prehistoric Tuberculosis in the Americas

Thomas Genn Cook. Koster: An Artifact Analysis of Two Archaic Phases in Westcentral Illinois

Lynn Gail Goldstein. Mississippian Mortuary Practices: A Case Study of Two Cemeteries in the Lower Illinois Valley

John Winfield Rick. Heat-Altered Cherts of the Lower Illinois Valley: An Experimental Study in Prehistoric Technology

Bonnie Whatley Styles. Faunal Exploitation and Resource Selection: Early Late Woodland Subsistence in the Lower Illinois Valley

Handbook on
Soil Resistivity Surveying

Handbook on
Soil Resistivity Surveying

INTERPRETATION OF DATA FROM
EARTHEN ARCHEOLOGICAL SITES

Christopher Carr
Department of Anthropology
University of Arkansas
Fayetteville, Arkansas 72701

CENTER FOR AMERICAN ARCHEOLOGY PRESS
Evanston, Illinois

Center for American Archeology
Research Series, Volume 2

COPYRIGHT © 1982, BY THE CENTER FOR AMERICAN ARCHEOLOGY
All Rights Reserved.
No part of this publication may be reproduced
in whole or in part
without permission in writing
from the publisher.

CENTER FOR AMERICAN ARCHEOLOGY
At Northwestern University
1911 Ridge Avenue
P.O. Box 1499
Evanston, Illinois 60201

Library of Congress Cataloging in Publication Data

Carr, Christopher. Date.
   Handbook on soil resistivity surveying.

   (Research series ; #2)
   Bibliography: p.
   Includes index.
1. Soil science in archeology.
2. Archaeological surveying.   3. Soils—Electric
properties.   I. Title.   II. Series: Research
series (Center for American Archeology
Press) ; 2.
CC79.S6C37   930.1'028   82-4277
ISBN 0–942118–13–8   AACR2

PRINTED IN THE UNITED STATES OF AMERICA

TO MOM AND DAD
*with all my love and thanks
for all you gave and gave up
so many years*

Contents

Foreword by Stuart Struever   *xv*
Preface   *xvii*
Acknowledgments   *xxiii*

1
Introduction   1

    Statement of Problem and Thesis   1
    The Nature of Resistivity Surveying Methods and Their Traditional
       Use in Archeology   4
    A Theoretical Reason for Taking the Use-Area as
       the Unit of Analysis When Interpreting Resistivity Data   10
    Operational Reasons for Taking the Use-Area as
       the Unit of Analysis When Interpreting Resistivity Data   14
    The Complementary Use of Controlled Surface Surveys
       and Resistivity Surveys   19
    Summary   21
    References   22

2
Mathematical Methods for Preparing Resistivity Data for Interpretation   29

    The Barnes Layer Method of Resistivity Data Interpretation   33
    Time-Series Analysis and Running Filter Functions   37
    Conclusion   43
    References   44

viii   Contents

3
Factors Affecting the Electrical Conductivity of Soils   47

The Formation Factor and Soil Structure	48
Soil Porosity, Pore-Size Distribution, and Particle-Size Distribution	50
Soil Moisture Content as a Function of Soil Structure	51
Soil Moisture Content and Hysteresis	61
Soil Moisture Content as a Function of Drainage Conditions: Saturated Flow	62
Soil Moisture Content as a Function of Drainage Conditions: Unsaturated Flow	64
Capillary Conductance	66
Soil Organic Matter: Its Hygroscopic Capacity	68
Soil Organic Matter: Its Effects on Soil Aggregation	71
Soil Organic Matter: Its Effects on the Infiltration of Rainfall	77
Soil Organic Matter: Its Effects on Soil Chemistry and Flocculation of Soil Colloids	78
Soil Organic Matter: Its Effect on Soil Chemistry and the Osmotic Potential	79
Conductivity of Soil Water: Introductory Framework	79
The Conductivity of Soil Water as a Function of the Concentration of Strong Electrolytes and Solution Temperature	84
The Conductivity of Soil Water as a Function of the Concentration of Weak Electrolytes	91
The Conductivity of Soil Water as a Function of the Concentration of Colloidal Particles	92
Implications of the Relationships Governing Soil Water Conductivity on Resistivity Surveys	94
Interaction between the Physical and Chemical Properties of Organically Enriched Disturbed Soils and Natural Soils in Producing Contrast in Their Resistivities	97
Summary	101
References	101

4
Natural Processes Determining the Formation of Soil from Human Refuse and the Maintenance of Anthropic Soil Anomalies within Archeological Sites   108

Long-Term Soil Alterations Produced by Prehistoric Human Activities: Overview	108
Previously Documented Long-Term Alterations in Soil Ion Concentration and Organic Matter Content	109

Previously Documented Long-Term Alterations in Soil Moisture Equilibria and Structure	115
Maintenance of Soil Alterations on Archeological Sites, I: Rates of Decomposition of Organic Matter	116
Factors Affecting the Rates of Decomposition of Organic Matter	117
Rates of Decomposition of Organic Matter	119
Maintenance of Soil Alterations on Archeological Sites, II: Rates of Decomposition of Humus	127
The Stability of Humus and Its Residence Time within the Soil	127
Nutrients Stored within Humus Micelles	129
Maintenance of Soil Alterations on Archeological Sites, III: Nutrient Cycling	131
Cation Adsorption and Exchange and the Structure of Clays	136
Anion Adsorption and Exchange and the Structure of Clays	140
Ion Exchange on Hydrous Oxides and Humus Micelles	141
Fixation of Cations and Anions within Clays	143
Fixation of Cations and Anions within Surface Vegetation	145
The Nitrogen Cycle	146
The Sulfur Cycle	150
The Phosphorus Cycle	154
The Calcium, Potassium, Magnesium, and Sodium Cycles	159
Maintenance of Soil Alterations on Archeological Sites, IV: Interaction Effects among Byproducts of Human Activity	160
Temporal Alterations of Soils within Archeological Sites: A General Model	167
Summary	176
References	177

5

A Functional and Distributional Study of Surface Artifacts from the Crane Site   183

Introduction to the Crane Site	184
Nature of the Data Used to Define Activity and Depositional Areas within the Crane Site	191
A Model of the Nature of Organization of the Archeological Record Used in Analyzing the Crane Data	195
Methods Used to Define Use-Areas within the Crane Site	201
The Approach Used in Classifying the Crane Assemblage into Functional Types	212
Description of the Crane Assemblage and Assignment of Artifact Functions	218
Use-Areas within the Crane Site: Their Depositional and Functional Characteristics	308
Summary	341
References	342

6
Research Designs for Collecting Pedological and Resistivity Data from the Crane Site   353

    Soil and Resistivity Surveying along Transects   353
    Soil Analyses Performed on the Soil Samples Collected from Transects I–V   365
    Coding and Presentation of the Data Collected for the Transect Survey   371
    The Block Survey: A More Selective Soil and Resistivity Survey   373
    Supplementary Soil Surveys at the Crane Site   376
    Summary   383
    References   384

7
Natural Soil Variations and Soil Alterations Produced by Historic Human Activity at the Crane Site   387

    Overview of the Parent Materials and Soils at Crane Site   387
    Natural, Spatial Variation of Parent Materials at the Crane Site   392
    Spatial Variations in the Degree of Soil Development at the Crane Site   403
    Spatial Variations in Soil Profile Development Compared to Spatial Variations in Parent Material Composition   406
    Chemical Soil Variation at Crane Site Related to Historic Land-Use Patterns: Deforestation and Cultivation   407
    Physical Soil Disturbances at Crane Site Related to Historic Land Use: Stratigraphy   435
    Summary   436
    References   437

8
Soil Alterations Produced by Prehistoric Human Activities at the Crane Site   441

    Relationships between Use-Areas and Soil Chemistry at the Crane Site: Differentiation of Use-Areas by the Spectra of Their Natural Anomalies   442
    Relationships between Use-Areas and Soil Chemistry at the Crane Site: Differentiation of Use-Areas by the Magnitudes of Their Nutrient Anomalies   515
    Physical Alterations of the Soils at Crane Site: Pit Fills Compared to Their Natural Soil Matrices   524
    Physical Alterations of the Soils at the Crane Site: Midden Deposits Compared to Natural B Horizons   535

Physical Alterations of the Soils at the Crane Site: Comparison among Midden Deposits	541
Conclusion	548
References	548

9
Soil Resistivity as a Product of Multiple Physical and Chemical Soil Properties Illustrated at the Crane Site — 553

10
The Feasibility of Using the Barnes Layer Method and Spatial Filtering Techniques to Isolate Archeologically Significant Soil Resistivity Variation Illustrated at the Crane Site — 563

Tests of the Capability of the Barnes Layer Method	564
Use-Areas Differentiated by the Mean Resistivity of Their Soils	566
Use-Areas Differentiated by the Variance of the Resistivity of Their Soils	571
Use of Spatial Filtering Techniques to Remove Low Frequency, Nonarcheological Variation from Resistivity Series	576
Use of Spectral Analysis and Spatial Filtering Techniques to Remove Periodic and High-Frequency Nonarcheological Variation from Resistivity Series	583
Use of Spatial Filtering Techniques to Partition Archeological Variation into Components of Differing Scales	589
Use of Spatial Filtering Techniques to Define and Differentiate Use-Areas: An Outline of Specific Procedures	591
Use of Spatial Filtering Techniques to Locate and Differentiate Use-Areas: Illustration	598
Conclusion	621
References	622

11
Boundary Conditions for the Proposed Methods, and Conclusions — 623

Behavioral and Archeological Conditions	624
Soil Conditions Prior to Occupation and Environmental Conditions during Occupation	626
Weathering and Aging of Sites	628

Conditions of Survey 628
Survey Efficiency and Time Limitations 629
Economic Considerations 633
Boundary Conditions for the Use of the Barnes Layer Method 635
Boundary Conditions for the Use of Spatial Filtering Techniques 636
Conclusion 637
References 638

Appendix I
Computer Programs for Calculating Barnes Layer Resistivity Values and for Spatial Filtering Resistivity Data. 639

Appendix II
Operational Procedures of Defining Depositional Sets within the Crane Assemblage 655

Appendix III
Additional Attributes of Retouched and Unretouched Chipped Stone Tools from the Crane site

Appendix IV
Artifact Densities in the Controlled Surface Survey Units at the Crane Site*

Appendix V
Resistivity and Soil Data Collected from the Transect Survey at the Crane Site*

    Table 1. Grid Locations of Resistivity Survey Stations Along Transects I–V.
    Table 2. Barnes Layer Resistivity Values (Ohm–cm) Calculated from Resistance Data from Transects I–V.
    Table 3. Physical Characteristics of Soils That Were Sampled Along Transects I–V.
    Table 4. Chemical Characteristics of Soils That Were Sampled Along Transects I–V.

Appendix VI
Resistivity and Soil Data Collected from the Block Survey at the Crane Site*

> Table 1. Barnes Layer Resistivity Values (Ohm-cm) Calculated from Resistance Data from the Block Survey
> Table 2. Physical Characteristics of Soils That Were Sampled in the Block Survey.
> Table 3. Chemical Characteristics of Soils That Were Sampled in the Block Survey.

Appendix VII
Physical and Chemical Properties of Natural Soils, Midden Deposits, and Pit Fills, and Their Matrices at the Crane Site*

> Table 1. Physical Characteristics of Soils in Those Locations Sampled for Soil Texture and Pore-Size Distribution But Not Described in the Field.
> Table 2. Chemical Characteristics of Soils in Those Locations Sampled for Soil Texture and Pore-Size Distribution But Not Described in the Field.
> Table 3. Physical Characteristics of Pit Fills and Pit Matrices from Various Locations at the Crane Site.
> Table 4. Chemical Characteristics of Pit Fills and Pit Matrices from Various Locations at the Crane Site.
> Table 5. Observations Made on Soil Profiles at Several Locations within and around the Crane Site.
> Table 6. Physical Characteristics of Soil Profiles That Were Described in the Field for Observable Traits.
> Table 7. Chemical Characteristics of Soil Profiles That Were Described in the Field for Observable Traits.

Subject Index 669

Erratum 677

---

*Appendices IV–VII are on the microfiche inserted inside the back cover of this book.

Center for American Archeology
At Northwestern University

To realize its potential, archeology faces the formidable task of devising new approaches to research that will increase its capacity to interpret the remains of extinct cultures. Archeology today attempts to reconstruct, from excavated remains alone, important elements of three prehistoric systems—culture, environment, and human biology. It seeks to develop and to test hypotheses that explain observed relationships within and between aspects of the three systems. All have left traces in the soil.

But the most brilliant archeologists cannot possibly control the range of knowledge necessary to interpret the archeological remains of the three extinct systems. It will take an interdisciplinary effort among specialists who together possess the necessary expertise.

In recent years, there have been major advances in the physical and biological sciences. These have an enormous potential for increasing the archeologist's capacity for interpreting the prehistoric record of culture, environment, and human biology. The difficulty has been that many of the new methods in physics, chemistry, mathematics, mining engineering, and medicine, to mention a few, have not been regularly applied to archeological problem solving because they require expertise that is unavailable and capital expenditures that are not feasible in conventional archeological research efforts.

Most of these interfaces between archeology and other sciences entail the development of new research methods by specialists using expensive instrumentation in a framework of long-term research. Application of these methods to archeological problem solving can occur only when institutional frameworks within which archeology is conducted have themselves evolved to a level capable of sustaining the diverse specialities that today offer so much potential for the growth of the discipline.

In 1964, archeologists at Northwestern University began an attempt to create an institutional framework capable of sustaining the diverse research expertise and substantial facilities and budgetary levels required to conduct long-term, regional research of the kind archeology today increasingly defines for itself. From the beginning, this research effort carried the name "Northwestern Archeological Program" (NAP). In June, 1981, the *Center for American Archeology* (CAA) was founded as a nonprofit research institution affiliated with Northwestern University. Headquarters of the new institution is Northwestern's Evanston, Illinois, campus.

The geographic focus of the Center and the antecedent NAP has been a 64 × 112-km (40 × 70-mile) area of west–central Illinois situated immediately north of the confluence of the Illinois and Mississippi Rivers. The human record here, preserved in the more than 2000 known archeological sites, is at least 12,000 years long. Three major river valleys (Mississippi, Illinois, and Missouri) converge here, a factor partially responsible for the complex culture history that is beginning to emerge from recent archeological investigations. The Mississippi–Illinois–Missouri confluence area appears to have been a major center of prehistoric cultural development for thousands of years, a factor relating to the rich and stable resources of the river floodplains.

The intensity of focus that the CAA and NAP have given this area since 1964 is unusual in North American archeology. There are both practical and philosophical reasons for the lack of long-term regional research projects of this kind. It is believed, nonetheless, that such an intense focus can produce a data base that enhances observation of the relationships between cultural variables and environmental and human biological factors, enhancing the opportunity for archeologists to generate explanatory models.

By 1981, Center archeologists and their colleagues from various other universities, have excavated 16 prehistoric habitation and mortuary sites in the 7200-km^2 research universe surrounding the Kampsville Archeological Center. These projects have focused on delineating economic patterns, population changes, and various interdependent cultural and biological processes that characterize the Archaic, Woodland, and Mississippian periods of this region.

Only time will tell whether this geographically focused archeological research effort will produce a significantly higher level of interpretation. What is clear is that the Center for American Archeology is attempting to develop an organizational structure that can accommodate the specialized expertise backed by facilities and supported by an institutional budget adequate to capitalize and to sustain the research program. In this, it represents an unusual experiment in modern archeology.

The Lower Illinois Valley as a Field Laboratory

There are practical reasons why the lower Illinois River Valley is an ideal field laboratory for development of new methods and approaches in archeology:

First, today this region is sparsely populated farmland, its southern limits separated from the St. Louis metropolitan area by the Mississippi River which has acted as an effective barrier to the northward spread of the city. Thus the lower Illinois Valley has escaped the widespread destruction that has greatly reduced the archeological research potential of many comparable river valleys throughout the United States. Second, the cost of performing research in the lower Illinois Valley is as low as can be expected anywhere in the world. Third, extensive information on climate, landforms, soils, water resources, plants, and animals is available for this region. These data provide an essential baseline for the cultural–ecological studies which are a major focus of CAA research.

The cost of doing archeological research has been too long overlooked as a significant factor affecting both the limitations archeologists impose on the goals of their research and the outcomes of that research.

As long as archeological dollars are scarce, cost and accessibility are key factors when selecting an area for long-term archeological research. The factors to be considered are (a) cost of transportation between the university and the field; (b) cost of maintaining and operating a field headquarters; (c) accessibility to cooperating natural scientists; (d) accessibility to supplies and equipment; (e) cost of transporting and maintaining students in the field; (f) ability to maintain continuity in the field program in terms of political and economic factors; (g) the prior availability of information on the natural environment (past and present); and (h) the ease of transporting artifacts to the home institution and technical specialists to the field. With respect to all these factors, the lower Illinois Valley is an ideal location for testing new archeological methods.

The Kampsville Archeological Center

To serve the research and teaching efforts of CAA, a permanent field center is under development at Kampsville. Kampsville is ideally suited for this purpose, since it is located in a region rich in archeological sites, yet one in which real estate values are modest enough to make feasible the purchase and construction of the facilities required to operate the Center for American Archeology's research and teaching programs.

During the late 1950s, I directed the excavation of Hopewell sites on the Illinois River near the small village of Kampsville. This area could be reached from Chicago in six hours and from St. Louis Airport in two. In steamboat days, Kampsville had been a major river port but had steadily declined since the 1920s. The facilities needed to create CAA's archeological field campus might be obtained most economically by purchasing a series of contiguous store buildings and houses in what had been the central business district of nineteenth-century Kampsville.

The purchase in 1968 of the building that formerly housed the Getz Hardware Store began the development of what has, by 1981, become a 39-building teaching and research center.

The Kampsville Archeological Center includes eight private residences that have been converted into student dorm houses. Several former store buildings are now laboratories for archeologists, zoologists, botanists, geologists, and other specialists. The 15,000-volume library at Kampsville represents an important resource, as does the PDP 11-34 computer. A small museum has been established to interpret the results of the Kampsville-based research for the public.

Research Publications

During the 1960s, and early 1970s, the Northwestern Archeological Program published research monographs in cooperation with the Illinois State Museum. The growth of archeological programs in both institutions made this cooperative arrangement less practical over the years, and beginning in 1976 Northwestern began to publish under its own imprint the research of its in-house faculty and students, as well as that of cooperating scholars from other institutions.

The Center for American Archeology will carry this publication program forward. The present volume, by Christopher Carr, is the seventeenth in a continuing series of monographs and collected works that represent the research results of the Center and the Northwestern Archeological Program before it.

Stuart Struever
EVANSTON, ILLINOIS

Preface

This volume introduces a new approach for using resistivity surveying techniques to investigate archeological sites. Traditionally, resistivity methods have been used to locate and delimit individual features. Surveys have been successful where features are large or linear and contrast greatly from their matrices in conductivity. On earthen archeological sites, however, where features such as pits and house trenches contrast little from their matrices and are small or nonlinear, natural and agriculturally caused soil variation can mask out or be confused with archeologically significant soil variation, as manifested in resistivity data sets. Individual anomalies within resistivity signatures from earthen sites often are unreliable indicators of prehistoric disturbance. It is suggested that in such circumstances, broad-scale use-areas are more appropriate targets than individual features. Resistivity data may be interpreted statistically by examining the mean resistivity and variance in resistivity of broad geographic zones in order to locate use-areas and to characterize and differentiate them by these attributes. The spatial scales of such use-areas distinguished in their means and variances also may be used to characterize and differentiate them. That use-areas of different kinds are distinguishable in their mean resistivity is based on the assumptions that: (a) different activities produce residues of different chemical compositions; (b) these various residues alter in different ways and to different degrees the physical and chemical properties of soils within the areas where they are deposited; and (c) the different soil changes within different use-areas result in their having different soil resistivities. That use-areas of different kinds are distinguishable in the variance of their resistivity signatures is based on the assumption that different activities may require different sizes or numbers/areal densities of facilities, or generate residue deposits in different spatial configurations and areal densities. These factors will determine the degree of homogeneity of the soils within each use-area, and thus, the variability of soil resistivity within each use-area. It also is assumed that different kinds of activities may require different amounts of space. Use-areas defined within resistivity data sets by statistical and geographic analysis, along with spatial patterns found in surface artifact distributions, may be employed as the spatial strata of excavation sampling designs in order to obtain an unbiased sample of both the artifacts and features within a site.

The soil alterations produced by humans on archeological sites, as direct causes of the resistivity variations found in sites, are of primary concern to this study. Therefore, detailed discussions are given of the natural processes by which residues of prehistoric activity are incorporated within soils and alter the physical and chemical states of soils. The natural processes by which the resultant anthropic soil anomalies are maintained over time also are described. Spatial overlap of activities and deposition of different kinds of refuse in a single location is shown to affect the maintenance of anthropic soil anomalies, as well. Finally, a general model of the changes that occur in archeological soils from the time of site abandonment to the time of archeological investigation is offered. Changes in anomalous soil organic matter content and in both the magnitude and ratios of anomalous ion concentrations are considered. All of these discussions serve as a basis for understanding the nature of archeological soils and soil resistivity data.

Mathematical techniques are introduced for examining resistivity data sets from a geographic, statistical perspective. Variability of resistivity data sets may be partitioned along the dimensions of both space and depth, isolating archeologically significant information, and removing noise due to natural soil variations and contemporary farming practices. Both spatial filtering techniques and formulas describing the resistances of stratified sediments as if they were resistors hooked in parallel circuitry (the Barnes Layer method) may be utilized for this purpose. Spatial filtering techniques may be used also to analyze the isolated archeological component of resistivity variation, itself, in order to locate use-areas within it and to characterize them by their dimension, local mean resistivity, and local variability in resistivity. Models for interpreting the restructured data sets are provided, using theory from physics, chemistry, and soil science, and empirically derived relationships between pedological and archeological variables. The empirical relationships are constructed from archeological and pedological data collected at the Crane site, a Middle Woodland base camp in the Illinois River drainage.

The feasibility of interpreting resistivity data from a geographic, statistical perspective in order to define use-areas within sites is illustrated with data from Crane. Use-areas in which different amounts and/or kinds of activities have occurred and in which different amounts and/or kinds of refuse materials have been deposited are shown to be differentiable in the mean resistivities of their soils. Use-areas having different areal densities of pit features are shown to be distinguishable in the variance of the resistivity of their soils. The use of spatial filtering techniques to isolate archeologically significant variation within a complex resistivity series and to locate, characterize, and differentiate use-areas is illustrated. Also, the Barnes Layer method is shown to be capable of removing the noise-producing effects of surface soil horizons from resistivity data and to be accurate in estimating the resistivity of buried archeological horizons. Boundary conditions for the application of the proposed analytical design are provided.

Acknowledgments

Throughout the course of work leading to this volume, I have paused many times in private, thankful of the faith that so many people have had in the project, and of the support that they have given in their time and energy, personality, and spirit. I now wish to express my gratitude publicly. Each of the persons listed below has been critical to the completion of this study. Each has given more than was required—some to the point of personal stress and discomfort at crucial times of need, and others through their continuous efforts over many years. To these individuals, for their personal interest and involvement in this project, I extend by deepest thanks:

Sarah Anderson
David L. Asch
Karen Atwell
Leland Berkwits
Judie Birdsall
Charles Boast
  University of Illinois at Urbana
James R. Boyle
  The University of Michigan
James A. Brown
  Northwestern University
Susan Brown
Ellen Bruce
Margaret Burright
Peter Carlson
Anton G. Carr
Betty Carr
Judith Chase

George Chick
Frank Coates
Roy Coleman
Donald Crane
Page Danley
John W. Delano
Kathy Denham
Nancy Dietz
Charles Downy
Robert Etheridge
Kenneth B. Farnsworth
Richard I. Ford
  The University of Michigan
Lawrence Fox
Andrew Getz
James Gigliotti
Bruce G. Gladfelter
  University of Illinois at Chicago Circle

Leslie Glick
Daniel Haas
Robert L. Hall
   University of Illinois at Chicago Circle
Michael Hambacher
Alan D. Harn
Janice Hill
Cheryl Hurzwurm
Karl L. Hutterer
James Johnson
Lauren Kelley
Beth Kerr
Mark B. King
Leonard Kotin
Linda Krakker
Leslie Landefeld
Judeth Larson
Miles S. Linnabery
Joseph Manning
Julie Martin
Vincent P. McCarren
Andrew Moore
Robert Murowchick
Linda Myers
Ann Myhre
Charles E. Olson, Jr.
   The University of Michigan
Samuel Outcalt
   The University of Michigan
Corinne Pearson
Richard Rossier
Alan Rostoker
Gary R. Rotramel
Malisa Rotramel
Cecile M. Sands
Kent A. Schneider
Timothy Shaker
Wayne F. Shields
David M. Smith
John D. Speth
   The University of Michigan
Glenn Stone
Judy Stone
Linda Stone
Stuart Struever
   Northwestern University
David Sun
Jan Surkamp
Simone Taylor
Sue Tituskin
Lisanne Traxler
Seymour Turner
Margaret Van Bolt
Ira Vogel
Timothy Wendt
Darrel Wheaton
Susan Wiedenbeck
Jonathan Wool

    Of all these important people, some have been especially helpful and deserve special recognition. First, I want to thank Richard I. Ford and Robert L. Hall, my graduate and undergraduate advisors, for their understanding of my personal interest in multiple fields and for the liberal academic attitudes they expressed while allowing me the freedom to investigate a problem that lies outside the heart of anthropology. At the same time, they continuously helped me to focus my work toward ends that would be useful to archeologists. The balance between broad-based and focused research, which they encouraged in my studies, directly influenced the form this volume has taken.

    To Stuart Struever, Ann Myhre, Kent Schneider, Charles Boast, and Judie Birdsall, my heart cannot open deeply enough in appreciation and respect. Theirs were truly labors and acts of faith. Stuart Struever, through the Center for American Archeology, granted the initial financial support that carried this project to the field. Since then, he has continued to finance portions of the analyses and writing through his own exhaustive fund-raising activities. His commitment to this project reflects a strong and more general belief that he has in education and research, and in encouraging young people to try out their ideas. (At the beginning of this project, I was 21, anxious, but unexperienced.) I will always admire Stuart Struever for his philosophy and be thankful for the strong support he has given. In a similar vein, Ann Myhre, Kent Schneider, and Charles Boast freely gave tremendous amounts of their time performing most of the chemical and physical soil analyses reported here, simply from their general good will and interest in the project. I particularly want to thank Ann Myhre for seeing through the chemical soil analyses despite the numerous personal and administrative frustrations she had to endure. Judie Birdsall undertook the typing of this manuscript free of compensation, again out of personal kindness and good spirit. Without her help, I could have not afforded to complete this document at this time.

A large number of individuals have contributed to the ideas and analyses presented here. Most important of whom are my doctoral committee members: Richard I. Ford, John D. Speth, Karl L. Hutterer, and James R. Boyle. They have worked hard to give me the foundations in lithic tool analysis, in the formation of the archeological record, in the archeology of the eastern United States, and in pedology, which are integral aspects of this project. They also have offered many constructive criticisms of my analyses and have spent much time and energy editing this lengthy document. Samuel Outcalt and Robert Whallon have had major influences on my use of mathematical and statistical methods as conceptual frameworks and analytical techniques. James. A. Brown and David L. Asch helped me to phrase my research questions in testable forms and provided enormous help in designing the field research. For their more general contributions in teaching me about the scientific process and how to do scientific research, I will always be indebted to my father, Roy Coleman, James A. Brown, and Robert Whallon.

The research reported here could not have been realized without the invaluable support of Kenneth B. Farnsworth. Over the course of two hard field seasons at the Crane site, Ken and his crews collected the controlled surface survey data and excavation data that are analyzed here, and generously made these fruits of his labor available to me. Ken's initial impressions of the spatial organization of the Crane site from his preliminary surface surveys of it were instrumental to me in designing the transect surveys. I also appreciate the more general understanding of Middle Woodland archeology of the lower Illinois Valley with which Ken provided me. This information was of immense help to me in determining the functions of various artifact classes from Crane and the nature of the site.

In helping me to collect the resistivity and soil data for this work, I owe special thanks to my assistant field supervisors, Robert Murowchick, James R. Johnson, and Page Danley. Without their continuous efforts to insure a high level of accuracy in the gathering and recording of the field data, this project would not have been successful.

I extend deep thanks to the numerous institutions and foundations that granted support. Primary financial aid was given by the Center for American Archaeology (multiple grants) and the National Science Foundation (NSF–SOS Grant No. EPP75-08645; Doctoral Dissertation Improvement grant;

xxvi   *Acknowledgments*

Graduate Fellowship). Funds for computer analyses were provided by the Museum of Anthropology and the Departments of Anthropology at Northwestern University and the University of Arkansas, and Geography at the University of Michigan, by the Department of Anthropology at Northwestern University, and by the University of Arkansas. Facilities for performing soil analyses were provided free of charge by the North Carolina Department of Agriculture, Agronomic Divison at Raleigh and by the Department of Agronomy at the University of Illinois at Urbana. Field equipment was lent by the Center for American Archeology, the Departments of Anthropology at Northwestern University and the University of Illinois at Chicago Circle, the Department of Agronomy at the University of Illinois at Urbana, and the Department of Pharmacy and School of Natural Resources at the University of Michigan. The Department of Research and Sponsored Programs and the Department of Anthropology at the University of Arkansas typed the final revisions of this manuscript.

Finally, with all my heart, I thank my father and mother for the continual training and encouragement in scholarship that they gave me; for developing in me a love of learning and a desire to make the world a whole; and for raising me with the belief that giving knowledge is one way we may contribute what we owe.

1
Introduction

Statement of Problem and Thesis

Archeologists have investigated, and still do investigate, the remains of past peoples for a variety of academic reasons: (*a*) to reconstruct the lifeways of those peoples (e.g., Braidwood 1959:79; Taylor 1948; McCartney 1980), including their economy, social and community structure (Chang 1968; Longacre 1968) and ideology (Hall 1976, 1977); (*b*) to trace their histories (e.g., Kroeber 1937:163; Rouse 1965:2; Aikens 1966); and (*c*) to investigate more general problems on the nature of cultural, behavioral and cognitive (choice) processes (e.g., Willey and Phillips 1958:5; Binford 1968; Jochim 1976; Limp 1978).

In each kind of study, however, and from a broad perspective, the categories of data used to achieve the particular goal are similar. Minimally, they include: (*a*) the tools and debris manufactured by the people; (*b*) nonmobile, permanent features such as house foundations, storage pits, moats; (*c*) intrasite spatial relations among artifacts and features of the same or different classes, whether the classes be functional or stylistic in nature; (*d*) the relative frequencies, within sites, of artifacts and features of different classes; and (*e*) spatial relations among archeological sites of the same or different types, as defined by their artifact and feature inventories.

Whatever the particular goals of the archeologist, information on these several categories of archeological phenomena must be collected—by survey or excavation—in such a way that the *samples* of artifacts, features, their relative proportions, and their spatial patterning obtained for study are unbiased and *representative of the total state of the archeological record within the universe of study*. The appropriate *design* of survey or excavation for collecting such a sample for study, however, may be unclear when the spatial structure of the archeological entities to be investigated is unknown. Different kinds of sampling design are more or less efficient, depending on the spatial arrangement of the entities (Berry and Baker 1968; Cochran 1963; Plog 1976; Read 1975).

To circumvent this problem, archeologists have developed a number of field techniques for obtaining an initial idea of the structure of the archeological record and for guiding them in creating reasonable collection designs. Among these are aerial photography and remote sensing (Aston and Rowley 1974; Bradford 1957; Lyons and Hitchcock 1977), systematic surface survey (Binford 1964; Binford *et al.* 1970; Redman and Watson 1970), soil survey (Hurley and Heidenreich 1971; Provan 1971), and resistivity and magnetometry survey (Aitken 1961; Atkinson 1963; Clark 1963). These different techniques are useful in guiding the collection of different classes of archeological data, and complement each other. For example, systematic surface survey techniques are appropriate for guiding the design of excavations that representatively sample the artifact populations within sites; soil surveys and resistivity and magnetometry surveys are appropriate for guiding the design of excavations that representatively sample the feature populations within sites; and aerial photography can be used for guiding the design of regional archeological surveys that inventory sites.

The potential of resistivity survey methods as an aid in designing excavations that representatively sample buried archeological features within sites has not been fully recognized by archeologists. Accordingly, in this volume, previous modes of application of the technique, will be reviewed, and new ones will be proposed and described. In particular, it will be argued that resistivity survey methods need not be restricted to detecting and delimiting *individual* archeological features. They also may be used to locate, characterize, and distinguish different kinds of "use-areas" or "activity areas" as *whole geographic units* within sites, *without* determining the location, morphology, and function of each feature within the use-areas. It is suggested that different kinds of use-areas can be characterized and differentiated statistically by their mean soil resistivity values and by the magnitude of the variance of their soil resistivity values. The scales of use-areas also may be determined from their resistivity responses and used to distinguish them. Use-areas defined in this manner then can be used as sampling strata within geographic excavation designs.

That use-areas of different kinds are distinguishable in their *mean* resistivity is based on three assumptions:

1. Different activities yield residues of different natures and nutrient compositions.
2. These various residues alter the physical and chemical properties of soils in different ways and amounts.
3. The different soil changes within different use-areas result in their having different soil resistivities.

That use-areas of different kinds are distinguishable in the *variance* of their resistivity signatures is based on the assumption that different activities may require different sizes or numbers (and areal densities) of facilities, or generate residue deposits of different spatial configurations and areal densities. These factors will determine the degree of homogeneity of the soils within each use-area, and thus, the variability of soil resistivity within each use-area. Finally, it is assumed that different kinds of activities may require different amounts of *space*.

This new approach to the interpretation of resistivity data may be used on sites where features contrast well from their matrices in their resistivity properties (e.g., masonry walls buried in soil) or on sites where the contrast between features and their matrices is low (e.g., earthen features within soil). It is tailored particularly, however, for use on sites of the latter kind.

In operationalizing, testing, and illustrating this thesis, the chapters of this volume follow a specific, logical sequence. In the remainder of this chapter, the nature of resistivity survey methods and their traditional modes of use to isolate individual features are described. Theoretical and operational reasons then are given for interpreting resistivity data from an alternative, statistical, and geographic perspective, in which case the definition of use-areas is the goal. The means by which the alternative resistivity method and surface collection methods may be used in a complementary fashion to define the spatial organization of sites and to guide archeologists in designing representative excavation samples also is discussed. Chapter 2 describes the particular mathematical methods by which resistivity data may be restructured within a statistical and geographic framework, so as to allow the location, characterization, and differentiation of use-areas within sites. Chapters 3 and 4 provide various models useful in the interpretation of resistivity data once it has been treated by the appropriate mathematical methods. Mathematical and qualitative relationships beween soil resistivity and the physical and chemical soil attributes that determine it are described. The ways in which such determinants of soil resistivity may be altered by the deposition of human refuse on archeological sites are described and the relative degrees to which such alterations are maintained within archeological soils by natural pedological and ecological processes from the time of deposition to the present are discussed.

The second half of this work tests and/or illustrates the various aspects of the thesis with archeological, pedological, and resistivity data collected at an earthen site (Crane) on silt loam soils in southern Illinois. Chapter 5 describes the site, and the research design and analytical methods used to investigate it archeologically. Use-areas within the site are reconstructed. The designs for collecting soil and resistivity data from the Crane site are presented in Chapter 6. In Chapter 7, soil variations of natural causes and resulting from spatially nonuniform historic land-use patterns are reconstructed in order to separate them later from other soil variations, which presumably are of prehistoric, anthropic cause. In Chapter 8, the relationship between prehistoric activity and soil alteration proposed in this work is tested. The physical and chemical soil alterations that one would expect to occur in the different use-areas at the Crane site are compared with those of prehistoric origin that actually do occur in them. The degree to which different kinds of physical and chemical soil alterations of prehistoric, anthropic origin have been maintained relative to each other within the Crane soils also is examined. Chapter 9 tests that portion of the thesis relating soil physical and chemical properties to soil resistivity. Variation in the resistivity responses of different areas of the Crane site are related to spatial variations in the individual soil properties that determine soil resistivity. In Chapter 10, the mathematical methods presented in Chapter 2 for restructuring resistivity data sets within a statistical, geographic framework are illustrated and evaluated for their effectiveness in allowing the location, characterization, and differentiation of use-areas within sites. The final chapter summarizes the boundary conditions within which the proposed methods and models are applicable.

To follow the logical flow of this work in detail, despite its length, I urge the reader to refer back to the table of contents continuously. The headings and subheadings of sections that are listed in the table have been worded carefully to reflect their position within the logical fabric of the study.

The Nature of Resistivity Surveying Methods and Their Traditional Use in Archeology

Electrical resistivity surveying is a technique for locating and mapping buried structural or chemical anomalies within consolidated and unconsolidated sediments and within soils. In geology, where the method was first applied and developed (see Van Nostrand and Cook 1966), detectable anomalies of interest may include buried ore deposits, infilled channels and sinks, vertical stratigraphic variation, faults, and areas of different water quality. In archeology, the buried anomalies may include stone walls, infilled ditches and pits, or midden accumulations.

Resistivity survey methods were applied in archeology first by R. J. Atkinson in 1946 to locate the infilled ditches of Neolithic henge monuments at Dorchester, Oxfordshire (Atkinson 1952). Since then, and particularly during the 1950s and 1960s, the technique has been applied with enthusiasm to locate and delimit many different types of features (Table 1).

The application of resistivity survey techniques to archeology relies on the fact that the anomalies of interest may differ from their matrices in properties that determine their capacity to conduct an electric current. If different electric potentials are applied to the earth at two locations with a pair of electrodes, a current will be generated between them. Positive ions dissolved within the water

TABLE 1
Archeological features of various types which have or have not been identifiable within resistivity data sets

Feature and Matrix	Success or Failure in Differentiating Feature	Reference
Features with Fill Concentrating from their Matrices:		
void chambers within masonry pyramids	horizontal location defined	Lerci (1959); Lerci et al. (1959)
void prehistoric well buried below soil and gravel	horizontal location defined	Chalabi (1965:133)
void mine shafts and galleries in limestone, below stone rubble	horizontal location defined	Dabrowski (1963:85)
mining shafts in chalk filled with loose chalk rubble or soil; underlying a plowzone	spatial distribution of shafts defined	Sieveking, et.al (1973)
loose pile of stones (?) of either a Bronze Age burial or a recent archaeological excavation; buried below soil fill of a mound	location and length defined	Palmer (1960:73)
Roman age masonry walls or loose piles of stone	location defined	Palmer (1960:74)
Roman age masonry walls, rubble, and paving in soil	spatial distribution and alignments defined	Rees and Wright (1969)
prehistoric compact gravel road buried within soil	horizontal location and width defined	Chalabi (1965:133)
prehistoric gravel causeways buried below plowed soil	horizontal location and width defined	Atkinson (1963:23)
Roman age road of limestone rubble bound with stiff clay and dressed with pebbles; buried below clayey soil	horizontal location defined	Atkinson (1963:20)
Roman age stone roadway below plowed soil	location and width defined	Palmer (1960:74)

TABLE 1 (cont.)

Feature and Matrix	Success or Failure in Differentiating Feature	Reference
masonry wall buried in soil	horizontal location defined	Aitken (1961:70)
Roman age masonry walls below soil	horizontal location and orientation defined	Schwarz (1961:68)
Roman age wall buried in soil	horizontal location defined	Atkinson (1963:25)
Roman age masonry wall buried in clayey soil	horizontal location defined	Clark (1963:574)
Roman age masonry walls buried in clayey soil	horizontal location defined	Dunk (1962)
prehistoric masonry wall backed with a bank of clay and turf; buried under a clay soil	horizontal location defined; wall and backing distinguished	Atkinson (1963:22)
Roman age masonry wall within earth and rubble matrix overlaid with bricks	horizontal location and width determined	Linington (1968)
masonry remains of Roman age dwellings and smoke houses; burial conditions not stated	horizontal locations and sizes determined	Dabrowski (1963: 86-87)
masonry walls and building foundations; burial conditions not stated	horizontal locations defined	Annable (1958)
18th and 19th century masonry foundations buried in well-drained sandy soils	horizontal location and orientation determined	Leith et al. (1977: 6,8)
18th century masonry walls buried in clayey soil	horizontal location and orientation determined	Leith et al. (1977:8)
iron slag and earthen furnaces; burial conditions not stated	horizontal locations defined	Dabrowski (1963:85)
Etruscan age, earth-filled tomb cut into limestone and filled with soil; burial conditions not specified	horizontal location defined	Linington (1963: (63,66,69)

TABLE 1 (cont.)

Feature and Matrix	Success or Failure in Differentiating Feature	Reference
Roman age ditches cut into soft sandstone; filled with sand, clay, and loam; buried below pebbly sand	horizontal locations defined	Chalabi and Rees (1962)
ditches of Roman age filled with soil, or soil and chalk silt; carved into chalk; buried by soil; plowed	horizontal locations defined	Atkinson (1963:21)
Neolithic age ditches dug into chalk, filled with loose chalk rubble; and buried below plowed soil	horizontal location, orientation, and shape determined	Linington (1967)
prehistoric ditch dug in natural gravel, filled with soil; burial conditions not specified	horizontal location and shape determined	Atkinson (1963:26)
Neolithic ditches filled with loose rubble and loam, dug within a gravelly loam, buried below plowed soil	horizontal locations defined	Linington (1966)
Roman age ditch filled with stones, buried below undisturbed soil	horizontal location defined	Schwarz (1961:68)
Roman age ditch filled with soil and gravel, buried below plowed soil	horizontal location defined	Palmer (1960:74,75)

Features with Fill not Contrasting with their Matrices*:

soil-filled pits within a heterogeneous sandy loam	only some pits with anomalous resistances, both positive and negative anomalies without pattern. Pedological anomalies with resistances similar to pits	Ford (1964)
soil-filled pits in soil matrices	only some pits with anomalous resistances; pedological anomalies with resistances similar to pits	Gansfuss (1975: personal communication)

TABLE 1 (cont.)

Feature and Matrix	Success or Failure in Differentiating Feature	Reference
soil-filled pits in soil matrices	only some pits with anomalous resistances; pedological anomalies with resistances similar to pits	Kent A. Schneider (personal communication)
soil-filled pits in clayey loam	only some pits with anomalous resistances; pedological anomalies with resistances similar to pits	Richard Leary, Illinois State Museum (personal communication)
Roman age pits and baking ovens filled with soil; burial conditions not stated	anomalous resistances typically not associated	Dabrowski (1963: 86-87)
2 prehistoric ditches and a prehistoric pit, filled with soil, dug in a soil matrix; burial conditions not stated	unpredictable results, depending on moisture conditions. Ditches somewhat more definable than pit under suboptimal moisture conditions	Atkinson (1963:24)
ditches and pits filled with soil in soil matrices	unpredictable results, depending upon moisture conditions	Aitken (1961:72)
Neolithic age ditches dug in loam and clay with flints; filled with soil	undetectable in most instances	Clark (1963:575)
Roman age ditch filled with soil and buried below plowed soil	horizontal location defined	Palmer (1960:76)
soil-filled trench in soil matrix	orientation determined	Gansfuss (1975)
organically-enriched mound cap buried below sandy loam	horizontal extent determined	Ford and Keslin (1969)
recently dug pits within organically-enriched mound fills	horizontal locations determined	Ford and Keslin (1969)

*Fewer examples of resistivity surveys over features of this nature are reported than surveys over features contrasting phenomenally from their matrices. This is in part presumably related to the discouraging results which have been obtained under the former conditions and the reluctance to report negative results.

held by the ground will migrate toward the location of relative negative potential, and ions and conductive colloidal solids (clays) will migrate toward the location of relative positive potential.[1] The ability of the earth to conduct the current will depend on the concentration of ions within the groundwater, the total moisture content of the earth, and the geometric arrangement of the moisture-holding pores within it. Local anomalies in these properties—either of natural origin or as a result of human disturbance—will produce localized differences in the conductivity of the ground.

The differences in ground conductivity can be determined using various geometric arrangements of four electrodes (e.g., Figure 3a). Two "current" electrodes apply a standard potential difference to the ground and generate a current of a standard amperage throughout the ground. Another two electrodes ("potential" electrodes), which are placed somewhere within the electric field, measure the variable drop in potential (voltage) that occurs within that part of the electric field between them, as a result of the resistance of the ground between them to electric flow. Using Ohm's law (voltage = amperage × resistance) and equations specific to the geometry of the electrode system (see Van Nostrand and Cook 1966), the total resistance of the ground between the potential electrodes, and its resistivity or conductivity,[2] may be calculated from the observed potential drop, the standard amperage, and the standard distances between the electrodes.

Importantly, anomalies that are deeply buried as well as those at the ground surface can be detected by this method without inserting the electrode into the ground more than is necessary to make electrical contact. The paths of flow of the current that is generated by the current electrodes penetrate the ground in all directions and (at least theoretically) to infinite depth, rather than follow a direct line between the electrodes. The observed potential drop between potential electrodes will reflect the resistivity conditions of the ground between them to that depth effectively penetrated by the current (e.g., Figure 3b). By increasing the distance separating the current electrodes, the effective depth of penetration of the current and the depth of investigation of the metering system may be increased (e.g., Figure 4). Reviews of the general principles of resistivity surveying and the equipment used in surveys are given by Aitken (1961), Atkinson (1963), Chalabi (1965), and Palmer (1960). One particular method of survey is described in detail in Chapter 3 of this work.

Because resistivity methods can be used to detect *buried* anomalies within the ground as well as those at the ground surface, they may be applied to an

---

1. The precise mode of current flow is "electron transfer," rather than "electron migration." Electrons are displaced *short distances*, from one location of rest to an adjacent location of rest within the ground in a manner similar to the way energy is transferred by falling dominos. *Long-distance* migration of individual ions, as that which is pictured to occur in an idealized electric cell, is not a primary mechanism of current flow within the ground.
2. The "resistivity" of a substance is the "resistance," per unit volume of it, to electric flow. "Conductivity" is the reciprocal of resistivity.

advantage in archeology to *systematically* locate and delimit buried archeological features that in other ways are not observable from the surface and that might be missed by excavation. Resistivity survey methods may help archeologists focus their excavation efforts on those portions of an archaeological site that are of most interest to them in relation to the kinds, densities, and arrangements of features they contain. Also, since the costs in time, labor, and supplies of using resistivity surveys as a method for prospecting for buried features are considerably less than those of archeological excavation, resistivity methods are a double asset to the archeologist.

Resistivity survey methods have been applied most commonly in archeology to locate and delimit *individual* features (Table 1). They also have been used to determine the *spatial organization* of features within site structure (e.g., Dabrowski 1963; Dunk 1962; Ford 1964; Rees and Wright 1969). In the latter studies, however, determination of the spatial organization of features has been attempted by examining spatial patterning of *individual* resistivity anomalies and by trying to associate individual anomalies with individual features. A statistical geographic approach, in which use-areas having different kinds and densities of archeological residues and facilities would be identified *directly* as *whole areal units* by the *statistical* attributes of their resistivity signature has not been attempted.

A Theoretical Reason for Taking the Use-Area as the Unit of Analysis When Interpreting Resistivity Data

Resistivity surveys approached from a statistical, geographic perspective, and focusing on use areas, in addition to resistivity surveys focusing on the location and delimiting of individual features, have a place in archeology. Two reasons are clear. One is of a theoretical nature, related to the goals of archeology and the units of analysis that are appropriate for achieving those goals. The second is of a practical nature, concerned with how readily apparent use-areas as opposed to individual features may be within resistivity data sets. Let us consider first the theoretical reason for taking the use–area as the unit of analysis.

Despite the differences that archeologists may have in their theoretical orientation and research objectives, a primary aspect of archeological research that commonly is necessary for archeologists to achieve their stated goals is the reconstruction of the *activities* of past peoples. Toward this end, the determination of *spatial associations* between artifacts and features and the definition of areas of different use within archeological sites is critical. The activity, as evidenced by spatial associations between artifacts and features within use-areas, is a fundametal *unit of inference* in archeology.

The importance of the activity and spatial associations within use-areas has repeatedly been recognized as important by archeologists having different goals. Let us consider several of the more common approaches to archeology, and how concern with activities and spatial associations within use-areas is basic to the achievement of their goals.

One of the commonly stated goals of archeology is the reconstruction of the "lifeways" of past peoples (Vaillant 1930:9; Taylor 1948). By the very nature of this goal, the activities of those people are of primary concern.

A major spokesman for this goal of archeology is Walter Taylor (1948), who advocated that in attempting to achieve it, a "conjunctive approach" to the interpretation of the archeological record should be taken. In this approach, the "relation of item to item, the associations and relationships, . . . the affinities within the manifestation under investigation" are of major interest, being useful as evidence of past activities and lifeways (Taylor 1948:95–96). Although Taylor's conjunctive approach embraces the study of a number of forms of relationships at different levels of organization of material culture, including for example the association of attributes among artifacts, the relative proportions of artifacts within sites, and the relationship of sites to their natural environmental context and to each other in their artifact contents, it also involves and emphasizes the use of the *spatial relationships and associations* between artifacts and features (pp. 111–112).

A second stated goal of archeologists is the reconstruction of the histories of past peoples. This approach may involve simply the ordering of artifact types and sites within a temporal–spatial framework (e.g., Kidder 1924; Kroeber 1937:163), that is, "area synthesis" (Willey and Sabloff 1980:110), or it may be concerned with descriptive changes in the lifeways of people within a region (e.g., Braidwood 1960; MacNeish 1964). When areal synthesis is the goal, cultural historians depend heavily on stratigraphic excavation of multicomponent sites and seriation techniques (Ford 1962). Primary concern is with the vertical relationships between entities within different components of sites— the changes in frequencies or presence–absence states of entities between components of sites—rather than the internal arrangement of sites. When description of changing lifeways is the goal, cultural historians again emphasize excavation of multicomponent sites, but attention is given to the study of contextual and spatial relationships within components, as well as changes in entities between components, in order to document time-discrete lifeways and their chronological evolution.

Today, most cultural historians realize that intrasite structure cannot be ignored, even when chronological ordering of artifact types and sites is the primary objective. Successful seriation of types and sites requires that the sample of each component, which is compared to others, be representative of the component *as a whole*, and that compared components be of similar function. Both requirements must be met if variations among samples are to reflect only time differences. The representativeness of a given sample of a component and the function of a component can be assessed only if the archeologist has some understanding of the arrangement of use-areas within the site.

K.C. Chang also has discussed the importance of interpreting the archeological record with respect to the activities of the people who produced it, and the utility of spatial associations found between artifacts and features in making those interpretations. Chang sees as a primary goal of archeology the reconstruction of social and community organizations and the development of social structural models like those of social anthropology (Chang 1958:324; 1968). To achieve this, he has suggested (1967:231–232) that archeologists characterize the archeological remains of the prehistoric social groups they study by the activities they imply, and that:

> instead of conceiving society as having a social structure [archeologists should] conceive of social behavior as being structured by participation in given activities within which behavioral choices are regular and redictable (quoting Howard 1963:410). *[Such] activities can be physically indicated by loci and instruments* ... (emphasis added).

that is, by the spatial aggregation of tools or facilities used in the activities.

Archeologists who employ a "human ecological" perspective likewise stress the reconstruction of prehistoric activity and the use of spatial relations among artifacts and features as a step toward achieving their goals. The ecological perspective is a broad one, concerned with the relationships between the demographic, biological, and cultural aspects of local populations and their environments. The perspective encompasses a number of variant approaches (Orlove 1980) presently used by archeologists. Most current are the "neofunctionalist" and "processual ecological"approaches.

The neofunctionalist approach stems from the work of Vayda and Rappaport (1967) and is concerned with (*a*) social organization and cultural practices as mechanisms that allow *local* human populations to adapt to their natural and social environments, and (*b*) the structure, maintenance, and evolution of *whole ecosystems*. The approach focuses on the population as the unit of analysis and explains ecosystemic equilibria and trajectories using structural principles at the ecosystem level. Works by Flannery (1972) and by Ford (1974) represent this school.

The processual ecological approach has developed as part of the recent trend in anthropology and biology toward the use of actor-based models and decision making models (Orlove 1980:246, 248). It is concerned with individual and group activities as the outcome of decision-making processes that allow local populations to adapt to their environments. The approach focuses upon the individual faced with alternative choices and a hierarchy of goals as the unit of analysis. It explains ecosystemic organization, maintenance, and evolution, as the product of individual decisions curtailed by natural selection rather than by evoking principles at the ecosystemic level. Works by Jochim (1976), Limp

(1978), Reidhead (1979), and Keene (1979) exemplify this approach in archeology.

Importantly, the ecological approach, including both neofunctionalism and processual ecology, is a *behavioral* one, concerned with the *practices and activities* of human populations. As Vayda and Rappaport have stressed, human behavior is "studied as animal behavior and interpreted in the same way as the behavior (or part of the behavior) of any species, for instance, in its adaptive aspects and consequent interaction with natural selection (Vayda and Rappaport 1967, quoting Simpson 1962:106; Rappaport 1971:244). The activities studied may be biologically based, technological, economic, social, or ritualistic in nature.

In operationalizing the ecological perspective, archeologists have defined and made use of the concept of the "activity area": portions of sites where different sets of tools, debris, and features are found in association and where different activities presumably[3] occurred (Brown and Freeman 1964; Freeman and Butzer 1966; Hill 1970; Struever 1968; Yellen 1972). Definition of such areas within a site and reconstruction of the activities they presumably reflect are seen as critical steps in determining the nature of the social unit that occupied the site, the function of the site within the subsistence–settlement system of the occupants (see Struever 1968), and how the occupants interacted with their natural and social environments as part of a whole ecosystem (Binford 1964). Although the concept of the activity area has been applied by archeologists for purposes other than strictly ecological ones (e.g., Chang 1968; Schiffer 1976), its original definition, formalization, and popularization as an operational aid to archeological inference occured among archeologists who were concerned with ecological problems (e.g., Binford 1968, Struever 1968).

Thus, despite the diversity of goals and theoretical perspectives found among archeologists, concern with the activities of past peoples and with spatial associations of artifacts and features within use-areas as products of those activities is common and basic to the approaches of many archeologists. The use-area, as a *whole, bounded, geographic zone* encompassing *multiple artifacts and features*, is an important *unit of analysis and interpretation* in archeology in general.

Accordingly, it is critical that use-areas—the spatial associations of artifacts and features of one or more types—as well as the individual types of

---

3. The concepts, "activity area" and "activity set," are too restrictive in the archeological phenomena they describe to be applicable to all kinds of artifact aggregation and to all locations of artifact aggregation. They do not acknowledge the wide variety of archeological formation processes which can be responsible for artifact aggregation. In this work, the broader terms, "use-area," "depositional area," and "depositional set," in addition to "activity area," and "activity set," are used in reference to the various possible kinds of artifact aggregations and locations of artifact aggregation that may occur. These terms are defined and discussed in the section of Chapter 5, "A Model of the Nature of Organization of the Archeological Record."

artifacts, and features, themselves, be sampled representatively with respect to their diversity when excavating a site. One means of doing so is by using one or more of the guiding techniques mentioned previously—such as resistivity surveying or controlled surface surveying—to locate and define prior to excavation the approximate boundaries of areas having different kinds of artifact associations, feature associations, and uses, and to design excavations that sample these areas adequately and representatively. Consequently, from a theoretical standpoint, there is good reason to interpret resistivity data with respect to information it contains on use-areas and spatial associations of features as well as information it contains on individual features. A geographic, statistical mode of interpretation of resistivity survey data can be recommended from this theoretical perspective.

Operational Reasons for Taking the Use-Area as the Unit of Analysis when Interpreting Resistivity Data

Theoretical considerations suggest an advantage in interpreting resistivity data from a geographic, statistical perspective to define use-areas within sites, rather than attempting to isolate individual features. Which analytical framework is selected, however, will also depend on the particular problem at hand and the nature of field conditions. For example, an archeologist might be interested in determining the exact, unknown boundaries of previously excavated areas within a site, so that excavation efforts might be concentrated on undisturbed deposits. In this case, a feature-oriented approach, in which the excavated areas are the features of interest, would be more appropriate. Under some field conditions, (e.g., physical and chemical states of features) analysis at the scale of the individual feature may be successful, whereas during other conditions, it may be completely unrewarding and a statistical geographic mode of analysis may be the *only* available option for interpretation.

A review of the successes and failures of resistivity surveys using traditional, feature-oriented means of interpretation will point out the limitations of this perspective and the circumstances under which a geographical, statistical perspective may be applied with better results. Electrical resistivity surveying methods have been used with *consistent* success numerous times by archeologists to locate individual *masonry* structures and *hollow* chambers (e.g., walls and tombs), particularly in Europe, but also America (see Table 1). In these cases, the features of interest differ phenomenally in their resistance properties from their matrices. Traditional methods of data interpretation in which outstanding local maxima and minima within the data are isolated are adequate to define such features. However, attempts to relate specific anomalies within resistivity data sets to specific anthropic *soil* disturbances within soil matrices more often than not have met with failure or with results that are unreliable for guiding excavation (Table 1). Exceptions include the attempts that have been made to locate and define recently infilled earthen features such as old archeological test pits and large-scale earthen features such as mound caps (Ford and Kelsin 1969).

Reasons for the failure of soil resistivity surveys are varied, but there are some reoccurring factors. First, the anomalies of interest tend to have low contrast profiles. Soils tend to show great variability in those physical and chemical properties that determine their electrical resistivities. Consequently, the mean of the distribution of resistivity values describing an earthen feature often differs insignificantly from that of its soil matrix compared to the variances of the two resistivity distributions. This circumstance may be considered the *spatial* aspect of the problem of "high noise-to-signal ratios."

A second situation that introduces noise into resistivity measurements and prevents the location of individual features is the masking effect of highly variable layers of soil above the subterranean target. Such layers are included within the volume of soil measured for its resistivity when surveying for subterranean features. The sources of surface variation are multiple. Over natural landscapes, these include spatial variation in: effective precipitation, infiltration, and evaporation; small-scale topographic and drainage regimes; the kind and density of surface vegetation; soil water movements, ion contents, and ion concentrations within root zones; and temperature. On cultivated lands, additional factors may make plowzone the most variable horizon with respect to its physical, chemical, and electrical attributes. Among these are spatially nonuniform plowing, liming, and fertilizing practices. The well-aerated loose structure of freshly plowed soil also facilitates high and variable contact resistances between the soil and the electrode systems used in making soil resistivity measurements (Clark 1963:574). All these surface factors, man-made and natural, may mask deeper variations of archeological interest from the resistivity surveyor. They contribute to a depth-dependent aspect of the problem of high noise-to-signal ratios.

Third, the depth-dependent aspect of high noise-to-signal ratios is enhanced by the fact that the effect of any volume of soil on the resistivity of the total volume of soil measured decreases with the depth of the contributing volume. Thus, in a typical situation where earthen features lay immediately below plowzone and the total volume of soil to be measured includes plowzone plus the anomalous feature and its matrix, it is the plowzone and the zone of natural surface variation that most affects the resistivity value. Surface noise is magnified by the geometry of the resistivity method.

Fourth, in the Eastern Woodlands of North America, at least, many of the features of interest are too small to be easily detectable, within economic limits, by resistivity equipment. Given the low-contrast profiles of many earthen archeological features and the high levels of background noise found in soils, several measurements of the resistivity of an anomaly of interest and its matrix may have to be taken to distinguish the feature. If the average size of pits on a site to be surveyed is, for the sake of argument, 50 cm, the spacing between adjacent stations at which resistivity readings are taken would have to be 25 or 12 cm in order to discriminate those pits. (Even the latter spacing would not result in four measurements of pit fill alone; rather, the measurements would include the matrix as well, within the volume of soil affecting the resistivity reading.) *In a 5 x 5-m square, 400 to 1600 resistivity measurements would be required in order to discriminate a pit.* Clearly, the minimum sampling spacing necessary to distinguish most pits by the resistivity method is uneconomical.

Fifth, features of round shapes, the most common form of pits and many types of houses in the Eastern Woodlands, are not as well distinguished by those electrode arrangements used in most resistivity surveys (four, collinear electrodes) as are features of linear and square shapes (Clark 1963:570). Consider a graph (Figure 1b) of standardized resistivity values against the location of the center of a Wenner electrode array (four, collinear, evenly spaced electrodes) as the array passes over a vertical electrical discontinuity of infinite depth and horizontal expanse (Figure 1a). As the array crosses the discontinuity, the graph will trend toward the mean resistivity of the new medium being entered. In the midst of this trend, however, the graph will "jut" in the opposite direction. These juts are called "subsidiary peaks" and are useful in determing the precise location of the discontinuity.

The clearness of the graph in indicating the location of a discontinuity depends on the magnitude of the difference in the resistivities of the two media on either side of the discontinuity, on the electrode configuration and separation, and on the angle that the array makes with the discontinuity (Chalabi 1969). As the angle that the array makes with the discontinuity deviates from 90 degrees (Figure 1a), the distance between the array and the discontinuity at which the graph starts to rise and fall increases and the magnitude of the diagnostic subsidiary peaks decreases. If a rounded, vertical discontinuity and a linear vertical discontinuity are both approached at the same angle, as described by the line of the array and the tangent to the perimeter, the *average* angle of approach with respect to the whole perimeter will be less for the circular discontinuity than the linear discontinuity. Consequently, circular, vertical discontinuities such as pits will produce less distinctive resistivity graphs than will linear, vertical discontinuities such as trenches, when traversed with resistivity equipment.

Finally, there are numerous agricultural and natural features which cannot be distinguished from archeological features in their effect on electric currents. Such features include infilled depressions from uprooted trees, large roots, large animal burrows, natural stones, localized clay pans, drainage tiles, banded and other uneven distributions of fertilizers and lime.

In conclusion, there are very clear, statistical, physical, chemical, and geometric for expecting negative results for resistivity surveys that are performed on archeological sites having small, nonlinear, earthen features and that have as their goal the isolation of individual features using economic

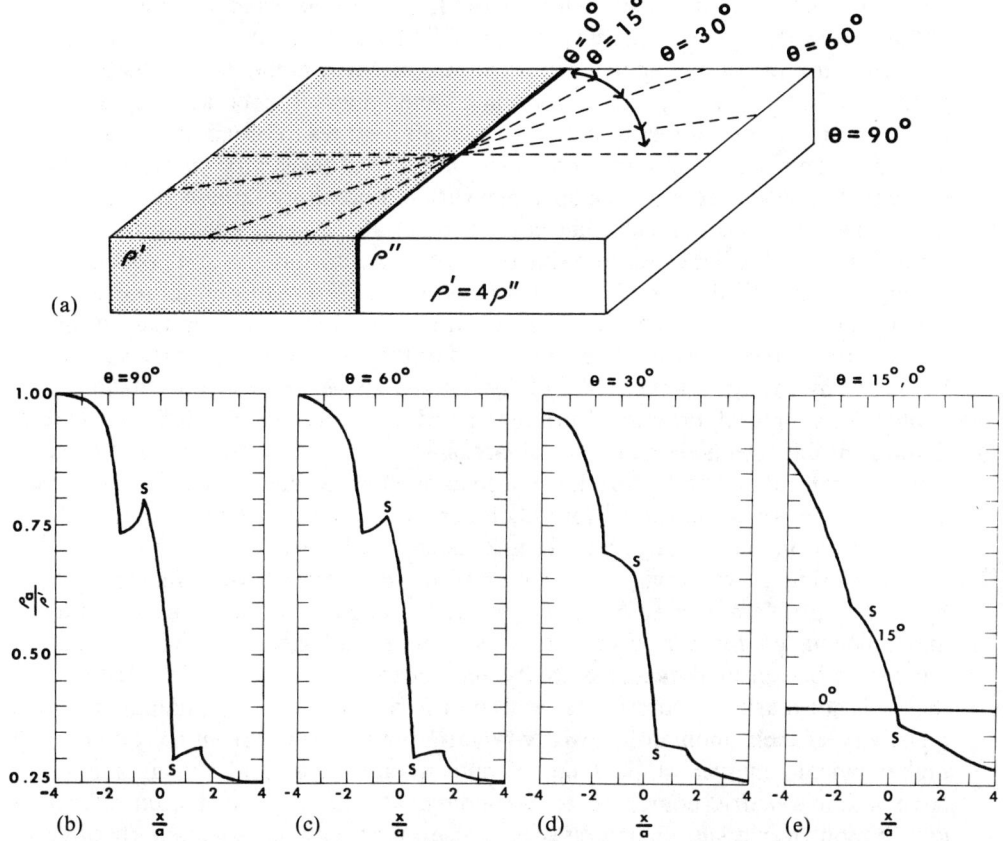

FIGURE 1. *(a) A vertical discontinuity of infinite depth and horizontal expanse crossed at various angles by a resistivity electrode array. (b–d) Theoretical resistivity values as a Wenner electrode array crosses a vertical discontinuity of infinite depth and horizontal expanse. Traverses are made at angles of 90° (b), 60° (c), 30° (d), and 15° and 0° (e), with respect to the discontinuity. $\rho'/\rho'' = 4$. ρ_a = apparent resistivity. x = the distance of the center of the array from the discontinuity. a = the electrode spacing, s = subsidiary peaks. Adapted from Van Nostrand and Cook (1966:120).*

sampling spacings between measurements. It is suggested that under such circumstances, resistivity data generally will be more interpretable and reliable when examined from a geographical, statistical viewpoint and used to differentiate broad use-zones within sites than when used to locate and delimit individual earthen features.

This last statement can be logically supported. The various problems involved in isolating individual earthen features within soil matrices can be summarized by three general factors: (*a*) the high variability in the resistivity of both natural and human-disturbed soils, compared to the slight differences in their mean resistivity; (*b*) the uneconomic distance between adjacent resistivity

18   1. Introduction

readings (large sample size), which would have to be used to resolve the difference in the mean resistivity of small, individual features and their soil matrices; and (c) the indistinguishable nature of some archeological features from some natural and agriculturally related anomalies. By increasing the *scale* of the anomalies of interest from that of smaller features to broad use-areas, all three problems can be circumvented. The number of resistivity measurements that can be made at economical spacings within a broad use-area is larger than the number of measurements that can be made at the same spacing within a small feature. This larger sample size of measurements may make it possible to distinguish statistically the disturbed soils within a use-area from natural soils, when it might not be possible to do so for a smaller feature despite the similar low contrast of both disturbances compared to the natural soil (problems *a* and *b*). Also, by increasing the scale of the anomalies of interest, the chance of confusing a natural or agriculturally caused anomaly for an archeological feature (problem *c*) is decreased. Use-areas are of a larger scale than the most common natural and agriculturally caused anomalies. Mathematical techniques are available for separating overlaid, broad-scale and small-scale resistivity variations from each other, and are introduced in Chapter 2.

Considering the resistivity data from a *statistical* perspective is another way by which problem *c* may be circumvented. Where natural or agricultural anomalies have a *uniform distribution* over an archeological site, they produce a constant background mean resistivity and background variances in resistivity that will not mask out differences between use-areas in their anomalous mean resistivity or their anomalous resistivity variation that result from the different archeological remains within them. Where natural or agricultural anomalies have a *low density,* compared to archeological features, but a *nonuniform* distribution over a site, their presence in use-areas may not significantly alter the estimated anomalous means and variances of resistivity values found in different use-areas; and their differential distribution among use-areas may not significantly alter the difference in anomalous means and variances of resistivity values otherwise occurring between different use-areas.

In summary, resistivity survey data can be used for two different purposes, and interpreted in two different ways, under different surveying conditions.

1. Individual features can be isolated using deterministic traditional methods of data analysis most successfully when the features are large and/or linear, and when they have high contrast profiles. These conditions are most common on archeological sites that have masonry features. On earthen archeological sites, they are less common, but may arise when certain problems are addressed. These include locating areas recently dug and infilled with unconsolidated earth (e.g., locations of previous excavation); following out the directions and limits of linear, ditchlike features that have already been partially revealed through excavation; and defining the areal limits of sites.

2. Broader-scale use-areas can be defined and differentiated, using a statistical geographic approach, in both earthen and masonry sites. This

approach is particularly useful, however, on earthern sites where features have low-contrast profiles, are small, and/or not linear, and where feature-oriented deterministic analyses would give less reliable indications of the spatial organization of the site.

The Complementary Use of Controlled Surface Surveys and Resistivity Surveys

Over the past 15 years, archeologists have become more aware of the fact that in order to evaluate correctly the variety and proportion of activities that occurred on a site and to assess site function, artifacts, features, and spatial relationships among them must be sampled representatively. Toward this goal, controlled surface surveying techniques and geographic sampling designs have been applied (Baker 1978; Binford 1964; Binford et al. 1970; Lewarch and O'Brien 1981; Redman 1970; Roper 1976; Struever 1968; Trubowitz 1981) The spatial distribution of artifacts at the surface of sites has been taken to indicate areas of different prehistoric use, each of which must be sampled by excavation to determine its artifact and feature inventory.

In light of these developments, two questions arise:

1. To what extent are controlled surface surveys capable of accurately defining use-areas within sites?

2. To what extent do they make superfluous those resistivity surveys that are of a statistical, geographic nature and that, like surface surveys, have as their aim the definition of use-areas?

The view taken here is that there are definite problems in, or limitations to, using the controlled surface survey as a means for guiding the archeologists in designing excavations that representatively sample artifacts, features, and their spatial relations within sites. Resistivity surveys approached from a geographic, statistical perspective are seen as a means for overcoming some of these problems and limitations, and consequently, are considered a necessary complement to surface surveys in the development of appropriate sampling designs. Let us consider some of the problems in using controlled surface data and how resistivity data may be used to correct for them.

First, and most important, controlled surface survey data at best indicate the spatial distribution of only artifacts, and need not reflect the locations of different kinds of features (Binford et al. 1970; Redman 1970). For example, earth ovens for baking food may have been swept clean periodically, leaving debris and artifacts aggregated adjacent to the activity area rather than within it. In such circumstances, surface surveying techniques will not be adequate for quiding the archeologists in designing excavation samples that optimally reveal subsurface features. This in itself is not a problem if the artifacts found on the surface of the site are diagnostic of the specific kinds of activities that occurred there, and if one is interested in reconstructing only the range of activities represented in the site and not their spatial patterning. The surface distribution of artifacts in such a case can be used to design excavation samples that will increase the sample size of the artifact collection from the site and allow a better estimate to be made of the relative frequencies of different artifact classes within the site and the relative intensities with which various activities took place on the site. Often, however, artifacts are not interpretable as to their specific function, and data on subsurface features are required to determine the specific activities that occurred on the site. Moreover, the spatial distribution of portable, scatterable artifacts often can not be used to examine community structure and the nature of the social segment that occupied the site as easily as can stationary subsurface features. For example, the size and variation in size of shelter or house remains within a site, their number, and their pattern of spatial clustering or dispersement can be used to estimate the size of the community that occupied the site and something of its social organization (Cook and Heizer 1968; Naroll 1962; Wiessner 1974). The size, density, and number of surface artifact clusters found on a site are more ambiguous clues to community size and organization (Yellen 1974). Thus, in some cases, both subsurface features and artifacts must be adequately sampled by excavation to make a more reliable assessment of the site. In these circumstances, geographic resistivity survey data indicating portions of sites having different kinds and densities of buried features, in addition to controlled surface survey data indicating portions of sites having different kinds and densities of artifacts, are required if excavations are to be designed to yield a representative sample of features, artifacts, and the spatial associations among them.

A second problem with surface survey data is that they may be biased by local geomorphological processes, stratigraphic circumstances, and contemporary farming practices. Where alluvium or colluvium partially cap archeological remains, the extent of surface artifact litter need not represent the true areal extent of prehistoric activity. Excavations designed on the basis of surface data alone and situated only within the area of surface artifact scatter could miss part of the range of variability of artifacts and features occurring at the site and could lead to a biased reconstruction of site function. Sedimentation and erosion, in combination with artifact reuse, also may result in the disproportionate occurrence of large artifacts at the surface of sites (Baker 1978). Activities represented by larger artifacts thus may be better indicated by surface survey data than are activities represented by smaller artifacts. Excavations guided by surface survey data, alone, could systematically overlook certain kinds of use-areas represented by smaller kinds of artifacts. In these cases, geographic resistivity surveys indicating the distribution of use-areas within sites on the basis of undisturbed, buried archeological deposits could be helpful in alerting the archeologist to the biases in his surface survey data and in adjusting excavation sampling schemes based on those data.

Archeological sites under cultivation offer a wide range of problems. Where archeological deposits have been differentially capped with sterile deposits or partially deflated by wind or water and then reworked by plowing, the density or artifacts incorporated within the plow zone and revealed at the site surface may not reflect their density in undisturbed deposits below the plowzone. Also, artifacts may be displaced through plowing (see Roper 1976). Plowing will displace artifacts randomly and enlarge the dimensions of artifact concentrations in all directions if the direction of cultivation varies from year to year, or may elongate artifact concentration if cultivation consistently occurs along one axis. In such cases, where the densities and spatial distributions of surface artifacts do not accurately reflect subsurface artifact densities and locations of concentration, surface survey data can not be used to design excavation strategies that optimally sample buried artifacts, buried features, or associations of these within use-areas. Data from geographic resistivity surveys may be more reliable for assessing the location and organization of use-areas within sites and for guiding archeologists in the designing of excavations.

Finally, if an archeological site has deposits of some depth, surface artifacts will indicate locations of different activities and use only for the upper levels having their artifacts brought to the surface, either naturally by soil fauna, erosion, etc., or by cultivation. Data from a geographic resistivity survey are necessary to assess the nature and spatial organization of use-areas within the more deeply buried archeological horizons and to design excavations that appropriately sample them.

The use of controlled surface survey data to indicate the spatial distribution of use-areas within an archeological site and to guide the designing of optimal excavation strategies that disclose the full range of activities found at individual sites thus has serious problems. Resistivity survey data interpreted from a geographic perspective can be used to detect and overcome some of these problems and to aid in designing more appropriate excavation samples. Resistivity surveys and controlled surface surveys have complementary roles in archeology.

Summary

Resistivity surveying techniques traditionally have been used by archeologists to locate and delimit individual features so that excavation efforts can be focused on them. Resistivity techniques also, however, may be used to locate, characterize, and differentiate use-areas within sites, and to assess intrasite structure. The latter information may be used, along with controlled surface survey data, to guide archeologists in designing excavations that representatively sample the different kinds of artifacts, features, and the spatial associations of these (i.e., use-areas), that are contained in a site. Inasmuch as the use-area is an important unit of archeological analysis and inference, the latter geographic approach to the application of resistivity methods in archeology is relevant. It may also may be the only possible approach to the use of resistivity methods under some difficult but common field conditions in which individual archeological features are not easily detectable. This is particularly true for sites having features that are earthen and buried in a soil matrix.

Use-areas may be defined and characterized within resistivity data sets by interpreting the data within a statistical and geographic framework. Rather than seeking one-to-one correspondences between features and resistivity anomalies and attempting to determine the location, morphology, and functional nature of *each* feature and the absolute densities of features within use-areas in order to define them, use-areas may be defined as *whole* geographic zones by their mean resistivity, the variance of their resistivity values, and their scale. This possibility is based on the assumption that different kinds of human activities may leave residues of different natures and in different amounts, causing different kinds and magnitudes of physical and chemical alterations to natural soil profiles of different use-areas. These various alterations, in turn, presumably are reflected in the different mean resistivity responses of the soils in different use-areas. It also is assumed that different activities may require different sizes or numbers/areal densities of facilities, and may generate residue deposits of different spatial configurations and areal densities. These factors presumably are reflected in the different variances of soil resistivity responses within different use-areas. Finally, it is assumed that different activities may require different amounts of space.

References

Anthropological and Archeological Framework

Ascher, Robert
    1968   Times arrow and the archaeology of a contemporary community. In *Settlement archaeology*, K.C. Chang (ed.), pp. 43–52. National Press Books, Palo Alto, Cal.
Aikens, C. Melvin
    1966   Fremont–promontory–plains relationships. *University of Utah, Anthropological Papers* 82.
Aston, Michael, and Trevor Rowley
    1974   *Landscape archaeology*. David and Charles, London.
Baker, Charles M.
    1978   The size effect: An explanation of variability in surface artifact assemblage content. *American Antiquity* 43(2):288–293.
Berry, Brian J. L., and Alan M. Baker
    1968   Geographic sampling. In *Spatial analysis: A reader in statistical geography*, B. J. L. Berry and D. F. Marble (eds.), pp. 91–100. Prentice-Hall, Englewood Cliffs, N.J.

Binford, Lewis R.
- 1964 A consideration of archaeological research design. *American Antiquity* 29:425–441.
- 1965 Archaeological systematics and the study of cultural process. *American Antiquity* 31(2, Part I):203–210.
- 1968 Archaeological perspectives. In *New perspectives in archaeology*, S. R. Binford and L. R. Binford, (eds.), pp. 5–32. Aldine, Chicago.
- 1974 Forty-seven trips. In *Contributions to anthropology: The interior peoples of northern Alaska. Archaeological Survey of Canada, Paper* 49:299–381.

Binford, Lewis R., Sally R. Binford, Robert Whallon, and Margaret Ann Hardin
- 1970 Archaeology at Hatchery West. *Society for American Archaeology, Memoir* 24.

Bradford, John
- 1957 *Ancient landscapes*. G. Bell and Sons, London.

Braidwood, Robert J.
- 1959 Archaeology and evolutionary theory. In *Evolution and anthropology: A centennial appraisal*, pp. 76–89. Anthropological Society of Washington, Washington, D.C.
- 1960 The agricultural revolution. *Scientific American* 203:130–148.

Brown, James A., and Leslie Freeman
- 1964 A UNIVAC analysis of sherd frequencies from the Carter Ranch Pueblo, Eastern Arizona. *American Antiquity* 31:203–210.

Chang, K. C.
- 1958 Study of the neolithic social grouping: Examples from the New World. *American Anthropologist* 60(2):298–334.
- 1967 Major aspects of the interrelationship of archaeology and ethnology. *Current Anthropology. 8(3):227–243.*
- *1968 Toward a science of prehistoric society*. In *Settlement archaeology*, K. C. Chang (ed.), pp. 1–9. National Press Books, Palo Alto, Cal.

Cochran, William G.
- 1963 *Sampling techniques*. John Wiley, New York.

Cook, S. F., and R. F. Heizer
- 1968 Relationships among houses, settlement areas, and population in aboriginal California. In *Settlement archaeology*, K. C. Chang (ed.), pp. 79–116. National Press Books, Palo Alto, Cal.

Ford, James A.
- 1954 The type concept revisited. *American Anthropologist* 56 (1):42–57.
- 1962 A quantitative method for deriving cultural chronology. Pan American Union Department of Social Affairs, *Technical Manual* 1. Washington, D. C.

Flannery, Kent V.
- 1972 The cultural evolution of civilizations. *Annual Review of Ecology and Systematics* 3:399–425.

Ford, Richard I.
- 1974 Northeastern archaeology: Past and future directions. *Annual Review of Anthropology* 3:385–413.

Freeman, Leslie, and Karl W. Butzer
- 1966 The Acheulean station of Torralba (Spain): A progress report. *Quaternia* 8:9–21.

Griffin, James B.
- 1943 *The Fort Ancient aspect: Its cultural and chronological position in Mississippi Valley archaeology*. The University of Michigan Press, Ann Arbor.

Hall, Robert L.
- 1976 Ghosts, water barriers, corn and sacred enclosures in the Eastern Woodlands. *American Antiquity* 41(3):360–363.
- 1977 An anthropocentric perspective for eastern United States prehistory. *American Antiquity* 42(4):499–517.

Hill, James N.

1970 Broken K. pueblo: Prehistorical social organization in the American Southwest. *University of Arizona, Anthropological Paper* 18.

Howard, Alan
1963 Land, activity systems, and decision making models in Rotuma. *Ethnology* 2:407–440.

Hurley, W. M., and C. C. E. Heidenreich
1971 *Paleoecology and Ontario prehistory*. University of Toronto Department of Anthropology, *Research Report* 2.

Jochim, Michael A.
1976 *Hunter–gatherer subsistence and settlement: A predictive model*. Academic Press, New York.

Keene, Arthur S.
1979 Prehistoric hunter-gatherers of the deciduous forest: A linear programming approach to Late Archaic subsistence in the Saginaw valley (Michigan). Unpublished Ph.D dissertation, Department of Anthropology, University of Michigan.

Kidder, Alfred V.
1924 *An introduction to the study of Southwestern archaeology*. Papers of the Southwestern Expedition, Phillips Academy, 1. Yale University Press, New Haven.

Kroeber, A. L.
1937 Archaeology. In *Encyclopedia of the social sciences*, E. R. A. Seligman and A. Johnson (eds.), pp. 163–166. MacMillan, New York.

Lewarch, Dennis E. and Michael J. O'Brien
1981 Effects of short term tillage on aggregate provenience surface pattern. In *Plowzone archaeology: contributions to theory and technique*, M. J. O'Brien and D. E. Lewarch (eds.). *Vanderbuilt University, Papers in Anthropology*. In press.

Limp, Fredrick
1978 Optimization theory and subistence change: Implications for prehistoric settlement location analysis. Unpublished paper presented at the 43rd Annual Meetings of the Society for American Archaeology, Tucson.

Longacre, William
1968 Some aspects of prehistoric society in east-central Arizona. In *New perspectives in archaeology*, Sally R. Binford and Lewis R. Binford (eds.), pp. 89–102. Aldine, Chicago.

Lyons, Thomas R., and Robert K. Hitchcock (eds.)
1977 Aerial remote sensing techniques in archaeology. *Chaco Center Reports* 2. National Park Service, Albuquerque, N.M.

MacNeish, Richard S.
1964 The origins of New World civilization. *Scientific American*. 211(5):29–37.

McCarthy, Allen P.
1980 Archaeological whale bone: A northern resource. *University of Arkansas, Anthropological Papers* 1.

Naroll, R.
1962 Floor area and settlement population. *American Antiquity* 27(4):587–589.

Orlove, Benjamin S.
1980 Ecological anthropology. *Annual Review of Anthropology* 9:235–273.

Plog, Stephen
1976 Relative efficiencies of sampling techniques for archaeological surveys. In *The early Mesoamerican village*, K. V. Flannery (ed.), pp. 135–158. Academic Press, New York.

Provan, Donald M. J.
1971 Soil phosphate analysis as a tool in archaeology. *Norwegian Archaeological Review* 4:27–50.

Rappaport, Roy
1971 Nature, culture, and ecological anthropology. In *Man, culture, and society*, H. Shapiro (ed.), pp. 237–267. Oxford University Press, New York.

Read, Dwight
 1975 Regional sampling. In *Sampling in archaeology,* J. W. Mueller (ed.), pp. 45–60. University of Arizona Press, Tucson.
Reidhead, Van
 1979 Linear programming models in archaeology. *Annual Review of Anthropology* 8:543–578.
Redman, Charles L., and Patty Jo Watson
 1970 Systematic, intensive surface collection. *American Antiquity* 35(3):279–291.
Roper, Donna C.
 1976 Lateral displacement of artifacts due to plowing. *American Antiquity* 41(3):372–375.
Rouse, Irving
 1965 The place of people in prehistoric research. *Journal of the Royal Anthropological Institute* 95:1–15.
Shiffer, Michael B.
 1972 Archaeological context and systematic context. *American Antiquity* 37(2):156–165.
 1976 *Behavioral archaeology.* Academic Press, New York.
Simpson, George Gaylord
 1962 Comments on cultural evolution. In *Evolution and man's progress*, H. H. and R. W. Burhoe (eds.), Columbia University Press, New York.
Struever, Stuart
 1968 Woodland subsistence–settlement systems in the lower Illinois Valley. In *New perspectives in archaeology,* S. R. Binford and L. R. Binford (eds.), pp. 285–312. Aldine, Chicago.
Taylor, Walter W.
 1948 A study of archaeology. *American Anthropological Association, Memoir* 69.
Trubowitz, Neil
 1981 Settlement pattern survival on plowed northeastern sites. Paper presented at the Annual Meetings of the Society for American Archaeology, San Diego.
Vaillant, G. C.
 1930 Excavations at Zacatenco. American Museum of Natural History, *Anthropological Papers* 32(1).
Vayda, Andrew P., and Roy A. Rappaport
 1967 Ecology, cultural and noncultural. In *Introduction to cultural anthropology,* J. Clifton (ed.), pp. 477–497. Houghton-Mifflin, Boston.
Wiessner, Polly
 1974 A functional estimator of population from floor area. *American Antiquity* 39(2):343–349.
Willey, Gordon R., and Phillip Phillips
 1958 *Method and theory in American archaeology.* University of Chicago Press, Chicago.
Willey, Gordon R. and Jeremy A. Sabloff
 1980 *A history of American archaeology.* W. H. Freeman, San Francisco.
Winters, Howard D.
 1969 The Riverton culture. *Illinois State Museum, Reports of Investigation* 13.
Yellen, John E.
 1974 The !Kung settlement pattern: An archaeological perspective. Unpublished Ph. D. dissertation, Department of Anthropology, Harvard University.

Studies in Resistivity Surveying

Aitken, M. J.
 1961 *Physics and archaeology.* Willey(Interscience), New York.

1. Introduction

    1962   Physics applied to archaeology, Part II. *Contemporary Physics* 3(5):333–351.
Annable, F. K.
    1958   Excavation and field-work in Wiltshire. *Wiltshire Archaeological Natual History Magazine* 207:233.
Atkinson, R. J. C.
    1946   *Field archaeology*. Methuen, London.
    1952   Methodes electrique de prospection en archeologie. In *La decouverte du passé*, A. Laming (ed.), pp. 59–70. Picard, Paris.
    1963   Resistivity surveying in archaeology. In *The scientist and archaeology*, E. Pyddoke (ed.), pp. 1–30. Phoenix House, London.
Carr, Christopher
    1975   Resistivity surveying on earthen archaeological sites. Unpublished *Final Technical Report* to the National Science Foundation for Grant Project EPP75-08645, on file in The University of Michigan Museum of Anthropology, Ann Arbor.
    1976   Soil resistivity surveying and regional archaeological planning, lower Illinois Valley. In *Abstract Reports of Student Originated Studies Projects*, 1975. National *Science Foundation*, Washington, D. C.
    1977   A new role and analytical design for the use of resistivity surveying in archaeology. *Mid-Continental Journal of Archaeology* 2(2):161–193.
Chalabi, Mahboub al-
    1965   Applications of geo-electrical methods in archaeology. *Sumer* 21(1, 2).
    1969   Theoretical resistivity anomalies across a single vertical discontinuity. *Geophysical Prospecting* 17(1):63–81.
Chalabi, M. M. al- and A. I. Rees
    1962   An experiment on the effect of rainfall on electrical resistivity anomalies in and near surface. *Bonner Jährbucher* 162:266–271.
Clark, Anthony
    1963   Resistivity surveying. In *Science in archaeology*, D. Browell and E. Higgs (eds.), pp. 569–581. Basic Books, New York.
Dabrowski, Krzysztof
    1963   Geophysical methods to archaeological research in Poland. *Archaeometry* 6:83–88.
Dabrowski, K., and W. Stopinski
    1962   The application of the electric resistivity method to archaeological investigations illustrated on the example of an early mediaeval hillfort in Kaliez. *Archaeologia Polona* 5:21–30.
Dunk, A. J.
    1962   An electrical resistance survey over a Romano–British site. *Bonner Jährbucher* 162:272–276.
Ford, Richard I.
    1964   A preliminary report of the 1964 resistivity survey at the Schulz Site 20 SA 2. *Michigan Archaeologist* 10(3):54–58.
Ford, Richard I., and Richard O. Kelsin
    1969   A resistivity survey at the Norton mound group, 20 KT 1, Kent County, Michigan. *Michigan Archaeologist* 15(3):86–92.
Gansfuss, John E.
    1975   The application of geophysical techniques to archaeological exploration. In *National Science Foundation, Student Originated Studies Projects, 1975 Abstract Reports*.
Hesse, A.
    1962   Geophysical prospecting for archaeology in France. *Archaeometry* 5:123–125.
    1966   *Prospections, géophysiques à faible profondeur, applications à l'archaeologie*. Dumed, Paris.
Lerci, C. M.
    1959   Periscope on the Etruscan past. *National Geographic Magazine* 32.

Lerci, C. M., R. Bartoccini, and M. Moretti
  1959 Necropoli di Tarquina. *Fondazione Ing. C. M. Lerci, Publication* 15. Milan.
Leith, C. J., K. A. Schneider, and C. Carr
  1977 Geophysical investigations of archaeological sites. *International Association of Engineering Geology, Bulletin* 14:123-128.
Linington, R. E.
  1963 The application of geophysics to archaeology. *American Scientist* 51:48-70.
  1966 Etude de la résistivité des sols. Appendix to Les Camps Neolithique des Matignon à Juilla-le-Coq, by C. Burnez and H. Case. *Gallia Prehistòria* 9:198-200.
  1967 An electrical resistivity survey at Les Matignons. *Prospezioni Archeologiche* 2:91-93.
  1968 The search for the supposed obelisk under Via Giustiniana, Rome. *Prospezioni Archeologiche* 3:77-81.
Martinaud, M., and G. Colmont
  1971 Interêt de l'étude des sols par mesure de resistivite et carottages mecanique. *Prospezioni Archaeologiche* 6:53-60.
Palmer, L. S.
  1959 Examples of geoelectric surveys. *Institution of Electrical Engineers*, Proceedings 106A: 231-239.
  1960 Geoelectrical surveying of archaeological sites. *Prehistoric Society, Proceedings* 26:64-75.
Rainey, Froelich, and Elizabeth K. Ralph
  1966 Archaeology and its new technology. *Science* 153(3743):1481-1491.
Rees, A. I., and A. E. Wright
  1969 Resistivity surveys at Barnsley Park. *Prospezioni Archaeologiche* 5:121-124.
Schwartz, G. Theodor
  1961 The "Zirkelsone": A new technique for resistivity surveying. *Archaeometry* 4:67-70.
Sieveking, G., I. H. Longworth, M. J. Hughes, A. J. Clark, and A. Millett.
  1973 A new survey of Grime's grave—First Report. *Prehistoric Society, Proceedings* 39:182-218.
Thompson, R.
  1978 Resistivity investigation of an infilled kettle hole. *Quarternary Research* 9(2):231-237.
Tite, M. S.
  1972 *Methods of physical examination in archaeology.* Seminar Press, London.
Tite, M. S., and C. Mullins
  1970 Electromagnetic prospecting on archaeological sites using a soil conductivity meter. *Archaeometry* 12:97-104.
Van Nostrand, Robert G., and Kenneth L. Cook
  1966 Interpretation resistivity data. *United States Geological Survey, Professional Paper* 499.

2
Mathematical Methods for Preparing Resistivity Data for Interpretation

Resistivity survey data from archeological sites can be used in two ways. Individual features may be isolated or broad scale use-areas may be defined, depending on the nature of the features and the goals of the archeologist. The mathematical methods by which resistivity data should be prepared for interpretation vary in accord with which one of these two approaches is used.

When archeological features contrast well with their matrices and are large compared to economical spacings between resistivity measurements and when it is the goal to isolate them individually, standard techniques may be used that repeatedly have been shown to be successful. The references given in Table 1 document these methods in detail, and only a brief overview of them will be given here. Under optimal circumstances of feature definition, soil resistivity soundings can be made over the area to be surveyed using a single depth of investigation that encompasses the buried level of archeological features plus the soil above them. Surface resistivity variations will not mask the high variability of the archeological horizon, so that the resistivity response of the archeological horizon alone can be estimated by the resistivity response of all soil from the surface through the archeological horizon. Soil resistivity values can be plotted as graphs against distance for transect surveys or as contour maps for areal surveys. Outstanding local maxima or minima representing the features of interest can be defined directly by visual means. In many studies, previous knowledge of the approximate location, size, shape, and orientation of the features has been important in this definition process. Investigations where the buried portions of partially excavated ditches and moats have been followed out with resistivity equipment are cases in point (Atkinson 1963:20–27; Dabrowski and Stopinski 1962:84; Linington 1967, 1970; Palmer 1960:71–75; Rees and Wright 1969:121).

In somewhat less optimal conditions, where background noise competes with resistivity variations produced by the features of interest, additional techniques have been used to isolate features. Dot density maps, in which areas around observation points are differentially shared with randomly placed dots in proportion to resistance values, may be used instead of contour mapping techniques (Scollar 1969; Scollar and Krückeberg 1966; Wilcock 1970). This method of representing resistivity data utilizes the innate, pattern-finding abilities of the human eye to filter random background noise from meaningful patterns of high and low resistivity caused by the features of interest. Less subjective methods for defining significant resistivity anomalies include trend surface analysis (Leith *et al.* 1977) and spatial filtering (Linington 1970).

When archeological features contrast little with their matrices, as in the case of earthen features in soil matrices, and when they are small and nonlinear, interpretive frameworks and mathematical methods of preparing resistivity data for interpretation different from those used traditionally are more appropriate (Carr 1972, 1976, 1977). First, consider the general interpretive framework.

In Chapter 1, it was suggested that when interpreting resistivity data from earthen sites, a geographical and statistical approach should be taken, focusing on use-zones within archeological sites and differentiating them by their mean soil resistivity values, the magnitudes of their variances, and their scales. In such an approach, soil resistivity values across a landscape or over depth can be envisioned as a single-dimensional summary series, or palimpsest, reflecting multiple soil properties. Variation in some of the soil properties, such as texture, might reflect largely natural phenomena; variation in others such as organic matter content might be caused largely by human activity; and variation in still others such as the availability of certain ions might reflect both natural and anthropic phenomena. Depending on the spatial patterns and magnitudes of variation of the several soil properties, they will either augment, partially obscure, or completely negate each other's effect on the conductivity of the soil; the anthropic, natural, or agricultural phenomena may or may not be reflected and distinguished in the summary resistivity series.

Two summation processes are possible. In the first case, a single phenomenon may produce two different kinds of soil alterations that counteract or augment each other in their effect on the resistivity of the soil. Because the alterations are produced by a single, underlying variable, they will be synchronous in their spatial variation. The extent to which they augment or cancel each other's effects on soil resistivity will be constant over space; the two effects will be indistinguishable in the summary resistivity series. For example, anthropic enrichments of soil organic matter content will both increase the number of organic colloidal particles that have water-attracting properties and increase soil granulation and drainage conditions. The first condition will increase a soil's water-holding capacity and potential for conducting electricity. The second will decrease a soil's water-holding capacity and potential for conducting electricity. The net effect of the added organic matter may be null, an increase in soil conductivity, or a decrease in soil conductivity, depending on the relative degree to which the number of micelles and degree of granulation have been augmented to the soil. The relative importance of these two factors in their effect on soil resistivity will be constant over space, however, and they will be indistinguishable from each other in the summary resistivity series.

The second kind of summation process occurs when two different kinds of soil alterations are produced by two *different*, independent phenomena. Again, soil alterations might interfere "destructively" or "constructively" with each other in their effects on soil resistivity. However, because the alterations are caused by independent phenomena, they may not vary in synchrony over space and the degree to which they interfere destructively or constructively with each other in their effect on soil resistivity may vary over space. Should this be the case, the two kinds of soil alterations will be distinguishable in the summary resistivity series. On the other hand, the two soil alterations might happen to vary in sychrony, in which case they will not be distinguishable in the summary resistivity series, just as two alterations produced by a single underlying variable would not.

An example of a situation where the two alterations would be distinguishable is as follows. Over an archeological site, large-scale (low spatial frequency) variation in the natural clay content of the soils and smaller-scale (high spatial frequency) variations in the thickness of midden deposits might both occur. A decrease in clay content would cause a decrease in the cation exchange capacity and moisture holding capacity of the soils and alone would result in higher resistivity values. Increases in midden thickness, alone, would result in lower soil resistivity values; midden deposits often have greater moisture-holding capacities and higher cation exchange capacities compared to natural soils, as a result of their higher organic matter contents. The combined variation of soil clay content and midden depth over the site might cause constructive and destructive patterns of interference over space in their effects upon the resistivity of the soil, but both variable components would be distinguishable in the summary resistivity series. The natural variation in soil clay content would be manifested by large-scale, gentle trends within the resistivity series, whereas variations in midden thickness would be indicated by smaller-scale, local variation in resistivity superimposed on the broader-scale trends.

Thus, a series of soil resistivity values over a landscape can be envisioned as the summation of the effects of multiple, physical and chemical soil variables, that in turn reflect different prehistoric human, natural, or agricultural phenomena. This summation process can be described more precisely with the aid of time-series analysis—a method that is derived from applied mathematics and that makes it possible to dissect a summary resistivity series into discrete sets of information about its causative variables. Any linear series of data points can be envisioned as the sum of three components (Figure 2): (*a*) a trend, that can be obtained by fitting a curve to the data points using least-squares methods or smoothing methods; (*b*) a periodic component, which can be isolated through

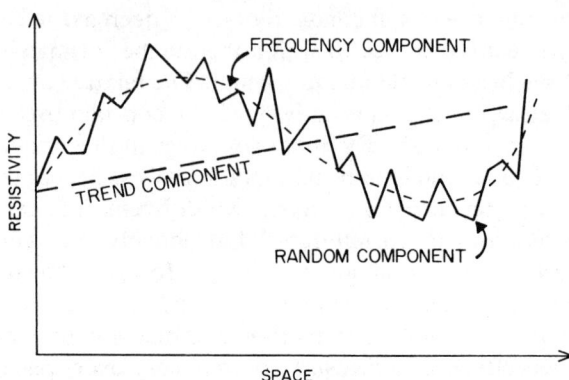

FIGURE 2. Three components of variability within a hypothetical soil resistivity data series. Adapted from Rich (1973:216).

spectral analysis and Fourier series analysis; and (c) a random component, which can be described by Markovian matrices of probability (Davis 1973; Rich 1973). The periodic component, itself, may be considered the sum of multiple waves having different frequencies, amplitudes, and phase angles.

Resistivity data series may contain all three kinds of information: trends, periodicities, and random components. Natural soil variation, such as decreasing particle size with increasing distance from a stream, may be apparent in a major trend of decreasing resistivity away from the stream. Band fertilization and plowing may manifest themselves in the periodic components of the data series. Contact resistance between the electrodes and the soil (which raises resistivity values above normal), near-surface sources of noise, and measurement errors will appear as random and semirandom fluctuations in the data. If resistivity series are seen in this interpretive framework, they may be subjected to the descriptive and analytical techniques of time-series analysis. Archeologically significant soil variability of one scale and form can be isolated from other sources of soil variation of other scales and forms, to allow the definition of use-areas.

Importantly, however, information on these different kinds of soil variations may be distinguishable not only in their *spatial* patterning, but also over *depth*. The effects of agricultural and near-surface natural phenomena will be most apparent in the upper soil horizons, or plowzone, if present. Archeologically significant soil variation may be isolated within lower levels encompassing midden deposits and features or natual soil horizons that have been chemically enriched by organic residues. Natural soil variations related to soil parent material variations and to the degree of profile development may be mapped in still lower horizons. The separation, however, will seldom be perfect. For example, the effects of fertilizer leachates and near-surface natural phenomena such as differential infiltration of rainfall may extend into the levels of archeological and natural character. Nevertheless, the structuring of different kinds of information over depth as well as over space should be recognized.

This circumstance can be used to an advantage in isolating archeologically significant components of variation within soil resistivity data sets from natural or agriculturally caused components of variation. Multiple resistivity soundings may be made at each location within the area to be surveyed, successively encompassing deeper and deeper horizons—first the agricultural horizon alone, then the agricultural and archeological, and finally the agricultural, archeological, and natural horizons. The multilevel resistivity data may be used to calculate the resistivity of the *individual* horizons and to isolate the resistivity response of the archeological horizon from the masking responses of the horizons above it. This method of estimating the resistivities of buried layers is called the *Barnes Layer method*. It contrasts with what will be termed here the *Whole Volume method* of layer estimation, where only one resistivity sounding from the surface through the archeological horizon of interest is made at each location within the area to be surveyed, and where the resistivity response of the archeological horizon is estimated by the response of *all soil* from the surface through the archeological horizon, and includes surficial noise. The Whole Volume method is the approach that traditionally has been used in estimating resistivity variation within archeological horizons. It has worked reasonably well on sites where features contrast greatly from their matrices in their resistivity. On earthen sites, however, use of the Barnes Layer method rather than the Whole Volume method is recommended.

The Barnes Layer Method of Resistivity Data Interpretation

Having introduced the general perspectives with which resistivity data may be viewed in order to define use-areas on archeological sites, we now can examine the particular analytical techniques allowing the isolation of archeologically significant variation within resistivity data sets. Two techniques will be introduced, one concerned with variation in soil resistivity over space; the other concerned with variation in soil resistivity over depth. The aim of the techniques is to partition the mean and variance of resistivity palimpsets along *both* dimensions in order to maximize the quantity and clarity of information obtained for each class of variables affecting the composite soil resistivity measurements.

Division of resistivity data sets into components of variability and information should first be done along the dimension of depth in order to remove or reduce the effect of surface and near-surface noise and natural, subsoil variation on the resistivity data, and to isolate the response of those layers in which archeological deposits occur. This may be achieved with the Barnes Layer method of resistivity estimation, which was invented by H. E. Barnes (1952,

1954, n.d.) and extensively tested and confirmed by the Michigan State Highway Department (Malott 1964, 1965, 1967, 1968).

The theory of the Barnes Layer method is based on Wenner's (1915) equation for determining the average apparent resistivity of a homogenous medium. Using a collinear array of four, equidistant electrodes, the outer two supplying current and the inner two measuring the potential drop across the medium between them, the apparent resistivity of the medium is defined by

$$\rho_a = \frac{4\pi AR}{1 + [2A/(A^2 + 4B^2)^{1/2}] - [A/(A^2 + B^2)^{1/2}]} \quad (2.1)$$

where ρ_a is the average, apparent resistivity in ohm-centimeters for a volume of the medium extending to a depth of approximately "A centimeters," where A is the equidistant electrode spacing in centimeters, where B is the penetration distance of the electrodes into the medium, and where R is the observed apparent resistance in ohms. The volume of the medium that affects the resistivity measurement theoretically has infinite lateral and vertical dimensions, lying between two hemispherical, equipotential bowls, as shown in Figure 3a. Operationally, however, the volume is smaller, that portion of the medium having the most effect on the resistivity measurement extending only to a depth of approximately A centimeters, when the depth of penetration of the electrodes is very small compared to the electrode spacing (Figure 3b). When the depth of penetration of the electrodes is significant compared to the electrode spacing, the approximate depth of the effectively measured volume will be $A + B$.

Assuming a layered model with no lateral changes within strata, the hypothesis then is made that the apparent resistances of layers behave in a manner analogous to parallel electrical resistors. A number of resistance measurements with successively greater depths of investigation may be made at one location by expanding the electrode array around a single point. If the electrode separations are $A_1, A_2, A_3, \ldots, A_n$ and the approximate depth of the measured volumes of the medium are $A_1, A_2, A_3, \ldots, A_n$, then the measured apparent resistance, $R_1, R_2, R_3, \ldots, R_n$ will decrease such that:

$$1/R_L = (1/R_N) - (1/R_{N-1}) \quad (2.2)$$

where R_N is the apparent resistance, in ohms, of a volume extending between the surface and a depth A_N; where R_{N-1} is the apparent resistance, in ohms, of a volume extending between the surface and a depth A_{N-1}; and where R_L is the apparent resistance, in ohms, of a layer of the medium, between depths A_{N-1} and A_N. The value of R_L, the layer thickness, and the average of the electrode separations and the average of the electrode penetrations that were used in generating the two measured volumes, may be used in Wenner's equation in order to calculate apparent layer resistivity values. The layer thickness ($A_N -$

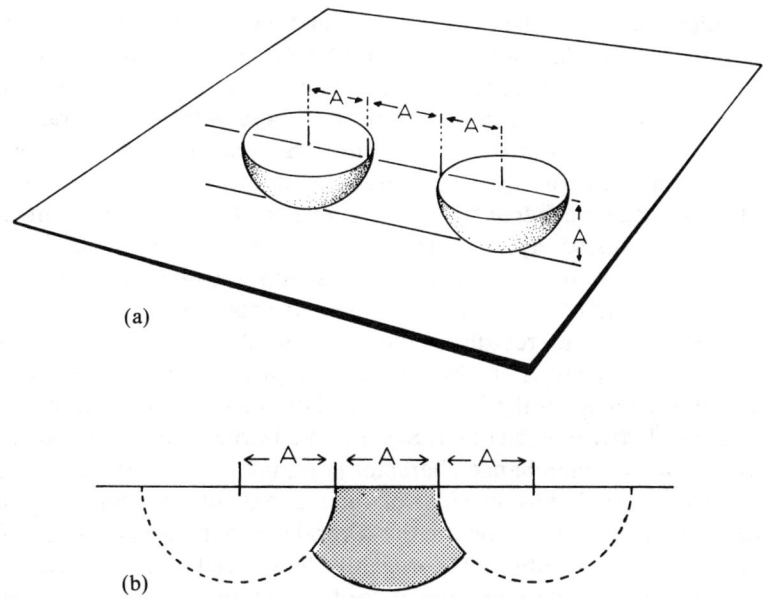

FIGURE 3. *(a) Equipotential bowls formed when measuring the resistivity of a homogeneous medium with a Wenner electrode configuration. (b) That volume of a conducting medium having the most effect upon the resistivity measurement (stippled).*

A_{N-1}) is used as A in the denominator of the equation; the average electrode separations and average electrode penetrations are used as A and B in the denominator of the equation. The "layer" of influence that is generated by such a mathematical operation is shown in Figure 4.

Thus, by using the Barnes Layer method, it is possible to find the resistivity of soil horizons that are largely agricultural, archeological, or natural in their soil

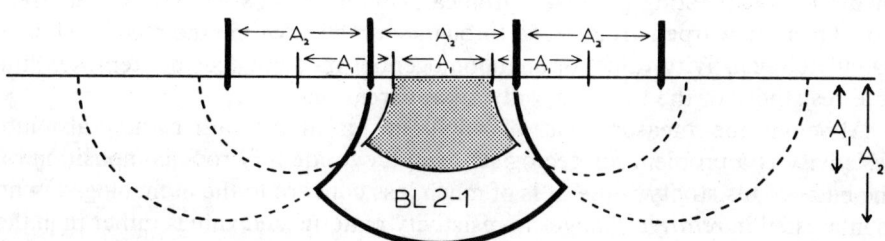

FIGURE 4. *That layer of a conducting medium whose resistance may be calculated with the Barnes Layer method from two resistance measurements. The measurements are made with Wenner electrode arrays which have different spacings but which are both centered around one point.*

variation. Optimal use of the Barnes Layer method for this purpose, however, requires some knowledge of the stratigraphy of the archeological site to be surveyed. Electrode separations must be chosen so as to allow the generation of Barnes Layers that coincide in depth and thickness with agricultural, archeological, and natural soil horizons of the site. This knowledge may be obtained quickly with a limited soil coring survey of the site.

It should be made clear that the Barnes Layer technique is only a method of *interpretation* of resistivity, that is, it allows the calculation of "apparent resistivity" values rather than absolute resistivity values. There are three reasons for this. First, the analogy between a profile of rock strata or soil horizons and a group of resistors hooked in parallel is not perfect. Parallel resistors are not contiguous and do not allow the flow of current between each other, whereas depositional strata and soil horizons are continuous and do conduct current between each other. Second, the Barnes Layer method assumes that (*a*) the strata do not change character horizontally, and (*b*) the resistance-affecting materials added to the sides of a measured volume when an electrode array is expanded (Figure 4) contain the several strata in the same proportion in which they occurred within the originally measured volume. These later conditions are assumed to be true in order that the change in resistances between one measurement and a measurement of deeper investigation be attributable to only the material added *below* the orginally measured volume, and not that added to its sides. In practice, however, neither of these assumptions are true. The process of subtracting the effect of overlying deposits [Eq. (2.2)] therefore, is only approximate. Barnes Layer volumes, consequently do reflect, to some extent, the resistivities of the strata above that being investigated. Finally, the Wenner equation, itself, is derived with the assumption that the conducting medium is totally homogeneous, and the current flow lines follow a predictable, ellipsoid pattern. This assumption allows the calculation of the expected potential drop (and, therefore, resistance, by Ohm's law) between the two inner (potential) electrodes, given a specified amperage between the current (outer) electrodes and a specific resistivity of the conducting medium. The application of the Wenner equation to inhomogenous media for prospecting purposes (ironically) violates this assumption: Current flow lines are warped around and through anomalies within the media. Thus, a calculated resistivity value for an inhomogenous medium does not represent the true resistivity of the medium, only an apparent one.

Although the measurement of apparent resistivity rather than absolute resistivity is a problem for geologists, who try to identify rock composition on the basis of resistivity value, it is of much less concern to the archeologist, who is interested in *relative* changes in resistivity patterns and values rather than the absolute resistivity of the medium he is investigating.

Time-Series Analysis and Running Filter Functions

Once a resistivity data set has been partitioned into stratigraphically semi-independent data series that reflect primarily agricultural, archeological, or natural pedological variation, each series can be analyzed further using the methods of time-series analysis. A standard time-series approach (e.g., Rich 1973) involves:

1. Isolation of broad-scale trends with polynomial least squares methods
2. Removal of such trends from the data series by subtracting from each observation the value that would be expected there based on the polynomial regression alone
3. Performance of Fourier analyses and spectral analyses (Jenkins 1961; Jenkins and Watts 1968) on the residual series to determine those frequencies that account for the greatest variability of the data series
4. Removal of such periodicities by subtracting from each residual observation the value which would be expected, there, if only the frequencies accounting for the greatest variance of the data series were present within it
5. Examination of the secondary residuals for significant attributes.

With respect to soil resistivity data sets from archeological sites, it might be thought on an a priori basis that these steps could be taken in a mechanical way to isolate archeologically significant soil variability. Steps 1 and 2 might be used to identify and remove from the data gentle spatial variations in natural soil properties such as parent material or topographically related variations in soil profile development. Steps 3 and 4 might be used to identify and remove from resistivity series periodic components that are a result of agricultural practices such as band fertilization or plowing. The residual data series left after the removal of trends and periodicities from it presumably would reflect archeologically significant soil variability along with small-scale natural soil anomalies.

Such a standard time-series approach is not sufficient, however, for the analysis of resistivity data sets that have been collected to define use-areas on archeological sites; other approaches are more appropriate. There are at least two reasons. First, the patterns of data variation within resistivity series obtained from earthen archeological sites can be quite complex with respect to broad-scale, natural or nonprehistoric human-caused variations. Spatial changes in parent material, topography and drainage, land use (present or historic), and surface vegetation may combine to produce broad-scale soil resistivity variations that are so intricate that they can be completely and accurately described by polynomial equations only in a cumbersome manner, using a large number of terms. Least squares, polynomial methods are more appropriate for generating *generalized, predictive* models which account for a *high* percentage of the broad-scale variations that can be expected to occur in a phenomenon that *repeats itself* in a number of different data sets than they are appropriate for generating *data-specific, descriptive* models which account for 100% of the broad-scale variations that occur in a *unique* data set. Other methods are more appropriate for isolating and removing broad-scale variations completely from data series.

Second, when viewing resistivity variation over an archeological site as the summation of the effects of multiple phenomena having different spatial scales, it is desirable to dissect such variation into its differently caused components using techniques that allow one control over the *spatial scale* of variability being removed from the data. Use of least-squares methods to describe broad-scale trends within a data series for the purpose of later removing them from the series does not explicitly allow one this control. The technique considers the *percentage* of variation accounted for by the polynomial model rather than the *scale* of variation that the model encompasses.

In place of the use of the standard time-series analytic design to investigate resistivity data series, I would suggest the use of running filter functions (Holloway 1958), which are designed to remove variation of *specific spatial scales* as completely as is possible. In this approach, a resistivity data series still is envisioned as being composed of gentle, nonperiodic trends, periodic components, and semirandom and random components, but each component is interpreted within the *Fourier (wave)* domain rather than the *spatial* domain. The entire data series is envisioned as the sum of multiple waves having different wavelengths, amplitudes, and phase angles. Naturally caused broad-scale variations are envisioned as the sum of low-frequency waves, or portions of low-frequency waves running through the data. Archeologically significant variations are envisioned as the sum of intermediate-frequency waves. Variations resulting from measurement errors; from small-scale natural pedological, topographic, or vegetational anomalies; or from periodic agricultural phenomena are envisioned as high-frequency and ultra-high-frequency waves. Running filter functions sensitive to particular scales of variability then can be designed to isolate for study such scale-specific, phenomenon-specific variations within a resistivity series. Running filter functions also may be used to analyze the isolated frequency-specific components of variation. Frequency-specific resistivity series may be transformed with running filter functions so as to emphasize areas of high or low local mean resistivity or local variance in resistivity. When applied to data series containing variations largely of the scale of use-areas, they may be used to locate and to characterize use-areas.

The general stepwise procedures by which running filter functions can be applied to a resistivity data series to dissect it into its components of variation and to analyze the archeological component are described in the following paragraphs. Specific details and illustrations of the procedures are given in Chapter 9.

As a first step in the analysis of a resistivity data series, those observations that are too high or low to represent information on archeological, natural, or agriculturally caused soil variation and that most probably represent noise due to erratic probe contact resistance or instrument-reading errors should be replaced by the average of their adjacent observations. The major problem at this point is to define limiting values for "too high" and "too low," which are stringent yet do not call for the removal of meaningful observations. One might define a single pair of upper and lower thresholds with respect to the whole data series, such as three standard deviations above and below the series mean. Such an operation, however, is not recommended. If low frequency trends run through the series, this approach will inconsistently remove or pass by noise, and in some locations may remove meaningful observations, depending on the *local* mean (Figure 5). This situation can be corrected somewhat if the 3σ thresholds are defined in reference to local means. Running operators that find the mean and standard deviation of n-number of adjacent points for all sets of n-adjacent points along the series and that define locally specific thresholds in relation to these means and standard deviations can be used for this purpose.

A problem, however, still remains. It will be recalled from Chapter 1 that when a Wenner electrode array crosses a vertical discontinuity separating two media of different resistivities or when it crosses a local anomaly, subsidiary peaks are produced. These peaks are diagnostic of inhomogeneities and should not be removed from data series. Such peaks, however, can be much higher or lower than the mean resistivity of either the anomaly of interest or its matrix

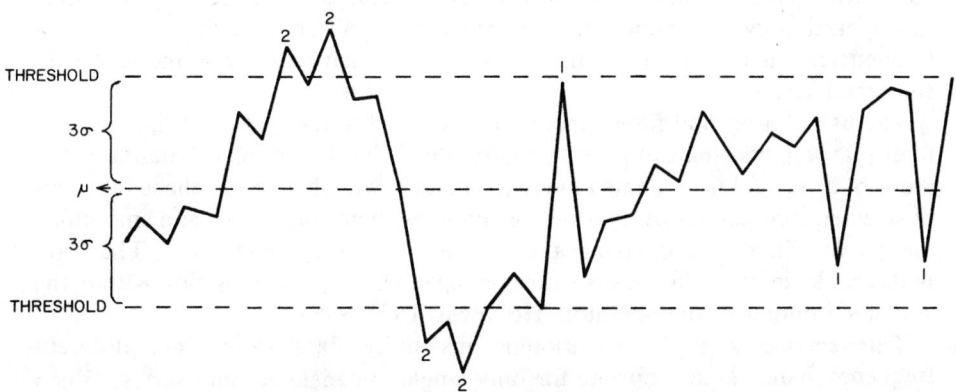

FIGURE 5. *Noise may be inconsistently removed or passed by and meaningful observations may be removed when serieswide threshold levels are applied to a data series having low frequency trends. (1) Noise not removed by serieswide thresholds. (2) Meaningful observations removed by serieswide thresholds.*

(Van Nostrand and Cook 1966: 137, 141, 148, 153–157, 219, 221, 227) and could be removed by a 3σ threshold, even if this threshold is defined locally. To avoid this mistatke, the expected deviations of subsidiary peaks above or below local mean resistivity values should be calculated and applied as a threshold, if greater in value than the more arbitrary 3σ threshold. The expected magnitude of subsidiary peaks can be calculated if one knows or assumes the following: (*a*) the electrode spacing of the Wenner array; (*b*) the approximate size, shape, and depth of burial of the anomaly of interest; (*c*) the approximate factor by which the anomaly of interest is greater or lesser in resistivity than its matrix; and (*d*) the electrode array crosses the anomaly perpendicularly (in which case subsidiary peaks would be at their maximum magnitudes). Specific formulas for calculating the magnitudes of subsidiary peaks under variable circumstances are given in Van Nostrand and Cook (1966).

It should also be noted that when making an initial screening of noise from a resistivity series, the observations defined as noise should be replaced with an average value representing not more than one tier of surrounding points. If a larger smoothing interval is used, "undesirable" ripples (called *polarity reversals*) may be introduced in the data.

The second step of the proposed analytical design requires that the data series be partitioned into a number of bands of different frequencies and widths, representing variabiity of different spatial scales and possibly different causes. This can be done best with a "normal" smoothing filter function that replaces each data point in the series by a weighted average of the points surround it, the weighting values being the ordinates of the normal curve. The smoothed series derived in this manner then is used to obtain high-frequency components of the raw data series by subtracting the smoothed (low-frequency) series from the raw data series. Intermediate bands of frequencies can be obtained by performing the operation twice, using filtering intervals of different widths, or "cutoff frequencies," and then subtracting the well-smoothed series from the less well-smoothed series.

The use of a normal filter function is preferred to a simple, running average filter that weights adjacent points equally. Both filter functions will maintain the mean of a series, but a running average operator will change the phase of waves of selected frequencies by 180° while changing their amplitudes and smoothing the curve. Such phase changes are called "polarity reversals." They are undesirable in that they cause misleading high-frequency ripples within the smoothed output of the operator (Holloway 1958:358).

This second step for partitioning resistivity data series into different frequency bands is appropriate for only single-dimensional data series. When observations have been made over a gridded area, it is possible to design filters that smooth in two dimensions by replacing each observation with an average of the points surrounding them within a ring of a given radius. High-frequency components can be obtained by subtracting the smoothed surface from the raw data surface. Surfaces reflecting intermediate frequency bands can be found by filtering the raw data observations with ring operators of different radii and by subtracting the well-smoothed surface from the less-well-smoothed surface. Such procedures, however, give responses that are poorly defined in directions oblique to the principle axes of the gridded data. When dealing with two-dimensional data, it is more appropriate to filter them in the Fourier domain than in the spatial domain. Scollar (1970) has discussed in detail the appropriate techniques for designing filters to be used in the Fourier domain.

Once a resistivity data series has been partitioned into multiple series of different frequency bands pertaining to resistivity variability of different spatial scales and of different origins, it should be possible to examine each of these component series for subportions of the archeological site that differ in their local mean resistivity and variability in resistivity. These zones may correspond to areas of different parent material, prehistoric use, or to natural or agricultural soil anomalies, depending on the scale of variability isolated within the frequency band. Whether meaningful interpretations and optimal differentiation of agricultural, archeological, and natural phenomena can be achieved at this stage of the analysis will depend in part on whether the raw data series has been partitioned into frequency bands of appropriate widths and ranges. This can be achieved only if the cutoff frequencies of the filters used in defining the band have been chosen intelligently, in light of the expected spatial scale of the phenomena of interest. For example, at the archeological site to be examined in this study (cf. Chapter 5), use-areas such as discrete clusters of pits and multipurpose open work areas in one excavated protion of the site ranged from 6.5–8.5 m in diameter. Optimal isolation and differentiation of these use-areas with resistivity data would require that bands of only the 6.5–8.5-m spatial frequency be removed from the raw data series and examined. Wider bands would include frequencies pertaining to other phenomena of different spatial scales in addition to those pertaining to use-areas. As composites of frequencies relating to multiple phenomena, they would be less readily interpretable. Ideally, one would like to partition raw resistivity series into bands that are of the range of frequencies in which the particular phenomena of interest will be manifested.

To aid in the process of designing filters that optimally segregate agricultural variability from archeological and natural variability within a resistivity series, the technique of spectral analysis may be used (Davis 1973:256–272; Jenkins 1961). This technique allows one to calculate those frequencies within a resistivity series that are the predominant *periodic* components and within which periodic, agriculturally caused resistivity variations are manifested. To apply the technique, first the resistivity series should be made stationary—that is, made to have broad-scale means that do not change over the course of the series. This may be achieved by smoothing the series with a normal filter function, which averages points over a broad interval, and by subtracting the smoothed series from the original series. The resulting series of residuals will be stationary with respect to all local means assessed at scales larger than the smoothing interval of the running filter. Next, the autocorrelation function of the stationary series can be used to determine the percentage of variance contributed by each frequency composing the residual series (those having wavelengths less than the smoothing interval used to derive the stationary series). The frequencies found to contribute most to the total series variance may be interpreted as those that are the primary periodic components within the series and those to which periodic, agricultural variation is restricted. When designing normal filters to segregate archeologically significant variation from agricultural and natural variation within the *original* resistivity series, this information may be used to choose cutoff frequencies that are optimal for this purpose.

Once archeologically significant variation has been isolated within a resistivity series, it, too, may be dissected again using running filter functions. If it is suspected that several differnt kinds of use-areas occur on an archeological site and that they are distinguishable by their dimensions, a number of component series, each containing waves of a particular spatial scale pertaining to use-areas of a particular expected dimension, can be derived from the resistivity series containing archeologically significant variation. Each of the component series then can be examined for locations that are anomalous in their mean resistivity or variance in resistivity in order to define use-areas that are characterized by these resistivity attributes *plus their scale*, that is, the frequency of the band in which they are found.

To locate and characterize use-areas spatially within a particular band series of archeologically significant variation by their anomalous mean resistivity and/or anomalous variance in resistivity, spatial filtering techniques may be used. Running filters may be designed that replace each observation with the mean of the points surrounding it or the standard deviation of the points surrounding it, within a set interval. The interval over which local means and standard deviations are assessed again must be chosen intelligently, in light of the expected spatial scale of the phenomenon of interest. For example, if the series to be filtered is a band that contains waves with periods of 8–10 m and in which use-areas of these dimensions are expected to manifest themselves, local means should be assessed over an interval ranging somewhere between 8–10 m and equivalent to the precise dimension of the use-areas to be located. The derived series then may be examined for positions at which local minima in mean resistivity or local maxima in resistivity variability occur. Such positions may correspond to the centers of prehistoric use-area—low mean resistivity indicating, for example, anthropic enrichments of ions (carriers of electric current) to the soil, and high resistivity variability indicating, for example, the inclusion of a large variety and/or density of features within the use-areas.

In stepping through these procedures, it is useful to examine the resistivity signatures of *each* surveyed Barnes Layer for agricultural, archeological, and natural pedological variation that may be reflected within them. One first should examine the Barnes Layer resistivity series, which has been generated in order to segregate a particular agricultural, archeological, or natural horizon of soil variability for indications of *that* particular source of variation. For example, resistivity series from Barnes Layers encompassing surficial horizons should be investigated for periodic patterns due to agricultural practices; series from Barnes Layers encompassing midden deposits and natural soils with intrusive features should be investigated for varations of the scale of use-areas. Resistivity patterns due to an agricultural, archeological, or natural phenomenon should appear strongest in the Barnes Layer encompassing the horizon that is primarily of that nature. Since the segregation of agricultural, archeological, and natural pedological variation over depth is never perfect, however, each Barnes Layer resistivity series may manifest variations due to phenomena other than those that they are intended to monitor. Patterns found in the resistivity series of Barnes Layers encompassing archeological deposits, thus, should be compared to those found in the resistivity series of Barnes Layers encompassing agricultural and natural soil horizons to determine whether they are, in fact, archeological in nature or whether they are agricultural or natural in origin. Similar comparisons should be made between the resistivity series of Barnes Layers encompassing other combinations of different horizons. By making comparisons between the resistivity series of Barnes Layers encompassing different kinds of horizons, it should be possible to understand interhorizon phenomena that might otherwise confuse the analyses of horizons examined individually. Such interhorizon phenomena include the leaching of fertilizers from a plowzone into archeological and natural subsoil horizons; the leaching of nutrient and organic enrichments within archeological horizons into subsoil horizons; and the mixing of soil from all horizons by soil megafauna and plant roots.

In summary, by designing spatial operators that can isolate specific bands of frequencies within resistivity data sets and that can emphasize particular attributes of those frequency bands, it is possible (*a*) to discriminate objectively agricultural, archeological, and natural sources of soil variation within a complex resistivity series; and (*b*) to locate areas within archeological soil horizons that have different soil attributes and that may have been used prehistorically for different purposes. The extent of success of this approach, however, will depend on the degree to which one can deduce a set of phenomena that are the *probable* causes of the soil resistivity variations on the basis of one's general knowledge of the site being investigated, and the degree to which these phenomena are understood. Optimal spatial filters can be designed only in reference to *expectable* circumstances.

Conclusion

Agricultural, archeological, and natural soil variations, as reflected in soil resistivity series, can be segregated over two dimensions, space and depth, by using both filter operations and the Barnes Layer technique. Although both of these techniques of data manipulation can be applied toward this common goal, they are by no means equivalent or redundant in their effects. It might be thought that when resistivity data series are interpreted by Wenner's equation for volumes of soil extending from the surface through agricultural horizons and archeological horizons and into the subsoil, and are filtered with spatial operators, these components of soil variation can be isolated as effectively as when spatial operators are used on individual Barnes Layer resistivity series. This is not true. It was noted before that within resistivity survey data that are collected with sampling intervals appropriate to the resolution of use-areas (.5–1.0 m), archeologically significant information occurs in the highest frequencies. These are the same frequencies in which random noise from surface and near-surface natural variations occur. If spatial filtering methods rather than the Barnes Layer method are used to remove surficial, high-frequency noise from resistivity soundings extending into archeological deposits, high-frequency information about human-caused soil variability will also be removed in the same process and lost from the analysis. Thus, the Barnes Layer method and spatial filtering are complementary mathematical techniques. When used together, they allow the extraction of much more archeologically significant information from a resistivity palimpsest than can the use of spatial filtering on resistivity data interpreted by Wenner's equation alone.

References

Atkinson, R.J.C.
    1963   Resistivity surveying in archeology. In *The scientist and archeology,* Edward Pyddoke (ed.), pp. 1–30. Phoenix House, London.
Barnes, H. E.
    1952   Mapping and subsurface exploration for engineering purposes. *Michigan State Highway Department, Research Board Bulletin* 65.
    1954   Electrical subsurface exploration simplified. *Roads and Streets* 97 (5):81–84.
    n.d.   Barnes Layer method of earth resistivity interpretation. Unpublished manuscript. Michigan State Highway Department, Ann Arbor.
Carr, Christopher
    1972   An adaptation of the Barnes Layer method of earth resistivity data interpretation to very shallow lateral surveying for archeology. Unpublished manuscript on file, Department of Anthropology, University of Illinois at Chicago Circle.
    1976   Soil resistivity surveying and regional archeological planning, lower Illinois Valley. In *Abstract Reports of Student Originated Studies Projects, 1975.* National Science Foundation, Washington, D.C.
    1977   A new role and analytical design for the use of resistivity surveying in archaeology. *Mid-Continental Journal of Archaeology* 2(2):161–193.
Dabrowski, K., and W. Stopinski
    1962   The application of the electric resistivity method to archeological investigation illustrated on the example of an early mediaeval hillfort in Kaliez. *Archaeologia Polona* 5:21–30.

Davis, John C.
  1973 *Statistics and data analysis in geology.* John Wiley, New York.
Griffin, W. R.
  1949 Residual gravity in theory and practice. *Geophysics* 14:39–56.
Holloway, J. Leith
  1958 Smoothing and filtering of time-series and space fields. *Advances in Geophysics* 4:351–389.
Jenkins, G. M.
  1961 General considerations in the analysis of spectra. *Technometrics* 3:133–166.
Jenkins, G. M., and D. G. Watts
  1968 *Spectral analysis and its application.* Holden-Day, San Francisco.
Leith, C. J., K. A. Schneider, and C. Carr
  1977 Geophysical investigation of archeological sites. *International Association of Engineering Geology, Bulletin* 14:123–128.
Linington, R. E.
  1967 An electrical resistivity survey at Les Matignens. *Prospezioni Archaeologiche* 2:91–93.
  1970 A first use of linear filtering techniques on archeological prospecting results. *Prospezioni Archeologiche* 2:43–54.
Malott, Donald F.
  1964 The application of geophysics to highway engineering by the Michigan State Highway Department. Unpublished manuscript. Michigan State Highway Department, Ann Arbor.
  1965 The application of geophysics to highway engineering in Michigan. *Michigan State Highway Department, Highway Research Record* 81.
  1967 Shallow geophysical exploration by the Michigan Department of State Highways. *Michigan State Highway Department, Report* TG-16.
  1968 Uses of geophysics in subsurface surveying. *Michigan State Highway Department, Report* TG-17.
Palmer, L. S.
  1960 Geoelectrical surveying of archeological sites. *Prehistoric Society, Proceedings* 26:64–75.
Rees, A. I., and A. E. Wright
  1969 Resistivity surveys at Barnsely Park. *Prospezioni Archaeologiche* 5:121–124.
Rich, Linvil G.
  1973 *Environmental systems engineering,* pp. 214–253. McGraw-Hill, New York.
Scollar, Irwin
  1969 Some Techniques for the evaluation of archeological magnetometer surveys. *World Archaeology* 1(1):77–89.
  1970 Fourier transform methods for the evaluation of magnetic maps. *Prospezioni Archaeologiche* 15:9–40.
Schollar, I., and F. Krückeberg
  1966 Computer treatment of magnetic measurements from archeological sites. *Archaeometry* 9:61–71.
Van Nostrand, Robert G., and Kenneth L. Cook
  1966 Interpretation of resistivity data. *United States Geological Survey, Professional Paper* 499.
Wenner, Frank
  1915 A method of measuring earth resistivity. *United States Bureau of Standards, Bulletin* 12, *Scientific Paper* 258:469–478.
Wilcock, J. D.
  1970 Some developments in the portrayal of magnetic anomalies by digital incremental plotter. *Prospezioni Archeologiche* 5:55–58.

3
Factors Affecting the Electrical Conductivity of Soils

A number of mathematical techniques for preparing resistivity survey data for interpretation within a geographic, statistical framework have been offered in Chapter 2. The particular operations used, the strategy with which they are employed, and interpretation of the manipulated data, however, are not mechanical procedures. Reasonable data manipulations and interpretations can be made only when the multiple phenomena causing soil variations and electrical resources are understood. This chapter and Chapter 4 provide the archeologist with the necessary understanding of these phenomena.

In this chapter, the physical and chemical variables and phenomena that are directly involved in the conduction of electric currents through soils (e.g., influence of soil structure on the paths taken by electric currents) will be discussed. Variables that are not directly involved in conduction but that have important effects on the primary phenomena (e.g., the influence of soil organic matter content on soil structure) also will be considered. In Chapter 4, the processes by which the natural states of these soil variables are altered by the occupants of archeological sites from natural soil states and the processes by which such altered soil states are maintained or not maintained over time will be presented. The models presented in these chapters will serve to operationalize the thesis that different kinds of prehistoric use-areas are characterized by different kinds of soil alterations and resistivity values.

This chapter is organized into two halves. The first half discusses the relationships between physical soil properties and soil resistivity; the second half presents relationships between the chemical properties of soils and their resistivity. When presenting the effects of alteration of the physical properties of a soil on its resistivity, those chemical alterations that often accompany the physical on archeological sites and that have an effect on a soil's resistivity are excluded from consideration, as if they do not occur. Likewise, when discussing the effects of alternations of the chemical properties of a soil on its resistivity, the physical alterations that often accompany the chemical and that have an effect on a soil's resistivity are ignored. This strategy is good in that it allows a more complete modeling of the individual relationships responsible for the resistivity of soils. It also is a necessary strategy, because the relative importance and combined effect of factors that change a soil's resistivity and that covary in archeological sites has not yet been systematically investigated.

Because modeling is partial in each half of the chapter, however, it is possible to speak of only the changes in soil resistivity *favored* by particular chemical and physical alterations. This qualification has been emphasized in footnotes when it is critical to the modeling process. The combined effects of physical and chemical soil changes on soil resistivity are modeled in Chapter 9 with the aid of empirical data collected from the Crane site.

The Formation Factor and Soil Structure

Unlike the electrodynamics of metallic conductors, the electrical phenomena that occur in porous semiconductors (e.g., soil) are poorly understood (Schopper 1966). Theoretical relationships between the electrical conductance of porous media and the characteristics of such a media have yet to be formulated in concise mathematical models. Consequently, it is not possible to present *summarizing* models of the relationships that must be understood in order to interpret resistivity data accurately and to operationalize the thesis of this study. Instead, a number of *separate* electric and electrolytic phenomena will be discussed in detail, linking them together and to this study by qualitative logic. The variables to be considered include: soil structure and texture, organic matter content, water content, the concentration and kinds of ions present in the soil water solution, and soil temperature.

The conductivity of soil, in the simplest of models, depends on two variables: the physical structure of the soil matrix,[1] and the conductivity of the liquid filling the interstitial soil pores. The conductivity of the matrix, itself, is insignificant. This relationship can be summarized by the expression

$$\rho_0 = F\rho_w \qquad (3.1)$$

where ρ_0 is the "apparent resistivity" of the soil saturated with a conductive liquid (soil water) of resistivity ρ_w, and where F is a constant—the formation

---

1. Throughout this study the term soil structure will be used to indicate the organization and packing of constituent soil particles, regardless of whether that organization is a result of the particle *size* distribution or particle *aggregation*. I will not use the term "structure" to indicate only the degree of aggregation of a soil, as often is done in soil science.

factor—which assesses the physical arrangement of pores that are filled with the liquid and that conduct current (Patnode and Wyllie 1950). The formation factor varies for different media with the tortuosity of the path that the current must follow through the soil pores and capillaries when flowing from one point to another; that is, it varies with the ratio of the fictitious average distance followed by the current flowing between the two points in the medium to the shortest distance between the two points (Brown 1971). The formation factor normally is defined for porous media saturated with water, but the concept can be extended to media of any *constant* moisture content. When a porous medium is not saturated with water, there are simply fewer conductive pores, the path of current flow is more tortuous, and ρ_0 is greater.

The relationship between the structures of porous media and their apparent resistivities as given in Eq. (3.1) is an empirical one, requiring the experimental assessment of structure (determination of F) for each unique medium under examination. As such, its predictive value is limited. An examination of the values of F taken by various saturated porous media, however, is helpful in illustrating the effect of the structure of a medium on its resistivity and the contribution which structure makes to its resistivity relative to the resistivity of its interstitial waters. Table 2 lists the values of the formation factor that have been observed in various staturated, consolidated and unconsolidated sediments These values represent the multifold increase in the resistivity of natural water structured within the sediments compared to the resistivity of water itself, alone and in an unstructured state. The factors of increase in resistivity are very significant in absolute terms—between five and several thousand times. The resistivity of any *single* consolidated or unconsolidated sediment (or soil) thus is highly dependent on its structure.

If one compares *variation* in the formation factor of a saturated, consolidated or unconsolidated sediment (or soil) across a landscape relative to spatial variation in the resistivities of its interstitial waters, however, in general the latter will be larger. The role of the formation factor in determining spatial variation in the apparent resistivities of saturated, consolidated or unconsolidated sediments usually is secondary to that of the resistivity of the interstitial waters (Hackett 1956). It is this fact that has permitted geophysicists to use resistivity surveying methods to map variations in the ionic concentrations and portability of ground waters within geological formations—even when they show lateral and vertical structural inhomogeneity (e.g., Merkel 1972).

Where the interstitial waters of soils and sediments have unnaturally low resistivities, however, structural variation may become the dominant factor affecting their apparent resistivities. This can be true of the soils on archeological sites, where the ionic concentrations of soil waters have been augmented through the deposition and decay of human refuse and soil water resistivity is consequently low (see the last section of this chapter).

TABLE 2
The effect of the physical structures of porous media upon their total resistivity at 100% saturation

Medium	Formation Factor	Total Resistivity (ohm-cm)
Limestones	10 - 1000s*	ca. 500
Sandstones	5 - 100s**	ca. 2000
Unconsolidated Gravels and Sands	5 - 100s***	ca. 5000

* Patnode et al. (1950)
** Hackett (1956:16)
*** Biddle (1970:33)

Soil Porosity, Pore-Size Distribution, and Particle-Size Distribution

The value of the formation factor in itself tells little about the nature of structural organization of porous media, other than the *summary* effect of structure on the path followed by electric currents when the media have a given moisture content. The formation factor is only an empirical constant. It would be helpful in evaluating resistivity survey data if theoretical models were available that relate the resistivity of porous media to variables describing, or associated with, the geometric nature of structural organization itself. Such variables might include soil texture, porosity, pore-size distribution, or the specific surface area of a soil.

Although each of these variables summarizes the states of more primary variables governing electric transport phenomena within porous media (Schopper 1966), no universally applicable mathematical expression relating them to apparent resistivity has been found empirically or derived theoretically. The problems that have retarded progress in this area are the same as those that have stymied all attempts to model the effects of the structure of porous media on their hydraulic conductivity (even though electrical conductivity and hydraulic conductivity are determined by different primary physical and chemical variables). These problems have been summarized by Childs (1969). The simplest approaches have sought correlations between conductivity (electric or fluid) and porosity alone (Jacob 1946; Franzini 1951). This approach is too simplistic. The porosity of a medium describes only the *volume* of void space within it capable of conducting fluids or currents; it does not measure the diameter, continuity, shape, or tortuosity of the conducting channels. Fluid and electrical conductivity depend on all of these additional factors.

Where attempts have been made to consider the *structural nature* of void space and its effects on fluid or electrical conductivity, in addition to the *total quantity* of void space, the models that have been developed are too idealized. Most are based on the flow of fluids or currents between parallel planes, or within a set of cylindrical tubes that parallel each other and have a given density per cross-sectional area (porosity evaluated in one plane). Such models are highly anisotropic (i.e., the electric or hydraulic conductivity along one axis does not equal that along other axes) and are not applicable to media such as soils, which contain more randomly oriented capillaries and allow the flow of currents and fluids in all directions (Childs 1969:179).

Other researchers have approached the problem of predicting the conductivity of porous media using the particle size distributions of the media. The major problem with this approach is the indeterminant relationship between particle size distribution and mode or structure of particle packing (Scheiddegger 1957:18–19). Even in the simplest of a priori models, where all particles are considered to be spheres and are of one diameter, multiple methods of packing are possible, yielding capillaries with different geometries, total conductive cross-sectional areas (porosities), and capabilities for conducting fluids or currents. Moreover, for any one mode of packing, the porosity of the medium will be independent of the size of the spheres (Hrubišek 1941). Consideration of other geometric variations found in natural soils presents even more complex problems to model building. Particles in natural soils are always of multiple sizes, creating situations where smaller particles fill the spaces between larger ones, reduce the porosity and conductive cross-sectional area of the soil and increase the tortuosity of flow lines. In addition, particles within natural soils tend to be angular rather than rounded and permit bridging to occur, resulting in increases in porosity and conductive cross-sectional area. These circumstances have made it impossible to derive any theoretical relationship between the textural distribution of porous media and their hydraulic or electrical conductivity.

The most promising approach to the problem of relating the conductivity of soils to some primary structural variable involves the use of their pore-size distributions (Hillel 1971:97). The major problem here has been in determining an appropriate numerical way to characterize the distribution as a whole. To date, little progress has been made along this line.

Soil Moisture Content as a Function of Soil Structure

Little success has been achieved in quantitatively or qualitatively describing the relationship of the structural organization of porous media to their apparent resistivities *when holding moisture content constant.* Although it would be theoretically elegant and practically useful if such a relationship were known from a practical standpoint, alone, an alternative approach to the problem can be taken. The resistivity of a soil is dependent on both the structure of a soil and its moisture content, but more so on moisture content. Under natural conditions, these variables tend to covary, the structure of the soil determining its equilibrium moisture content under specific drying conditions. Consequently, *the differences in the resistivities of the two soils attributable to their different structures can be examined largely in terms of the differences in the moisture contents that their structures produce.* It is not necessary to hold moisture content constant in determining the effect of soil structure on soil resistivity, as has been done in the previous section. Let us first examine the relationship between structure and moisture content, and then the joint relationship of these two variables to apparent resistivity.

The moisture content of a soil (natural or anthropically disturbed) at any point in time is influenced by the soil's structure in two ways. First, the structure of a soil will determine potential energy relations, which govern how strongly water is held within it. Second, the structure of a soil will affect the rate at which water can flow through the soil and be removed from it during drying conditions.

With regard to energy relations, the equilibrium moisture content of a soil can be seen as a balance of four energy potentials: matric, osmotic, gravitational, and environmental. The matric potential of soil water is that energy which occurs in potential form as the attraction of water to soil solids (matrix) and itself. It is the result of two forces: those of adsorption and those of surface tension. Adsorptive forces include both the adhesive forces that bind water dipoles directly to electrostatically charged soil surfaces and exchangeable cations, and the cohesive forces that bind other water dipoles to those already joined to the matrix by adhesion (Brady 1974:164–166). The surface tension of soil water is the net inward force that it experiences at its interface with soil air. It is the result of the greater attraction that water dipoles have for each other than for the air. The matric potential of soil water is always negative, indicating the reduction of free energy that adsorption and surface tension cause to soil water.

The two forces, adsorption and surface tension, which determine the matric potential of soil water, are observable in the phenomenon of capillary rise (e.g., the rise of water in a straw above the level of water in a glass). They are described by the equation,

$$h = 2T/rdg \qquad (3.2)$$

where h is the height of capillary rise equivalent to the matric potential, T is the surface tension of the fluid under consideration, r is the radius of the capillary, d is the density of the fluid, and g is the acceleration of gravity. For soil water, this reduces approximately to

$$h = .15/r \qquad (3.3)$$

depending on the temperature and ionic concentration of the soil water.

The second soil water potential, the osmotic potential, results from the attraction of solutes within the soil water to the water dipoles. Its major effect is on the ease of vaporization of soil water. As more solutes are added to water, its vapor pressure decreases and the energy required for vaporization increases. The osmotic potential, like the matric potential, is negative.

A third, form of potential energy attributable to soil water is the gravitational potential. This potential is the energy that would be released in kinetic form if the water within a soil drained from it under the force of gravity. It is related to the height of the water above a reference elevation (usually the water table).

$$P_g = Gh \quad (3.4)$$

where P_g is the gravitational potential, G is the acceleration of gravity, and h is the height of the water above the reference level. The gravitational potential of soil water above the reference elevation is is positive. The potential of that below the reference elevation is negative.

The sum total of the matric, osmotic, and gravitational potentials of a soil water defines its total potential:

$$P_t = P_m + P_o + P_g \quad (3.5)$$

The total potential of a soil water may be thought of as the energy that would have to be expended to begin to draw water to the surface of the soil and evaporate it.

Opposing the matric, osmotic, and gravitational potentials are environmental forces and conditions at the soil surface that can draw water from the soil and dry it: solar heat, air movement, air humidity, etc. These forces and conditions constitute what is termed here an "environmental potential."[2] When a soil is in the process of drying, the environmental potential is greater than the total potential of the soil:

$$P_e > P_m + P_o + P_g \quad (3.6)$$

When the moisture content of a soil holds constant over time, the environmental potential is less than or equal to the total soil potential:

$$P_e \leq P_m + P_o + P_g \quad (3.7)$$

The environmental potential is positive and can be expressed as a suction or tension (measured in bars or atmospheres), or as the height in centimeters of a unit water column, the weight of which equals the suction under consideration.

In relating the structure of a soil to its equilibrium moisture content, the most critical potentials are the matric, environmental, and gravitational potentials. As Eq. (3.3) indicates, the force with which water is held within a soil capillary or pore against gravity or environmental forces is a direct function of its size.

2. The term "environmental potential" is not used in soil science. It is a logical extension of the concept of "soil potentials," used here to aid the reader in envisioning the forces acting on soil water under specific weather conditions.

The average force with which water is held within a soil as a whole thus is a function of its pore-size distribution. The equilibrium moisture content of the soil will depend on the particular balance found between its total matric forces, as determined by its pore-size distribution, and the particular environmental and gravitational suction to which it is subjected. Soils with different structures and pore-size distributions will retain different amounts of water under the same environmental conditions (total porosity held constant). Those having a greater number of large pores will retain less water; those having a greater number of small pores will retain more water, for a given environmental and gravitational suction.

This relationship between the pore-size distribution of a soil and its moisture content has direct implications on the effects of both the particle-size distribution and the degree of aggregation or compaction of particles within a soil on its moisture content and resistivity, and consequently, on the interpretation of resistivity data from archeological sites. First let us consider particle-size distribution.

Finer textured soils have smaller pores whereas coarser textured soils have larger pores. Because of this difference in pore sizes, if both a finer textured soil and a coarser textured soil are saturated with water and then subjected to the same total environmental and gravitational suction, the latter will retain less water (Figure 6). Table 3 summarizes the approximate moisture contents that can be expected to be held by soils of different textural classes at field capacity (greater than or equal to ca. .1–.2 bar tension) and at the primary wilting point (greater or equal to ca. 15 bar tension). As a result of these differences in moisture equilibria, soils of different textural classes will have different conductive cross-sectional areas available for current flow at any single environmental suction and will have different resistivities. Natural textural variation within the soils of an archeological site therefore must be taken into consideration when interpreting resistivity data.

Over a whole period of drying, as environmental suction increases, soils of different textures will achieve their maximum distinctness from each other in moisture content and in resistivity when suctions are intermediate between those at field capacity and those at the primary wilting point (Russell 1939). This is a result of the different suctions at which different textured soils loose most of their water. Coarse-textured soils, having a high frequency of large pores, will loose their moisture rapidly at first under low suctions and then will retain what little moisture is left until much higher suctions are attained. More finely textured soils, on the other hand, have more uniform pore-size distributions, and will loose their moisture more gradually and uniformly over a full range of suctions while drying. Thus, when attempting to use resistivity survey methods to map anthropically caused structural variation within the soils of archeological sites (e.g., aggregation, compaction) rather than natural structural variation associated with textural changes, better results may be achieved if the survey is performed during rainfall–evaporation regimes when

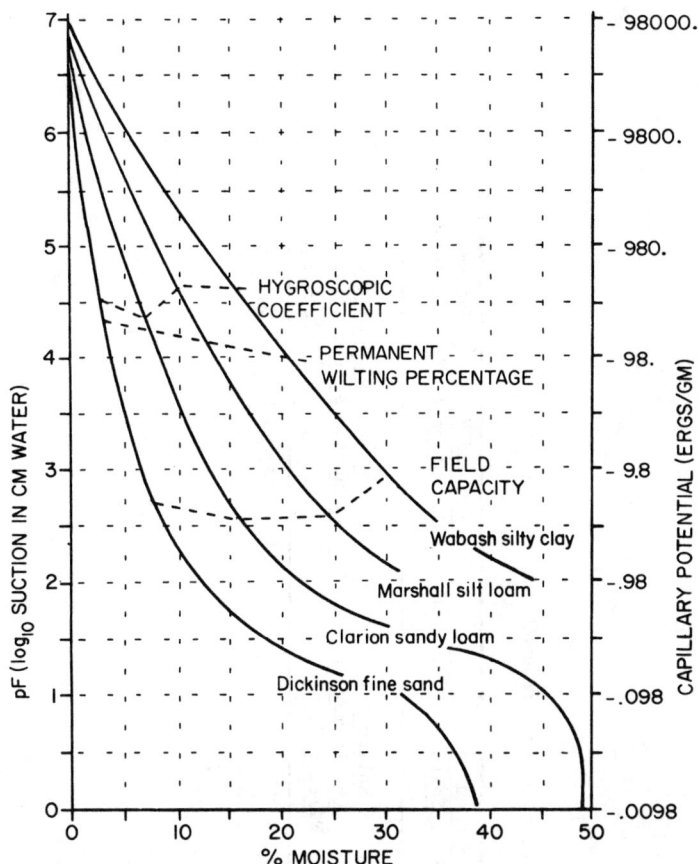

FIGURE 6. *The effect of soil particle size distribution upon the moisture retentive potential of soils. Adapted from Russell (1939).*

suctions are not within the intermediate range and are not optimal for differentiating soils of different textures.

The degree of aggregation or compaction of particles within a soil, like its particle-size distribution, has an effect on its pore-size distribution and moisture equilibria. The relationships between these variables usually is considered as a part of the larger topic of the effects of organic matter on moisture retention. Organic matter helps to maintain water in a soil both directly, through its hygroscopic powers, and indirectly, by facilitating soil aggregation. A detailed discussion of the relationship between organic matter, soil aggregation, pore-size distribution, and moisture retention therefore, will be reserved for the next section, where it will be considered in itself, as a unit. Only a general overview of the effects of aggregation and compaction on soil pore-size distribution and soil–moisture equilibria will be given here.

TABLE 3
*Approximate moisture contents of soils of various textual classes expected to be found at different suctions**

Textual Class	International System of Classification of Soil Textures		British and U.S.D.A. System of Classification of Soil Textures	
	% Moisture at Field Capacity	% Moisture at the Primary Wilting Point	% Moisture at Field Capacity	% Moisture at the Primary Wilting Point
Sands				
coarse sand	8 ⎫			
sand	14 ⎬	4	8-11	4
fine sand	19 ⎬			
very fine sand	20 ⎭			
Loamy sands				
loamy coarse sand	13 ⎫			
loamy sand	18 ⎬	7	13-14	5
loamy fine sand	22 ⎬			
loamy very fine sand	25 ⎭			

			16-24
Sandy loams			
coarse sandy loam	19		
sandy loam	26		
fine sandy loam	28		
very fine sandy loam	28	9	7
Loam	30	13	27
Silty loam	34	12	--
Silt loam	39	16	35
Silt	--	--	41
Sandy clay loam	26	15	21
Clay loam	34	18	31
Silty clay loam	43	20	41
Sandy clay	29	19	26
Silty clay	47	25	41
Clay	42	25	37

*Data are from Salter and Williams (1969)

On archeological sites, variation in the pore-size distribution of a soil that is not attributable to variation in particle-size distribution may result from either granulation (aggregation) or physical compaction. Granulation may relate either to enrichment of the soil with organic matter or to the physical churning of the soil by the occupants of the site, or to both. A soil that has become more granulated will, as a whole unit, have a greater number of both large pores and small pores than the natural soil from which it is derived. The increase in number of large pores will be greater than the increase in number of small pores. Considering soil void space on a volume-standardized basis, the granulated soil consequently will have a greater total porosity, more void space within the large pore-size classes per unit volume of soil, and *less*[3] void space within the small pore size clases per unit volume than does its natural counterpart. On the other hand, a compacted soil, as a whole unit, will have fewer large pores and perhaps fewer small pores than the natural soil from which it is derived. The decrease in number of large pores will be greater than the decrease in number of small pores. On a volumetric basis, the compacted soil thus will have a smaller total porosity, less void space within the large pore-size classes per unit volume of soil, and *more* void space within the small pore-size classes per unit volume than does its natural counterpart.

Because of these differences in the proportion of pores of given radii found in the granulated, natural, and compacted soils, they will retain different amounts of water per unit volume at a given suction. These differences in moisture content equilibria will result in their having different conductive cross-sectional areas available for current flow, and will favor their having different resistivities. The different pore-size distributions of the three kinds of soils and the different proportions of air-filled as opposed to water-filled pores they may have at a given suction will result also in their having paths available for current flow that differ in their structural organization and tortuosity. The resultant resistivities of the granulated, natural, and compacted soils will depend on the particular balance they offer, at a given suction, between their conductive cross-sectional areas and the tortuosity of their paths of current flow (chemical differences between the soils held constant). In general, however, the former property will be more important; the soils with greater moisture contents per unit volume and greater conductive cross-sectional areas will have the lower resistivities.

Whether it is the granulated, natural, or compacted soil that holds more water, has a greater conductive cross-sectional area, and has a lower resistivity will depend on the particular environmental suction at the time of observation. During periods of low environmental suctions, when both large and small pores are filled with water, the granulated soil will have a greater volumetric moisture

---

3. Note that the reduction in void space within the small pore-size classes does not reflect the process of aggregation, itself, which actually inreases the absolute number of small pores within the soil as a whole, but rather, reflects standardization of void space by the volume of soil under consideration.

content than the natural soil from which it is derived; the natural soil will have a greater volumetric moisture content than its compacted counterpart. These conditions will favor an *increase* in the resistivities of the soils through the series: granulated soil, natural soil, compacted soil. During periods of higher environmental suctions, when only small pores are filled with water, the granulated soil will have a lower volumetric moisture content than the natural soil; the natural soil will have a lower volumetric mosture content than the compacted. These conditions will favor a *decrease* in the resistivities of the the three soil types through the series: granulated soil, natural soil, compacted soil.[4]

The degree to which different kinds of disturbed and undisturbed soils are distinguishable in their equilibrium moisture contents and electrical resistivities will vary over the full range of suctions to which they are subjected. Two soils may share a common relative frequency of pores that are less than a given diameter and that hold water at one suction, but may differ in their relative frequencies of pores that are less than other given diameters and that hold water at other suctions. This situation can be visualized most easily by making a comparison between the "cumulative moisture characteristic curves" of the soils under consideration (Figure 7). A cumulative moisture characteristic curve is simply a graph of the various moisture contents at which a soil will equilibrate at a number of different suctions, and reflects the cumulative pore-size distribution of a soil.

Because different soils can be more or less similar in their moisture contents and resistivities at different suctions, the timing of resistivity surveys on archeological sites must be planned carefully. During some rainfall–evaporation regimes, anthropically disturbed soils and feature fills may be distinguishable from natural, undisturbed soils in their volumetric moisture contents and resistivities. During other regimes, they may not. The particular moisture conditions optimal for differentiating disturbed and undisturbed soils will vary from site to site and soil to soil.

In general, however, differentiation of a granulated or compacted soil from its natural counterpart in moisture content and resistivity will be greatest at very low and at very high environmental suctions.[5] As the different soil types dry

---

4. This model will be qualified in the following to take into consideration the effects of the hygroscopic powers of organic matter upon soil moisture retention at high suctions.

Also, whether the decrease in resistivity through the series of three soil types that is *favored* by their physical conditions is *realized* will depend on the balance achieved between the effects of their physical properties and the effects of their chemical properties on their resistivities. This topic is discussed in the last section of this chapter.

5. Of the two extreme environmental suctions—low and high—at which disturbed and undisturbed soils are most distinct in their moisture contents and conductive cross-sectional areas, it is during the latter that the two soil types will be most distinguished in their resistivities. This fact reflects a relationship between the size of capillaries filled with water, capillary conductance, and total soil conductivity (see following). Again, chemical soil properties are being held constant, and the effects of variation in only physical soil properties are being considered.

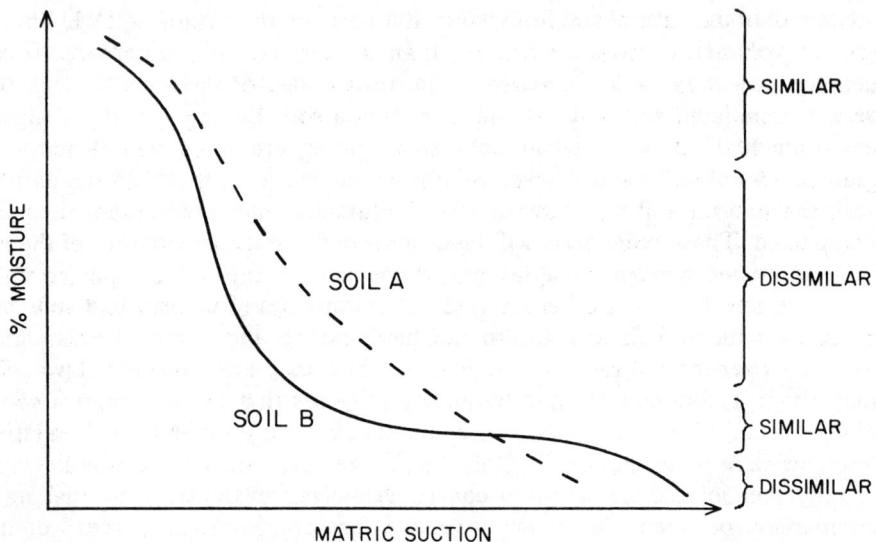

FIGURE 7. *The moisture characteristic curves of two soils distinguishable in their moisture contents at some environmental suctions but not at others.*

from saturation (0 bar suction) to the point at which only hygroscopic waters remain within them (ca. 31 bars), the difference in their volumetric moisture contents and conductive cross-sectional areas will at first decrease to nil, and then increase. At the particular intermediate suction at which the conductive cross-sectional areas of the disturbed and undisturbed soils are equivalent, the difference in their resistivities will be minimal. At this point, the difference will depend only on differences in the tortuosity of the paths they offer for current flow, and not on differences in the number of paths they offer.[6] Resistivity surveys for archeological purposes should be planned for rainfall–evaporation regimes other than when environmental suctions are in this intermediate range, at times when differences between disturbed and undisturbed soils in their conductive cross-sectional areas are maximized.

Note also, that during the periods of intermediate suction, not only is the effect of aggregation or compaction on differences in the moisture balances of disturbed and natural soils minimal, but the effect of textural differences between the soils on differences in their moisture balances is maximal. Resistivity surveys performed during regimes of intermediate environmental suctions thus will have doubly a hard time mapping anthropic soil differences and distinguishing them from natural soil variations.

---

6. Differences between granulated, natural, and compacted soils in their chemical properties such as their cation exchange capacities, availabilities of nutrients, and concentrations of ions within their soil waters are not being considered here. See the last section of this chapter.

Soil Moisture Content and Hysteresis

Up to this point, the impression has been given that for any particular environmental suction, the moisture content of a soil will equilibriate at a unique value. This is not true. The value taken will depend on whether the soil is in the process of drying or being wetted. It will be greater in the former case (Figure 8). This phenomenon, called hysteresis, is the result of a number of factors. First, there is the difference between the diameters of the pores within the soil and the diameters of the capillaries connecting them. In the drying process, it is the narrow radius of the channels connecting pores that determines the matric potential (high) and the force required to empty the pores and to obtain a specific moisture content. In the wetting process, it is the larger diameter of the pores that determines the matric potential (low) and the force required to fill the pores and to obtain that same moisture content. The desorption portion of a hysteresis curve indicates the relative amount of pore space within a soil connected by *capillaries* of particular diameters, whereas the sorption portion of a hysteresis curve indicates the relative amount of pore space within a soil composed of *pores* of particular diameters. Second, the contact angle and radius of curvature of an advancing meniscus within a capillary is greater than that of a receding meniscus. As a result, at a given water content, a drying soil will

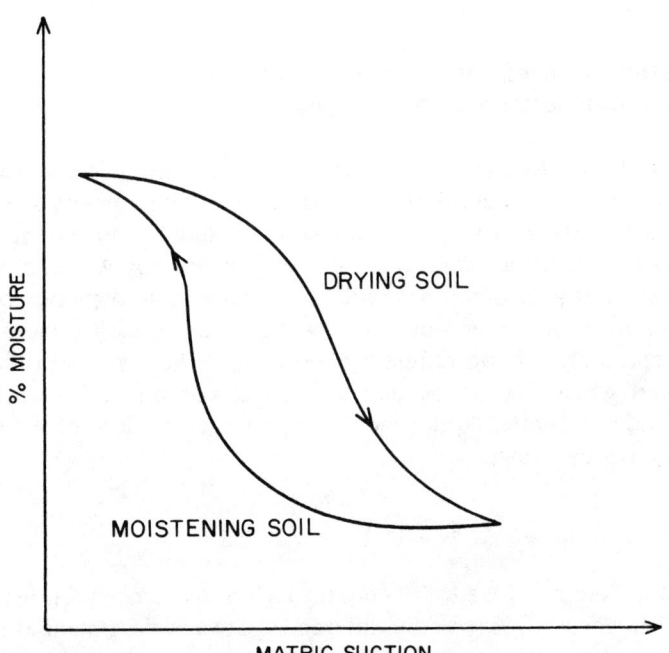

FIGURE 8. *Variations in the moisture content of a soil under a range of suctions, during both wetting and drying processes.*

exhibit a greater negative potential than a moistening soil. Third, when a soil is wetted, air bubbles may become trapped within capillaries and prevent water from filling pores that normally would be filled at the suction of investigation. This does not occur as a soil dries. Finally, as a soil dries, it may shrink, and as it moistens, it may swell. The history of structural changes associated with previous contractions and expansions within a soil may result in a hysteresis effect (Hillel 1971:65–67).

Hysteresis is of relevance to this study in that it may affect the degree to which soils of different structures (e.g., disturbed and undisturbed soils) are distinguishable by their moisture contents and electrical resistivities. In the simplest examples, two soils might have similar total porosities and pores of similar size distributions but differ in the diameters of the capillaries connecting pores. As a result of the first factor causing hysteresis, the two soils would be found to have similar moisture characteristic curves on wetting, but on drying, the soil with the smaller capillaries would consistently be found to be more moist. In other instances, two soils having similar moisture characteristic curves during drying might be found to have different moisture characteristic curves during wetting. Such results generally will be found most pronounced in coarser-textured soils in the low suction ranges, where pores may empty at much larger suctions than those at which they fill (Hillel 1971:67).

Soil Moisture Content as a Function of Drainage Conditions: Saturated Flow

The effect of soil structure on potential energy relations of soil water is one manner in which soil structure determines the moisture content of a soil. The structure of a soil also determines soil moisture content by influencing the rate at which water can infiltrate, drain through, or flow laterally within a soil profile under the forces of gravitation or gradients in matric and osmotic potentials.

The dependence of water flow rates within soils on soil structure is best described separately for two different conditions: When the soil is completely saturated and when it is moistened to only a portion of its water-holding capacity. Under saturated soil conditions, the rate of flow of water can be described by Darcy's law:

$$q = K\left(\frac{\Delta H}{L}\right) \qquad (3.8)$$

where q is the volume of water flowing through a unit cross-sectional area per unit time (flux density), ΔH is the difference between the potential of the soil water at the source of flow and the potential of the soil water at the sink, L is the length of soil through which the water is passing, and K is the hydraulic conductivity constant. The value of K and its effect on water flow rates depends on both the total porosity of the soil under investigation and the pore-size distribution. The greater the porosity of the soil, the more passages there will be available for conducting water per unit cross-sectional area of the soil and the greater will be the value of K and the rate of flow. The larger the conducting pores, the greater will be the value of K and the rate of flow. Within the ranges of porosities and pore-size distributions usually found in soils, the effect of pore size is more important than the effect of total porosity in determining hydraulic conductivity. A sandy soil with a high frequency of large pores will conduct water more rapidly than will a clayey soil having a high frequency of small pores—even though the sandy soil will have a much lower porosity than the clayey soil (Brady 1974:192). Likewise, a well-aggregated soil having a high frequency of larger pores will conduct water more rapidly than will a poorly aggregated soil having a high frequency of smaller pores.

As mentioned at the beginnning of this chapter, a universally applicable, precise, quantitative expression relating the hydraulic conductivity constant or the rate of flow of water through a soil to variables that measure aspects of soil structural organization has yet to be found. The best approximation to such a function presently available is Kozeny's theoretically derived equation:

$$K = \frac{f^3}{ca^2(1-f)^2} \tag{3.9}$$

where K is the hydraulic conductivity constant, f is soil porosity, a is the specific surface area exposed to the soil water, and c is a constant representing particle shape. Although Kozeny's equation is supposedly applicable to any natural, porous material—no matter how complicated its pore space—in fact, it falls short of this. There are two primary reasons. First, the derivation of the equation is based on an anisotropic parallel-tube model, the effects of which are hidden in the scalar quantities of surface area and porosity. Second, the equation considers only that aspect of the structural organization of pore space which is attributable to the texture of the particles that compose the medium, and ignores that aspect of structural organization due to aggregation. The equation fails, for example, to describe the conductivity of fissured clays, where the structural fissures contribute negligibly to both the porosity and specific surface area of the clays, yet are major contributors of their permeabilities (Childs 1969:180–183).

Despite these drawbacks, Kozeny's equation is useful. It provides a *minimum* estimate of the hydraulic conductivity of a soil and may be used, along with Darcy's law and equations for nonsaturated hydraulic flow (following), to determine the approximate length of time after a rainfall that moisture flow within a soil has essentially ceased and soil moisture content has equilibriated. The latter condition is desirable when performing a resistivity survey. If soil moisture content has not stabilized, changes in it over time while surveying over space may lead to complex results, which are difficult to interpret. The typical length of time required for moisture flow within the top several feet of a soil to slow after a rain to the point where a resistivity survey is practical is 2 to 3 days.

Soil Moisture Content as a Function of Drainage Conditions: Unsaturated Flow

Flow of water through unsaturated soils differs from the flow in saturated soils; the rate of flow is a function of the moisture content of the soil as well as the force of the driving potential gradient. As a soil drains and becomes unsaturated, pores begin to fill with air—the larger first and then the smaller. Once filled with air, pores will no longer conduct water (unless a wetting regime is activated) and consequently reduce the conductive cross-sectional area on which flow rates directly depend [Eq. (3.4)]. They also provide obstacles and increase the tortuosity of the path of flow. Thus, decreases in the moisture content of a soil as it drains reduces the hydraulic conductivity of the soil and the rate of flow of water through it. The transition from saturated to unsaturated flow generally results in a huge drop in hydraulic conductivity, by several orders of magnitude (Hillel 1971:105).

Unsaturated flow within a soil differs from saturated flow also in the effect of the soils' pore-size distribution on it. At saturation, the most conductive soils are those with large pores (coarse-textured). At less than saturation, smaller-pored soils (fine-textured) are more conductive. At any given suction below saturation, and holding total soil porosity constant, a smaller-pored soil will hold more water than a larger-pored soil, have a greater conductive cross-sectional area, a less tortuous path of flow, and consequently, a greater hydraulic conductivity.

Quantitative descriptions of unsaturated flow require the use of Darcy's law [Eq. (3.8)] plus empirically derived formulas that estimate the hydraulic conductivity constant from the moisture content or matric potential of the soil. The two equations used most frequently for estimting the hydraulic conductivity constant are

$$K = a/p^m \qquad K = a/(b + p^m) \qquad (3.10)$$

where p is the matric potential and a, b, and m are empirical constants. Of these latter parameters, m is the most critical. It usually varies between 2 or less for clayey soils and 4 or more for sandy soils (Gardner 1960).

In regard to practical applications, it was indicated previously that resistivity surveys should be timed with respect to the last rainfall and the period necessary for the soils under investigation to drain and reach a semistable equilibrium content. Equations (3.8) through (3.10) can be used to estimate this period.

Equations (3.8) through (3.10) also can be used to estimate the *pattern of rates of change* in the distinguishability of structurally disturbed and undisturbed soils from each other in moisture content and in resistivity as they dry and move from one moisture equilibrium state to another. As discussed earlier, anthropically disturbed soils may be more granulated, have higher total porosities, and have higher relative frequencies of large pores than do the natural soils from which they are derived. They also may be more compacted, have lower porosities, and have lower relative frequencies of large pores than the natural soils from which they derived. The degree of distinction between granulated, natural, and compacted soils in their moisture contents and resistivities *at times of moisture equilibrium* will depend on static relations among soil and environmental potentials, as influenced by pore-size distribution, total porosity, and organic matter content.[7] In general, at low environmental suctions, granulated soils will retain more moisture than the natural soils from which they are derived and compacted soils will retain less. At high environmental suctions, the ordering of soils by their moisture contents will be reversed. During period of *drying and disequilibrium,* the distinctiveness of granulated, natural, and compacted soils will depend on (*a*) their distinctiveness at the initial moisture equilibrium state; (*b*) their distinctiveness at the final moisture equilibrium state to be achieved; and (*c*) the different rates at which moisture is capable of moving in the several soil types and they are able to reach their final equilibrium states. When environmental suction changes from a low equilibrium point to an intermediate equilibrium point, the moisture content of a granulated soil will adjust to the new equilibrium faster than does that of its natural counterpart and the moisture content of the natural soil will adjust faster than that of its compacted counterpart. This results from the greater number of large pores and the greater conductive cross-sectional area in the granulated soil than the natural soil, and in the natural soil than the compacted. The initially larger losses of moisture from the better granulated soil will quickly reduce the excessive moisture it originally holds compared to its natural counterpart. Likewise, the initially larger losses of moisture from the natural soil compared to those of the compacted will quickly reduce the excessive moisture held by the natural soil. Thus, over a single period of drying from low to intermediate suction, as the moisture contents and favored resistivities of the granulated, natural, and compacted soils change from those of graeter distinctiveness to those of lesser distinctiveness, their distinctiveness, their distinctiveness will decrease most rapidly at the beginning of the period of change and more slowly at the end of the period of change. At higher environmental suctions, the pattern of rates of change in the distinctiveness of granulated, natural, and compacted soils is analogous to that at lower suctions. When environmental suction changes form an intermediate equilibrium point to a high equilibrium point, the moisture content of a compacted soil will adjust to the new equilibrium more quickly than does that of its natural counterpart, and the moisture content of the

---

7. Again, differences between the granulated, natural, and compacted soils in their cation exchange capacities, availabilities of nutrients, and concentration of ions within their soil waters are not being considered here. See the last section of this chapter.

natural soil will adjust more quickly than that of the granulated soil. This pattern results from the greater moisture content of the fine-pored soils, their greater conductive cross-sectional area, and the less tortuous shapes of their paths of water flow. Consequently, as the moisture contents and favored resistivities of the compacted, natural, and granulated soils change from those of less distinctiveness to those of greater distinctiveness, their distinctiveness will increase most rapidly at the beginning of the period of change and more slowly at the end of the period of change.

Capillary Conductance

Having examined how the structural organization of a soil will determine to a great extent its moisture content, let us now consider how these two variables together can affect a soil's resistivity. As a soil dries, its electrical resistivity increases (e.g., Figure 9). This increase is a function of three factors. First, as pores within a soil dry and loose their capacity to conduct electrical current, there is a reduction in the conductive cross-sectional area of the soil and the number of paths available for current flow. Second, as pores dry, the paths of flow become more tortuous. Third, over a drying period, as environmental suction increases, the radius of the pores and capillaries retaining water decreases. As the electrical conductance of smaller pores and capillaries is less than that of larger pores and capillaries, the conductivity of a soil will decrease both per unit volume holding water and in total, as it dries.

This last factor requires explanation. The water filling a soil capillary may be visualized as having two components—a thin layer around the perimeter of the capillary and an interior cylinder of water. The concentration of ions within the peripheral water of a capillary is always greater than that within the interior water of a capillary. Electrostatically charged clay and humus particles, which compose the walls of soil capillaries, attract both cations and anions to their surfaces (see Chapter 4) and increase the concentration of ions within peripheral soil water. The concentration of the ions is highest in the immediate vicinity of the capillary wall, and decreases away from it—at first very rapidly, and then asymptotically—to the concentration of ions found in the interior soil water. This phenomenon is called the Gouy effect, and is common to electrolytic solutions bounded by solid surfaces in general—electrostatically charged or neutral in charge (Gouy 1910; Wilklander 1964:166–170). It is analogous to the more familiar osmotic effects that occur when one electrolyte is separated from another by a semipermeable membrane.

As a result of the Gouy effect, the conductivity[8] of peripheral capillary water

---

8. The electrical conductivity of a material is its *ability* per unit volume to conduct a current. The electrical conductance of a material is the *amount* of current flowing across it cross-sectional area per unit time.

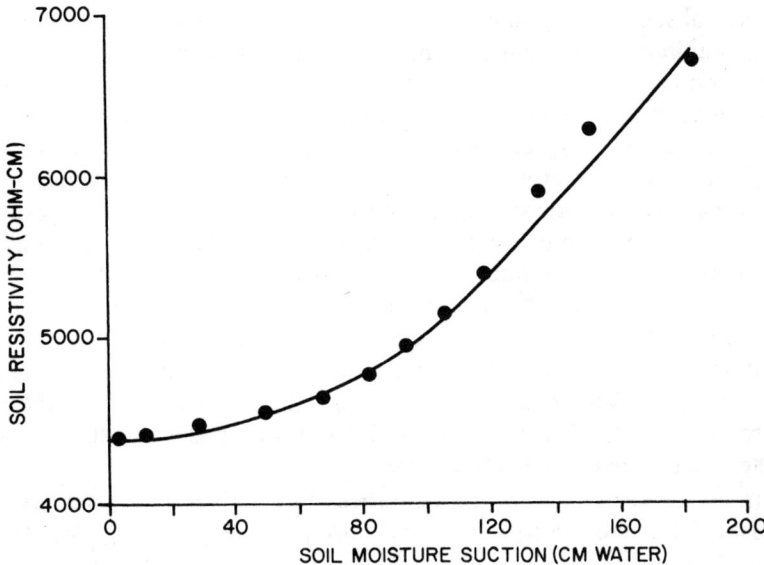

FIGURE 9. *Electrical resistivity of a sandy loam as a function of soil moisture content. Adapted from Shea and Lutkin (1941:374).*

is greater than that of interior capillary water: There are a greater number of electric carriers (ions) within peripheral capillary water than in interior capillary water. The electrical conductance of a whole, water-filled soil capillary is a function of the conductivities of both its interior and peripheral waters and its radius, as given by the expression:

$$C_T = C_i + C_P = (r^2/L)K_i + (2\pi r/L)K_P \tag{3.11}$$

where C_t is the total capillary conductance, C_i is the interior water conductance, C_p is the peripheral water conductance, K_i is the interior water conductivity, K_p is the peripheral water conductivity, r is the radius of the capillary, and L is the length of the capillary (van Olpen 1963:190). Thus, as a soil dries and the capillaries still retaining water and capable of conducting electricity decrease in radius, their conductances decrease, and the total capillary conductance of the soil—its "conductivity"—decreases. This decrease in conductivity is additional to that caused directly by the decrease in soil moisture content and the conductive cross-sectional area available for current flow; it is a result of the *interaction* between the structural organization of a soil and its moisture content.

The effect of capillary size and conductance on total soil conductivity complements the effect of soil moisture content on soil conductivity in differentiating disturbed and undisturbed soils in their resistivities.[9] At low

---

9. Differences between the disturbed and undisturbed soils in the cation exchange capacity, availability of nutrients, and concentration of ions within their soil waters are not being considered here. (See the last section of this chapter.)

environmental suctions, a granulated soil will have more water-filled large pores with high capillary conductances and less water-filled small pores with low capillary conductances than the natural soil from which it was derived. Consequently, it will have a higher total capillary conductance and conductivity, as well as a higher conductive cross-sectional area, than its natural counterpart; these two factors will conjoin constructively to give the granulated soil a lower resistivity. A natural, undisturbed soil will have a lower resistivity than its compacted counterpart for the same double reason. At high environmental suctions, the granulated soil will have less water-filled small pores than the natural soil from which it was derived. It will have a lower total capillary conductance and conductivity, as well as a lower conductive cross-sectional area than its natural counterpart; these two factors will combine constructively to give the granulated soil a higher resistivity. For the same double reason, a natural, undisturbed soil will have a higher resistivity than its compacted counterpart at high suctions.

The moisture regime at which disturbed and undisturbed soils are most distinguishable in their resistivities and during which resistivity surveys are optimal is dependent on the effect of capillary size on capillary conductance and total soil conductivity. It was noted earlier that disturbed soils and their natural counterparts are most distinguished in their volumetric moisture contents and conductive cross-sectional areas at low and high suctions, and least distinguished in these attributes at some intermediate suction. If the differences between the two soil types in their volumetric moisture contents are the same at a low suction and a high suction on either side of the intermediate-suction suboptimum, however, the difference in their resistivities will be greater at the *high* suction. This circumstance results from the fact that capillary resistance (the reciprocal of capillary conductance) increases at a faster rate at high environmental suctions than at low ones per unit decrease in the average radii of the capillaries still holding water, or approximately per unit increase in suction [Eq. (3.11) Figure 9; Edlefsen and Anderson 1941; Shea and Lutkin 1961].

Soil Organic Matter: Its Hygroscopic Capacity

Most of the changes in structural organization and moisture equilibrium states that occur in the soils of archeological sites and that favor those soils being distinguished from natural soils in their resistivities are the result of the deposition of organic residues by human occupants.[10] The following four sections will consider the various pathways by which organic enrichments can cause these changes (Figure 10).

The most direct way in which human-deposited organic residues can alter soils on archeological sites so as to distinguish them in their resistivity is by increasing their content of moisture-holding humus colloids. Through the

---

10. Physical compaction is an exception.

Soil Organic Matter: Its Hygroscopic Capacity   69

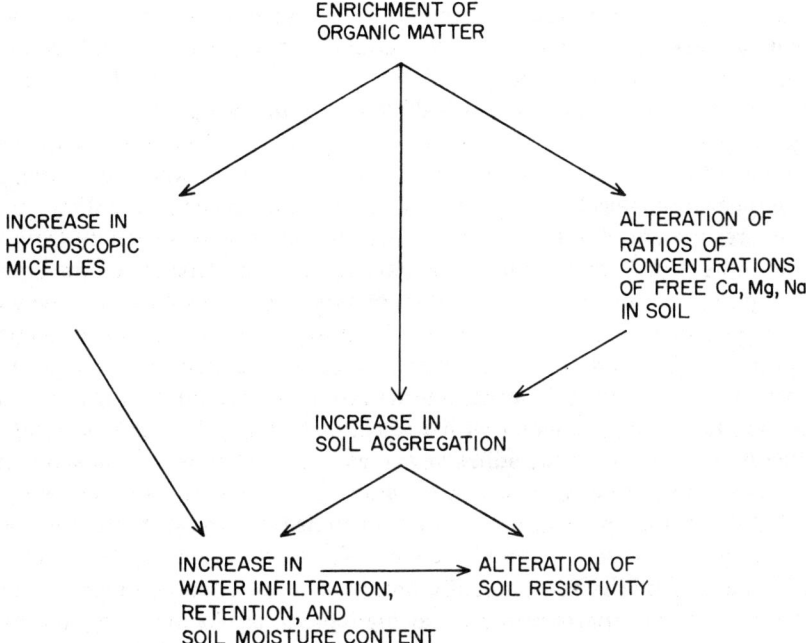

FIGURE 10. *Various manners in which enrichment of organic matter to a soil may augment its water retentive capacities.*

biochemical processes of decomposition and polycondensation, organic residues are transformed into highly reactive humic structures (Chapter 4) that have a great affinity for water. The hydrophilic nature of humus colloids, in part, is a result of the great density of oxygen-containing functional groups that occur on their surfaces and are capable of bonding water molecules. Among these functional groups are COOH, phenolic-OH, aliphatic-OH, enolic-OH, and C–O-reactive sites (Stevenson 1972:644). The geometry of humus micelles also is significant. Humus molecules have a loose, "spongy" structure with a large number of internal spaces and a great surface area, which gives them considerable absorptive and adsorptive capacity (Brady 1974:148; Kononova 1961:69).

The quantity of water that humus micelles can adsorb depends on their particular structure, which in turn is related to the soil environment in which they formed. Humus micelles that form in podzolic soils and other soils that are consistently moistened by rainfall tend to have an aliphatic structure (linear, with many side chains), and have a pronounced hydrophilic character. Humus micelles that form in chernozems and other soils that periodically suffer moisture deficits tend to be more spheroidal, and have lower moisture-holding capacities (Kononova 1961:69, 88). In general, however, all forms of humus micelles are much more hydrophilic than other soil components, and an increase in their content will increase the moisture-holding capacity of the soil. A humus micelle can adsorb an amount of water equivalent to about 80–90% of its own weight from a saturated atmosphere. A clay colloid, on the other hand, may adsorb an amount of water only 15–20% of its own weight.

The hygroscopic capacities of humus micelles outlined previously are of significance to this study in that they affect the degree to which soils that have been disturbed by enrichment of their organic matter contents are distinguished in moisture content, conductive cross-sectional area, and resistivity from their natural undisturbed counterparts. The particular effect depends on the environmental suction to which the soils under consideration are subjected. The water-retentive property of organic micelles will most influence the moisture content of a soil at high suctions rather than low ones. At high suctions, a greater proportion of the water still remaining within a soil will be held within smaller pores, will be in direct contact with soil solids forming the walls of pores, and will be directly subject to the adhesive forces of those solids as opposed to other moisture-retaining forces (cohesion of water molecules, the surface tension of water). Also, at higher suctions, a greater proportion of the still-remaining water will be absorbed *within* humus micelles. Thus, the hygroscopic property of organic matter will favor organically enriched soils retaining more water than their natural counterparts primarily at high suctions. At low ones, the hygroscopic property of organic matter will not significantly influence the relative amounts of water retained by disturbed and undisturbed soils.

When the effects of the hygroscopic nature of organic matter on soil moisture balance are combined with the effects of organic matter on soil aggregation and its influence on soil moisture equilibria, anthropically disturbed soils and natural soils will differ in their moisture contents and the resistivites[11] favored by those moisture contents in the following pattern. At low environmental suctions, the greater degree of aggregation of an anthropically disturbed soil than the natural soil from which it is derived will result in the disturbed soil tending to have a greater volumetric moisture content than the natural soil, and consequently, a greater conductive cross-sectional area and lower resistivity, as just described. The greater total hygroscopic capacity of the organic matter within the disturbed soil than the natural soil, as a result of the greater concentration of organic matter in the disturbed soil than the undisturbed soil will not significantly augment this tendency. As environmental suction increases through the intermediate range, and again as a result of the difference between the disturbed and natural soils in only their degree of aggregation and not the total hygroscopic capacity of their humus micelles, the moisture content of the disturbed soil will approach that of its natural counterpart, and eventually become less than that of its natural counterpart. The tendency for the disturbed soil to retain less water than the natural soil as a result of differences in their pore-size distribution, however, will increasingly be offset as environmental suctions continue to rise, by the greater organic matter content and the greater

---

11. Differences between the disturbed and natural soils in cation exchange capacity, availability of nutrients, and concentration of ions within their soil waters are not being considered here.

total hygroscopic capacity of the organic matter within the disturbed soil. Eventually, a suction may be reached where the difference in the organic matter content and total hygroscopic capacity of the humus micelles in the disturbed and natural soils is as important as the difference in the degree of aggregation in the two types of soils in determining their equilibrium moisture contents. At this point, the two soil types will again approach a common moisture content and a similar resistivity will be favored.

Thus, over an entire range of increasing environmental suctions, the distinctiveness of an anthropically disturbed soil and its natural counterpart in their moisture contents and resistivities will at first decrease, then increase, and finally decrease again. The initial decrease and subsequent rise in distinctiveness reflect the effects of organic matter on soil moisture through its influence on soil aggregation. The final decrease in distinctiveness reflects the effect of organic matter on soil moisture equilibria, both indirectly, through its influence on soil aggregation, and directly, through its hygroscopic nature.

This model of the contrast found between an organically enriched soil and its natural counterpart in their moisture contents, and of the relative importance of aggregation and of the hygroscopic capacity of humus micelles in determining the contrast, is applicable only to soils with sandy loam or finer textures. An increase in the organic matter content of a sand or loamy sand soil will increase its moisture-holding capacity at any suction. This results from the fact that a sandy soil has almost no fine capillaries and can hold almost no water at moderately low through high suctions. Organic enrichments to a sandy soil will increase the amount of moisture it can hold at low suctions by developing its structure, and will increase the amount of moisture it can hold at moderately low through high suctions by increasing its total hygroscopic moisture capacity. Structural developments, which cause losses of moisture from soils of finer texture at intermediate and high suctions, cannot possibly cause such losses in sandy soils, which retain almost no moisture at such suctions to begin with.

Finally, it should be noted that the moisture-holding nature of humus micelles is significant, not only to helping to *retain* water within soils, but also in the capture of rainfall. Organic matter on the surface of a soil may hold water there long enough for it to seep into the soil rather than run off (Wischmeier and Mannering 1965). This phenomenon will tend to give the soils in those areas of archeological sites where many organic residues have been deposited greater moisture contents than the soils in areas of less organic enrichment—regardless of the environmental suction.

Soil Organic Matter: Its Effects on Soil Aggregation

The second way in which anthropically deposited organic materials can alter soils on archeological sites so as to distinguish them in their resistivities is by indirectly facilitating moisture retention or loss through the development of soil aggregation. First let us consider the process of aggregation itself and then the effects of aggregation on soil moisture balance.

3. Factors Affecting Soil Conductivity

A *soil aggregate* is a group of particles in which the forces holding them together are much stronger than the forces between adjacent aggregates (Allison 1973:316). Two separate processes are required to produce this situation: (*a*) aggregate formation, in which the particles are brought into contact by their physical displacement; and (*b*) aggregate stabilization, in which the abutting particles are bonded together by some chemical means. Physical grouping of soil particles may result from a number of circumstances. On freezing, soil water will expand and can produce high pressures that displace particles. Slow freezing will result in large but infrequently spaced ice crystals and large units of aggregation, whereas rapid freezing will yield smaller, closely spaced ice crystals and smaller units of aggregation. The burrowing action of earthworms and other soil megafauna, and the penetration of rootlets through soil also can cause local compaction and aggregate formation. Rootlets also cause stresses and strains that may move particles within soils by removing water in some localized areas and not in others. Macroscale drying and wetting of a soil is of some value in aggregate formation, in that it causes 2:1 expandable clays (see Chapter 4) to shrink and expand and results in horizontal and vertical shearing movements of soil particles. Such movements can reorient soil particles so as to form dense masses with a minimum of pore space. In general, however, soil wetting and drying cycles favor the disruption of stabilized aggregates more than they do aggregate formation (Allison 1973: 316–324).

The primary role of organic matter is in aggregate *stabilization* rather than aggregate *formation*. The mechanisms are multiple, and depend on the form and complexity of the organic matter. First, complex humic and fulvic acids—the final, high-molecular weight products of decomposition (Chapter 4)—may precipitate and envelop soil aggregates as the soil dries. If the humus colloids, themselves, dry or freeze, they will denature irreversibly (at which point they are called humin) and form a permanent, chemically very stable coating over the aggregates (Allison 1973:325). Humic substances that stabilize aggregates in this matter can participate only once in the development of soil structure (Kononova 1961:70). Second, the fatty and waxy byproducts of decomposition of organic residues may reside within soil aggregates. These substances are hydrophobic (i.e., they prevent moisture from entering soil aggregates) and thereby minimize aggregate disintegration during drying and wetting cycles (Allison 1973:329). Third, low-weight, linear, organic polymers may link two or more clay particles together by bonding to them chemically. The organic substances primarily involved include polysaccharides, polyuronides, and cellulose—all of which are products of decomposition.

Bonding may be of several varieties. Hydrogen bonds may form between hydroxyl groups on the polymers and oxygen atoms on the exposed surfaces of the clay crystals (Chapter 4). Anion exchange reactions between polymer functional groups and reactive sites on clay colloids also may bind these micelles (Harris *et al.* 1966). Water dipoles may serve as binding agents, linking organic polymers and clay particles together. The primary means by which humic materials and clays are linked, however, is by ionic bonding involving polyvalent cations (Ca, Mg, Fe, Al) or oxides (Brady 1974:62; Greenland 1965a, 1965b).

The particular structural alterations that occur in a soil as it becomes better aggregated through the effects of organic matter enrichments and the impact of these changes on soil water balance have been mentioned briefly in the previous section. Now let us examine them in detail. As a soil's organic matter content is augmented and aggregation develops, its total porosity and maximum moisture-holding capacity increases. These increases are a function of increases in the numbers of *both* larger and smaller pores over the 0 through 15 bar (or more) suction range, but more so in the numbers of large pores. This particular pattern of alteration of the porosity, pore-size distribution, and moisture-holding characteristics of soils has been found repeatedly in experimental studies where various forms of organic matter have been added to soils of a wide range of textural classes (Table 4). Finer-textured soils exhibit smaller increases in total porosity than do coarser-textured soils per unit of organic matter added (anonymous 1955; Bouyoucos 1939), particularly within the small pore-size range (Coile 1939). Table 5 will give the reader an idea of the magnitudes of change in soil structure and water balance that organic matter enrichments can yield in soils of various textural classes.

The precise manner in which soil aggregation augments the total porosity of a soil and changes its pore-size distribution may be understood if two types of pore space and their pore size distributions are considered: that within aggregates and that between aggregates. First let us examine changes in total porosity. As the size of aggregates within a soil increases (i.e., as aggregation increases), the corresponding increase in total soil porosity is primarily a result of an increase in intra-aggregate porosity (Figure 11; Ameniya 1965; Wiltmuss and Mazurak 1958). Intra-aggregate porosity increases with aggregate size because the volume-to-surface area ratio of an aggregate increases as it grows in

TABLE 4
Experimental studies suggesting that organic additions to soils increase total soil porosity, the number of larger pores, the number of small pores, but more so the larger pores

Reference	Kind of Organic Matter Added	Textural Class of Soils Examined
Bouyoucos (1939)	muck, manure, peat	clays to sands
Coile (1939)	?	fine, coarse-textured
Gingrich and Stauffer (1955)	manure, crop stubble	silt loam
Jamison (1953)	manure	clays to sands
Junker and Madison (1967)	peat	sand
Kute and Jacobs (1950)	manure	silt loam
Salter et al. (1967)	manure	?
Salter and Haworth (1961)	manure	sandy loam

3. Factors Affecting Soil Conductivity

TABLE 5
Changes in soil structure and water balance in soils of various textural classes with various amounts of muck added

% Muck Added	Moisture Content at Field Capacity (ca. .1-.2 bar)	Moisture Content at Primary Wilting Point (ca. 15 bars)
Sand		
0	9.4	1.70
2	13.9	2.69
4	16.5	4.08
6	18.8	5.60
8	20.1	6.55
10	22.3	7.58
12	24.1	8.50
Increase, 0-12%	14.7	6.80
Sandy Loam		
0	13.3	7.61
2	15.7	8.65
4	18.7	9.44
6	20.4	9.90
8	23.9	10.36
10	25.3	11.28
12	27.0	12.03
Increase, 0-12%	13.7	4.42
Silt Loam		
0	21.7	7.80
2	24.3	8.50
4	26.0	9.30
6	27.2	10.1
8	28.0	10.6
10	31.5	11.4

size, and the frequency with which pores are located within its interior rather than at its surface and constitute interaggregate pore space increases. Interaggregate porosity, on the other hand, will vary indeterminantly with aggregate size—in the same way that there is an indeterminant relationship between the textural distribution of a soil and its porosity (Hrubíšek 1941).

TABLE 5 (cont.)

% Muck Added	Moisture Content at Field Capacity (ca. .1-.2 bar)	Moisture Content at Primary Wilting Point (ca. 15 bars)
Silt Loam (cont.)		
12	34.2	12.1
Increase, 0-12%	12.5	4.3
Clay Loam		
0	34.2	16.6
2	35.0	17.2
4	35.8	18.0
6	37.5	18.8
8	39.4	19.8
10	41.7	20.4
12	43.2	21.5
Increase, 0-12%	8.0	4.9
Clay		
0	39.8	19.4
2	42.5	19.8
4	43.7	20.9
6	44.7	21.2
8	45.4	22.4
10	46.3	23.4
12	47.6	23.6
Increase, 0-12%	7.8	4.2

Ordering of textural classes by their increments in moisture holding capacity per unit of organic matter added, for the range 0-12% organic matter content:
.1-.2 bar section: sand > sandy loam > silt loam >> clay loam ≃ clay
15 bar section: sand > sandy loam ≃ silt loam ≃ clay loam ≃ clay

 The change in the total pore-size distribution of a soil as its aggregation increases is a result of alterations in the pore size distributions of both interaggregate and intra-aggregate void space. As the size of aggregates within a soil increases, both the volume of large pores and the volume of small pores within aggregates will increase, but more so the volume of large-sized pores

(Tamboli *et al.* 1964:105). This tendency for large pore-size void space to be increased more than small pore-size void space within aggregates is complemented by similar changes that occur in the pore-size distribution of interaggregate void space as aggregation increases. At the interaggregate level, the volume of large pores increases and the volume of small pores decreases. The changes in pore-size distribution at the interaggregate level are analogous to the greater volume of large pores and the lesser volume of small pores that a coarse-textured soil has compared to a fine-textured soil (see page 54).

With respect to the interpretation of resistivity data from archeological sites, these patterns of structural alteration are critical. Where soils have been enriched with organic matter by human occupation, they will hold more water per unit volume, favoring their lower resistivity, than their natural soil counterparts, when environmental suctions are low and when both large pores and small pores are water filled.[12] At moderately high suctions, when only small pores are water-filled, organically enriched soils will hold less water per unit volume,[13] which will favor them having higher resistivities than their natural counterparts. At very high suctions, the hygroscopic nature of organic matter will tend to offset the pattern found at moderately high suctions, decreasing the contrast between the disturbed and natural soils in their moisture contents and in the resistivities favored by their moisture contents.

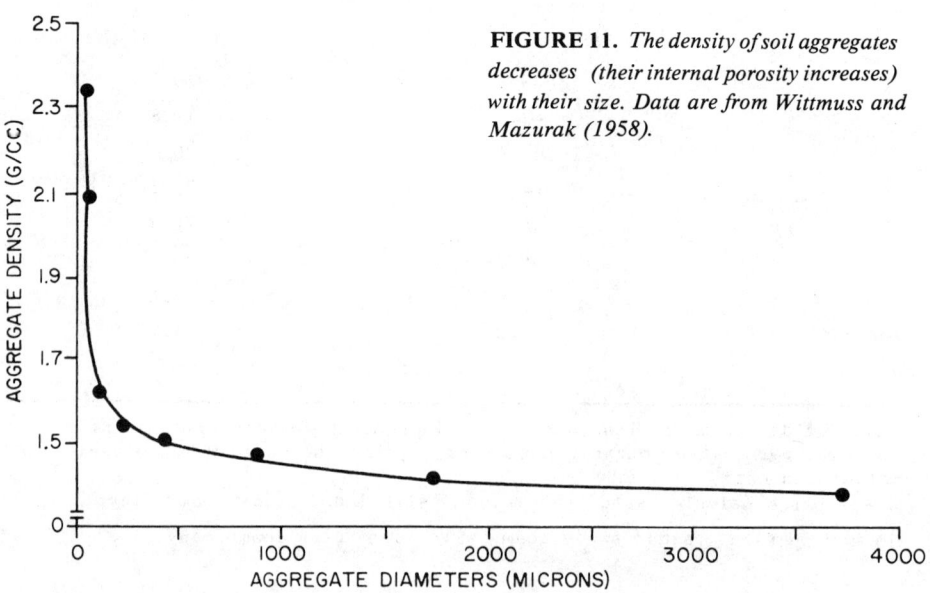

FIGURE 11. *The density of soil aggregates decreases (their internal porosity increases) with their size. Data are from Wittmuss and Mazurak (1958).*

12. Differences between the disturbed and natural soils in cation exchange capacity, availability of nutrients, and concentration of ions within their soil waters are not being considered here.

13. Applicable to soils having sandy loam or finer textures only. An increase in the organic matter content of a sandy or loamy sandy soil will increase its moisture-holding capacity for any suction. See the previous section.

Soil Organic Matter: Its Effects on the Infiltration of Rainfall

The aggregation produced by organic enrichment to a soil will affect a soil's water balance and resistivity not only by increasing the amount of water it is capable of retaining at particular environmental suctions, but also by increasing the amount of water it is capable of capturing during rainfall. Large capillaries produced within a soil by the aggregation process permit rainwater to seep into the soil and to distribute itself throughout the soil more rapidly, and consequently, prevent loss of rain water by surface runoff.

That this infiltration effect can be very significant in determining a soil's moisture content is demonstrated by a study of Wischmeier and Mannering (1965). They examined infiltration and runoff rates as related to a large number of soil variables and simulated rainfall patterns for 44 soils, ranging in texture from sandy loams through clay loams (most within the silt loam and loam classes). In regressing runoff and infiltration against pedological variables they found that between 36 and 43% of the variance in runoff and infiltration could be accounted for by the organic matter contents of the soils and presumably the degree of aggregation of the soils. The percentage of the variance that could be accounted for increased as the amount of rainfall was increased (longer showers at the same rate of precipitation).

Thus, soils that have been disturbed by the addition of anthropically derived organic matter and that show greater degrees of aggregation should be capable of capturing greater amounts of water than their natural counterparts. During periods of low environmental suction, this circumstance will complement that relationship between soil organic matter content, aggregation, and soil moisture content; it will tend to give the disturbed soil a higher moisture content, and will augment the contrast between the disturbed and undisturbed soils in their moisture balances and resisitivities. During periods of high environmental suction, however, when a disturbed soil will tend to have less water than its natural counterpart,[14] the greater ability of the disturbed soil to capture precipitation will diminish the contrast between the two soil types.

Soil Organic Matter: Its Effects on Soil Chemistry and Flocculation of Soil Colloids

Additions of organic matter to a soil may also alter its moisture equilibria and resisitivity by changing the ratio of the amounts of different kinds of free ions that are found within the soil and that may facilitate or retard soil aggregation. When an organic residue is introduced to a soil and undergoes decomposition, some of the ions forming the residue are released to the soil in ionic form (see Chapter 4). In general, Ca and Mg occur in organic matters in greater

---

14. Sandy loam or finer textured soils, only.

proportions than does Na (Table 67) and they will be released to the soil in greater amounts than will Na. The resulting increase in the concentration of Ca and Mg within the soil, both in absolute quantities and relative to the concentration of Na, will favor the development of soil aggregates, and thus, moisture retention. At least two mechanisms are involved. First, Ca and Mg, along with other polyvalent cations, encourage soil colloids (both clays and humus micelles) to flocculate (i.e., clump into aggregates). Na, on the other hand, encourages soil colloids to maintain a dispersed state (Brady 1974:108–109; Kemper 1958; Puri and Mahajan 1962). By altering the balance of Ca, Mg, and Na in favor of Ca and Mg, the introduction of organic residues to a soil facilitates flocculation. Flocculated soil colloids do not constitute stabilized soil aggregates, but the addition of Ca and Mg to a soil (particularly Ca) will encourage aggregate stabilization, as well. Chemical bonding between polyvalent cations and clay micelles and between polyvalent cations and organic polymers are the primary means by which aggregates of colloids are stablized within soils (Allison 1973:325; Russell 1961).

The degree to which organic residues will facilitate the formation and stabilization of aggregates within a soil, of course, will depend on the ratios of Ca, Mg, and Na found within the residues. As a consequence, different areas of an archeological site where different kinds of activities have been performed and different kinds of refuse materials have been deposited will differ in the degree to which their soils have undergone aggregation. The degree of aggregation found in the soils of different use-areas also will depend, however, on the amount of refuse deposited within them.

Soil Organic Matter: Its Effect on
Soil Chemistry and the Osmotic Potential

A final matter in which additions of organic matter to a soil may alter its moisture equilibria and resistivity is by increasing the osmotic potential and reducing the free energy of its soil water. Ions that are released to the soil water solution when organic amendments decompose are attracted to and bind together the dipolar molecules of water within the solution. This attractive force, or "osmotic suction," reduces the capacity of the soil water to be lost from the soil in two ways: (a) It lowers the vapor pressure of the water and prevents it from evaporating within the soil or at the soil surface; (b) it reduces the ability of plant roots to absorb water that would subsequently by lost be transpiration (Brady 1974:172, 196).

The effect of changes in the osmotic potential of a soil water on its retention, compared to the effects of changes in soil structure and changes in the total hygroscopic capacity of organic colloids on soil water retention, is relatively small per measure of organic matter added to the soil (Brady 1974:196). For purposes of modeling the changes in moisture equilibria and resistivity that occur in a soil after organic matter has been added to it, this factor may be disregarded.

Conductivity of Soil Water: Introductory Framework

The structure of soil water, as determined by the structural organization of a soil, its moisture content, and indirectly, its organic matter content, is one summary variable that influences soil resistivity. The second summary variable determining soil resistivity is the electrical conductivity that soil water would exhibit in an unstructured state outside the soil [Eq. (3.1)]. The electrical conductivity of soil water, as one form of aqueous solution, is determined by a number of electrolytic phenomena, most of which may be subsumed under and accounted for by four operational variables: (*a*) the kinds of ions present in the water, (*b*) their concentration within the water, (*c*) the concentration of conductive colloidal particles, and (*d*) the temperature of the water. In the following sections, I will discuss the relationship of each of these variables, and their associated phenomena, to the conductivity of soil water.

If one examines pure water for its electrical conductivity, one will find it to be very low—a few hundredths of a micromho per centimeter at 25°C (Hem 1959). As soluble compounds are added to water and the concentration of dissolved ions (electric carriers) within the water increases, the conductivity of the aqueous solution will increase. It also will increase as the concentration of conductive collodial particles (also electric carriers) increases and the temperature of the solution rises. For aqueous solutions composed of uni–uni salts (salts formed with a cationic species having a valence of 1 and an anionic species having a valence of 1) in *very* low concentration, the increase in conductivity with ionic concentration or temperature is linear (Figures 12, 13). The relationship between the conductivity of a liquid and the concentration of conductive colloidal particles within it also shows a linear relationship at very low concentrations (Figure 14; Hem 1959).

Unfortunately, linear models of these types are very restricted in the range of circumstances to which they are applicable. As the concentration of ions or dissolved colloidal particles increases any appreciable amount in an aqueous solution and as the number of different kinds of ions dissolved within an aqueous solution increases, the relationship between conductivity and concentration becomes more complex; so does the relationship between conductivity and temperature. In order to present the models of electrolytic chemistry that do describe these more complex relationships, it is best if some basic concepts in chemistry are explained first.

Most important, it is necessary to understand that the conductivity of a substance is the ability *per unit volume* of that substance to conduct electric current. Likewise, the resisitivity of a substance (the reciprocal of conductivity) is the resistance *per unit volume* of that substance to electric current flow. This is not apparent, at first, for the units in which conductivity and resistivity are measured do not imply standardization by volume. The unit measure of resistivity is the ohm-cm. The unit measures of conductivity are the ohm^{-1} −cm^{-1} or the mho/cm (10^5 micromhos/cm). If we examine the physical model on which the definition of conductivity and resistivity are based, however, their volumetric nature and the rationale for the particular units attached to them will become obvious.

Suppose a conductive parallelepiped, either solid or electrolytic, has a length L, a cross-sectional area a, and is divided into 1-cm-thick slices. Also imagine that each slice is composed of a cubes, each 1 cm on the side and having a resistance ρ, measured in ohms (Figures 15a, 15b). This resistance, by definition, is called the resistivity of the substance; that is, resistivity by definition is resistance per unit volume. The unit-measure given to resistivity, the ohm-cm, however, does not imply this standardization by volume. Rather, it is designed to keep it and other unit measures *consistent* with each other within basic equations of physics describing the resistance of more complex, multi-resistor systems such as that shown in Figure 15. This may be seen in the following derivation. The resistance of the whole parallelepiped shown in Figure 15 may be determined if each slice is thought of as a resistor hooked in series with other slices and if each 1-cm^3 cube within a slice is thought of as a resistor hooked in *parallel* with other cubes within the same slice. The resistance, r, of each slice is described by the relation:

$$\frac{1}{r} = \frac{1}{\rho} + \frac{1}{\rho} + \frac{1}{\rho} + \frac{1}{\rho} + \cdots$$

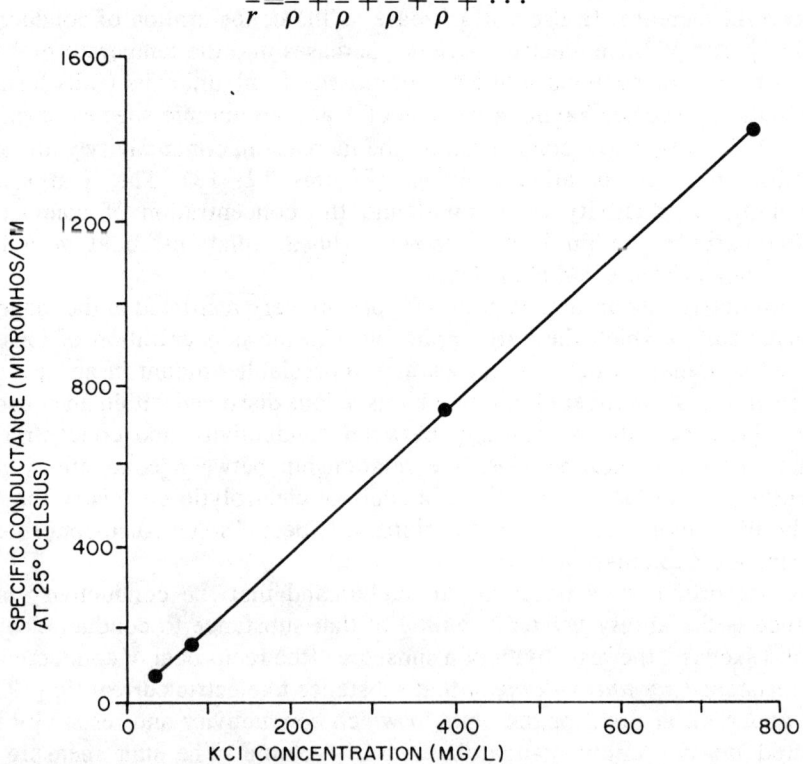

FIGURE 12. *Variation in the conductivity of an aqueous solution of KCl at several very low concentrations. Adapted from Hem (1959:97).*

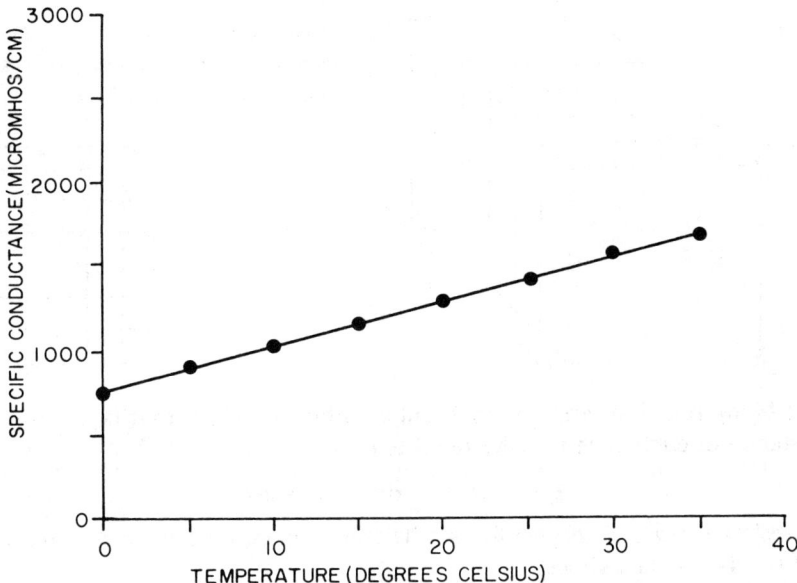

FIGURE 13. *Variation in the conductivity of an aqueous solution of KCl (.10 molar concentration, very low) at several temperatures. Adapted from Hem (1959:98).*

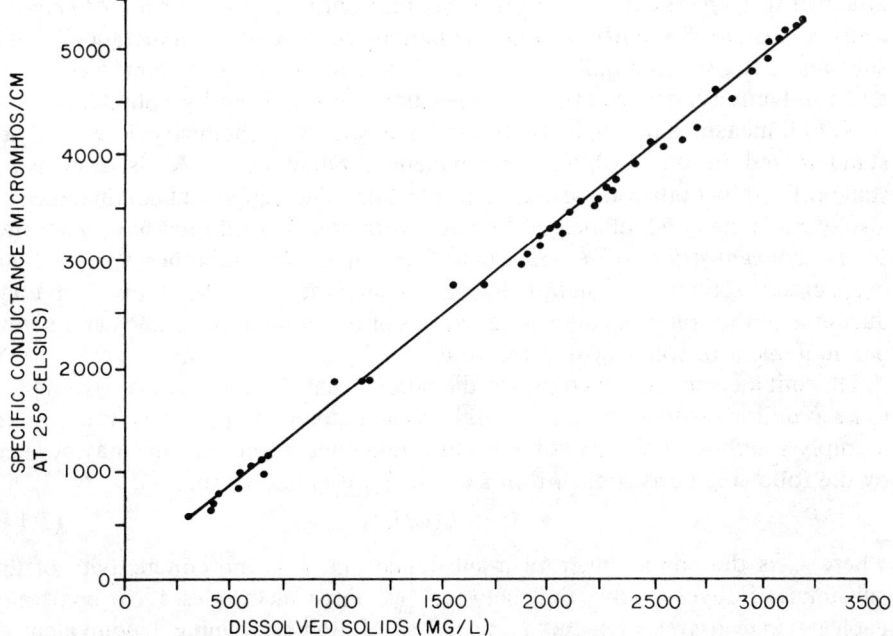

FIGURE 14. *Variation in the conductivity of water samples from the Gila River at Bylas, Arizona, as the concentration of conductive colloidal particles changes. Adapted from Hem (1959:100).*

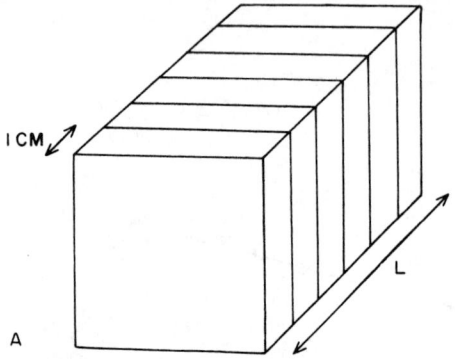
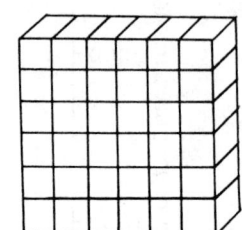

FIGURE 15. *A conductive parallelepiped thought of as a group of resistors hooked in a series and in parallel (see text).*

there being one $1/\rho$ term for each cube within the slice, that is a terms. The resistance of each slice may be rewritten:

$$1/r = a/\rho \quad \text{or} \quad r = \rho/a$$

The resistance, R, of the whole parallelepiped is equivalent to the sum of all L of the 1-cm-thick slices:

$$R = Lr = L\rho/a \tag{3.13}$$

(Glasstone 1942). As R is measured in ohms and $1/a$ has the unit cm^{-1} attached to it, ρ (resistivity) is *given* the unit ohm-cm *for the sake of keeping units consistent*. Similarly, K, the conductivity or "specific conductance"[15] of a substance, is given the units $ohm^{-1}-cm^{-1}$ to keep units consistent. Resistivity and conductivity, nevertheless, are measures standardized by volume.

Not all measures of conductivity used in electrolytic chemistry, however, are standardized in this fashion. "Equivalent conductance" (Λ) is a measure standardized by both volume and concentration. The equivalent conductance of a solution is the conductance of *1 cubic centimeter* of a solution having *solutes in the concentration of 1 equivalent*[16] *per liter*. Or, in other terms, if an unspecified volume of solution has a conductance, C, the equivalent conductance of the solution is the conductance of the volume per cubic centimeter, per equivalent of solute within the volume.

The unit measure attached to Λ is the $ohm^{-1}-cm^{-2}/equiv$. It, too, is designed to keep unit measures consistent within basic equations of physics, rather than to imply standardization by volume and solute concentration. This may be seen by the following derivation. From Eq. (3.13), it is known that

$$C = k(a/L), \tag{3.14}$$

where C is the conductance of a substance and k is the conductivity of the substance. Suppose, now, that between two large electrodes 1 cm apart and each having an area of v cm² is placed a solution containing 1 equivalent of solute. By Eq. (3.14), the conductance C of the solution is

$$C = k(V/1) = Kv = \Lambda,$$

---

15. These terms may be used interchangeably.
16. The equivalent weight of a substance and other measures of weight and concentration used in this section are defined in Table 6.

TABLE 6
Definition of basic concepts used in electrolytic chemistry

Measures of Weight and Mass

   Molar Weight. One mole of an element is equal to one gram-atomic weight of that element. One mole of a compound is equal to one gram-molecular weight of that compound. The gram atomic weight of an element may be found in the Periodic Table of Elements.

   Example 1. The atomic weight of Mg is 24.33 amu. The atomic weight of Cl is 35.457 amu. What are the weights of 1 mole of Mg, 1 mole of Cl, and 1 mole of $MgCl_2$?

$$1 \text{ mole of } Mg = 24.32 \text{ g}$$
$$1 \text{ mole of } Cl = 35.457 \text{ g}$$
$$1 \text{ mole of } MgCl_2 = 24.32 \text{ g} + 2(35.457)\text{g} = 95.234 \text{ g}$$

   Equivalent Weight. One equivalent of an element is equal to the gram-atomic weight of that element divided by its valence. One equivalent of a compound is equal to the gram-molecular weight of the compound divided by the total positive (or negative) valence of the compound. The gram equivalent weight and valence of an element may be found in the Periodic Table of Elements.

   Example 2. What is the equivalent weight of one equivalent of Mg, Cl, and $MgCl_2$

$$1 \text{ equiv. of } Mg = \frac{24.32}{2} \text{ g} = 12.16 \text{ g}$$
$$1 \text{ equiv. of } Cl = \frac{35.45}{1} \text{ g} = 35.457 \text{ g}$$
$$1 \text{ equiv. of } MgCl_2 = \frac{24.32 \text{ g} + 2(35.45)\text{g}}{2}$$
$$= \frac{95.234}{2} \text{ g} = 47.667 \text{ g}$$

   Conversion. The molar weight of an element or compound is equal to the equivalent weight of an element or compound divided by its valence.

Measures of Solution Concentration

   Molar Concentration. A one molar solution of a compound is equal to one gram-molecular weight of the compound dissolved in a solvent so as to make one liter of solution.

   Example 3. A one molar solution of $MgCl_2$ (written 1M $MgCl_2$) is 95.234 g of $MgCl_2$ dissolved in nearly one liter of solvent. In this solution are 1 mole of Mg ions and 2 moles of Cl ions.

   Molal Solution. A one molal solution of a compound is equal to one gram-molecular weight of the compound dissolved in 1000 g of solvent. One milliliter (1 cc) of water at 3.98°C has a weight of 1 gram, so at 3.98°C, a one molal aqueous solution of a compound is equal to one gram-molecular weight of the compound dissolved in 1 liter of water.

   Example 4. At 3.98°C, a one molal aqueous solution of $MgCl_2$ is 95.234 g of $MgCl_2$ dissolved in 1 full liter of water.

   Equivalent Solution. A one equivalent solution (1 normal solution) of a compound is equal to 1 equivalent weight of the compound dissolved in a solvent so as to make one liter of solution.

TABLE 6 (cont.)

Example 5. A one normal solution of $MgCl_2$ (written 1N $MgCl_2$) is 47.667 g of $MgCl_2$ dissolved in <u>nearly</u> one liter of solvent. In this solution are one equivalent of Mg ions and one equivalent of Cl ions.

<u>Conversion</u>. To convert the concentration of a solution from one standard to another, always first find the gram weight of the element or compound found in 1 liter <u>of the solution</u>, and then apply the above definitions to those data.

---

where C by definition is the equivalent conductance Λ of the solution. Here, v is not only the area of the electrodes, but also the volume containing 1 equivalent of solute, that is, the dilution of the solution in cm³/equiv. The reciprocal of v, then, is the concentration of the solution in equivalent cm³. If c is defined as the concentration of the solution in equivalents per liter, then $v = 1000/c$ and

$$\Lambda = k(1000/c) \tag{3.15}$$

Since k has the units ohm⁻¹–cm⁻¹ and c has the units equiv/l, Λ is given the units ohm⁻¹–cm⁻²/equiv for the sake of consistency. This should not confuse the reader, however, as to the nature of equivalent conductance with respect to solution volume and concentration.

Although the relationships we wish to investigate are those between various chemical characteristics of soil water solutions and the conductivity (k) of soil water solutions, it is necessary that these relationships be presented in terms of Λ rather than k. It is in terms of Λ that physical chemists have chosen to derive the mathematical models that describe the determinants of the electrical properties of solutions.

The Conductivity of Soil Water as a Function of the Concentration of Strong Electrolytes and Solution Temperature

Having reviewed some basic chemistry, let us now consider how the content and concentration of ions within an aqueous solution and the temperature of an aqueous solution affects its equivalent conductance. If the ions of any solution are subjected to the external force of an electric field, two electrokinetic phenomena arise, the *electrophoretic* effect and the *relaxation* effect; each determines the equivalent conductance of the solution and is affected by the concentration of ions in the solution. The electrophoretic effect occurs when the ions moving through a viscous solvent toward an oppositely charged electrode drag along with them molecules of solvent in their immediate vicinity, such that neighboring ions heading in the opposite direction toward the other electrode must move against the carried stream of solvent (Robinson and Stokes 1955). As the concentration of the ions increases and the distance between the ions in the solution decreases, each ion must move against a greater resistant force of the streaming solvent. The mobility of the ions in the solution and their ability to migrate to the oppositely charged electrodes, therefore, is reduced, and the solution as a whole has a lower equivalent conductance. This effect is

TABLE 7a
Limiting equivalent conductivities, $\lambda°$, of inorganic ions in water at various temperatures*

Ion	0° C	5° C	15° C	18° C	25° C	34° C	45° C	55° C	100° C
H+	225.0	250.1	300.6	315.0	349.8	397.0	441.4	483.1	630
OH-	105.0	---	---	171.0	198.6	---	---	---	450
Li+	19.4	22.7	30.2	32.8	38.6	48.0	58.0	68.7	115
Na+	26.5	30.3	39.7	42.8	50.10	61.5	73.7	86.8	145
K+	40.7	46.7	59.6	63.9	73.50	88.2	103.4	119.2	195
Rb+	43.9	50.1	63.4	66.5	77.8	92.9	108.5	124.2	---
Cs+	44.0	50.0	63.1	67.0	77.2	92.1	107.5	123.6	---
Ag+	33.1	---	---	53.5	61.9	---	---	---	175
NH_4^+	40.2	---	---	63.9	73.5	---	---	---	180
F-	---	---	---	47.3	55.4	---	---	---	---
Cl-	41.0	47.5	61.4	66.0	76.35	92.2	108.9	126.4	212
Br-	42.6	49.2	63.1	68.0	78.1	94.0	110.6	127.8	---
I-	41.4	48.5	62.1	66.5	76.8	92.3	108.6	125.4	---
NO_3^-	40.0	---	---	62.3	71.46	---	---	---	195
ClO_4^-	36.9	---	---	58.8	67.3	---	---	---	185
Mg++	28.9	---	---	44.9	53.0	---	---	---	165
Sr++	31.0	---	---	50.9	59.4	---	---	---	---
Ca++	31.2	---	46.9	50.7	59.50	73.2	88.2	---	180
Ba++	34.0	---	---	54.6	63.6	---	---	---	195
Cd++	---	---	---	44.8	---	---	---	---	---
La+++	34.0	---	---	59.5	69.7	---	---	---	215
SO_4^{--}	41	---	---	68.4	80.0	---	---	---	260
Viscosity of water (centipoises)	1.792	1.519	1.140	1.056	0.8937	0.7725	0.5988	0.5064	0.2838

*Taken from Robinson and Stokes (1955:454)

particularly active in aqueous solutions, for water dipole molecules have a strong tendency to follow neighboring ions.

The relaxation effect occurs when an electrical asymmetry in the distribution of the charged ions in a solution forms as a result of the motion of the ions under the external force of an electric field. In a solution at equilibrium, outside of an electric field, every ion on a time-average basis is surrounded by a sphere of other ions of both positive and negative charges that are in balance and do not exert any force on that ion. When the ion is subjected to an electric field and is moved away from the center position of its neutral sphere (its "ionic atmosphere"), it experiences a restoring force that decreases as the surrounding ions rearrange to accommodate themselves to the new position of the ion (i.e., the ionic atmosphere "relaxes"). As the concentration of the ions in a solution increases and the distance between ions decreases, the restoring force becomes greater (in agreement with Coulomb's law), the velocity of each ion decreases, and the equivalent conductance of the solution decreases. In sum, the equivalent conductance of a solution decreases as its concentration increases because the ions in the solution interfere with each other's ability to move and "carry an electric current." The equivalent conductance of a solution is greatest when the solution is most *dilute*.

It should be apparent from this description of electrokinetic phenomena that in order to model the relationship between the concentration of solutes in a

TABLE 7b
Limiting equivalent conductances, $\lambda°$, of some organic ions in water at 25° C*

Cations		Anions		Anions (cont.)	
i-Butylammonium	38.0	Acetate	40.9	Lactate	38.8
n-Decylpyridinium	29.5	p-Anisate	29.0	Malonate	63.5
Diethylammonium	42.0	Azelate	40.6	Methyl Sulfonate	48.8
Dimethylammonium	51.5	Benzoate	32.4	C_2O_4	74.2
Dipropylammonium	30.1	Bromobenzoate	30.0	Octyl sulfonate	29.0
n-Dodecylammonium	23.8	n-Butyrate	32.6	Phenyl acetate	30.6
Ethylammonium	47.2	Chloroacetate	39.7	Picrate	30.2
Ethyltrimethylammonium	40.5	Chlorobenzoate	33.0	Propionate	35.8
Methylammonium	58.3	Citrate	70.2	Propyl sulfonate	37.1
Histadyl	23.0	n-Crotonate	33.2	Salicylate	36.0
Piperidinium	37.2	Cyanoacetate	41.8	Suberate	36.0
Propylammonium	40.8	Cyclohexane carboxylate	28.7	Succinate	58.8
Pyrilammonium	24.3	Cyclopropane-1, 1-dicarboxylate	53.4	Sulfonate	43.1
Tetra-n-butylammonium	19.1	Decyl sulfonate	26.0	Tartrate	64.0
Tetramethylammonium $N(CH_3)_4$	44.9	Dichloroacetate	38.3	Trichloroacetate	36.6
Tetraethylammonium $N(C_2H_5)_4$	32.6	Diethyl barbiturate	26.3		
$N(C_3H_7)_4$	23.4	Dihydrogen citrate	30.0		
$N(C_4H_9)_4$	19.4	Dimethyl malonate	49.4		
$N(C_5H_{11})_4$	17.4	3,5-Dinitrobenzoate	28.3		

Tetra-n-propylammonium	23.5	Dodecyl sulfonate	24.0
Triethylammonium	34.3	Ethyl malonate	49.3
Triethylsulfonium	36.1	Ethyl sulfonate	39.6
Trimethylammonium	46.6	Fluorobenzoate	33.0
Trimethylsulfonium	51.4	Formate	54.6
Tripropylammonium	26.1	HC_2O_4	40.2

*A temperature coefficient of .02 deg^{-1} is applicable to the cations and anions. Extracted from Dean (1973:6-30, 6-31).

TABLE 7c
Limiting equivalent conductances for H^+ and OH^-

Temperature	H^+	OH^-	Temperature	H^+	OH^-
0°C	229.0	118	25°C	350.0	196
10°C	275.6	149	35°C	399.6	228
15°C	300.4	164.5	40°C	421.4	244
18°C	315.2	174	50°C	464.3	276

Taken from National Research Council (1929)

solution and its equivalent conductance, two quantities must be considered: (a) the equivalent conductance of the solution at concentrations approaching infinite dilution, when solute and solvent ions do not interfere with each other; and (b) the magnitude of loss in the equivalent conductance of the solution when the concentration of solute ions is increased above that of infinite dilution. The first quantity can be determined using Kohlrausch's law of independent migration of ions. This law states that when a solution is composed of one or more compounds, its equivalent conductance near infinite dilution is the sum of the limiting equivalent conductances of the separate ions:

$$\Lambda_0 = (\lambda^0_{+1} + \lambda^0_{-1}) + (\lambda^0_{+2} + \lambda^0_{-2}) + (\lambda^0_{+3} + \lambda^0_{-3}) + \cdots \qquad (3.16)$$

limiting equivalent conductance of the cation constituent of compound n, and λ^0_{+n} is the limiting equivalence conductance of the anion constituent of compound n.

The limiting equivalent conductances of separate ions (also called ionic conductance or ion conductance) are constants which have been determined for each ionic species (Tables 7a, 7b, 7c). Each is dependent primarily on the charge (valence) of the ion and secondarily on the radius of the ion. The ionic conductance of an ion does not depend on the other species of ions found within the solution under consideration or the other species of ion which was a constituent of the salt from which the ion was freed upon dissolution. The value of λ^0 does depend, however, on the temperature of the solution, an increase in temperature resulting in an increase in ionic conductance. The relationship is summarized by the expression:

$$\lambda_t = \lambda^0_{25}[1 + a(t - 25)] \qquad (3.17)$$

where λ^0_t is the ionic conductance of the species at infinite dilution at temperature t, λ^0_{25} is the ionic conductance of the species at 25°C, t is the new temperature, and where a is a constant, approximately equal to .02 for most ions, but equal to .0142 for H^+ and .016 for OH^-. Ion conductance, then, increases by about 2% per degree Centigrade in the vicinity of 25°C.

Kohlrausch's law and Eq. (3.17) constitute one portion of a general model allowing the calculation of the equivalent conductance of a solution from the concentration of solutes within it. The remaining relationship necessary to complete the model is found in an equation derived by Debye, Hückel, and Onsanger (Onsanger, 1926, 1927). If a solution is composed of only two kinds of ions derived from a single salt, the loss of equivalent ionic conductance of either one of the ionic species as a result of an increase in concentration of the salt above infinite dilution (and an increase in the retarding effects of electrophoretic and relaxation phenomena) is given by the expression:

$$\Delta\lambda_i = \lambda_i - \lambda^0_i = -\left(\frac{29.15z_i}{(DT)^{1/2}\eta} + \frac{9.90 \times 10^5}{(DT)^{3/2}}\lambda^0_i w\right)(\mathcal{C}_+z^2_+ + \mathcal{C}_-z^2_-)^{1/2}$$

or
$$\qquad (3.18)$$

$$\Delta\lambda_i = \lambda_i - \lambda^0_i = -\left(\frac{29.15z_i}{(DT)^{1/2}\eta} + \frac{9.90 \times 10^5}{(DT)^{3/2}}\lambda^0_i w\right)(c_+z_+ + c_-z_-)^{1/2}$$

where $\Delta\lambda_i$ is the loss of equivalent conductance of the ionic species of interest at a given concentration, λ_i is the equivalent conductance of the ionic species at the given concentration, λ_i^0 is the equivalent conductance of the ion at infinite dilution, T is the temperature of the solution in degrees Kelvin, D is the dielectric constant for the solvent at temperature T (78.5 for water at 25°C), η is the viscosity of the solvent at temperature T (8.95×10^{-3} for water at 25°C), $\mathcal{C}_+$ and $\mathcal{C}_-$ are the respective concentrations in moles/liter of the positive and negative ions that are found in the solution and derived from the same salt (i.e., $\mathcal{C}_+$ and $\mathcal{C}_-$ are equivalent), c_+ and c_- are the same as $\mathcal{C}_+$ and $\mathcal{C}_-$ but given in equivalents per liter, z_+ and z_- are the respective valences of the positive and negative ions excluding their sign, and where

$$w = z_- z_+ + \frac{2q}{1 + q^{1/2}} \qquad q = \left(\frac{z_+ z_-}{z_+ + z_-}\right)\left(\frac{\lambda_-^0 + \lambda_+^0}{z_+\lambda_-^0 + z_-\lambda_+^0}\right),$$

λ_+ and λ_- representing the equivalent conductance of the constituent positive and negative ions of the salt dissolved at infinite dilution. Either λ_+ or λ_- is equivalent to λ^0; depending on which is the ion of interest. The equivalent conductance of a solution composed of ions from a single salt in a known concentration thus can be found by summing the limiting ionic conductances of the cationic and anionic specids (λ_+^0, λ_-^0) according to Kohlrauch's law, and by subtracting from this total the losses in equivalent conductance encumbered by both ionic species ($\Delta\lambda_+, \Delta\lambda_-$) as described by the Onsanger equation.

For solutions of uniunivalent salts, biunivalent salts, and unibivalent salts having concentrations of up to .002 equivalents per liter, the Onsanger equation predicts the loss of equivalent conductances of solutions with increasing concentration very well (Glasstone 1942; Table 8). It gives a reasonably good approximation of the losses of equivalent conductances of solutions having solute concentrations of several hundreds of an equivalent per liter. The equation is not very accurate at predicting the losses of equivalent conductances of solutions of bi–bivalent solutes, but this is not a serious drawback for this study. Soil waters are never composed solely of bivalent ions.

The Onsanger equation does offer a problem, however, in that it accommodates only *two* species of ions, which are assumed to originate from a single salt. Soil water solutions, on the other hand, are composed of multiple ionic species. To circumvent this problem, it is possible to generalize the Onsanger equation. If it is assumed that in solutions composed of a number of ionic species, the effect on a given ionic species of all the other ionic species at various concentrations is equivalent to the effect of *one* ionic species having (*a*) a concentration equal to the total concentration of all the other ions, (*b*) a valence that is an average of the valences of all the other ions, weighted according to their relative concentrations, and (*c*) a limiting ionic conductance that is an average of the limiting ionic conductances of all the other ions, weighted according to their concentrations, then the Onsanger equation may be rewritten as:

$$\Delta\lambda_i = \lambda_i - \lambda_i^0 = -\left(\frac{29.15}{(DT)^{1/2}\eta} + \frac{9.90 \times 10^5}{(DT)^{3/2}}\lambda_1^0 w\right)\left(\sum_j^j \mathcal{C} z_j^2\right)^{1/2}$$

TABLE 8
Comparison of observed and calculated Onsager slopes in aqueous solutions at 25°C*

Electrolyte	Observed Slope	Calculated Slope
LiCl	81.1	72.7
NaNO$_3$	82.4	74.3
KBR	87.9	80.2
KCNS	76.5	77.8
CsCl	76.0	80.5
MgCl$_2$	144.1	145.6
Ba(NO$_3$)$_2$	160.7	150.5
K$_2$SO$_4$	140.3	159.5

*Taken from Glasstone (1952:92)

or

$$\Delta\lambda_i = \lambda_i, \lambda_i^0, D = -\left(\frac{29.15}{(DT)^{1/2}\eta} + \frac{9.90 \times 10^5}{(DT)^{3/2}}\lambda_i^0 w\right)\left(\sum_j c_j z_j\right)^{1/2} \quad (3.19)$$

where λ_i, λ_i^0, D, T, and η are as before; c_j and $¢_j$ are the concentration of a given ionic species, j; z_j is the unsigned valence of the given ionic species; where j ranges from 2 (the limiting Osanger case) to the number of ionic species in the solution; and where

$$w = z_i \bar{z} \frac{2q}{1 + q^{1/2}}$$

and

$$q = \left(\frac{z_i \bar{z}}{z_i + \bar{z}}\right)\left(\frac{(\lambda_i^0 + \bar{\lambda}_i^0)}{z_i \bar{\lambda}_0 + \bar{z}\lambda_i^0}\right),$$

z_i representing the unsigned valence of the ionic species of interest, $\bar{z}$ representing the unsigned weighted average of the valences of the other ionic species, and $\bar{\lambda}_0$ representing the weighted average of the limiting ionic conductances of the other ionic species in the solution.[17]

[17] One will not find this modified version of the Osanger equation in the literature of chemistry—it is my own solution to the problem at hand. The reliability of the equation, however, appears satisfactory, for at least two reasons. First, for multi-ion solutions having ionic concentrations and temperatures similar to those found in soil waters, I have found the modified Onsanger equation to

Summarizing Eqs. (3.15)–(3.19), the conductivity of a soil water solution may be calculated by the expression:

$$K = \sum_{j=1}^{n} c_j \lambda_{i_j}/1000 \qquad (3.20)$$

where K is the conductivity of the solution and c_j is the equivalent concentration of an ionic species j having an equivalent ionic conductance λ_{i_j} as adjusted for temperature [Eq. (3.17)] and solution concentration [Eq. (3.19)].

The Conductivity of Soil Water as a Function of the Concentration of Weak Electrolytes

The model presented above for calculating the equivalent conductance of a solution has been described as if it were applicable to solutions of all kinds of electrolytes. This is not true. Electrolytes fall into two classes: strong and weak.[18] The equivalent conductance of solutions of strong electrolytes behaves according to the previously outlined model. At very low concentrations, the equivalent conductance of the solution increases approximately with a decrease in the square root of concentration. Weak electrolytes behave very differently. At low concentrations, the equivalent conductance of solutions of weak electrolytes increases very rapidly as solute concentration decreases (Figure 16). This difference may be explained as follows. Suppose water is continuously added to a solution of strong electrolytes. As infinite dilution is approached, very little change will occur in the equivalent conductance of the solution with concentration decreases: (*a*) the number of ions (electric carriers) within the solution remains the same, and (*b*) the relaxation and electrophoretic effects remain approximately the same (negligible). For solutions of weak electrolytes, on the other hand, as infinite dilution is approached, the number of ions (electron carriers) increases dramatically. This phenomenon is a result of the incomplete dissociation of weak electrolyte molecules at higher solute concen-

---

predict losses of equivalent ionic conductance such that total-solution equivalent conductances may be estimated within 10% of their true value. This is only a slightly lower level of accuracy than that of the original Onsanger equation. Second, my use of the expression $(\Sigma_j c_j z_j)$ and the weighted average values λ_i^0 and z_j have some theoretical basis. If one examines the derivation of Onsanger's equation (Glasstone 1942: 87–89), one will see that the quantity $(\Sigma_j c_j z_j)$, indeed, is used up through the final steps of the derivation, at which point the quantity is simplified to $(c_+ z_+ + c_- z_-)$ for solutions of only two ionic species. Moreover, in modeling the electrophoretic and relaxation effects of an ionic atmosphere upon a single ion, Onsanger replaced all ions within the ionic atmosphere by a single charge placed away from the ion of interest at a distance that would yield the same effect as the ionic atmosphere. This modeling procedure is similar (although not equivalent) to the weighting procedure by which I have modified Onsanger's equation.

18. Ionic species which compose strong electrolytes and which are found in soil waters include NO_2, NO_3, K, Ca, Mg, Na, SO_4, and Cl. Ionic species that compose weak electrolytes and that are found in soil water include NH_4 (of the ammonium hydroxide equilibrium system), HCO_3 (of the carbonic equilibrium system) and HPO_4 and H_2PO_4 (of the phosphoric equilibrium system).

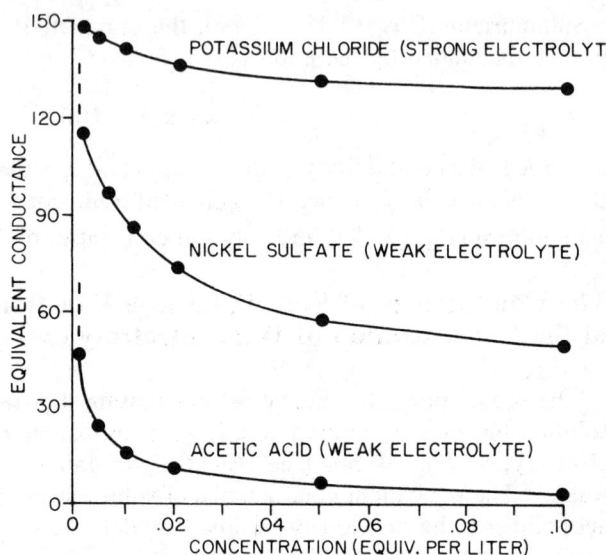

FIGURE 16. *Solutions of strong and weak electrolytes differ in their relationship between concentration and equivalent conductance. Adapted from Glasstone (1952:91).*

trations, followed by more complete dissociation when more solvent is added (Daniels 1944). Consequently, the Onsanger equation is not applicable to solutions of weak electrolytes.

No single, general model exists that relates the equivalent conductivity of a solution composed of both strong and weak electrolytes to the concentration of the ionic species found within the solution; nor is a model available for weak electrolytes alone. As a general approximation, however, a log–log relationship between the equivalent conductance (or resistivity) of a solution of a weak electrolyte and solute concentration appears appropriate (Figure 17; Merkel 1972), equivalent conductance decreasing as concentration increases.

The Conductivity of Soil Water as a Function of the Concentration of Colloidal Particles

A final factor that affects the conductivity of soil water solutions is the concentration of conductive clay colloidal particles within the solutions. Consider a solution composed of free ions plus clay particles that have exchangeable cations and anions adsorbed to their surface (see Chapter 4). When an electric field is applied to such a solution, the positive free ions will move toward the negative electrode and the negative free ions and clay colloids will move toward the positive electrode. Clay particles have an electrostatic negative charge (see Chapter 4) that accounts for their mobility in this manner within an electric field. As each clay particle migrates, a shearing plane develops between its solid phase and its adsorbed ions within the liquid phase

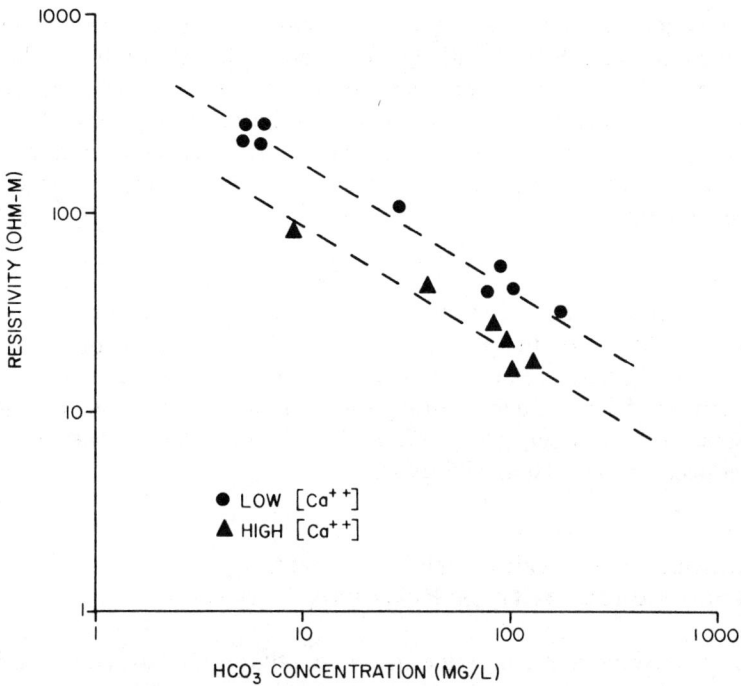

FIGURE 17. *Relationship between the resistivity of a solution and its concentration for the weak electrolytic species HCO_3^-, with both low and high Ca concentrations. Adapted from Merkel (1972:318).*

surrounding the particles. Adsorbed ions are continually pulled away from the clay particle at its back end (with respect to its direction of motion), and other ions are added and adsorbed to the front end. Adsorbed ions between the front and back of the particle shift to the rear during this process. This continual loss and replacement of adsorbed ions within the liquid phase surrounding a clay particle subjects it to a relaxation force that retards its motion. By the same token, the motions of the free ions in the liquid phase are retarded. The free ions and colloidal particles also are subjected to retarding electrophoretic forces that develop when the free ions and clay particles have to move against the stream of water dipoles bonded to oppositely moving particles or ions (van Olpen 1963:192). Thus, theoretically, the equivalent conductance of a solution having a high concentration of both clay particles and free ions that interfere with each other's mobilities should be less than the sum of the equivalent conductances of two solutions, one with only clay particles in a concentration equivalent to that of the original solution but lacking the interfering free ions, and one with only free ions in a concentration equivalent to that of the original solution but lacking the interfering clay particles.

Empirically, however, Patnode and Wyllie (1950:48) have found that the conductivity (and therefore the equivalent conductances) of aqueous solutions of salts and clay colloids is essentially equivalent to the sum of the conductivities contributed by the ions, alone, and the clay particles, alone. The contribution of the ions to the conductivity of the complex solution may be found using the modified Onsanger and Kohlrausch equations. The contributions of all conducting colloidal particles to the conductivity of the solution may be found using the relationship

$$K = AS \qquad (3.21)$$

where K is the specific conductance of the solution contributed by the conducting colloidal particles; where S is the concentration of the colloidal particles in milligrams per liter; where A is a constant of proportionality which ranges between .54 and .96 in natural ground water and stream water, but, more often takes values between .55 and .75 and has higher values in waters with high sulfate concentrations (Hem 1959:99).

Implications of the Relationships Governing Soil Water Conductivity on Resistivity Surveys

The relationships that determine the conductivity of unstructured soil water and that are discussed previously have direct bearing on this study in at least two ways. First, they serve to operationalize and give theoretical credibility to our primary thesis that the different alterations produced in the soils within different kinds of use-areas of archeological sites are distinguishable in the soil resistivity signatures of those areas. One of the manners in which the soils within different kinds of use-areas may be distinguished is in the relative proportions of the various cation and anion species that have been added to them through human refuse deposition and that occur in their soil waters (see Chapters 4 and 8; Tables 9 and 67). These differences almost always will yield differences in soil resistivity because the various ionic species occuring in different proportions have different limiting conductances. Even if the soils of two different kinds of use-areas are augmented with the same total number of ions by human activity, their resistivities will differ.

The second way in which these relationships are important is that they suggest the archeological and environmental conditions which are optimal for soil resistivity surveys. To begin, given the different limiting ionic conductances of various ionic species and the different spectra of ionic species enriched within the soils of different use-areas, some types of use-areas will be more distinguishable than others. The ions and radicals that occur most commonly in soil waters and that also are most likely to be of anthropic origin in the soils of archeological sites (i.e., those occurring in major proportions within organic materials that might be deposited there) include NH_4, NO_3, HPO_4, H_2PO_4, K, Ca, Mg, Na, SO_4, Cl, and HCO_3 (Table 11). These have a wide range of

TABLE 9
Ionic species occurring most commonly in soil waters and in organic materials likely to be deposited on archeological sites

Ionic Species in Soil Waters*				Elements in Organic Materials	
Acid Soils		Basic Soils			
Ionic Species	Concentration Moles/l x10	Ionic Species	Concentration Moles/l x10	Element	Typical Relative Concentrations:Rank**
NH_4, NO_3	12.1	Na	29.0	N	1
Ca	3.4	S	24.0	K	2
Mg	1.9	Cl	20.0	P	3
Cl	1.1	Ca	14.0	Ca	4
Na	1.0	NH_4, NO_3	13.0	Mg	5
K	.07	Mg	7.0	Na	5
S	.05	K	1.0	S	5
HPO_4, H_2PO_4	.007	HPO_4, H_2PO_4	.03	Cl	5
HCO_3	Trace				

*Representative data, as given by Fried and Broeshart (1967)
**Representative organic materials from which ranks are determined are the same as those in Table 67.

limiting ionic conductances, varying between 33 and 80 mho-cm^2/equiv at 25°C (Table 7a). Different kinds of use-areas on archeological sites will tend to be more distinguishable in their resistivities when: (*a*) the residues that have been deposited within them differ in the relative proportions of their constituent nutrient species (e.g., see Table 67, Chapter 8); (*b*) when those differences in nutrient spectra have been maintained until the present within the soil and soil waters (see Chapters 4, 8); and (*c*) when the nutrient species exhibiting the greatest differences in their relative proportions within the soil water also are those with large limiting ionic conductances, such as NH_4 or K.[20] Of course, where different kinds of use-areas are distinguished in the total amount of refuse deposited within them, and in the total concentration of nutrient ions within their soil waters, these differences may be as important or more important than differences in nutrient spectra and the limiting ionic conductances of nutrients in distinguishing use-areas in their resistivities.

With respect to the environmental conditions during which resistivity surveys would be most optimal, the relationships just described between the conductivity of solutions and their temperatures favor warmer periods. The equivalent conductances of different ionic species are most distinct at higher temperatures (Table 7a). Consequently, soil water solutions in different kinds of use-areas and having different spectra of ionic species dissolved within them will be most distinguished in their conductivities at higher temperatures. The gains in distinguishability of use-areas obtained at greater temperatures will be minor, however, compared to differences between them that are a function of the ionic concentrations within their soil waters.

Finally, the relationships between the conductivity of solutions, solution concentration, and solution temperature emphasize how important it is that weather conditions remain *constant* during resistivity surveys if all observations are to be comparable. An increase in the temperature of a soil by 1°C will increase ionic conductances by approximately 2% within the vicinity of 25°C and will cause a decrease in soil resistivity of a similar amount (Hesse 1962). A decrease in the moisture content of a soil will cause the concentration of ions dissolved within its soil water to increase, and will increase the conductivity of the soil water solution, partially offsetting the decrease in soil conductivity resulting from the increased tortuosity of the paths of current flow and the loss of conductive cross-sectional area associated with moisture loss. If such changes in soil temperature and soil water solution concentration occur during the course of a resistivity survey, temporal variation in soil resistivity will

---

20. NO_3, SO_4, and Cl also have high limiting ionic conductances. However, SO_4 and Cl usually are not the nutrient species exhibiting the greatest differences in their relative proportions within organic residues (see Table 67). Moreover, NO_3, SO_4, and Cl are easily leached from soils and usually will not be the primary nutrient species distinguishing the soils of different kinds of use-areas even when they occur in widely different proportions within the different kinds of residues deposited in those areas.

be compounded with spatial variation, making an interpretation of geographic variation in soils difficult.

One might think it is possible to remove such unwanted temporal variations in a soil resistivity data series by mathematical procedures. For changes in soil temperature, this may be so. If soil temperature variations at the depth of investigation have been monitored, a correction factor of 2% decrease in soil resistivity per degree centigrade gained above some base line temperature may be applied to the data to make them internally comparable with respect to soil temperature. Temporal variations in soils moisture content, however, can never be removed from resistivity data obtained during nonuniform moisture regimes. Unlike changes in soil temperature, the alterations in soil water conductivity that accompany alterations in soil moisture represent not only changes in the *states* of the variables determining soil resistivity, but also the *relative importance of the variables, themselves,* in determining soil resistivity.

A change in soil moisture content will change not only the conductivity of the total soil water solution, but also the relative contributions made to the conductivity of the soil water solution by strong electrolytes, weak electrolytes, and conductive colloidal particles. This circumstance arises because each of the three different types of electric carriers exhibit different functional relationships between their concentration in a solution and solution conductivity (square root, log–log, and linear). Thus, it is critical that soil resistivity data be obtained under as similar moisture conditions as is possible.

Interaction between the Physical and Chemical Properties of Organically Enriched, Disturbed Soils and Natural Soils in Producing Contrast in Their Resistivities

Throughout the discussion of the effects of anthropic alterations in the physical properties of soils on their resistivities, the chemical properties were held constant. Similarily, when discussing alterations in the chemical properties of soils, the physical properties were held constant. On archeological sites, however, in circumstances where the form of pedological alteration involves the incorporation of organic residues of anthropic origin within the soil, physical and chemical changes in the soil occur *jointly*.[21] Consequently, any changes that occur in the resistivity of the soil reflect (*a*) whether the effects of modification of its physical properties and the effects of modification of its chemical properties on its resistivity combine in a constructive or destructive *pattern of interference;* and (*b*) the relative *magnitudes* of the effects of the physical and chemical alterations. Let us examine these two factors and how they determine

---

21. Other forms of pedological disturbance where this is not true include simple physical churning or compaction of the soil, or the deposition of dissolved or soluble inorganic compounds (e.g., salts).

the contrast in resistivity between an organically enriched, disturbed soil and its natural counterpart in their resistivities (Figure 18).

When organic matter is introduced to the soil, its primary effects on the *chemistry* of the soil always favor the lowering of soil resistivity. The cation exchange capacity of the soil, the availability of various ions and radicals, and their concentration within the soil water solution all are increased. The magnitude of the changes in these properties per unit weight of organic matter added to the soil will vary with (*a*) the nutrient contents of the organic matter and the amount of ions and radicals it frees to the soil and soil water during the process of decomposition; and (*b*) whether the decompositional environment in which humic molecules are formed from the organic residues favors the humic molecules having complex *aromatic* structures (spherical) or more simple, *aliphatic* structures (linear, with many side chains). Humic molecules with an aromatic structure tend to have higher cation exchange capacities than those with aliphatic structures (see Chapter 4; Kononova 1961). Their presence in a soil will produce higher concentrations of ions and radicals within soil water and lower soil water resistivities than will humic molecules with aliphatic structures. Whatever the magnitudes of the changes caused in the chemical properties of a soil on the incorporation of organic matter within it, however, the directions of the changes will *always* favor a *reduction* in soil resistivity.

The primary effects of organic enrichments on the *physical* properties of a soil, on the other hand, may favor either a reduction or an increase in soil resistivity, depending on the environmental suction at the time of observation. An increase in the organic matter content of a soil will facilitate an increase in its degree of aggregation, total porosity, and relative frequency of large pores, but a decrease in its relative frequency of small pores. These circumstances will allow the disturbed soil to hold more water and to have a greater conductive cross-sectional area at low suctions than its natural counterpart—favoring a decrease in its resistivity. The increase in number of large, water-filled pores having high capillary conductances with the disturbed soil also will favor a reduction in its resistivity. At moderate to high environmental suctions, the disturbed soil will hold less water,[22] have a lower conductive cross-sectional area, and have fewer small, water-filled pores and a lower total capillary conductance than the natural soil from which it was derived, favoring an increase in its resistivity. At very high suctions, the conditions found at moderate to high suctions and favoring an increase in the resistivity of the disturbed soil will be offset somewhat, or perhaps even be exceeded, by the effects of the hygroscopic capacity of organic matter. The tendency for the disturbed soil to have a lower moisture content and conductive cross-sectional area than the natural soil from which it was derived and to have a higher resistivity will be decreased.

As in the case of the effect of organic enrichments on the chemical properties of a soil, the effect of organic enrichments on the physical properties of a soil

---

22. Sandy loam or finer textured soils, only.

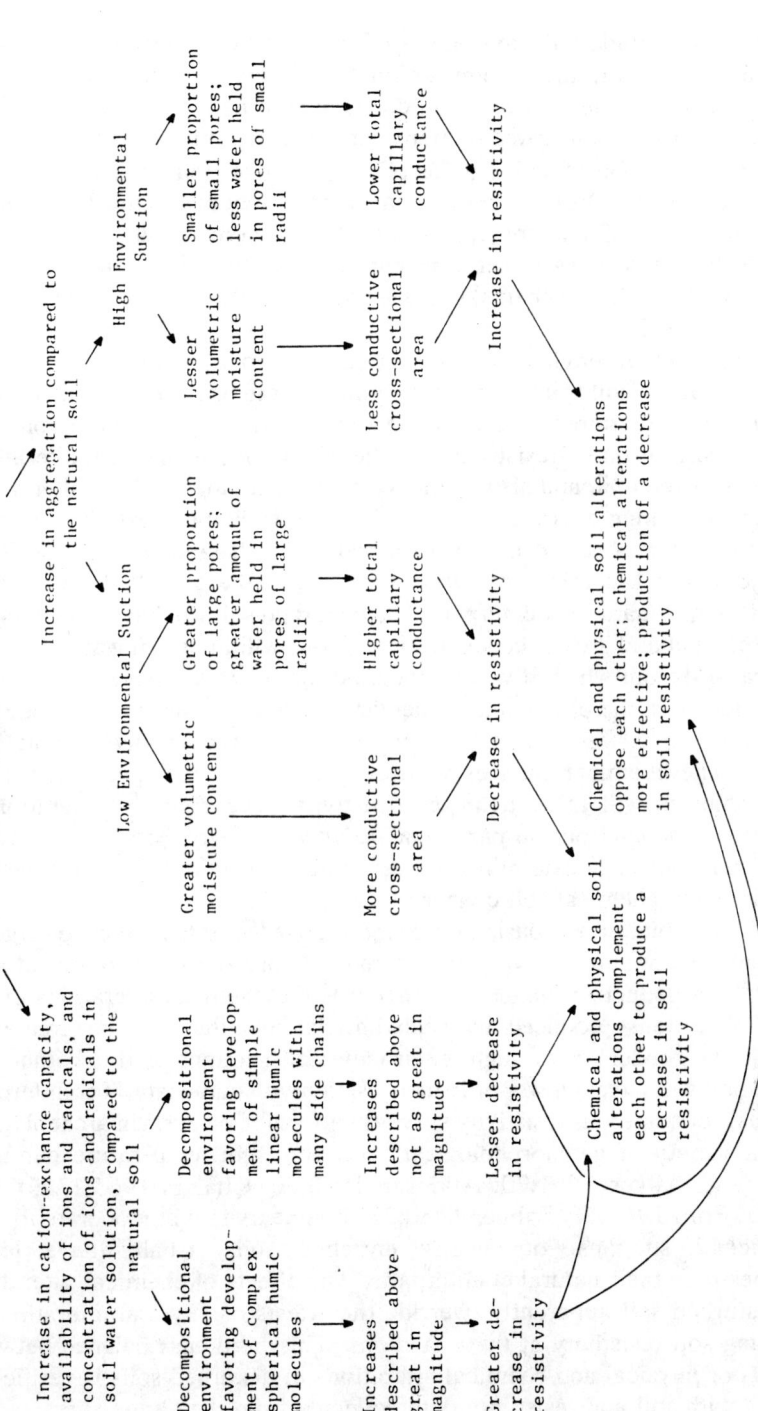

FIGURE 18. *Alterations in the resistivity of an organically enriched soil as a result of both physical and chemical changes within it.*

will vary in magnitude with the nature of the organic materials incorporated within the soil and with the decompositional environment. The development of soil aggregation and the changes it produces in soil moisture equilibria will be facilitated when (*a*) the organic matter contains a high proportion of nutrients promoting flocculation of soil colloids (Ca, Mg) rather than dispersion of soil colloids (Na); and (*b*) the decompositional environment favors the development of humic molecules of a complex, aromatic structure (spherical) that are easily coagulated by electrolytes rather than humic molecules of a simple, aliphatic structure (linear, with side chains) that are not so easily coagulated (see Chapter 4; Kononova 1961).

At any particular environmental suction, the contrast of an organically enriched, disturbed soil from its natural counterpart in its resistivity will depend on the *pattern* of interference between the effects of the physical alterations that have occurred in it on its resistivity and the effects of the chemical alterations that have occurred in it, and also on the relative magnitudes of these alterations. At low and very high environmental suctions, both the physical and chemical changes that have occurred in the disturbed soil will favor a decrease in its resistivity; they will combine *constructively* to give the disturbed soil a lower resistivity than its natural counterpart. The disturbed soil will have both a higher moisture content and greater concentration of ions within its soil water than will the natural soil from which it was derived. At moderate to high environmental suctions, however, the physical and chemical changes within the disturbed soil will combine in a *destructive* pattern of interference in determining soil resistivity. The chemical changes will favor a lower resistivity, whereas the physical changes will favor a higher resistivity. The resultant alteration in resistivity will depend on the particular *balance* achieved between these two opposing tendencies, as determined by the relative magnitudes of the effects of the physical and chemical soil changes.

It is not possible, on the basis of present knowledge of the effects of organic enrichments on the nutrient availabilities and the pore-size distributions of soils to predict in a deductive manner whether it is the chemical alterations or the physical alterations produced by such enrichments that will be greater in magnitude at moderate to high environmental suctions and whether the disturbed soil will have a lower or higher resistivity than its natural counterpart. On the basis of numerous resistivity surveys performed on earthen archeological sites in a variety of geomorphological settings within temperature climates, however (e.g., Aitken 1961:70; Atkinson 1963:20; Chalabi 1965:132; Clark 1963:575; Ford 1964:56; Palmer 1960:73) it appears that at moderate to high environmental suctions, organically enriched soils usually have lower resistivities than their natural counterparts. The effects of chemical alterations in the disturbed soil apparently override the effects of physical alterations in determining soil resistivity at these suctions. This particular balance between the effects of physical and chemical alterations in disturbed soils is verified in Chapter 9 with soil and resistivity data collected from the Crane site.

Finally, the environmental suctions at which organically enriched soils will be found to contrast most from their natural counterparts in their resistivities can be predicted deductively to be those in the very low and very high ranges. At these times, the physical and chemical determinants of soil resistivity differentiating a disturbed soil from its natural counterpart will combine constructively to give the disturbed soil its lower resistivity. At moderate to high suctions, the lower resistivity of the disturbed soil favored by its chemical characteristics will be offset somewhat by the higher resistivity favored by its physical characteristics, reducing its contrast with its natural soil counterpart. The accuracy of these predictions is supported data presented by Malott (1964:10) for natural soils of various kinds, and is systematically tested and verified in Chapter 9 of this work.

Summary

Throughout this chapter, the relationships of numerous physical and chemical variables to soil resistivity have been discussed. Environmental and pedological conditions under which the soils in different kinds of use-areas of archeological sites are most distinguishable in their resistivity have been pointed out in the process. In some cases, the conditions that would be optimal with respect to the state of one determinant variable of soil resistivity are suboptimal with respect to the states of other determinant variables. If all the determinants of soil resistivity are considered together, it will be found that under most circumstances, optimal differentiation of disturbed and undisturbed soils can be obtained when soil moisture contents are very low or very high, and when soil temperatures are high. The primary exceptions to this are when soils are extremely sandy and are practically incapable of conducting electric currents when low in moisture, and when soil porosity at the surface is so great that probe contact resistance becomes significant during dry periods (as in newly plowed fields upon which rain has not yet fallen).

References

Factors Directly Determining the Electrical
Resistivity of Porous Media: Soil Texture,
Soil Structure, Soil Water Solution Conductivity,
Capillary Conductance.

Aitken, M. J.
   1961 *Physics and archaeology.* Wiley (Interscience), New York.
Atkinson, R. J. C.
   1963 Resistivity surveying in archaelogy. In *The scientist and archaelogy,* E. Pyddoke (ed.), pp. 1–30. Phoenix House, London.

Biddle, James G., and Company
  1970 *Getting down-to-earth*. James G. Biddle Co., Plymouth Meeting, Penn.
Brown, Donald C.
  1971 Techniques for quality-of-water interpretation from calibrated geophysical logs, Atlantic Coastal Area. *Ground Water* 9(4):25–38.
Chalabi, Mahboub al-
  1965 Applications of geo-electrical methods in archaeology. *Summer*, 21(1,2).
Chalabi, Mahboub al- and A.I. Rees
  1962 An experiment on the effect of rainfall on electrical resistivity anomalies in the near surface. *Bonner Jahrbucher* 162:266–271.
Childs, E. C.
  1969 *Soil water phenomena*. Wiley, New York.
Clark, Anthony
  1963 Resistivity surveying. In *Science in archaeology*, D. Brothwell and E. Higgs (eds.), pp. 569–581. Basic Books, New York.
Dean, John A. (ed.)
  1973 *Lange's handbook of chemistry, 11th edition*. McGraw-Hill, New York.
Edlefsen, N. E., and A. B. C. Anderson
  1941 The four electrode resistance method for measuring soil moisture content under field conditions. *Soil Science* 51:367–376.
Ford, Richard I.
  1964 A preliminary report of the 1964 resistivity survey at the Schulz Site (20 SA 2). *Michigan Archaeologist* 10(3):54–58.
Franzini, J. B.
  1951 Porosity factor for case of laminar flow through granular media. *American Geophysical Union, Transactions* 32:443–446.
Fried, Maurice, and Hans Broeshart
  1967 *Soil–plant system*. Academic Press, New York.
Getmna, Frederick H., and Farrington Daniels
  1944 *Outlines of physical chemistry*. Wiley, New York.
Glasstone, Samuel
  1942 *An introduction to electrochemistry*. Van Nostrand, Toronto.
Gouy, C.
  1910 Sur la constitution de la charge electrique a la surface d'un electrolyte. *Journal de Phys.* 9(4):457–463.
Gupta, S. C., and R. J. Hanks
  1972 Influence of water content on electrical conductivity of the soil. *Soil Science Society of America, Proceedings* 36:855–857.
Hackett, James E.
  1956 Relation between earth resistivity and glacial deposits. *Illinois State Geological Survey, Circular* 223.
Hem, John D.
  1959 Study and interpretation of the chemical characteristics of natural water. *United States Geological Survey, Water Supply Paper* 1473.
Hesse, A.
  1962 Geophysical prospecting for archaeology in France. *Archaeometry* 5:123–125.
Hillel, Daniel
  1971 *Soil and water*. Academic Press, New York.
Hrubišek, J.
  1941 In *Kolloid-Beiheft* 53:385.
Jacob, C. E.
  1946 Radial flow in a leaky artesian aquifer. *American Geogphysical Union, Transactions* 27:198.

Keller, G. V., and F. C. Frischknecht
  1966 *Electrical methods in geophysical prospecting.* Pergamon, Oxford.
Kirkham, D., and G. S. Taylor
  1950 Some tests on a four-electrode probe for soil moisture measurement. *Soil Science Society of America, Proceedings* 14:42–46.
Merkel, R. H.
  1972 The use of resistivity techniques to delineate acid mine drainage in ground water. *Ground Water* 10(5):38–48.
National Research Council
  1929 *International Critical Tables* 6. McGraw-Hill, New York.
Onsanger, Lars
  1926 Theory of electrolytes, part I. *Physikalische Zeitschrift*, 27:388–392.
  1927 Theory of electrolytes, part II. *Physikalische Zeitschrift*, 28:277–298.
Palmer, L. S.
  1960 Geoelectrical surveying of archaeological sites. *Proceedings of the Prehistoric Society* 5:64–75.
Patnode, H. W., and M. R. Wyllie
  1950 The presence of conductive solids in reservoir rocks as a factor in electric log interpretation. *American Institute of Mining and Metallurgical Engineers, Transactions* 189:47–52.
Rhoades, J. D., and R. D. Ingvalson
  1971 Determining salinity in field soils with soil resistance measurements. *Soil Science Society of America, Proceedings* 35:54–60.
Robinson, R. A., and R. H. Stokes
  1955 *Electrolyte solutions.* Butterworths, London.
Scheidegger, A. E.
  1957 *The Physics of flow through porous media.* MacMillan, New York.
Schopper, Jurgen R.
  1966 A theoretical investigation on the formation factor/permeability/porosity relationship using a network model. *Geophysical Prospecting* 14(3):301–341.
Shea, P. F., and J. N. Lutkin
  1961 An investigation of the use of the four-electrode probe for measuring soil salinity *in situ*. *Soil Science* 92:331–339.
Van Olpen, H.
  1963 *Clay colloid chemistry.* Wiley (Interscience), New York.
Wilklander, Lambert
  1964 Cation and anion exchange phenomena. In *Chemistry of the soil*, F. E. Bear (ed.), pp. 163–205. Reinhold, New York.

Indirect Determinants of Soil Resistivity: Relationships between the Organic Matter Content, Texture, Structure, and Moisture Content of Soils.

Allison, F. E.
  1973 *Soil organic matter and its role in crop production.* Elsevier New York.
Amemiya, M.
  1965 Influence of aggregate size on soil moisture content capillary conductivity. *Soil Science Society of America, Proceedings* 2:744–748.
Amer, Fathi
  1960 Relation of laboratory hydraulic conductivity to texture, aggregation and soluble calcium plus magnesium percentage. *Soil Science.* 89:45–48.

Anonymous
1955 Soil aggregation in relation to texture and organic matter, by R. Heinonen, Review. *Soil Science* 80:336.
Baver, L. D., Walter H. Gardner, and Wilford R. Gardner.
1972 *Soil physics.* Wiley, New York.
Bouyoucos, G. J.
1939 Effect of organic matter on the water-holding capacity and wilting points of mineral soils, *Soil Science* 47:377–383.
Brady, Nyle C.
1974 *The nature and property of soils,* Macmillan, New York.
Childs, E. C.
1969 *Soil water phenomena.* Wiley, New York.
Coile, T. S.
1939 Effect of incorporated organic matter on the moisture equivalent and wilting percentage values of soil. *Soil Science Society of America, Proceedings* 3:43.
Gardner, W. R.
1960 Dynamic aspects of water availability to plants. *Soil Science* 89:63–73.
Gingrich, J. K., and R. S. Stauffer
1955 Effect of long-time soil treatments on some physical properties of several Illinois soils. *Soil Science Society of America, Proceedings* 19:257–260.
Greenland, D. J.
1965a Interaction between clay and organic compounds in soils. Part I. Mechanisms of interaction beteen clays and defined organic compounds. *Soils and Fertilizers* 28:415–425.
1965b Interaction between clay and organic compounds in soils. Part II. Adsorption of soil organic compounds and its effect on soil properties. *Soils and Fertilizers* 28:521–532.
Harris, R. F., G. Chesters, and O. N. Allen
1966 Dynamics of soil aggregation. *Advances in Agronomy* 18:107–169.
Hillel, Daniel
1971 *Soil and water.* Academic Press, New York.
Jamison, V. C.
1953 Changes in air–water relationships due to structural improvement of soils. *Soil Science* 76:143–157.
Jamison, V. C., and E. M. Kroth
1958 Available moisture storage capacity in relation to textural composition and organic matter content of several Missouri soils. *Soil Science Society of America, Proceedings* 22:189–192.
Juncker, P. H., and J. J. Madison
1967 Soil moisture characteristics of sand–peat mixes. *Soil Science Society of America, Proceedings* 31:5–8.
Kemper, W. D.
1958 Structural implications of moisture retention in clay-size soil materials. *Soil Science Society of America, Proceedings* 22:5–8.
Klute, A., and W. C. Jacob
1950 Physical properties of sassafras silt loam as affected by long term organic matter additions. *Soil Society of America, Proceedings* 14:24–28.
Kononova, M. M.
1961 *Soil organic matter.* Pergamon Press, New York.
Malott, Donald F.
1964 The application of geophysics to highway engineering by the Michigan State Highway Department. Unpublished manuscript. Michigan State Highway Department, Ann Arbor.

Pillsbury, A. F.
  1950 Effects of particle size and temperature on the permeability of sand to water. *Soil Science* 70:299.
Puri, B. R., and O. P. Mahajan
  1962 Effect of adding charcoal on the moisture-retension capacity of soil. *Soil Science* 94:162–167.
Reeve, R. C.
  1957 Factors which affect permeability. In *Drainage of Agricultural Land*, pp. 404–414. American Society of Agronomy, *Monograph*, 7.
Russel, M. B.
  1939 Soil moisture system curves for four Iowa Soils. *Soil Science Society of America, Proceedings* 4:51–54.
  1961 *Soil conditions and plant growth.* Longmans, New York.
Salter, P. J., and G. Berry, and J. B. Williams
  1967 Effect of farmyard manure on matric suctions prevailing in a sandy loam soil. *Journal of Soil Science* 18:318–328.
Salter, P. J., and F. Haworth
  1961 The available water capacity of a sandy loam, 2. The effects of farmyard manure and different primary cultigens. *Journal of Soil Science* 12:335–342.
Salter, P. J., and J. B. Williams
  1963 The effect of farmyard manure on the moisture characteristic of a sandy loam. *Journal of Soil Science* 14:73–81.
  1969 Influence of texture on the moisture characteristics of soil: Relationships between particle-size composition and moisture contents at the upper and lower limits of available water. *Journal of Soil Science* 20:126–131.
Shaykewich, C. F., and B. P. Workentin
  1970 Effect of clay content and aggregate-size on availability of soil water to tomato plants. *Canadian Journal of Soil Science* 50:205–217.
Smalley, R. R.
  1962 Effects of four amendments on soil physical properties on yield and quality of putting greens. *Agronomic Journal* 54:393–395.
Stevenson, F. J.
  1972 Role and function of humus on soil with emphasis on adsorption of herbicides and relation of micronutrients. *Bioscience* 22(11):643–650.
Tamboli, P. M., W. E. Larson, and M. Amemiya
  1964 Influence of aggregate size on soil moisture retension. *Iowa Academy of Sciences, Proceedings* 71:103–108.
Thijeel, A. A., and J. R. Burford
  1975 Effect of the application of cow slurry to grassland on nitrate levels in soil and soil water content. *Journal of Science Food and Agriculture* 26:1203–1213.
Thomas, Moyer D., and Karl Harris
  1926 The moisture equivalent of soils. *Soil Science* 21(6): 411–424.
Visser, W. C.
  1966 Progress in the knowledge about the effect of soil moisture content on plant production. *Institute of Land Water Management, Wageningen, Netherlands, Technical Bulletin* 45.
Wiltmuss, H. D., and A. P. Mazurak
  1958 Physical and chemical properties of soil aggregates in a Bruizerm soil. *Soil Science Society of America, Proceedings* 22(1):1–5.
Wischmeier, W. H., and J. V. Mannering
  1965 Effect of organic matter content of the soil on infiltration. *Journal of Soil and Water Conservation* 20:150–152.

4
Natural Processes Determining the Formation of Soil from Human Refuse and the Maintenance of Anthropic Soil Anomalies within Archeological Sites

Thus far, the functional relationships between particular soil variables (e.g., the concentration of ions found in soil solutions) and soil resistivity values have been described. Consideration has been given to variables directly determining electrical conductivity, such as temperature and ion concentration, and to variables having a less direct effect, such as organic matter content as a moisture-holding agent. The values taken by some of these pedological variables, in turn, are determined by the kinds and amounts of refuse deposited on an archeological site by prehistoric human activity, as well as by natural soil processes that *alter* the refuse and *incorporate* its components within the soil, and by larger soil–vegetational ecosystemic processes that *maintain* its effects on the soil over extended periods of time. Although the kinds and quantity of refuse deposited on an archeological site are specific to the site in question, the natural soil and ecosystemic processes are of a more general character, and may be used in a wider set of circumstances to understand soil variation and soil conductivity variation within archeological sites. These general natural processes are the focus of this chapter.

Long-Term Soil Alterations Produced by Prehistoric Human Activities: Overview

In considering soil variation on archeological sites within the context of resistivity surveying, several questions are pertinent:

1. Which of the multiple variables determining soil conductivity are affected by the prehistoric activities that can occur on earthen archeological sites?

2. Are the soil alterations that are produced by such activities maintained over time?
3. If the alterations are maintained, what natural processes or mechanisms are responsible?

The first two questions may be answered through a review of soil analyses previously performed on archeological sites and by reference to common-knowledge field observations. The following section is concerned with these topics, whereas the remainder of the chapter considers answers to the third question.

Of the several variables that determine soil conductivity, some are not affected to any significant degree by the kinds of activities that occur on earthen archeological sites and are nonessential to the topic of this chapter. The textural distribution of the mineral phase of soils is one such variable. Most materials deposited on earthen archeological sites where prehistoric activity has had some impact on its soils are of an organic nature rather than mineral. Consequently, there is little alteration of soil texture on earthen archeological sites. Exceptions include (a) sites into which exotic clays have been brought, in order to build mounds and earthworks; (b) kitchen middens where shell (largely $CaCO_3$) has been deposited in large quantities; (c) use-areas where the mineral component of bone (apatite) has been deposited in large amounts; and (d) areas that have been regularly trampled and where small-scale but continuous deflation of artifacts has occurred.

The temperature of soils on archeological sites may be somewhat altered from those expected under natural conditions. This is achieved through the accumulation of organic matter, which in turn increases or decreases the equilibrium moisture content and specific heat of the soil and yields lower or higher soil temperatures, respectively. The temperature of the top several inches of a wet soil may be 6 to 12°F lower than that of a moist to dry soil (Brady 1974:271). From the perspective of the resistivity surveyor, however, such soil temperature differences are insignificant. They will be limited largely to the top several inches of the soil, whereas the soil encompassed in a resistivity survey usually will extend to much greater depths, if for measurement limitations alone. The volume of soil that exhibits temeprature distinction between disturbed and undisturbed soils and that is included in a resistivity measurement usually will be small compared to the total volume of soil measured.

Several soil variables that can be sufficiently altered by prehistoric activities so as to cause significant soil conductivity anomalies include (a) the concentration of nutrient ions and radicals[1] in the soil water solution, (b) organic matter content, (c) soil moisture content, and (d) the degree of development of

---

1. Throughout this work, reference often will be made to ions and radicals within the soil using the term *nutrients*, as is done in agronomy. The term will encompass NH_4, NO_3, NO_2, HPO_4, H_2PO_4, H_3PO_4, K, Ca, Mg, Na, SO_4, and Cl. In precise terms, Na is not a nutrient for it is not required by plants for growth, but for convenience in this study it will be called a nutrient.

soil structure. The latter two variables are directly dependent on organic matter content. The concentration of nutrient ions and radicals found within soil water is related to both their concentration in the refuse deposited on the archeological site and to organic matter content as a factor controlling the cation exchange capacity of soils.

Previously Documented Long-Term Alterations in Soil Ion Concentration and Organic Matter Content

Literature on the analysis of soils from eathen archeological sites clearly demonstrates that both the concentrations of ions (total or available[2]) and the amount of organic matter within the soils of sites can be increased by prehistoric human activity, and that these soil alterations can be maintained over long periods of time. Let us review this literature by topic and level of generalization, considering first those studies concerned with how sites *as whole units* may be distinguished from their natural surroundings in their soil chemistry, and then those studies concerned with more specific relationships between activity and soil chemistry alteration. The latter include studies of variations in soil chemistry related to the *amounts* of activity that occurred on sites and/or the length of site occupation, and to the *kinds* of activities that occurred on sites.

At the broadest level of generalization, a number of analyses of soils from archeological sites have demonstrated that the refuse materials deposited by prehistoric occupants of sites can significantly increase the concentrations of specific elements and percentages of organic matter within their soils above the levels occurring naturally in surrounding, undistributed soils. The particular soil attributes examined and the bibliographic references to the studies are summarized in Table 10. The greatest amount of effort has been focused on total inorganic P concentrations, and secondarily on organic matter content. Other attributes that have been used successfully to differentiate on-site from off-site locations include total N concentrations, total Ca, organically fixed P concentrations, and the availabilities of exchangeable Ca, Mg, K, and PO_4. Most studies have been carried out on villages having appreciable midden accumulations and ranging in age from 300 years (Cook and Heizer 1965; Hurley and Heidenreich 1971) to over 6000 years (Sawbridge and Bell 1972). It should be emphasized that the successful results reported in the previously listed studies are for the most part from sites located on natural soils where the accumulation and maintenance of anomalously high levels of nutrients and organic matter is least expected. Most of the studied sites occur on acidic forest

---

2. *Available nutrients* are those that can be extracted from the soil with a standard leachate using a standard procedure. The *availability* of a particular nutrient is the amount of the nutrient extracted per unit volume of soil. It approximates the amount of the nutrient held on exchange sites of clay and humus colloids per unti volume of soil, and the maximum number of ions of that type available for electric current transfer per unit volume.

TABLE 10
Analyses of the soils from archeological sites demonstrating various effects of prehistoric refuse deposition upon them

Soil Attribute Found to be Enriched in On-Site Locations Compared to Off-Site Locations	References
Total inorganic P concentration	Arrhenius (1929, 1931a, 1931b, 1934, 1935, 1938, 1950, 1955a, 1955b, 1963)
	Cook and Heizer (1965)
	Dietze (1957)
	Eddy and Dregne (1964)
	Hurley and Heidenreich (1971)
	Sawbridge and Bell (1972)
	Solecki (1950)
	Weide (1966)
	Zeiner (1946)
Total N concentration	Lutz (1951)
	Cook and Heizer (1965)
	Sawbridge and Bell (1972)
Total Ca concentration	Cook and Heizer (1965)
Organically-fixed P	Hurley and Heidenreich (1971)
Exchangeable Ca, Mg, K and PO_4	Hurley and Heidenreich (1971)
	Lutz (1951)
	Sawbridge and Bell (1972)
	Solecki (1950)
Organic matter content (organic carbon)	Cook and Heizer (1965)
	Hurley and Heidenreich (1971)
	Sawbridge and Bell (1972)

soils—some highly leached—where nutrient ions and humus micelles are most mobile and most likely to be lost from the soil.

Although these analyses are critical to this study in that they document an association between prehistoric human activity and long-preserved soil alterations on archeological sites, this association is at a much more general level than the argument proposed here—that different kinds of use-areas within an archeological site can be distinguished by regularly recurring patterns of soil alterations. Several classes of studies that have investigated more specifically the relationship between prehistoric activity and soil chemical alterations on sites, however, *approach* the documentation of this proposition and support it.

To begin are those analyses that *relate the magnitude of soil anomalies to intensity of activity or length of occupation* of the archeological site. These are of several varieties. First, there are studies showing a distinct relationship between the *thickness of midden deposits* on a site and the concentration of particular elements. The University of Birmingham's Department of Archaeology (Dauncey 1952:35) was able to demonstrate this for total soil phosphate content at Crococalana, a Roman posting station. Likewise, Provan (1971:43) found a good agreement between the thickness of refuse layers inside different Norwegian farm houses (A.D. 350–550) and the levels of total phosphate within these deposits. Wiede (1966:161) was able to show correlations between soil pH and the degree of midden accumulation within a Chumash site in California. In these three studies, the variation in total phosphate content can be attributed only to the intensity or duration of occupation: The factors of microenvironmental setting and the age of the deposits, which can also explain soil chemistry variability, were held constant by examining deposits within *single* sites. Similar in intent, but not holding locational–microenvironmental factors as constant, is Weide's (1966:159–160) comparison of the soils of two Chumash sites. One site was a village site, having houses, burials, and midden deposits, whereas the second was a more ephemeral occupation lacking these features. The two sites could be differentiated by soil pH, the more intensely occupied village site having higher pH anomalies above the natural soil reaction than the more ephemerally occupied site.

A second variety of studies relating the intensity or duration of prehistoric activity to the magnitude of soil alterations on archeological sites uses *position within a settlement,* rather than midden thickness, as the indicator of the intensity or duration of occupation. Arrhenius (1929, 1931a), one of the first to investigate the soil changes produced in archeological sites by their prehistoric occupants, repeatedly was able to show a significant correlation between the magnitude of total phosphate concentrations at specific locations within Swedish rural villages and the distance of the locations from the centers of the sites. Central areas were found to have higher phosphate concentration than peripheral areas. Likewise, Hurley and Heidenreich (1971:185) found on several prehistoric Huron sites in Ontario, Canada, that soil pH values were systematically higher at the centers of the sites than at their peripheries. They attributed this pattern to the greater amounts of garbage and ash found in the central portions of the sites. Cook and Heizer (1965:58) also have been able to distinguish the peripheries of archeological sites from their centers of activity. They report that total soil C, N, P, and Ca all showed higher magnitudes at the center of a prehistoric mound–village site in the Napa Valley, California.

A final variety of studies concerned with intensity or duration of activity is represented by Deetz and Dethlefsen's (1963:243) examination of pH *differences between distinct occupational strata* of a single early historic Indian site in Santa Barbara County, California. Two occupational strata (zone 2; zones 3 and 4 combined) within the site were defined, one thicker than the other. In this particular case, the thicknesses of the two strata can reasonably be taken to be primarily a function of intensity of activity rather than the kinds of activities performed. Both occupations are represented by a similar range of subsistence and maintenance activities. Deetz and Dethlefsen found that the thicker stratum of the more intense or longer-term occupation had higher pH values than the thinner stratum of the less intense or shorter-term occupation. Whereas soil pH indicates only the relative balance of cations and hydrogen within the soil, and not the absolute concentration of cations, higher pH is what one would expect to find in the soils of a more intensely occupied site where cation concentration had been augmented to a greater extent than in the soils of a less intensely occupied site. Unfortunately, it is not possible to conclude from Deetz and Dethlefsen's study that the higher pH values found for the thicker stratum are attributable to the greater degree of soil disturbance represented by its thickness. The thicker stratum underlays the thinner one. Soil pH often increases naturally with depth as a result of the downward leaching of cations, and the observed differences in pH between strata might reflect only this natural variation.

Approaching more closely the documentation of the proposition that different kinds of use-areas within an archeological site can be distinguished by regularly recurring patterns of soil alteration is a second class of studies that make specific reference to variation in kinds of activities which occurred at sites. At the simplest level are those studies that use the relative *concentration of single elements to differentiate either sites or subareas of sites where different kinds of activity occurred*. The earliest steps in this direction were made by Arrhenius (1929, 1931a). Arrhenius found that total phosphate enrichments in the soils of Stone Age camps are generally of a higher magnitude than enrichments in the soils of Neolithic sites. This difference was explained as being a result of a change in subsistence patterns and diet over time; from a diet based largely on fish and game and yielding phosphate-rich refuse to one where plants and cultigens played a greater role and phosphate-depleted refuse were generated. Lorsch (1940) continued to document the pattern that Arrenius had noted, using Swedish Stone Age and Neolithic sites, and elaborated Arrhenius's explanation. Lorsch claimed that the lower total phosphate concentrations found in agricultural settlements were a result of not only the lower phosphate content of the food refuse deposited on such sites, but also a change in depositional patterns. He suggested that in Neolithic sites, organic wastes were spread over wider areas through the agency of livestock and deliberate manuring.

On a similar note, variation in the total P content in the soils of 48 prehistoric sites in northern California was examined by Cook and Heizer (1965) for the purpose of distinguishing sites where vertebrates constituted a major food resource and where midden deposits were composed primarily of bone from sites where mollusks constituted the major food resource and where midden deposits were composed primarily of shell. They found that when holding constant the age of the sites, the intensity of occupation, and the parent materials on which the sites occur, the concentration of P was significantly greater within the middens of sites where primarily bone was deposited. This is an expectable result, considering that bone is composed mainly of Ca and P, whereas mollusk shell is largely lacking in P and composed mainly of $CaCO_3$ (see Table 67).

Oriented toward different goals but still using differences in the magnitude of *single* elements as indicators of different *kinds* of activities are a number of studies that defined the spatial organization of activities *within sites* using soil phosphate concentration. Dietz (1957:405) in examining the total phosphate contents of soils on a prehistoric habitation site in Wisconsin, was able to distinguish the locations of domiciles that were swept clean (low phosphate content) from areas that served as dumping locations for the sweepings (high phosphate content). Winter hearth cleanings, in particular, were specified as the major source of phosphate. A similar study will be presented in Chapter 8 of this work for a prehistoric site in Illinois. On a slightly broader scale of intrasite activity organization, Cook and Heizer (1965:70–79) successfully differentiated inhabited areas from uninhabited ceremonial precints within the site of Cuicuilco, Mexico. Both total P and total C concentrations were systematically higher in the soils of the inhabited areas of the site.

A number of successful studies of the internal organization of Norwegian sties using total phosphate assays have been reported by Provan (1971). Within a farm settlement dating to A.D. 350–550, he was able to distinguish between a portion of the farm house, encompassing a hearth area with very high total phosphate levels in the soil; a portion of the farm house used as a barn or byre area and having lower levels of phosphate; refuse areas outside the house having fairly high phosphate concentrations; and an enclosed garden area used for cultivating wheat and oats, and having low phosphate concentrations. Presumably, the crops grown in the garden had been depleting the soil of phosphates. At a second Norwegian site, dating between 400 B.C. and the time of Christ, Provan was able to segregate a house location from an associated refuse dump located downslope from it. Both areas had anomalously high total phosphate concentrations relative to those found in the surrounding natural soils, but the trash area had 10 times the concentration of phosphates found within the house. Finally, on a Mesolithic site, Provan (1971:47) was able to distinguish subportions of the site where concentrations of bones of seal and birds had accumulated from subportions lacking bone material. Areas of bone accumulation had up to 36 times the amount of total phosphate found in areas lacking bone material. Similar success in differentiating areas of archeological sites where bone had or had not been deposited was obtained by Eddy and Dregne (1964) who examined the pH of soils within several Indian sites in northwestern New Mexico.

Another instance of the use of soil pH to distinguish different kinds of use-areas within archeological sites is the work of Weide (1966:159, Figure 3). Weide investigated the areal distribution of soil pH in a Chumash village site in California and was able to distinguish spatially dispersed trash midden from burial structures and a pit house. The trash areas caused high pH values over a widespread area whereas the burial structures and pit house were associated with more localized high pH anomalies.

To date, the most complex and detailed studies that have begun to document that different kinds of use-areas within earthen archeological sites can be distinguished by regularly recurring patterns of soil alteration are those that have examined the *different spectra of elements in archeological deposits*. Hurley and Heidenriech (1971:189) asked whether it might be possible to distinguish midden deposits where different food refuses had been dumped, using elemental ratios. Upon examining the soils from a 300-year-old Huron Indian village in Ontario, Canada, they found (1971:212) that different, spatially discrete midden deposits had different Ca:Mg ratios. Although they make no attempt to infer the specific kinds of food refuse that generated this pattern, they do attribute this soil variation to the different kinds of food refuses that were deposited at different locales. They note (1971:209) that corn, among other substances, is high in Ca. Additionally, Hurley and Heidenreich (1971:205) noted that the spatial distributions of exchangeable Mg anomalies and total P anomalies differed within the Huron site. High Mg anomalies were confined to the center of the village whereas P occurred in high concentrations throughout the village. The distribution of Mg anomalies was taken to indicate the greater concentrations of ash within the center of the village and the locations of fireplaces within long houses. The generally high levels of P around the village were attributed to the general scattering of mixed refuse throughout the village area.

A second study of the spectra of elements within the soils of archeological sites is the analyses performed by Cook and Heizer (1965) on a late prehistoric Indian site in the Napa Valley, California. Cook and Heizer (1965:20) suggest that the ratios of Ca, P, and CO_2 in soils might be used to distinguish soils upon which primarily molluskan shell was deposited from soils upon which primarily bone was deposited. In soils where the pH is above 7 and the carbonate radical is immobile and not subject to leaching, locations of bone deposition should have high Ca and P levels approximately in the ratio found in bone (Ca : P = 2.4) and should have low carbonate concentrations, whereas locations of mollusk shell deposition should have high Ca and CO_3 levels approximately in the ratio found in shell ($Ca:CO_2 = .91$) and should have low phosphate concentrations. On examining their target site, Cook and Heizer (1965:34) found it possible to differentiate middens formed primarily with bone from middens formed primarily with shell using Ca:P ratios.

Although the previous studies of the soil chemistry of archeological sites reviewed here clearly demonstrate that a number of different kinds of ions of anthropic origin are maintained within the soils of archeological sites for long periods of time, it should be pointed out that most of these studies are concerned with *total* ion concentrations rather than *exchangeable* ion concentrations. Total elemental concentrations include both insoluble, fixed forms of the element and soluble, exchangeable forms of the element. Only the latter species are in short-term dynamic equilibrium with the concentration of ions and radicals within the soil water solution; only they are capable of directly influencing soil conductivity values. Fixed, insoluble forms of an element may over long time periods, be in equilibrium with readily exchangeable ionic forms and ions in solution (Brady 1974:453, 467–468, 477–478) but it is not clear that such fixed forms provide any major contribution to the ions and radicals represented in soil water solutions. Since many of the previous studies are concerned with total ion concentrations that may include a high proportion of fixed ions, it can not be concluded that the full range of archeological phenomena found distinguishable with soil chemistry data—including use-areas—also are distinguishable with resistivity methods that depend on soluble ionic forms. This will be demonstrated in Chapter 8, where a detailed analysis is given of the spectra of exchangeable ions and radicals found in different use-areas of an archeological site.

Previously Documented Long-Term Alterations in Soil Moisture Equilibria and Structure

Little systematic work has been done to relate soil moisture content and degree of development of soil structure to prehistoric human activity. It is common knowledge to field archeologists, however, that human-disturbed soils and the fills of earthen archeological features are often less consolidated and more easily cut by the trowel than undisturbed, surrounding soils (Cornwall 1958:60). They also often tend to remain moist longer than undisturbed soils when exposed to the sun.

It is probable that both of these soil attributes—the degree of development of crumblike soil structure and moisture retention—do vary consistently with the kind of uses to which different areas of an archeological site are put. Both attributes are directly related to the amount of organic matter contained within a soil, and there is good documentation in Cook and Heizer's (1965:45–61) study that organic matter content (indicated by total C) does vary with site function as well as with the duration of occupation and the age of a site. A direct and systematic study of the relationship between prehistoric activities on an archeological site and the kinds of changes in soil structure and moisture retention which they produce is given in Chapter 8.

Maintenance of Soil Alterations on Archeological Sites, I: Rates of Decomposition of Organic Matter

It is apparent from a review of the literature on chemical analyses of the soils from archeological sites that some of the soil attributes which affect soil conductivity can be altered by human activity and that such alterations can be maintained in the soils of archeological sites for long periods of time. The literature hints that it may very well be true that different kinds of use-areas on archeological sites are characterized by different spectra and/or amounts of available nutrients, different organic matter contents, different soil moisture equilibria, and different degrees of development of soil structure. For the moment, and until a thorough study is made of this proposition (Chapter 8), let us assume it to be true.

Given that the soils in different use-areas within an archeological site can be differentiated by their physical and chemical attributes, we may ask by what mechanisms or processes these differences can come down to us from the past. Once the different kinds of refuse deposited in different use-areas of an archeological site have been incorporated within its soils and have altered the physical and chemical characteristics of those soils, how are those alterations preserved? Are they, in fact, preserved, or are the new, anthropically generated soil characteristics subject to changes over time that might obscure the differences originally occurring in the soils of different kinds of use-areas? For example, do the ratios of different elements within the soil of a use-area change over time with the effects of leaching and natural inputs of nutrients into that soil? To answer these questions, the remainder of this chapter will describe the natural processes that help to maintain or obscure differences in the soil characteristics of different kinds of use-areas.

A first matter of concern is the rates at which the organic refuse deposited on archeological sites are decomposed from their fixed (insoluble) organic forms and (*a*) are mineralized to soluble ionic constituents that change the conductivity of the soil, and (*b*) are resynthesized into stable but reactive humic materials that facilitate the development of soil structure and alter soil conductivity. Are the rates of mineralization and humus formation *slow* enough that they, alone, could provide a steady input of mineralized ions to the soil water and a steady input of humus to the soil *over an extended period of time*? If so, then the postulated lengthy process of decomposition, itself, would serve as a mechanism for maintaining differences in the soil chemistry and physics of different use-areas. The following review of the literature on rates of mineralization and humification of organic debris under natural conditions and the factors affecting these rates suggests that this mechanism is inadequate to explain the chemical and physical differencs found between different use-areas of archeological sites.

Factors Affecting the Rates of Decomposition of Organic Matter

A great complexity of factors influence the rates at which organic residues decompose, mineralize, and/or are resynthesized. These factors offer multiple opportunities for the inhibition of the decomposition process and the extension of time over which mineralization and humification occur. Most of the determinant factors are concerned with the living requirements of microbial decomposers. The complexity of organic compounds composing the residue and their various proportions are of primary importance. Fractionation of more complex compounds with more diverse chemical bonds requires more kinds of chemical reactions, more kinds of enzymes that facilitate these reactions, and a greater number of species of microbes each capable of only certain enzymatic reactions. As the mulitiple reactions usually must proceed sequentially and because there are time lags between the bloom stages of the different microbial species responsible for the different sequential reactions (Bell 1974; Jensen 1974; Kononova 1961:134), rates of decomposition decrease as the complexity of the compound increases. All else being equal, the rate of decomposition of organic compounds decreases in the following order: sugars, starches, simple proteins, crude proteins, hemicellulose, cellulose, lignins, fats, waxes (Brady 1974:140). In general, plant tissues will decompose more slowly than animal tissues because the former possess higher proportions of complex compounds, particularly hemicelluloses, celluloses, and lignins, which they posses uniquely. More fibrous or woody, lignin-rich plant tissues decompose more slowly than succulent, nonwoody plant tissue (see Heath et al., 1966; Lockett 1937).

A second factor influencing the rate of mineralization and humification of organic materials is the ratio of nutrients found within them. Particularly critical as a limiting factor is the ratio of carbon to nitrogen. When the C:N ratio is less than ca. 25, which occurs in plants that are not very fibrous or lignin-rich, decomposition can proceed at the maximum rate possible under environmental conditions (Allison 1973:104). Nitrogen for the manufacture of microbial protoplasm occurs in sufficient quantity compared to the energy available in carbon bonds for microbial metabolism so as not be be a limiting factor in microbial reproduction. When the C:N ratio within an organic residue is greater than ca. 25, however, the nitrogen required for building microbial tissues becomes a limiting resource, and decomposition is slowed. Similarly, organic materials having a high content of the ash-element nutrients are more readily decomposed than refuse low in these nutrients (Jensen 1974:80). Large quantities of calcium are particularly important (Brady 1974:130).

Additional properties of organic residues that affect their rates of decay are the percentage of water-soluble nitrogen and bases and water-soluble organic compounds they contain. Such readily available nutrients that do not require the enzymatic activities of microbes to procure them hasten the expansion of microbial numbers, and consequently, decomposition rates. The percentage of soluble organic substances is most important in the initial phases of bacterial decomposition, whereas the percentage of soluble bases becomes important in later stages (Jensen 1974:98–99).

Two factors with a very significant effect on rates of decomposition, but not usually given much notice, are the mechanical breakdown of organic residues by soil megafuna and the inhibiting effect that the polyphenol content of these residues have on soil megafauna activity. Mechanical comminution considerably quickens the rate of this decomposition (Heath *et al.* 1966). Disintegration increases the surface area of the residue relative to its volume and provides microorganisms greater access to the material. At the same time, mechanical disintegration improves aeration and the moisture-holding capacity of the residue (Jensen 1974:94). Whether or not organic matter deposited in soil is reworked and comminuted by soil fauna depends on its palatability. The texture of the material apparently is of small significance but the concentration of distasteful polyphenols within the residue does have a major effect on the rate at which it is disintegrated (Jensen 1974:93–94, 99; King and Heath 1967). Often the residue must undergo a period of weathering when polyphenols are leached from them before soil fauna will act upon them.

In addition to the characteristics of the residues, themselves, a number of soil conditions affect the rates at which microbes can metabolize and reproduce, and thus influence decomposition rates. Among these are soil moisture content and aeration, pH, and temperature. Optimum moisture contents of soils for the growth of bacteria are wide, ranging from 50–100% of moisture-holding capacity, or between .001 and 3 bars water tension, depending on the texture of the soil (Allison 1973:105; Dickinson 1974:640). High moisture contents of soil will delay the decay of organic matter for several reasons. First, such conditions limit the supply of gaseous oxygen available for microflora metabolism. Decomposition can be continued under anaerobic conditions by microbes that utilize oxygen combined with iron and manganese for their respiration (Brady 1974:260), but rates of decomposition are slower. High soil moisture contents also create conditions nonoptional for microbial growth and the decomposition of organic residues by preventing the escape of the carbon dioxide produced by microbial metabolism. Accumulation of carbon dioxide within soils decreases soil pH and provides a less suitable environment for soil microflora (Dickinson 1974:640; Williams and Gray 1974:619). With respect to soil pH, the decomposition of organic residues proceeds most quickly when soil reaction is between 6 and 8. Under these conditions, bacteria will grow faster than fungi will. Generally, bacteria are capable of breaking down organic residues more quickly than fungi, and environmental conditions that facilitate bacterial growth over fungal growth tend to increase decomposition rates (Brady 1974:130; Allison 1973:107). Soil temperature has a major effect on the rates of decomposition of organic residues (Williams and Gray 1974:615–618). In general, within the range of temperatures tolerable by any given species of microflora, the rate of metabolism and growth of that species will increase two or three times for each 10°C rise in temperature (Allison 1973:106). Most soil microflora are mesophiles, with optimum temperature requirements of between 25 and 37°C and effective temperatures for activity of between 5 and 10°C.

The rates of decomposition of organic residues also are affected by whether they occur on the ground surface or are buried within the mineral soil. Organic matter will decompose most quickly on a bare soil surface, less quickly within the litter layer of a woodland soil, and even more slowly if it is incorporated within the A1 horizon (Williams and Gray 1974:613). This is particularly true where the mineral soil contains a fair amount of clay (Allison 1973:130). Proteins and other complex organic molecules are capable of being absorbed between crystal planes of 2:1 expandable clays and protected from further microbial attack (Brady 1974:117; Toth 1964:153). Also, ammonia-containing organic compounds within a residue may participate in cation exchange reactions with clay micelles (Toth 1964:153) and become less susceptible to microbial attack.

Rates of Decomposition of Organic Matter

The factors that influence the rates of decomposition of organic materials deposited within soils and the possibilities that these multiple factors offer for inhibiting decomposition and extending the time over which mineralization and humification might occur are complex, indeed. In most instances, however, decomposition of organic residues is rapid compared to the lengths of time dealt with by archeologists. Except in environments of extreme cold or dryness, decomposition, mineralization, and humification typically are completed within a matter of a few months to a few years. A number of studies document these rates. The variables used as measures of decomposition rates include the visual disappearance of residues, microbial population size, percentage weight loss of the residue, and percentage weight loss of ash elements and soluble organic ions. Studies using these variables will be reviewed, one by one.

Visual disappearance of leaves of a number of tree species as well as ash and lime have been monitored by Heath et al. (1966) for temperate woodland soils. The various materials were buried in the soil 2.5 cm below surface within nylon bags having holes large enough to permit the entrance of soil megafauna. A summary of their findings is shown in Figure 19. Within a year and 2 months, all tested materials had disappeared by more than 50%, and most had disappeared completely. Bobcock et al. (1960) found that leaves of ash (*Fraxinus* sp.) lying on the surface of forest soils with mull humus disappeared with equal rapidity, only a few midribs remaining after 6 months.

The length of time required for buried organic refuse to sustain a maximum microbial population has consistently been measured as very short—whether or

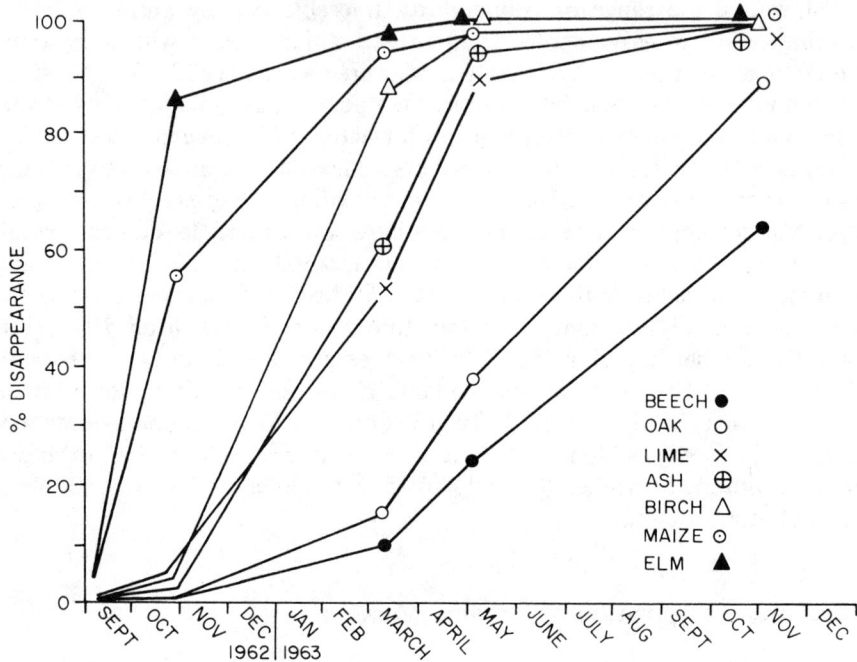

FIGURE 19. *Rate of visual disappearance of the leaves of several tree species, as well as lime and wood ash, buried within a temperate woodland soil. Data from Heath et al. (1966:8).*

not residues are at the ground surface or buried several inches within the ground. Stevenson (1962) buried several kinds of cultigens 2 in. below the surface of greenhouse soils and found the highest microbial activity to occur between 7 and 14 days after burial (Table 11). Allison and Klein (1962) added wheat straw to a sandy loam and noted that maximum microbial weight was achieved within 20 days after burial and constituted 1.7 times the original weight of the straw. Witkamp (1960) and Minderman and Daniels (1967) found that forest leaf fall on the surface of a calcareous mull soil in Holland achieved its maximum bacterial population within a few weeks, followed by a gradual decrease in microbial numbers. Similar results were obtained for the surface litter of a mixed oak forest in Belgium (Remacle 1970, 1971) and for the litter of beech (*Fagus* sp.) in Germany (Meyer 1960) and Denmark (Holm and Jensen 1972). Longer periods of time are required for wood to develop a maximum microbial feeding population than are noted for the more succulent organic materials mentioned previously. Allison *et al.* (1963) observed the rate of invasion of microbes on the sawdust of 19 softwood and 9 hardwood species buried in soil. Maximum microbial populations were attained within 40 days for 10 of the tree species whereas the remaining 18 tree species required 80 to 160 days.

TABLE 11
*Length of time required for several kinds of organic refuse to sustain maximum microbial populations**

Test	Crop	Sampling Time (Days)								
		0	1	3	5	7	10	14	18	25	66
Numbers of bacteria ($\times 10^6$/g soil)	Control	48	64	54	45	46	64	45	64	26	48
	Red clover		53	85	270	300	291	75	85	113	140
	Alfalfa		39	71	350	370	223	175	174	104	76
	Timothy		77	73	86	132	263	201	144	123	133
	Flax		48	65	400	485	354	402	296	133	75
	Oats		56	72	124	480	479	522	320	136	156
	Wheat		64	89	83	340	291	450	158	155	84

*Data from Stevenson (1962:503)

The several plant types shown were buried 2" below surface in greenhouse conditions.

4. Soil Formation from Human Refuse and Maintenance of Anomalies

A number of studies have monitored the total loss of weight of various organic materials. These weight losses largely reflect the release of carbon, hydrogen, and oxygen—the major constituents of organic matter—rather than nitgrogen or the ash elements, which are found in much lower proportions. Tables 12 and 13 illustrate typical rates of weight loss for a number of different kinds of tree leaves after their fall to the ground. Rates differ systematically with the base content of the leaf type and with soil conditions, but all are rapid from our perspective. A detailed documentation of the weight losses found for different organic residues, showing the changing trajectory of weight losses over time, is given in Figure 20. Differences among the several kinds of residues in their changing rates of decomposition over time are largely related to the surface area available for microbial attack compared to residue volume (Allison 1973:42).

When estimating the total length of time of mineralization and humification from the weight-loss data just presented, it should be remembered that completed mineralization and humification does not occur over a time period involving 100% weight loss, but rather, over a much shorter time span. Stable humus end products may account for 50–70% of the original total weight of the residues (Kononova 1961:136), and the length of time over which mineralized ions are being released and humus is being synthesized may be only that required for weight losses of 30–50%. Kononova (1961) has documented the lengths of time required for clover leaves and roots, hazel leaves, and scots-pine needles to humify completely under natural conditions. One hundred and eighty days or less were necessary, with weight losses of only 30–50%.

Complementing studies of the amounts of weight lost in specific time intervals by organic residues introduced to the soil are studies of the percentage weight loss over time of ash elements, nitrogen, and trace elements from residues. Again, rapid rates of mineralization are the pattern. A compilation of the

TABLE 12
Weight losses observed for the leaves of several tree species during the first six months after fall, as related to soil conditions

Species	Loss in Dry Weight (%) in Six Months	
	Mull Humus	Moder Humus
Fraxinus*	95.0	50.0
Betula verrucosa** (ex limestone)	82.9	26.3
Succulent greens***	70.–80.	---
Tilia cordata**	55.6	22.6
Quercus petraea	26.2	23.2
Quercus robur	17.4	17.0

TABLE 13
Weight loss observed for the leaves of several tree species during the first six months after fall, as related to leaf nutrient content*

Species	Excess Base Content	% Decomposition in 6 Months
Group 1		
75 mequiv excess base per 100 g		
Pinus strobus	42	17
Pinus rigida	44	21
Fagus grandifolia	70	17
Platanus occidentalis L.	72	21
Group 2		
75-110 mequiv excess base per 100 g		
Quercus alba	80	28
Quercus velutina Lam.	88	29
Juniperus virginiana L.	99	40
Acer saccharum	109	33
Group 3		
>110 mequiv excess base per 100 g		
Liriodendron tulipifera L.	121	55
Robinia pseudo-acacia	145	47
Cornus florida L.	152	43
Aesculus hippocastanum L.	174	30

*Data from Broadfoot and Pierre (1939)

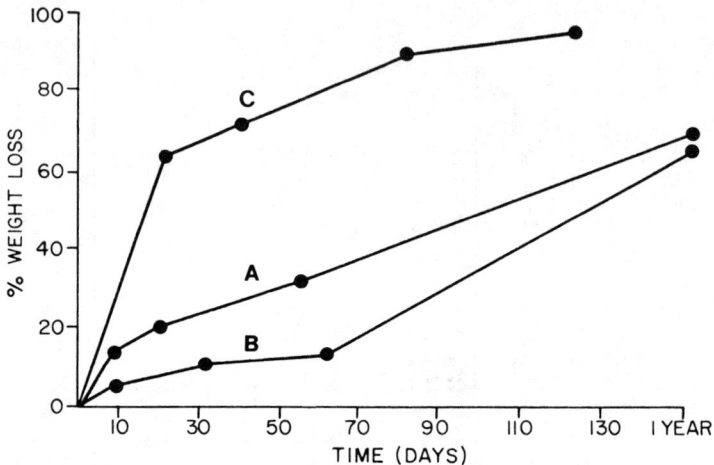

FIGURE 20. Changes over time in the rates of decomposition of the leaves of several tree species: (A) Melandrium, (B) Festuca sylvatica, (C) Vaccinium myrtillus. Data from Mangenot (1966).

amounts of nitrogen lost after 148 days from the leaf litters of various tree species and related to soil types (Table 14) is given by Bobcock (1964). Remezov (1961) examined the amounts of various elements lost from the leaf litter of oak (*Quercus* sp.) during the first 2 months after leaf fall and found the following losses: K, 54%; P, 47%; N, 35%; Mg, 35%; Ca, 28%; Si, 17%; Al, 10%; Fe, 0%. From 8 to 90% of the initial amount of K in the leaf litter was lost in the first year, and often between the time of leaf fall and early the next spring. Similar sequences and percentages of nutrient losses from leaf litters are documented by Burges (1967) and Attiwell (1968). Burges found that well over half the Na and K within leaf litter was lost in only a few weeks after its fall. Phosphorous and Mg also were mobilized fairly rapidly (greater than 50% loss) during the same period, but approximately half the Ca remained in the plant material. Attiwell observed a 90% reduction in the amounts of Na and K in leaf litter that had weathered 1 year, and only a 50% reduction in the amounts of Ca, Mg, and P after 2 years.

Much of these elemental losses result from the leaching of soluble forms of the elements rather than from biologically directed mineralization. The ordered sequences of different elements lost to greater or lesser degrees reflect the relative proportions of the elements that occur in soluble form within the particular kinds of organic residues examined. In general, Na and K occur in the greatest percentages in soluble form in plant tissues, whereas N occurs in the least percentages in soluble form (Table 15).

Most of the work done on rates of decay of organic residues considers leaf litter, which is relatively succulent and easily decomposed compared to more lignin-rich materials, such as wood. Some data on the rates of decomposition of

TABLE 14
Amounts of nitrogen lost after 148 days from the leaf litters of various species of trees*

Species	Initial Nitrogen (% Dry Wt.)	Percentage of Initial Nitrogen Left After 148 Days** on Mull	on Moder
Alnus glutinosa	3.06	10	55
Acer pseudoplatanus	2.52	20	50
Fraxinus excelsior	1.55	25	45
Tilia cordata	1.45	40	80
Corylus avellana	1.39	73	76
Salix sp.	1.38	45	60
Ulmus glabra	1.32	20	15
Fagus sylvatica	1.17	65***	80
Betula pubescens Ehrh.	1.13	42	75
Betula pendula	1.07	20	70
Aesculus hippocastanum	0.89	75	70
Quercus robur	0.79	40***	60
Quercus petraea	0.77	50***	65
Castanea sativa Mill.	0.69	75	70

*Data from Bocock (1964)

**Approximate values from figures.

***After 267 days.

wood are available, though, and indicate the relatively short amount of time—from our perspective—which is required for its mineralization and humification. It was mentioned previously that the lengths of time required for buried sawdust of various species of trees to attain maximum numbers of microbes required from only 40 to 160 days (Allison *et al.* 1963). Allison and Murphy (1962) and Allison *et al.* (1963) have summarized the percentages of C oxidized after 60 days in wood and bark tissues that were pulverized and buried within the litters of woodland soils. Substantial losses were evident (Table 16). Kaärik (1974:168) reports that in northern, humid climates, debarked posts standing in the soil may reach an advanced stage of decay after only 4 years.

TABLE 15
*Percentage of elements in mature plants soluble in water**

	K	Na	Mg	Ca	P	N
barley	65	52	45	40	36	2
wheat	54	41	46	34	33	7
oats	36	23	45	40	33	2
apple leaves	36	20	12	9	32	8
potato vines	18	22	12	6	25	3

*Data are from Bear (1951:18)

In conclusion, the lengths of time required for the mineralization of and formation of stable humus compounds from organic materials that have been deposited on soils are very short compared to the ages of archeological sites with which archeologists usually are concerned. This is true even for those organic residues containing high proportions of complex organic molecules that resist decay. It also is true when the factors governing decomposition rates are suboptimal, excluding extremely dry or frigid conditions. Consequently, slow decomposition rates, themselves, typically cannot account for the preservation of differences in the concentrations of mineralized ions and radicals and the percentages of humic materials that can occur in different use-areas of archeological sites. Except in extremely young archeological sites, unaltered organic residues deposited on them during the period of occupation do not serve for any significant length of time as the storehouses of ions that later are released and

TABLE 16.
*Percentages of carbon oxidized after 60 days in wood and bark tissues which were pulverized and buried within the litters of woodland soils**

	% C oxidized After 60 Days (Average Values)	
	Wood	Bark
hardwoods	30.3	22.4
pines	16.2	8.7

*Data from Allison and Murphy (1962, 1963)

found in mineralized form within the soils of such sites at the time of their investigation by archeologists. Other mechanisms must be evoked to explain the differencs in soil chemistry and physics found within different use-areas of archeological sites.

Maintenance of Soil Alterations on Archeological Sites, II: Rates of Decomposition of Humus

Microbial decomposition of organic matter yields two kinds of end products: mineralized compounds, such as CO_2, H_2O, and NH_3, and humic substances. The former are either gaseous or water soluble, and can be lost quickly from soil, either to the atmosphere, or to ground water by leaching. Humic substances, however, are not water soluble and are not easily lost from soils, although they can be eluviated in acidic or humid environments. Humus also is relatively resistant to microbial attack and mineralizes at only very slow rates. Can it be that humus, as a relatively immobile, slowly mineralizing substance, is one means by which the chemical and physical differences found between different use-areas of an archeological site are maintained over time? Does humus, as an agent in the *production* of different degrees of development of soil structure found in different use-areas of an archeological site also *remain* in the soil long enough to maintain those differences over extended periods of time? And does humus, as a relatively stable, yet slowly mineralizing substance, serve as both (*a*) a *reservoir* for maintaining within the soil the different kinds and amounts of nutrients that have been deposited within different use-areas of an archeological site and (*b*) a *mechanism* for the release of these stored nutrients into the soil water at the time of archeological investigation?

The Stability of Humus and Its Residence Time within the Soil

In answering these questions, a first matter of concern is the length of time over which humic substances are stable and reside within the soil. The chemical stability and residence times of humic substances within soils can be estimated by at least two kinds of data. First, the annual rate at which nitrogen—which forms an integral part of humus molecules—is released from them, can be used to assess the speed of decomposition of humus molecules. Rates of nitrogen release from humic components of soils range between 3 and 4.7% per annum in most soils (Allison 1973:117, 148). The half lives of humus molecules calculated from these rates of decay would be between 14.7 and 23 years.

The stability of humus molecules, as estimated by rates of release of nitrogen, is not very great compared to the ages of sites with which archeologists often are concerned and would not suggest humus to be a mechanism for maintaining chemical and physical differences between the use-areas of archeological sites. Estimates of the rates of decomposition of humus molecules from nitrogen-release rates, however, may be much too great, and misleading. Such rates of release are not determined directly on purified extracts of humic substances, but rather, are estimated from various field parameters. As a result, the estimates can include nitrogen released from incompletely humified organic matter that is decomposing at a much quicker rate as well as nitrogen being released from more slowly decomposing humus molecules.

A second, more direct and reliable method for estimating the stability of humus molecules within the soil is to determine the average age of carbon atoms within the humus molecule of a soil, as estimated by radiocarbon dating. This method gives a combined measure of both the chemical resistance of humus molecules to decomposition, mineralization, and loss from the soil, as well as the resistance of undecomposed humus molecules to eluviation and loss. The dates obtained by this method represent an average age of all humus molecules, including those deposited in the soil many centuries ago (if still present) and those just recently added to the soil in recycled plant litter. If the soil–vegetational system for which estimates are being made can be presumed to be in equilibrium, with the rate of humus formation equivalent to the rate of humus decomposition and eluviation, then the average determined age of humus molecules will reflect their rates of replacement. A number of age determinations under such circumstances have been made. The age of humus in the A horizons of prairie soils in Iowa and North Dakota have been estimated at 210 ± 130 and 440 ± 120 B.P. Correcting for the Suess effect would add 200 years to these dates (Broecker et al. 1956; Smith et al. 1960). In Iowa, a woodland soil was found to have an A1 horizon 610 years old and an A2 horizon 1040 years old (corrected dates). Podzols in Georgia and Holland have been dated at 1350 and 1140 years old. In Wyoming, the humus within A1 horizons of grassland soils have been found to date between 1174 and 3280 B.P., and the humus fixed within the clay fraction of the subsoil horizons of these soils have been dated to 6690 B.P. (Allison 1973:157). All of these estimates of humus stability are much greater than those estimated with rates of release of nitrogen and are much more in line with the degree of stability that would be required of humus molecules if they were to function as a mechanism for maintaining the chemical and physical differences found between different kinds of use-areas on archeological sites.

In summary, the estimates of the average age of carbon within the humic components of natural soils suggests that humic substances are chemically stable and relatively immobile compounds. Their time of residence within soils is long enough in many cases for them to be responsible for the maintenance of structural and chemical differences within the soils of different use-areas of archeological sites.

Explanations for the greater stability of humic substances within soils compared to the more transient nature of undecomposed organic matter are multiple but all concern the ease with which microorganisms can utilize the carbon and nutrients within humus. First, the large sizes of the molecular compounds forming humus make them less vulnerable to attack by microorganisms (Allison 1973:148). Humic and fulvic acids—the primary constituents of soil humus—have molecular weights of 1000–50,000 and ca. 300, respectively.

A second factor rendering humus less easily utilized by microbes is the diversity of the chemical units comprising humic acids and fulvic acids and the amorphous structuring of those units. Humic fulvic acids are complex polymers consisting of many different kinds of amino acids, phenol groups, and quinone units. These functional groups do not occur in any specified locations or order within the humus molecule (Allison 1973:153, 159). Biochemical fractionation of humus molecules consequently requires a large number of enzymes that *cannot* be applied in any particular successional sequence. Or from the perspective of microbial populations, a greater diversity of microbial species are required, acting sporadically rather than successionally or in unison on the humus molecule. Such a situation is not possible biologically, however, for microbes, like all living organisms, require a *continuous* supply of energy and nutrients rather than a *sporadic*, unreliable one. Consequently, humus molecules are not subject to large scale microbial attack, despite the large pool of energy and nutrients they offer. A third attribute of humus molecules rendering them less easily decomposed by microbes is the ring structure of some of their functional units (Allison 1973:149). The core constituents of humic and fulvic acids are largely aromatic ring compounds (organic compounds containing one or more benzene rings), particularly quinones and polyphenolic compounds oxidized to quinones (Kononova 1961:114, 157). To be assimilated as energy sources by microbes, these compounds must be fissioned. This reaction cannot be achieved by most microbial species,[3] and consequently the humus molecule has a biochemically very sound structure.

Nutrients Stored within Humus Molecules

The stability and long residence times of humic substances within soils suggests that humus not only is an agent in the production of differences in the degrees of development of soil structure within different use-areas of archeological sites, but also is capable of maintaining those differences over extended periods of time. Does the humus molecule also act as a mechanism for maintaining differences in the chemistry of soils of different use-areas? Are nutrient ions integrated within the core structure[4] of humus molecules and

---

3. Exceptions include *Penicilium glaucum*, which can fission phenols, and *Asperillus niger* and *Citromyces glaber*, which can utilize a number of kinds of polyphenols.

4. Here, reference is being made to the functional groups, themselves, comprising humic acid and fulvic acid, rather than exchangeable cations and anions adsorbed to humus molecules.

stabilized in the soil over time, these ions being released to the soil solution only slowly as the humus molecules decompose and affecting soil conductivity continuously? To answer these questions, we must examine in greater detail the chemical composition and structure of humus molecules.

Historically, soil humus has been studied in three components: humic acid, fulvic acid, and humin. *Humic acid* is defined as the organic material that is extracted from soil by alkaline solutions and that is precipitated from these extracting solutions by acidification. *Fulvic acid* is the material remaining in the extracting solution after acidification. That organic component which cannot be leached from soil by either acid or basic extracting solutions is defined as *humin* (Stevenson 1972:649). The differences that these three components of humus exhibit in their reaction to acids and bases reflect differences in both their reactive side groups and in their core constitutents. Humic acids are more complex in structure than fulvic acids. They are more oxidized and hydrated than fulvic acids, and have a more spherical structure, whereas fulvic acids have a more aliphatic structure (linear, with many side chains). Humin is a form of humic acid that has been irreversibly denatured and bonded with the mineral components of soil as a result of desiccation or freezing (Kononova 1961).

Although humic acids, fulvic acids, and humin do differ in many important ways in their chemical structure, and although these differences have a determinant effect on such soil properties as water-holding capacity, C.E.C., and soil dispersion, all three components of humus are basically similar with respect to their incorporation of nutrients. The primary constituents of humus are carbon, hydrogen, and oxygen, with smaller amounts (3–5%) of nitrogen. These elements occur in fairly consistent proportions within the humic components of all soil types (Tables 17a, 17b) so as to form the basic building blocks of humus: aromatic rings of the phenolic and quinonic types (Kononova 1961:114, 157). In addition, humus contains small amounts of sulfur that, along with –O–, –NH–, and –N– linkages, bridge the aromatic rings together into a single structure (Kononova 1961:66; Stevenson 1972:645). The ash elements other than sulfur are not part of the primary structure of humus.

Although most of the nutrients that are found in soil water (see Chapter 3) and determine soil conductivity are not an integral part of the structure of humic acid, fulvic acid, or humin, it would be a mistake to conclude that humus does not serve as a reservoir for many of these nutrients. Many of the compounds that compose the primary structure of humic and fulvic acid, as well as side chains of these acids, are capable of tightly fixing bivalent and polyvalent cations such as Mg, Ca, and Fe, and incorporating them within the integral structure of humus. The process by which this occurs is called *chelation*—the formation of heterocyclic rings (organic, closed-chain or ring compounds containing atoms other than C) by attachment of multiple compounds to a common, central metal ion. Humic and fulvic acids contain a number of core structural groups and side groups that present a variety of sites where such rings can form. Included among them are phenolic hydroxyl groups; weakly acidic carboxyls; more stable, strongly acidic carboxyls; poysaccharides; as well as amino and imino groups (Kononova 1961:73; Mortensen 1963:181). Metal ions incorporated within the structure of humus in this manner cannot be replaced by more reactive ions, as they normally would in simple ion-exchange, and as does occur with cations adsorbed to exchange sites on humus and clay micelles (Mortensen 1963). For the chelated metals to be released, integral parts of the humus molecule must be decomposed. Consequently, humus molecules do serve as a mechanism for maintaining bivalent and polyvalent cations within the soil over extended periods of time, in addition to N and S.

It should be pointed out that although chelation is a chemical phenomenon defined simply with respect to the pattern of chemical bonds, it usually is discussed in reference to means by which nutrient ions are kept in available, water-soluble form within the soil (Stevenson 1972). There are a number of water-soluble compounds within soil waters—including amino acids, phytic acid, adenosine phosphates, aliphatic acids, keto acids, chlorophyll, catechin, and simple sugars—which are capable of chelating metal ions just as humus moleules can. Metals chelated by these water-soluble compounds are kept in solution even under soil conditions in which they normally would be precipitated (see Brady 1974:493–495) and consequently are subject to leaching and loss from the soil. Metals chelated by humus molecules, however, are not subject to loss, but rather, are stabilized within the soil, for humus molecules are not water-soluble and are relatively immobile in soils.

In summary, soil humus does serve as a mechanism whereby both physical and chemical differences between the soils within different use-areas of an archeological site are maintained over time. Humus molecules facilitate the development of soil structure, and have long enough residence times within some soils to explain the maintenance, over extended periods of time, of differences in the degree of development of soil structure in different use-areas. Soil humus also incorporates N, S, Mg, and Ca within its primary chemical structure and can serve as reservoir of anomalous concentrations of these nutrient ions within the soils of use-areas, releasing them to the soil water solution slowing over time as it decomposes.

Maintenance of Soil Alterations on Archeological Sites, III: Nutrient Cycling

Although some of the nutrient ions deposited as refuse on the soils of archeological sites are maintained in the soils as an integral part of humus micelles, most are drawn into the pool of nutrients that naturally occur within the local ecosystem, and are recycled along with them between the vegetational and pedological component of the system. The various nutrient pools that occur within the ecosystem include the vegetational component, soil microbial tissue,

TABLE 17a
The elementary composition of humic acids from different soils*

Soils From Which Humic Acids Were Isolated	C	H	N	O	C:N	C:H	O:H	Author
Podzolic soil	52.39	4.82	3.74	39.05	14.0	10.9	8.1	
Rendzina	54.90	4.36	4.07	36.67	13.5	12.6	8.4	
Degraded chernozem	56.34	3.54	3.58	36.65	15.7	15.9	10.3	Tischenko and Rydalevskaya (1936)
Deep chernozem	57.47	3.38	3.78	35.37	15.2	17.0	10.4	
Ordinary chernozem	58.37	3.26	3.70	34.67	15.7	17.9	10.6	
Chestnut soil	58.56	3.40	4.09	33.95	14.3	17.2	10.0	
Podzolic soil	56.67	4.79	5.14	33.40	11.0	11.8	7.0	Natkina (1940)
Ordinary chernozem	62.55	2.78	3.32	31.35	18.8	22.5	11.3	
Columnar solonets	59.21	3.83	4.28	32.68	13.8	15.4	8.5	Kononova (1943)

*Data from Kononova (1961:53)

TABLE 17b
Elementary composition of fulvic acids from different soils*

Fulvic Acids	C	H	O	N	Absorption Capacity (m eq/100 g Substance)
From chernozem (Tyurin, 1940)	44.35	5.94	44.20	5.52	324
From podzolic soil (Tyurin, 1940)	44.82	5.77	43.66	5.75	318
From humus-illuvial horizon (Ponomareva, 1947)	45.3	5.0	48.6	1.1	300

*Data from Kononova (1961:79)

insoluble inorganic compounds, and exchangeable ions adsorbed on the surfaces of clay and humus colloids. These pools, as well as the natural processes by which nutrients are interchanged between them, comprise the major means by which the nutrients added to the soils of an archeological site are maintained within them. Additionally, nutrient inputs from precipitation and the weathering of parent materials help replace the nutrients lost to the ground water by leaching and to the atmosphere via biochemical transformations. The following section provides a schematic overview of the nutrient cycles, generalizing the paths taken by metallic ions, and then discusses in detail the particulars of the biochemical cycles of individual nutrients. Only those elements that occur in large proportions within soil water solutions (see Chapter 3) and that have a significant effect on soil conductivity will be considered.

Local natural ecosystems, as pools of nutrients, are composed of five subsystems: the soil and surface flora and fauna, which together form the major recycling components of the system, and the atmosphere and parent material, which are compartments of both nutrient input and output (Figure 21). The soil, itself, is composed of living and nonliving matter. The nonliving portion is a complex of inorganic particles formed by the physical, chemical, and biological weathering of the parent material, as well as organic colloids (humus) formed by the decomposition of plant and animal matter derived from the surface biological community. The inorganic particles of clay size, including crystalline clays and amorphous hydrous oxides of iron and aluminum, as well as the humus colloids, have net negative charges. They consequently are chemically reactive, and are capable of adsorbing to their surfaces H^+ cations and base nutrients such as NH_4^+, K^+, Na^+, Ca^{2+}, and Mg^{2+}, which are released to the soil by decaying organic matter, by decomposing humus micelles, by weathering parent materials, or by atmospheric precipitation. The interactions of clay and humus micelles with the remainder of the solid and liquid phases of the soil subsystem (including materials of anthropic origin) and also with the local vegetational community, soil organisms, the atmosphere, and the parent material constitute the major pathways of the generalized nutrient cycle (Figure 21).

The reactive sites on clay and humus micelles, under normal soil conditions, are filled with both hydrogen ions and base nutrients. All of these can dissociate from the clay and humus micelles, and may react with molecules dissolved in the surrounding soil water. They also exchange places (base exchange, base replacement) with ions dissolved in the soil water so as to maintain an equilibrium between (*a*) the relative concentrations of base nutrients and hydrogen ions adsorbed on the micelle and the relative concentrations of base nutrients and hydrogen ions in the soil water solution, and (*b*) the relative concentrations of different base nutrients adsorbed on the micelles and the relative concentrations of different base nutrients within the soil water solution (Wilklander 1964). For the purpose of illustration, here, only the balance of hydrogen and base nutrients will be considered; that is, that aspect of the

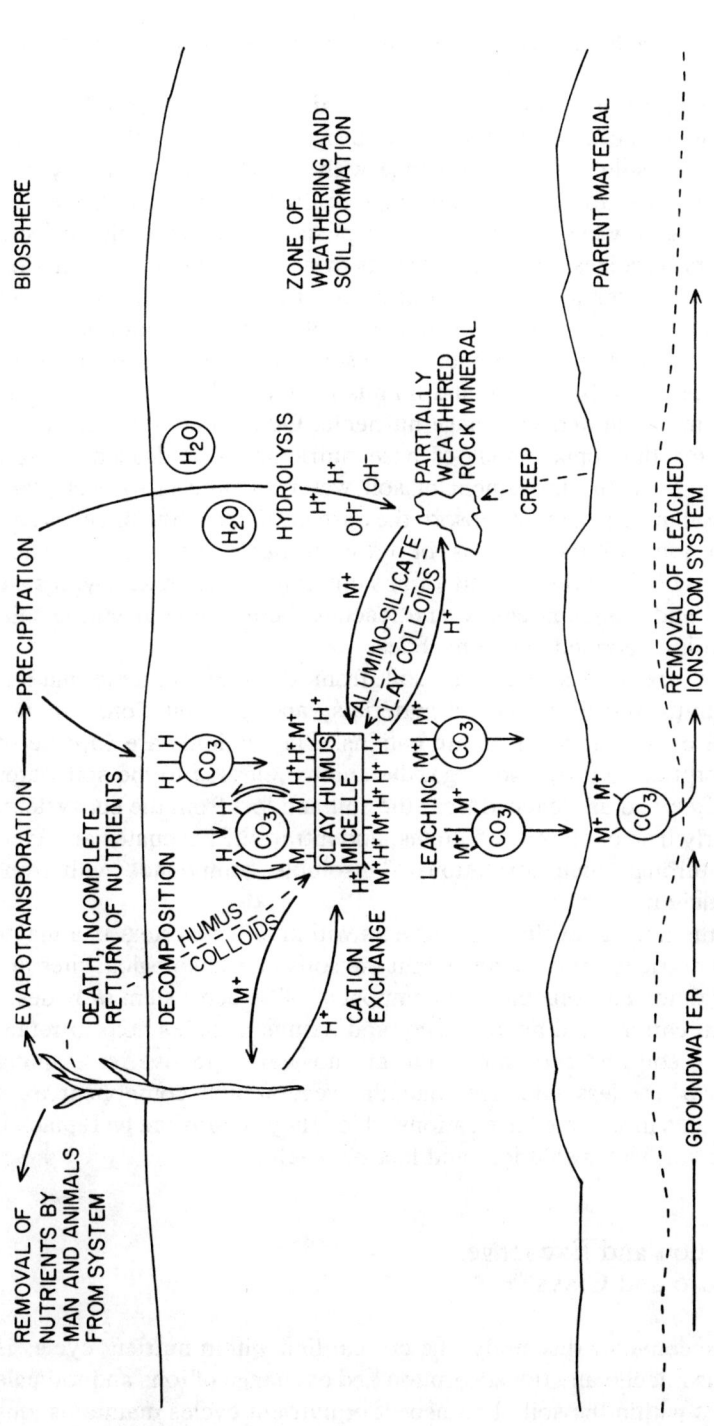

FIGURE 21. *A schematic representation of the nutrient cycles of metallic ions (M^+). Adapted from Dr. Bruce Gladfelter (personal communication) and Eyer (1968).*

generalized nutrient cycle concerned with maintenance of the pH balance of a soil.

When the concentration of base nutrients within the soil water solution is increased relative to the concentration of hydrogen ions, the base nutrients within the soil water solution may exchange with hydrogen ions on clay and humus micelles in order to establish chemical equilibrium between these two phases. This may occur when organic matter deposited on an archeological site decomposes and mineralizes, or when nutrients from atmospheric precipitation are added to the soil water solution. Hydrogen ions freed in such instances can then become involved in the chemical weathering of the parent material, which in turn may release more base nutrients to the soil water solution. If this were the only soil process involving clay and humus micelles, their exchange sites would soon become saturated with base nutrients. Other processes, however, are responsible for the replacement of base nutrients by hydrogen on soil micelles. Through both the movement of soil water and the growth of plant rootlets near to soil micelles, plants absorb the base nutrients of the micelles and return hydrogen ions to them. Loss of base nutrients also occurs when rainwater, which is dilute carbonic acid, drains through the soil, reacts with base nutrients on clay and humus micelles, and leaches the nutrients from the soil, leaving hydrogen ions bonded to the micelles.

Anions, which are released to the soil from decaying organic matter, decomposing humus micelles, parent materials, and precipitation, are not readily adsorbed by soil micelles as are cations. Clay colloids are capable of some anion adsorption, but by and large, the anions released to the soil water solution remain free and are leached from the soil and lost from the ecosystem. This is particularly true of chloride, sulfates, and nitrates. Phosphates are fixed within the soil through their formation of insoluble compounds with iron, aluminum, or calcium.

In summary, the natural cycling of nutrients within an ecosystem is a major means by which nutrient enrichments within the soils of archeological sites are maintained over time. The vegetational component of an ecosystem provides a reservoir for both cations and anions. Clay and humus micelles help to retain cations within the soil and ecosystems but are not very effective in retaining anions. They also are less effective than the vegetational component as a reservoir of nutrients in that the base cations which they adsorb can be replaced, released into the soil water solution, and lost be leaching.

Cation Adsorption and Exchange, and the Structure and Clays

From the perspective of this study, the critical link within nutrient cycles is the clay or humus micelle and the adsorption and exchange of ions and radicals that it encourages within the soil. This aspect of nutrient cycles maintains ions and radical within the soil, but in a soluble rather than fixed form such that they are capable of affecting the conductivity of the soil. To understand these phenomena more fully, it is necessary to consider the chemical nature of clays and humus.

Clay colloids are crystalline substances, having two structural components: silica tetrahedra and aluminum octahedra. Silica tetrahedra may be arranged in an interlocking manner, sharing oxygen atoms, so as to form a sheetlike "tetrahedral layer." Similarly, aluminum octahedra may be bound to each other through the sharing of O atoms so as to form a sheetlike "octahedral layer." These two different layers may be bound together in different orders so as to provide a number of different kinds of crystalline clays, including kaolinite with alternating tetrahedral and octahedral layers (1:1 clay structure), and montmorillonite, vermiculite, and illite, with an octahedral layer sandwiched between two tetrahedral layers (2:1 clay structure) (Brady 1974:71–83).

The net negative charge of clay micelles and their ability to adsorb cations to their surface are attributable to various kinds of anomalies within the crystal lattice of those micelles. Among clays with a 1:1 crystal structure, unsatisfied valences at the broken edges of silica tetrahedral and aluminum octahedral sheets act as negatively charged sites. Such charged sites hold hydrogen ions by covalent bonding. When the soil water solution has a pH of less than 6, the hydrogen ions are tightly adsorbed and are not easily exchanged for nutrient cations. When the soil water solution has a pH above 6, however, the hydrogen of hydroxyl groups dissociates slightly and can be replaced by other cations, particularly Ca and Mg. These nutrients tend to dominate exchange sites in neutral and alkaline soils.

Among clays with a 2:1 crystal lattice structure, another circumstance is responsible for their net negative charge: "isomorphous substitution." Silicon atoms within the centers of tetrahedra are subject to replacement by atoms of similar size. Aluminum is only slightly larger than Si and is able to substitute for Si. Although this replacement maintains the structural organization of atoms within the tetrahedral layer, it yields a net negative charge to the layer: Silicon has a valence of 4 whereas Al has a valence of only 3. In a similar manner, trivalent Al ions within the octahedral layers of clays may be replaced by bivalent Mg, Fe, and Zn ions of a similar size, and to a minor extent, by univalent K ions, leaving unsatisfied negative charges in the octahedral layer. Vermiculite and illite clays are distinguished by isomorphous substitution within the tetrahedral layer, whereas montmorillonite is characterized by replacement within the octrahedral layer (Brady 1974:76–81; Kardos 1964:386–389; Toth 1964).

Broken valences at crystal edges, hydrogen dissociation from hydroxyl groups, and isomorphous substitution may all determine the degree to which a clay colloid is negatively charged and its capacity to adsorb cations to its surface. They are general factors governing the maintenance of *all kinds* of cations within the soil. Factors governing which particular cations are adsorbed to a clay colloid and which cation species are better retained within the soil than others include: (*a*) the types of cations found in the soil water solution; (*b*) their relative concentrations in the soil water solution; (*c*) the degree of hydration of the cations; and (*d*) the nature of the anion species associated with the cations.

In general, the proportions of cations of different species found in clay colloids and the patterns of replacements of one cation by another as their concentrations within the soil water change can be predicted by the law of mass action. As the concentration of a given ionic species increases in the soil water, that species will replace cations of other species on clay micelles and increase its proportional representation on the micelles. Ions of higher valence have greater replacing power and are less easy to displace than ions of lower valence. An exception is hydrogen, which tends to behave as a divalent or trivalent ion. For ions of the same valence and concentration, replacing power tends to increase with the radius of the ion, taking into consideration its degree of hydration (Table 18). For unhydrated ions, the relative replacing powers of univalent species tends to increase in the same order as the lyotropic series of the periodic table: Li < Na < K < Rb < Cs. The relative replacing powers of bivalent species tends to increase in the same order as the alkaline earth series of the periodic table: Mg < Ca < Sr < Ba (Figure 22). Greater hydration of a given cation species compared to other cation species decreases the strength with which it is adsorbed by a clay and may lower its position within its replacement series. The relative replacement powers of cations also are affected by the anion species with which they are associated. The replacing power of a cation and its proportion on clay micelles will be increased if the associated anions are strongly adsorbed (see the next section on anion adsorption). If the associated anions do not tend to be adsorbed by the micelles, however, polyvalent cations may behave as monovalent ions and not be as strongly adsorbed to the micelles as expected; the accompanying anions have the effect of neutralizing the excess charges of the cation species. Anions found to function in this depreciating manner include OH^-, Cl^-, and NO_3^- (Kardos 1964; Toth 1964; Wilklander 1964).

Although the effects of each of these several factors can be documented and predicted when holding the others constant, their combined effect can not be generalized on and depends on the specific kinds of clay colloids that act as the exchanger medium. Holding the concentration and degree of hydration of cations constant, the manner of intergration of the lyotropic and alkaline earth replacement series has been found to be quite variable (Wilklander 1964:193–196). It can be noted, however, that in an ideal soil with the most probable relative concentrations of different cations species, the proportions of those species expected to be found on exchange sites of clay micelles would be 65% Ca, 20% H, 10% Mg, 5% K (Toth 1964:148–149).

The patterns of cation adsorption just described are significant to this study as they suggest that certain cation species are more easily maintained within the soil than others. Depending on the types of clay colloids found within the soils of

TABLE 18
Effective mean diameters of some hydrated and unhydrated ions, given in Angstrom units°

	Ion					
	H+	Mg++	Ca++	Na	K	NH$_4$
Hydrated diameter	9	8	6	4.3	3	?
Unhydrated diameter		1.3	1.98	1.9	2.66	2.96

*Data from Wilklander (1964:179) and Kardos (1964:191)

CATION	Li	Na	K	NH$_4$	Rb	Cs	Mg	Ca	Ba
UNHYDRATED DIAMETER (Å)	1.20	1.90	2.66	2.96	2.96	3.38	1.30	1.98	2.70

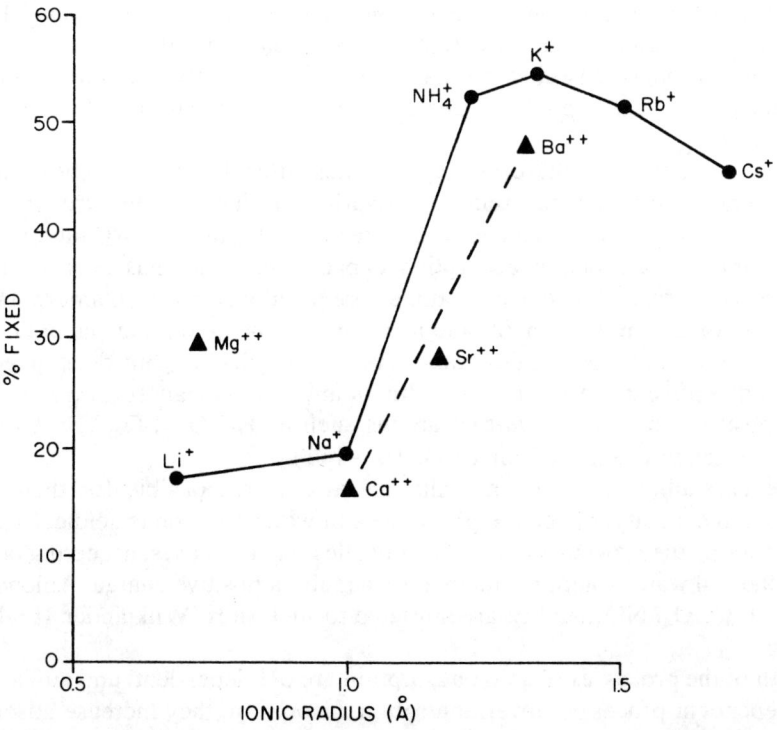

FIGURE 22. *Relationship between the unhydrated ionic sizes of various cations and their relative replacing powers, for the clay fraction of a Miami soil. The replacing powers of particular cation species are indicated by the percentage of cations fixed by the clay.*

an archeological site, certain nutrients deposited on the site will tend to be retained, whereas other nutrients will not. Those activities producing refuse having higher than average proportions of nutrients that tend to be better retained will be associated with use-areas having more distinct anomalies in their soil chemistry (intensity and duration of activity held constant). Activities producing refuse having higher than average proportions of nutrients that tend to be poorly retained will be associated with use-areas having less distinctive anomalies in their soil chemistry (intensity and duration of activity held constant).

Anion Adsorption and Exchange, and the Structure of Clays

Up to this point, the nutrient cycling and maintenance within ecosystems (and thereby, the soil) of only cations have been considered. Anions also are cycled within ecosystems, however, and like cations, are pooled within the vegetational and pedological components of the system. The primary mechanism for the cycling of anions within the soil subsystem—adsorption of anions to clay–humus colloids—is similar to that for cations, but not nearly as effective. Anions tend to be less strongly adsorbed to clay–humus micelles than cations, and consequently, they are subject to greater degrees of leaching and loss from the ecosystem.

The "amphoteric" nature of clay micelles—that is, their ability to adsorb both cations and anions—relates primarily to their mixed structure. The crystalline core of clay micelles, which has a net negative charge and which is responsible for the cation adsorption capacity of clays, has been described previously. Surrounding this crystalline core usually is found an amorphous gel of precipitated iron, aluminum, and manganese. The amorphous layer does not decrease the cation adsorption capacity of clay micelles, but does give them anion adsorptive powers. The iron, aluminum, and manganese within the layer have positive charges and attract anions such as H_2PO_4, SO_4, and Cl to the surface of clay micelles (Toth 1964: 146–151).

The crystalline core of clay minerals also is responsible for their anion adsorption capacity. When the pH of the soil water solution is acidic, hydroxyl groups along the broken edges of clay micelles react as bases, accepting protons from the soil water solution, and thereby acquire a positive charge. Anions such as H_2PO_4, SO_4, NO_3, and Cl are attracted to such sites (Wilklander 1964:164, 200).

Both of the processes of anion adsorption are pH dependent, and unlike those pH-dependent processes governing cation adsorption, they increase adsorptive capacity as the soil water solution becomes more acidic (Wiklander 1964:198). At lower pH values, more Fe, Al, and Mn on the surface of clay colloids becomes soluble and capable of reacting with anions. Activation of hydroxyl groups on the broken edges of clay micelles also increases with decreasing pH. Empirical studies on the anion adsorption capacities of several soils (Table 19) indicate that these pH-dependent processes occur in both 1:1 and 2:1 clays—as would be expected. These studies also indicate how weakly anions are adsorbed by clay micelles. Chloride ions and sulfate radicals apparently are not adsorbed at all when soil water solutions have a pH above 7, and they may not be retained even at lower pHs.

Systematic data on the relative replacement powers of different anion species are retained within the soil are scarce (Table 19). A commonly observed replacement series, however, is $Cl < SO_4 << PO_4$ (Wilklander 1964:198).

Ion Exchange on Hydrous Oxides and Humus Micelles

Silicate clays are one form of colloidal micelle that adsorb ions and maintain them within the soil in a form which is soluble and has an effect on soil conductivity. Two other adsorptive, colloidal forms are hydrous oxides and humus. Together these three colloidal substances account for the vast majority of the cation exchange capacities of soils.

Hydrous oxides of aluminum and iron, like the silicate clays, are crystalline compounds with net negative charges. With respect to their chemical reactivity, however, they are amphoteric, and are capable of both cation and anion exchange. The phenomena responsible for their adsorptive powers are a subset

TABLE 19
Anion adsorption capacities of kaolinitic colloids from a Nipe soil and montmorillonitic colloids from a Sharkey soil[*]

	Nipe				Sharkey	
pH	meq Sorbed per 100 g Colloid			pH	meq Sorbed per 100 g Colloid	
	Cl	SO_4	PO_4		Cl	PO_4
7.2	0.0	0.0	31.2	6.8	0.0	22
6.7	0.3	2.0	41.2	5.6	0.0	36.5
6.1	1.1	5.5	46.5	4.0	0.05	47.4
5.8	2.4	7.1	50.8	3.2	0.1	64.0
5.0	4.4	10.5	66.1	3.0	0.1	73.5
4.0	6.0	---	88.2	2.8	0.4	100

*Data from Wilklander (1964:198)

of those responsible for the exchange capacities of clays. They include the pH-dependent reactions, the hydrogen dissociation and protonation. In the former reaction, the hydrogen ions of hydroxyl groups that are bonded to Al or Fe at the surface of hydrous oxides dissociate, leaving the micelles with net negative charges. This occurs at higher pH values. In the latter reaction, the reverse occurs, and hydrogen ions associate with surface hydroxyl groups, giving the micelles net positive charges. This reaction occurs as the pH of the soil water solution is lowered, and is responsible for the anion exchange capacity of hydrous oxides. An additional reaction by which anions are exchanged on hydrous oxides is the replacement of surficial hydroxyl groups by the monovalent species of phosphate H_2PO_4. This reaction also occurs primarily when the pH of the soil water solution is acidic (Brady 1974:91–92).

Humus colloids are extremely reactive compared to silicate clay colloids and hydrous oxide colloids, and play a very significant role in ion adsorption phenomena. Even though the humus content of soils typically constitutes only 3–5% of the total soil mass, it usually accounts for 20–50% of a soil's cation exchange capacity (Toth 1964:154). Humus does not, however, have the anion adsorptive capacities that the silicate and hydrous oxide clays do, and consequenty, both forms of colloids are important in nutrient cycling.

Humus is an amorphous, noncrystalline, organic substance (see above), and the mechanisms responsible for its cation adsorptive capacities differ from those of crystalline clays and hydrous oxides. The reactive nature of humus owes to its high content of oxygen-containing functional groups, including COOH, phenolic-OH, aliphatic-OH, enolic-OH, and C=O structures (Stevenson 1972:644). Carboxyl, phenolic, and enolic hydroxyl groups predominate on humus colloids and are the primary groups responsible for the cation exchange capacity of humus micelles (Kononova 1961:51–54; Toth 1964:155). Other functional groups that questionably participate in cation exchange include amino, heterocyclic amino, imino, and sulhydral groups (Stevenson 1972:644).

Because many of the reactive sites on humus micelles are hydroxyl groups, the charge and cation adsorption capacity of humus micelles are pH-dependent. Replacement reactions on phenolic hydroxyl sites occur primarily under alkaline conditions. Carboxyl groups participate in exchange reactions when the soil water solution is neutral or alkaline in pH, but not as often when it is acidic. Consequently, as the pH of a soil system increases, the cation exchange capacity of humus micelles increases (Brady 1974:95; Kononova 1961:55).

Humus, hydrous oxides, and silicate clays are all significant to this particular study in that they are complexes which naturally occur within all soils and are universal mechanisms by which nutrients deposited on earthen archeological sites can be cycled and maintained in the soil and the larger ecosystem. Hydrous oxides and humus have an added significance in that their quantities within soils are enriched by human occupation, whereas the quantity of silicate clays usually is not augmented.[5] Enrichments of humus within the soils of archeological sites are a direct result of the organic debris deposited there, and are extremely common. Enrichments of hydrous oxides are less ubiquitous, but can occur in limited areas of archeological sites where limonite (yellow ochre) or red ochre were used, or where red ochre was produced from limonite (Winters 1969:26). In both cases, the adsorption capacity of the soil may be significantly increased, augmenting its capability to cycle and maintain anthropically derived cations and anions within it and within the ecosystem, at large.

Fixation of Cations and Anions within Clays

In contrast to nutrient ions and radicals that occur in exchangeable, soluble forms within the soils of archeological sites are those that are fixed in inorganic, insoluble forms. These nutrients comprise another of the several pools of nutrients that are cycled and recycled within the soil and the ecosystem. Many of the processes by which such nutrients are fixed are specific to particular elements, and they will be discussed when I describe the cycles of individual elements, one by one, in the following pages. One group of fixation processes that have a more general character and concern multiple nutrients, however, are those that involve the structural integration of nutrients within the crystal lattices of clays.

The primary process by which nutrients are fixed by clay micelles is isomorphous substitution—the same process responsible for the negative charge of clay micelles. The nutrients most commonly involved in this process are Mg, and in a much more minor way K, which are similar in size to aluminum ions and which can replace Al ions within the crystal lattices of 2:1 clays. Phosphorous also may become involved in several kinds of isomorphous substitutions. First, whole phosphate tetrahedra may replace whole silicate tetrahedra within the tetrahedral layers of clays. Second, phosphate tetrahedra may replace hydroxyl ions found peripherally on the cleavage planes between crystal sheets. Finally, as the degree of isomorphous substitution becomes excessive and a clay crystal lattice can no longer accommodate misfitted phosphate tetrahedra within its original form of organization, the lattice may recrystallize, incorporating the phosphate tetrahedra more stably within some new form of organization (Kardos 1964:372–374).

A second process by which nutrients are fixed by clay micelles does not occur among all types of clay minerals and is specific to those with 2:1 lattice structures of the illite type. In this case, K and NH_4 ions may be tightly fixed

---

[5] Clays brought onto sites to cap mounds or to make puddled floors are some possible exceptions.

within interlattice spaces between the adjacent exposed surfaces of three-layered tetrahedral–octahedral–tetrahedral crystal units. The exposed surfaces of tetrahedral layers consists of oxygen ions arranged hexagonally, with openings within the hexagons. The openings are 2.8 Å in diameter—the approximate diameter of unhydrated K ions (2.66 Å) and unhydrated NH_4 ions (2.96 Å). K and NH_4 ions can partially fit within these openings. The bonding of K and NH_4 ions to tetrahedral layers serves to balance the negative electrical charges that the layers develop through isomorphous substitutions of the type described previously (Kardos 1964:386–387), and is one means by which K and NH_4 are fixed within illite-like clays. The strength of K and NH_4 fixation is increased, however, by a second circumstance. The bound K and NH_4 ions between adjacent trilayered crystal units draw these units together and lock themselves within the clay crystal. This prevents them from being readily exchanged with other ions in the soil water solution and helps to maintain them within the soil. Over extended periods of time, however, K and NH_4 ions fixed in this manner can be displaced by Na, Ca, or Mg ions, if the concentrations of the latter within the soil water solution are great enough. An Na:K ratio of 3:2 or a Ca:K ratio of 4:1 apparently is sufficient for K replacement (Kardos 1964:386–387).

In all of these reactions, nutrients incorporated within the clay lattices have their source in the soil water solution and *remain in equilibrium* with nutrients of the same species within the soil water solution after their incorporation. In this respect, they resemble simple cation exchange reactions on the surface of clays. The length of time over which such equilibria are established, however, is much longer than that required by cation exchange reactions. In one study of kaolinitic clay colloids (Kardos 1964:374), it was found that phosphate adsorption reactions required only 1½–3 hours to reach equilibrium. Replacement of silicate tetrahedra by phosphate tetrahedra, however, did not begin until 8 days after adequate phosphate concentrations were provided in the soil water solution. Substitution then continued for 35 days, at which point the clay began to recrystallize and incorporate phosphates in a chemically more stable manner.

The delayed equilibrium involved in the processes of fixation and release of nutrients within clay crystal lattices is particularly important from the viewpoint of this study. First, the equilibrium times of such reactions are still *short* enough that Mg, K, P, and NH_4, which are released to the soils of archeological sites during the decomposition of anthropically deposited organic matter can be incorporated within clay crystals and be protected there from leaching and loss from those soils. Second, the equilibrium times of such reactions are *long* enough that the reactions do not occur with day-to-day fluctuations in soil moisture content and the concentration of nutrients within the soil water. Consequently, the fixed nutrients are not released to the soil water and lost when, for example, the soil is flushed after a rain and the concentration of phosphates within the soil water is low. This situation contrasts with that of nutrient ions that are adsorbed to the surfaces of clay micelles. The latter are in a more dynamic equilibrium with the nutrients in the soil water solution and are subject to exchange reactions and loss during short-term depletions of nutrients within the soil water solution. Finally, although on a short-term basis the nutrient ions that are incorporated within clay lattices are not available to affect the conductivity of soil water solutions and the soil, they are effective during periods of relatively minor soil moisture alterations when equilibrium conditions between those nutrients incorporated within clay micelles and those within the soil water solution can be approximately achieved.

The latter consideration, in particular, has bearing on the times at which resistivity surveys might optimally be performed on archeological sites. The use of resistivity methods to differentiate areas having soils with anomalously high nutrient concentrations from other areas within archeological sites can probably be achieved best when the moisture contents of the soils have reached an approximate equilibrium and sufficient time has passed for nutrients fixed within clays to establish an equilibrium with nutrient concentrations in the soil water solution.

Fixation of Cations and Anions with Surface Vegetation

A final general aspect of nutrient cycling that pertains to more than one nutrient and that is responsible for the maintenance of nutrient anomalies within the soils of archeological sites is the uptake and fixation of nutrients by surface vegetation. Natural vegetation cover over an archeological site may constitute a significant stabilized pool of nutrients within the local soil–vegetational subsystem. This would not be significant to our problem at hand if the amounts of nutrients adsorbed by the surface vegetation over archeological sites were only those that are required by the vegetation and would naturally be retained in the ecosystem. Many plants, however, will adsorb *more* nutrients than they require (Bear 1951:12). Consequently, the excesses of nutrients deposited on the soils of archeological sites in the form of refuse may in part be stabilized in the local ecosystem through their uptake by vegetation. The excess nutrients stabilized within the vegetation, of course, are recycled to the soil and help to maintain anomalously high nutrient concentrations there are net losses of nutrients are encumbered by the soil.

The phenomenon of excessive nutrient uptake by plants is called "luxury consumption" (Brady 1974:474). It has been documented for a number of nutrients, including N, P, and K (Bennett *et al*. 1953; Dumenil 1961; Fletcher and Kurtz 1964; Tyner 1946), but appears to be most pronounced for K (Brady 1974:474). Much of the work in this area has centered on crops rather than natural vegetation, so it is not possible to state the kinds of natural vegetational and ecosystemic contexts for which luxury consumption is most significant as a mechanism in the retention of anomalously high nutrient concentrations.

The Nitrogen Cycle

An overview has been given of some of the general processes involved in nutrient cycling and the overall manner by which it can help to maintain anomalously high nutrient concentrations within the soils of archeological sites. Having established this general conceptual framework, it is now possible to consider more specific processes of nutrient cycling unique to one or a few elements.

While considering these element-specific processes within individual nutrient cycles, it should be recognized at this point that they not only *maintain* nutrients and anthropic nutrient anomalies within soils, but also modify over time the magnitudes and ratios of those anomalies. Anomalous nutrient ratios are modified because the different cycles of the different elements encompasses different pathways of loss, gain, and transformation, and differ in their effectiveness in maintaining these elements within the soil. The last section of this chapter considers this idea in greater detail, but the reader should keep this broader perspective in mind while the various nutrient cycles are discussed individually. Let us begin with the nitrogen cycle.[6]

Within the soil, nitrogen is found in three major forms: as organically fixed (insoluble) N, as ammonium ions on exchange sites on clay–humus micelles, and as ammonium and nitrate ions dissolved within the soil water solution. Organically fixed nitrogen accounts for at least 60% of all nitrogen within the soil. Most of this fraction occurs in the form of stable humus micelles ranging in N content from 3–5%. Nitrogen within humus micelles occurs in several functional positions: (*a*) as a constituent of the cyclic chemical structures that are the primary building blocks of the humus molecule; (*b*) as –NH– and –N– linkages between such building blocks; and (*c*) as NH_2 reactive groups on peripheral chains of the micelle. Exchangeable ammonium radicals on the surface of clay–humus micelles usually account for less than 40% of the total N within a soil. Dissolved ammonium and nitrate radicals usually only comprise 1–2% of the total N within a soil.

The transformations that elemental nitrogen undergoes as it is cycled within the soil and between the soil, vegetational, and atmospheric components of an ecosystem are summarized in Figure 23 and Table 20. Most of the processes internal to the soil system are biochemical rather than chemical. They involve the use of nitrogen compounds by soil microbes as either: (*a*) raw materials for building their own protoplasm, or (*b*) sources of energy for their metabolism. The capture of energy is achieved by their oxidizing or reducing the various nitrogenous compounds (Delwiche 1970:138–139).

The cycling of nitrogen within the soil component of an ecosystem may be summarized as follows. When organic debris is added to the soil, it is subject to

---

[6] This survey of the nitrogen cycle is abstracted and synthesized from discussions by Allison (1973), Bormann and Liken (1970), Brady (1974), Deevey (1970), Delwicke (1970), Kardos (1964), Kormondy (1969).

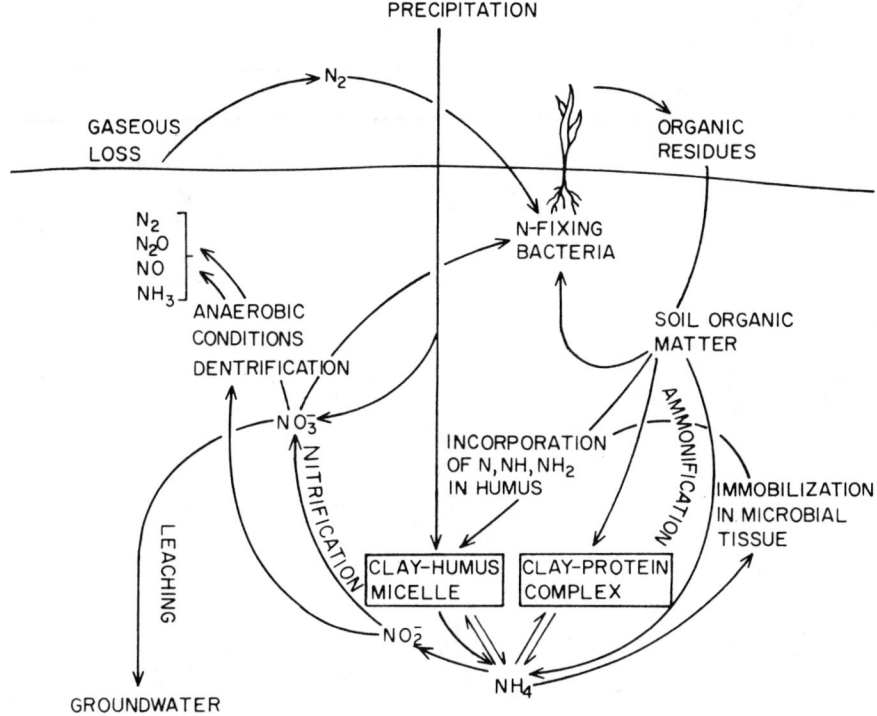

FIGURE 23. *The nitrogen cycle. Adapted from Brady (1974:424).*

a number of successive microbial transformations (see the section on decomposition, p. 117 this chapter). The ultimate end products of such decomposition include humus micelles, which can hold NH_4 in exchangeable or fixed forms, and NH_4. Much of the NH_4 released by microbial decomposition is reutilized by successive microbial populations to build their tissues or as a source of energy. The remainder may be utilized by higher plants or held in exchangeable or fixed forms within clay and humus micelles. Little remains within the soil water solution and is subject to leaching, compared to NO_3. The NH_4, which is released to clay micelles, can be fixed under moist conditions if the clay is illite or vermiculite, but requires drying–rewetting cycles if the clay is montmorillonite. The specific processes of fixation have been described already. NH_4 may also become fixed in a semistable form within humus micelles by reacting with their aromatic and quinonic components. A final pathway for the nitrogen of decomposing organic matter results in its incorporation within semistable protein–clay complexes. Clays with expandable lattices facilitate the development of such complexes.

Of the NH_4 that is released and then reutilized by soil microbial populations, some is incorporated directly into microbial protoplasm and some is utilized as

TABLE 20
Biochemical reactions within the nitrogen cycle

Name of Reaction	Chemical Process
Ammoniafication	$\underset{\text{amino combination}}{R-NH_2} + HOH \underset{\text{enzymatic hydrolysis}}{\longrightarrow} R-OH + NH_3 + \text{energy}$
	$2NH_3 + H_2CO_3 \longrightarrow (NH_4)_2CO_3 \rightleftharpoons 2NH_4^+ + CO_3^-$
Nitrification	$2NH_4^+ + 3O_2 \underset{\text{enzymatic oxidation}}{\longrightarrow} 2NO_2^- + 2H_2O + 4H^+ + \text{energy}$
	$2NO_2^- + O_2 \underset{\text{enzymatic oxidation}}{\longrightarrow} 2NO_3^- + \text{energy}$
Denitrification	$NO_3^- \underset{-2[O]}{\rightarrow} 2NH_2O \underset{\substack{-H_2O \\ -2[O]}}{\rightarrow} N_2O \underset{-O}{\rightarrow} N_2 \underset{\substack{-H_2O \\ =O}}{\rightarrow} 2NO$
Nitrogen fixation	$N_2 \underset{\text{enzymatic}}{\longrightarrow} \underset{\text{amino combination}}{R-NH_2}$

$R-NH_2$ = organically fixed nitrogen
NH_3 = ammonia
NH_4^+ = ammonium radical
NO_3^- = nitrate radical
NO_2^- = nitrite radical
N_2 = gaseous nitrogen

a source of energy for microbial metabolic processes. In the latter case, the hydrogen and nitrogen of ammonium radicals are broken apart via enzymatic oxidation so as to release bonded energy. Nitrogen-containing nitrites and nitrates, as well as water and free hydrogen, are released.

From the perspective of nutrient cycling, the important part of this transformation is the formation of nitrites and nitrates, rather than the release of energy. This process—nitrification—occurs sequentially, in two steps, as shown in Table 20. First, ammonium radicals are oxidized to nitrites and then the nitrites are oxidized to nitrates. Under typical soil conditions, the second step immediately follows the first, and nitrites do not accumulate within the soil. Under very alkaline conditions, however, the transformation of nitrites to nitrates is retarded and nitrites do accumulate. The dependence of nitrite accumulation on pH results from the fact that the two steps of nitrification are carried out by different microbial species with different pH requirements. Transformation of ammonium radicals to nitrates is achieved by *Nitrosomonas* sp., which have wide pH tolerances. Oxidation of nitrites to nitrates is carried out by *Nitrobacter* sp., which do not propagate well under very alkaline conditions.

A soil's potential for sustaining the nitrification process depends on a number of environmental requirements that the nitrifying soil microflora have. Many of these requirements have been discussed in a previous section of this chapter for soil microflora (pp. 117–119), in general, but an additional factor, specific to the nitrifying microflora is whether or not the soil occurs under a deciduous forest. Such vegetational communities apparently can inhibit microbial production of nitrites and nitrates from NH_4 (Bormann and Liken 1970:98–99). The precise mechanism of inhibition is not known, but is viewed as an adaptive means for maintaining nitrogen and other nutrients within forest ecosystems. Nitrate is more easily leached from soils than is NH_4, and inhibition of the nitrification process is a good strategy for minimizing the amount of nitrogen lost by leaching. Additionally, by slowing nitrification, this inhibitory mechanism curbs the production of hydrogen ions that can replace exchangeable cations on colloidal micelles. The latter would be subject to leaching upon their replacement by hydrogen and dissolution within the soil water.

When nitrates are released to the soil water solution, they follow several pathways within the nutrient cycle: (*a*) utilization by microbes as a source of nitrogen to build their tissues, (*b*) adsorption by higher plants, (*c*) leaching, and (*d*) conversion to N_2 gaseous forms, which escape to the atmosphere. The degree to which NO_3 is utilized by microbes as a nutrient or is adsorbed by higher plants depends on the availability of carbon within the soil. If the $C:NO_3$ ratio of a soil is high, nitrates within the soil water solution will be rapidly incorporated and fixed within microbial tissue and made unavailable to plants.

Loss of nitrogen from soils and ecosystems occurs by both the leaching of nitrates into ground waters and the conversion of nitrates to gaseous forms. In well-drained soils of humic regions, leaching is the primary pathway of nitrogen loss. Nitrates, unlike NH_4 radicals, are not adsorbed or fixed strongly by clay or humus micelles. Those nitrate radicals not utilized by microbes or plants tend to remain in the soil water solution and consequently can be flushed from the soil when it rains. Gaseous losses of N—either seasonal or permanent—occur in anaerobic soil conditions when microbes reduce nitrates to NO, N_2, and N_2O for metabolic purposes. These gases escape to the atmosphere and, in well drained soils of humic regions, may account for a loss of 10–20% of the annual input of nitrogen. In poorly drained soils, the loss can be substantially more.

Inputs of N to soils and ecosystems occur by both precipitation and the fixation of atmospheric nitrogen into organic N by bacteria. Rainwater contains both ammonium and nitrate forms of nitrogen. In temperate regions, the amount of ammonium added tends to be greater than the amount of nitrate added, average annual enrichments being 5.0 kg/ha and 1.7 kg/ha, respectively (Brady 1974:441). Much larger inputs (28.7 kg/ha NH_4, 19.0 kg/ha NO_3) have been reported for the New England area (Likens et al. 1977).

The fixation of atmospheric nitrogen by bacteria is the primary process by which N is incremented to most natural soils. Atmospheric nitrogen is utilized by various kinds of bacteria and released to the soil in organic form. The manner of release may be direct or indirect, depending on whether or not the bacteria involved maintain symbiotic relationships with plants. In the case of the nonsymbiotic microbes (e.g., *Azotobacter, Clostridium*, and blue-green algae in the soils of the Midwest United States), nitrogen is incorporated within their tissues and released directly to the soil when they die and decompose. In the case of the symbiotic bacteria (*Rhizobium* sp.), nitrogen is released first to the plant in which the bacteria reside, and is freed to the soil only when the plant dies and decomposes. The amount of nitrogen with which symbiotic bacteria yearly enrich the soil under natural conditions depends partially on the density plant species in the ecosystem that are capable of maintaining symbiotic relationships with them. In temperate forest ecosystems, this tends to be low compared to the amount of nitrogen fixed by nonsymbiotic microflora. The latter has been estimated at 62 kg/ha/yr, on the average, for temperate forests in general (James Boyle, personal communication).

Nitrogen added to the soil by precipitations and microbial fixation helps to maintain within the soils of archeological sites anomalously high concentrations of N, which otherwise would be rapidly dissipated by leaching and gaseous losses. At the same time, however, microbial fixation of N can tend to reduce the *differences* in the N content of soils in different use-areas of archeological sites. Nitrogen fixation rates are reduced in soils when NH_4 or NO_3 occur in them in high concentrations. Consequently, the nitrogen within the soils of use-areas where N concentrations are high is not replenished by microbial fixation processes as much as is the nitrogen within the soils of use-areas where N concentrations are lower. Differences in the N content of such areas thus can be decreased over time.

The Sulfur Cycle

The cycling of sulfur within soils and within ecosystems at large is similar to the cycling of nitrogen. In each case, (*a*) the atmosphere is the most important source of input of the element into the ecosystem; (*b*) within the soil, the organic combinations of the element comprise the major form in which it occurs; and (*c*) within the soil, cycling encompasses a number of transformations of mineralized forms of the element, with oxidation and reduction reactions under microbial control being more important than strictly chemical reactions.

The Sulfur Cycle

The major form of sulfur and the biochemical transformations involved in the sulfur cycle are summarized in Figure 24 and Table 21. Considering first the pathways of sulfur within the soil, itself, we can enter the cycle at the point where fresh organic matter has just been added to the soil. Successive populations of microbial species decompose the organic matter, yielding two kinds of sulfur-containing end products. First are stable humus micelles that include sulfur as an integral part of their structure. Sulfur serves to link the various monomers that comprise the humus polymer. The second kind of end product of decomposition is a set of various mineralized forms of S. Included among these are sulfides, plus incompletely oxidized compounds such as elemental sulfur, thiosulfates, and polythionates. All of these mineralized substances are freed to the soil water solution and are subject to rapid leaching from the soil. They are not adsorbed or fixed by clay or humus micelles. In this respect, they differ from the mineralized nitrogenous end product of decomposition, NH_4, which can be held in exchangeable and fixed forms by clay and humus micelles.

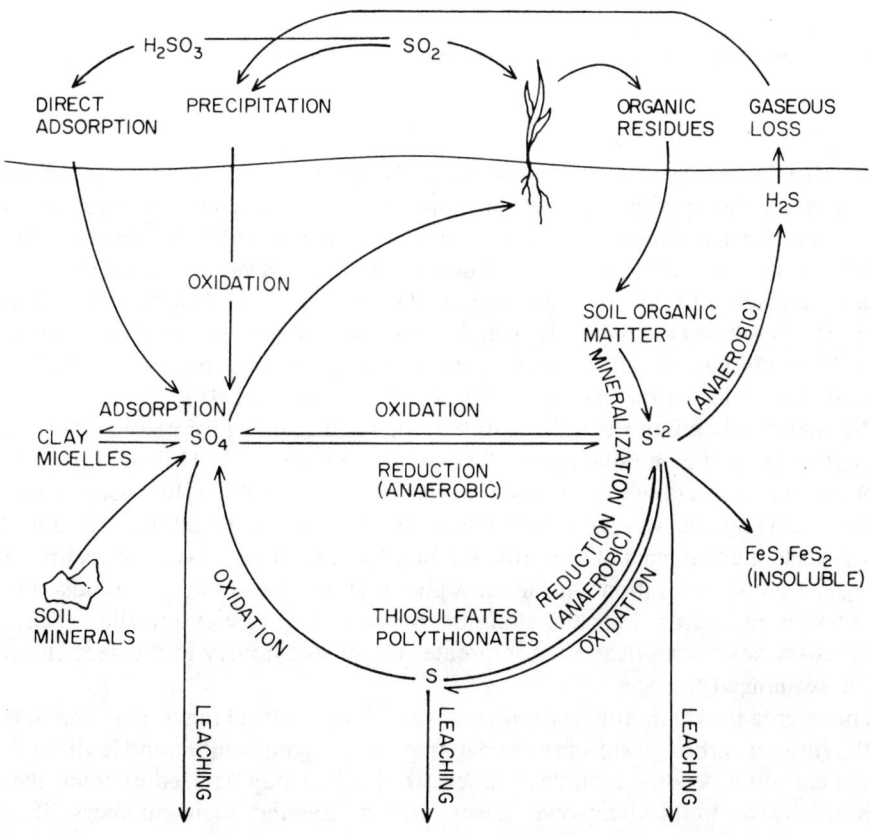

FIGURE 24. *The sulfur cycle. Adapted from Brady (1974:450).*

TABLE 21
Forms of sulfur within the soil

Form of Sulfur	Name of Form
--	proteins and other organic combinations
S	elemental sulfur
H_2S	hydrogen sulfide
FeS, FeS_2	ferrous, ferric sulfide
S^{-2}	sulfide radical
SO_3^{-2}	sulfite radical
SO_4^{-2}	sulfate radical
$S_2O_3^{-2}$	thiosulfate radical
SO_2	sulfur dioxide

As with the initial nitrogenous products of decomposition, those initial sulfur products of decomposition that are incompletely oxidized are subject to oxidation. Some sulfur products, such as sulfites and sulfides, are oxidized by purely chemical reactions. Most sulfur oxidation, however, is biochemical in nature, carried out by various species of the bacteria *Thiobacillus*. The oxidation processes are means by which these bacteria free energy for metabolic uses. The ultimate oxidized form yielded by these reactions is the sulfate radical. Its analog in the nitrogen cycle is the nitrate radical.

Production of sulfate radicals from the other mineralized forms of sulfur is a critical step in the maintenance of sulfur within the soil. Unlike the other mineralized products of decomposition, the sulfate radical can be adsorbed (to some extent) by clay micelles and hydrous oxides and protected from leaching. The precise mechanisms of retention include the attractive powers of positively charged Fe and Al cations in the amorphous outer layer of clay colloids, and protonation of hydroxyl groups along the broken edges of clay micelles. These mechanism have been described in greater detail, previously in the section on anion exchange (p. 140).

The degree to which mineralized sulfur is subject to leaching in part depends on the ratio of carbon to sulfur in the decomposing organic matter and in the soil. When carbon is readily available, mineralized sulfur may be used by microbial populations to build their own tissue and to expand their numbers. The mineralized sulfur becomes fixed in an organic form, again, and is protected from leaching. When the C:S ratio is low, however, the mineralized products of decomposition are not reincorporated, as much, into the pool of organic sulfur within the soil; sulfur then is lost more readily from the ecosystem. This is particularly true when sulfate formation can not keep in step with the formation of the initial species of mineralized sulfur, which are more easily leached than sulfates.

Oxidized forms of sulfur, like oxidized forms of nitrogen, are subject to reduction in anaerobic soil conditions. Five species of the bacteria *Desulfovibro* and three species of the species of the bacteria *Desulfatomaculum* are capable of using the oxygen combined in sulfates, sulfites, and thiosulfites to oxidize organic materials. The sulfur end products of such reactions are sulfide ions, which immediately are precipitated by free iron in the soil to form ferrous sulfide. The precipitate may be oxidized to ferric sulfide if the soil becomes better aerated. Both forms of iron sulfate are highly insoluble and constitute reservoirs of sulfur within soils.

Although the sulfur cycle resembles the nitrogen cycle with respect to the transformations that occur within the soil system, there are some differences between them in the way nutrients flow into and out of a soil system from *other components* in an ecosystem. With respect to inputs, a considerable quantity of S (ca. 1 kg/ha/yr elemental S) enters the soil *directly* from the atmosphere as H_2SO_2 and is oxidized by microbes to SO_4. There is no analog to this process in the nitrogen cycle. Second, sulfur is added to the soil indirectly through the adsorption of atmospheric SO_2 by plants and the incorporation of those plants within the soil after death. The closest analog to this in the nitrogen cycle is N fixation. In this case, bacteria rather than plants function as the fixing agent. Third, in some ecosystems, the parent material serves as a significant source of sulfur input to the soil system. Gypsum ($CaSO_4 : 2H_2O$) is a primary sulfur-bearing mineral that occurs commonly in parent materials. In contrast, N is very scarce within parent materials, and the lithosphere in general (Deevey 1970:151). One form of input of S into the soil system, which is the same as in the nitrogen cycle, is precipitation. Sulfur dioxide in the atmosphere reacts with water and water vapor to form sulfurous acid (H_2SO_3), which is added to the soil in rain. Until the recent industrial revolution and the augmenting of atmospheric sulfur with the gaseous sulfur products of coal combustion, the amount of sulfur added to the soil in precipitation was about 1 kg/ha/yr. Presently, in industrial areas, this figure may approach 110 kg/ha/yr.

With respect to losses of S from the soil system, leaching and gaseous escape constitute the primary depleting process, as in the N cycle. The two cycles are similar in that gaseous loss is facilitated by anaerobic soil conditions. Leaching of sulfur from the soil, however, is much more intensive than leaching of nitrogen. None of the mineralized forms of sulfur that occur in the soil are strongly adsorbed or fixed by clay or humus micelles. Ammonium radicals, on the other hand, are held tightly within the soil system by clay and humus adsorption and fixation.

The Phosphorus Cycle[7]

Unlike the nitrogen and sulfur cycles, the nutrient cycles of P, K, Ca, Mg, and Na involve largely chemical reactions rather than biochemical transformations within the soil system. The status of the soil system with respect to the stability and availability of these nutrients depends mainly on multiphase chemical equilibrium phenomena and, in some cases, on soil pH. The nutrient cycles of P, K, Ca, Mg, and Na also differ from the nitrogen and sulfur cycles in that the atmosphere is not a major reservoir of them.

Phosphorus occurs in five forms within soils: (*a*) organic P, (*b*) inorganic phosphate precipitates of Ca, (*c*) inorganic phosphate precipitates of Fe, Al, and Mn, (*d*) inorganic phosphates fixed by hydrous oxides of Fe and Al and by silicate clays, and (*e*) exchangeable and dissolved inorganic phosphates and organic P-bearing compounds within the soil water solution. Most P in soils is bound in the first three, unavailable forms. Organic forms of P often account for 20–50% of the total soil P. Of this fraction, usually ca. 35% occurs in inositol phosphates, 2% as nucleic acids, 1% as phospholipids, and small amounts of phytin and phytin derivatives. The remaining organic phosphorus-bearing compounds are of complex, unknown natures (Allison 1973:150). Phytin and nucleic acids are water soluble whereas most of the organic P in soils is insoluble.

With respect to inorganic, dissolved phosphates and inorganic phosphate precipitates, soil pH plays a critical role in determining the form in which they occur in the soil. As described in Chapter 3, soluble phosphate radicals occur in several species—PO_4, HPO_4, H_2PO_4, and H_3PO_4—which along with H and OH ions, define a reversible equilibrium system. The relative concentrations of the several phosphate species in aqueous solutions varies with the pH of the solution, as shown in Figure 25 and Table 22. In soil water solutions of typical pHs, HPO_4^{2-} and $H_2PO_4^{-}$ are the main species found. The monovalent species predominates in soil solutions with pHs of less than 6.71 whereas the divalent species predominates in soil water with pHs greater than this (Kardos 1964:370).

The relationship between pH and the form in which inorganic phosphate precipitates occur in the soil is shown in Figure 26. In more acidic soils, particularly those with a pH of less than 6, phosphates released to the soil are rendered insoluble by dissolved Fe, Al, and Mn. The reaction with Al exemplifies this:

$$Al^{+3} + H_2PO_4^{-} + 2H_2O \rightleftharpoons Al(OH)_2H_2PO_4$$
hydroxy phosphate

In soils with pHs below 5, the concentration of dissolved Fe and Al exceeds that of H_2PO_4, as determined by the solubility product constants of reactions

---

[7] This survey of the phosphorus cycles is abstracted from Allison (1973), Brady (1974), Deevey (1970), Kardos (1964), Kormondy (1969), and Wilklander (1964).

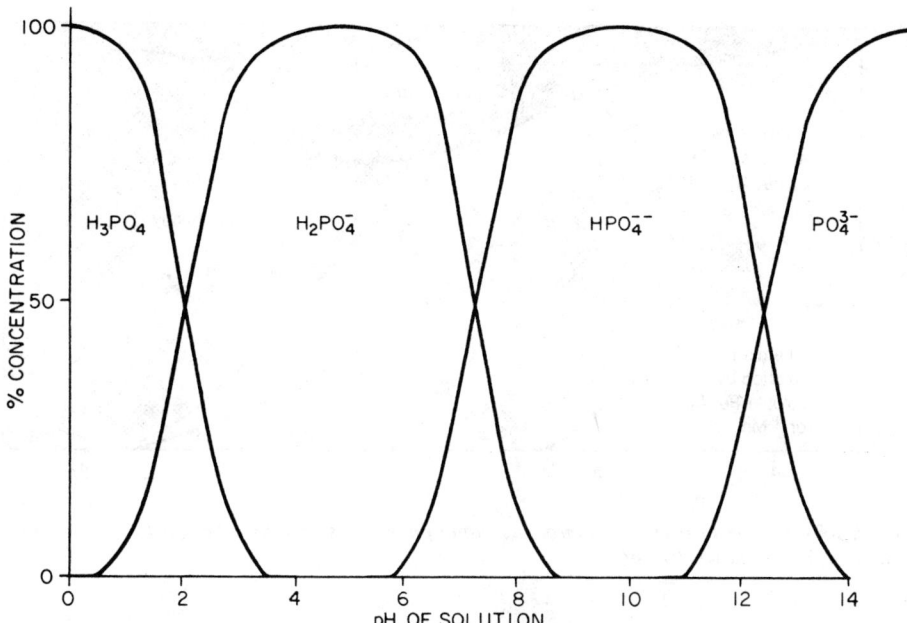

FIGURE 25. *Relative concentrations of various phosphate ionic species within aqueous solutions of various hydrogen ion concentrations. Adapted from Brady (1974:461).*

TABLE 22
*Concentration of phosphate ions in solution at various hydrogen ion concentrations**

pH	(H^+)	(H_3PO_4) (moles/l)	$(H_2PO_4^-)$ (moles/l)	(HPO_4^{2-}) (moles/l)	(PO_4^{3-}) (moles/l)
3	10^{-3}	8.6×10^{-7}	9.5×10^{-6}	1.9×10^{-8}	6.8×10^{-19}
4	10^{-4}	9.3×10^{-8}	1.02×10^{-5}	2.05×10^{-8}	7.4×10^{-17}
5	10^{-5}	9.1×10^{-9}	1.01×10^{-5}	2.02×10^{-7}	7.3×10^{-15}
6	10^{-6}	7.8×10^{-10}	8.2×10^{-6}	1.71×10^{-6}	6.2×10^{-13}
7	10^{-7}	3.1×10^{-11}	3.4×10^{-6}	6.9×10^{-6}	2.47×10^{-11}
8	10^{-8}	4.5×10^{-13}	4.9×10^{-7}	9.8×10^{-6}	3.53×10^{-10}
9	10^{-9}	4.6×10^{-15}	5.1×10^{-8}	1.02×10^{-5}	3.7×10^{-9}
10	10^{-10}	4.7×10^{-17}	5.2×10^{-10}	1.03×10^{-5}	3.7×10^{-7}

*Data from Kardos (1964:370)

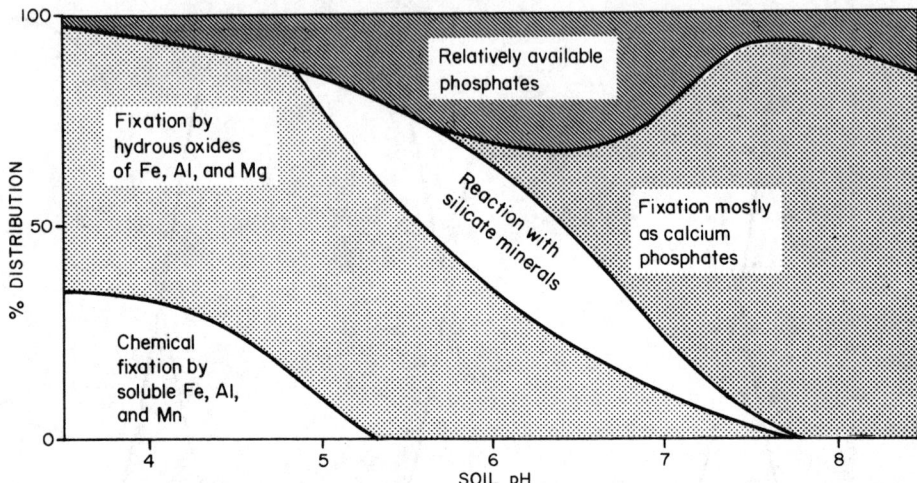

FIGURE 26. *The form of fixation of inorganic phosphates in soils as a function of soil pH. Adapted from Brady (1974:462).*

involving these ions, and almost all H_2PO_4 is rendered insoluble by these metals.

In soils with pHs above 6.5, nearly all inorganic phosphates are fixed by Ca, as shown here:

$$2H_2PO_4 + 3Ca^{2+} \rightleftharpoons Ca_3(PO_4)_2$$

$$2H_2PO_4 + Ca^{2+} + 2CaCO_3 \rightleftharpoons Ca_3(PO_4)_2 + 2CO_2 + 2H_2O$$

The tricalcium phosphate formed in these reactions is very insoluble. It usually reacts with other compounds, however, to form even less-soluble precipitates, including oxy apatite, hydroxy apatite, carbonate apatite, and fluor apatite. The particular balance of these several forms of calcium phosphate depend on the pH and carbonate content of the soil, and on the solubility product equilibrium constants and free energy constants of the reactions between their constituents (Figure 27, Table 23).

Phosphates that have been precipitated by either calcium or iron, aluminum, or manganese are insoluble and not subject to loss by leaching. Nevertheless, they can be uptaken by plants through the direct contact of rootlets with the precipitated particles and remain within the nutrient cycle. After several years, however, precipitated phosphates become unavailable even to plants. The crystal sizes of the precipitated phosphates increase and the surface area of exposed phosphate radicals decreases compared to their volume. Additionally, phosphates held by $CaCO_3$, iron oxide, and aluminum oxide particles penetrate more deeply into them and become less available for plant uptake. These latter processes help to maintain phosphates within the soil over extended periods of

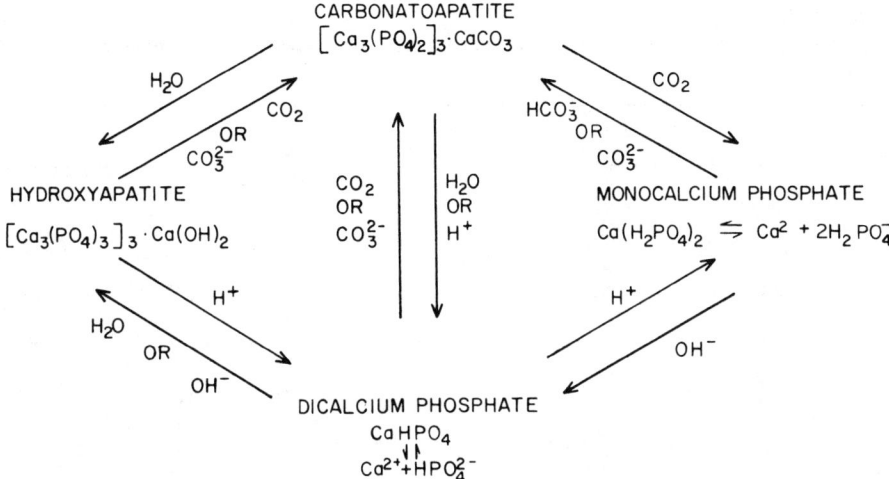

FIGURE 27. *Several forms of calcium phosphate which occur within soils and the reactions which can transform one form to another.*

time, but not in a way that allows them also to have an effect on the conductivity of soil water solutions.

Another form of phosphate that occurs within the soil is that fixed by hydrous oxides of iron and aluminum and by silicate clays. The mechanisms by which fixation can occur have been discussed in detail previously and include (*a*) replacement of hydroxyl groups by the monovalent phosphate species; (*b*) replacement of silicate tetrahedra by phosphate tetrahedra in silicate clays; and (*c*) incorporation of phosphate tetrahedra within recrystallizing silicate clays. In each case, the fixed phosphate radicals remain in a delayed equilibrium with phosphates dissolved in the soil water solution. They are both maintained within the soil over long periods of time and capable of affecting soil conductivity.

Fixation of phosphates by hydrous oxides, in general, is the major means by which P is held within soils in unavailable forms. Hydrous oxides occur in large quantities in most soils. The importance of phosphate fixation by silicate clays is more variable, depending on the textural distribution of the soil. On some archeological sites, hydrous oxides may be especially important in the retention of phosphates. The quantity of hydrous oxides in the soil and its phosphate retention capacity may be augmented by yellow ochre (limonite) and red ochre (oxidized limonite), which are added to the soil by the occupants of the sites.

Exchangeable and dissolved forms of phosphate occur in only small percentages within soils, maximum availability occurring when the soil pH is within the range of 6 and 6.5 (Figure 26). Exchangeable phosphates are adsorbed by clay micelles and hydrous oxides via mechanisms previously discussed for anions in general. These mechanisms apply to soluble organic

TABLE 23
Solubility product equilibrium constants and standard free energy constants of possible soil phosphate reactions*

	Reaction	K_{eq}	ΔF
1.	$H_3PO_4 = H^+ + H_2PO_4^-$	1.1×10^{-2}	2,670
2.	$H_3PO_4 + HCO_3^- = H_2CO_3 + H_2PO_4^-$	3.14×10^4	-6,140
3.	$H_3PO_4 + OH^- = H_2PO_4^- + H_2O$	1.12×10^{12}	-16,450
4.	$2H_3PO_4 + CO_3^{-2} = H_2CO_3 + 2H_2PO_4^-$	6.4×10^{12}	-17,500
5.	$H_2PO_4^- + HOH = H_3PO_4 + OH^-$	9.15×10^{-13}	16,430
6.	$H_2PO_4^- + HCO_3^- = H_3PO_4 + CO_3^{-2}$	4.91×10^{-9}	11,340
7.	$H_2PO_4^- = H^+ + HPO_4^{-2}$	1.95×10^{-7}	8,940
8.	$H_2PO_4^- + H_2CO_3 = H_3PO_4 + HCO_3^-$	3.18×10^{-5}	6,140
9.	$H_2PO_4^- + HCO_3^- = HPO_4^{-2} + H_2CO_3$	5.57×10^{-1}	-347
10.	$H_2PO_4^- + OH^- = HPO_4^{-2} + H_2O$	1.94×10^7	-10,000
11.	$HPO_4^{-2} = H^+ + PO_4^{-3}$	3.6×10^{-13}	16,600
12.	$HPO_4^{-2} + HOH = H_2PO_4^- + OH^-$	5.16×10^{-8}	10,000
13.	$HPO_4^{-2} + HCO_3^- = PO_4^{-3} + H_2CO_3$	1.03×10^{-6}	8,170
14.	$HPO_4^{-2} + HCO_3^- = H_2PO_4^- + CO_3^{-2}$	2.77×10^{-4}	4,860
15.	$HPO_4^{-2} + CO_3^{-2} = PO_4^{-3} + HCO_3^-$	6.67×10^{-3}	2,970
16.	$PO_4^{-3} + HOH = HPO_4^{-2} + OH^-$	2.79×10^{-2}	2,120
17.	$PO_4^{-3} + HCO_3^- = HPO_4^{-2} + CO_3^{-2}$	6.67×10^3	-2,970
18.	$PO_4^{-3} + H_2CO_3 = HPO_4^{-2} + HCO_3^-$	9.73×10^5	-8,170

*Taken from Kardos (1964:380)

phosphates as well as to soluble inorganic phosphates. Soluble organic phosphates also are subject to precipitation by Fe, Al, and Ca under the same pH conditions that inorganic phosphates are.

Both the adsorption of phosphates on clay micelles and hydrous oxides and the fixation of phosphates within the crystal lattices of silicate clays are decreased by the enrichment of soils with humus (Kardos 1964:514). Humus micelles, having a net negative charge, compete with the phosphate ion and other anions for exchange sites on clay micelles and hydrous oxides. They also can be attracted between the expandable layers of 2:1 silicate clays and can block phosphates from gaining access to the interior surfaces of clay micelles, where phosphate fixation can occur. These interactions between humus, clays, and phosphates are of direct relevance to this study in that they determine, in part, the amount of phosphorous of anthropic origins maintained within the soils of archeological sites and within different use-areas of archeological sites. Given two different kinds of use-areas where the same amounts of phosphorus have been added to the soil but the ratios of augmented humus to augmented phosphorous differ (due to the kinds of organic debris deposited), more phosphorous will be maintained in the soil in the use-areas where the humus:phosphorus ratio is less.

The Calcium, Potassium, Magnesium, and Sodium Cycles

Calcium, potassium, and magnesium are similar in the manner in which they cycle through an ecosystem in several respects:

1. Their cycles are not linked to the atmosphere.
2. Mineral weathering is the primary process by which they enter the soil system.
3. All of them occur in only one available form within the soil—ionic.
4. They are not utilized as energy sources by microbes and subject to biochemical transformations as are N and S.
5. All three nutrients are leached from the soil more quickly than are P and N.

Sodium is similar to Ca, Mg, and K in all these respects, but stands alone in that it is not a nutrient requirement of plants (Brady 1974:21–23). It is incorporated within and cycled through the vegetational components of ecosystem, but not as a result of active, enzymatically controlled, selective absorption by plant roots. Sodium is drawn into plants from soil water solutions simply as a result of concentration gradients that may occur between the soil water solution and plant roots.

Inputs of Ca, K, Mg, and Na to the soil system occur in the form of both organic litter from surficial vegetation and ions released from weathered parent materials. In natural ecosystems, parent materials are the ultimate source of these nutrients. Plagioclase feldspars, limestone, and gypsum are commonly occurring sources of calcium. Potassium is obtained from the weathering of orthoclase feldspars and biotite and muscovite micas, and Mg from dolomite and biotite mica. Plagioclase feldspars and simple salts are the most commonly occurring sources of Na to soils.

Once these nutrients have been released to the soil water solution, they are subject to various reactions that help to maintain them within the soil. These include adsorption onto the surfaces of clay and humus micelles (all four nutrients), chelation by humus micelles (Mg, Ca), fixation within the crystal lattices of silicate clays (Mg, K), and precipitation (Ca, Mg). The details of most of these phenomena have been discussed previously. With respect to precipitation, Ca and Mg are rendered insoluble by phosphate radicals when the soil water solution has a pH of 6.5 or greater (Figure 26) and by carbonate and bicarbonate radicals when the soil water solution has a pH of 7 or greater. Below a pH of 6.5, phosphate radicals are fixed largely by Fe and Al ions and are not available in soluble form to precipitate Ca. Below a pH of 7, calcium carbonate and calcium bicarbonate are highly soluble and do not precipitate, given their natural concentrations within the soil.

In terms of total quantities of nutrients found within soils, K is generally more abundant than any other element in the soil. Most (90–95%) K within the soil, however, is held in unavailable forms—feldspars and micas—which are resistant to weathering. The amount of exchangeable K occurring within soils usually is quite small. Calcium, on the other hand, is usually the most plentiful element among those held in exchangeable forms.

All four nutrients—K, Mg, Ca, and Na—are subject to loss from the soil system by leaching. In general, sodium is the most mobile of these four nutrients within the soil system and potassium the most stable. All four are more readily leached from soils than are P and N. As a result, maintenance of K, Mg, Ca, and Na within ecosystems and soils depends to a greater degree on nutrient cycling and the temporary fixing of them within the vegetational components of ecosystem than does maintenance of P and N.

Maintenance of Soil Alterations on Archeological Sites, IV: Interaction Effects among Byproducts of Human Activity

Throughout the previous discussions of nutrient cycles, a number of specific mechanisms have been enumerated that either (*a*) directly maintain nutrients and anthropic nutrient anomalies within soils (e.g., preciptiation of calcium phosphates) or (*b*) indirectly facilitate the maintenance of nutrients and anthropic nutrient anomalies within soils by comprising a portion of a nutrient cycle (e.g., biochemical transformations of N, S). In all cases, the mechanisms have been described from the standpoint of their effect on individual nutrients.

In the natural world, however, nutrients recycled to the soil in the form of organic matter enter the soil in *sets* rather than individually. The same is true of the nutrients within residues deposited on archeological sites. Moreover, on archeological sites, there may be areal overlap in the spatial distribution of activities and in the locations where several different kinds of refuse having *different nutrient spectra* are deposited. In these circumstances, the *combination* of nutrients that are released together to the soil in high concentrations, along with the natural processes that act upon them, may constitute a means by which they are maintained or lost from the soil. The mechanism of nutrient maintenance is, in fact, an "interaction effect" between the nutrients concerned. This perspective can be broadened to include the effects of interaction of soil properties other than the kind of nutrient, as well, such as soil pH and humus content.

For the archeological context, some of the most direct kinds of interactions between nutrients or other soil properties and the effects of those interactions are summarized in Table 24. The first two columns of this table give pairs of anthropic soil changes for which an interaction effect may occur. The second two columns give the actual effect yielded by the interaction between the two kinds of soil changes. The last column explains the natural soil processes responsible for the interaction effect.

Most of the entries in the table are self-explanatory, and the natural processes involved have been documented in detail, above, when considering maintenance mechanisms of individual nutrients. An exception is the effect of the common deposition of limonite or red ochre and residues having high concentrations of P (Table 24, row 5). Limonite and red ochre could fix phosphates in insoluble form and maintain them within the soil by two different mechanisms. First, yellow and red ochre might serve as a source of dissolved iron that would precipitate phosphates. Second, yellow and red ochre might fix phosphates directly, by the replacement of their hydroxyl groups by H_2PO_4. That these reactions do occur within areas where the deposition of limonite or red ochre has been *great enough* to fix significant quantities of phosphate, however, is a moot point. Both reactions require acidic soil conditions. The principle context in which one would expect to find *large* amounts of limonite and red ochre deposited along with phosphorous-rich residues on the other hand, is alkaline: locations where red ochre has been produced from limonite by firing (Winters 1969:26) and where ash would occur. One archeological context where the proposed mechanism for the maintenance of P within soils might be of significance is burials. Solecki (1950) found high phosphate contents with

TABLE 24
Interactions between soil properties affected by human activity and their implications upon the maintenance of anthropic soil anomalies in archeological sites

Soil Property #1 Affected	Soil Property #2 Affected	Effect of Interaction upon Soil Property #1	Effect of Interaction upon Soil Property #2	Reason for Interaction Effect
increase total N	increase pH, base status	maintain N	--	liberation of gaseous forms of N when NO_2 reacts with ammonium salts, simple amenes of urea, or non-nitrogenous sulfur compounds occurs under slightly acidic conditions
increase total N	increase in exchangeable Ca	maintain N	--	high amounts of exchangeable Ca favor N-fixation
increase in NH_4, K fixed between plates of 2:1 clays	increase in exchangeable Na, Mg	loss of fixed NH_4, K	--	Na and Mg can overcome the lattice-contrasting forces of K and NH_4 and displace them by simple cation exchange
increase total SO_4	increase soil pH	loss of SO_4	--	sulfates adsorbed to hydrous oxides are released when soil pH is increased
increase S^{-2} ion, PO_4	increase quantity of hydrous oxides (limonite and red ochre)	maintain S^{-2}, PO_4	--	sulfide and phosphate fixation capacity of soil increased by addition of limonite and red ochre as (cont.)

increase total P	increase total Ca	maintain Ca	maintain PO_4	fixing agents. See Solecki (1950) for archaeological verification. precipitation of calcium phosphates, when soil pH>6.5
increase total P	increase humus	loss of PO_4	--	negatively charged humus micelles compete with PO_4 in anion exchange phenomena; attracted between plates of 2:1 silicate clays and block phosphates' access to interior surfaces of clays where fixation can occur
increase total K	increase Ca	maintain K	--	K fixation by 2:1 silicate clays increased by Ca enrichments. Precise mechanisms unknown.
increase total Ca	increase total CO_3	maintain Ca	maintain CO_3	precipitation of calcium carbonates when soil pH> 7
increase total Ca, Mg	increase soluble organic compounds capable of chelation	loss of Ca, Mg	loss of PO_4 contained in some chelating compounds (Mortenson 1963)	chelation of Ca, Mg

TABLE 24 (cont.)

Soil Property #1 Affected	Soil Property #2 Affected	Effect of Interaction upon Soil Property #1	Effect of Interaction upon Soil Property #2	Reason for Interaction Effect
increase total nutrients in general	increase humus content	maintain nutrients in general	--	increase in number of humus micelles, resulting in: 1. fixation of N,S,P in humus micelles 2. increase in cation, anion adsorption capacity 3. development of soil structure, providing better environment for microbes, fixation of nutrients in microbial tissue, and microbial fixation of atmospheric N.
increase total nutrients in general	greater increase in exchangeable Ca, Mg, than Na	maintains nutrients in general	--	predominance of exchangeable Ca, Mg over Na allows soil colloids to flocculate and encourages the development of soil aggregates and soil structure. The latter provides a better environment for microbes, fixation of nutrients in microbial tissue, and microbial fixation of atmospheric N.

increase total nutrients in general	greater increase in exchangeable Na than Ca, Mg	loss of nutrients in general	--	predominance of exchangeable Na over Ca, Mg causes soil colloids to disperse and discourages the development of soil aggregates and soil structure. Microbial propagation, fixation of nutrients in microbial tissue, and microbial fixation of atmospheric N are thereby discouraged
increase total nutrients in general	increase soil pH	maintain nutrients in general	maintain higher soil pH	increases in soil pH decreases the mobility and loss of humus micelles; provides a better environment for microbes, fixation of nutrients in microbial tissue, and microbial fixation of atmospheric N; maintains $CaCO_3$ in precipitated form
increase in cations strongly adsorbed to soil colloids	increase in cations less strongly adsorbed to soil colloids	maintains strongly adsorbed cations	loss of less strongly adsorbed cations	cation replacement, base exchange

Adena burial pits that contained red ochre and where skeletal material had disintegrated under acidic soil conditions.

Temporal Alterations of Soils within Archeological Sites: A General Model

In the previous sections of this chapter, the viewpoint has been taken that (*a*) organic matter deposited on archeological sites is rather quickly altered and incorporated within the soil as nutrient ions (fixed or soluble) and humus, and that (*b*) these enrichments then are *maintained* over time within the soil via various natural processes. Stress has been placed on the maintenance of the *presence* of anthropic enrichments over time rather than upon maintenance of the absolute *magnitudes* of those enrichments or *ratios of the magnitudes* of nutrient enrichments.

If one considers the magnitudes of humus or nutrient enrichments or ratios of the magnitudes of nutrient anomalies on archeological sites, rather than their presence or absence, however, an appropriate analytical framework would recognize that the natural processes discussed previously *modify* anthropic enrichments over time. Differences that occur in the amounts of nutrient and humus enrichment or the ratio of nutrient enrichments in the soils of different kinds of use-areas immediately after site abandonment and after subsequent incorporation of refuse within soils are not necessarily those that can be expected to occur hundreds of years later on archeological investigation. The following section provides a general framework for viewing within a time perspective the changes that occur in the absolute and relative magnitude of anthropic soil anomalies on archeological sites.

Prior to human occupation of an archeological site, the soil in the locale of the site is a part of a natural ecosystem. Providing that the ecosystem is approximately in a steady-state equilibrium[8] the soil will have attributes that reflect the state of the system as a whole. The ratios and amounts of nutrients and the amounts of humus within the soil will reflect a number of system parameters, including requirements of the surface vegetation and the soil

---

8 Johnson (1971) has argued that parent materials and the atmosphere should not be considered components of local ecosystems as *systems* capable of achieving *steady-state* equilibria with respect to the ratios and amounts of nutrients incorporated within them. Since chemical weathering reactions and atmospheric precipitation of nutrients on a local scale are noncyclic, nonequilibrium processes, parent materials and the atmosphere should be considered exterior to ecosystems—as sources of nutrient inputs or as sinks for nutrient losses. Johnson would define an ecosystem by only the soil, surface vegetational, and surface faunal components, which together are capable of achieving steady-state equilibria with respect to nutrient ratios and amounts. In this study, the more traditional definition of ecosystems (see Ricklefs 1979) will be used, including the atmosphere and parent material, as well as the soil and surface biota within the system. The state of an ecosystem with respect to nutrient ratios and amounts will be considered a product of all these components, despite the fact that the state may not reflect only equilibrating processes. Ecosystems will be considered

microbial populations; nutrient inputs from the parent material, and precipitation; nutrient outputs via leaching and gaseous escape; and climate. When man intrudes on the locale and occupies it for a period of time, the steady state of the local ecosystem is disrupted. The surface vegetation and vegetational requirements may be altered; erosion and leaching may increase; and rates of replenishment of nutrients and humus within the soil may be augmented by the deposition of refuse. A complete description of the disruption would entail many more variables, but the point to be made is that their net effect is to increase the nutrient and humus content of the soil. The various studies reviewed earlier on the chemistry of soils on archeological sites document this fact.

After abandonment of an archeological site, the nutrient and humus content within the local soil is not maintained unaltered over time, but decreases. Studies of the soil of archeological sites document this well (Arrhenius 1929, 1931a; Cook and Heizer 1965). These changes reflect a readjustment process of the total ecosystem toward a more stable, steady-state equilibrium. The equilibrium sought may be the same as that of the ecosystem prior to human intrusion, if occupation has not been too disruptive and the previous ecosystem can reestablish itself, or a new equilibrium defined by a differently structured ecosystem. In either case, *the trajectory of soil changes over time is from an anthropically altered state of disequilibrium to a natural steady-state. Unnatural excesses of nutrients and humus micelles are removed from the system.* Differences in the soil characteristics of different use-areas of an archeological site are slowly eradicated over time.

The rates at which individual soil attributes are altered toward their natural steady-state differ with the attribute concerned. Different soil nutrients follow separate cycles that encompass different pathways of loss, gain, and transformation and which vary in their effectiveness in maintaining those elements within the soil and ecosystem, at large. Consequently, as time progresses, the *ratios of nutrient anomalies that differentiate various use-areas of an archeological site from each other, as well as their magnitudes, change.* Differences in the ratios of nutrients that might be expected in the soils of different use-areas just after site abandonment, based on the relative proportions of nutrients within the kinds of refuse deposited in those different areas, can not necessarily be expected at later dates. The particular soil differences found within different kinds of use-areas will depend (in part) on the length of time that has progressed between site abandonment and the time at which investigation is initiated—the point in time at which the processes of rebound to natural soil conditons is interrupted.

With this overview of the general model of changes that occur within the soils

---

capable of achieving steady-state with respect to nutrients in as much as nutrient inputs from and losses to the atmosphere and parent material are held constant. For analytical purposes, however, attention will sometimes be focused on soil–vegetational subsystems, alone, which can achieve truly cyclic equilibrium states with respect to nutrients.

of sites during occupation and after abandonment, let us now examine in greater detail those aspects of it concerned with the rebound of disturbed soils toward natural equilibrium states.

Replacement of Nutrient Ions and Humus Micelles within Soils through Cycling Processes

One critical aspect of this model of temporal variation in the soils of archeological sites is that the nutrients and humus found in anomalously high concentrations within them are subject to *replacement* over time as part of natural cycling processes. The nutrient ions and humus micelles found in excess within the soils of archeological sites at the time of their investigation must not be considered the *same* ions and humus micelles that were augmented to the soil at the time of occupation centuries ago, minus those lost by leaching or gaseous escape, and decomposition. Rather, the anomalous concentrations should be seen as a pool, part of the larger pool of nutrients and micelles naturally occurring within the soil, continuously drawn on and replenished, with the rate of replenishment less than the rate of loss. This generalization applies to nutrients in nearly all forms—those readily exchangeable and soluble, and those more strongly fixed by clays and humus. It also applies to the semistable humus micelles. It is less true of anthropically enriched phospate ions, however. Phosphate ions found in highly stable, fixed, inorganic forms within the soils of archeological sites at the time of their investigation may be the same phosphate ions that were released to the soils at the time of occupation of the site.

It is precisely because nearly all forms of nutrients and humus micelles of anthropic origin ultimately are subject to cycling and replacement that net losses in their concentration and alterations of nutrient ratios within the soil of archeological sites can occur and the changes in the soils of an archeological site toward a natural steady-state do happen. This point may seem trivial after the preceding, detailed discussions of nutrient cycling, but it must be stressed.

Rates of nutrient replacement compared to the length of time with which archeologists often deal are impressive and underline the point that nutrient anomalies on archeological sites must be considered in a dynamic framework. Kline and Jordan (1972) have calculated turn-around times for calcium within oak–ash, mixed oak, and northern hardwood ecosystems. Cycles ranged from 60 to 200 years. The rates of replacement of humus micelles are much slower (see the preceding section, on the decomposition of humus) ranging from 400 to 3300 years for a variety of grassland and woodland ecosystems. These rates are still fast enough, however, that for most prehistoric archeological sites in North America, the humus micelles presently found in anomalous concentrations within their soils are largely replacements rather than those initially produced from the decomposition of organic residues deposited prehistorically on the sites. In summary, nutrient anomalies and humus anomalies within archeological sites represent replacement sets and it is through replacement that changes in those sets may occur.

Soil–Vegetational Subsystems Approach Equilibria in Nutrient Ratios and Concentrations

A second critical feature of the temporal model just outlined is the nature of the equilibria which anthropically disturbed ecosystems approach over time. Is it justifiable to say that definite *steady-state* equilibria in the concentration and ratios of nutrients within the soil subsystem and the soil–vegetational subsystem of an ecosystem are approached over time after the disturbance of an ecosystem by prehistoric human occupation? If so, what are the primary parameters determining the specific equilibrium value approached? Note that in both questions, the units that are of interest with respect to their states are only the soil and soil–vegetational components of the ecosystem—not the ecosystem as a whole.

The time trajectory of anthropically altered soils toward a natural steady state was recognized by Cook and Heizer (1965:18) with respect to N and C enrichments, but no support of this interpretation was given, other than the decrease of nutrient concentrations over time on archeological sites. Stronger empirical evidence is available, however, that soil systems do approach steady-state equilibria and that these equilibria may be reachieved after disruptions. Parsons (1962; Parsons, Schoter, and Riecken 1962) has shown that soil horizon development, including the depletion of exchangeable bases, clays, and humus from A1 horizons does not proceed indefinitely, but, rather, eventually attains a dynamic equilibrium condition. He examined the A1 horizons that had developed on a number of Indian mounds dating between 100 B.C. and A.D. 1400 and that occurred under oak–hickory forest cover. The parent materials from which the mounds were constructed and upon which the A1 horizons developed were similar in all cases. Regardless of the ages of the A1 horizons, all were found to have similar amounts of exchangeable bases, organic carbon (humus content) and presumably similar base exchange capacities (clay and humus content). They also were found to be similar in these attributes to the A1 horizons of natural Fayette silt loams occurring nearby under a similar vegetational cover and dating to ca. 12,000 B.C. The conclusion was drawn by Parsons that all the A1 horizons had developed quickly (in less time than the youngest mound, 600 years old) and that they had nearly achieved equilibrium states that could last at least 13,000 years under stable environmental conditions.

A study indicating that soil and vegetational components of ecosystems *together* have natural equilibrium states with respect to their nutrient concentrations and that they may reachieve their natural equilibria after disruption is a series of simulations designed by Kline and Jordan (1972). The flow of Ca, K, Mg, and Na within a Puerto Rican tropical rainforest–soil system was simulated using empirical data on the nutrient inputs and outputs of the various components of these ecosystems. The simulated systems were found to have steady-state qualities with respect to the cycling and replacement of nutrients within the soil systems and within the soil–vegetational systems as wholes. When the simulated systems were disrupted by minor nutrient additions or losses, all were found to rebound quickly to their natural, steady-state nutrient equilibria.

The studies of Parsons, Kline, and Jordan are concerned with the existence of steady-state equilibria in the *concentration* of nutrients with natural soils. Other data indicate that the *ratios* of nutrients within soils also are maintained at specific equilibrium states. In most soils, the ratios of C:N:P:S within organic fractions are constant, approximately 110:9:1:(1.125 to .75) (Allison 1973:159). These ratios are held fixed largely by processes of decomposition and humus formation that incorporate nutrients within humus micelles in rather specific ratios. For example, when organic matter with a very high N content compared to carbon (e.g., greens) is deposited within the soil, as decomposition proceeds, considerable amounts of N are released in leachable form (NH_4) rather than fixed in more stable humic compounds. If organic matter deficient in N compared to C (e.g., wheat straw) is deposited within the soil, all the N may be incorporated within humic substances, as well as N from outside sources (e.g., precipitation). Consequently, the relative concentrations of C and N within soils are held constant over time. Some other mechanisms maintaining the ratios of nutrients within natural soils at equilibrium levels include the relative replacing power of different kinds of nutrient ions on clay and humus micelles, the relative strengths with which different nutrients are fixed by clays or humus, the ratios in which nutrients are incorporated within the vegetational and soil microbial components of the ecosystem, and the relative water solubility (and hence, leachability) of nutrients comprising the surface vegetation of an ecosystem.

In summary, then, a number of specific empirical data, as well as what is known about the natural processes of nutrient cycling in general, suggest that soil systems and soil–vegetational systems tend to approach definite equilibrium states with respect to the magnitudes and ratios of nutrients they incorporate. These equilibria can be reestablished after minor systemic perturbations. The decrease in mangitude that nutrient and humus anomalies within the soils of archeological sites are known to be subject to over time thus may be interpreted to reflect a rebound process in which natural nutrient and humus concentrations and ratios are being approached.

Primary Parameters Determining the Nutrient Ratio Concentration Equilibria Approached by Soil and Soil–Vegetational Systems

Given that it is appropriate to interpret temporal changes in the nutrient status of the soils of archeological sites within a systems-equilibrium framework, what are the primary parameters that determine the specific equilibrium values that are approached? Are the determinants of *concentration* equilibria the same as the determinants of *ratio* equilibria? With respect to the maintenance of specific concentrations of nutrients and humus within soils, all of the mechanisms described in this chapter for stabilizing nutrients and humus within soils help to prevent their losses through leaching, gaseous escape, or decomposition, and help to maintain their concentrations at equilibrium levels. The equilibrium concentrations of nutrients and humus may be considered the cumulative effect of parameters governing rates of nutrient inputs and outputs (e.g., precipitation rates affecting both mineral weathering and leaching; climate conditions affecting microbial propagation), filtered by the various retentive mechanisms mentioned in this chapter (e.g., the clay content and cation exchange capacity of a soil).

With respect to the maintenance of specific *ratios* of nutrients within soils and soil–plant systems, a number of mechanisms have been mentioned, including the relative replacement powers and strengths of fixation of nutrients within clays, the ratios in which nutrients tend to be integrated within the structure of humus micelles, the ratios in which nutrients are incorporated within the surficial vegetational and soil microbial components of the ecosystem, and the relative water solubility of nutrients comprising the surface vegetation. One parameter in particular, however, has a primary role in determining nutrient ratio equilibria in soils and soil–plant systems: the nutrient requirements and composition of the surficial vegetation.

The primary role of the vegetational component in determining nutrient ratio equilibria within soils and soil–plant systems is not surprising in some instances, such as the tropical rainforest system (Kline and Jordan 1972). In this case, the majority of the biologically cycled nutrients within the soil–plant system are held within the surface vegetation. Soils are heavily leached of nutrients and those present tend to be the immediate products of decomposition of litter and deadwood from the vegetational component. Consequently, the ratios of nutrients held within the soil system reflect the composition and nutrient requirements of the surface vegetation.

In many other ecosystems, however, the primary role of the surface vegetation in determining nutrient ratios within the soil and soil–plant system can not be deduced so simply; the surface vegetational component does not hold the majority of biologically cycled nutrients within the soil–plant system. For example, Rolfe *et al.* (1978:127) have summarized the distribution of nutrients within different components of oak–hickory forests covering loessic parent materials in southern Illinois (Table 25). Eighty percent of the total nutrient pools for such soil–vegetational systems is held within the soil and only 19% is held within the forest stands. Ninety percent of the total N and Mg within the soil–vegetational system occurs within the soil. These circumstances suggest that soil processes might be primary in determining soil nutrient ratios. In particular, the nutrient requirements and composition of soil organisms could have a major determinant role. In oak–hickory forests similar to the one previously mentioned, soil organisms may have a total biomass of 79,900 kg/ha (Brady 1974:115) and account for 33% of the total biomass of such systems (237,000–243,000 kg/ha; Rolfe *et al.* 1978). The percentage of nutrients incorporated within soil organisms rivals that incorporated in the surface vegetational component and would suggest fixation of nutrients within microbial tissue as a possible determinant of soil nutrient ratio equilibria.

Despite the relatively small proportion of biologically cycled nutrients held within the surface vegetational components of some soil–plant systems, it would appear that the nutrient requirements and composition of this component have a major role in determining the ratios of nutrients found within soil–plant systems and within soil systems. This can be inferred by comparing the relative proportions of nutrients found within soil–plant systems and soil systems to (*a*) their relative proportions in the various components of the soil–plant systems that are suspected of playing a determinant role, and (*b*) the relative strengths with which they are held in the soil by various processes suspected of playing a determinant role. Such a comparison was made for the oak–hickory forest–soil systems just examined and is summarized in Table 26. Data on the nutrient requirements and compositions of soil organisms were not available for comparison to the nutrient compositions of the soil and soil–forest systems, but all other factors suspected of having a determinant role were compared. It is apparent that for these systems, at least, the relative proportions of various nutrients found within an entire oak–hickory forest–soil system and within its soil subsystem, alone, are most similar to the proportions of nutrients found within the *surficial vegetational* component of the system. The nutrient requirements and composition of the surficial vegetational component can be identified as a major phenomenon determining the equilibria ratios of nutrients found within such soil–plant systems and soil systems, if not the primary one.

It is possible that many soil–plant systems are similar to the oak–hickory soil–forest systems just examined with respect to the factors determining their nutrient ratio equilibria, but certainly not all are. Where input of nutrients through mineral weathering and precipitation greatly exceed the annual nutrient requirements of the surficial vegetation, the ratios of nutrients within the soil–plant system and the soil will not be controlled mainly by the nutrient requirements of the vegetation; the processes governing the amounts of excess in which nutrients are supplied to be system also will be determinant factors (Kline and Jordan 1972).

TABLE 25
*Distribution of nutrients within different component of oak-hickory forests covering loessic parent materials in southern Illinois**

Component	N		P		K		Ca		Mg		Total	
	kg/ha	%	kg/ha	%	kg/ha	%	kg/ha	%	kg/ha	%	kg/ha	%
Soil	6800	90.7	160	76.6	880	67.5	4440	67.4	1250	90.0	13,530	79.6
Aerial biomass	478	6.4	29	13.9	310	23.7	1603	24.3	105	7.6	2,525	14.9
Litter	69	0.9	8	3.8	7	0.5	118	1.8	7	0.5	209	1.2
Roots	153	2.0	12	5.7	109	8.3	426	6.5	26	1.9	726	4.3
Total	7500	100.0	209	100.0	1306	100.0	6587	100.0	1388	100.0	16,990	100.0

*Taken from Rolfe et al. (1978:127)

TABLE 26
*Testing various phenomena for their possible role in determining ratios of nutrients found in an oak-hickory forest and loessic soil system in southern Illinois**

Nutrient	N	P	K	Ca	Mg	Total
Percentage of nutrient in the entire oak-hickory vegetational soil system*	44.1%	1.23%	7.69%	38.8%	8.17%	99.99%
		N ≃ Ca > Mg ≃ K > P				
Percentage of nutrient in the soil system	50.25	1.18	6.50	32.82	9.24	99.99
		N ≃ Ca > Mg ≃ K > P				
Percentage of nutrient in the aerial portions and roots of the vegetational component of the system*	19.4	1.26	12.9	62.4	4.03	99.99
		Ca > N > K > Mg ≃ P				
Percentage of nutrient in precipitation on the ecosystem*	19.0	2.0	42.0	31.0	5.0	99.
		K ≃ Ca > N > Mg ≃ P				
Relative insolubility and immobility of nutrient from plant tissue**		K > Mg > Na > Ca > P > N				
Percentage of nutrient in humic acids***	91.67	.0833	0	0	0	100
		N >> P > Ca ≃ Mg ≃ K				
Immobility of nutrient within soil system (approximate)		P > K > NH$_4$ > Mg > Ca > Na > NO$_3$ > Cl				
primary factors governing mobility		phosphate fixation	clay fixation	clay and humus exchange phenomena		

*Rolfe et al. (1978)
**Attiwell (1968:143), Bear (1951:18), Gray and Williams (1974), Reikerk (1971), Remezov (1961)
***Allison (1973:150)

In summary, it would appear that for at least some soil–plant systems and soil systems, nutrient ratio equilibria are determined to a major degree by the nutrient requirements and composition of their surface vegetational component. An archeological site located in such natural contexts can be expected to have the nutrient ratios that originally characterize and distinguish its different kinds of use-areas altered over time and made more similar, approaching the single, natural set of nutrient equilibrium ratios determined by and characteristic of the specific kind of vegetational community that covers it. The equilibrium ratios that are approached will vary from site to site with the kind of surface vegetation. The specific nutrient ratios distinguishing the soils in different kinds of use-areas of a site at the time of its archeological investigation will depend on:

(a) the chemical nature of the refuse incorporated within the soil; (b) the age of the site and length of time its soils have been approaching the natural nutrient ratio equilibria; and (c) the nutrient requirements of the surface vegetation.

Time Required for Distributed Soils to Reach Natural Equilibria in Nutrient Ratios and Concentrations

A third critical feature of the temporal model of soil alternations on archeological sites is the *time* over which the magnitude of nutrient enrichments and differences in the ratios of nutrient anomalies within the soils of different use-areas are diminished. The major points to be made are: (a) different nutrients within the soils of archeological sites will be lost at different rates and will achieve natural concentrations over different lengths of time; and (b) consequently, the nutrient ratios specifically found to distinguish the soils of different kinds of use-areas will depend greatly on the age of the site and the length of time over which its soils have been approaching natural nutrient concentrations.

The length of time required for nutrients within the soils of archeological sites to reach natural concentrations largely depends on (a) their mobility within the ecosystem (average ecocycle time), and (b) whether or not their rates of supply to and loss from the system are great. Nutrients having long ecocycles and low input or output rates will be maintained in anomalous concentrations on archeological sites longer than will nutrients having shorter ecocycles and high rates of input and output. This generalization is supported by studies of Kline and Jordan (1972), who simulated several natural ecocsystems and determined rebound times for them when minor perturbations (nutrient losses or gains) were introduced to them.

The mobility of a nutrient within an ecosystem, in turn, depends on a number of phenomena, including those that help to retain the nutrient within a particular component of the system, and those that facilitate cycling. The various mechanisms discussed previously as means by which nutrients are maintained within soils (e.g., incorporation within humus, fixation by clays, chemical precipitation) are examples of the former. Those phenomena that faciliate nutrient cycling include: (a) chelation as a means by which nutrients that would normally be precipitated within the soil are maintained in solution for plant uptake; (b) the water solubility and leachability of a portion of the nutrients within the surface vegetational component of an ecosystem; and (c) rates of flow of water through the system.

The second factor affecting the length of time required for nutrients within the soils of archeological sites to reach natural concentrations—rates of nutrient input and output for the ecosystem—depends on a number of obvious parameters. Among these are rates of weathering of parent materials, the amount of the nutrient supplied to the system by precipitation, and the amount of the nutrient lost through leaching and gaseous escape.

In view of the large number of processes that govern the rates of cycling and flow of nutrients through ecosystems, and the variation of these processes under different kinds of environmental settings, it is not possible to establish a single universal ordering of nutrients reflecting the times required by them to regain natural concentrations within soils after having been augmented or depleted. Likewise, and as a result of this, no single model of temporal alterations in the spectra of nutrients characterizing the soils of different use-areas within archeological sites can be offered. It can be warned, however, that the different ratios of nutrients found successful in characterizing and distinguishing different use-areas on one archeological site of one age within one environmental setting must not be used to interpret soil anomaly ratios found in other archeological sites of different ages within other environmental settings. Studies of use-areas within archeological sites using soil chemistry data must be made in reference to *local* soil conditions and processes that govern the temporal alterations of nutrient concentrations and ratios, as well as in reference to the specific archeological context.

Summary

A number of studies of the chemistry of soils within archeological sites suggest that the refuse deposited on archeological sites enrich the concentration of nutrients and humus and alter the ratios of nutrients within their soils from natural concentrations and ratio equilibria. In most cases, these alterations probably also affect soil structure and soil moisture equilibria. The magnitude of nutrient anomalies has been related to the age and intensity of occupation of sites, and ratios of nutrient enrichments have been used to distinguish broadly differing kinds of use-areas within sites.

Soil anomalies within archeological sites are maintained over extended periods of time by a number of natural processes, including the slow rates of decomposition of humus micelles, and a large variety of phenomena encompassed within the cycles of nutrients. The magnitudes of these anomalies decrease over time toward nutrient concentration equilibria that are natural for the local soil and ecosystem. Likewise, ratios of nutrient enrichments distinguishing different kinds of use-areas are altered over time toward natural nutrient ratio equilibria. Rates of loss of nutrients from the soils of archeological sites are not constant for all nutrients nor are the rates of alteration of ratios of nutrient enrichments constant for all ratios. Consequently, the particular spectra of nutrients distinguishing different kinds of use-areas depend on the age of the site and the local environmental context.

References

Analysis of Soils on Archeological Sites

Arrhenius, Olaf
- 1929 Die Bodenanalyses im Dienst der Archaologie. *Zeits. für Pflanzenernährung, Düngung und Bodenkunde*, Teil A 10:185.
- 1931a In *Zeits. für Pflanzenernährung, Düngung und Bodenkunde*, Teil B 10:427–439.
- 1931b Markanalysen i Arkeologiens Tjanst. *Geol. fören Forbandlingar*, Stockholm.
- 1934 Fosfathalten i Skanska Jordar. *Sveriges Geol Undersokning*, Series C., No. 383.
- 1935 Markundersökning och Arkeologi, *Fornvännen* 30:65–76.
- 1938 Den Gotländska Akerjordens Fosfathalt. *Sveriges Geol. Undersokning*, Series C, No. 412.
- 1950 Forhistoriek Bebyggelse Antydd Genom Kemisk Analys. *Fornvänner* 45:59–62.
- 1955a Akermarkens Urgarmla Havd. *Fornvännen* 50:80–87.
- 1955b The Iron Age Settmenents in Gotland and the Nature of the Soil. In *Vallhagar* II, edited by M. Stenberger, pp. 1053–1064. Copenhagen: Munksgaard.
- 1963 Investigation of soil from old Indian sites. *Ethos* 2–4:122–136.

Baker, Charles M.
- 1975 Experimentation with soil phosphate analysis. In Arkansas Eastman Archaeological Project, C. M. Baker (ed.), pp. 67–82. *Arkanses Archaeological Survey, Research Report 6*.

Cook, S. F., and R. F. Heizer
- 1965 Chemical analysis of archaeological sites. *University of California, Publications in Anthropology* 2.

Cornwall, I. W.
- 1958 *Soils for the archaeologist*. Phoenix House, London.

Dauncey, K. D.
- 1952 Phosphate content of soils on archaeological sites. *Advancement of Science* 9:33–36.

Deetz, J., and E. Dethlefsen
- 1963 Soil pH as a tool in archaeological site interpretation. *American Antiquity* 29:242–243.

Dietz, Eugene F.
- 1957 Phosphorous accumulations in soil of an Indian site. *American Antiquity* 22(4):404–409.

Eddy, F. W., and H. E. Dregne
- 1964 Soil tests on alluvial archaeological deposits. Navajo Reservoir district. *El Palacio* 71(4):5–21.

Eidt, Robert C.
- 1973 A rapid chemical field test for archaeological site surveying. *American Antiquity* 38:206–210.

Hurley, W. M., and C. E. Heidenreich
- 1971 Paleoecology and Ontario prehistory. *University of Toronto Department of Anthropology, Research Report* 2.

Li, L. C.
- 1943 Rate of soil development as indicated by profile studies of Indian mounds. Unpublished Ph.D. thesis, University of Illinois. Champagne.

Lorsch, W.
- 1940 In *Naturwissenschaften* 40/41(28):633–640.

Lutz, H.J.
    1951   The concentration of certain chemical elements in the soils of Alaskan archaeological sites. *American Journal of Science* 249:925–928.

Mattingly, G. E. G., and R. J. B. Williams
    1962   A note on the chemical analysis of a soil buried since Roman times. *Journal of Soil Science* 13:254–258.

Parsons, R. B.
    1962   Indian mounds of northeast Iowa as soil genesis benchmarks. *Journal of the Iowa Archaeological Society* 12:1–70.

Parsons, R. B., W. H. Scholtes, and F. F. Riecken
    1962   Soils of Indian mounds in northeastern Iowa as benchmarks for studies of soil genesis, *Soil Science Society of America, Proceedings* 26:491–496.

Provan, Donald M.J.
    1971   Soil phosphate analysis as a tool in archaeology. *Norwegian Archaeological Review* 4:37–50.

Sawbridge, D. F., and M. A. Bell
    1972   Vegetation and soils of shell middens on the coast of British Columbia, *Ecology* 53:840–849.

Solecki, Ralph S.
    1950   Notes on soil analysis and archaeology. *American Antiquity* 16(2):254–256.

Weide, D. L.
    1966   Soil pH as a guide to archaeological investigation. *University of California (Los Angeles) Archaeological Survey*. Annual Report 8:155–163.

Wildesen, Leslie E.
    1974   A functional, factorial approach to archaeological site development. Unpublished manuscript on file, University of California archaeological research unit.

Zeiner, H. M.
    1946   Botanical survey of the Angel Mound site, Evansville, Indiana. *American Journal of Botany* 33:83–90 Brooklyn.

Decomposition and Humus Formation

Allison, F. E.
    1973   *Soil organic matter and its role in crop production*. Elsevier, New York.

Allison, F. E., and C. J. Klein
    1962   Rates of immobilization and release of nitrogen following additions of carbonaceous material and nitrogen to soils. *Soil Science* 93:383–386.

Allison, F. E., and R. M. Murphy
    1962   Comparative rates of decomposition in soil and wood and bark particles of several hardwood species. *Soil Science Society of America, Proceedings* 26:463–466.

Allison, F. E., R. M. Murphy, and C. I. Klein
    1963   Nitrogen requirements for the decomposition of various kinds of finely ground woods in soil. *Soil Science* 96:187–190.

Attiwell, P. M.
    1968   The loss of elements from decomposing litter. *Ecology* 49:142–145.

Bear, Firman
    1951   Soils and fertilizers. John Wiley, New York.

Bell, Mary K.
    1974   Decompostion of herbaceous litter. In *Biology of plant litter decomposition*, C. H. Dickinson and G. J. F. Pugh (eds.), pp 37–68. Academic Press, New York.

Bocock, K. L.
    1963   Changes in the amount of nitrogen in decomposing leaf litter of sessile oak (Quercus petraea). *Journal of Ecology* 51:555–566.

1964 Changes in the amounts of dry matter, nitrogen, carbon, and energy in decomposing woodland leaf litter in relation to the activities of soil fauna. *Journal of Ecology* 52:273–284.

Bocock, K. L., and O. J. W. Gilbert
1957 The disappearance of leaf litter under different woodland conditions. *Plant and Soil* 9(2):179–185.

Bocock, K. L., O. J. W. Gilbert, C. K. Capstick, D. C. Twinn, J. S. Ward, and M. J. Woodman
1960 Changes in leaf litter when placed on the surface of soils with contrasting humus types. *Journal of Soil Science* 11:1–9.

Brady, Nyle C.
1974 *The nature and properties of soils.* Macmillan, New York.

Broadfoot, W. M., and W. H. Pierce
1939 Forest soil studies: I. Relation of rate of decomposition of tree leaves to their acid-base balance and other chemical properties. *Soil Science* 48:329–348.

Broecker, W. S., J. L. Kulp, and C. S. Tucek
1956 Lamont natural radiocarbon measurements, III. *Science* 124:154–165.

Burges, A.
1956 The release of cations during the decomposition of forest litter. *6th International Congress of Soil Science, Paris 1956,* B:741–745.
1967 The decomposition of organic matter in the soil. In *Soil biology* A. Burges and F. Raw (eds.), pp 479–492. Academic Press, New York.

Dickinson, C. H.
1974 Decomposition of litter in the soil. In *Biology of plant litter decomposition,* C. H. Dickinson and G. J. F. Pugh (eds.), 633–658. Academic Press, New York.

Flaig, W., H. Beutelspacher, and E. Rietz
1975 Chemical composition and physical properties of humic substances. In *Soil components,* J. E. Gieseking (ed.). Springer-Verlag, New York.

Gieseking, John E. (ed.)
1975 *Soil components vol 1: Organic components.* Springer-Verlag, New York.

Gray, K. R., and A. J. Biddlestone
1974 Decomposition of urban waste. In *Biology of plant litter decomposition,* C. J. Dickinson and G. J. F. Pugh (eds.), pp. 611–631. Academic Press, New York.

Heath, G. W., M. K. Arnold, and C. A. Edwards
1966 Studies in leaf litter breakdown, I. Breakdown rates of leaves of different species. *Pedobiologia* 6:1–12.

Holm, E., and V. Jensen
1972 Aerobic chemoorganotrophic bacteria of a Danish beech forest. *Oikos* 23:248–260.

Jensen, V.
1974 Decomposition of angiosperm tree leaf litter. In *Biology of plant litter decomposition,* C. H. Dickinson and G. J. F. Pugh (eds.), pp 69–104. Academic Press, New York.

Käärik, Aino A.
1974 Decomposition of wood. In *Biology of plant litter decomposition,* C. H. Dickinson and G. J. F. Pugh (eds.), pp 129–174.

King, H. G. C., and G. W. Heath
1967 The chemical analysis of small samples of leaf material and the relationship between the disappearance and composition of leaves. *Pedobiologia* 7:192–197.

Kononova, M. M.
1961 *Soil organic matter.* Pergamon Press, New York.

Lockett, J. L.
1937 Microbiological aspects of decomposition of clover and rye plants at different growth stages. *Soil Science,* 44:425–439.

Mangenot, F.
1966 In *Bulletin of Ecol. Nat. Sup. Agronomy, Nancy* 8:113–125.

Meyer, F. H.
   1960   In *Arch. Mikrobiol.* 35:340–360.
Minderman, G., and L. Daniels
   1967   In *Progress in soil biology*, O. Graff and J. E. Satchell (eds.), pp 3–9. North-Holland, Amsterdam.
Mortensen, J. L.
   1963   Complexing of metals by soil organic matter. *Soil Science Society of America, Proceedings* 27:179–186.
Remezov, N. P.
   1961   Decomposition of forest litter and the cycle of elements in an oak forest. *Pocvoved*, 1961 (7):1–12. Trans. *Soviet Soil Science* 1961(7):703–711.
Remacle, J.
   1970   Le microflore des litères. *Bull. Soc. Roy. Bot. Belg.* 103:83–96.
   1971   Succession in the oak litter microflora in forests at Mesnil-Eglise (Ferage), Belgium. *Oikos* 22:411–413.
Romell, L. G.
   1932   Mull and duff as biotic equilibria. *Soil Science* 34:161–188.
Smith, G. D. et al.
   1960   In *Soil classification: A comprehensive system, 7th approximation*. United States Department of Agriculture, Washington, D.C.
Stevenson, I. L.
   1962   The effect of decomposition of various crop plants on the metabolic activity of the soil microflora. *Canadian Journal of Microbiology* 8:501–509.
Toth, Stephen, J.
   1964   Colloid chemistry of soils. In *Chemistry of the soil*, E. Bear (ed.), pp. 85–106. Reinhold, New York.
Waksman, Selman T.
   1938   *Humus*. Williams and Wilkins, Baltimore.
Witkamp, M.
   1960   In *Meded. Inst. Toegepast, Bio. Onderz. Nat.* 46:1–51.
Williams, S. T., and T. R. G. Gray
   1974   Decomposition of litter on the soil surface. In *Biology of plant litter decomposition*, C. H. Dickinson and G. J. F. Pugh (eds.). Academic Press, New York.

Nutrient Cycling and Fixation

Allison, F. E.
   1973   *Soil organic matter and its role in crop production:* Elsevier, New York.
Attiwell, P. M.
   1968   The loss of elements from decomposition litter. *Ecology* 49:142–145.
Bear, Firman
   1951   *Soils and fertilizers*. John Wiley, New York.
   1964   *Chemistry of the soil*. Reinhold, New York.
Bennett, W. F., G. Stanford, and L. Dumeil
   1953   Nitrogen, phosphorous, and potassium content of corn leaf and grain related to nitrogen fetilization and yield. *Soil Science Society of America, Proceedings* 17:252–258.
Bormann, Herbert F. and Gene E. Likens
   1970   The Nitrogen cycles of an ecosystem. *Scientific American* 223:92–101.
Brady, Nyle C.
   1974   *The nature and properties of soils*. Macmillan, New York.

Deevey, Edward S.
   1970 Mineral cycles. *Scientific American* 223:148–158.
Delwiche, C. C.
   1970 The nitrogen cycle. *Scientific American* 223:137–146.
Dumenil, Lloyd
   1961 Nitrogen and phosphorous composition of corn leaves and corn yield in relation to critical levels and nutrient balance, *Soil Science Society of America, Proceedings* 25(4):295–298.
Eyer, S. R.
   1968 *Vegetation and soils.* Edward Arnold, London.
Fletcher, H. F., and L. T. Kurtz
   1964 Differential effects of phosphorous fertility on soybean varieties. Soil Science Society of America, Proceedings 28(2):225–228.
Johnson, Noye M.
   1971 Mineral equilibria in ecosystem geochemistry. *Ecology* 52(3):529–531.
Kardos, Louis T.
   1964 Soil fixation of plant nutrients. In *Chemistry of the soil*, F. E. Bear (ed.), pp. 177–199. Reinhold, New York.
Kononova, M. M.
   1961 *Soil organic matter.* Pergamon Press, New York.
Kormondy, H. J.
   1969 *Concepts of ecology.* Prentice-Hall, Englewood Cliffs, NJ.
Kline, Jersy R., and Carl F. Jordan
   1972 Mineral cycling: Some basic concepts and their application in a tropical rain forest. *Annual review of ecology and systematics*, pp. 33–50.
Liken, Gene, E., F. Herbert Bormann, Robert S. Pierce, John S. Eaton, and Noye M. Johnson
   1977 *Biogeochemistry of a forest ecosystem.* Springer-Verlag. New York.
Remezov, N. P.
   1961 Decomposition of forest litter and the cycle of elements in an oak forest. *Pocvoved*, 1961(7):1–12. Trans. *Soviet Soil Science* 1961(7):703–711.
Ricklefs, Robert E.
   1979 *Ecology.* Chiron Press, New York.
Riekerk, Haus
   1971 The mobility of phosphorous, potassium, and calcium in a forest soil. *Soil Science Society of America, Proceedings* 35(2):350–356.
Rolfe, G. L., M. A. Artar, and L. E. Arnold
   1978 Nutrient distribution and flux in a mature Oak–Hickory forest. *Forest Science* 24(1):122–130.
Stevenson, F. J.
   1972 Role and function of humus in soil with emphasis on adsorption of herbicides and chelation of micronutrients, *Bioscience* 22(11):643–650.
Toth, Stephen J.
   1964 Colloid chemistry of soils. In *Chemistry of the soil*, F. E. Bear (ed.), pp 85–106. Reinhold, New York.
Tyner, E. H.
   1946 The relation of corn yields to leaf nitrogen, phosphorus, and potassium content. *Soil Science Society of America, Proceedings* 11:317–232.
Wallace, A.
   1963 Role of chelating agents on the availability of nutrients to plants. *Soil Science Society of America, Proceedings*, 27:176–179.

Wildesen, Leslie E.
   1974 A functional, factorial approach to archaeological site development. Unpublished manuscript on file at the University of California Archaeological Research Unit, Riverside.
Wilklander, Lambert
   1964 Cation and anion exchange phenomena. In *Chemistry of the Soil*, F. E. Bear (ed.), pp. 107–148. Reinhold, New York.
Winters, Howard
   1969 The Riverton culture. *Illinois State Museum, Reports of Investigation* 13.

5
A Functional and Distributional Study of Surface Artifacts from the Crane Site

To interpret resistivity data from earthen sites in terms of the different kinds of use-areas they may reflect, it is necessary to understand two groups of relationships: (*a*) those describing how different kinds of human activities can cause different kinds of physical and chemical alterations to natural soils, and (*b*) those describing how these various changes in soils affect their resistivity. It is possible to describe the relationships between soil properties and soil resistivity by drawing on a number of a priori models and established empirical models from physics, chemistry, and soil science. This has been done in Chapter 3. Models specifying the relationships between human activities and the soil alterations they produce, however, have yet to be formulated in detail. Some general concepts from soil science may be utilized (e.g., the effect of organic matter enrichments on soil structure), and these have been introduced in Chapter 3; but the *specific alterations* produced by specific activities remain to be modeled. Chapters 5 through 8 of this study are presented for this purpose. Archeological and pedological data from an earthen site will be examined in order to formulate *empirical* models relating specific human activities to specific soil changes. Also, the more general models provided by soil science will be illustrated with the data.

The data for formulating and illustrating these models come from the Crane site, a Middle Woodland habitation in the Illinois River drainage (Macoupin Valley). I surveyed the site at a number of locations for the chemical and physical properties of its soils, and the resistivity of its soils. It was concurrently surface collected and extensively excavated (during two field seasons) by Kenneth Farnsworth as part of a regional Middle Woodland subsistence–settlement study (cf. Farnsworth 1973). Farnsworth's work included sample excavations in areas of the site where soil and resistivity surveys were made. He made his controlled surface collection data, together with preliminary observations about the distribution of subsurface structural remains at the site, available for comparison to and evaluation of the results of the soil and resistivity surveys.

In this chapter, the particular kinds of uses that different areas of the Crane site served, will be reconstructed from artifacts found at the site surface, so that later (Chapter 8) a comparison can be made between use-areas and the soil alterations within them. Surface artifacts will be described and analyzed with respect to their functions. The functions of the artifacts found at given locations then will be used to hypothesize the activities that occurred at them and the kinds of chemical residues that probably were deposited at them. This study has been made independent from that by Farnsworth. Next, to determine whether the activities and residue deposition hypothesized to have occurred in given areas actually occurred in them, or whether the areas were simply locations of *disposal* of artifacts and not locations of activity and residue deposition, primary and secondary archeological deposits (Schiffer 1972) will be distinguished. This distinction is important, for correlations between particular activities and the soil alterations they produce can be expected to be found only in *primary* depositional contexts, where *both* artifactual evidence of activity and residues of activity have been deposited. Inasmuch as it is necessary to understand archeological formation processes in order to reconstruct use-areas and to distinguish primary and secondary deposits on archeological sites, these processes and the nature of organization of the archeological record will be discussed.

Introduction to the Crane Site

The Crane site is a multicomponent site located in Green County, Illinois, on the banks of precanalized Macoupin Creek, a tributary of the lower Illinois River. The site is bean-shaped (Figure 28), covering 6.5 acres and having maximal dimensions of 276 × 174 m. Most of the surface artifacts from the site are concentrated on two knolls (centers at 210R90, 120R60) having a meter or less relief (Figure 29) within the areas encompassed by the concentrations. The two concentrations are separated by a small gully (ca. 1 m deep) that leads into the bed of precanalized Macoupin Creek and that apparently was present during the time of site occupation. Smaller concentrations of artifacts surround the two major concentrations and occur on the planar topography that slopes gently away from the two knolls.

Two Middle Woodland occupations (Bedford and Steuben phases; as defined by Griffin *et al.* 1970) and two late Woodland occupations (White Hall–Weaver and Jersey Bluff; as defined by Struever 1964, 1968b:169; Houart 1971) spanning the period from the time of Christ to A.D. 1200 are responsible

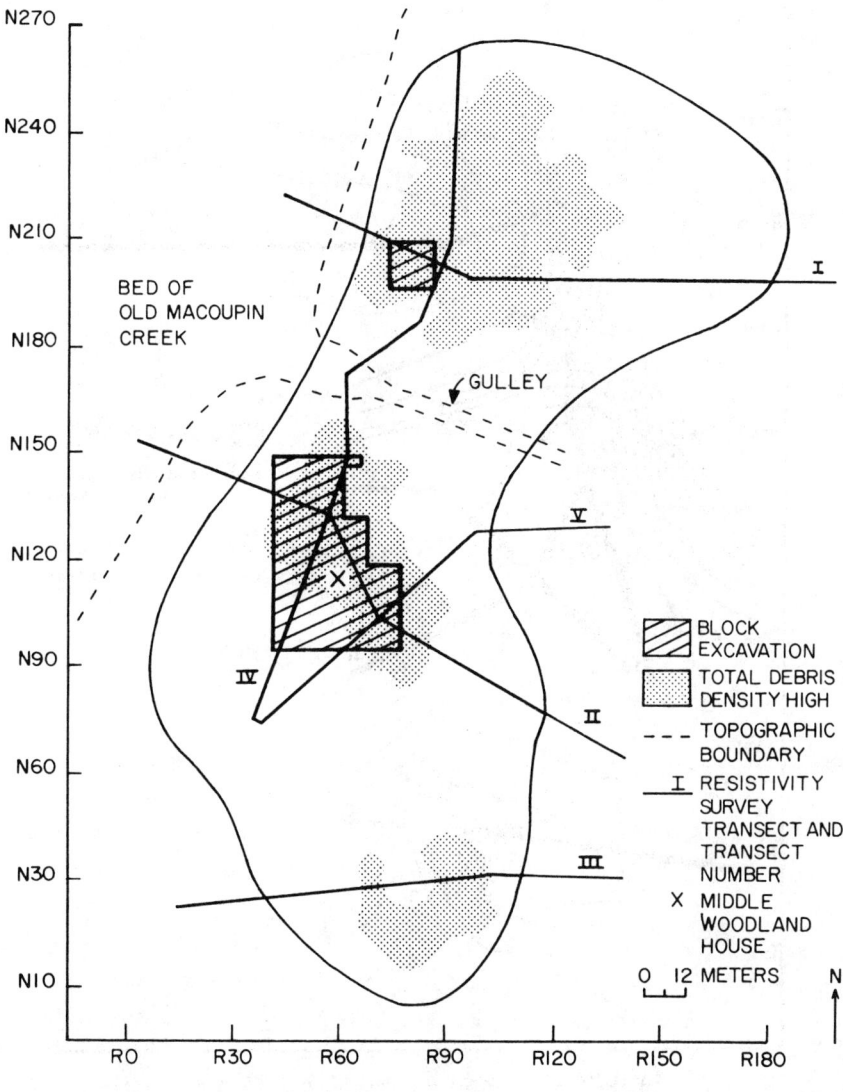

FIGURE 28. *The Crane site.*

for the formation of the Crane assemblage. It would appear that most of the materials are attributable to the Bedford phase (A.D. 1–100 or 150) of the Middle Woodland. This conclusion is based, first, on the relative surface densities of pottery types diagnostic of the four different time periods (Table 27). Havana pottery sherds (Griffin 1952), diagnostic of the Bedford phase of the Middle Woodland, occur in overwhelming frequencies compared to Pike and

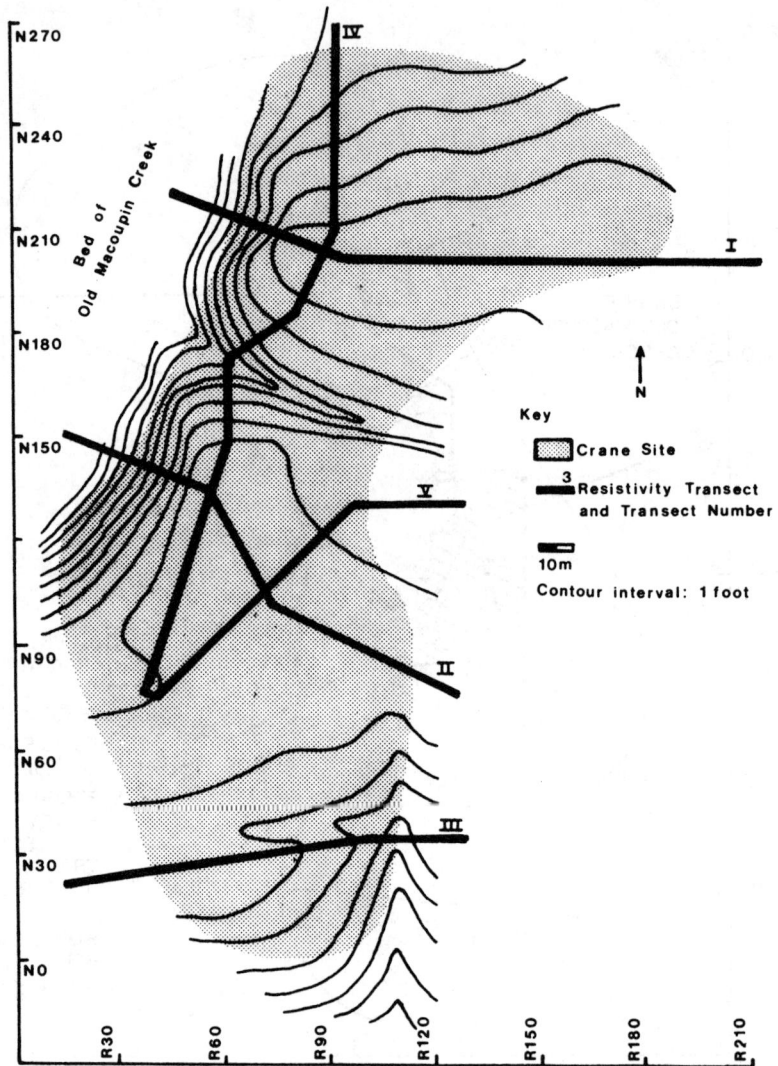

FIGURE 29. *Topographic map of the Crane site.*

White Hall–Weaver sherds (Struever 1964, 1968b) and Jersey Bluff sherds (Harn 1971; Munson 1971; Titterington 1935, 1942) from the later periods represented at Crane.[1] Second, the total spatial structuring of surface artifact conentrations at Crane site appears to be explainable only with respect to Bedford phase activities. Temporally diagnostic pottery and projectile points

---

1. The assumption is being made here that the ratio of pottery sherds to items of other artifacts types is similar for all occupations and that temporally undiagnostic artifact types are not disproportionately represented in high densities for the non-Bedford occupations.

TABLE 27
Densities and spatial coverage of pottery sherds from the several components at Crane site

Statistic	Bedford Phase (Havana Ware)	Steuben Phase (Pike Ware)	Weaver-Whitehall	Jersey Bluff	Late Woodland Type Indeterminant
Total number of sherds	12,476	289	38	496	110
Number of 6x6 m units having sherds present	662	139	29	232	66
Number of sherds per unit when present:					
Mean	18.846	2.0778	1.3103	2.1374	1.7652
Range	1-117	1-13	1-3	1-9	1-7

from the Bedford phase occur in *every* local surface concentration of artifacts. Pottery and point type time indicators of the other three phases at Crane are restricted in their spatial extent to certain portions of the site and do not occur in all local concentrations of artifacts at Crane (Figure 30). Details of the methods and assumptions involved in this analysis are given in Carr (1977). Third, the predominance of Bedford-age archeological deposits relative to deposits of the later phases represented at Crane is suggested by the number and kinds of subsurface features attributable to the several components at Crane. Of the 315 pits that were excavated at the Crane site, nearly all were assignable to the Bedford phase on the basis of the relative frequencies of Havana sherds that they contained. Finally, the single dwelling (Farnsworth, personal communication) excavated at Crane is assignable to the Bedford phase, based on the Havana storage vessels found in its interior. In sum, for all practical purposes, the Crane site may be treated as a single-component, Bedford-age site.

The function of the Crane site within the subsistence–settlement system of its occupants during the Bedford phase is reflected by the activities that probably occurred at the site; these will be documented in detail in the following pages. As an overview, however, the following may be noted. During the Bedford phase, the Crane site served as a base camp. It was occupied by the members of a single household, as evidenced by the remains of one dwelling, located at the crest of the southern knoll (116R56). The size of the social unit probably ranged between 6 and 17 persons, based on the floor area of the dwelling and cross-cultural regressions between floor area and population (Cook and Heizer 1968; Naroll 1962). Most of the activities reflected by the Crane assemblage are maintenance activities (Binford and Binford 1966): precooking processing of nuts and seeds, butchering, cooking, dressing of hides, woodworking and bone working, flint knapping, extraction of plant fibers used in making textiles and cordage, etc. Extractive activities are represented by curated items (e.g., projectile points) assumed not to have been used on the site. Exceptions include those tools used to obtain wood and bark from the periphery of the site, as well as hoes used in gardening.

Although extractive activities are not well represented at the Crane site, the maintenance activities performed at Crane and the faunal and floral debris yielded by these activities reflect the natural environment around Crane and how it was exploited. Crane falls within the oak–hickory forests in the uplands, beyond the main trench of the Illinois Valley (Zawacki and Hausfater 1969). The primary food resources here include nuts and deer. Tools for cracking, pounding, and pulverizing nuts are common at Crane site, as are scrapers and knives used in dressing hides. Of the faunal remains that were found in test excavations distributed randomly over Crane site and that were identifiable to class, over 90% (by count) were from mammals; most of these were deer. The

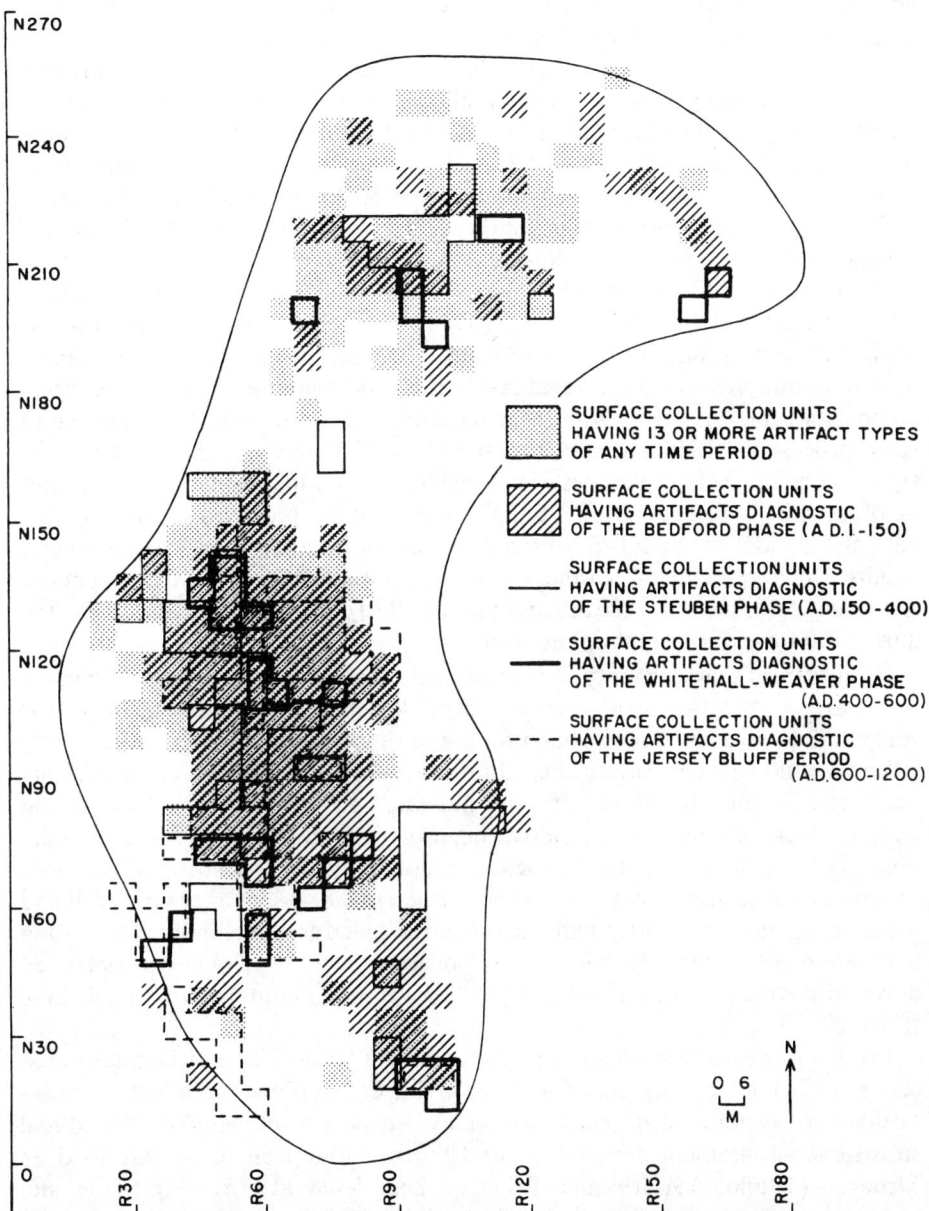

FIGURE 30. *Spatial arrangement of concentrations of all artifact types found on the surface of the Crane site compared to the arrangement of those surface artifacts diagnostic of the four components present at the Crane Site.*

nonmammal remains were composed largely of turtle.[2] Mollusks may have been a significant second-line animal resource, however, based on the excavation of a large quantity of shells found within a refuse dump apparently adjacent to a mollusk processing station. The mollusk remains cannot be compared quantitatively to the mammalian remains to determine more specifically their relative importance to the diet of the Crane occupants, for they were found in an area outside the random excavation sample. Carbonized nuts, primarily hickory, were unearthed in large numbers (nut weight relative to total charcoal weight)—slightly greater than those noted in the Archaic horizons, 6 and 8, at the Koster site within the Illinois Valley (David Asch, personal communication, Asch and Asch 1977; Asch et al. 1972). The relative importance of seeds compared to nuts, based on plant remains, is debatable because of the problem of differential preservation (David Asch, personal communication). The larger number of grinding stones with wear patterns and a morphology suggestive of seed processing (Riddell and Pritchard 1971), however, may reflect the significance of seeds to the diet of the residents at Crane. A shallow backwater of Macoupin Creek (Jackson "Lake") which likely provided an appropriate habitat for pioneer annuals during some years of flooding, was located several hundred meters west of Crane during the nineteenth and early twentieth centuries (Andreas et al. 1873; Anonymous 1909) and in all probability existed during the time of occupation, as well.

With respect to seasonality, it is most likely that Crane was occupied most of the year, if not year-round, except when "logistics" trips were made to the surrounding environment to obtain food and raw materials. Summer and early fall occupations seem certain, based on the excavated remains of cormorants and nuts. Spring occupation may be indicated by the remains of geese and mallard duck, and also by the cultivating tools, and extensive amounts of hide-dressing tools found at the site. Ethnographically, in the Northeast United States, hide dressing commonly was performed in the spring, after major fall and winter deer hunts and after hides had been allowed to decompose, facilitating hair removal (Mason 1889:567). It is uncertain whether Crane was occupied during the winter; males may have gone on extended hunting trips at this time (Carr 1977).

From a more general, regional perspective, the Crane site may be interpreted (Carr 1977) as having functioned as a base camp within a subsistence–settlement system *analogous* in many respects to Middle Woodland subsistence–settlement systems in the Illinois Valley trench, as described by Struever (1968a, 1968b) and Whatley and Asch (1975). First, the site apparently was occupied nearly year-round. Second, it was situated so as to be conveniently located to as *many* food resources as possible within its upland setting. Nuts, fruits and berries, and deer were located in the forest immediately

---

2. The analysis of faunal remains from the Crane site has not been finalized. The approximate relative frequencies of animal classes presented here, however, is substantiated by analyses to date.

surrounding the site. Seed-bearing annuals were obtainable from the backwater adjacent to the site during those years when it formed. Mollusks and turtle were available in small but significant amounts within Macoupin Creek.

Compared to the subsistence systems of the Bedford phase residents of the Illinois Valley trench, however, that of the occupants at Crane accommodated a greater deal of unreliability. There were fewer food resources and greater seasonal variations in food supply in the uplands surrounding the Illinois Valley trench than in the trench itself. This unreliability was manifested in the settlement system of the Crane occupants. The population of the Crane settlement was at least half that of known base camps in the Illinois Valley (Apple Creek site, Struever 1968b; Pool site, McGregor 1958). The residential mobility of the Crane occupants also appears to have been less than that of populations living in the Illinois Valley, based on several studies (Carr 1977), which later should be reappraised when more complete subsurface excavation data become available for analysis. The estimated length of occupation of Crane site is only 3 to 6 years (Carr 1977:164–169a). Additionally, the greater unreliability of the subsistence system of the Crane occupants may have been manifested in intersite, regional exchange that could have involved food resources. Deer hides appear to have been dressed in excess to the needs of the occupants of Crane. Hides and possibly surpluses of deer meat may have been give to populations within the Illinois Valley during seasons or years of plenty in return for other food resources that were not available to the Crane occupants but were available to Illinois Valley populations during the more lean periods for the Crane residents (Carr 1977:186a).

Nature of the Data Used to Define Activity and Depositional Areas within the Crane Site

In order to investigate at Crane the particular kinds of soil alterations produced in soils by particular kinds of activities and depositional behavior, it is necessary to partition the Crane site into multiple spatial units within which different kinds of activities were performed or different kinds of materials were deposited. These multiple units then might be compared to each other with respect to those of their pedological characteristics that are anomalous from the attributes of natural soils around the Crane site, in order to generalize the effects that particular activities and depositional patterns have on soils.

The data used to divide the Crane site into multiple use-areas consist largely of those artifacts found on the surface of Crane during an intensive systematic survey. The methods of collection that were used are similar to those described by Redman and Watson (1970). Crane, which was plowed, disked, and thoroughly rain-washed just prior to the surface survey, yielded a large number of artifacts exposed at its surface. All observed items, regardless of size, were collected from the entire surface of the site within 6 × 6-m collection units gridded over it. These units were large enough to make a total surface collection practicable but small enough to be consistent with the apparent scale of activity patterning at Crane. The latter is a requirement for the analysis of spatial associations between artifact types and for the definition of sets of tools and debris types that repeatedly were used together or deposited together (Whallon 1973).

That the scale of the surface collection units at Crane is consistent with the scale of activity patterning at Crane is based on the known diameters of discrete clusters of pits and open work areas that surrounded the single house, there. These activity loci averaged 8 m in diameter and ranged from 6.5–8.5 m in diameter ($N = 14$). If such work areas were kept somewhat clear of debris and were swept periodically, the tools and debris used and produced in such areas would have been deposited within and somewhat beyond the average 8-m areas. The 6 × 6-m dimension of the units used to collect the artifacts on the surface of Crane site is encompassed easily by the dimension of the known work areas at Crane. The collection units, therefore, are probably relatively homogeneous with respect to the number of distinct activity or depositional loci encompassed within them.

It is true that this conclusion is based on only a small area of the site exhibiting a limited range of activities (primarily household tasks probably performed by women; Mason 1899), and that other activities elsewhere on the site may be expected to have varied in the amount of space they required. Flint knapping, for example, does not require the large amounts of space that stretching and drying hides does. It may be expected, however—based on ethnographic observations on the internal structure of hunter–gatherer camps—that the sample of household activity loci used in estimating the dimension of activity areas in general at Crane is conservative. Other activity loci elsewhere on the site probably are larger, on the average, than those around the house, and still encompass a single 6 × 6-m collection unit. Among the Alyawara Aborigines of Australia (O'Connell 1977, 1979) and the !Kung Bushmen of Africa (Yellen 1971), more spacious, messy, and time-consuming activities occur away from the huts, toward the periphery of the camps, where work space is not at a premium, as it is around the huts. Work areas around huts are used daily for multiple, spatially and temporally restricted activities. This pattern appears to be true of the distribution of activities having different spatial requirements at Crane, too (Carr 1977:151–152, Table 16). Thus, the size of activity areas estimated for the Crane site, based on the size of activity areas situated around the house at Crane, appears to be conservative; the scale of the grid units within which surface artifacts were collected would appear to be small enough to allow those units to encompass relatively few activity loci or depositional loci. A meaningful analysis of the spatial associations between artifacts of different types, therefore, probably can be made.

Data Used to Define Activity and Depositional Areas 193

Materials from all four temporal components of the Crane site were collected from its surface. In no portion of the site were components buried so deeply that their midden deposits were not upturned by the plow. Havana pottery from the Bedford component at Crane—the earliest occupation of Crane—occurred in almost all surface collection units and very nearly always in densities greater than the densities of pottery from later components.

The surface collection from Crane covers the entire area of the site and provides an excellent opportunity for studying, in detail, the spatial arrangement of particular kinds of tools diagnostic of particular kinds of activities. Nevertheless, the data set contains certain biases that can distort a reconstruction of the manner in which different areas of the site were used. First, material items brought to the surface by the plow come from only the top 20 cm of archeological deposits. Although this depth encompasses all four components at Crane, it does not include all of the earliest Bedford phase deposits, some of which extend to 65 cm below the surface. If activities at the site changed in location over the time during which such deposits were formed, only those activities represented by the uppermost portion of the Bedford deposits would be included within the plowzone and the sample of items brought to the surface of Crane. Activities represented by deeper deposits would not be indicated, or be underrepresented, by the surface materials. This possible circumstance could be confusing when comparing the activities or depositional patterns that material items indicate as having occurred in given areas to the physical and chemical characteristics of the soils in those areas.

Excavation data are not available, at present, to monitor whether particular archeological deposits at Crane change over depth in the activities represented by their artifact inclusions. It only can be assumed that the archeological deposits in any one area really are homogeneous over depth with respect to the activities that generated them. This assumption does not appear to be unrealistic. The spatial organization of activities at the Crane site do not appear to have changed much over time. Only one house was built on Crane, and it was not moved during the entire Bedford occupation of the site. Around the house, pits form discrete clusters and rings, and do not intrude on each other with any great frequency, suggesting a highly structured organization of work space that did not change much during the Bedford occupation. An exception to this pattern is one area to the northwest of the dwelling, where pit densities are high, and several clusters appear to be superimposed (surface collection units 349, 350, 363, 364; Figure 37a). It is possible, however, that the activity overlap indicated by this pattern represents the use of the area by the Steuben phase occupants of Crane in addition to the Bedford phase occupants, rather than spatial movement of activities during the Bedford phase. The area has a relatively large number of artifact types diagnostic of the Steuben phase (Figure 30). Finally, the relatively short time (3–6 years) over which the Crane site was occupied by its Bedford phase residents decreases the possibility that the spatial arrangement of work space at Crane was reorganized much. Thus, it is likely that the archeological deposits in any one area of Crane are relatively homogeneous over depth with respect to the activities that produced them.

Several factors other than change in the arrangements of activities over time that could possibly bias the content of artifacts appearing on the surface of Crane site and distort a reconstruction of the manner in which different areas of the site were utilized prehistorically appear to be more critical. Most important, relic collecting by local residents over the last 150 years has probably lowered the proportions of finely made or large items within the zone of deposits brought to the surface by repeated plowing. Such items include point forms used for woodworking or in hunting, cache blades possibly used in defleshing hides, celts used for scraping wood and bone and for obtaining bark from trees, hoes used in gardening, finely decorated Hopewell ware used possibly as serving bowls, figurines, grooved and ungrooved polished axes used in felling or girdling trees and in removing slabs of wood and bark from trees, hammerstones used for flint knapping, and choppers used to dismember game and to rough out forms in wood. A second important biasing factor is the poor preservation of bone within the plowzone at Crane. Not one bone artifact was recovered on the surface of the site.

Finally, plowing, disking, and other agricultural operations at Crane are responsible for the lateral displacement of artifacts. The importance of this factor can be estimated using the results of investigations by Roper (1976) for an Archaic site in Illinois that had been plowed for 20 to 30 years. Roper identified excavated pieces of bifaces that belonged to single items and tabulated the distances separating the pieces of those items. A mean displacement of only 4.12 m was determined. This figure is below the 6-m dimension of the surface collection units used at Crane. If applicable to Crane, it would suggest that a good estimate of the artifacts that orginally occurred in a given collection unit prior to plowing could be obtained from the inventory of artifacts found in that unit. For that portion of the Crane site north of the gully the 4.12-m figure probably is a reliable maximal estimate of the average displacement of artifacts that has occurred. This portion of Crane had been cultivated for only 6 years immediately prior to the investigation of the site in 1974. We may assume that the inventory of artifacts found in a collection unit in this part of the site is a good approximation of the undisturbed artifact content of that unit. That portion of Crane lying south of the gully, however, has been subjected to cultivation continuously, beginning sometime between 1819 and 1873 (Andreas et al. 1873; Continental Historical Co. of Springfield 1885; Donnelley et al. 1879) and extending to the present. If degree of artifact displacement is related to years of plowing in a direct, linear manner, it is clear that agricultural operations in the southern portion of Crane have "smeared" the archeological record there significantly. The inventory of artifacts found in collection units in this portion of Crane might not be a good approximation of the undisturbed artifact contents of those units.

In sum, the surface collection from the Crane site provides an excellent opportunity for reconstructing in detail the kinds of activities and depositional patterns that occurred in particular locations of the site, both in the total nature of its areal coverage and in the appropriate size of the collection units with respect to the known sizes of some activity areas at Crane. The collection, nonetheless, constitutes a biased sample of the artifacts that were deposited in particular locations by the occupants of Crane site, some artifact types tending to be missing from the surface more than others, and all artifact types being displaced from their location of prehistoric deposition to a greater or lesser degree. These biases will be taken into consideration when defining empirical relationships between the activities indicated by the artifacts occurring in particular collection units and the alterations in the soils occurring in those units.

A Model of the Nature of Organization of the Archeological Record Used in Analyzing the Crane Data

To investigate the relationship between activities and the soil alterations they can produce, it is necessary to partition the Crane site into discrete areas that were utilized in different ways by its prehistoric occupants. From a simplistic viewpoint, it might be said that we wish to subdivide Crane site into different *activity areas*—locations in which particular types of tasks such as cooking, butchering, or woodworking occurred (Watson *et al.* 1971:119)—and then to determine the different kinds of soil alterations associated with the different kinds of activity areas. Activity areas might be identified within the surface collection data from Crane by spatial clusters of artifacts of different functions that tend to be correlated or repeatedly associated with each other in their spatial arrangements over the site as a whole and which represent *tool kits*. An appropriate analytical approach to the problem at hand, however, requires a model of activity, activity areas, and tool kits that is more sophisticated than that just implied, and that takes into consideration archeological formation processes. Thus, as a prelude to reconstructing use-areas at the Crane site, the nature of organization of the archeological record will be discussed and the analytical techniques appropriate for describing that organization will be introduced.

Archeological deposits may be viewed best as the end product of processes or *mapping relations* (Ammerman and Feldman 1974) between *two domains* (Schiffer 1972). On the one hand, tools and debris are used in or produced by the activities of the occupants, themselves, of a site. This may be called the *behavioral* domain. The sets of tool types repeatedly used in performing a particular task and the debris resulting from that task may be called an *activity set*. The area in which the work occurred comprises an *activity area*. On the other hand, the tools and debris left as a remainder of activities which occurred in the past constitute the *archeological* domain. The tool and debris types that repeatedly are found together in the archeological record may be termed in the broadest sense (see the following), *depositional sets,* and occur in *depositional areas.*

This distinction of domains, sets, and areas is necessary because the patterns of co-occurrence and spatial arrangement of different kinds of tools and debris may differ between the two domains. In the behavioral domain, the tools and debris that are associated are those actually produced and/or used together. In the archeological domain, the tools and debris found together could represent a number of behavioral phenomena. They might represent all the tools and debris produced and used together in one kind of task by the previous occupants of the site, or perhaps only a portion of them, if some were saved for use in other activities at a later time. Such an association has been called *primary refuse* by Schiffer (1972, 1975a). An association also could represent tools and debris that were thrown away together in a formalized dumping location. Associations of this kind have been called *secondary refuse* by Schiffer (1972, 1975a). Other possible kinds of artifact aggregations include items that were stored together for later use (caches)—a special kind of primary refuse—or items that were used in a number of independent tasks which occurred at different times but happened to overlap spatially. An association of artifacts also might reflect a particular social context rather than some common task in which the artifacts were put to use (Yellen 1974:204, 207). For example, among the Alyawara Aborigines (O'Connell 1979), the Western Desert Aborigines (Gould 1971) and the !Kung Bushmen (Yellen 1974), a large group of activities occur within the context of the family around the hearth. The remains from such activities are overlayed and mixed together in a single area. Co-occurrences between different artifact types in this situation do not reflect a common activity in which the artifacts were used, but rather, a common social context in which they occur together.

Adding further complexity, an aggregation of artifacts may not reflect past human behavioral processes at all, but rather, postdepositional processes of natural origin or contemporary human origin. Erosion, solifluction, rodent activity, and contemporary farming are examples of such processes.

Similarly, an *area* in which several kinds of tools and debris are found together on an archeological site does not necessarily correspond to an "activity area" in the behavioral domain. Other possibilities include a trash dump or a storage area. A local aggregate of tools and debris also may be a mixed composite of items from multiple activities that were performed in a common social context. The area in which a group of artifacts is found in association also might represent simply the final resting place of the artifacts, that have been removed and transported from different primary depositional contexts by geological or other natural processes.

Finally, it must be realized that although every unique activity performance in the behavioral domain is ipso facto associated with a location or an activity area, it is not necessarily associated with a depositional area in the archeological domain. Some activities do tend to be grounded to single locales, such as drying meat on racks anchored in the soil. Other activities are unrestricted in the locations in which they can occur, and the debris they produce may occur ubiquitously over an archeological site. Whittling, for example, is often done in hunter–gatherer camps on an impromptu basis, whenever and wherever people socialize.

The terms *activity set* and *activity area* thus are too restrictive in the archeological phenomena they describe to be applicable to all forms of artifact aggregations or to all locations of artifact aggregation occurring in the archeological record. The terms *depositional set* and *depositional area* may be used to describe a group of artifact types that repeatedly are found together and the locations of that group in the broadest sense, without specifying the processes by which the aggregates were formed and placed in space. If natural or contemporary human postdepositional disturbances do not appear to be responsible for the aggregations of artifacts, the more specific term *use-area* may be applied to their locations, implying that the areas were used prehistorically for artifact manufacture, use, storage, or disposal, but not specifying which of these. The repeatedly associated artifact types *cannot* be called, correspondingly, a "use-set," for they were not necessarily used together. They might have been manufactured, stored, or disposed of together. The associated types still are termed a depositional set. The terms activity set and activity area are reserved for those repeated associations of artifact types and their locations for which it can be shown that the artifacts were *used* together and *expediently* deposited together.

A more precise and formal statement of the nature of the archeological record that will be assumed in this analysis of the Crane surface collection may be with the aid of concepts taken from set theory. In set theory, an organization of entities may be described by using four basic concepts: (*a*) *sets*—groups of entities; (*b*) *members* or *elements of sets*—the entities grouped together; (*c*) *attributes*—the character states that the entities possess; and (*d*) the *list of attributes* that most entities in a set share or that all entities in a set are required to have to belong to it. Applying these concepts to an archeological situation, suppose that the archeological deposits within a site can be classified into several kinds, according to the functional types of artifacts they contain. The several deposits (entities) of one kind comprise a set; they always or often contain certain artifact types (attributes). The several *artifact types* held in common or tending to be held in common by the deposits comprise a list of attributes, and are what have been previously termed an activity set or depositional set.[3]

---

[3]. The use of the term *set* here is unfortunate, for a depositional set is not a set at all, but rather a list of attributes, in set-theoretical terms. However, as the term *activity set* prevails in archeological literature, and depositional sets are analogous to activity sets, I will continue to use this archeological terminology.

Sets and the list of attributes that characterize their members may be described as *overlapping* or *nonoverlapping* in nature, and *monothetic* or *polythetic* in nature. Sets are said to be overlapping when members of different sets share some of the character states that are required of them for admittance into their respective sets or that often are found among them. Sets are said to be nonoverlapping when members of different sets *do not have in common* any of the character states that are required of or often found among them (Sibson 1968; Sneath and Sokal 1973:207–208). Likewise, the various lists of attributes that are required of or often found among the members of several sets may be termed overlapping if some of the attributes in the several lists are the same, whereas they may be termed nonoverlapping if none of the attributes in the several lists is the same.[4] Two depositional sets (two lists of artifact types always or often found among members of two sets of deposits) would be considered overlapping if some of the artifact types comprising each depositional set were the same. The depositional sets would be considered nonoverlapping if none of the artifact types comprising each depositional set were the same.

The distinction between overlapping and nonoverlapping sets and attribute lists refers to the *external* organization of sets. The distinction between monothetic and polythetic sets, and between monothetic and polythetic attribute lists, refers to the *internal* organization of sets. In a monothetic set, the elements of the set all share the very same character states, all of which are essential to group membership. In a polythetic set, the elements share a large number of character states, but no single state is essential to group membership (Sneath and Sokal 1973:21; Clarke 1968:37). Similarly, if the attributes possessed by the members of a set at large *all* are possessed by *each* member of the set, the list of attributes may be said to be monothetic, or monothetically distributed. If some of the attributes possessed by the members of a set shared by only some members of the set and no one attribute is required for membership in the set, then the list of attributes may be said to be polythetic, or polythetically distributed. A depositional set (list of artifact types) would be monothetic if all deposits having one of the artifact types in the list also had every other artifact type on that list. A depositional set (list of artifact types) would be polythetic if the deposits having one of the artifact types in the list did not necessarily have some of the other artifact types in the list, and those having the other artifact types did not necessarily have the first.

In the model of the archeological record that will be assumed in this study, depositional sets are considered to be overlapping and polythetic in nature. These characteristics reflect assumptions about the nature of organization of material items in the behavioral domain and about the nature of the mapping

---

4. Set theoreticians use the adjectives *overlapping*, *nonoverlapping*, *monothetic*, and *polythetic* to describe only sets, not attribute lists. The use of these terms has been extended to attribute lists in this discussion to avoid confusions that might arise from the misuse of the term *set* (as in *activity set*) in the archeological literature.

relations that link the behavioral domain to the archeological domain. The overlapping nature of depositional sets mirrors the overlapping nature of activity sets in the behavioral domain. Some tools used for one activity may be used in other activities, as well. Among the Eskimo, for example, "women's knives" are used both in kitchen activities involving the cutting of meat and vegetables, and in sewing, to cut leather (Mason 1889).

The polythetic organization of depositional sets derives from a number of factors (Binford 1974; Schiffer 1972, 1973, 1975a, 1975b, 1976), some of which have already been mentioned briefly.

1. The artifact types comprising an activity set in the behavioral domain may enter the archeological domain as subsets separated in different locations. The types comprising an activity set may be deposited in the locations of their manufacture, use, storage, or discard, none of which need coincide.
2. If the artifact types within an activity set are *curated,* that it, stored for reuse at a later time (Binford 1974), and if the activity is not performed repeatedly in the same location, differential wear rates and breakage rates may lead to different subsets of the activity set being deposited in different locales. The degree to which the artifact types within an activity set are curated and not always deposited with each other depends on the labor invested in manufacturing them, their cultural importance, the ease with which they can be moved, the distance to the next site to be occupied in the annual round of the community, the availability of the types (or the raw materials from which they can be made) at the next site, and the degree of mobility of the community. The number of classes of tools that are curated by a social group tends to increase with the residential stability of that social group (Binford 1974:42). The great impact which curation can have on the organization of the archeological record is illustrated by Binford's work among the Nunamiut Eskimo. Binford recorded the number and kinds of items that were taken by the Nunamiut on 47 logistics trips away from their base camps. Of the 647 trip-items carried, 99 were totally consumed in the course of their use, most being food items. The remaining

> five hundred and forty-eight of the trip-items carried were visible in that there were tangible by-products from their use or no destruction occurred during their use. Of these items, fifty-three, or only 9.67 percent of the total visible items were not returned to the village.... Of these fifty-three (53) trip-items, thirty-six (36) are items which are disposable byproducts in their context of their use, including the peanut butter jar, sardine can.... Of the remaining eighteen (18) items not returned to the village, fourteen (14) or 26.3% of the total were cached in the field for future use. Of the remaining four items, three were unintentionally lost on the trail and *only one was discarded, broken at the location where it was used.* This is not, however, the only item broken during the course of the forty-seven trips. Twelve additional items were broken, but returned to the village for repair. [Binford 1974:35-36].

3. The polythetic organization of depositional sets also may be caused by the reworking of an artifact of one type used or produced in one activity into a different type used in another activity. Each such single-item, compound-type artifact can be deposited with the members of only one of the activity sets in which it participated.
4. When sites are abandoned over an extended period of time, material items in abandoned areas may be "mined" by the residual occupants and reused for the same or different purposes in other areas of the site, creating deposits in the abandoned area that form polythetic sets when analyzed. The same phenomenon can also occur as a site shifts gradually in location without loss of membership to the social group (Ascher 1968). Previously occupied areas may be mined of tools and debris for use in the newly occupied part of the site. The artifacts mined are not always tools, but also debris items and junk. Some debris items are recycled immediately, but some are cached for raw material at a later time and might never be reused (James O'Connell, personal communication). Schiffer also has noted the occurrence of "mining" at a smaller scale. As households expand and contract in size during their lifetime, new rooms or huts are built and abandoned. The abandoned ones may be mined for materials by the members of the household still remaining.
5. Multipurpose, single-type tools (as opposed to compound-type tools just mentioned), which are used in several spatially segregated activities, can be deposited with the material items involved in only *one* of the activities. Lamellar blades, for example, can be used for both woodworking (Sollberger 1969), butchering, or shaving the scalp (Crabtree 1968).
6. Activity sets with several alternative tool types accomplishing the same ends are polythetically organized. For example, the Nunamiut use both saws and knives to cut meat. In any particular activity performance, one or the other might be used, but not necessarily both.

The methods of data collection and analysis used by an archeologist may artificially cause the depositional sets with which he is working to be organized polythetically.

7. When recovery of artifacts is not complete, as is the case with surface survey data or when screening is not used during excavation, the depositional sets will be polythetic. This is true of the data analyzed in this work.
8. When tool and debris classifications are based on stylistic rather than functional attributes, items that are functionally equivalent and that belong to the same activity set may be classified into separate types. This is the case with classic lithic tool typologies (e.g., Balout 1967; Bordes 1961, 1969; de Heinzelin 1962; Tixier 1963) in which attention is given to flake shape and size, and retouch patterns rather than more functional attributes such as the angle and wear of the working edge of the tool (e.g., Ahler 1971; Wilmsen 1970). Attempts to define depositional sets using stylistically based artifact typologies will yield sets that are more polythetic than would be the case if a functional typology were used.

9. An overly divisive typology will also yield depositional sets of a polythetic nature. Care must be taken not to overclassify artifacts, particularly with the assemblages of mobile populations who seemingly tend to be more opportunistic in the tools they use to accomplish their ends. As Gould *et al.* (1971:154) have pointed out: It would be

> a mistake to overclassify the ethnographic adzes (i.e., scrapers) of the Western Desert Aborigines. Ethnographic observations over an extended period of time and in a variety of situations lead, instead, to an appreciation of the casualness and opportunism of present day Aborigine stone chipping. To these people, the primary aim is to perform a task involving either cutting (of sinew, flesh, vegetable fibers, *etc.*) or scraping (of wood) with little interest in the shape of the tool except for the angle of the working edge relative to the particular task involved.

James O'Connell (personal communication) estimates that there are only about 10 functional types of artifacts in Alyawara assemblages—quite a small number compared to the elaborateness of some typologies of tools of mobile groups that have been designed and are supposed to be functional. For example, Binford and Binford (1966:251) used 40 tool types in examining the Mousterian of levallois facies.

10. Finally, misclassification of items as to the functional category to which they belong will yield a greater degree of polythetic organization to the defined depositional sets than would otherwise be the case.

In summary, the model of the archeological record that will be assumed when analyzing the surface materials from Crane site differs to a great extent from the simplistic viewpoint that there is a direct relationship between activity sets or activity areas in the behavioral domain and the content of the archeological record. Tools are not necessarily deposited in the locations where they were used in the behavioral domain nor with the other kinds of tools with which they were used in the behavioral domain. Deposition of tools and debris on archeological sites is not necessarily an "expedient" process (Binford 1974). These assumptions restrict the methods that can be used to analyze the Crane surface collection.

Methods Used to Define Use-Areas within the Crane Site

In the pioneering studies of the spatial arrangement of activities within sites (e.g., Brown and Freeman 1964; Freeman and Butzer 1966; Hill 1970; Struever 1968a), activity sets typically were defined by means of correlation analysis and factor analysis. Activity areas were believed to be definable by plotting the spatial distribution of factor scores of individual collection units, or by plotting the spatial arrangement of artifact types within the same set and examining the arrangement for significant clusters. These techniques are consistent with or imply certain assumptions that were made about the nature of the archeological record. Among these assumptions are the following:

1. All member artifact types of an activity set are deposited together in an "expedient" manner (Binford 1974) in their locations of use.
2. Each activity area in the behavioral domain is segregated from all others and relates to a unique depositional area in the archeological domain.
3. Both activity sets and depositional sets have a monothetic, nonoverlapping organization.

It was suggested previously, however, that the archeological record is more complex in its organization and in its relationship to the behavioral domain than these three assumptions admit. Consequently, the correlation and factor analytic techniques that classically have been used to define "activity sets" and "activity areas" would not appear to be appropriate to their task. Speth and Johnson (1976) have stressed this incongruity between correlation techniques and data structure, as well.

Methods Applicable When the Functions of All Artifact Types Are Known

In considering how the Crane site and archeological sites in general might best be partitioned into different kinds of use-areas, the following position has been taken. First, it is clear that one means of avoiding the incongruity between analytical technique and data structure in defining different kinds of use-areas on an archeological site would be (*a*) to bypass those seps in the analysis concerned with the *sitewide correlations or associations* of different artifact types and with the definition of so-called tool kits; and (*b*) to proceed to the definition of the functional nature of different *individual* areas within the site directly, using information on the function of the *individual artifacts occurring within the specific subportions of the site* that are of interest. For example, if a particular collection unit contained knives thought to have been used to deflesh hides and scrapers thought to have been used to grain hides, hideworking would be indicated by the artifacts in the particular area, *regardless of whether the types of knives and scrapers found in association in that unit also tend to be found in association in other units with each other or with other artifact types involved in hide working.* Stress, here, is being placed on the function of the artifacts found in *a particular area* rather than on the *total site patterning* of association of artifact types.

To apply this procedure, of course, the function of all artifacts present within the spatial unit of interest must be known. It also is desirable that each artifact be single-purpose rather than multipurpose, so that the precise activity or activities associated with the unit may be specified, rather than a number of alternative activities. If multipurpose tools are present within a spatial unit, however, the range of activities they indicate might be narrowed by examining whether single-purpose artifacts used for those activities are present or absent from the unit.

Once the functions of the artifacts occurring in a particular unit have been inventoried, the kind of depositional process reponsible for the association must be considered: Does the association reflect primary refuse from *in situ* activity, a cache, secondary refuse, or a mixture of primary refuse from multiple activities that overlapped spatially and were performed at different times? Numerous approaches might be taken to determine the depositional nature of the archeological remains within a particular subsegment of a site, depending on the site's stratigraphic nature, the topographical constraints it placed on human behavior, and even the total site population and population density (Schiffer 1972:161–162). The methods that will be used at Crane site largely depend on an analogy between the spatial organization of different kinds of activities and depositional processes within hunter–gatherer camps and the spatial distribution of various archeological phenomena (e.g., midden depths, available work space) at Crane.

In this study, determination of whether the archeological deposits in a particular collection unit represent primary refuse, a cache, or secondary refuse is extremely important. An activity may produce two kinds of byproducts: organic refuse (e.g., the body fluids of a butchered animal) that may be incorporated into the soil almost simultaneously with the activity and at the location of activity, and more permanent artifactual refuse (e.g., stone tools) that are movable. If the artifactual refuse is moved from the location of activity to a secondary dump, the location of the artifacts will not be the location of soil change. This situation will confuse an analysis attempting to find correspondences between activities, as evidenced by artifact distributions, and soil alterations unless the secondary nature of the artifact deposits is realized.

In summary, when it is possible to identify the function of particular artifact types, an appropriate method for determining the manner in which different areas of an archeological site were used is to (*a*) consider the functions of the individual artifacts occurring within a particular area, and (*b*) determine whether the artifacts within that area represent primary refuse, a cache, secondary refuse, or mixed primary refuse from multiple, spatially overlapping activities. By considering the functional and depositional nature of *individual* areas *directly* rather than via the definition of sitewide tool associations, one may circumvent incongruities between the analytical techniques classically used to define use-areas and the nature of organization of the archeological record.

Methods Applicable When the Functions of Only Some Artifact Types Are Known

The procedures described previously for defining use-areas within archeological sites assume that every artifact type, as defined by certain functional attributes, can be identified with a particular task or set of tasks in which it participated. Operationaly, however, this sometimes is not feasible, and a given set of functional attributes characterizing a particular class of artifacts may suggest a number of alternative tasks in which the artifacts participated. In such a situation, an ancillary procedure is required for determining the probable function of the ambiguous artifact class.

One means for determining the probable function of an ambiguous artifact type is to consider the known functions of the tools with which it often associates across the site of interest at large. For example, a particular type of scraper might possess functional attributes suggesting its use for either graining hides or scraping wood, but no attributes allowing a finer specification of function. If, however, that scraper type also is known to occur often with other tool types used in dressing hides, then one might hypothesize that the ambiguous type of scraper belonged to a hide-dressing tool kit, and that it was used to grain hides rather than to work wood.

Determining the function of such ambiguous artifact types, of course, is important not with respect to those provenience units where the type co-occurs with other types which were used in the same task and have known functions. In these units, a direct reconstruction of the locations' use is possible. Rather, identification of the function of ambiguous types is important with respect to those units where the type occurs isolated from its associates and where reconstruction of activities would not be possible. Such circumstances may be expected, given the polythetic nature of depositional sets.

Thus, although the definition of sitewide depositional sets is not a prerequisite to the identification of the prehistoric use of each particular area on an archeological site, their definition may be useful in an ancillary manner, allowing one to narrow down the function of an ambiguous artifact type and the activities that it might indicate to have occurred in its resting place. In the preceding section, however, it was suggested that the analytical techniques classically used to define depositional sets imply certain assumptions about the nature of their organization that are untenable. What alternative methods might be used which take into consideration the complex form of organization of depositional sets, as modeled?

One possible method is a nearest-neighbor technique, which I have designed in another paper (Carr 1977). This method and its logic may be summarized as follows. In considering the nature of organization of the archeological record in mathematical terms, the pattern, covariation, is not the most appropriate model of the relationships holding between different artifact types that tend to be found together. Although tool and debris types used and produced together may covary in their frequencies in the behavioral domain, their frequencies may be independent in the archeological domain. Two general classes of factors are responsible for the inappropriateness of covariation as a model of artifact relationship: (*a*) one class concerned with the extent of spatial mingling of the artifacts generated and discarded by different activities, and (*b*) a second concerned with the processes governing the polythetic organization of depositional sets.

Consider the first class.

1. If the activity sets of several activities have some artifact types in common (i.e., are overlapping activity sets) and if the activities intermingle spatially, the relative frequencies of shared artifact types and the relative frequencies of shared to nonshared types within the refuse generated in such areas of overlap will *not* be equivalent to the relative frequencies of those artifact types in areas where the activities are spatially discrete and produce separate refuse deposits. The relative frequencies of the types will be a function not only of the *proportions* of them generated by each activity that contributes refuse to the area, but also the relative *amount* of refuse generated by each activity.
2. If the activity sets of several activities have some artifacts types in common and if they are not deposited in their respective areas of use, but rather, are translocated and deposited in a common secondary trash area, again the relative frequencies of shared to nonshared types in the dump will not be equivalent to the proportions in which those types are generated in the discrete areas of use, before translocation. The proportions of such types in the trash area will be a function of the proportions in which they are generated in the discrete areas of use *and* the relative amount of refuse generated by each activity using them.
3. Numerous postdepositional smearing and blending processes, such as the walking of inhabitants of a site over abandoned portions (Ascher 1968) or natural disturbances, may have the same effect as dumping the different activity sets of several activities in a common location.

The second class of factors leading to the inappropriateness of covariation as a model of artifact relationships in the archeological domain are the numerous processes leading to the polythetic organization of depositional sets. Consider an activity that generates a number of different kinds of tools and debris at different rates, that is, the discard rates of different types are different. If the activity is performed numerous times at a location and no polythetic-causing factors other than curation (differential discard rates) operate, then the relative frequencies of artifact types deposited in that location will stabilize over time to constant values approaching the ratios of the discard rates of those types. As the number or intensity of polythetic-causing processes involved in the formation of the deposits is increased and stochasticity is introduced, however, the tendency of the relative frequencies of the deposited types to approach the ratios of their discarded rates will decrease.

No analytical method seeking to define sets of artifacts that are manufactured, used, or stored separately can overcome the problems resulting from the first class of factors concerned with the final spatial mingling of initially discrete sets. These factors, when carried to their extreme, can destroy virtually all evidence of the forms of organization sought by the archeologist. On the other hand, it *is* possible to circumvent the problems resulting from the second class of factors by using analytical techniques other than correlation, which are congruent with the nature of organization of depositional sets.

Various forms of association analysis are a step in the right direction. When an activity that generates a number of different kinds of tools and debris at different rates is repeated numerous times at a single location, and when polythetic-causing processes are operative, the relative frequencies of the deposited artifact types may not stabilize over time, but the presence states of those types will.

Several alternative methods defining patterns of association, as opposed to covariation, among artifact types can be used to find sets of artifact types that repeatedly were manufactured, used, stored, or discarded together. In the realm of statistical tests of significant association, the chi-square test is used most commonly. Variable pairs (here, artifact types) may be tested for their independence at a given level of confidence. In the realm of nonstatistical procedures, any of a number of "association coefficients" can be calculated as a measure of the degree of association between pairs of artifact types (Sneath and Sokal 1973:129–137; Cole 1949). The coefficients differ in the weight they attach to the a, b, c, and d cells of a two-way contingency table. For example, the simple matching coefficient weights all matches and mismatches equally in determining the degree of association between two variable pairs, whereas the Jaccard coefficient omits consideration of negative matches (Sneath and Sokal 1973:131–133). The latter coefficient is useful particularly when many of the artifact types being related are rare. and the matrix of type-counts per observation unit contains many zeros. In this case, matches in the absence states of pairs of artifact types at many locations may not be a meaningful indicator of association. Once a matrix of association coefficients for all possible pairs of artifact types has been defined, it can be subjected to overlapping hierarchical clustering algorithms (Cole and Wishart 1970; Jardine and Sibson 1968) in order to isolate overlapping, interassociated sets of artifact types.

Association, as a measure of the strength of relationship holding between different artifact types, is more congruent with the nature of organization of the archeological record proposed in this study than is covariation. Nonetheless, it does not completely reflect the polythetic nature of depositional sets. Let us return to the example of the activity that involves a number of artifact types, but results in their discard at different rates. One cannot expect an activity of this kind to be performed, necessarily, at *every* location of use *enough times* that the more highly curated artifact types are deposited (present) in each location of use. Similarly, one can not expect compound tools or multipurpose tools participating in several activities to be deposited at every location of their use. Nor can one expect methods of data collection and typology to be accurate enough that they do not cause some absences of certain artifact types from deposits where they really do occur. These forms of absences of an artifact type from a cluster of several types in which we might otherwise expect them to occur do not indicate that the missing types are not part of the depositional set represented by the types in the cluster. Nevertheless, in association analysis, these forms of absence states count toward the dissociation of the missing type from the depositional set as if they did indicate dissociation. We now may see how association analysis is inappropriate for defining depositional sets. Currently used association coefficients do not distinguish between two possible explanations of the absence of a type from an aggregation of artifacts: (*a*) the actual dissociation of an artifact type from the depositional set represented by other types found in the aggregation, and (*b*) the polythetic organization of types among depositional sets. Mismatches (counts in the b and c cells) in fourfold contingency tables always are considered a measure of dissociation (Sokal and Sneath 1963:128–135).

One way to circumvent this problem and to take into consideration the polythetic distribution of artifact types among archeological deposits is to rephrase the question of association in terms of the geographic distances between items of different types. First, let us consider a measure of monothetic association within this framework. A simple statistic that can be calculated in comparing the distribution of items of two artifact types is the average absolute distance between items of one type and their nearest neighbors of the second type. A base type and reference type are chosen. For each item of the base type, the Euclidean distances at which surrounding items of the reference type occur are compared until the nearest neighbor of the reference type is found. The same procedure is then repeated, this time using the items of the reference type as base points and the items of the base type as the satellite reference points. The average intertype distance can be computed by:

$$\text{DIST}_{AB} = \frac{\sum_{1}^{n} \overline{AB} + \sum_{1}^{m} \overline{BA}}{n + m}$$

where n is the number of items of type A, m is the number of items of type B, $\overline{AB}$ is the distance from a given base point of type A to its nearest neighbor of type B, and $\overline{BA}$ is the distance from a given base point of type B to its nearest neighbor of type A. Note that the sum of the distances $\overline{AB}$ need not be equivalent to the sum of the distances $\overline{BA}$. This is so, both because the number of items of the two types need not be the same, and because the base item of one type to which the nearest neighbor of the opposite type is referred need not be the nearest neighbor of that reference item. Figure 31a illustrates the latter factor.

This statistic provides a monothetic measure of the association of the items of two different types. It is assumed that if an item of one type is not present at a "reasonable distance" (to be defined later) from an item of a second type, then the case constitutes dissociation of the two types. If this is the case on the average, considering all items of both types, then the two types can be said to be dissociated. It was shown previously, however, that numerous processes of artifact use and deposition in the behavioral domain can cause two functional types used in the same activity to be deposited separately in many cases. Figure 31b illustrates this. In this diagram, type X often pairs with type O, but not in all cases. Perhaps their rates of discard differ slightly, or perhaps type X has multiple purposes and is deposited at times with the members of a second activity set in which it occurs. A measure of polythetic association should reflect the strength of association of the two types where they do occur together and ignore those particular cases in which pairing does not occur.

Although it is possible to judge which cases of pairing and nonpairing should

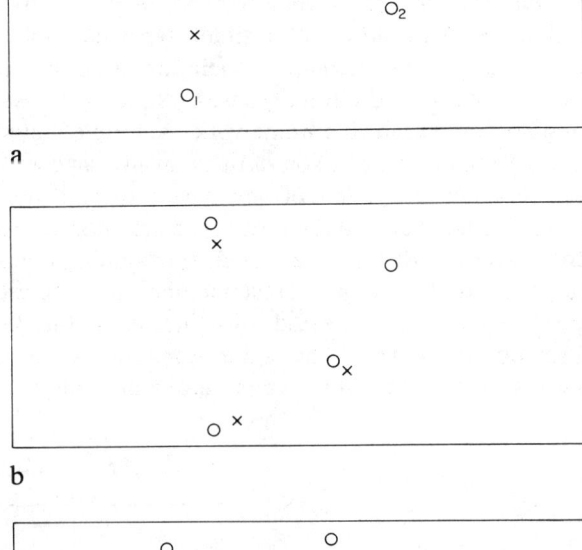

FIGURE 31. *(a) Asymmetry in nearest neighbors. X is O_2s nearest neighbor of the opposite type, but O_1 is Xs nearest neighbor of the opposite type. (b) An arrangement of items of two artifact types, X and O, associated in a polythetic nature. (c) A more ambiguous polythetic association of items of two artifact types, X and O.*

be included in a measure of polythetic association for the clear spatial arrangement shown in Figure 31b, more ambiguous arrangements may arise, as shown in Figure 31c. An algorithm for measuring polythetic association between types can be designed such that ambiguous arrangements like the one in Figure 31c can be evaluated in a *consistent* way. As before, the two sums of distances, $\overline{AB}$ and $\overline{BA}$, are calculated. Rather than summing these, however, and standardizing by the total number of base items considered in the joint sum, the individual sums $\overline{AB}$ and $\overline{BA}$ are standardized separately.

$$\text{AVDIST1} = \frac{\sum_{1}^{n} \overline{AB}}{n} \qquad \text{AVDIST2} = \frac{\sum_{1}^{m} \overline{BA}}{m}$$

The *minimum* average distance is then chosen as the measure of the strength of association of the two artifact types, *ignoring the effects of the incomplete pairing of types which resulted from the various processes by which the archeological record was formed.* In Figure 31c, for example, the average distance from type X to type O is small, because pairing between the two types is complete from the perspective of type X. The average distance from type O to type X, however, is larger, because pairing is incomplete from the perspective of type O. The smaller distance would be chosen as the measure of association.

By defining the strength of association between two types in this manner, two *causes* for the absence of pairing of artifact types have been separated: (*a*) the *actual dissociation* of the types from different depositional sets and possibly different activity sets; and (*b*) the *processes* by which activity sets are mapped from the behavioral domain into the archeological domain so as to produce polythetic depositional sets. The effects of incomplete pairing of different artifact types that result from the various processes by which an archeological record was formed do not increase the value of the dissociation statistic, AVDIST. Dissociation of two artifact types will be indicated only when the items of the two types do not tend to pair at all—either completely or incompletely.

Once a matrix of AVDIST coefficients for all possible pairwise combinations of types has been calculated, depositional sets may be defined using an overlapping, hierarchical clustering algorithm and some threshold value specifying the level of the hierarchy at which grouping is significant. The value of the threshold used in defining clusters should be consistent with the expectable scale of patterning of artifact type of co-occurrence. For example, at Crane site, a threshold of 16 m would seem appropriate. This value is equivalent to the known average diameter—8 m—of open work spaces and discrete clusters of pits surrounding the single house at Crane, plus another 8 m. The additional 8 m take into consideration the spreading of material items outward from the activity areas in which they were generated, as a result of both sweeping and clearing of the activity areas that may have occurred in the behavioral domain, as well as historic farming practices.

The method, which has just been summarized, for defining depositional sets of a polythetic nature requires grid coordinate data for the computation of geographic distances. For some data sets, however, such as the Crane surface collection, it is feasible to convert grid count data into a form consistent with the proposed methodology. If the archeological site from which the data are taken is large enough compared to the size of the collection units, the centers of collection units may be used to approximate the actual provenience of the artifacts. Multiple observations of the same artifact types in the same collection unit may be considered equivalent to a single artifact, for it is the *co-occurrence* of artifact types rather than the *covariation* of the frequencies of artifact types that is the appropriate measure of whether or not those types belong to the same depositional set.

Artifact types that occur ubiquitously across a site in various densities can be handled in either of two ways. First, in being present in all or most collection units, they might be considered constants in the analysis of depositional sets and dropped from the analysis. Second, if the grid cells with high counts appear to show spatial patterning, the artifact type might be considered present in only those units where counts are higher than locally defined threshold values that bring out the observable pattern. Multiple, locally defined thresholds, rather than a single, sitewide threshold must be used because the amount of activity of one type that occurred in different areas of a site can vary from area to area and the level of background counts in unused areas can vary from place to place. No single threshold may be capable of distinguishing activity areas of *all* levels of use from the unused areas surrounding them. To define local threshold values, the spatial filtering methods introduced in Chapter 2 may be used.

The proposed method for defining depositional sets is considered to be more consistent with their nature of organization than other techniques presently used in archeology. Nevertheless, there are at least two circumstances in which the technique may be incapable of resolving depositional sets that correspond to groups of artifacts which were manufactured, used, or stored together. The first case is when the refuse from different kinds of activities consistently is deposited together, as a result of either repeated spatial overlap of the activities and their primary refuse, or repeated secondary deposition of the refuse of the activities in common trash areas. The activities that occur around a house might exemplify these circumstances. As the degree of common deposition of the refuse of distinct activities increases, the capability of the method to resolve the different kinds of refuse will decrease. AVDIST1 and AVDIST2 both will decrease for pairwise comparisons of the artifact types in the different depositional sets, until the sets appear to be one. This problem does not reflect an inadequacy of the method, but rather, a loss of patterning of the relationships between artifacts as they are mapped from the behavioral domain to the archeological domain, and the inability of any technique to retrieve that lost patterning.

A second circumstance in which the technique may be incapable of resolving a set of artifacts that were manufactured, used, or stored together is when both the base and reference types of artifacts of *each* pair of types within the set are polythetically distributed with respect to each other (e.g., Figure 32). Such an arrangement of artifacts can occur when multiple processes leading to the polythetic organization of activity sets are operative, different processes having operated upon different types. In such cases, for any given pair of types, both AVDIST1 and AVDIST2 will have high values. Choosing the minimum of these two statistics for the average, absolute nearest neighbor statistic will not help to distinguish absence-states resulting from the actual dissociation of types and absence-states resulting from the processes leading to the polythetic organization of the sets. An arrangement of two artifact types from the same depositional set in which both base and reference types are polythetically distributed with respect to each other may yield an average absolute nearest neighbor statistic that has a value similar to that for an arrangement of two types that belong to different depositional sets but overlap spatially to a slight degree (Figures 32, 33). I do not have a good solution to this problem at present. However, it can be noted that if the overlapping, hierarchical clustering algorithm used to define multitype artifact sets is also an average linkage routine, the process by which types are clustered will tend to compensate for the inadequacy of the AVDIST statistic. The effectiveness of average linking in overriding the inadequacy of the AVDIST statistic will depend on whether the percentage of polythetically caused absence states of types with respect to each other is extreme or not and on the particular threshold chosen for defining clusters.

Thus, in considering the relationships between artifact types on the Crane site and their organization into depositional sets, the methodology outlined here will be used rather than those analytical techniques that classically have been used to define activity sets. The proposed methodology appears to be more consistent with the nature of organization of the archeological record. The specific operational procedures that were followed are described in Appendix 1.

To summarize this section, the use to which particular locations of Crane site were put during its prehistoric occupation will be defined along two dimensions. First, the *activities indicated by the functions of the individual artifact types present within individual collection units* will be tabulated. The function of

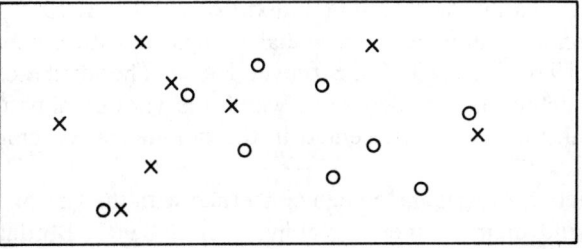

FIGURE 32. *Artifact type X is associated polythetically with respect to type O, and type O is associated polythetically with respect to type X.*

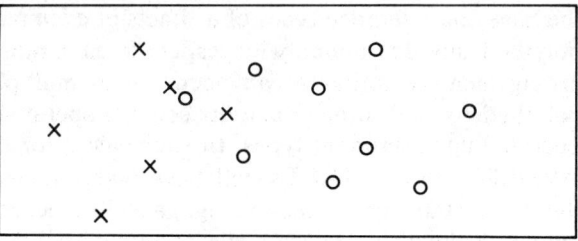

FIGURE 33. *Two artifact types which belong to different depositional sets but overlap spatially to a slight degree. Other artifact types included in the depositional sets are not shown.*

particular artifact types will be based primarily upon the morphological attributes characterizing the types. Where these are ambiguous, functions will be assigned on the basis of the function of other artifact types with which the ambiguous type tends to be associated most closely over the site as a whole. The degree of association between different artifact types will be measured with the algorithm that has just been described. The second dimension that will be used to define the way in which different locations of Crane site were utilized is the *means by which the artifacts present in single collection units were deposited:* whether the archeological deposits within a particular collection unit are a result of primary deposition from *in situ* activity or caching; secondary deposition; or the mixture of primary refuse from multiple, spatially overlapping episodes of activity. Both the function of the artifacts present in single collection units and the means by which they came to be deposited in those units must be known in order to determine the correspondences that may occur between particular activities and particular soil alterations.

The Approach Used in Classifying the Crane Assemblage into Functional Types

As a first step in defining use-areas within the Crane site, it was necessary to determine the functions of the various artifacts found on the site's surface. This is not a straightforward procedure. The artifact remains of past social groups carry multiple dimensions of information about those groups, including the tasks they performed (Binford and Binford 1966; Cook 1973), the status and importance of those activities within their social context (Wobst 1974, 1975), and the degree of intercommunication between social groups or within sub-segments of groups (Brose 1970; Leone 1968; Struever 1965). The attributes chosen to classify artifacts determine the degree to which the various dimensions of behavior encoded in them are represented in the typological scheme constructed.

Traditionally, lithic artifact typologies have been concerned with the size and outline shapes of artifacts and their manner of flaking (Balout 1967, Bordaz 1959; Bordes 1961; de Heinzelin 1962; Tixier 1963). Bordes (1961:8), for example, places emphasis on the shape of the flake scars resulting from retouching an edge. Balout (1967:716) stresses the method of fabrication, noting not only the shape of the flake scar but also other pertinent attributes such as the diffuseness of the bulb of percussion and ripple marks indicated within the flake scars. These attributes of artifacts do, to some extent, reflect the use to which the artifacts were put, but also reflect their social context and their relative importance within it. For example, tools exchanged between groups tend to be more elaborate and finely flaked than those used to accomplish "quick and dirty" tasks within groups. Men's tools tend to show more elaboration and less variation in morphology than women's tools, as it is men who are involved in the intergroup exchange of artifacts (Wobst 1974, 1975). In the same fashion, Bordes (1961:8) recognizes different cultural groups and/or traditions on the basis of the form of retouch used in making elaborate tools: *retouche en écaille* is thought diagnostic of Typical Mousterian, scalariform retouch is thought diagnostic of Quina Mousterian, and parallel retouch is thought indicative of the Solutrean period.

On the other hand, typologies can be constructed using attributes that reflect more directly the use of the artifact in accomplishing various tasks. These attributes include the nature of the raw material from which the artifact is made (its elasticity and strength); the angle of the functional edge of the artifact; the kind and location of wear along the edge; the shape of the edge; in the case of lithic artifacts, the presence or lack of retouch along the functional portion of the edge; the overall size or weight of the artifact; and the frequency and placement of breakage of the artifact category.

In this study of the artifacts from the surface of the Crane site, concern is with the kinds of tasks they represent. The attributes chosen to classify the artifacts from Crane, therefore, had to be functional ones with respect to both tools and debris. Initially, the Crane assemblage was sorted into broad classes, some of which had precise and narrow functional meaning, but others of which contained a mixture of functional classes and were homogeneous primarily along temporal dimensions. This initial sorting was done by Kenneth Farnsworth during the field season (1974) in an attempt to sketch out the spatial distributions of the different components at Crane and broadly different activity zones, to aid in the designing of an excavation sample. At a later time (1977) I reanalyzed the initial classes of materials and subdivided them into more homogeneous functional groups. This secondary sorting procedure was restricted to the chipped stone and ground stone tools. The debris and pottery from Crane site were too voluminous for me to resort into finer classes by myself. The following describes the general approach used by Farnsworth and me in classifying these materials.

Pottery. Pottery could not be sorted into true functional classes for several reasons. First, all of the pottery from the surface of Crane was broken into small sherds by repeated plowing. Ericson and De Atley (1976) have designed detailed procedures for estimating vessel shapes and volumes from sherd data, but the samples from Crane were too small to apply these procedures with any accuracy. Ericson and Stickel (1973:364–366) have shown how the weight of pottery sherds can be used to estimate the total ceramic capacity of a site, and the same methods could be applied to subareas within a site in order to differentiate their functions. Estimating total volume from total ceramic weight, however, requires that a unique regression between these variables be determined for the specific locale and time period under study. Local and temporal variability in the density of the ceramics and vessel thickness must be taken into consideration. The data for building the kind of regression model proposed by Ericson and Stickel were not available for Middle Woodland sites in the Macoupin Valley. Finally, the subsorting of pottery into true functional classes was prohibited by the quantity of sherds that would have had to have been handled, even if sampling procedures had been used.

Despite these problems, the classification scheme by which pottery from Crane was sorted does have some functional implications. Although sherds were categorized into traditional pottery types which primarily emphasize chronological variation, the distinctions between Havana and Hopewell wares, and Pike and Baehr wares also are functional ones.

Debris. Debris from the surface of Crane were classified so as to consider a number of attributes: (*a*) the nature of the raw material (e.g., burnt clay, bone, stone); (*b*) the means by which the raw material could be worked (e.g., chert, which can be chipped versus igneous rock, which must be pecked and polished); (*c*) the stage of manufacture, if any, implied by the debris (e.g., decortication material, fine chipping debris, hoe resharpening flakes); and (*d*) incidental modification indicating use (e.g., burning of bone, heat fracturing of igneous rock); (*e*) Social context also was considered through the identification of rare or exotic raw materials (e.g., galena, nonlocal cherts) and figurines.

Chipped and Ground Stone Tools. Neither the chipped nor ground stone tools of Middle Woodland assemblages in Illinois have previously been studied systematically and classified with respect to their functional attributes. Montet-White (1963, 1968) has proposed a typology for chipped stone tools for Middle Woodland assemblages along the central and lower Illinois River Valley, but the classification system considers largely the method of manufacture of the chipped artifacts and their morphology rather than their function. Therefore, I devised my own typology of chipped and ground stone tools to conform with the particular goals of this study.

Lithic tools were treated in one of two ways, according to their elaborateness. Elaborately finished artifacts such as polished stone celts, hoes, and point forms, which have standard morphological definitions in the literature on the Illinois Middle Woodland were placed into the categories defined in the literature. The members of these classes, however, were examined for the same set of functional attributes that less elaborate items were, in order to determine how they were used.

Less elaborately finished items, such as impromptu-made scrapers or knives, were examined for and classified using the following functional attributes: (*a*) shape of the functional edge, (*b*) the angle of the functional edge, (*c*) kind of edge wear, (*d*) the shape of the face(s) of the artifact bearing the wear marks, (*e*) the intensity of the wear, (*f*) the knapping technique by which the functional edge was shaped, (*g*) whether the artifact was shaped bifacially or unifacially, (*h*) size of the whole tool, and (*i*) the raw material from which the artifact was made. Not all of these attributes are applicable to the ground stone tools. The various states that can be taken by these variables for the chipped stone industry are shown in Table 28.

The particular manner in which wear patterns, among other attributes, were used to classify the lithic industry from the surface of Crane site and to identify artifact function requires some general remarks and qualifications. In examining the lithic items from Crane, hand lenses or magnifying scopes were not used. The scale of wear noted was very large compared to that used by Semenov (1964) and others (e.g., Ahler 1971, Hayden and Kamminga 1973; Keeley 1977; Tringham 1972; Wilmsen 1970). Such differences in scale of analysis radically influence the degree to which wear patterns can effectively be used to infer artifact function, particularly with respect to the *kind* of wear appearing on tools, as opposed to the *placement* of wear with respect to the functional edge. Keeley (1977) has noted that different kinds of wear, which are diagnostic of different kinds of materials that were worked with a tool, become differentiable only at magnifications of 100 to 400 diameters. Lower scales of analysis do not provide sufficient resolution to distinguish the particular morphology of individual wear marks that suggest the kind of material that was worked with a tool. This conclusion also was suggested by Tringham's (1972) program of study of artifact wear. Tringham used magnifications of less than 100 diameters. Despite her thorough work, she was able to make only a minimal differentiation of tools with respect to the materials they were used to work: between those tools used to work harder materials (bone, antler, wood) and those used to work softer materials (meat, hides, and nonwoody plant material). Finer distinctions were not possible. Consequently, the kinds of wear that were found on the Crane artifacts with no magnification whatsoever cannot be expected to yield detailed information about the kinds of materials worked by these artifacts and artifact function. Edge angle, placement of wear, plan-view shape of the functional edge, and retouch patterns are macroscopic attributes that are more appropriate for inferring the kind of material worked by a particular tool.

A second consequence of not having used magnifying scopes to observe wear on the artifacts from Crane site was an inability to identify consistently smoothing wear (polish) on artifacts. Smoothing wear could be identified in only a highly developed state; lesser degrees of smoothing wear could not be seen.

TABLE 28
States taken by variables used in classifying the less-elaborate chipped stone tools at Crane site

I. Shape of functional edge
    (a) straight
    (b) convex outward
    (c) concave outward

II. Angle of the functional edge
    (a) continuous measurement, with error levels of approximately ±2° for unifaces, ±4° for bifaces. It was possible to measure steep edge angles more accurately than fine edge angles.

III. Kind of wear
    (a) chipping (CH): small, hemispherical pocks usually less than 1/8 inch in diameter, differentiated from pressure retouch by their discontinuous distribution over the functional edge.

    (b) step fracture (ST)

    (c) thinning wear (TH): very shallow, wide chips ending in a step fracture, not to be confused with thinning flakes which are longer than wide and feather out at their ends.

    (d) smoothing (SM): polishing of the functional edge not systematically noted like the previous categories.

    (e) no wear visible

IV. Placement of wear

    Unifaces:

    (a) on the face which was shaped by chipping (D) (usually corresponding to the dorsal face of the flake, i.e., that opposite the bulb of percussion).

    (b) on the face which was left unchipped (V) (usually corresponding to the ventral face of the flake, i.e. that with the bulb of percussion).

    Bifaces of plano-convex cross-section:

    (a) on the convex face (D)

    (b) on the planar face (V)

TABLE 28 (cont.)

        Bifaces of wedge shape or lenticular shape:

           (a) on one face (1F)

           (b) on both faces (BF)

V.   Intensity of wear (not systematically noted)

           (a) slight

           (b) heavy

VI. Knapping technique, by which the functional edge was shaped (Crabtree, 1972; Ellis, 1940).

           (a) no retouch, only the coarse, direct-percussion, hard-hammer flaking used to shape the biface or uniface (DPS).

           (b) retouch with finer, direct percussion methods, perhaps soft hammer flaking. Flake scars are more shallow than in (a) and have a greater length/width ratio (DP).

           (c) retouch by pressure. Small elongated flake scars less than 1/8 inch long, or small deep chips similar to chipping wear but occurring continuously across the functional edge (P).

VII. Shaping

           (a) bifacial

           (b) unifacial

VIII. Size of the whole tool. This was used in a loose manner to distinguish heavy-duty tools having great weight and inertia from lighter-duty tools.

           (a) small

           (b) massive

IX. Raw Materials. Not systematically used in sorting through the chipped stone assemblage, but helped in the definition of a few classes (particularly heavy duty tools on large primary flakes).

Consequently, in suggesting functions for artifacts that bore only a few chips or no visible wear along their functional edges, it was not possible to distinguish between: (*a*) whether they had been used on softer materials less than a very long time and thus had not developed any visible smoothing wear (Semenov 1964), or (*b*) whether they had been used on harder materials, but only for a short time—not long enough to have developed chipping wear patterns diagnostic of this alternative use. When occurring with other functional attributes suggesting the use of the artifacts to work softer materials, however, lack of wear or only slight chipping wear was taken as evidence substantiating this use inferred from the other attributes of the artifacts.

In the process of sorting the unelaborate chipped stone tools from Crane, the functional significance of many of the created categories was tested informally. Sorting initially occurred on the basis of variables dealing with the shape of the functional edge, placement and kind of wear, whether shaping was done bifacially or unifacially, the knapping technique used to prepare the functional edge, size, and raw material. The items falling into each homogeneous group defined by these dimensions then were measured for their edge angles. In nearly all cases, edge angles clustered within an approximately 10–15 degree range, constituting an independent verification that artifacts had been sorted into functionally homogeneous classes. In several cases, however, bimodal frequency distributions of the edge angles were discovered. Items in these classes were subdivided into groups having homogeneous edge angles.

Once the chipped and ground stone tools had been divided into classes on the basis of the functional attributes they possessed, the spatial provenience of each item in every functional class was recorded. Single items having more than one type of functional edge (i.e., compound tools) generally were scored for the collection unit from which they came as a count for each of the several kinds of working edges they possessed. This procedure was followed for all chipped-stone items, based on the assumption that the several different kinds of edges occurring on single items were not related in a functional manner and used in single, tightly interrelated sets of activities. This assumption has been shown to be the case, elsewhere (Carr 1977:155–158).

One set of compound artifacts for which the chosen method of scoring was not followed consists of hammer–grinding stones (V60, V61, and V62). It is probable that the hammering and grinding surfaces occurring on single items were used in an integrated manner within one activity or a set of closely interrelated activities. This is documented in the following. Thus, the hammering and grinding surfaces of single items were not given separate counts for the collection units in which such items were found. A tool with both hammering and grinding surfaces on it was considered one functional item, scored once for the collection unit.

Description of the Crane Assemblage and Assignment of Artifact Functions

In this section, the functional attributes characterizing each of 121 classes of artifacts from the surface of the Crane site will be described. For classes specific to this study and not previously described in archeological literature, all functional attributes observed will be documented. Classes previously described in detail will be referenced and described for only those attributes used directly in inferring function.

One or more functions will be assigned to each artifact class, based on (*a*) ethnographic reports of tool use in North America and elsewhere, (*b*) experimental studies on artifact function, and (*c*) the morphology and wear patterns characterizing the artifacts. When these three kinds of information are not sufficient for deducing use and when several alternative uses seem possible, the functions of the classes of artifacts with which the ambiguous class tends to co-occur spatially will be considered in assessing artifact function. Classes of artifacts that tend to co-occur spatially and that comprise depositional sets, as defined by the model of the archeological record and the algorithm presented previously, are shown in Table 29. The average distances between artifact classes in each depositional set are presented in Table 30. Table 31 lists the final, deduced functions of each of the artifact classes. Of all 119 classes and subclasses listed in Table 31, 19% had attributes that made their functions ambiguous and required the use of patterns of spatial co-occurrence among classes to determine their function (V3, 4, 9, 19, 25, 26, 33, 36a, 36b, 51, 52, 53, 54, 55, 56, 59, 61, 66, 67, 68, 81, 82, 106).

Finally, when known, the chronological positioning of the artifact types will be given. Temporal placement is important to this study in that it will be used, along with other criteria, in judging the nature of the artifact associations and deposits within particular surface collection units: whether they are the result of *in situ* activity, caching, secondary refuse deposition, or the spatial overlap of multiple activities of the different occupants of the Crane site.

Pottery and Clay Items and Related Classes

V1. Havana Pottery. Formal descriptions of this pottery type are given by Griffin (1952) and Struever (1968b). In the lower Illinois Valley, vessels of this kind date to the period A.D. 1–150 (Griffin *et al.* 1970), and at Crane site they are attributable to the Bedford phase component. Havana vessels are utilitarian jar forms, having wide mouths, elongated bodies, and conoidal to more rounded bases. They exhibit this form with consistency over all sizes, and suggest a single generalized functional class (Braun 1977; McGregor 1959).

Havana vessels are known to have been used for cooking and other heating tasks, based on the smudging and caking of charcoal on their bottoms (Struever 1968b). They were placed directly upon coals rather than suspended over flames, for they have neither lugs nor everted rims for suspension. Both boiling and indirect-heat roasting (in which the food to be cooked is placed in the pot without water) could have been accomplished with Havana vessels. The later form of cooking is documented by Waugh (1916:139) for the historic Iroquois, who would roast insects and meat in dry pots. Boiling served many purposes for

TABLE 29
Artifact types composing depositional sets 1–30

Set 1	V26	Hematite
	V29	Galena
Set 2	V9	Hoes
	V49	Denticulates
Set 3	V42	Unretouched cache blades (bifaces)
	V62	Hammer-grinders with flat-face wear, edge wear, and end wear
	V99	Unretouched celts (bifaces)
Set 4	V77	Unifacially-retouched, straight-edged scrapers
	V78	Unifacially-retouched, straight-edged scrapers
	V89	Unifacially-retouched, concave-edged scrapers
Set 5	V3	Sherdlets
	V4	Clay blobs
	V17	Fire-cracked igneous rocks
	V23	Sandstone
	V24	Tempering material
Set 6	V2	Hopewell pottery
	V103	Bifacially-retouched endscrapers
	V108	Bifacially-retouched, round-edged knives
Set 7	V1	Havana pottery
	V25	Red ochre
	V30	Hornstone
	V38	Unifacially-retouched endscrapers
	V41	Bifacially-retouched, round-edged cache blades
	V83	Unifacially-retouched, straight-edged knives
	V84	Unifacially-retouched, straight-edged knives
Set 8	V28	Unburnt bone
	V43	Bifacially-retouched, round-edged cache blades
	V68	V-profile, slotted abraders
Set 9	V114	Small hammerstones, less than 2¼" in diameter
Set 10	V47	Open-angled notches
	V59	Gouges
Set 11	V10	Hoe chips
	V51	Spurs
	V75	Unifacially retouched, straight-edged scrapers
	V80	Unretouched, primary-flake unifaces
Set 12	V7	Snyders points, Snyders/Norton points

Description of the Crane Assemblage   221

TABLE 29 (cont.)

Set 12 (cont)	V33	"Flint Ridge" (?) chert
	V95	Chisels
Set 13	V20	Figurines
	V55	Perforators
Set 14	V21	Polished stone celts
	V73	Unifacially-retouched, straight-edged scrapers
	V100	Bifacially-retouched, rectangular celts
	V106	Bifacially-retouched, straight-edged scrapers
Set 15	V14	Decortication material
	V15	Chipping debris
	V46	Notches
	V60	Hammer-grinders with end wear only
	V67	Convex abraders
	V113	Hammerstones of all size classes, in general
Set 16	V12	Chert nodules
	V45	Notches
	V52	Spurs
	V115	Intermediate size hammerstones, 2½-3" in diameter

peripheral to set:
V74   Unifacially-retouched, straight-edged scrapers

Set 17	V39	Unifacially-retouched endscrapers
	V76	Unifacially-retouched, straight-edged scrapers
	V88	Unifacially-retouched, concave-edged scrapers
Set 18	V65	Nutting-grinding stones
	V70	Trough and open-angled abraders
Set 19	V18	Grinding stones without hammering wear
	V79	Unifacially-retouched straight-edged scrapers
	V86	Unifacially-retouched straight-edged knives
	V92	Choppers
Set 20	V11	Unretouched lamellar blades
	V44	Lamellar cores
	V54	Drills
	V71	Turtleback scrapers
	V102	Specialized planing tools
Set 21	V16	Igneous cobbles
	V40	Unifacially-retouched, round-edged cache blades
	V48	Notches
	V61	Hammer-grinders with edge wear and end wear
	V64	Nutting-hammering-grinding stones

TABLE 29 (cont.)

Set 22	V8	Gibson points
	V13	Flake cores
	V58	Proximal ends of lamellar blades
	V72	Turtleback scrapers
	V91	Choppers
	peripheral to set:	
	V69	Abraders with trough and open-angle profiles
Set 23	V19	Mauls
	V22	Grooved, polished axes
	V31	Missouri chert
	V50	Denticulates
	V98	Unretouched celts (bifaces)
Set 24	V36	Unifacially retouched endscrapers
	V105	Bifacially retouched, straight-edged scrapers
Set 25	V27	Burnt bone
	V97	Unretouched, round-edged celts (bifaces)
Set 26	V37	Unifacially retouched, endscrapers
	V82	Unifacially retouched, round-edged scrapers
Set 27	V85	Unifacially retouched, straight-edged knives
	V101	Unretouched, slightly round-edged celts (bifaces)
	V104	Unretouched, straight-edged celts (bifaces)
Set 28	V32	Kaolin chert
	V81	Unifacially retouched, round-edged scrapers
	V87	Unifacially retouched, round-edged knives
	V90	Bifacially retouched endscrapers
Set 29	V35	Unifacially retouched endscrapers
	V116	Large hammerstones, greater than 3" in diameter
	peripheral to set:	
	V6	Norton points
	V94	Wedges
Set 30	V5	Belknap points
	V34	Unifacially retouched endscrapers
	V53	Drills
	V56	Chert hammers
	V57	Saws
	V66	Flat Abraders

TABLE 30
Depositional sets and the statistics defining them

Variables	Average Absolute Nearest Neighbor Distance	Standard Deviation in Nearest Neighbor Distances	N_1	N_2
Set 1				
26,29	10.829	6.097	7	13
Set 2				
9,49	11.704	5.808	14	8
Set 3				
42,62	4.243	6.000	2	31
42,99	10.243	2.485	2	6
62,99	7.398	7.531	31	6
Set 4				
77,78	7.243	1.757	8	2
77,89	9.887	7.947	8	16
78,89	3.000	4.243	2	16
Set 5				
3,4	1.435	3.503	12	41
3,17	0.	0.	12	111
3,23	4.931	4.251	12	32
3,24	6.893	4.909	12	27
4,17	.585	1.802	41	111
4,23	5.090	4.883	41	32
4,24	8.468	6.025	41	27
17,23	1.953	2.973	111	32
17,24	6.028	9.171	111	27
23,24	10.421	7.022	32	27
Set 6				
2,103	13.355	7.644	12	8
2,108	17.491	8.491	12	4
103,108	19.690	4.801	8	4
Set 7				
1,25	6.34	10.608	70	17
1,30	8.307	6.807	70	65
1,38	6.975	5.579	70	4

TABLE 30 (cont.)

Variables	Average Absolute Nearest Neighbor Distance	Standard Deviation in Nearest Neighbor Distances	N_1	N_2
Set 7 (cont)				
1,41	0.	0.	70	2
1,83	12.885	14.277	70	17
1,84	7.030	9.786	70	20
25,30	8.565	7.404	17	65
25,38	9.149	14.574	17	4
25,41	10.243	2.485	17	2
25,83	15.276	5.153	17	17
25,84	13.207	11.388	17	20
30,38	3.621	4.303	65	4
30,41	6.000	0.000	65	2
30,83	13.434	12.884	65	17
30,84	8.802	12.613	65	20
38,41	13.817	11.054	4	2
38,83	13.884	5.567	4	17
38,84	11.365	5.638	4	20
41,83	8.485	12.000	2	17
41,84	6.708	9.487	2	20
83,84	12.749	9.099	17	20
Set 8				
28,43	11.580	11.374	66	5
28,68	2.828	4.899	66	3
43,68	10.472	3.937	5	3
Set 9				
114	---	---	--	--
Set 10				
47,59	0.	undefined	7	1
Set 11				
10,51	17.29	8.204	28	5
10,75	12.178	7.030	28	17
10,80	0.0	0.0	28	2
51,75	16.097	13.420	5	17
51,80	12.708	1.002	5	2
75,80	10.243	2.485	17	2

TABLE 30 (cont.)

Variables	Average Absolute Nearest Neighbor Distance	Standard Deviation in Nearest Neighbor Distances	N_1	N_2
Set 12				
7,33	13.612	8.722	37	12
7,95	4.472	7.746	37	3
33,95	11.301	2.539	12	3
Set 13				
20,55	8.485	12.000	6	2
Set 14				
21,73	19.489	10.435	8	7
21,100	17.507	5.504	8	25
21,106	6.000	0.0	8	2
73,100	10.530	8.639	7	25
73,106	13.729	7.416	7	2
100,106	6.000	0.0	25	2
Set 15				
14,15	10.478	5.447	41	41
14,46	9.847	5.583	41	23
14,60	10.955	6.369	41	22
14,67	6.000	8.485	41	2
14,113	12.069	7.194	41	34
15,46	9.900	6.390	41	23
15,60	12.157	6.950	41	22
15,67	12.487	9.174	41	2
15,113	7.731	4.642	41	34
46,60	14.742	7.544	23	22
46,67	11.485	7.757	23	2
46,113	16.358	8.819	23	34
60,67	10.951	3.487	22	2
60,113	11.219	6.687	22	34
67,113	15.059	9.297	2	34
Set 16				
12,45	12.053	7.496	17	11
12,52	20.215	14.010	17	13
12,74	22.427	8.575	17	3
12,115	13.470	7.637	17	7
45,52	19.229	10.744	11	13
45,74	21.032	19.961	11	3

TABLE 30 (cont.)

Variables	Average Absolute Nearest Neighbor Distance	Standard Deviation in Nearest Neighbor Distances	N_1	N_2
Set 16 (cont)				
45,115	15.787	5.765	11	7
52,74	19.073	18.339	13	3
52,115	12.422	5.999	13	7
74,115	23.799	30.829	17	7
Set 17				
39,76	9.487	13.416	2	13
39,88	3.000	4.243	2	8
76,88	12.070	12.753	13	8
Set 18				
65,70	8.576	5.190	13	6
Set 19				
18,79	11.301	2.539	31	3
18,86	9.256	7.456	31	6
18,92	12.166	6.966	31	16
79,86	26.537	31.267	3	6
79,92	7.657	1.435	3	16
86,92	14.207	11.806	6	16
Set 20				
11,44	7.207	8.925	53	11
11,54	9.641	8.400	53	11
11,71	11.213	8.389	53	20
11,102	0.	0.	53	2
44,54	13.239	11.765	11	11
44,71	12.425	8.233	11	20
44,102	6.000	8.485	11	2
54,71	7.161	5.137	11	20
54,102	10.243	2.485	11	2
71,102	7.243	1.757	20	2
Set 21				
16,40	10.884	9.059	27	4
16,48	3.000	4.243	27	2
16,61	11.482	10.879	27	28
16,64	7.169	8.959	27	11

TABLE 30 (cont.)

Variables	Average Absolute Nearest Neighbor Distance	Standard Deviation in Nearest Neighbor Distances	N_1	N_2
Set 21 (cont)				
40,48	15.059	9.297	4	2
40,61	4.243	8.485	4	28
40,64	2.121	4.243	4	11
48,61	13.243	6.728	2	28
48,64	15.059	9.297	2	11
61,64	4.718	6.349	28	11
Set 22				
8,13	0.	0.	2	53
8,58	9.708	5.244	2	19
8,69	26.972	7.550	2	2
8,72	9.708	5.244	2	16
8,91	19.155	18.605	2	5
13,58	7.463	6.099	53	19
13,69	9.000	12.728	53	2
13,72	8.545	3.965	53	16
13,91	7.200	5.020	53	5
58,69	17.659	12.974	19	2
58,72	14.805	11.756	19	16
58,91	9.624	10.011	19	5
69,72	19.416	10.488	2	16
69,91	19.416	10.488	2	5
72,91	11.864	8.722	16	5
Set 23				
19,22	9.000	12.728	3	2
19,31	2.000	3.464	3	45
19,50	8.472	4.282	3	14
19,98	10.944	4.282	3	10
22,31	9.708	5.244	2	45
22,50	7.243	1.757	2	14
22,98	9.708	5.244	2	10
31,50	13.210	11.900	45	14
31,98	7.243	3.117	45	10
50,98	15.411	11.318	45	10
Set 24				
36,105	5.475	6.635	14	4

TABLE 30 (cont.)

Variables	Average Absolute Nearest Neighbor Distance	Standard Deviation in Nearest Neighbor Distances	N_1	N_2
Set 25				
27,97	6.708	9.487	27	2
Set 26				
37,82	11.285	6.449	6	15
Set 27				
85,101	13.766	13.005	11	8
85,104	12.728	6.000	11	2
101,104	4.243	6.000	8	2
Set 28				
32,81	9.458	6.645	20	39
32,87	9.606	8.323	20	6
32,90	4.472	7.746	20	3
81,87	11.301	4.422	39	6
81,90	4.000	3.464	39	3
87,90	8.944	7.746	6	3
Set 29				
6,35	11.211	10.838	9	3
6,94	16.921	14.138	9	10
6,116	18.000	undefined	9	1
35,94	14.000	9.165	3	10
35,116	0.000	undefined	3	1
94,116	24.000	undefined	10	1
Set 30				
5,34	16.909	6.190	41	10
5,53	14.801	7.001	41	39
5,56	15.278	10.883	41	18
5,57	11.986	4.944	41	4
5,66	18.992	13.265	41	16
34,53	12.866	8.251	10	39
34,56	19.256	8.485	10	18
34,57	20.052	8.558	10	4
34,66	14.404	9.026	10	16
56,57	8.208	6.494	18	4
56,66	19.582	12.709	18	16
57,66	15.184	8.038	10	16

TABLE 31
Possible functions of artifacts of various classes found on the surface of Crane site

Variable Number	Kind of Artifact	Possible Function or Activity Implied
V1	Havana pottery	boil meat and vegetables, seeds
		boil nuts for oil
		simmer hides with brains or plain, warm water
		soak hides in salt water, in soapy water made with roots, in soapy water made with dissolved deer brains, in water with bark, in water with organic dyes, in water with ash (lye), or in urine
		boil fibrous plants to break up parenchymatic tissue
		boil bone to soften it, prior to working it
		roast insects, meat, in dry pot
		storage
		boil textiles with berries, vegetable dyes, inorganic pigments
		boil crushed bone for marrow, tallow, and grease
V2	Hopewell pottery	not for cooking
		serving bowls?
V3	sherdlets	areas of heavy trampling
		see V1 Havana pottery
V4	clay blobs	hearth liner (if no tempering material present in collection unit)
		manufacture of pottery (if tempering material present in collection unit)
V5	Belknap points	function determined by attributes specific to individual specimen (see Table 42)
V6	Norton points	function determined by attributes specific to individual specimen (see Table 42)

TABLE 31 (cont.)

Variable Number	Kind of Artifact	Possible Function or Activity Implied
V7	Snyders, Snyders/Norton points	function determined by attributes specific to individual specimen (see Table 42)
V8	Gibson points	function determined by attributes specific to individual specimen (see Table 42)
V9	hoes	cultivation of ground
		used secondarily for thinning and graining hides, if other hide-working tools present
V10	hoe chips	resharpening hoes
V11	unretouched lamellar blades	knives used to whittle wood, bone
		cut soft materials: meat, hides, vegetable fibers
V12	chert nodules	manufacture of stone artifacts: decortication
V13	flake cores	production of flakes used as knives to cut meat, hides, vegetable fibers
		whittle wood, bone
V14	decortication flakes	manufacture of stone artifacts: decortication
V15	chipping debris	manufacture of stone artifacts: secondary flaking
V16	igneous cobbles	used within earth ovens or as the walls of hearths
		shaping and smoothing insides of pots -- smaller specimens, only
V17	fire-cracked igneous rock	used within earth ovens or as the walls of hearths
		hearth dumpings
V18	grindingstones without hammering wear	hulling as well as grinding small, delicate seeds such as those of Chenopods
		grind red ochre
V19	mauls	removal of wood slabs and bark from trees
		possibly pounding bark into cloth

TABLE 31 (cont.)

Variable Number	Kind of Artifact	Possible Function or Activity Implied
V20	figurines	ceremonial uses
V21	polished celts	shaping hard wood
		felling or girdling trees?
		removing layers of wood and bark from trees?
V22	grooved, polished axes	felling or girdling trees
		removing layers of wood and bark from trees
V23	sandstone	used as the walls of hearths
		hearth cleanings
V24	tempering material	manufacture of pottery
V25	red ochre	combined with oil for paint
		ingredient in bread made from bitter acorns
		color hides
V26	hematite	ceremonial uses
		ground into red ochre
V27	burnt bone	roasting meat, baking meat
		fuel for fire
		boil bone to obtain tallow, grease
V28	unburnt bone	butchering of animals
		boiling meat
		manufacture of bone tools used in sewing, leather work, weaving, basket-making
V29	galena	ceremonial uses
V30	hornstone	manufacture of stone artifacts
V31	"Missouri" chert	manufacture of stone artifacts, particularly lamellar blades
V32	kaolin chert	manufacture of stone artifacts, often hoes
V33	"Flint Ridge" chert	manufacture of stone artifacts
V34	unifacial endscrapers	function determined by attributes specific to individual specimen (See Table 42)

TABLE 31 (cont.)

Variable Number	Kind of Artifact	Possible Function or Activity Implied
V35	unifacial endscrapers	working bone
V36	unifacial endscrapers	working hides
V37	unifacial endscrapers	working soft woods
V38	unifacial endscrapers	working soft or hard woods
V39	unifacial endscrapers	deflesh hides
V40	cache blades	deflesh hides
V41	cache blades	dismemberment
V42	cache blades	deflesh hides
V43	cache blades	grain hides
V44	lamellar cores	production of lamellar blades used to whittle wood, bone; to cut soft materials such as meat, hides, vegetable fibers
V45	notches	smoothing hard wooden shafts of ¼" in diameter
V46	notches	smoothing hard wooden shafts of ¼"-1/16" in diameter
V47	notches	working wood: scraping angular edges (e.g., edges of bowls, atlatls, shuttles)
V48	notches	initial rough shaping of wooden shafts shaping shafts of bone
V49	denticulates	extract plant fibers deflesh and/or grain hides
V50	denticulates	extract plant fibers deflesh and/or grain hides
V51	spurs	working hard woods, bone
V52	spurs	working hard woods, bone grooving hard wood shafts
V53	drills	drilling soft woods, hard woods, bone
V54	drills	drilling soft woods
V55	perforator	?
V56	chert hammers/crushers	crush bones to extract marrow and prepare it for boiling for tallow and grease

TABLE 31 (cont.)

Variable Number	Kind of Artifact	Possible Function or Activity Implied
V56 (cont.)	chert hammers/crushers (cont.)	pound bark into cloth and basketry
		pulverize dried fruits, bulbs, roots, rhizomes, meat
		pound fibrous plants prior to extraction of fibers with denticulates
V57	saws	working wood and bone
V58	lamellar blades, proximal ends	manufacture of blade edges used to groove wood, bone
V59	gouge	working wood
V60	hammer-grinders	hammerstones used in manufacture of stone tools without preparation of platforms
		grind acorns, hazelnuts
		pound and hull seeds in wooden mortar
		grind hulled seeds along with V18 grinders
V61	hammer-grinders	pound nuts and sunflower seeds in shell
		grind acorns, hazelnuts
V62	hammer-grinders	anvil on which nuts, in general, and sunflower seeds pounded
		anvil on which bark pounded
		grind acorn or hazelnut meats
		possibly anvils on which roots, rhizomes, bulbs, fruits, or bone powdered
V64	nutting/grinding/hammerstones	grind seeds
		pound nuts
V65	nutting/grinding hammerstones	grind seeds
		pound nuts?
V66	abraders, flat	sanding wood
V67	abraders, convex	sanding wood
V68	abraders, slotted	sharpen pointed bone and possibly wooden implements
V69	abraders, slotted	smooth shafts

TABLE 31 (cont.)

Variable Number	Kind of Artifact	Possible Function or Activity Implied
V70	abraders, trough, open angle	sanding wood or bone polish stone axes
V71	turtleback scrapers	scraping soft woods, hard woods
V72	turtleback scrapers	scraping hard woods, bone
V73	unifacial, straight-edged scrapers	scraping hard woods
V74	unifacial, straight-edged scrapers	scraping hard woods, bone
V75	unifacial, straight-edged scrapers	scraping soft woods, hard woods
V76	unifacial, straight-edged scrapers	scraping hard woods, bone
V77	unifacial, straight-edged scrapers	scraping soft woods, hard woods
V78	unifacial, straight-edged scrapers	scraping hard woods, bone
V79a	unifacial, straight-edged scrapers	chopping and roughing out forms in hard woods
V79b	unifacial, straight-edged scrapers	scraping and roughing out forms in hard woods
V80	unifacial, straight-edged scrapers	scraping bone, coarser stages
V81	unifacial, round-edged scrapers	hide graining
V82	unifacial, round-edged scrapers	hide graining or soft wood working
V83	unifacial, straight-edged knives	dehair hides
V84	unifacial, straight-edged knives	whittle wood, bone, horn
V85	unifacial, straight-edged knives	dehair hides
V86	unifacial, straight-edged knives	shredding plant fibers and animal sinew
V87a	unifacial, round-edged knives	cut meat, skin, sinew

TABLE 31 (cont.)

Variable Number	Kind of Artifact	Possible Function or Activity Implied
V87b	unifacial, round-edged knives	deflesh hides
V88	unifacial, concave-edged knives	dehair hides
V89a	unifacial, concave-edged scraper	scrape hard woods, bone
V89b	unifacial, concave-edged scraper	grain hides
V89c	unifacial, concave-edged scraper	grain hides
V90	bifacial endscrapers	grain hides
V91	choppers	dismemberment
V92	choppers	rough out forms in soft and hard wood
V94	wedges	splitting wood or bone
V95	chisels	work soft woods
V97	bifacial, round-edged scrapers	grain hides
V98	celts	procure bark, soft wood slabs? scraping soft wood
V99	celts	procure bark scraping hard woods or bone
V100	rectangular celts	scraping bone
V101	celts	thin and grain hides
V102	specialized planing tools	wood working
V103	bifacial endscrapers	grain hides
V104	bifacial, straight-edged scrapers	scraping hard wood?
V105	bifacial, straight-edged scrapers	pound bark and fibrous plants?
V106	bifacial, straight-edged scrapers	working bone
V107a	bifacial, round-edged scrapers	grain hides
V107b	bifacial straight-edged scrapers	scraping bone or pounding bark or fibrous plants

TABLE 31 (cont.)

Variable	Kind of Artifact	Possible Function or Activity Implied
V108	bifacial round-edged knives	grain hides
V113	hammerstones of all sizes	flint knapping in general
V114	hammerstones: small	flint knapping: secondary flaking
		crushing bone to be boiled for grease, tallow?
V115	hammerstones: medium	flint knapping: decortication, secondary flaking
V116	hammerstones: large	flint knapping: decortication, secondary flaking
V117	Pike pottery	see V1, Havana pottery
V118	Baehr pottery	see V2, Hopewell pottery
V119	White Hall-Weaver pottery	see V1, Havana pottery
		serving bowls
		storage vessels?
V120	Jersey Bluff pottery	see V1, Havana pottery
		serving bowls
		storage of solids
		storage of liquids?
		salt evaporation?
V121	Steuben points	function determined by attributes specific to individual specimen. (See Table 42)

the historic North American Indians, both in food preparation and a variety of other maintenance tasks (Table 32). Havana vessels, as boiling utensils, may have been used for some or all of these purposes. Their use at Crane to boil meat is suggested by the co-occurrence of Havana sherds with V41 cache blades (depositional set 7) that were used in dismembering animals. Their use in extracting oil from nuts to make paint may be indicated by the association of Havana sherds with the pigment, red ochre (depositional set 7).

In addition to heating tasks, Havana vessels also may have served as storage vessels. Reconstructable Havana pots as large as 60 cm in height are known in the lower Illinois Valley (Farnsworth, personal communication). These may have been too large and cumbersome to have been used as cooking vessels placed on coals, but could have been used as the liners of storage pits. Large Havana vessels have been found placed within pits bearing no evidence of

TABLE 32
Tasks accomplished by historic Indians by boiling (or soaking) materials in containers functionally analogous to Havana vessels

Task	Reference
Boiling was the most common means of cooking meat in the prairie and eastern United States	Driver (1961:89)
Cook seed and vegetables	Waugh (1916)
Extract oil from nuts such as hickory, walnut, and butternut, which are not easily shelled, to make paints, and for bodily anointment	Battle (1922), Waugh (1916:123)
Render fat for tallow and grease used in making pemmican and in cold packing procedures	Leechman (1951), Mason (1895:28), Peale (1981), Wheat (1972:113)
Break down the parenchymatic tissue of fibrous plants prior to scraping them in order to extract fibers	Osborne (1965)
Color textile and cordage by boiling with vegetable dyes and inorganic pigments	Gilmore (1931:96,99) Swanton (1946:245)
Soften bone prior to working it	Semenov (1964:159)
Loosen the hairs on hides from their follicles, as one step in dressing them, by simmering them in warm water, warm water with brains	Mason (1889:564)
Soak hides in salt water; soapy water made with roots, or dissolved brains; water with ash, urine, bark, or organic dyes, during various stages of dressing them	Mason (1889)

burning (Braun, personal communication). Solids (dried foods), and possibly liquids (water, sap), could have been contained by them, although the wide mouths of Havana vessels would make them less appropriate for the latter.

V2. Hopewell Pottery. Formal descriptions of this ware are given by Griffin (1952) and Struever (1968b). Hopewell pottery dates to the period A.D.1–150, and at Crane it can be attributed to the Bedford phase component. Hopewell vessels differ from Havana vessels in several respects that suggest they were used in a different manner than Havana vessels. Hopewell ware is constructed with much more care than Havana ware. It is thinner walled and more finely decorated. Vessels lack caking of charcoal on their bottoms.

Hopewell vessels occur in two forms—bowls and jars. The jars are small compared to Havana jars. In Middle Woodland villages, Hopewell sherds typically represent only small percentages of combined Havana–Hopewell total sherd counts (see Farnsworth 1973 for a summary tabulation). Bowl forms are particularly rare (McGregor 1959:25). In sum, the size, forms, execution, and frequency of Hopewell vessels suggest their use in a manner different from Havana vessels—possibly as serving vessels; their use for culinary purposes may be ruled out.

Hopewell vessels served as status markers in burial contexts. It once was thought that Hopewell ware was restricted to this context and function, but vessels of this kind have since been found in midden deposits of a number of Middle Woodland villages (McGregor 1952, 1958: Struever 1968b; Wray and MacNeish 1961). They were not used solely as ceremonial ware.

V117, V118. Pike, Baehr Pottery. Formal descriptions of these wares have been given by Griffin (1952) and Struever (1965:219, 1968b). These vessel forms date to the Steuben phase of the Middle Woodland, and span the period from ca. A.D. 150–400 (Griffin *et al.* 1970). The distinction between Pike and Baehr vessels is analogous to the distinction between Havana and Hopewell vessels with respect to vessel size, form, execution, relative frequency, and function. Havana ware evolved into Pike ware over time, and Hopewell ware evolve into Baehr ware over time (Griffin *et al.* 1970). The distinction between Havana and Pike vessels is made largely on the basis of decorative designs, the earlier vessel types having more finely executed and often more complex designs. The earlier wares also are thicker than the later wares (Braun 1977:138–173). The transition from Havana to Pike vessels furthermore includes a change in tempering material from grit (quartz and feldspar) to crushed limestone (Struever 1968b).

As with Hopewell vessels, Baehr vessels occur in burial contexts in addition to fulfilling more utilitarian purposes (McGregor 1958; Struever 1968b; Wray and MacNeish 1961).

V118. White Hall–Weaver Pottery. Vessels of this form and design have been defined by Struever (1964:93–94, 1968b:169–172), Griffin (1952:121–122) and Wray and MacNeish (1961:55–59). They represent a style cline along the Illinois River Valley that includes Weaver ware in the central Valley and White Hall ware in the lower Valley. White Hall pottery dates from A.D. 350–400 to A.D. 600 or 700 (Griffin *et al.* 1970:10; Struever 1968b:172), and is considered by Struever to mark the beginning of the Late Woodland period.

Like Havana and Pike vessels, Weaver vessels have wide mouths, elongated globular bodies, and conoidal bases (Wray and MacNeish 1961:57). Some have vertical rather than incurving rims, but none have everted rims or lugs for suspension. Weaver vessels exhibit this form with consistency over all size classes (Wray and MacNeish 1961:57). This might suggest that White Hall–

Weaver jars form a single, generalized functional class used in manners similar to Havana and Pike jars, but further considerations would suggest otherwise. Unlike circumstances in the Middle Woodland period, in which two different kinds of jars were used contemporaneously (Havana and Hopewell, or Pike and Baehr), in the Early Late Woodland period, only one type of jar was utilized—that represented by White Hall–Weaver pottery. It might be suggested that this single type of jar served in capacities including the functions of *both* Havana–Pike and Hopewell–Baehr jars. Several data support this inference. First, the size range of White Hall–Weaver jars—10–30 cm in diameter (Wray and MacNeish 1961:57)—encompasses the size ranges of both Havana–Pike and Hopewell–Baehr jars. Moreover, the decorative elements and spatial organization of the design units that occur on White Hall–Weaver pottery reflect some carry over of *both* Pike and Baehr modes of decoration (Struever 1968b). Third, in the subsequent Jersey Bluff period, jars are of a singular form resembling Havana, Pike, and White Hall–Weaver jars, but fall into two distinct size classes that have size ranges similar to those of Hopewell–Baehr jars, on the one hand, and Havana–Pike jars, on the other. Assuming continuity of vessel function from the Middle Woodland through the Late Woodland periods, we might expect the same bimodal distribution of jar sizes to characterize White Hall–Weaver jars, indicating their use for the same set of tasks in which both Havana–Pike and Hopewell–Baehr jars were used. Unfortunately, the shape of the frequency distribution of White Hall–Weaver jar sizes is not documented in the literature to test this expectation.

If White Hall–Weaver vessels were used for some of the same purposes as Havana–Pike vessels and Hopewell–Baehr vessels, at the same time, they may not have been used for *all* the tasks in which Havana–Pike and Hopewell–Baehr vessels were involved. Very large White Hall–Weaver jars that would have been more appropriate for storage than heating–soaking tasks have not been found. The largest White Hall–Weaver jars could have been used for storage purposes, but this cannot be verified on the basis of vessel size, or the contexts in which they have been found. Second, White Hall–Weaver vessels are all of jar forms, and do not include the rarer bowl forms that are noted among Hopewell and Baehr vessels. White Hall–Weaver vessels may not reflect as wide a range of serving functions or other possible functions encompassed by Hopewell and Baehr bowls. These "lost" functions could have been provided by bowls made of perishable materials, such as wood and bark, as were many of the containers among historic Indians in the eastern United States (Waugh 1916).

Finally, it should be mentioned that some White Hall–Weaver jars exhibit caking and smudging of charcoal on their surfaces, which may be used along with the similarity of their form to Havana and Pike vessels to infer their use in heating tasks.

V119. Jersey Bluff Pottery. The term Jersey Bluff originally was used by Titterington (1935, 1942) and Shalkop (1949) in describing pottery styles in

Jersey County, Illinois spanning the broad Late Woodland period from A.D. 300–1100, but largely attributable to the latter portion of this period. Here, however, the term is used in the more restrictive manner occurring in recent literature (Farnsworth 1973; Houart 1971) for pottery styles dating between A.D. 600 and 1200 in the lower Illinois Valley. It includes the pottery styles similar to those designated "Early Bluff" (A.D. 600–900) and "Late Bluff" (A.D. 900–1100) as described by Munson (1971) and Harn (1971) for the American Bottoms area around East St. Louis, Illinois.

Formal descriptions and illustrations of Jersey Bluff and similar wares have been given by Titterington (1935, 1942) for the lower Illinois Valley, by Harn (1971), Munson (1971), O'Brien (1972:47–50), and Vogel (1964:Fig. 18–24). for the American Bottoms region, by Maxwell (1951:213, 231) for the Dillinger Focus in the Carbondale area, and by Munson and Anderson (1973:37–39) for the central Illinois Valley. During Early Bluff times, wares in the broader southern Illinois area were largely of a single, generalized jar form having a wide mouth, elongated body, and conoidal to round base, similar in form to the utilitarian jars described for earlier periods. Some variability occurred in the rim shape of these vessels (straight, incurving, or slightly everted), but all had rims that did not allow them to be suspended, as was the case for earlier jar forms. Early Bluff jars occur in sizes ranging from ca. 8 to 30+ cm in their orifice diameters, encompassing the range of jar sizes found among both Havana–Pike and Hopewell–Baehr jars. Differentiation of Early Bluff jars along functional dimensions similar to those distinguishing Havana–Pike and Hopewell–Baehr jars in the utilitarian sphere is suggested by the distribution of Early Bluff jar sizes into two modes, one ranging from 8 to ca. 20 cm, the other from ca. 20 to 30+ cm (O'Brien 1972). The Early Bluff jars of larger sizes could have functioned in manners analogous to the ways in which Havana–Pike jars were used—for heating and soaking tasks, and possibly for storage. Those of smaller sizes, which are of a volume similar to Hopewell and Baehr jars, might have been used as serving ware.

During Late Bluff times, and particularly at the end of this period, ceramic vessels were manufactured in a much larger variety of forms used in a wider range of tasks. Maxwell (1951) notes four basic shapes of vessels for the Dillinger focus around Carbondale, Illinois: salt pans, bowls, globular jars, and vases. Late Bluff jars also are differentiated into those that have lugs and could have been suspended and those that lack lugs. This may indicate a differentiation of jars used for heating tasks from those used for storage.[5]

Additional evidence for the functional differentiation of Late Bluff jar forms is

---

5. In the Titterington Collection of Late Bluff pottery from Jersey and Green Counties, Illinois, which are housed in the Museum of Anthropology, University of Michigan, many lugged jars bear charcoal and smudging on their exteriors but none of the unlugged jars show such signs of their having been used in heating tasks. This fact is *consistent* with the hypothesis that lugged jars were used for heating tasks whereas jars lacking lugs were used for other tasks, perhaps storage. It does not *verify* the hypothesis, however, for the sample of jars is small (<20) and perhaps biased.

their division into distinct size classes. O'Brien (1972) has noted that Late Bluff jars in the Cahokia region fall into two size classes, having modal orifice diameters of 16 cm and 34 cm. During the earlier Late Bluff and prior to the elaboration of Late Bluff ceramics into multiple, functionally specific forms, this distinction in jar sizes could have reflected differences in the use of jars paralleling those reflected by Havana–Pike and Hopewell–Baehr wares. By the end of the Late Bluff period, however, specialized bowl and plate forms had evolved and probably served in the utilitarian capacities that are reflected by Hopewell and Baehr wares.

The Jersey Bluff pottery from the Crane site is all later material, probably within the A.D. 1000–1200 range. It thus may be taken to indicate a wide variety of functions, including not only those that Havana–Pike and Hopewell–Baehr vessel served—heating and soaking tasks, serving, and storage of solids—but also more specialized tasks including the storage of water or other liquids, and possibly salt formation.

V3. Sherdlets. These are pieces of broken pottery which are smaller than a dime, without decoration, and unidentifiable to type. Most are grit tempered and probably come from Havana vessels, but they also could come from White Hall–Weaver and Jersey Bluff vessels used by the Late Woodland occupants of Crane. The overwhelming number of identifiable Havana sherds compared to identifiable Late Woodland sherds (Table 2) found at Crane suggests their likely origin in Havana vessels.

The activities that might be reflected by sherdlets include all those described for Havana pottery. The spatial association of high densities of sherdlets with hearth dumping material (e.g., clay hearth liner, sandstone, fire-cracked igneous rock; V4, V23, V17, depositional set 5), however, suggest that the vessels from which the sherdlets were derived more likely were involved in tasks requiring heat. Soaking and storage tasks are less probable. This inference is supported by the fact that high densities of small, broken sherds may indicate locations in which a large amount of trampling occurred (Brose 1968:286; McPherron 1967), such as firesides as foci of activity. High densities of small, broken sherds also, however, could reflect refuse dumps containing the sweepings from such areas of heavy foot traffic.

V4. Clay Blobs. These are amorphous pieces of clay, usually less than one inch in diameter, without tempering. They could be pieces of hearth liner. Two clay-lined hearths are known at the Crane site. They might also represent clay used for pottery manufactured prior to the mixing in of tempering material. The spatial association of clay blobs with other debris classes (depositional set 5) does not help to resolve this question. The co-occurrence of clay blobs with high densities of hearth dumping materials would suggest they are pieces of hearth liner, whereas the co-occurrence of clay blobs with tempering material would suggest they were to be used in the manufacture of pottery.

V24. Tempering Material. These are very small pieces of pink plagioclase feldspar. This raw material was used to temper Havana ware but not Hopewell ware or vessels of either the Steuben Phase of the Middle Woodland or the following Late Woodland in the lower Illinois Valley (Struever 1968b). It may be used as an indicator of the manufacture of Havana vessels or the heavy trampling and disintegration of Havana sherds.

V20. Figurines. This category contained the head of only one crudely made ceramic figurine and several ceramic cylinders that could be appendages of figurines. Middle Woodland figurines elsewhere in Illinois are known from both burial contexts and village sites. In Mound 8 of the Knight mound group (McKern et al. 1945: Deuel 1952), five realistically modeled, painted clay figurines were found associated with a male in its central chamber. They appear to depict real people of varying characteristics. The single head from Crane is much less elaborate, having fingernail impression slots for eyes, a poorly applied clay lump nose, no headdress, no mouth, and no ears. In these characteristics, it resembles the figurine found at Whitnah Village, Fulton County, Illinois (Cole and Deuel 1937: Plate 34). A figurine of such simple workmanship also was found with an infant buried in Knight Mound 8. The cylinders of baked clay at Crane might be considered pottery coils rather than appendages of figurines. Havana and Hopewell pottery vessels were made by the coil and paddle technique (Griffin 1952a). However, the lack of tempering of all but one of these cylinders does not support this interpretation.

Bone Items

V27. Burnt Bone These items may be taken as indicators of roasting or baking meat as opposed to boiling it. The burnt nature of the bone might also indicate its use as a fuel, however, or the accidental burning of midden material below a hearth.

V28. Unburnt Bone. These items may be used as indicators of the boiling of meat, butchering of animals prior to cooking, or the working of bone into tools. Evidence for the latter activity can be found in the spatial co-occurrence of V-profile slotted abraders with high densities of unburnt bone (depositional set 8).

Lithic Raw Materials and Debris Classes

V12. Chert Nodules. These are unmodified, natural chert nodules, probably from the Burlington limestone formation within the lower Illinois Valley (Meyers 1970). They do not occur naturally on the Crane site, itself, and were brought to the site most likely as raw materials from which stone tools were to be knapped. The close spatial proximity of chert nodules to V115 hammerstones, on the average (depositional set 16), supports this contention.

V14. Decortication Material. Primary flakes of chert bearing some of the cortex of the nodules from which they were struck are included in this class. Together with chert nodules, these items indicate the initial stages of flint knapping, including the production of preforms.

V15. Chipping Debris. This category includes debris from only secondary stages of flint knapping. It does not contain large primary flakes or flakes bearing cortex. Nor does it contain any flakes showing use-wear or modification that might indicate activities other than knapping (e.g., use of flakes as knives to whittle wood). Most of the debris is probably local Burlington chert.

V16,17. Igneous Cobbles, Fire-Cracked Igneous Rock. Items from both of these classes could have been used in stone boiling. However, among the historic Indian tribes of the eastern United States and prairie states, where pottery was available, almost all boiling was done using direct heat (Driver 1961:89) rather than indirect heat. The use of these items in stone boiling is thus less likely. Alternately, the items in these classes might have been used in the construction of earth ovens or the retaining walls of hearths. Earth ovens occurred historically over most of the United States (Driver 1961:89) and are known in Late Woodland contexts in southern Illinois at the site of Hatchery West (Binford *et al.* 1970:7,92). At Hatchery West, both cobbles and amorphous crystalline rock were used in constructing earth ovens. The spatial association of fire-cracked igneous rocks with hearth dumping materials (depositional set 5) at Crane suggests this interpretation of the function of these classes, as well.

The smoothest and smaller igneous cobbles at the Crane site may have been used in pottery manufacture. Historic Indians of the southeastern United States used smooth pebbles to shape the insides of pots (Swanton 1946:243, 529).

V23. Sandstone. This raw material was used in making abraders at Crane that functioned in several activities (see V66–69). This is not the primary function of this class, however, for sandstone occurs in great quantities across Crane, whereas abraders are not common. The primary use of sandstone at Crane was in the construction of hearths or ovens in a manner similar to the way in which burnt igneous rock may have been used. During the excavation of pits at Crane, sandstone often was found in association with charcoal and ash that had been dumped in them, presumably while cleaning out hearths.

V25. Red Ochre. This material was used historically in the Eastern Woodlands as a pigment. It was combined with oil (Battle 1922) to make paint for body adornment. Its use in this manner during the Middle Woodland in Illinois as evidenced by one of the figurines at the Knight mound group (a warrior) that was painted red on its body, white on its face. It also was used to paint pottery vessels (Griffin 1952b). At the Crane site, the use of red ochre as a paint might be suggested by its spatial association with Havana pottery (depositional set 7). Havana pottery could have been used to boil nuts in order to derive oil with which red ochre might have been combined to form a paint.

A second manner in which red ochre may have been used is as an ingredient in acorn bread. Bitter acorns contain tannic acid, which gives them a poor taste and a poisonous character if eaten in great quantity (Driver 1961:91). If red ochre or ferruginous earth are combined with acorn meal prior to baking it into bread, any tannin in the meal will be converted to an insoluble form, making it indigestible and safe to consume. This practice is recorded for the historic Indians of California (Chestnut 1902; Gifford 1936:90; Merriam 1918:130). It is not documented for the Indians of the eastern United States, although it may well have been practiced here, for bitter acorns were exploited by the Eastern Woodland Indians. The documented means by which the Eastern Woodlands Indians exploited bitter acorns include boiling them whole, with or without ash, and pouring water through acorn meal held in a porous basket. At the Crane site, the spatial association of red ochre with Havana pottery (depositional set 7) might reflect the use of red ochre in processing bitter acorns. Whole acorn meats might have been boiled to remove most of the tannic acid they contained and to extract oil from them. The boiled meats then might have been ground to meal, to which red ochre was added, to make acorn bread.

A third manner in which red ochre might have been used at the Crane site is in dressing hides to give them a bright color. This practice is documented for the Tuski Eskimo (Mason 1889), and might be inferred for the occupants of Crane site by the spatial association of red ochre with tools used in dressing hides (V83 knives; depositional set 7).

Red ochre also might have been used in ceremonial contexts at Crane. In Illinois during the Middle Woodland period, red ochre is found in burials, sprinkled over corpses (Deuel 1952:173). Although no burials were found during the excavation of Crane and although this specific use cannot be documented there, broader, ceremonial uses of red ochre at Crane are still a possibility.

A final and unlikely use of red ochre at the Crane site is as an abrasive for polishing stones. This is a feasible function physically, but has not been documented ethnographically (Witthoft 1967:384–385).

V26. Hematite. This natural form of iron was used prehistorically in the Eastern Woodlands to make hammers, grooved axes, celts, atlatl weights, grooved plummets and sinkers, rounded burnishers, gorgets, pendants, and gaming disks. In Jersey and Calhoun counties, Illinois, plummets were the main type of artifact made of hematite. Hematite artifacts occur in both utilitarian, village contexts and in ceremonial, burial contexts (Moorehead 1912). The same is true of unmodified hematite nodules. At Crane, the spatial association of hematite with galena (depositional set 1), a raw material that occurs in burial and ceremonial contexts but which was not used for utilitarian purposes suggests that hematite was used there in the sphere of ceremonial activities rather than the mundane.

An alternative manner in which hematite may have been used is as the raw material from which red ochre was derived, by grinding. This would apply only to the more weathered specimens.

V29. Galena. This material is found in burial contexts during the Middle Woodland Period in Illinois (Griffin 1967) but is also known from village sites other than Crane (Struever 1968b). It was not used as a raw material from which utilitarian artifacts were manufactured, and can be designated clearly to the ceremonial sphere. Galena was one of the raw materials traded throughout the Midwest during Middle Woodland times. Its source was northwestern Illinois (Griffin 1967:184; Walthal 1978).

V30. Hornstone (Dongola Chert, Harrison County Flint, "Turkey-Tail" Chert). This is a dark gray chert with concentric lighter colored bands (Cook 1973:58). It has a very distinctive pattern of fracture, tending to break tabularly rather than conchoidally as most flints do (Shepherd 1972:37).

Hornstone is one of the raw materials that are traded over broad areas of the Midwest during the Middle Woodland period (Griffin 1967:184). Source locations of Dongola chert are widespread through southern Indiana, southern Illinois, and western Kentucky. The items included in this class at Crane, however, may not all be of nonlocal Dongola chert. A formation of Chouteau limestone bearing dark gray chert resembling Dongola chert occurs in the lower Illinois Valley trench (Cook 1973, Chapter 2). Some of the items identified as Dongola chert may actually be Chouteau chert.

Hornstone was made into a number of different kinds of tools at Crane; it does not appear to have been reserved for the manufacture of a restricted range of tool forms. Nor does hornstone co-occur with other artifact classes that can be allocated to the ceremonial sphere. The generalized, mundane activity of artifact manufacture is the most appropriate description of the use to which hornstone was put at Crane.

V31. "Missouri" Chert. This is an arbitrary name given to a pink chert peppered with dark red speckles. The chert resembles Burlington chert that has been heat treated. (Rick, personal communication). Farnsworth (personal communication) and Struever and Houart (1972), however, believe that Missouri chert is not a heat-treated variant of Burlington chert, but rather a nonlocal chert that was traded or brought into the lower Illinois Valley. This may be true, as Missouri chert does not contain the fossiliferous anomalies that most Burlington chert does.

The view taken here is that Missouri chert is a heat-treated, local variety of Burlington chert, based on the kinds of tools manufactured from Missouri chert. At Crane, Missouri chert was made primarily into lamellar blade knives but was not used to make tools designed to withstand twisting and shock, such as drills, perforator, and scrapers. This pattern of raw material usage is understandable in terms of heat treatment. Heat treatment facilitates the development of long fracture planes in chert when it is knapped. It consequently is one means by which a chert of poor knapping quality such as Burlington chert, can be upgraded to allow the removal of longer blades than otherwise would be possible. By the same token, heat treatment increases the brittleness of chert, and is not an appropriate preparatory step in the manufacture of tools to be used under great pressure or in a twisting manner. Thus, the kinds of tools made and not made from Missouri chert is understandable if Missouri chert were a heat treated variant of a local, lower grade chert.

The activity reflected by Missouri chert is the manufacture of stone tools—particularly lamellar blades.

V32. Kaolin Chert. This stone is milk-colored when examined under a light but when held in front of a light, it exhibits a translucent quality that reveals thin, interlaced yellow lines through it. Its source is believed to be the Mill Creek area of southern Illinois (Howard Winters, personal communication). Kaolin chert is most often made into hoes, but the percentage of hoes made out of Kaolin chert is much less than 50%. The occurrence of this raw material in an area may be taken to indicate artifact manufacturing and probably the manufacturing of hoes.

V33. "Flint Ridge" Chert. This is a marblized chert of yellow, red, brown, and white. It resembles descriptions of the chert from Flint Ridge, Ohio, but no chemical assays of the examples from Crane have been made to verify this identification. It is more probable that chert of this kind is not exotic, for the artifacts manufactured from it are not elaborate, finely retouched tools, nor ceremonial in nature. Mundane, unelaborate tools such as choppers were made from the raw material.

The occurrence of pieces of "Flint Ridge" chert in an area may be taken to indicate simply stone artifact manufacture, in general.

Igneous and Sandstone Items

V21,22. Polished, Ungrooved Celts and Grooved Axes of Metamorphic Rocks. The grooved axes at the Crane site are similar to those fully grooved specimens described by Griffin (1955). The polished celts at Crane have straight, expanding sides and rounded bases similar to those described by McGregor (1958:94) for the Pool village and by Wray and MacNeish

(1961:43) for the Weaver village. Also like the specimens at Pool site, those at Crane are almost all broken segments of celts rather than complete specimens. The ungrooved celts at Crane do not resemble the specimens of a bell shape with sharply expanding proximal ends found at the Liverpool mound group (Cole and Deuel 1937:Plate 31).

With respect to their ages, ungrooved celts usually are seen from a regional perspective (that of the eastern United States) as replacing grooved axes, around 500 B.C. (Griffin 1955:41; Struever 1964:91). On the local level, however, grooved axes have been found in clearly Middle Woodland contexts. In Illinois, a three-quarters-grooved axe was found at the Snyders village and a fully grooved axe was found at the Manker village (Griffin 1955:38), in both cases in Middle Woodland deposits. Witthoft and Miller (1952:83), based on excavation and surface surveys in Pennsylvania, also concluded that the three-quarters-grooved axe extended forward into Middle Woodland times. Nonetheless, grooved axes are rare compared to ungrooved celts in Middle Woodland contexts in Illinois, and the same is true at Crane. Griffin (1955:33) explains their occurrence in Middle Woodland contexts as an example of mining (Ascher 1968) of Archaic-made axes by Middle Woodland peoples.

The range of tasks in which the grooved axes and ungrooved celts at Crane site could feasibly have been used include the chopping of wood to girdle trees, to fell trees or to obtain slabs of wood or bark; the shaping of wood; and the thinning of hides. In the latter activity, they would have been used in an adze-like manner; in the former activities, as axes or adzes.

The use of ground stone celts in an adze-like manner to thin hides has been described for the North American Indians by Mason (1889). Semenov (1964:93) also notes the use of adze-like blades to dress skins. Their use in shaping wood has been documented by Miles (1973:70) for the North American Indians. At the Crane site, however, neither grooved axes nor ungrooved celts were used as adzes, for these activities or others. The Crane specimens all were axes. Several lines of evidence suggest this conclusion. First, none of the ungrooved celts at Crane site that I examined had wear of a kind typical for adzes: occurring on one face only, and in a uniform distribution over the total width of the working edge (Semenov 1964:123–131). All specimens had wear on both faces, and more wear at the top of their blades than their bottoms—a pattern found on axes. Secondly, in the prehistoric Midwest, there is good evidence that ungrooved celts were used as axes rather than adzes. Ungrooved celts have been found in bogs in Pennsylvania with their orginal wooden handles, showing that they were mounted as axes rather than adzes (Witthoft 1955:16).

As axes, the grooved and ungrooved polished blades at Crane site could have been used for several tasks, as summarized in Table 33. The use of grooved axes at Crane to fell tress or to obtain wide pieces of bark by the girdling and pounding strategy described in this table is suggested by the spatial association grooved axes with mauls on the site (depositional set 23). Also, both grooved

TABLE 33
Functions which stone axes (grooved or ungrooved) can serve

Function	Reference
Felling trees, by cutting clear through them	Ethnographic documentation: Mitchell (1959:192). Australian Aborigines. Experimental documentation: Cusance (1968)
Felling trees by girdling them. Strategy: cut 2 grooves around trunk, several feet apart; bruise layers of wood with the butt end of axes (or hammerstones or mauls); remove layer of wood; repeat	Driver (1961:80). Slash and burn agriculturalists of North America. Mason (1895:213). California tribes of the 1850's.
Obtain wide pieces of bark to be pounded into cloth. Strategy: the girdling method described above	
Obtaining strips of bark to be used in making bark containers. Strategy: cut into trunk at an acute angle	Waugh (1916:55). Historic Iroquois. Mitchell (1959:191). Australian Aborigines.
Cut out slabs of wood. Strategy: cut into truck at an acute angle	Experimental documentation: Semenov (1964:129)
Work and shape wood	

and ungrooved blades may have been used at Crane to obtain slabs of wood or strips of bark from trees by cutting into trunks at an acute angle; some specimens exhibit more intensive wear on one face than the other.

The distinction between the grooved and ungrooved varieties of polished axes found at Crane may reflect whether they functioned in heavier duty tasks such as felling trees and obtaining slabs of wood, or whether they were used in lighter duty tasks such as working and shaping wood. Grooved axes may have been mounted in heavier-duty hafts appropriate for rigorous tasks, whereas ungrooved axes may have been mounted in lighter-duty hafts or held in the hand, in line with their use in less demanding tasks. Among the Delaware Indians, grooved axes were hafted, presumably for heavy duty work, whereas ungrooved axes were only handheld and useful only in light duty tasks such as shaping wooden forms. This possible pattern of differential use at Crane can be given some support by the spatial distributions of these two classes of axes and the different artifact types with which they associate. At Crane, grooved axes consistently were found near the periphery of the site, which was probably forest edge and where wood could have been obtained and where trees could have been felled. Ungrooved celts, on the other hand, were found closer to the house. (An analogous distribution may occur in Pennsylvania, as well, where grooved axes often are found isolated in fields which produce no other Indian artifacts.) Moreover, grooved axes at Crane co-occur spatially with mauls (depositional set 27)—an association that could indicate their joint use in felling trees or obtaining slabs of wood. Ungrooved axes, on the other hand, do not co-occur with tools that might be used in these ways, but rather, co-occur with scrapers that were used to work hard woods (V73, depositional set 21). Thus, ungrooved celts and grooved axes may be functionally differentiated in their specific use on wood.

Whether or not ungrooved celts and grooved axes were functionally differentiated in the domestic domain, they do differ with respect to their burial contexts. Grooved axes rarely were placed in Late Archaic, Early Woodland, or Middle Woodland graves, whereas ungrooved celts commonly occur in Middle Woodland graves (Griffin 1955:34).

V19. Mauls (Grooved Hammerstones). The items in this class were hafted hammerstones. Some have a flat, pecked surface for pounding similar to those described by Wray and MacNeish (1961:43, Fig. 11) at the Weaver site. Others at Crane have rounded pounding surfaces similar to those at the Pool and Havana sites (McGregror 1958:95; 1952).

Mauls were used for multiple purposes by the North American Indians, as shown in Table 34. Some or all of these functions may be assigned to the mauls from Crane. Their use in wood and bark procurement and felling of trees probably is suggested by their spatial association with polished axes (V22, depositional set 23).

V113,114,115,116. Hammerstones. These classes are composed of spherical items and oblong igneous cobbles having abrasion, crushing, and pecking marks, which tend to occur at their ends. They have no traces of parallel straitions indicative of grinding (see Grindingstones, following).

Among the historic Indians of North America, hammerstones were used for a large number of purposes, including flint knapping, crushing bones to extract marrow, pounding dried meat into meal to make pemmican, driving down pegs and driving wedges, beating the hides of animals to make them pliable, and breaking up dry wood for fires (Mason 1895:53). They also were used in the American Southwest to peck at grinding slabs to roughen their surfaces. The small size of the hammerstones at Crane (most are less than 3 in. long in their longest dimension) suggest that they were not used in the heavier pounding tasks mentioned previously, for they do not have the required weight. They do, however, fall within the range of sizes of hammerstones used in flint knapping (1–4 in. in diameter; Crabtree 1967:61), which requires control as well as force. Many of the scrapers, knives, denticulates, and saws from Crane were made on flakes derived by hard hammer methods.

TABLE 34
Functions which mauls can serve

Function	Reference
As "pemmican pounders," to crush choke cherries	Miles (1973:48). Plains Indians
Crushing bones which were to be boiled to extract the products of marrow	Mason (1895:28). North American Indians, in general
Quarrying chert and the primary knapping stage of breaking up chert nodules, removing cortex, and preforming	Experimental documentation: Crabtree (1940, 1967), Ellis (1940)
Collecting bark, wood. Strategy: pound tree trunk with a flat-surfaced mallet so as to separate layers of bark and wood; cut out the layers with an axe or knife	Waugh (1916). Iroquois
Pounding bark and wood which subsequently cut into strips to make baskets	Waugh (1916). Iroquois
Felling trees by girdling them Strategy: see Table 33	Waugh (1916). Iroquois
Breaking up dry wood in forests, to be brought back to camp in faggots	Mason (1895:29). North American Indians, in general

The several classes of hammerstones at Crane reflect size variations. Variable 113 includes hammerstone of all sizes, whereas V114–116 include particular size classes that together form only a subsample of all hammerstones recorded in V113. (Only a portion of all hammerstones found at the surface of Crane were available for inspection of their size when the surface collection from Crane was analyzed.) Variables 114, 115 and 116 respectively have maximum dimensions of less than 2¼ in., between 2¼ and 3 in., and greater than 3 in. These classes were defined on the basis of the clustering of their members when cross-plotting the longest axis of each item against the second longest axis (Figure 34). If the hammerstones at Crane were used largely for knapping flint, the clusters may be interpreted as hammerstones used in different stages of knapping, the smaller ones for finer flaking. It is not possible, however, to relate the particular size classes to particular tasks in knapping, for the size of hammerstone appropriate for the removal of a particular size and shape of flake also depends upon the density and resiliency of the material being worked (Crabtree 1972:9).

It can be stated, however, that if the hammerstones at Crane were used

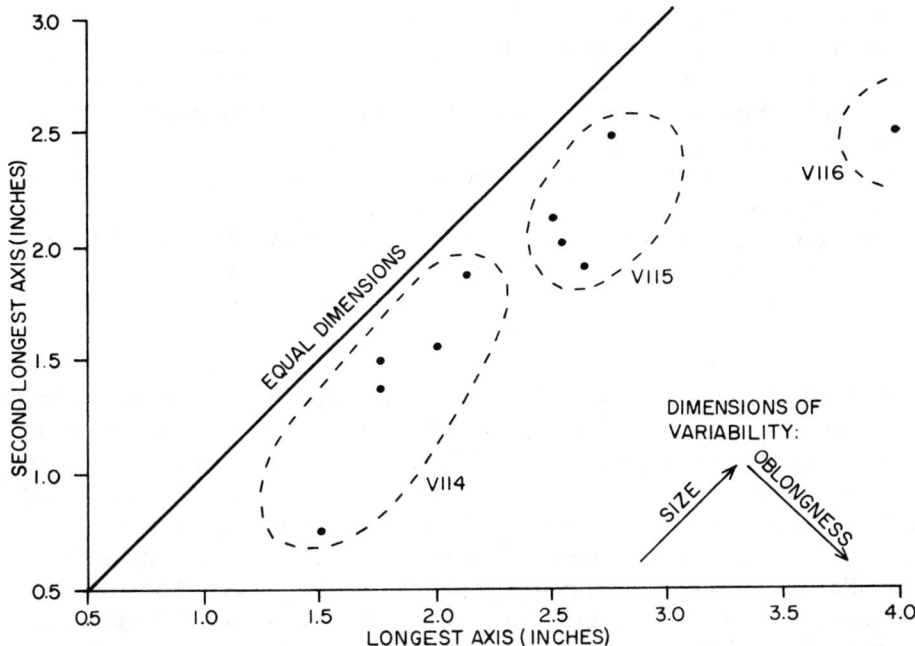

FIGURE 34. *A crossplot of the longest axis against the second longest axis of hammerstones at the Crane site. Three size classes are defined, V114, V115, and V116.*

primarily for knapping, they represent removal of flakes without prepared platforms. Crabtree (1967:60) notes that knapping without platform preparation is best accomplished by hammerstones with a convex or pointed surface. Hammerstones most useful in removing flakes with prepared platforms have a uniform, flat surface. Oblong cobbles are particularly useful (Crabtree and Swanson 1968). The cobble is held with its long side at a slight angle from the vertical and is brought down over the protruding platform. This allows the worker a level of accuracy in striking the platform at a particular angle, which is not possible when using a round hammerstone. Both the rounder shapes of the hammerstones at Crane and the wear at their ends rather than along their edges suggest their inappropriateness for knapping with platforms and their more probable use in removing flakes without prepared platforms.

One implication of this conclusion is that the hammerstones at Crane probably were not used in removing lamellar blades from prismatic cores. The lamellar blades at Crane have prepared platforms. Crabtree (1968; with Swanson 1968) has shown how lamellar blades, specifically, can be removed more easily by using the edge of oblong cobbles, indirect percussion, or pressure with the aid of a crutch, as opposed to using the ends of oblong cobbles or spherical hammerstones. The lamellar blades at Crane were removed from prismatic cores probably using pressure or indirect percussion methods.

The hammerstones at Crane might have been used to remove flakes without platforms from more amorphous flake cores, but Montet-White (1968:29) believes that during the Middle Woodland in Illinois, such flakes were produced by soft-hammer (Cobden) techniques rather than hard-hammer direct-percussion methods. Bifacial preforms, however, may have been shaped by direct-percussion, hard-hammer techniques at Crane and at Middle Woodland sites in Illinois in general (Montet-White 1968:31). Additionally, at Crane, less formal tools (e.g., scrapers, choppers) have flake scars indicating that they, too, were manufactured by direct percussion methods (see the following). Hammerstones in classes V114, V115, and/or V116 at Crane probably were used in shaping these preforms and tools.

Support of the hypothesis that the several size classes of hammerstone reflect functional differentiation only with respect to the stage of flint knapping in which they were used, and not with respect to the other possible activities in which they might have participated, cannot be found in the spatial data from Crane. Hammerstones in general (V113) associate spatially with chipping debris and decortication material at Crane (V15, V14, respectively; depositional set 15), but the individual size classes of hammerstones do not associate with the specific subclasses of knapping debris one might expect them to. This circumstance may, in part, result from the fact that the individual size classes of hammerstones include only a small portion of all the hammerstones that actually belong to them, the others having been unavailable for study. The incomplete nature of the classes may have unduly distorted the analysis of spatial associations.

V18. Grindingstones; V60, V61, V62. Hammer–Grinders; V64, V65. Nuttingstones. The items in these categories are igneous rocks and cobbles that have grinding striations on smooth, flat surfaces and/or abrasion–crushing wear resulting either from their use in hammering tasks or perhaps from intentional pecking. The morphological and wear attributes distinguishing each class are summarized in Table 35. The range of functions that they might have served, based on ethnographic analogy or experimental studies, is shown in Table 36.

The particular tasks for which the items in the several class of grindingstones, hammer–grinders, and "nuttingstones" were used can be specified more closely, in part, by considering their morphological attributes with respect to the results of experimental studies that have been made with such artifacts. Ethnographic analogy and the patterns of spatial covariation at Crane between these several classes of grinders, hammer–grinders, and nuttingstones and other artifact classes also can be used to assess function. First, and of considerable importance, is a study made by Riddell and Pritchard (1971) on the function of grinding manos of different morphologies. A major distinction these authors make is between grindingstones that are appropriate for hulling (threshing) small, delicate, hard-shelled seeds prior to winnowing, and those that can be used to crack, smash, and pulverize nuts, large hard seeds, and dried bulbs and

TABLE 35
Morphological and wear attributes of classes of grindingstones, hammer-grinders, and nuttingstones at the Crane site

Variable	Name	Diagnostic Attributes
V18	Grindingstones	Irregularly shaped igneous rocks averaging 5" in diameter and several inches thick Possess only grinding wear, on a smooth, flat surface; no hammering wear
V60	Hammer-grinders	Oblong igneous cobbles, 2 1/4 - 4 1/4" long in their longest dimension Exhibit both hammering and grinding wear, but on separate surfaces Grinding wear occurs on a smooth, flat surface Hammering wear occurs on their ends, alone
V61	Hammer-grinders	Oblong igneous cobbles, 2 1/4 - 4 1/4" long in their longest dimension Exhibit both hammering and grinding wear, but on separate surfaces Grinding wear occurs on a smooth flat surface Hammering wear occurs along their length, optionally at their ends
V62	Hammer-grinders	Oblong igneous cobbles, 2 1/4 - 4 1/4" long in their longest dimension Exhibit both hammering and grinding wear, but on separate surfaces Grinding wear occurs on a smooth flat surface Hammering wear occurs over their flatter surfaces, optionally at their ends
V64	Nuttingstones	Round to oblong cobbles, 2 1/4 - 3 3/4" long in their longest dimension Grinding striations on a smooth, flat surface with a shallow pit in the center of the surface (pecked, or abrasion/crushing wear) Hammering wear on one or both ends
V65	Nuttingstones	Round to oblong cobbles, 2 1/4 - 3 3/4" long in their longest dimension Grinding striations on a smooth, flat surface with a shallow pit in the center of the surface (pecked, or abrasion/crushing wear) No hammering wear

TABLE 36
Functions which hammering and grinding tools can serve

Function	Reference
Grinders:	
Hull small seeds, pulverize seeds	Ethnographic documentation: Kraybill (1977). Cross-cultural Experimental documention: Riddell and Pritchard (1971)
Grind nuts	Experimental documentation: Riddell and Pritchard (1971)
Grind pigments	Moorehead (1912). North American Indians
Hammers, Pounders:	
Mash/pulverize roots, rhizomes, bulbs, fruits, and larger seeds in preparation for cooking	Kraybill (1977). North American Indians
Crack nuts with hammerstones prior to boiling them for oil	Battle (1922), Swanton (1946), Waugh (1916:123). Eastern United States Indians
Pound nut meats into meal	<u>Ibid</u>.
Crack and finely fragment bones to extract marrow, tallow, grease	Leechman (1951), Peale (1871), Wheat (1972:113). North American Indians
Pulverize dried meat, fish to make pemmican, or for storage	Driver (1961:93), Miles (1973:44), Wheat (1972:117). North American Indians
Pound bark to make broad, flexible expanses used for cloth and narrow ribbons used in weaving bark baskets	McCarthy (1967:51), Waugh (1916:61). Australian Aborigines, Iroquois
Pound fibrous plants prior to shredding them to obtain fibers for making cordage	Miles (1973:94), Waugh (1916:61). North American Indians
Direct percussion flaking and bipolar flaking with cobbles	Experimental documentation: Crabtree (1967)

roots. Riddell and Pritchard argue that in order to hull fragile seeds such as those of grasses or *Chenopodium,* a tool is needed which lightly cracks the shell but does not mash the soft, meaty interior. Small, lightweight manos that may be held in one hand and that have a flat to slightly convex grinding surface are said to provide the amount of touch-control required to ensure the tool is not pressed down too hard. The grinding slabs used with a mano of such morphology are flat. In contrast, when hard, lumpy material must be reduced to a meal, considerable force is required. Manos appropriate to this task are heavier so that less muscle power is required, and larger so that both hands can be used side by side on the mano to exert more force. They also have a convex grinding facet that allows the worker to rock the mano and thus maintain a continuous, heavy, downward force of the mano as it is pushed across the grinding surface.

A second study which is of interest is the experimental work that has been done by Crabtree (1967:60) and Crabtree and Swanson (1968). As just discussed, these authors have concluded that hammerstones with convex or pointed surfaces are appropriate for direct-percussion knapping without platform preparation. The slightly convex edges of oblong cobbles, on the other hand, are particularly useful for the thinning of bifaces or the removal of lamellar blades by direct percussion. Consequently, *if* the grinders and nuttingstones at Crane site bearing hammering wear were multipurpose tools and used in flint knapping as well as food preparation, those with end wear would have been appropriate for knapping without prepared platforms, whereas those with edge wear would have been appropriate for knapping with prepared platforms.

Finally, a crucial ethnographic observation is that recorded by Waugh (1916:59) with respect to the manner of use of "pitted hammerstones." Waugh notes that the Iroquois use two flattened cobbles, one of which is shallowly pitted (intentionally pecked) to grind corn into meal. The pitted cobble was placed in a wooden bowl and corn kernels were ground between the pitted cobble and a second cobble on top of it. The bowl served to catch the meal. The depression in the lower cobble served to hold the kernels. The same couplet of stones, however, also was used in cracking nuts (Waugh 1916:59, 122) prior to their being boiled for oil or pounded into meal. McCarthy (1967:55) describes a similar set of tools called *kulki,* used by the Australian Aborigines in both hammering and grinding tasks, presumably in food preparation.

These experimental and ethnographic observations can be used, along with relationships specific to the Crane assemblage, to assign functions to the several classes of grinders, hammer–grinders, and nuttingstones found at Crane Site. V18 grindingstones may be considered metates upon which small, delicate seeds were hulled prior to winnowing, and then ground afterward. Their large size, thickness, and irregular form suggest their use as basal elements of couplets of grindingstones rather than manos, and their flat grinding surfaces suggest the kind of material which was ground. Additionally, some of these metates were used to grind red ochre. This function is indicated by red ochre stains that occur in the striations of a few of the specimens. The items that served as the mano elements, used in conjunction with V18 grindingstones could belong to any of the cobble-grinders considered here (V60–V65), but more probably belong to those in classes V64 and V60. The cobbles in class V64 occur, on the average, in greater proximity to V18 grindingstones (AVDIST = 9.756 m) than do any of the other classes of cobbles with grinding surfaces considered here. Moreover, the items in both V64 and V60 can be assigned tasks concerned with seed processing on the basis of attributes other than simply the presence of a flat grinding surface on them, whereas the other classes of cobble–grinders considered here cannot be linked to seed processing with multiple, independent considerations. (The items in class V64 resemble the pitted hammerstones used by the Iroquois to grind corn kernels. Those artifacts in class V60 could have been used to hull and pulverize seeds in wooden mortars.)

It may be that not all of the items in class V18 are grinders; some may be simply glacially worn rocks misidentified as grindingstones. The criteria that were used to identify grindingstones from glacial debris are some of those suggested by Howard Winters (personal communication). First, stones with patina on all but their ground surfaces were considered grindingstones. Second, stones with parallel striations in multiple directions were considered grindingstones whereas those with striations in one direction only were considered artifacts of glacial action. This later criterion was applicable because the grinders at Crane do not have troughs or channels requiring and indicating single-directional use; they have perfectly flat grinding surfaces and can be used for grinding with multidirectional motions.

The cobbles in V60 could have served a number of functions equally well. The hammering wear on the ends of the items in this class could have been produced when they were used as hammerstones in flint knapping by direct percussion techniques, without platform preparation (Crabtree 1967:60). This function likewise is indicated by the close spatial association of V60 hammer-grinders with decortication flakes and chipping debris (depositional set 15). Moreover, the sizes of the items in class V60 are within the range that is expected of hammerstones used in direct percussion knapping. They overlap considerably in their size range with that of hammerstones V114, V115, and V116, which appear to have been used in flint knapping. All of the specimens fall within the upper size limit of hammerstones that Crabtree (1967:61) finds efficient for direct percussion knapping—4 in.

A second task to which V60 hammer–grinders are suited is cracking nuts prior to boiling them for oil or pounding or grinding their meats into meal. Both of these methods of processing nuts were used among the historic tribes in the eastern United States (Battle 1922; Swanton 1946; Waugh 1916:123), depending on the kind of nut utilized and whether or not its meat was easily extractable from its shell. The grinding surfaces on the same items in class V60 could have been used to mill easily extractable nut meats into meal after the nut shells were cracked. It is more probable, however, that the grinding surfaces were used to mill hulled seeds along with V18 grindingstones. Most of the nuts at Crane were hickory nuts, which, compared to halvable nuts such as pecans or butternuts, do not possess easily extractable meats, and are more efficiently utilized by boiling them for their oil.

The abrasion at the end of V60 hammer-grinders also might have been produced by pounding seeds in wooden mortars or on stone anvils in order to hull them prior to winnowing and grinding (Kraybill 1977:8, 10–11, 43). Wooden mortars probably would have been used if small, fragile seeds such as those of grasses or *Chenopodium* were to be possessed, in order to avoid mashing the meats along with the hulls (Kraybill 1977:11; Riddell and Pritchard 1971). Stone anvils could have been used if larger seeds were to be hulled. The items in V62 hammer-grinders could have been stone anvils associated with V60 hammer-grinders. The grinding surfaces on V60 hammer-grinders could have been used in grinding the seeds after they were pounded and hulled.

The cobbles in class V62 served primarily as anvils. They have abrasion wear and/or pitting on their flat faces similar to that illustrated by Spears (1975) for anvils that she used to crack nuts and knap flint in her experimental studies. Other items that feasibly could have been pounded on these items include large seeds, dried roots, rhizomes, bulbs, fruits, meat, bark, and bone. The hammer–grinders, V60, bearing hammering wear only at their ends, could have been used in conjunction with V62 anvils in pounding these materials. The anvils, themselves, also apparently served at times as the pounding element, for they bear hammering wear at their ends.

The most probable materials that were pounded on the V62 anvils are those that later were ground. Both V62 and V60 cobbles have grinding surfaces as well as hammering wear. Thus, nuts with readily extractable meats such as acorn and hazelnut, and large seeds, such as those of the sunflower, were the most likely items processed with V62 hammer-grinders. Nut shells could have been cracked and nut meats then could have been ground with V62 and V60 hammer–grinders. Small, delicate seeds probably were not hulled on V62 anvils, but rather in wooden mortars, in order to prevent their meats from being mashed (Kraybill 1977:10–11).

Another material that more probably was pounded on V62 anvils is bark, in the process of making cloth and basketry. This inference is based on the spatial association of V62 anvils with V99 celts, which may have been used in procuring bark (see the following).

Cobbles bearing abrasion wear on their edges and falling in class V61 were expected to have been used in the production of lamellar blades and thinning of bifaces, based on the experimental studies of Crabtree and Swanson (1968). The spatial distribution of this class of cobbles, however, suggests otherwise. V61 cobbles did not occur in close proximity to lamellar cores, unretouched lamellar blades, or the proximal ends of lamellar blades (AVDIST = 12.325 m,

17.532 m, and 17.066 m, respectively). In contrast, V61 cobbles did occur in close proximity to V64 "nuttingstones," which probably served as anvils used in grinding large seeds and cracking nuts. V61 cobbles could have served as the pounders and grinders used in conjunction with V64 "nuttingstones." The edge- and end-worn faces of V61 cobbles could have been used to crack nuts and thresh large seeds held in V64 anvils. The grinding facets on V61 cobbles could have been used in grinding extracted nut meats and threshed and winnowed seeds.

The items in classes V64 and V65 resemble the "pitted hammerstones" used by the historic Iroquois to grind large seeds and to crack nuts (Waugh 1916:59), and can be assigned one or both of these functions in all probability. The use of V64 cobbles for cracking nuts is more certain than the use of V65 cobbles for this task. V64 cobbles bear hammering wear on their ends and could have served as both anvils and hammerstones on different occasions when cracking nuts. The items in class V65 lack end hammering wear, and if used for cracking nuts, they served as anvils, only. It is possible that the specimens in class V65 were used only in grinding large seeds, for they do not, on the average, occur in close proximity with any artifacts that could have been used as hammerstones for cracking nuts.

Finally, on a more general level, an assumption should be made explicit which was used in assigning functions to each of the categories of hammer-grinders and nuttingstones considered previously. The grinding and hammering elements found on the items in each of these classes have been evaluated as if their joint association on single items is significant with respect to the kinds of tasks in which those items were used. This allowed the assignment of a narrow range of functions to each class. However, for some specimens, and perhaps for some whole classes of specimens, the association of grinding and hammering elements may be fortuitous, and a result of their having been recycled or used in multiple activities at one time. In this case, the tools may be been used for a wider range of tasks than those specified.

V66, V67, V68, V69, V70. Sandstone Abraders. These items were divided into five classes, on the basis of the shape of their grinding surfaces: (V66) flat, without striations; (V67) convex outward (diameter: ½–1 in.) without striations; (V68) narrow slotted abraders; in cross section, the grooves are V-shaped; in plan view, they are approximately ⅛ in. broad, widening slightly at one end; (V69) medium slotted abraders, with hemispherical grooves ca. ¼ in. diameter and parallel-sided the whole length of the abrader; (V70) wide slotted abraders; in cross section the grooves are trough-shaped or an open angle of ca. 100°, and are ⅝ in. wide.

Ethnographically, abraders with different contours were used for different purposes (Table 37), and this is likely true of the Crane specimens. Of the multiple functions that V66 flat abraders could have served (Table 37), wood-working is the most probable. Abraders having this face contour occur in close

TABLE 37
Functions which sandstone abraders of different morphologies can serve

Function	Reference
Flat Abraders:	
Deflesh and thin hides	Mason (1889:560,572-573; 1899: 78-79). Eskimo, Sioux, Pawnee
Polish ground stone celts	Battle (1922), Miles (1973:73), Swanton (1946:545), Witthoft and Miller (1952:84). North American Indians
Work shell or wood	Cook (1973). North American Indians
In pressure flaking and indirect percussion, to roughen platforms to keep the flaking tool from slipping; also to weaken the surface so as to induce fracture	Experimental documentation: Crabtree (1972b:7), Speth (1972)
Dull the sharp edges of artifacts where hafts were to be applied	Cook (1973:73). North American Indians
Abraders with a V-shape cross-section	
Resharpen bone awls	Winters (1969:60). North American Indians
Resharpen bone projectile points	Semenov (1964:141). Cross-cultural
Sharpen the tips of wooden shafts	
Abraders used for the above purposes often were used in pairs, held together while pressing the item to be sharpened inward and twisting it, wedging the abraders apart. This manner of use, however, will result in the abraders having round cross-sections rather than V-shaped cross-sections	Waugh (1916), Iroquois

TABLE 37 (cont.)

Hemispherical-slotted abraders	
Smooth shafts, single or paired abraders	Miles (1973:92). North American Indians
Straighten shafts. Strategy: First hold the green branch to be made into a shaft and straightened over a fire; bend it to a desired corrective curve and allow it to cool in this position. For more localized straightening, heat a grooved abrader; lay the shaft on the groove until hot; bend it locally to the desired corrective curve and allow it to cool in this position; repeat wherever a crook occurs in the shaft	Mason (1899). Panamint Indians of California

spatial proximity, on the average, to drills and saws (V53, V57; depositional set 30) that were used to work wood. V66 abraders at Crane were not used in the knapping and hafting tasks shown the the table. The abraders bear no striations to indicate their use on harder, angular materials like chert. V68 V-shaped abraders were more likely used to work bone than wood. Abraders of this morphology associate spatially with unburnt bone (V28, depositional set 8), but do not co-occur with any tools used in working wood. They also were manipulated singly rather than in pairs, for their cross-sections are V-shaped rather than round. The slotted abraders in class V69 would have been appropriate for smoothing shafts of the size of darts or arrows. None of them, however, show fire reddening, which would have occurred had they been used for shaft straighteners as well as smoothers. The spatial association of class V69 abraders with other artifact classes also suggests their use as shaft smoothers, but not as shaft straighteners at Crane. V69 abraders co-occur with tools usable in woodworking, such as lamellar blades, turtleback scrapers, and flake cores from which flake knives could have been produced (depositional set 22). They do not occur in close proximity to any of the artifact classes used in conjunction with fire.

The precise functions of the abraders in classes V67 and V70 cannot be determined. V67 abraders would have been appropriate for sanding concave surfaces of any of the materials mentioned previously including wood, bone, shell, and ground stone. The spatial association of V67 abraders with V46 notches that were used to smooth wooden shafts (depositional set 15) might indicate in a very general way the use of these abraders in woodworking tasks.

They are not, however, of the appropriate morphology for having been used specifically with V46 notches in shaft smoothing.

The abraders in class V70 could have been used to sand obtuse-angled surfaces of a number of different materials. The edges of ground stone celts might have been polished with these items.

Chipped Stone Tools of a Functionally Specific Nature

The tool classes described in this section all have distinctive shapes and/or wear patterns that suggest a limited range of functions. They contrast, in this respect, to items that usually fall under the headings of knives, scrapers, celts, or points and for which precise functions are more difficult to determine.

V9. Hoes. The items in this class are similar to the expanding-sided hoes described by Montet-White (1968:83–85). They often are made of Kaolin chert. Their primary function as hoes rather than celts is suggested by the hoe polish that has accumulated on their faces. Witthoft (1967:387) attributes this polish to silica, which presumably was deposited on the hoe faces by the prairie sod through which the hoes were used to dig in the course of agricultural tasks. His interpretation is based on: (*a*) the assumption that agriculture was an important aspect of Middle Woodland subsistence patterns—an interpretation accepted at the time of his writing during and prior to the mid 1960s (Griffin 1960, 1964:243), (*b*) an analogy to the cultivation methods of historic Americans, who planted corn in plots of newly turned-over sod by cutting holes in the clumps of sod with axes; and (*c*) the spatial distribution of hoes and hoe chips around the Snyders site in the Mississippi Valley, Illinois. There, they were found on the prairie of the first terrace of the Mississippi River floodplain. Middle Woodland hoes might have been used, however, simply as digging tools to obtain roots (Sonnenfield 1962:63), and need not have been used for agricultural purposes.

Once dulled, hoes were used secondarily in hide working (Semenov 1964) by the North American Indians, in general. Their use in this capacity is noted by Miles (1973:99) and Mason (1899). Among the Pawnee, hoes were used specifically in thinning and graining hides, which were stretched and water-soaked (Mason 1889:572–573).

At Crane, hoes apparently were used for both cultivation and hide-working tasks. The polish on the items that defines this class suggests the first function. The spatial association of hoes with V49 denticulates (depositional set 2), which could have been used to either deflesh or grain hides, suggests the secondary, reuse of hoes for hide dressing.

A major concern is whether the hoes at Crane can be attributed to the Middle Woodland components or whether they belong to the more agriculturally oriented Jersey Bluff occupants, or less likely, to the emphemeral White Hall–Weaver occupants of Crane (Chapter 6). Hoe resharpening flakes at Crane site largely concentrate in the area of deep Middle Woodland midden deposits surrounding the one known Middle Woodland house there and some of the whole hoes occur in this same area. This spatial pattern could indicate that dull hoes were resharpened in the vicinity of the house by the Bedford phase occupants of Crane. It also is possible, however, that this spatial pattern reflects the use of Middle Woodland midden heaps as favored locations for Jersey Bluff garden plots, and the resharpening of hoes in those fields immediately as they dulled. An analogy can be made to the Dugum Dani of New Guinea, who frequently move their compounds and reuse the soil-enriched areas of previous occupations for sweet potato gardens (Heider 1967). Binford *et al.* (1970) have suggested a similar successional sequence at the Hatchery West site in Illinois. The hypothesis that hoes are resharpened expediently in garden plots and can be used to identify ancient fields is a reasonable one.

Determination of which of these two circumstances mentioned is applicable to the Crane site, and specification of the precise age of the hoes and hoe chips at Crane, can be made on the basis of both the shape of the hoes and the raw material from which they were made. The hoes at Crane site are all of the expanding, straight-sided variety that are known from clearly Middle Woodland contexts in numerous sites in the lower and central Illinois Valley (Montet-White 1968). They are not of the oval shape characterizing Early Late Woodland hoes (Wray and MacNeish 1961:61, Fig. 16) or Late Bluff hoes (Munson and Anderson 1973:34, 43) in the central Illinois Valley. Moreover, the hoes at Crane site are not made of Mill Creek chert—the raw material from which hoes were manufactured during Late Bluff times in the central Illinois Valley and in the American Bottoms (Munson 1971:13; Munson and Anderson 1973:34, 43), and probably in the lower Illinois Valley, as well. These attributes of the Crane hoes suggest that they belong to the Middle Woodland component. Final word on this interpretation, however, must await thorough documentation of the shape and raw materials of Late Woodland hoes in the lower Illinois Valley.

V10. Hoe Chips. The items in this class are very thin flakes with diffuse bulbs of percussion and bearing hoe polish. They are taken to be flakes removed to resharpen hoes (Harn 1971; Munson 1971).

V45, V46, V47, V48. Notches. Items in these four classes differ in the size and shape of the notch and the size and nature of the flakes upon which they are made. V45 consists of thin, semicircular notches, all having diameters of approximately ¼ in. and occurring on lamellar blades that have been backed on the edge opposite that bearing the notch. The notch was made by unifacial pressure flaking techniques. V46 consists of thin, unifacially or bifacially pressure flaked semicircular notches between ¼–11/16 in. in diameter. The notches occur on small, thin flakes of an inch or less in diameter and ⅛–7/16-in. thick, or on point forms (Snyders, Belknap). V47 consists of notches which have a cross sectional shape like an obtuse angle. Their sides and bottoms are straight and form a true angle rather than round into each other. The notches are ½ to 1¼ in. from side to bottom and were made by finer direct percussion flaking methods on amorphous flakes. V48 consists of heavy-duty, thick, pressure flaked semicircular notches between ¼ and 11/16 in. in diameter and occurring on large, thick, amorphous flakes up to 3 in. long and 1 in. thick. Notches of all classes bear step fracture on one face. In the case of unifacially flaked notches, it is the face with the retouch.

The semicircular notches could have functioned as shaft smoothers used prior to slotted abraders for the coarser work. The different diameters could reflect the *size* of shaft being smoothed, the *stage* of whittling down the shaft to the desired diameter, or the *degree* of use and resharpening of the tools. The first interpretation is unlikely for the hemispherical slotted abraders found at Crane have only one size of slot, around ¼ in in diameter. V48 heavy-duty notches could have been used for the very initial stages of work, or else for work upon harder woods. John Speth (personal communication) has suggested that some semicircular notches may have been used in a rotational manner to cut through shafts, but the unifacial wear on the specimens at Crane does not suggest this use there. With respect to the kinds of materials worked with the notches in classes V45, V46, and V48, it should be noted that the edge angles of these notches are all around 90°. This suggests that they were used to smooth very hard wood or bone. The notches of Paleo-Indians that Wilmsen analyzed and considered useful for working wood and bone had edge angles of only 65–75°.

The artifact classes with which V45, V46, and V48 notches associate spatially do not, in general, reflect the function assigned to them. V45 notches, however, do occur in close proximity, on the average, to V74 scrapers (depositional set 16) that were used to work either hard wood or bone.

The function of the obtuse-angled notches in class V47 cannot be stated specifically. On the basis of their spatial association with V59 gouges (depositional set 10), their general use in working wood seems probable. They might have been used to smooth the edges of bowls made with the gouges. Other wooden items having angled surfaces and which might have been smoothed with V47 notches include atlatls, shuttles, and loom frames.

V49, V50. Denticulates. The items in these classes are flakes and blades bearing a series of adjacent notches (at least two) forming a very coarse, serrated edge. Class 49 consists of denticulates made on lamellar blades with unifacial pressure flaking techniques, whereas Class 50 consists of denticulates made on amorphous flakes with either unifacial pressure flaking techniques or coarse, unifacial, direct percussion techniques. For items in both classes, the individual notches are ⅜ in. in width or less and wear is slight (step fracture or chips) or nonexistent. Wear occurs on the face bearing the retouch. The distinction between V49 and V50 denticulates is not a functional one, but rather, segregates those items that belong to the lamellar blade industry specific to the Middle Woodland times from those that could have been manufactured during any of the several periods of occupation of Crane site.

Denticulates could have been used for several purposes at Crane. Among the North American Indians, they were used to deflesh hides (Miles 1973:93). Usually, however, fleshers of *bone* with a serrated edge (functionally equivalent) were used (Mason 1889:568), if any serrated-edged tool was used. Second, denticulates could have been used to separate plant fibers from leaves and stems for making cordage and textiles. The Indians of the American Southwest scraped serrated stone flakes over fibrous plants placed on a cobble or wood anvil to extract fibers. They also used the same serrated scrapers with which they defleshed hides for extracting fibers (Osborne 1965:47–48). It is possible that the stone denticulates at Crane had such dual purposes. A third manner in which denticulates can be used but which does not seem to apply to the Crane specimens is as rasp files, for working soft wood. Some of the denticulates at Crane site having finer and shallower teeth could have functioned in this way, but do not bear the chipping wear that would be expected to have developed on them if they had.

Whether the denticulates at Crane site were used to extract fibers or dress hides, or both, cannot be determined. Winters (1969) does not believe that the denticulates from the Late Archaic in the Wabash Valley were used to shred plants, based on the lack of silica polish on them. This same argument might be made for the Crane specimens. The use of this criterion in determining the function of denticulates, however, has not been demonstrated to be adequate.

V51, V52. Spurs (Gravers, Becs). The items in these two classes are flakes and blades that have been flaked laterally so as to form a point at their base. Class 51 contains spurs made on lamellar blades by unifacial pressure retouch. Class 52 contains spurs made on amorphous flakes by bifacial pressure retouch. As with the two classes of denticulates, the two classes of spurs do not reflect a functional distinction, but rather, a segregation of those items that can be attributed to the Middle Woodland period, specifically (V51), from those that cannot (V52). The length of the spurs range from ⅛ to 7/16 in. All of the items in these categories are true spurs rather than gravers or becs. Spurs are retouched to a *point* whereas gravers are retouched so as to leave 1/32–3/16 in. of the original *flat* base of the flake or blade (Cook 1973:77). Spurs are retouched from both edges, whereas becs are retouched from only one.

Spurs were used in aboriginal America for a number of incising and piercing tasks, as shown in Table 38. The spatial associations of the spurs at Crane with other tool types allows one to specify a narrower range of tasks for which they probably were used. V51 spurs occur in close proximity, on the average (depositional set 11), to scrapers that were used to work hardwood (V75) and bone (V80). V52 spurs likewise associate spatially with scrapers used to work

TABLE 38
Functions which spurs served for North American Indians

Function	Reference
Smooth arrow shafts with concave portions above the spur, then groove the shafts with the spur and fill the grooves with pigment with the spur	Winters (1969:54). Omaha
Make parallel slots in bones in order to obtain long splinters used in making needles and other bone tools, involved in sewing, weaving, and basket making	Semenov (1964)
Work antler by grooving and splintering	Clark and Thompson (1954)
Incising tasks: carve eyes in needles and shuttles, pierce holes in leather while sewing	de Heinzelin (1962:29), Mason (1899), Nero (1957), Winters (1969)

hardwood and bone (V74; depositional set 16). These associations suggest the use of the spurs at Crane for incising wood and bone rather than the piercing of leather during sewing tasks. More specifically, the spatial association of V52 spurs with V45 notches suggests the use of this class of spurs to groove atlatl shafts or arrow shafts.

V53, V54. Drills without Smoothing Wear. Class 53 contains drills made on lamellar blades and amorphous flakes. They have cross sections shaped like rectangles, parallelograms, trapezoids, and biconvex lenses, forming two or four biting edges. All are narrow, from 1/16 to ⅜ in., and have parallel sides except where the tip is approached. Class 54 contains drills likewise made on lamellar blades and amorphous flakes and having cross sections like parallelograms or biconvex lenses. They are wide, however, and have expanding sides rather than parallel sides. Some have been thinned at their wide tops for hafting. Both classes were produced by pressure retouch, either unifacial or bifacial, depending on the number and shape of biting edge desired. They show no smoothing wear and are approximately the same in length (⅝–1¾ in.).

The drills in these two classes could have been used in drilling bone, wood, or shell, or in piercing holes in hides for sewing purposes. The narrower ones could have been used to begin drilling in harder materials where the wider bit drills would have been inappropriate. None of these drills would have been appropriate for drilling in stone. In North America, this was accomplished with solid wood drills or hollow cane drills used with a sand abrasive (Rau 1869).

The spatial association of V53 and V54 drills with artifacts of other classes suggests the interpretation that they were used to work wood and bone. V53 drills occur in close proximity on the average (depositional set 30) to V57 saws, which could have been used to work either wood or bone, and V66 flat abraders, which were used to sand wood. V54 drills co-occur, on the average (depositional set 20), with V71 scrapers, which were used to work soft and hard wood.

V55. "Drills" with Smoothing Wear. This class is composed of "drills" made on amorphous flakes by unifacial chipping. They have rapidly expanding sides (tip angles of ca. 45°) and blunted round tips with smoothing wear. They were not haftable, suggesting a function other than drilling. One task in which they could have been used and which their morphology suggests is the punctating of Havana pots. The smoothing wear borne by these drills, in this case, would be an essential attribute of the tool rather than wear from use. This interpretation does not, however, appear to be correct. The items in class V55 do not occur in close proximity to any of the artifact classes involved in or possibly involved in pottery manufacture (V1 Havana pottery, V4 clay tools, V24 tempering material). At present, the function of the items in class V55 cannot be specified.

V57. Saws. These items consist of lamellar blades that have been retouched by pressure, bifacially, and in an alternate flaking pattern so as to produce a sinuous edge when viewed edge forward. In plan view, the working edge is straight. These items could have been used to work either wood or bone, as were similar items among the Australian Aborigines (McCarthy 1946:34). They probably were used mostly in "crosscutting" (going against the grain of the wood) rather than "ripping" (going with the grain) (Mason 1895:47; Semenov 1964:152). Few rip saws are known among the industries of modern primitives (Mason 1895:47). Spurs, gravers, and becs generally fulfilled this function.

One exception to this generalization is the use of saws in notching arrow or dart shafts parallel to the grain of the wood. To do this, however, requires more than the saw. After some cutting has occurred, the saw will wedge itself in the wood. At this point, a flake knife must be used to widen the sides of the notch so that sawing may continue. Such operations may require two to three saws per shaft (Sollberger 1969:238–239), and multiple knife blades. At the Crane site, there is no evidence for this specific use of saws, based on spatial associations of artifact classes.

V56. Chert "Hammers." The items in this category are faceted polyhedral nodules or primary flakes. They bear straight edges formed by faces joining in obtuse angles. The edges are badly crushed and step-fractured. The nodules range from 1¼–1¾ in. × 2–4 in. in width and length.

Cook (1973:79) attributes chert hammers the function of pecking and shaping ground stone tools. This does not seem appropriate to the Crane specimens, however. Pecking tends to reduce chert hammers used in this way to spheroids. None of the Crane specimens show profile reduction to this state. Alternately, they could have been used to pound, crush, and pulverize a number of the softer materials shown in Table 36.

The precise set of tasks in which V56 chert hammers were used cannot be determined with or clarified by the spatial data from Crane site. V56 chert hammers co-occur most closely with V53 drills used in woodworking and secondarily with other woodworking tools (depositional set 30). It might be suggested from this spatial association, alone, that the chert hammers at Crane were used in some wood- or bone-working task. Crabtree (1973) has noted the appropriateness of obtuse angles for removing spiral shavings from dry bone, and it might be postulated that the chert hammers at Crane functioned in this manner. The chert hammers from Crane, however, have much duller edges than the items with which Crabtree experimented, so this alternative interpretation seems highly unlikely. No conclusion can be reached as to which of the functions listed in Table 36 the polyhedral nodules at Crane might have served.

V59. Gouges. These items are characterized by an edge that is U-shaped in profile and straight in plan view. They were produced by cleaving prismatic cores crosswise at an angle from the horizontal and then retouching the fresh edge with pressure. Step fracture occurs on their undersurface where the blades previously had been removed during their use as a core and which would have contacted the material being worked during their use as a gouge. This placement of wear is consistent with that described by Semenov (1964:130) for gouges.

The gouges at Crane have a wide diameter bit that suggests their use in carving out gentle concavities such as those of wooden bowls. If used in this respect, they probably did not function to *cut* away the concavity: Their wide bits would have required a huge force to do this. They more probably were used in scraper fashion in conjuction with fire that was used to burn out the concavity, as is done among the Iroquois (Waugh 1916:58) and Maori (Mitchell 1949:9). The edge angles of these items (53–57°) also suggests their use in a scraper manner rather than for cutting (see page 280).

V95. Whole Chisels (Reamers) and Their Proximal Ends. This class contains long, narrow, and very thick items similar in dimensions to a finger. They have a biconvex cross section, parallel sides or one side slightly constricted, and rounded tips. Step fracture and crushing wear occur on the edges (both faces) in a few places but is more concentrated at the tips (one face).

Their absolute size is approximately 1⅛ × 3⅞ × ¾ in. and the edge angles of their tips and sides fall between 67 and 71°.

In overall morphology, they resemble what McGregor (1958:116, Fig. 39C) has identified as reamers or gouges. This function also is suggested for the Crane specimens, which have unifacial wear and steep edge angles appropriate to such tasks. Like the wider-diameter gouges at Crane (V59), they could have been used well in scraping out concavities with the help of fire. The concavities could have been of smaller diameter, however. Chisels are known to have been used among modern primitives for such purposes. They were handheld rather than hit with a mallet (Mason 1895:33, 47). The length of the Crane specimens and their slightly concave nature along one edge suggests their use in the hand. The edges of the Crane specimens also might have been used in scraping straight surfaces found on the item being gouged, and is suggested by the bifacial wear along the edges of the chisels.

V13, V44. Flake Cores, Lamellar Cores. These two classes of items have been described by Montet-White (1963, 1968) in relation to the mode of manufacture of flakes and blades in the Middle Woodland sites of the central and lower Illinois Valley. Middle Woodland chipped stone artifacts are divisible into two sets, according to the knapping techniques used: flake artifacts and blade artifacts. The flake artifacts constitute an inhomogeneous group that have (*a*) broad, flat striking platforms, (*b*) wide variation in the angle of percussion used in removing the flake (between 95 and 120° at Snyders site; 115 to 130° at Worthy-Merrigan), and (*c*) prominent bulbs of percussion (Montet-White 1963:12). Direct percussion techniques, both hard hammer and soft hammer with the aid of an anvil, were used in removing the flakes. Flake cores had both prepared and unprepared platforms. The name, "Cobden technique," has been given to those cores with prepared platforms and flaked with soft hammers (Montet-White 1968).

The name, "Fulton technique," has been given (Montet-White 1968:28) to the punch and hammer technique of removing blades from prismatic cores. The cores are true prismatic-blade cores (Crabtree 1968:447) with small platforms prepared for placing the punch.

At Crane, many of the lamellar cores and blades are made out of Missouri chert, which may be a heat-treated variety of local Burlington chert. McGregor (1958:120, Fig. 40) noted that the polyhedral cores at the Pool site, too, were a distinctive color—largely white but mottled with pink to red speckles. Cook (1973:50) considers Illinois Middle Woodland cores and blades to have been heat treated.

Middle Woodland cores from the Illinois Valley were not used only in the production of blades or flakes. When exhausted, they sometimes were used for other purposes, resulting in wear on their edges unrelated to platform preparation (Montet-White 1963:20). Core scrapers are known among the historic Indians of North America (Crabtree 1968:427; Miles 1973). Crabtree

(1973:48) notes that the ridges of prismatic cores could have been used as "obtuse angle" scrapers, likely for bone working. Neither of these functions can be demonstrated for the cores at Crane site; they do not bear the wear markings distinctive of such uses.

The activities that one would expect flake cores and lamellar cores to indicate include not only the production of flakes and blades, but also the tasks in which flakes and blades, themselves, were used. Among the historic Indians of North America, blades often were manufactured at the time and place of their use rather than in advance. Cores were carried from activity to activity rather than the blades or flakes. This circumstance relates to the very delicate, molecular-thin edges (Speth 1972) that freshly struck flakes and blades have and that are hard to maintain if the blades are carried around (Crabtree 1968:455). One consequently would expect exhausted cores to be dumped in locations where the flakes and blades obtained from them were used. Partially used cores, on the other hand, might occur in temporary storage locations which were used for neither flake–blade production nor the activities in which flakes and blades were involved.

Chipped Stone Tools: Retouched Knives, Scrapers, "Celts," "Cache Blades," and Point Forms (V5–8, 34–41, 71–78, 81–90, 94, 96, 100, 102, 103, 105–108, 117)

The items in these classes all performed various cutting, scraping, chopping, or wedging tasks, the specific nature of which is not directly apparent from their overall morphology or wear patterns. The number of classes that have been defined is large, and reflects an overly divisive taxonomy that is more meaningful for descriptive purposes than for defining discrete functional classes. The taxonomy was created by completely crosscutting the multiple functional attributes shown in Table 29, without regard for the significance of the resulting classes. While the *individual* attributes in Table 29 are all significant functional variables, it is not true that *all possible combinations* of the several states taken by these variables also are significant. Some combinations do not occur at all within the Crane assemblage, and others which do occur might share in a common function.

In order to consolidate the descriptive taxonomy into a smaller set of functionally nonredundant and discrete classes and to assign functions to these classes, a logical key was developed (Figure 35) for sorting and combining some of the descriptive classes into functional classes. The key considers only retouched chipped stone items and is not meant for sorting either weighty, unretouched bifaces or unifaces such as choppers, wedges, hoes, and disjointing tools, or unretouched, natural-edged scrapers and knives. These have been considered or will be treated in a later section. Many of the mechanical

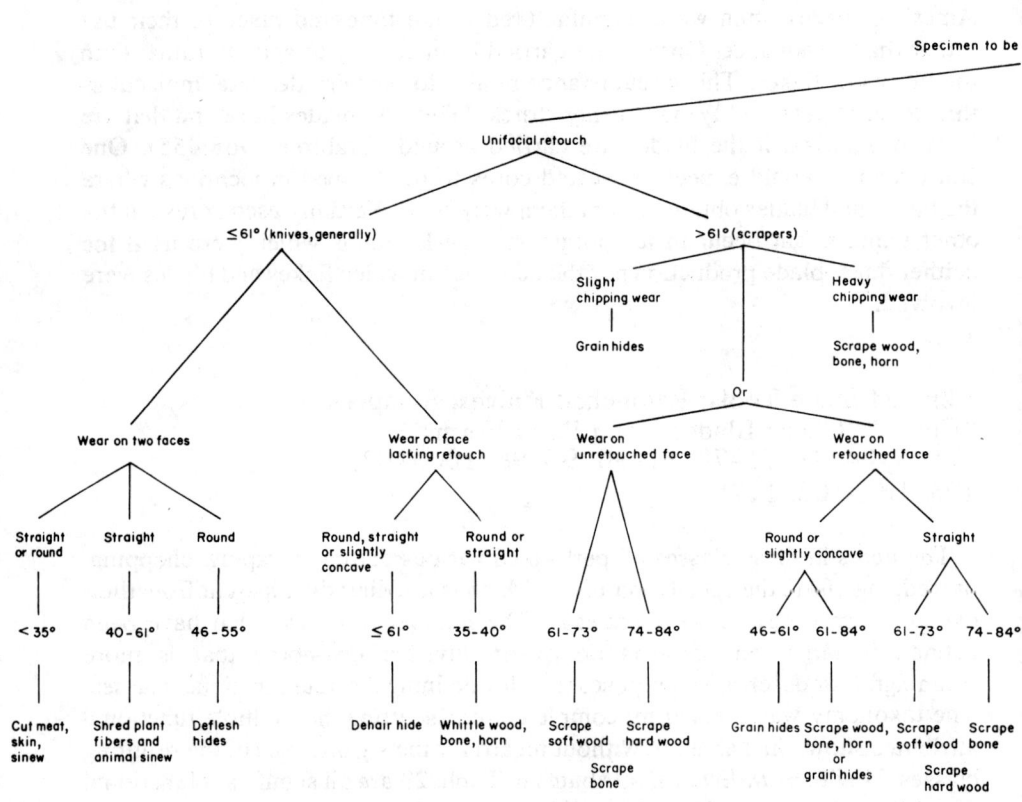

FIGURE 35. *A logical key for sorting unifacially and bifacially chipped stone tools.*

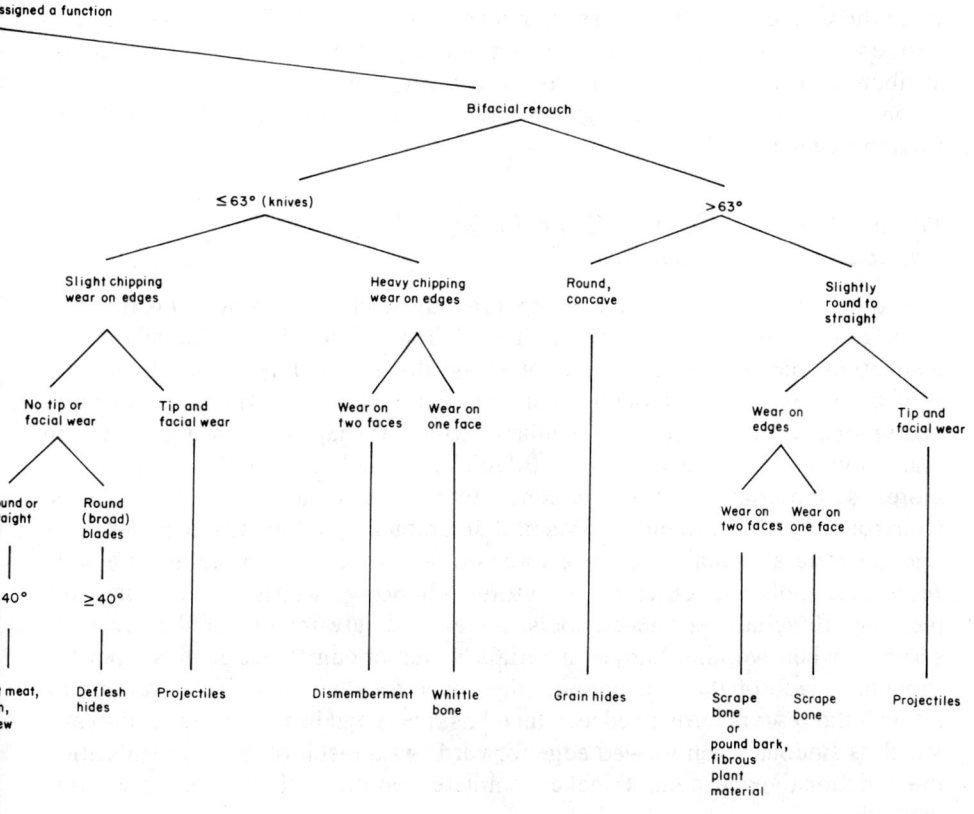

principles and other logical premises used in constructing the key are applicable to the sorting of unretouched items as well as the retouched items presently being considered. However, for organizational purposes, it will be clearer if the unretouched unifaces, bifaces, and other items are analyzed separately from the retouched items.

The key that will be presented was used to sort the descriptive *classes* of knives, scrapers, celts, etc. rather than the individual artifacts, themselves. The latter procedure would have been possible, had the key been avaiable at the time when the Crane assemblage was examined, but it was not. The key was built, in part, on associations that were found empirically among the multiple functional attributes only after the artifacts had been classified by the descriptive taxonomy. The following paragraphs present and document the logic and data used in developing the key.

Primary Sorting of Chipped Stone Tools: Unifacial versus Bifacial Retouch

The first bifurcation in the key separates those chipped stone items that had been retouched unifacially from those items retouched bifacially. This distinction allows the separation of tools along two dimensions: first, with respect to the distribution of forces on their functional edges when they do work; and second, with respect to the regular or irregular shape of the working edge in plan view and edge-forward view. Bifacially retouched tools, ipso facto, have more symmetrical cross sections than unifacially retouched items. Consequently, they are more efficient than unifacially retouched items for tasks that generate an equal amount of force on both faces. In particular, bifacially retouched tools are efficient at cutting, chopping, wedging, chiseling, and piercing. Bifacially retouched tools are exceedingly inefficient, however, as scrapers when working harder materials.[6] This circumstance results from the irregular shape of their functional edge, edge-forward view. The overlapping retouch flake-scars form an edge, which has ridges and lows in cross section and which is sinuous when viewed edge-forward. As a result of these irregularities, the functional edge cannot make complete contact with the material being worked.

The tasks for which unifacially retouched tools are most appropriate are those that generate an unequal amount of force on the faces forming their functional edge. In particular, they are efficient in scraping, whittling, and shaving. When scraping hard materials, unifacially retouched tools are more efficient than bifacially retouched tools, but less efficient than natural-fracture edges (e.g., lamellar blades) that make full contact with the material being worked (Sollberger 1969).

The significance of the distinction between unifacially retouched and

---

6. An exception is the use of bifacially-retouched tools to work bone (Ahler 1971:84).

bifacially retouched tools may be summarized in the following way. Both unifacially and bifacially retouched tools can function as knives for cutting materials. Scraping functions, however, are more likely to be assumed by unifacially retouched tools, while chopping, wedging, and piercing functions are more likely to be assumed by bifacially retouched tools.

Secondary Sorting of Chipped Stone Tools: Edge Angle

Edge angle, as a means for distinguishing between cutting and scraping functions among unifacially and bifacially retouched tools, and between cutting and chopping–wedging–piercing functions among bifacially retouched tools, is considered in the second level of bifurcation in the key. The differences between knives and scrapers will be discussed first.

The difference in the manner in which knives (unifacial or bifacial) and scrapers function is simply a kinematic one, that is, it concerns the motion of the tool with respect to the orientation of its working edge and is not concerned with the distribution of forces involved in tool use. Knives cut into a material, the motion of the tool being in the same plane defined by the cutting edge. Scrapers rake across a surface, the motion of the tool being at an angle to the functional edge (Mason 1897:727; McCarthy 1946:28; Sollberger 1969:236). One useful attribute for distinguishing knives from scrapers is edge angle. Only edges with a small angle can effectively force their way into a material and cut it. These same, thin edges cannot be used in a scraping motion, for they are delicate and would break.

The precise edge angle used to divide the chipped stone tools from Crane into knife and scraper categories could have been determined in two ways. It could have been determined in an a priori manner using the edge angle data and functional specifications defined by Gould *et al.* (1971) for Australian Aborigine assemblages, or by way of comparison to Wilmsen's (1968a, 1968b, 1970) analysis of Paleo-Indian tools. However, since the absolute range of edge angles characterizing knives and scrapers varies from assemblage to assemblage with the materials available in the natural environment for utilization (Wilmsen 1968b), this approach did not seem wise. Instead, patterns of co-occurrence between wear placement and range of edge angles were observed for the classes of items that have been defined on the basis of functional attributes other than edge angle. These patterns of co-occurrence were used to define the specific angle by which knives and scrapers could be distinguished.

Analysis of the *unifacially* retouched tools showed that items having edge angles of greater than 61° could bear wear markings on either their faces with retouch or their faces without retouch. Those unifacial items with edge angles below 61°, however, almost always had "chipping" or "thinning" wear (see Table 31 for definitions) on their unretouched faces. Experimental studies (Semenov 1964:20) demonstrate that knives of stone bearing wear on *one* side will consistently exhibit that wear on the *unretouched* side. Such knives are used in a whittling manner on harder materials such as wood and bone, or to shave hides (Mason 1889). Wear occurs on only the unretouched side because it is this side that must be drawn over the material being worked. The retouched side cannot be drawn over the material because it will not bite into the material as effectively (Sollberger 1969). In contrast, scrapers are used to work a material by drawing either their retouched or unretouched side over it, depending on the kind of material and task. Consequently, those items at Crane with edge angles above 61° and with wear on their retouched side could not have been knives, and could have been used only in a raking, scraperlike manner, whereas those items with edge angles below 61° and with wear on their unretouched sides were identifiable as knives, by analogy to the experimental studies. Thus, 61° was taken as the angle distinguishing unifacially retouched knives from unifacially retouched scrapers. Those unifacially retouched items with edge angles greater than 61° and with wear on their unretouched side were considered scrapers by way of analogy to unifacially retouched items with edge angles greater than 61° and with wear on their retouched side.

The particular division of unifacially retouched scrapers and knives which has been concluded is lent support by the form of the frequency distribution of the items considered scrapers as opposed to that of the items considered knives. Those items that belong to classes with edge angles of greater than 61° and that are considered to be scrapers form a bimodal distribution, with modal ranges of 61–73° and 74–84°. Those items that belong to classes with edge angles of less than 61° and that are considered to be knives form a unimodal distribution. This is the same pattern that occurs for the knives and scrapers used by the Western Desert Australian Aborigines and the Wonkonguru Aborigines (Gould *et al.* 1971), and the knives and scrapers identified by Wilmsen (1970) in Paleo-Indian assemblages (Table 39).

Importantly, this pattern seems to be a general one, independent of the specific kinds of stone from which knives and scrapers are made, and the specific raw materials upon which they are used. Wilmsen's data pertain to assemblages distributed over a wide range of environments in the United States, which differ from each other and from Australia in the stones and raw materials they offer. Thus, the consistency found between patterning of edge angle of tools identified as knives and scrapers at Crane and the patterning of edge angles of knives and scrapers used by the Australian Aborigines and Paleo-Indians is a significant one.

Finally, a comparison of the absolute edge angles of knives and scrapers used by the Australian Aborigines and Paleo-Indians to those of the items identified as knives and scrapers at Crane lends some support to the proposed classification. Table 39 shows the edge angles of knives and scrapers for each of these groups. Note the good correspondence between the edge angles of knives and scrapers at Crane and the angles of knives and scrapers used by the Australian Aborigines.

Thus, several lines of evidence suggest that unifacially retouched tools at

TABLE 39
Edge angles of knives and scrapers used by Australian Aborigines, Paleo-Indians, and the occupants of Crane

Group	Knives	Scrapers	
		Mode 1	Mode 2
Paleo-Indians*	20–35°	46–55°	66–75°
Western Desert Aborigines**	19–59°	50–59°	75–85°
Wongkonguree Aborigines**	19–59°	60–69°	80–89°
Crane Assemblage (unifacially retouched tools)	26–59°	60–73°	74–84°

*Wilmsen (1970)

**Gould et al. (1971)

Crane site can be divided into knives and scrapers on the basis of edge angle, those items having edge angles of more than 61° being scrapers; those items having edge angles of less than 61° being knives.

Just as unifacially retouched chipped stone tools can be classified into knives and scrapers, so bifacially retouched items can be classified into knives, on the one hand, and chopping, wedging, and piercing tools,[7] on the other. Again, the approach used is kinematic and uses the angle of the functional edge as the diagnostic attribute. It is argued that functional edges of only a small angle can force their way into a material and cut it, and that such thin edges are too fragile to be used in more heavy-duty tasks such as chopping, wedging, and chiseling.

To determine an edge angle for appropriately dividing bifacially retouched tools into groups representing knives, versus other tools, the frequency distribution of edge angles found on *all* classes of bifacially retouched items was plotted. The different descriptive classes of items were keyed differently in plotting the histogram so that their placement with respect to the total distribution could be observed (Figure 36). Two attributes of this plot are apparent, which allow the definition of a discriminant edge angle. First, the frequency distribution of edge angles possibly[8] is bimodal, with a cut point at 63°. This cut point is very close to the edge angle (61°) distinguishing knives from scrapers among unifacially retouched tools. By analogy, one might consider those bifacially retouched items with edge angles of less than 63° as knives, and suggest other functions for those items possessing edge angles of

---

7. And exceptionally, scraping tools, when working bone.

8. The shape of the frequency distribution is dependent on the width of the class intervals used in plotting the distribution.

5. Surface Artifacts from the Crane Site

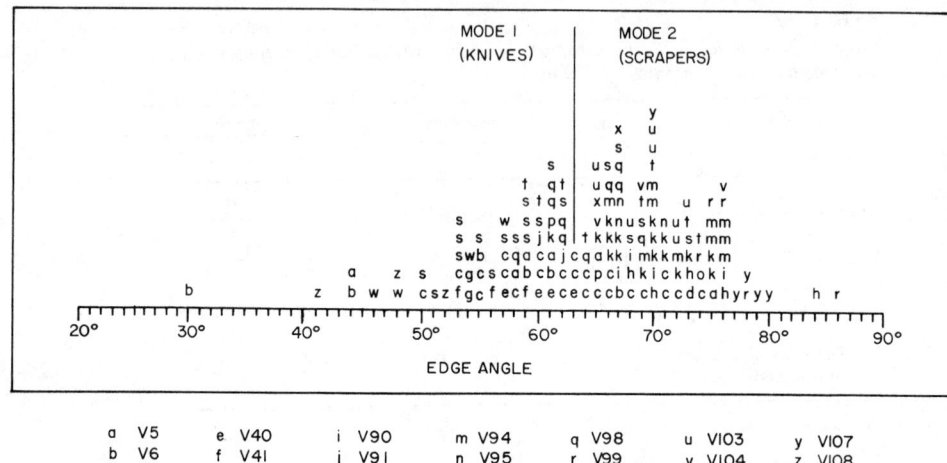

FIGURE 36. *The frequency distribution of edge angles of bifacially retouched tools found at the Crane site. Different descriptive classes of items have been keyed differently so that their placement with respect to the total distribution can be observed.*

more than 63°. This hypothesis is supported by the second characteristic of the plot—that the items within individual descriptive classes which were defined earlier by multiple functional attributes other than edge angle fall into one or the other of the two possible modes. The two different ranges of edge angles associate with *different restricted* sets of combinations of states taken by other functional attributes, suggesting that the possible bimodality both is real and is significant with respect to artifact function. On the basis of these attributes of the distribution of edge angles found on bifacially retouched items, and by way of analogy to the particular edge angle discriminating knives and scrapers among unifacially retouched items, an angle of 63° may be used to classify bifacially retouched items as knives or tools with other purposes.

To this point, the view has been taken that edge angle is an unambiguous indicatory of whether a stone tool was used in the manner of a knife or a scraper, and that the distributions of edge angles of knives and scrapers do not overlap. This is not the circumstance among contemporary hunter-gatherers, however, and there is some degree of overlap in the edge-angle distributions of items used as knives as opposed to scrapers. Gould *et al.* (1971:151, Table 1) shows that among the Western Desert Aborigines of Australia, knives and scrapers share common edge angles in the 40–59° range, and particularly in the 50–59° range. A similar situation occurs in the Crane data. A few unifacially retouched items (e.g., class V89c) bear step fracture wear on their retouched faces, alone—a diagnostic of scrapers—yet also have edge angles, which previously have been said to be diagnostic of knives. The items have edge angles in the 46–61° range.

Thus, the 61° or 63° threshold edge angles that are used in this study to define scrapers and knives are not absolute; some items with edge angles in the upper range of those defining knives actually may be scrapers. In some cases, it was possible to determine that an item classified as a knife on the basis of its edge angle actually functioned as a scraper, by noting the placement of wear on its edge. In other instances, however, classification errors may have gone unnoticed.

Tertiary Sorting of Chipped Stone Tools: Unifacially Retouched Knives

Having subdivided a chipped stone assemblage into unifacially retouched knives, unifacially retouched scrapers, bifacially retouched knives, and other bifacially retouched tools, it is possible to call on a number of subordinate attributes of functional edges to determine more precisely the tasks in which particular kinds of tools were used. I will trace these lower levels of the key beginning with unifacially retouched knives.

One attribute for determining the nature of the tasks performed by different kinds of unifacially retouched knives is whether wear is found on one face, on the side lacking retouch or on both faces of the artifact. Knives having wear on two faces most likely were used in cutting soft materials such as meat, sinew, leather, or plant fiber. Those bearing wear on only their unretouched side are likely to have been used in a whittling manner to work harder materials such as wood, bone, or horn. These conclusions were reached by Semenov (1964:20) on the basis of experimental replications of wear patterns as well as logical deduction. When stone knives are used to cut hard materials, only a thin paring can be produced, which does not mar the upper surface of the knife. Wear is created only on the face contacting the material being worked (Semenov 1964:109) The face that is held against the material and on which wear develops always lacks retouch. The knife cannot be used with its retouched face held against the material to be worked because this face bears ridges produced by overlapping retouch flake-scars and will prevent the tool from making complete contact with the material being worked. In contrast, knives used to cut soft material need not be used in a whittling manner, and may cut into material with a substantial amount of bulk on both sides of the knife. Consequently, both faces of the knife receive wear.

To Semenov's conclusions on the task for which stone knives may be used must be added one other. Unifacially retouched knives with wear on one face can be used in a shaving manner as well as for whittling. Mason (1889:562) has documented ethnographically the use of drawing knives (i.e., knives held in a whittling manner) to save the hair off hides in the process of dressing them. Either unifacially retouched or natural-edged knives would be appropriate for this task.

Finer specification of the function of the unifacially retouched knives having different wear patterns can be made using the angle of the functional edge and/or its plan-view shape. First will be discussed knives bearing wear on one face, and then knives bearing wear on two faces. With respect to knives with wear on one face, Semenov (1964:20) has found experimentally that when they are used to whittle harder materials such as wood, bone, or horn, they are most effective, yet still durable, when edge angles are between 35–40°. No such restriction on edge angle applies to them, however, when they are used to shave the hair from hides. Mason (1889) has noted ethnographically that both "sharp" knives (p. 573) and "dull" knives (pp. 562, 564, 575–576) are used in dehairing hides. Corresponding with the difference in the range of edge angles used for these different tasks should be differences in the intensity of wear found on the artifact. Intensive step-fracturing and crushing of the functional edge is more likely to be found among knives used to whittle wood or bone than among knives used in dehairing a soft hide. This secondary difference can be used as an independent check on the validity of the classes into which different items have been sorted using edge angle, and was done in this study.

Those unifacially retouched knives that bear wear on two faces and that were used to cut soft materials likewise can be subdivided into more specific functional categories, first using the plan-view shape of their edges. On the basis of ethnographic reports of hide-dressing practices in North America (Mason 1889, 1899), it can be concluded that in most instances, only knives with rounded contours were used in defleshing hides. Straight knives are equally efficient in removing flesh from hides, but are undesirable because their pointed ends can puncture the hide during the cleaning process. Puncture is less probable if rounded knives are used. On the other hand, when cutting slabs of meat or sinew while butchering, when cutting leather or hide while tailoring or butchering, or when cutting plant fibers, a straight knife with an angular tip is more efficient, although not a requisite. The angle of the functional edge of unifacially retouched, bifacially worn knives provides a second, independent, but equally important attribute for sorting them. Wilmsen (1970) has suggested for the Paleo-Indian tools he examined that unifacially retouched tools having edge angles of less than 35° could have been used efficiently for cutting meat or skin (p. 70) whereas those with angles of 46–55° could have been used for defleshing hides (pp. 70, 73). This segregation makes sense intuitively and ethnographically. One would not expect sharp knives to be used to deflesh hides, for fear of puncturing or slashing them, whereas sharp knives *would* be desirable in less restictive circumstances such as cutting slabs of meat during butchering or cutting leather while tailoring. Among the North American Indians, adzes, scrapers, and dull knives were the tools used more typically in defleshing hides—not sharp knives (Mason 1889).

The distinction between round, moderately dull knife edges and sharp knife edges (straight or round) for distinguishing hide defleshing tools from tools used to cut meat, sinew, skin, and plant fiber is a neat one. Empirically, however, the Crane assemblage also offered unifacially retouched, bifacially worn knives with straight, moderately dull edges. These items required explanation. Wilmsen (1970:70) suggests for his Paleo-Indian materials that large, unhafted tools with edge angles in the range of 46–55° may have been used to shred plant fibers and animal sinew. Such functions reasonably could be performd by tools also having straight edges and on which bifacial wear develops. Osborne (1968:48) has found that straight edges used in conjunction with an anvil are effective tools for shredding plant fibers. The same method might also be used to shred sinew. The experiments by Osborne produced marked wear on the tools he used to shred plant fibers, as a result of fortuitous contacts made between the shredder and the stone anvil. Although he does not report whether the wear occurred bifacially or on only the unretouched side of the tools he used, it is suspected that wear was bifacial, based on the steep angle with which Osborne held his tools to shred fibers. Thus, those unifacially retouched items having edge angles of 40–61°, straight edges, and wear on two faces might have been used for shredding fibers. I do not feel confident with this particular assignment of function, but I have no suggestions for alternative functions.

Tertiary Sorting of Chipped Stone Tools: Unifacially Retouched Scrapers

The next section of the key to be discussed is that concerned with unifacially retouched scrapers. The primary materials worked by scrapers include hides, wood, and bone. Hides require dehairing (sometimes), defleshing, and graining during the dressing process, but only graining was done with tools having edge angles that fall strictly into the range encompassed by the tools here defined as scrapers on a kinematic basis. Defleshing and dehairing implements are used to *cut* rather than *rasp*, and technically belong to the realm of dull knives rather than scrapers, although they have been called scrapers in the literature.

Unifacially retouched scrapers can be divided into functional classes, first, with respect to the amount of chipping wear and step fracture occurring on their edges. Slight amounts of chipping and step fracture wear can be used to indicate hide graining, whereas heavier amounts can be taken to indicate the working of wood, bone, or horn. The chipping and step fracture wear that occurs on those items used to grain hide is not a result of the contact the scrapers make with hide itself, but rather contact made with sand and stone fortuitously occurring on the hide (Keeley 1974:330; Odell 1975:230). Discrimination of the different functions of scrapers on the basis of the amount of roughening wear borne by their functional edges is suggested by several studies. Barnes (1932:53) worked oak, bone, and leather with steep-bitted scrapers, holding the tools with their retouched face against the material being worked. The scrapers that were used to work hardwood and bone developed step fracture on their retouched faces very quickly, whereas those used to grain leather developed only a few small step flakes after even a half hour of work effort. Crabtree and Davis (1968) also have experimented with the use of unifacially retouched scrapers on hardwoods (oak), but have held the scrapers both with their retouched face down, and with their unretouched, planar side down, using a rake angle of 45°. For both methods of use, they concluded that woodworking *quickly* consumes the functional edges of scrapers. That differences in amount of chipping wear result from working hides as opposed to harder materials also has been suggested by Wilmsen (1970). In his study of Paleo-Indian endscrapers, Wilmsen found two groups: those bearing "nibbling" (my "chipping") wear, and those generally devoid of nibbling wear but bearing polish. These wear groups also were homogenous and distinct with respect to their edge angles. Those items having more nibbling wear had edge angles clustering around 75° whereas those items with little nibbling wear and primarily bearing polish had edge angles clustering around 55°. If larger edge angles on scrapers can be taken to indicate their use on harder materials whereas smaller angles on scrapers may be taken to indicate their use on softer materials, then Wilmsen's data suggest that the amount of chipping wear occurring on scrapers is a function of the hardness of the material being worked. The patterning in Wilmsen's data independently corroborates the conclusions reached by Barnes, Crabtree, and Davis from their experimental studies. Thus, the amount of chipping wear occurring on a scraper might be used to distinguish those used to grain hides from those used to work wood.

One problem with using the amount of chipping wear occurring on a scraper edge as a criterion for sorting scrapers into functional classes is that it requires the assumption of equal use of all scrapers. In some cases, this might not be appropriate; slight amounts of chipping wear might derive from the shortness of time over which the scraper was used rather than the nature of the material worked. Consequently, although heavy chipping wear may be taken as an indication of the use of a unifacially retouched scraper for working wood, bone, or horn, slight amounts of chipping wear cannot be used as a certain diagnostic of the kind of material worked.

An alternative logic for assigning specific functions to unifacially retouched scrapers uses the criteria of placement of wear, edge contour in plan view, and edge angle. Placement of wear is the primary attribute for sorting, in this scheme, and allows the distinction between those items used to scrape wood, bone, or horn from those items that could have been used to scrape either hides or wood, bone, or horn. Ethnograhically, both hides and wood-bone-horn are worked with the retouched face of scrapers (Mason 1889, 1899; Semenov 1964:87), whereas only wood, bone, and horn are worked with the unretouched face of scrapers (Gould *et al.* 1970; Hayden and Kamminga 1973; White 1967). Wear on the unretouched face of a unifacially retouched scraper, thus, may be used to infer its use for working wood, bone, or horn, whereas wear on the retouched face of a unifacially retouched scraper is ambiguous as an indicator of artifact function.

An explanation for this circumstance can be found if one considers the mechanics of hide graining within the context of Sollberger's thesis on how unifacially retouched scrapers work. After a hide has been defleshed and thinned, it may be softened by repeatedly bending and stretching it. This may be done on a large scale, by stretching and twisting the hide as a whole, or on a small scale by chewing it or scraping the hide with a serrated edge (Mason 1889). Serrated edged, ethnographically, often were formed by notching the end of a long bone, but also were available in the overlapping flake-scars and ridges on the retouched face of unifacially retouched stone scrapers. By holding a unifacially retouched scraper face down and pulling it over a hide, the irregular retouched face of the scraper can stretch and twist the hide on a small scale, and grain or soften the hide. The same effect cannot be gotten by scraping a hide with the planar, unretouched face of a unifacially retouched scraper, for this face lacks protrusions. Consequently, if hide graining is done with a unifacially retouched scraper, the scraper must be held with its retouched side down. Only scrapers with wear on their retouched side could have been used to grain hides.

In contrast, there are no restrictions on the manner in which a unifacially retouched scraper can be used on wood. Such scrapers can be used with their retouched face down to rake the wood or with their unretouched face down, if a planing effect is desired (Crabtree and Davis 1968). Thus, scrapers that bear wear on their unretouched face and that were used with their unretouched face in contact with the material being worked could have been used to scrape only wood, bone, or horn. Scrapers that bear wear on their retouched face and that were used with their retouched face in contact with the material being worked could have been used to grain hides as well as scrape wood, bone, or horn.

Those scrapers bearing wear on their retouched faces and for which a precise function cannot be specified can be subdivided according to the plan-view contour of their functional edge in order to resolve partially the ambiguity of their function. Scrapers having edges with straight contours could have been used in working only wood, bone, or horn, whereas those having edges with round contours or slightly concave contours could have been used to grain hides, as well. This circumstance arises from mechanical constraints on the shape edges can assume if they are to be used for graining. When graining a hide over a soft surface, scrapers must have a round shape that follows the contour of the skin as it gives under the pressure of the scraper (Semenov 1964:89). This allows the protrusions that occur on the retouched face of the scrapers to bite into the hide and mechanically grain it. Alternately, the hide may be draped over a hard surface when being grained. In North America, Indians commonly draped hides over a log and grained them with a scraper having an edge which was slightly concave and of a diameter similar to that of the log (Mason 1889:571, 575). In this circumstance, too, the edge of the scraper follows the contour of the hide and bites into it. Straight-edged scrapers would be inappropriate for graining hides, either when they are placed on a soft surface or when they are draped over a log; the scraper edge would not follow the contour of the hide. Moreover, if a straight-edged scraper were used to grain a hide placed on a soft surface, the sharp ends of the scraper could pierce the skin accidentally, which would be undesirable. Thus, one would not expect grainers to have straight edges, and scrapers of this morphology can be designated woodworking, bone-working, or horn-working functions.

Those unifacially retouched scrapers that definitely can be assigned woodworking, bone-working, or horn-working functions may be subdivided approximately with respect to which of these several materials they were used on, with the aid of edge angle data. When harder materials are scraped, more force must be applied by the functional edge to perform work than when softer materials are scraped. This greater force requires that the end of the scraper be of a steeper angle if it is not to fracture (Semenov 1964; Wilmsen 1970).

Scrapers from the Crane site that definitely can be assigned woodworking, bone-working, or horn-working functions have edge angles that fall into two modes: 61–73° and 74–84°. These modes might reflect woodworking as opposed to bone–horn working, or might indicate soft wood versus hard wood–bone–horn working. It is possible to say, however, that scrapers having edge angles at the lower end of the range of the 61–73° mode probably were used to work soft woods, whereas those having edge angles at the upper end of the range of the 74–84° mode probably were used to work bone. The position of hard woodworking tools within either the upper end of the 61–73° mode or the lower end of the 74–84° mode is ambiguous.

Finally, a few comments should be made about hide working, in general, which do not concern the particular organization of the proposed key. Among the North American Indians, unifacially retouched scrapers were used both to deflesh and grain hides. Both kinds of scrapers had edges of a round contour in plan view but differed in the manner in which they were held. Those used in defleshing hides were held with their *unretouched* face down and *pushed* over the hide, whereas those used to grain hides were held with their *retouched* face down and *pulled* over the hide (Miles 1973:92; Semenov 1964:88). These two different actions would produce wear on different faces of the scraper—unretouched versus retouched side. The two different functions also are associated with different ranges of edge angles. Scrapers used in defleshing hides require sharper edges than those used to grain hides (Semenov 1964:87). It should be noticed that in the key presented, scrapers used to grain hides have been accounted for, but those used to deflesh hides have not. This reflects the fact that among the artifacts collected from the surface of Crane, no round scrapers bearing wear on only their unretouched side were found. All defleshing apparently was done with dull knives (unifacially or bifacially retouched) held in such a manner that bifacial wear was produced. This pattern of hide working that has been reconstructed for the Crane site apparently is not unusual, and has been documented by Mason (1889:563) for the Eskimo.

Tertiary Sorting of Chipped Stone Tools: Bifacially Retouched Knives

The second half of the key concerns the function of bifacially retouched tools. Let us begin with those items having edge angles of less than 63°, that is, the bifacial knives. One major source of information used in developing a key for sorting bifacially retouched knives into functional subcategories was Ahler's (1971) experiments with the use of point forms as part of his analysis of the "projectiles" from the Archaic site, Rodgers Rock Shelter, in Missouri. Ahler fashioned a number of points having different functional edge attributes and performed a number of different activities with them, including wood whittling, wood sawing, bone whittling, bone sawing, skinning a deer, meat carving, and disjointing. He noted the efficiency of the point forms in these several activities and also the wear patterns associated with them.

Ahler's study suggest some of the activities in which bifacially retouched knives might have been used. For point forms having edge angles around 40°, Ahler found that dismemberment and whittling bone were the only tasks among those he performed that could be accomplished efficiently. When used to whittle wood, saw wood, or carve meat, the functional edges of his knives quickly became clogged with fibers of the material being worked and became inefficent at their tasks. Bone sawing quickly dulled the edges of Ahler's knives and is a task to which bifacially retouched edges are likewise unsuited.

Ahler's study incorporated point forms having edge angles only around 40° and other sources must be considered when developing a list of activities to which bifacially retouched knives of other edge angle ranges might be suited. Wilmsen (1970) has suggested that knives (unifacially or bifacially retouched) having edge angles within the 25–35° range are appropriate for carving meat or cutting hide. Crabtree and Davis (1968) and Winters (1969) have suggested these functions for knives with bifacial retouch, explicitly. The serration allotted by the flake-scars of bifacially retouched knives is thought by Winters (1969:31) to be advantageous in cutting softer materials.

Defleshing of hides is another activity to which bifacial knives are suited. A particularly diagnostic form of knife used for this purpose is the Eskimo's "woman's knife" (Mason 1889, 1895). These are broad-bladed knives, either crescent-shaped or subtriangular to teardrop shaped, which were held by hand with a leather pad (Miles 1973:69–70) or with a backing of woody fiber (Mason 1895). The ovate and subtriangular "preforms" found on Middle Woodland sites in Illinois (White 1963:45) are of a similar morphology and some may have been used for defleshing purposes. Smoothing wear was noted by Rau (1873:402–403) to occur on those "disk-shaped preforms" found in a Middle Woodland mound near Fayetteville, Illinois.

Thus, the tasks for which bifacially retouched knives are appropriate and for which ethnographic or experimental documentation of use or efficiency are available include cutting of meat and leather, defleshing of hides, dismemberment, and whittling of bone. The key shown in Figure 35 for sorting bifacially retouched knives according to which of these several tasks they participated in was constructed on the following logic.

A primary attribute distinguishing the use to which a bifacial knife was put is the amount of chipping wear occurring on the functional edge. On the basis of analogy to the wear patterns found on scrapers of different functions (Barnes

1932; Crabtree and Davis 1968; Wilmsen 1970), as previously described, one would expect heavy chipping wear on knives used to work bone and only slight wear on knives used to cut softer materials such as meat and hide. This was precisely the pattern found by Ahler (1971:82–84) on the knives with which he worked different materials, excepting those knives used in dismemberment. Knives used in disjointing developed pronounced step flaking. This wear presumably did not develop as a result of the act of cutting sinew and tendons during the dismembering procedure, but rather as a result of repeated contact between the knife edge and the bones being separated. Use of the criterion intensity of chipping wear to subdivide knives into functional classes is subject to the assumption of their use for equal and long periods of time, as mentioned.

Those knives which bear pronounced chipping and step fracture wear on their functional edges and which were used either in dismemberment or in whittling bone can be subclassified with respect to their precise function on the basis of the placement of wear. Bifacially retouched knives used to dismember an animal are used in a sawlike manner, and will develop wear on both faces. Bone whittling, on the other hand, requires that only one face maintain contact with the material being worked; consequently, wear will occur on only one face.

Bifacially retouched knives used to deflesh hides and to cut meat or skin can be distinguished by their edge angle and partially by their shape. Wilmsen (1970) has suggested in general that knives with edge angles of 25–35° are appropriate for cutting meat or hide. No restriction is placed by Wilmsen on the shape of the functional edge or the overall size of the tool. In contrast, broad, flat, unhafted tools with edge angles of 46–55°, he suggests, might be used to deflesh hides. The restriction of a rounded edge contour also can be placed on bifacially retouched knives classified as defleshers, for the same reason that this was a requirement for unifacially retouched knives classified as defleshers: to prevent puncture and cutting of the hide being worked. While this mechanically based subdivision of bifacially retouched knives might reflect those most efficiently used for cutting meat or hide versus those most efficiently used for defleshing hides, ethnographically, the distinction between these two classes of knives is not always a clear one. Eskimo women used their crescent and ovate knives to cut leather and chop meat (Mason 1889; Miles 1973:69–70) as well as to deflesh hides (Mason 1889).

Tertiary Sorting of Chipped Stone Tools: Bifacially Retouched Items with Edge Angles Greater Than 63°

This portion of the key does not consider weighty, heavy-duty tools such as choppers, wedges, or hoes used for dismemberment, removal of slabs of wood and bark from trees, cultivation, or hide graining. Such tools typically are bifacially *shaped* but do not usually bear bifacial *retouch* by pressure flaking or finer direct-percussion methods. Only lighter weight, retouched tools are considered.

The activities to which bifacially retouched tools with edge angles of more than 63° are suited include scraping bone; graining hides; pounding, disintegrating, and/or tenderizing bark, and plants to be exploited for their fibers. Projectile points also fall in this class. The use of bifacially retouched, steep-bitted tools to scrape wood is documented by Semenov (1964:159). As opposed to the whittling of bone, where a knife is used at low rake angles, the scraping of bone requires a steep-bitted tool held nearly perpendicular to the surface. Scraping produces closely spaced, parallel ridges and troughs on the bone that presumably can then be scraped away with a unifacially retouched scraper more easily than could an unprepared smooth bone surface. Hide graining could have been achieved efficiently with steep-bitted, bifacially retouched tools. The serrated edge composed by the overlapping flake scars of a bifacially retouched scraper, like those of a unifacial scraper, are capable of stretching and twisting a hide on a small scale as the scraper is drawn over a hide, and thus can soften the hide. Mason (1889:Plate 71, #2, 3) shows historic Eskimo grainers made of bone and iron in a manner functionally equivalent to bifacially retouched stone tools that may have been used aboriginally in North America to grain hides. The pounding of bark into broad, flexible expanses used as cloth or into ribbons used in weaving bark baskets and the pounding of plant materials preliminary to shredding them to obtain fibers for making cordage has been recorded for the historic Indians of North America (Miles 1973:94; Waugh 1916:61) and elsewhere (Kraybill 1977). The important role of pounding in extracting fibers from plant material also has been documented experimentally (Kidder and Guernsey 1919; Osborne 1968). The primary tool used ethnographically and experimentally to pound bark and/or fibrous plant material is the hammerstone. Miles (1973:94) however, records that tools similar to the Eskimo "woman's knives," but which had dulled edges that crushed and did not cut, also were used in working bark and fibrous plants. Less formalized, bifacially retouched, steep-bitted, dull stone tools might have served in the same capacity.

The attributes by which steep-bitted bifacially retouched chipped stone tools can be sorted into the various functional categories just described are as follows: The plan-view shape of the functional edge is a primary attribute by which hide-graining tools can be separated from tools of other functions. As described previously for unifacially retouched grainers, tools used aboriginally to grain hides in North America had either well-rounded or slightly concave working edges, in order to follow the contour of skins placed on either a soft surface or over a log. Straight edges, on the other hand, were not used aboriginally in North America to grain hides, and are more appropriate for scraping bone or pounding bark and fibrous plant material, or as projectiles. A round shaped edge would accentuate the serrated nature of the functional edge and consequently would be inefficient for scraping bone (Sollberger 1969). Women's knives, which are similar to those tools that ethnographically were used to pound bark (Miles 1973:94), are broad blades with a contour slightly round to straight (ovate to subtriangular). Likewise, projectiles typically have straight to slightly convex edges. Consequently, the relatively tight curvature of the edges of hide grainers distinguishes them from bifacially retouched steep-bitted tools of other functions.

Among the straighter-edged, bifacially retouched steep-bitted tools, projectiles can be distinguished from bone scrapers and plant pounders by the lack of heavy chipping or step fracturing on their edges, the concentration of rounding and smoothing wear on their tips, and by light amounts of rounding and smoothing wear on the flake ridges adjacent to their tips (Ahler 1971:85–86). Bone scrapers and plant pounders, in contrast, will exhibit heavy wear along their edges alone. Differentiation of bone scrapers from plant pounders can be made only in part. Bone scrapers can bear chipping wear on either one face or two faces, depending on the motor habits of the worker. These tools are symmetrical and either face could be held in contact to the material being working. Tools used to pound bark or fibrous plants may be expected to exhibit wear on both faces, having been used vertically in a chopping manner. They cannot be distinguished from bone scrapers that bear wear on both faces.

Results of the Sorting Procedures: Functional Description of Classes and Assignment of Functions to Them

Of the attributes observed for each class of retouched, nonweighty, chipped stone tools found on the surface of Crane and into which the Crane assemblage was sorted (Table 29), only a limited number are required by the key formulated previously in order to assign functions. These include the plan-view shape of the functional edge, cross-sectional shape of the faces forming the edge, edge angle, placement of wear, kind of wear, knapping technique by which the edge was retouched, and whether retouched is unifacial or bifacial. Using these variables, functional descriptions of the classes into which the chipped stone tools from Crane were sorted can be made. Most descriptions are shown in Table 40.[9]

Other observations made for each class minimally include: (*a*) overall tool morphology and peculiarities in shape and size, and (*b*) whether the class is composed of complete tools or the edges of broken tools with unknown overall morphology. These attributes are presented in Appendix 3. Bibliographic references to previous descriptions of the morphology of elaborate tools (e.g., celts) also are given in this location.

The functions assigned to the various classes are shown in Table 31. In most cases, function was assigned on the basis of the attributes defining the classes (Table 40) and the placement of those attributes within the logical key. For a few classes, however, the attributes possessed by them were not diagnostic of their function, and suggested several possible functions. In these cases, the function of the ambiguous type was chosen on the basis of more information

---

9. Exceptions are the descriptions for point forms. V5–8, 117. These classes were defined on the basis of their outline shape, alone, and represent functionally mixed categories. Each specimen from these classes had to be considered separately as to its function, using the formal key. No general listing of functional attributes for these classes can be given. Point forms are considered later.

than that incorporated by the logical key. This additional information includes (a) the functions of the artifact types with which the ambiguous type co-occurs spatially and forms a depositional set; (b) ethnographic information additional to that given previously and specific to the class; or (c) morphological attributes peculiar to the class in their usefulness in defining function, but not generally useful in this regard. This additional information and logic not found within the key and Table 40 is presented in the following class descriptions.

V36a. Unifacial Endscrapers. The morphological attributes of this class of artifacts suggest their use in either hide graining or working of wood, bone, or horn. On the basis of their spatial association with V50 denticulates (depositional set 23), the items in this class are assigned the function of hide graining.

V36b. Unifacial Endscrapers. The functional characteristics of the items in this class key to either hide graining or working of wood, bone, or horn. On the basis of their spatial association with V50 denticulates (depositional set 23), the items are assigned the function of hide graining.

V71. Turtleback Scrapers. The morphological characteristics of the items in this class key to either working wood or graining hides. On the basis of their spatial association with V54 drills and V102 planing tools, the items in this class are assigned the function of working wood. A woodworking function is suggested also by the similarity of this class in overall morphology to items used in Australia to work wood. These items are called "horse hoof" scrapers and have been described by Mitchell (1949:192, 195) and McCarthy (1946, 1967). They are planar convex tools, having coarse flaking part way up their sides and cortex on their tops. The edge angles of those specimens pictured by McCarthy (1967:19, Fig. 4, #1) appear similar to the edge angles of the Crane specimens. The pattern of rejuvenation of the edges of the Australian horse hoof cores also is similar to that of the turtlebacks from the Crane site. With progressive resharpening of their edges, the lower portions of the convex faces of the scrapers eventually undercut the upper portion of the scraper. The upper portions of the scraper form an overhang that gives the scrapers a profile resembling a horse's hoof and lower foot (McCarthy 1946:47). The Australian specimens are used to do both rough woodworking with their retouched face against the wood, and finer, planing tasks, with their unretouched faces in contact with the wood (Mitchell 1949:192). Only the former use is indicated by the wear patterns on the specimens from Crane.

V72. Turtleback Scrapers. The items in this class are the same in all respects to those in class V71, excepting their larger edge angles. On the basis of the morphological attributes of the items in this class and also their similarity to the Australian horse hoof cores, the function of woodworking may be assigned to the items of this class. The large edge angles of these items suggest their use to work harder wood. Bone working also is a possibility.

TABLE 40
Functional attributes of retouched and unretouched chipped stone tools at Crane site

Variable Number	Plan View Shape of Edge	Cross-Sectional Shape	Edge Angle Range	Placement of Wear*	Kind of Wear*	Manner of Retouch*	Bifacial or Unifacial Retouch or Shaping
KNIVES							
83	straight	plano-convex	49-61°	V	TH, CH	DP, P	largely unifacial
84	straight	plano-convex	26-41°	V	TH, CH	DP, P	unifacial
85	straight	plano-convex	39-47°	V	TH, CH	P	unifacial
86	straight	plano-convex	53-60, 66°	D V	ST TH, ST	DP	unifacial
87a	convex	plano-convex	32-36°	D V	ST TH, CH	P	unifacial
87b	convex	plano-convex	46-48°	D V	ST TH, CH	P	unifacial
88	concave	plano-concave	47-62°	D	ST	P	unifacial
108	convex	bi-convex	41-52°	2F	slight CH	DP, P	bifacial
ENDSCRAPERS							
34	convex	plano-convex	57-72°	D	ST	P	unifacial
35	convex	plano-convex	80-89°	D	none or CH, ST	P	unifacial

TABLE 40 (cont.)

Variable Number	Plan View Shape of Edge	Cross-Sectional Shape	Edge Angle Range	Placement of Wear	Kind of Wear	Manner of Retouch	Bifacial or Unifacial Retouch or Shaping
ENDSCRAPERS (cont)							
36a	convex	plano-convex	73–85°	D	ST	P	unifacial
36b	convex	plano-convex	72–79°	D,V	ST	P	unifacial
37	convex	plano-convex	60–66°	V	ST,CH	DP,P	unifacial
38	convex	plano-convex	68°	D	heavy CH, some ST	DP,P	unifacial
39	convex	plano-convex	50–54°	D V	ST TH	P	unifacial
90	convex	bi-convex	67–76°	2F	ST	DP,P	bifacial
103	convex	plano-convex to bi-convex	65–73°	2F	slightly ST or none	DP,P	bifacial
CONVEX-EDGED SCRAPERS, CELTS							
71	convex	plano-convex	51,62–76°	D	ST	DP, (P rarely)	unifacial
72	convex	plano-convex	75–84°	D	ST	DP, (P rarely)	unifacial
81	convex	plano-convex	63–84°	D	ST	DP, (P rarely)	unifacial
82	convex	plano-convex	56–62°	D	ST	DP, (P rarely)	unifacial

TABLE 40 (cont.)

Variable Number	Plan View Shape of Edge	Cross-Sectional Shape	Edge Angle Range	Placement of Wear	Kind of Wear	Manner of Retouch	Bifacial or Unifacial Retouch or Shaping
CONVEX-EDGED SCRAPERS, CELTS (cont)							
91	convex	bi-convex	60-74°	2F	ST	DPS	bifacial
92	convex	bi-convex	61-75°	1F	ST	DPS	bifacial
95	convex	bi-convex	67-71°	1F	ST, crushing	DP	bifacial
97	convex	plano-convex	61,65°	D	slight CH	DPS	bifacial
98	convex/straight	plano-convex	58-69°	V	ST	DPS	bifacial
99	convex/straight	plano-convex	74-86°	V	ST	DPS	bifacial
101	slightly convex	bi-convex	59-74°	2F	ST	DPS	bifacial
107a	convex	bi-convex	70,80°	2F	slight CH	DP,P	bifacial
40	slightly convex	bi-convex	60-63°	2F	slight ST or none	P	unifacial
41	slightly convex	bi-convex	53-59°	2F	slight CH or none	P	bifacial
42	slightly convex	bi-convex	54°	2F	ST	DPS	bifacial
43a	convex, slightly convex	bi-convex	68-76,84°	2F	slight CH,ST	DP,P optional	bifacial

TABLE 40 (cont.)

Variable Number	Plan View Shape of Edge	Cross-Sectional Shape	Edge Angle Range	Placement of Wear	Kind of Wear	Manner of Retouch	Bifacial or Unifacial Retouch or Shaping
CONVEX-EDGED SCRAPERS, CELTS (cont)							
43b	convex, slightly convex	bi-convex	57-70, 75°	2F	slight CH, ST	DP, P optional	bifacial
STRAIGHT-EDGED SCRAPERS, CELTS							
73	straight	plano-convex	67-73°	D	ST	DP, P optional	unifacial
74	straight	plano-convex	76-84°	D	ST	DP, P optional	unifacial
75	straight	plano-convex	62-73°	D	ST	P	unifacial
76	straight	plano-convex	75-83°	D	ST	P	unifacial
77a	straight	plano-convex	64-68°	D V	CH TH, CH	DP, P	unifacial
77b	straight	plano-convex	65-67°	V	TH, CH	P	unifacial
78	straight	plano-convex	73-56°	D	ST	DP, P optional	unifacial
79a	straight	plano-convex	69-83°	D	ST	DP, P	unifacial
79b	straight	wedge	66°	1F	ST	DPS	bifacial
80	straight	plano-convex	81-83°	D	ST	DPS	unifacial

TABLE 40 (cont.)

Variable Number	Plan View Shape of Edge	Cross-Sectional Shape	Edge Angle Range	Placement of Wear	Kind of Wear	Manner of Retouch	Bifacial or Unifacial Retouch or Shaping
STRAIGHT-EDGED SCRAPERS, CELTS (cont)							
100	straight	bi-convex	50-70°	2F, end	ST	DP, P optional	bifacial
102	straight	plano-convex plano-concave	65°? 30°?	D V	ST TH	P none	bifacial --
104	straight	plano-convex	~68°	D V	ST TH	DPS	bifacial
105	straight	plano-convex or wedge	46-57, 74-80° on same edge	D,V	ST	DP,P	bifacial
106	straight	plano-convex or wedge	65-67°	D,V	ST	DP,P	bifacial
107b	straight	bi-convex or wedge	77-79°	2F	ST	DP,P	bifacial
SLIGHTLY CONCAVE-EDGED SCRAPERS							
89a	concave	plano-convex	71-99°	D	ST	DP,P optional	unifacial
89b	concave	plano-convex	68-82°	D	ST	DP,P optional	unifacial

TABLE 40 (cont.)

Variable Number	Plan View Shape of Edge	Cross-Sectional Shape	Edge Angle Range	Placement of Wear	Kind of Wear	Manner of Retouch	Bifacial or Unifacial Retouch or Shaping
89c	concave	bi-convex	64-67°	1F, 2F	slight ST, CH	DP, P optional	bifacial
WEDGES							
94	straight	bi-convex	63-78°	2F	ST	DPS, DP P	bifacial

*Abbreviations for the placement of wear, kind of wear, and manner of retouch are explained in Table 28.

V81. Unifacial, Round-Edged Scrapers. These items could have been used either to grain hides or work wood, bone, or horn, based on their morphology. Their spatial co-occurrence with V90 endscrapers and V87 knives (depositional set 28) suggests the function of hide graining.

V82. Unifacial, Round-Edged Scrapers. The morphological attributes of the items in this class, as well as the functions of the artifacts with which they associate spatially, suggest their use for either graining hide or working wood, bone, or horn.

V77a. Unifacial, Straight-Edged Scrapers. These are compound tools, used in both a planing and a raking manner to work soft and hard wood. The wear on the retouched faces of these items suggest their use in either scraping wood or graining hides. However, the additional wear on their unretouched side keys to only woodworking, and indicates this as their most likely function.

V100. Rectangular Celts. The morphological attributes used in the key suggest this class was appropriate for either scraping bone or pounding bark or fibrous plants. The moderate size of these items and the placement of wear on them, however, suggest they were not used in a pounding manner. To have been used as pounders, these items would have had to have been hafted, and intense wear would be expected to occur along only one edge, opposite the haft. These items, on the other hand, have wear on both their elongated edges and their ends. The alternative function of bone scraping is assigned, therefore, to this class. This function is not inconsistent with the wear patterns on the items in this class.

V105. Bifacial, Straight-Edged Scrapers. Each item in this class has a sinuous edge produced by the bimodal distribution of angles along its edge. The morphological attributes used in the key suggest either the scraping of bone or pounding of bark and fibrous plants. The sinuous nature of the edge is inappropriate for scraping bone; such as edge crushes easily. Thus, the alternate function, pounding and disintegrating bark and fibrous plants, is suggested.

V106. Bifacial, Straight-Edged Scrapers. The morphological attributes of artifacts in this class suggest either scraping bone or pounding bark or fibrous plants. On the basis of their spatial association with V100 rectangular celts (depositional set 14), the items in this class are assigned the function of bone scraping.

V94. Wedges. Half of the items in this class have their tips broken off. On the basis of this breakage pattern and their thick, almost conelike shape (Appendix

3), the items in this class may be assigned the function of wedges, used in splitting wood or bone.

Reliability of the Functions Assigned to Chipped Stone Tools

The reliability of the key in sorting and in assigning functions to chipped stone tools was not tested formally, but an informal evaluation of the key was made while it was used. The number of possible combinations of those functional attributes listed in Table 31 is huge compared to the restricted number of sets of co-occurring attributes that are implied by the formal key. If the logic behind the key were misguided, one consequence that we might expect is that some of the unifacially and bifacially retouched artifacts from Crane might have trait combinations different from any of the limited sets of combinations specified by the key. This seldom occurred. One exception is class V86 knives. These knives bear wear on both faces and have edge angles similar to knives used in defleshing hides, yet their edges are straight instead of curved. During the sorting process, it was necessary to construct a special branch of the key for them in order to accommodate them, and to assign to them the function of shredding plant fibers and animal sinew. I am not confident of this assignment of function. Analyses presented in Chapter 8 support the view that the assigned function is questionable.

Aside from this one problem, the key was successful in sorting the retouched classes at Crane site into discrete categories. This gives some credibility to its usefulness in assigning functions to the chipped stone artifacts of Crane, but does not *verify* the key or the functional assignments. First, the success of the key in sorting most of the retouched tools from Crane helps to evaluate directly only that aspect of it: its ability to *sort* artifacts into mutually exclusive classes, and not its ability to assign correct functions to those classes. It is true that since the patterns of co-occurrence of attribute states and the classes defined by the key are based on logical, ethnographically documented, and/or experimentally documented relationships between particular functions and particular attribute states, successful sorting also implies some degree of accuracy in the assignment of function. Nonetheless, the evaluation is an indirect one, and does not specifically test the hypothesized relationships and assignments, themselves. Second, the credibility of the key may be questioned because particular combinations of artifact traits encompassed by it could be indicative of more than one function, and the limited number of tasks considered by the key may not represent all the tasks in which the retouched tools from Crane site were used.

An independent check on the credibility of the key and the functions assigned by it might have been made by analyzing the kinds of wear occurring at the microscale on a sample of retouched tools from each functional class defined by the key, and determination of the functions implied by those wear patterns (Keeley 1977; Semenov 1964). Data for this checking procedure, however, are not available.

Point Forms

V5, V6, V7, V8, V117. Belknap, Norton, Snyders and Snyders/Norton, Gibson, and Steuben Points. Descriptions of and assignment of functions to these classes of retouched, nonweighty, chipped stone tools were not tabulated because they are distinguished largely on the basis of a traditionally recognized, stylistically and chronologically meaningful trait—outlined shape—rather than the functional attributes required by the key formulated. The classes were defined by Kenneth Farnsworth so that their spatial distributions might be plotted and the location of the different temporal components of Crane might be revealed.

The *classes* of point forms represent functionally mixed categories. For the purpose of this study, therefore, *individual* specimens rather than the classes had to be described and assigned functions. Only some specimens collected from the surface of Crane were available for study, however. For those points available, their particular defined functions were used in tabulating the activities reflected in the surface collection units in which they occurred. For the points unavailable for study, the full range of functions represented by their stylistic class—as determined by all those items that could be examined—were used in tabulating the activities possibly reflected in the surface collection units in which they occurred. The following generalities can be made about the items in each class as a whole.

With respect to their stylistic attributes, the various point forms included here have been described by Montet-White (1968) and Winters (1967:27). Chronologically, they belong to several different phases, as shown in Table 41. The temporal placement of the Norton, Snyders, and Gibson forms has been pinpointed by Montet-White (1968) through her examination of numerous Middle Woodland assemblages along the central and lower Illinois Valley. The chronological positions of Belknap and Steuben points are debatable. Rackerby (1973) believes that in the lower Illinois Valley, Belknap points are restricted to the Early Woodland period. This conclusion is based on his analysis of the spatial distributions of several temporal indicators and Belknap points at the Macoupin site. In the central Illinois Valley, however, Belknap points have been recovered from clearly Middle Woodland contexts—the Pond site, a single component, early Havana site. Winters (1967:36, 44) suggests that Belknap points extend from the Early Woodland into the Middle Woodland based on his survey of the Wabash Valley. Montet-White (1968:179) sees Belknap points as occurring mainly in the later portions of the Middle Woodland, but also notes (pp. 108, 109) the occurrence of similar contracting-stemmed forms in Early Woodland burials in the lower Illinois. At Crane site, Belknap points probably

TABLE 41
Pottery and projectile point types diagnostic of the individual components at Crane site

Type	Time Range	Component
Havana pottery, Bedford Phase	A.D. 1–150	Bedford
Hopewell pottery	A.D. 1–150	Bedford
Snyders points	250 B.C.–A.D. 100	Bedford
Norton points	A.D. 50–125	Bedford
Gibson points	A.D. 75–175	Bedford
Pike pottery	A.D. 150–400	Steuben
Baehr pottery	A.D. 150–400	Steuben
Steuben points	A.D. 125–250	Steuben
Whitehall-Weaver pottery	A.D. 400–600	Whitehall-Weaver
Jersey Bluff pottery	A.D. 600–1200	Jersey Bluff
Late Woodland points	A.D. 400–1200	Whitehall-Weaver and Jersey Bluff

can be assigned to the Middle Woodland rather than Early Woodland; no other Early Woodland diagnostics were recovered from Crane, nor are any Early Woodland occupations recorded in the Macoupin Valley, at large (Farnsworth 1973). If Winters is correct, they probably relate to the Bedford phase occupation at Crane, as do Snyders, Norton, and Gibson points.

The chronological positioning of Steuben points, likewise, is debatable. With respect to the lower Illinois Valley, Montet-White (1968) restricts Steuben points to the Middle Woodland Period. Johnson (1976) suggests a similar viewpoint in his discussion of sites in the lower Missouri Valley. Griffin (personal communication), however, has seen "Steuben-looking, Lowe flared base" points in Kentucky clearly associated with Late Woodland materials, and Maxwell (1951: 200, 245) pictures points with low flared bases for the Raymond Focus and Lewis Focus (Early Late Woodland) in southern Illinois. In this study, the viewpoint will be taken that the Steuben points at Crane belong to the latter portion for the Middle Woodland, alone, following the conclusions of Montet-White on those archeological assemblages most close to Crane in their geographic location.

With respect to function, the points within each of the several stylistic classes served multiple purposes that overlap greatly between groups. Table 42 shows the functional attributes of and the functions assigned to those specimens available for examination. The functional attributes of individual Snyders points keyed to scraping bone, scraping wood, dismemberment, defleshing hides, and graining hides. Norton points were found to be appropriate for scraping bone, scraping wood, cutting meat, and as projectiles. Belknap points were found to be suited to scraping bone and working wood. The Steuben points that I examined possessed attributes that suggested their use as projectiles and bone scrapers. In most cases, it was equivocal, on the basis of the attributes used in the formal key, whether individual specimens should be considered projectiles or scrapers. However, the high percentage of tip breakage among the items within this style class suggest that in general, Stueben points were used as projectiles.[10] Only two Gibson points were examined and both could be assigned the function of bone scrapers.

In addition to the multiple, overlapping functional nature of the several classes of points, two further generalizations can be made about their uses. The first concerns the functions of Snyders points compared to other point styles. Whereas the Norton and Steuben points that were examined indicate that sometimes these point styles were used as projectiles, none of the Snyders points that were examined exhibited attributes suggesting their use as projectiles. Of those attributes used by the formal key to assign function, most important is the unifacial pattern of retouch that tends to characterize Snyders points. Unifacial retouch, being asymmetrical and yielding an unequal distribution of forces on the two faces of the point, is less appropriate to projectiles as piercing tools than is bifacial retouch. The asymmetrical pattern of retouch occurring on Snyders points is carried even further in those points trimmed in a "rotary" fashion, that is, points with unifacial retouch from opposite faces on opposite edges. Points retouched in this manner are well suited to scraping tasks, allowing the use of both edges with a $180°$ twist of the item. Among the North American Indians, rotary point forms were used in working wood (Miles 1973:72). A second attribute that characterizes Snyders points and that makes them unsuitable projectiles is their large tip angle ($60°$ or more), and related to this, their rapidly expanding edges and wide dimensions. These characteristics are not appropriate to piercing tasks. In sum, from the limited sample of Snyders points that were examined from Crane site, it does not appear that Snyders points functioned as projectiles.

The second generalization of concern pertains to the function of Belknap points. Belknap points are thought by Winters (personal communication; Rackerby 1973) to have functioned as knives, typically, rather than projectiles. This inference is based on a distinction which Winters (1969:3) makes between projectiles and knives. Projectiles are thought to be bifacially retouched in such

---

10. Tip breakage on *individual* specimens cannot be used as an attribute diagnostic of their function as projectiles. All points examined come from the plowzone, where the probability of tip breakage is high. The greater percentage of tip breakage among Steuben points as a *whole class* compared to other classes, however, suggests their use as projectiles.

TABLE 42
Functional attributes and functions of some projectile points from Crane site

Variable Number	Point Style	Plan-View Shape of Edge	Cross-Sectional Shape of Edge	Edge Angle	Placement of Wear on Medial Section of Edge	Kind of Wear	Bifacial or Unifacial Retouch	Function to Which the Attributes Key
7	Snyders	slightly round	bi-convex	60°	2F	heavy ST, CH,TH	bifacial	scrape bone
7	Snyders	slightly round	bi-convex	62°	1F	ST,CH	bifacial	scrape bone
7	Snyders	slightly round	bi-convex	73°	2F	crushed	bifacial	scrape bone
7	Snyders	slightly round	bi-convex	73°	1F	ST	bifacial	scrape bone
7	Snyders	slightly round	bi-convex	68°	1F	CH	bifacial	scrape bone
7	Snyders	slightly round	bi-convex	64, 69°	2F	CH	bifacial	scrape bone
7	Snyders	slightly round	bi-convex	56°	--	none	bifacial	deflesh hides
7	Snyders	slightly round	bi-convex	57°	1F	CH	bifacial	deflesh hides
7	Snyders	slightly round	bi-convex	58°	2F	ST,CH	bifacial	dismember
7	Snyders	straight	bi-convex	54°	2F	ST	bifacial	dismember

TABLE 42 (cont.)

Variable Number	Point Style	Plan-View Shape of Edge	Cross-Sectional Shape of Edge	Edge Angle	Placement of Wear on Medial Section of Edge	Kind of Wear	Bifacial or Unifacial Retouch	Function to Which the Attributes Key
7	Snyders	slightly concave	bi-convex	71°	--	none	bifacial	grain hides
7	Snyders	slightly round	bi-convex	66°	D,V	CH	unifacial	scrape bone
7	Snyders	slightly round	plano-convex	68°	D, rotary	CH	unifacial	scrape wood or grain hides
7	Snyders	straight	plano-convex	66°	D, rotary	CH	unifacial	scrape wood or grain hides
7	Snyders	slightly round	bi-convex	65°	2F	CH	bifacial	scrape bone
7	Snyders	straight	plano-convex	51°	D	CH	unifacial	grain hides
6	Norton	straight	bi-convex	44,55°	slight CH,TH	2F	bifacial	projectile
6	Norton	straight	bi-convex	59°	CH	2F	bifacial	projectile
6	Norton	straight	bi-convex	<30°	none	2F	bifacial	projectile or cut meat, skin
6	Norton	straight	bi-convex	67°	heavy CH	1F	bifacial	scrape bone
6	Norton	straight	bi-convex	61° <20°	CH heavy CH	1F 2F	bifacial bifacial	scrape bone cut meat?
6	Norton	straight	plano-convex	67°	ST	D	unifacial	scrape wood

TABLE 42 (cont.)

5	Belknap	straight to slightly convex	bi-convex	64°	heavy CH, ST, TH	2F	bifacial	scrape bone
5	Belknap	straight to slightly convex	bi-convex	75°	ST	2F, 1F on different edges	bifacial	scrape bone
5	Belknap	straight to slightly convex	plano-convex	59°	ST	D	unifacial	scrape wood
8	Gibson	straight	bi-convex	67°	crushed	2F	bifacial	scrape bone
8	Gibson	straight	bi-convex	73°	ST,CH	1F	bifacial	scrape bone
12	Steuben	straight	bi-convex	57-74°	none	--	bifacial	projectile
12	Steuben	straight	bi-convex	51°	CH	2F, tip breakage	bifacial	projectile
12	Steuben	straight	bi-convex	67°	CH	1F, tip breakage	bifacial	scrape bone
12	Steuben	straight	bi-convex	59°	CH	2F, tip breakage	bifacial	scrape bone
121	Steuben	straight	bi-convex	65°	ST	2F, tip breakage	bifacial	scrape bone or projectile
121	Steuben	straight	bi-convex	64°	CH	2F	bifacial	scrape bone or projectile

a way that their edges are straight when viewed edge forward, whereas knives are thought to be bifacially retouched in an alternate flaking pattern so as to produce a wavy, sawlike edge efficient at cutting. Belknap points typically have the latter pattern of retouch. Whether or not Winters' *logic* is correct, Ahler's (1971) study of the point forms from the Missouri Archaic site, Rodgers Rock Shelter, supports Winters' *conclusion*. Broad-bladed points with contracting stems from this site were found to lack traits which typify projectiles—wear on the proximal portions of their faces and edges, and fracture by means of impact. Also, it can be noted that contemporary primitives use contracting stemmed points as hafted knives.

The Belknap points at Crane site support only a portion of the logic and conclusions reviewed here. There is no evidence to suggest that the Belknap points at Crane site functioned as projectiles. The specimens lack wear on the proximal portions of their faces and edges, and lack tip fracture. However, the alternative hypothesis, that they consistently served as knives, does not seem appropriate, either. The attributes possessed by Belknap points key to wood scraping and bone scraping.

In sum, in contrast to Norton and Steuben points, which appear to have been used as projectiles as well as tools of other functions, at Crane site, Snyders and Belknap points were not used as projectiles at all.

Chipped Stone Tools: Unretouched Knives, Scrapers "Celts," and "Cache" Blades (V11, 42, 58, 79b, 80, 97–99, 101, 104)

The items in nearly all of these classes are weighty tools that lack retouch and that have edges formed by only the flakes that have been removed with direct-percussion, hard-hammer techniques to give the tools their overall shape. Most are bifaces or unifaces, shaped in their entirety (faces and edges) by this knapping technique, but one class (V80) is composed of large, primary flakes, only the edges and a portion of their faces of which have been so shaped. Also to be discussed in this section are natural-edged knives made on lamellar blades. As with the retouched knives, scrapers, "celts," and "cache" blades, the overall morphology or wear patterns of the items in all these classes do not immediately suggest their function.

In assigning functions to these various classes of artifacts, use can be made of many of the same mechanical principles discussed when constructing the formal key for sorting retouched artifacts. For example, the distinction of knives and scrapers can be made on the basis of edge angle for unretouched unifaces and bifaces just as it can be for retouched tools. Not relevant, however, are those principles that pertain to the serrated nature of retouched edges and their efficiency in working different materials. For example, Ahler (1971) found that when bifacially retouched knives are used to carve meat or whittle wood, the protruding ridges along their functional edges quickly catch fibers of the material being worked, clog the edges, and make the tools inefficient at cutting. This circumstance could not be expected to characterize unretouched bifaces, however, for their edges lack closely spaced protrusions capable of being clogged.

Thus, the key that was used to assign functions to the retouched tools from Crane site cannot be applied directly to the unretouched unifaces and bifaces to determine the tasks in which they were used. In assigning functions to the tool classes considered in this section, notice will be given as to when the logic entailed in the formal key is being used and when other arguments are necessary. The required logic is presented here.

Descriptions of the functional attributes of the tool classes are summarized in Table 40. Other observations, including overall tool morphology and peculiarities in shape size, are given in Appendix 3. Bibliographic references to previous descriptions of the morphology of elaborate tools (e.g., celts) also are given in Appendix 3. The functions assigned to the tool classes discussed in this section are listed in Table 31.

As with some classes of retouched chipped stone tools, some classes of unretouched chipped stone tools possessed functional attributes that do not make clear the function of the tools. In these cases, the functions of the ambiguous types were chosen using information additional to that incorporated in the logical key: (*a*) the functions of the artifact types with which the ambiguous type co-occurs spatially and forms a depositional set; (*b*) ethnographic information additional to that used to construct the key and specific to the class; and (*c*) morphological attributes peculiar to the class in their usefulness in defining function, but not generally useful in this regard. This additional information is presented in the following, along with the logic used to assign functions to the classes.

V11, V58. Lamellar (Prismatic) Blades and Their Proximal Ends. On the basis of the formal key, these tools would be assigned the function of knives used in cutting meat, skin, and sinew during butchering and sewing activities. Experimental studies and ethnographic and archeological data concerned with natural-edged knives would suggest these functions, and also their use to cut vegetable fibers when weaving or making cordage, and to whittle wood, bone or horn. Gould (1971:154), Kenyon (1927), and McCarthy (1967:32) record the use of natural-edged knives by Australian Aborigines for cutting a wide range of soft materials, in general, including vegetable fibers. McCarthy (1949:10, 32) notes their use in Australia for the final, smoothing stages of woodworking, after shaping stages had been accomplished with heavy, natural-edged choppers and unifacially retouched scrapers. The possibility of using lamellar blades to work wood and bone has been described by Sollberger (1969). Lamellar blades have a distinct advantage over retouched knives and scrapers in working wood. A lamellar blade has a natural edge that is straight and unmarred, unlike unifacially and bifacially retouched edges, which have protruding ridges produced by the intersection of adjacent flake scars. The ridges of a retouched scraper striate the worked surface and never permit a completely smooth surface to be produced. The ridges of a retouched scraper also make them inefficient on hard substances, for they must be worn down before the scraper edge can make full contact with the material. A lamellar blade makes full contact with the material being worked from the start. The ridges of a scraper also collect fibers that clog the working edge. A lamellar blade, on the other hand, has a straight edge that does not collect fibers. The concave upper surface of a blade also provides room for shavings to rise with a minimum of friction; thus, the blade is self-cleaning. The convex lower surface of a lamellar blade allows control over the depth of cut simply by varying the working angle between the blade and the item being shaved. A retouched scraper has no attributes allowing depth control. Finally, because the base of a blade is convex, the cutting face can be oriented parallel to the forward motion of the blade, rather than in the perpendicular position in which a scraper is held. This allows thicker shavings to be made with less opportunity for tool wear.

The use of lamellar blades to work wood and bone is suggested by archeological evidence, as well. Montet-White (1968:95) has noted in Middle Woodland burials in Illinois that unretouched lamellar blades often are found in association with tools of bone. She infers from this that lamellar blades were used to manufacture bone tools. Green (1963) has described a cache of 17 lamellar blades found at the Clovis site (Paleo-Indian; New Mexico), which apparently had been carried in a skin bag on a hunting trip and left at the site. On the basis of the large amount of chipping wear that was restricted to the central portions of the blades, and the occurrence of accessory notches on two of the blades, Green (p. 156) suggests that the tools were used to prepare wooden shafts or bone foreshafts used while hunting at the Clovis site.

The pattern of wear on those blades at the Crane site bearing edge damage and their spatial association with other artifact classes also suggests that some lamellar blades, there, were used to work wood and bone. The lamellar blades at Crane that bear chipping wear usually exhibit it unifacially. This is the pattern that would be expected if the blades were used in a whittling manner. Spatially, the lamellar blades and portions of blades in classes V11 and V58 associate with tools that are used for woodworking. Class V11 lamellar blades co-occur, on the average, with the turtleback scrapers (V71) that have been assigned the function of soft wood scrapers, drills (V54) that were used to drill soft wood, and a specialized planing tool (V102) probably used in woodworking. Class V58 lamellar blade proximal ends are found in close proximity to turtleback scrapers (V72) used to scrape hard wood. The unretouched lamellar blades lacking visible wear, however, likely were used in cutting the softer materials mentioned previously.

The distinction between whole lamellar blades (V11) and the proximal ends of lamellar blades may reflect activity differentiation as well as a morphological distinction. The proximal ends of lamellar blades could indicate a purposeful modification of blades in order to produce a sharp, squared-off tip, such as that of a razor blade. An edge of this shape would be appropriate for making incisions, as when cutting meat. In contrast, whole lamellar blades might have been exploited for their lateral edges, which could be used to cut fibers or whittle wood or bone. The distal edges of whole lamellar blades, however, could also have been used in incising tasks.

V91, V92. Choppers. The items in this class have extremely sinuous edges (when viewed edge forward), formed by a coarse, direct-percussion, alternate-flaking technique. This attribute makes using the formal key inappropriate for assigning functions to these artifacts. It is necessary to rely on ethnographic analogies, spatial associations, and particular tool attributes to arrive at functional assignments for these choppers. Choppers had multiple uses among primitive peoples, as shown in Table 43. Within the limits of this set of functions, dismemberment is suggested by the functional attributes of the choppers in class V91, whereas woodworking is suggested by the functional attributes of the choppers in class V92. The items in both classes bear step fracture on their edges, which suggests their contact with harder materials, as when bones are disjointed or wood is scraped (Ahler 1971:84, 82). The items in class V91 possess wear on both faces, which is the expectable distribution for tools used to disjoint animal bones, whereas those in class V92 possess wear on only one face as would be expected for tools used to scrape wood. Neither class exhibits the marked crushed edges that would be expected to occur on tools used to pound and pulverize meat and vegetable materials.

The edge angles of the items in class V92 support the inference that they were used for working wood. Edge angles fall between 61 and 75°. The items in class V91 have edge angles that are larger than those suggested in the formal key for tools used in dismemberment. However, the tools with which the formal key is concerned are lighter-weight knives used in a cutting manner, whereas the tools in class V91 are heavier in weight and most likely were used in a vigorous, chopping manner. Larger edge angles would be expected in the latter case, where the force on the functional edge would be greater, and a stronger edge would be needed.

The functions of V91 and V92 choppers inferred from their morphological attributes and from ethnographic analogy are supported by the spatial associations of the items in these classes with artifacts of other classes. Class V91 choppers co-occur (depositional set 22) with lamellar blades (V11) and flake cores (V13), which were used, in part, in butchering tasks. Class V92 choppers occur in close proximity, on the average (depositional set 19), to V79 scrapers, which were used to rough-out forms in hard wood.

V97. Bifacial, Round-Edged Scrapers. The morphological attributes of the items in this class key them to hide graining. The lack of bifacial retouch on the edges of these items, the range of their edge angles, and the round, plan-view contour of their edges would suggest woodworking as an additional possible function. The slight amount of wear occurring on the edges of these tools, however, suggests that their use in graining hides is more probable.

TABLE 43
Functions which choppers can serve

Function	Reference
Crush meat and vegetable materials in a manner analogous to hammers (Table 36)	Cook (1973:9). North American Indians
Butchering and dismembering	Bordaz (1959), White (1952:338). Cross-cultural
Rough out forms in hard wood. Forms in hardwood cannot be roughed out easily with blades and whittling techniques as can forms in softwood, and require more heavy-duty tools	Ethnographic documentation: Horn and Aiston (1924), McCarthy (1949:10,12,15), Mitchell (1959:192). Australian Aborigines Experimental documentation: Crabtree (1968:426)

V98. "Celts." Although the items in this class are shaped bifacially, their cross-sectional shape is plano–convex, and they are more analogous to unifacially retouched tools than bifacially retouched tools in the distribution of forces on their functional edges. Taking this into consideration, the attributes of this class of items would key them to working soft woods. Additional information would suggest their use in scraping ribbons of bark and possibly wood from trees. The artifact types near which V98 bifaces are found most often at Crane include V22 grooved axes and V19 mauls. As already discussed, the latter two kinds of tools were used together, by the Iroquois (Waugh 1916:61) to remove layers of bark and wood from trees. Mallets were used to pound trees until layers of bark and wood separated from their interiors, and axes were used in the final removal of such layers. Bifaces of class V98 could have been used in this same procedure, replacing axes. Once bark of trees had been loosened by pounding, ribbons of it could have been scraped off with V98 bifaces. These artifacts are of appropriate weights for such operations. They also are long enough to allow a good grip with both hands, if they were pulled down over bark. Thus, on the basis of the morphology of these items and the functions of those artifacts with which they co-occur, it is suggested that V98 celts functioned in the procurement of bark and possibly wood.

V99. "Celts." As with V98 celts, the items in this class have a plano-convex cross-sectional shape and are more analogous to unifacially retouched tools than bifacially retouched tools in the distribution of forces on their functional edges. If this feature is taken into consideration, along with the other morphological attributes of these items, they would key to scraping hard wood and bone. Alternatively, V99 celts could have been used in the same manner as V98 celts, to procure bark and wood ribbons from trees. They are similar morphologically to V98 celts in all but their edge angles, which are larger. Spatially, they do not co-occur with wood and bark procurement tools, but do occur with hammer–grinders (V62) that could have served secondarily as anvils used in pounding bark prior to weaving baskets of this material (depositional set 3). This association lends some support to the use of V99 celts in activities involving bark.

On a more general level, the fact that the tools with which V99 celts occur are all used in women's activities also suggests the use of the celts to procure bark. Class V99 celts co-occur with cache blades (V42) probably used to deflesh hides, and hammer–grinders (V62) used primarily to process nuts and seeds. These activities were performed by women in aboriginal North America (Mason 1889). On the basis of the association of V99 celts with tools used in female activities, we can suggest their use, too, in a female activity. This would favor the interpretation of their use in procuring bark rather than scraping hard wood and bone. Procurement of bark ribbons for weaving baskets was a typically woman's job among the American Indians. On the other hand, working bone and manufacturing bone tools typically was a male activity (Mason 1899).

The one attribute of V99 celts that prevents complete rejection of the idea that they were not used in hardwood and bone scraping but rather in bark procurement is the large edge angles which they possess. The steep bits of V99 celts may be inappropriate for removal of long peelings of bark. No experimental data are available to judge this capability of the celts. Consequently, it can be concluded only that V99 celts may have been used for procuring bark or for scraping hard wood or bone.

V101. "Celts." The morphological attributes of the items in this class relate to graining hides, according to the formal key. This assignment of function is not unreasonable, despite the lack of fine retouch serrations along the edges of the items. In North America, graining and thinning of hides was done with unserrated blades resembling adzes in shape and manner of use, and with coarsely serrated tools, as well as with finely serrated tools (Mason 1889). Some of the tools used to grain hides have been called "hoes" (Mason 1889; Miles 1973:99)—a characterization that easily could describe the general morphology of items in class V101. The spatial association of V101 celts with V85 knives, which have functional attributes keying them to dehairing hides, supports the inference that the celts in class V101 were used in hide working.

V42. "Cache" Blades. The functional attributes of the items in this artifact class key to dismemberment. The lack of fine retouch along the edges of these items does not alter this assignment of function as a possibility, nor does it admit other functions as possibilities. The other attributes of these items still limit the set of tasks for which they are appropriate.

V79b. Unifacial, Straight-Edged Scrapers. The functional attributes of the items in this class key to scraping bone, but scraping of hardwood is more probable. In the formal key for retouched tools, scraping wood was not considered a possible function of bifacially retouched tools based on the studies of Sollberger (1969) and Ahler (1971). The protrusions along the edges of bifacially retouched items do not allow good contact between their functional edge and the material being worked, if it is of a harder nature and does not give under pressure. Moreover, the flake scars between the protrusions on bifacially retouched edges clog with wood after only short periods of use, and render the edges even more inefficient. These circumstances do not arise with the edges of unretouched bifaces and do not prohibit their use in scraping wood. Thus, V79b tools could have been used to scrape wood as well as bone. On the basis of the edge angles of the items in this class, and by way of analogy to unifacial scrapers, hard wood is more likely to have been worked. The weighty nature of these items suggests their use in the coarser stages of woodworking, such as roughing-out forms.

V80. Unifacial, Straight-Edged Scrapers. The functional attributes of these items key to scraping bone. Their heavy weight suggests their use in coarser stages of bone working.

V104. Bifacial, Straight-Edged Scrapers. These items are similar in shape to those in classes V105 and V106 but are larger, more crudely shaped, and lacking in retouch. The plano–convex cross-sectional shape of these items makes them more analogous to unifacially retouched tools than to bifacially retouched tools. On the basis of this consideration and the other functional attributes possessed by the items in this class, the function of scraping hardwood may be assigned to them.

Use-Areas within the Crane Site: Their Depositional and Functional Characteristics

The aim of this chapter is to determine areas of Crane site where different activities actually occurred, generated residues of different kinds, and possibly altered the soil in different ways. It was suggested that to do this, it is necessary to partition Crane into subareas along two dimensions of use: one concerned with the *kind of activity* indicated by the refuse which occurs in those subareas, and a second concerned with the *depositional nature* of the subareas—whether they actually served as locations of activity, or instead as locations of artifact disposal, storage, or caching.

The task of dividing Crane into use-areas along these two dimensions may be achieved if the surface collection unit is taken as the spatial unit of analysis and if the artifact inventories of each collection unit are examined. The first dimension of use of each collection unit can be specified by tabulating the classes of artifacts found in each collection unit at Crane and by considering the manner in which they functioned, as determined in the previous sections. Table 44 lists the artifact classes present or occurring in "significantly high densities" in each collection unit where soil and/or resistivity surveys were made. Also listed (by code, Table 45) are the activities implied by these classes. The spatial distribution of the collection units for which this information is listed is shown in Figure 37a and 37b.

In most cases, the activities listed for a given collection unit represent a complete tabulation of all the activities implied by all the artifacts found in that unit. An exception to this concerns tools that had multiple purposes at the Crane site (e.g., V11 lamellar blade knives) or tools for which it was not possible to specify a precise function and for which multiple, possible, alternative functions had to be postulated (e.g, V56 chert hammers), even after an analysis of artifact type co-occurrences had been made. In units having tools of this kind, not all of the alternative tasks for which the tools possibly could have been used were listed; only those tasks implied by the other tools present in the units were listed. For example, V72 turtleback scrapers at Crane could have been used to work either wood or bone. Woodworking functions were listed for those units in which they co-occurred with other wood-working tools, whereas bone-working functions were listed for those units in which they co-occurred with other bone-working tools. Where the artifacts occurring in a collection unit do not clarify the functional nature of an ambiguous item, all the alternative activities in which that item might have been used are listed (e.g., "activities 14 or 25").

It should be noted that the artifact classes tabulated to occur in the collection units at Crane and used to reconstruct the activities that may have occurred in them include all homogeneous artifact types defined for the surface materials (Table 31), regardless of their age. In contrast, only artifacts of the Bedford phase were used in determining spatial patterns of co-occurrence among artifact types (see Appendix 2). The different sets of artifact types used at these two different stages of analysis reflect the different goals and assumptions required at each stage. Patterns of co-occurrence among artifact types were analyzed to help determine the possible functions of ambiguous types. The analysis required that all items be synchronous in nature. Consequently, artifacts not belonging to the Bedford phase had to be excluded from the analysis; and it had to be assumed that all items which could not be dated probably are assignable to the Bedford phase. This assumption was checked at the beginning of this chapter and seems reasonable. On the other hand, when reconstructing the activities that may have occurred in the collection units at Crane, activities that occurred during any of the time phases represented at Crane are of concern. Activities that occurred during any of these phases may have caused soil alterations. Thus, artifact types of any age were used in reconstructing the use of space on the site.

The depositional nature of the archeological remains in each collection unit—

TABLE 44
Surface collection units at Crane site and their archeological characteristics

Unit Number+	Number of Artifact Classes Present*	Artifact Classes Present*	Activities Indicated by Artifact Classes**	Primary or Secondary Deposit	Criterion Used***
TI-141	1	5	(8 or 23)	p	1,7
TI-162	0		none	p	1,7
TI-142	3	56,71	4, (39 or 42 or 54 or 70)	p	1,7
TI-163	6	12,16,40,57, 61,64,113	31,46, (19a or 19b), (64,67,68)	s	1,7,8
TI-164	4	30,53,56,81	29,47, (17 or 18a or 18b), (39 or 42 or 70 or 71)	p	1,7
TI-185	4	13,31,32,34, 117	33,34, [[(3,4) or (22,23)] or (44 or 47 or 59)]	p	1,7
TI-165	6	7,13,30,53, 56,66,113	See TIV-165	s	1,7,8
TI-186	6	11,13,28,30, 31,38	See TIV-186	s	1,7,8
TI-187	8	12,24,30,34, 60,81,84,91, 113,117	47,49,61, [(3 or 22), (4 or 23)], (29,30,31), (28 or 44 or 73 or 75 or 76), [(63,65) or (67,68)]	s	1,7,8
TI-188	4	5,62,68,107b, 119	(23,26,28) [(67,68,75,76) or (70,74)],	p	1,7
TI-189	8	1,7,13,30,44, 60,69,84	(3,15), (29,33), [(63,65,75) or (67,68,76)]	s	1,7,8

TI-190	3	19,31,66	(2,14,33)	p	1,7
TI-191	8	2,10,14,30, 34,58,66,72	83, (8,14,36), (29,31,35)	s	1,7,8
TI-192	5	12,13,64,71, 81,118	31,47,83, (3,4,33), (64,66)	p	1,7
TI-193	5	1,26,50,81, 103,118	83, (44,46,47,48, or 52)	p	1,7
TI-194	1	29	86	p	1,7
TI-195	1	7	?	p	1,7
TI-196	2	26,31	33,86	p	1,7
TI-197	0		none	p	1,7
TI-198	0		none	p	1,7
TI-199	1	75	4	p	1,7
TI-200	2	81,83	(45,47)	p	1,7
TI-201	0		none	p	1,7
TII-318	0		none	p	1,7
TII-332	5	18,27,28,32, 37,84,117, 118	34,83, (3,6), (62,63,73), (78 or 79)	p	1,7

TABLE 44 (cont.)

Unit Number	Number of Artifact Classes Present*	Artifact Classes Present*	Activities Indicated by Artifact Classes**	Primary or Secondary Deposit	Criterion Used***
TII-333	3	27,65,71,117	4, (64,56,75,76), (78 or 79)	p	1,7
TII-334	12	1,4,7,12,17, 25,27,28,30, 31,70,86,117	(14 or 25 or 37), (29,31,33), [(38,40) or 41], (53 or 36), (55 or 73), (78 or 79), (80 or 82)	p or s	6,5,8
TII-349	12	1,3,4,7,11, 12,14,17,27, 28,49,54,84, 119	(3,17), [(21,22,28,43) or (55,56,57,73)], (31,33), (40 or 46 or 47), (78 or 79), (80,82)	p or s	6,5,8
TII-350	17	1,3,4,11,15, 16,17,23,25, 26,27,28,29, 30,61,62,64, 70,117,118,119	29, 83, [(((3,14) or 37),(55,56,57))or (21,22,25)] p or s [28 or 73 or (75,76)], (64,67,68,69), (78 or 79), (80,82)	p or s	6,5,8
TII-351	16	1,2,3,4,10, 11,16,17,23, 27,28,30,61, 62,64,65,94, 117	83, [(21,22,27) or (3,20, 55,56,57)], (29,35), [28 or 73 or (75,76)], (64,67,68), (78 or 79), (80,82)	s	6,5,8
TII-365	7	1,4,12,15, 17,23,27,28, 119,120	(29,31,32), (55 or 73), (78 or 79), (80,82)	p	4,3
TII-380	8	1,3,4,13,15, 17,27,28,56, 120	54, [(21,22,28) or (55,56,57,73)], (32,33), (78 or 79), (80,82)	p	4,3

312

TII-381	6	1,17,23,27 28,92	(5 or 7), [(55 or 73), (78 or 79)], (80,82)	p	4,3
TII-396	5	1,17,27,28 83,84	45, (3 or 22), (28 or 43 or 55 or 73), (78 or 79),(80,82)	largely p	4,3
TII-413	9	1,3,10,15, 17,27,28,72, 77,89a,119	8,83, (32,35), (55 or 73), (78 or 79), (80,82)	largely s	4,3
TII-414	11	1,3,4,11,15, 17,23,27,28, 78,81,82,120	32,47,83, [(21,22,23) or (3,8, 55,56,57)], (28 or 43 or 73), (78 or 79), (80,82)	largely s	4,3,2
TII-431	13	1,3,4,15,16, 17,23,25,27, 28,62,76,100, 113,117,118	23,83, (28 or 73 or 76), (30,32), (67,68,69), (78 or 79), (80,82)	s	6,5,8
TII-448	6	1,15,17,45 77,92,118	32,72,83, [4, (5 or 7), 10], (80,82)	p	4
TII-449	12	1,11,15,17, 23,27,44,57, 75,77,89a,117	77, (32,33), (19a,38), (78 or 79), (80,82)	s	6,5,8
TII-450	4	1,16,17,27, 53,117,120	77,83, (17 or 18a or 18b), (80,82)	p	1,7
TII-451	2	17,60	[(63,65) or (67,68)],(80,82)	p	1,7

TABLE 44 (cont.)

Unit Number	Number of Artifact Classes Present*	Artifact Classes Present*	Activities Indicated by Artifact Classes**	Primary or Secondary Deposit	Criterion Used***
TII-467	6	3,14,17,23, 27,46	11,31, (78 or 79), (80,82)	p	1,7
TII-468	6	13,16,23,31, 58,81,120	(33,36), (43,47,59), (80,82)	p	1,7
TII-469	4	17,61,62,66	14, (67,68), (80,82)	p	1,7
TII-470	4	17,62,82,89a	(4 or 23), [(67,68) or 70], (80,82)	p	1,7
TII-487	1	31	33	p	1,7
TII-488	0		none	p	1,7
TII-489	1	17	(80,82)	p	1,7
TII-490	1	13	33	p	1,7
TII-507	2	52,61,113	30, (12 or 24), (67,68)	p	1,7
TII-508	2	17,52,117,118	[(12-72) or (24,28)], (80,82)	p	1,7
TII-509	1	24	49	p	1,7
TIII-619	1	12	31	p	1,7
TIII-620	1	18	(52 or 62)	p	1,7
TIII-621	3	76,94,100	(23,27)	p	1,7
TIII-630	1	13	33	p	1,7
TIII-631	1	50	[40 or (46 or 47)]	p	1,7

TIII-632	4	17,18,84,89a, 75, (3,8), (62,63), (80,82) 120		p	1,7
TIII-633	1	107a	47	p	1,7
TIII-634	0		none	p	1,7
TIII-635	4	6,60,106, 107b	23, [(63,65) or (67,68)]	p	1,7
TIII-622	2	26,36	(32,47,48)	p	1,7
TIII-636	0	120	(72 or 84)	p	1,7
TIII-624	2	62,87a	(57 or 58 or 59), [(67,68) or 70]	p	1,7
TIII-625	1	99	(2 or 8 or 23)	p	1,7
TIII-626	1	2	83	p	1,7
TIII-627	3	7,10,103	35,47	p	1,7
TIII-614	1	65	64,66?	p	1,7
TIII-628	1	92	(5 or 7)	p	1,7
TIII-615	0		none	p	1,7
TIII-629	0		none	p	1,7
TIV-1	1	88	45	p	1,7

315

TABLE 44 (cont.)

Unit Number	Number of Artifact Classes Present*	Artifact Classes Present*	Activities Indicated by Artifact Classes**	Primary or Secondary Deposit	Criterion Used***
TIV-9	4	60,75,85,101	(6,8), (45,47), [(63,65) or (67,68)]	p	1,7
TIV-20	1	1	(72 or 84)	p	1,7
TIV-33	4	25,66,81,84	(3,14), (47,48)	p	1,7
TIV-49	1	66	14	p	1,7
TIV-66	3	14,27,46	11,31, (78 or 79)	p	1,7
TIV-85	3	11,13,14,27	31,33, (78 or 79)	p	1,7
TIV-105	5	15,27,28,30, 46,71,118	29,32,73,83, (8,11), (78 or 79)	p	1,7
TIV-125	5	7,21,27,36, 44,54	33,47, (3,7,17), (78 or 79)	p	1,7
TIV-145	3	15,27,54,118	17,32, (78,83)	p	1,7
TIV-166	4	1,20,27,28, 83	23, (21 or 55 or 73), (44,45), (78 or 79)	p	1,7
TIV-165	6	7,13,30,53, 56,66,113	(3 or 60), (29,30,33), [(17 or 18a),14], 39 or 42 or 70 or 71)	s	1,7,8
TIV-186	6	11,13,27,28 30,31,38	(3,4), [(21,22,28) or 55,56,57,73)], (29,33), (78 or 79)	s	1,7,8
TIV-185	4	13,31,32,34, 117	33, [(3 or 22),(4 or 23)) or (44,47,59)], (34,35)	p	1,7
TIV-206	3	7,58,81	(33,47,59)	s	1,7,8

TIV-225	2	7,31,118	83, (33,47,59)	p	1,7
TIV-224	4	14,32,33,113, 114,118	83, (29,30,31,32,34)	p	1,7
TIV-241	2	31,113,114	(30,32,33)	p	1,7
TIV-242	3	14,34,100, 118	23,31,83	p	1,7
TIV-240	1	30	29	p	1,7
TIV-254	3	58,72,81	8,47 [36, (12 or 59)]	p	1,7
TIV-253	0		none	p	1,7
TIV-252	0		none	p	1,7
TIV-267	0	118	83	p	1,7
TIV-281	2	5,45	8,10	p	1,7
TIV-282	0		none	p	1,7
TIV-295	6	1,17,27,28, 30,31,91,117	(29,33), (61,73,78), (78 or 79), (80,82)	p	1,7
TIV-296	1	17	(80,82)	p	1,7
TIV-309	4	30,31,76,77, 117	8, (29,33), (72 or 84)	p	1,7
TIV-310	2	30,43	(29,47)	p	1,7

TABLE 44 (cont.)

Unit Number	Number of Artifact Classes Present*	Artifact Classes Present*	Activities Indicated by Artifact Classes**	Primary or Secondary Deposit	Criterion Used***
TIV-322	8	1,4,11,17, 28,30,52, 108,118	29,33,47,83,[(21,22,24) or (3,12,55,56,57)], (28 or 43 or 73), (78 or 79), (80,82)	p	1,7
TIV-336	8	1,4,11,17 28,33,61,62, 117	[(21,22) or (55,56,57)], (29,33), (28 or 73 or 76), (67,68), (78 or 79), (80,82)	s	2
TIV-350	17	1,3,4,11,15, 16,17,23,25, 26,28,29,30, 61,62,64,70, 117,118,119	see TII-350	p or s	6,5,8
TIV-351	16	1,2,3,4,10, 11,16,17,23, 28,30,61,62, 64,65,94,117	see TII-351	s	6,5,8
TIV-364	15	1,3,4,5,10, 11,14,15,16, 17,23,27,28, 60,84,117,118, 119,121	83, [(3,55,56,57) or (21,22)], (31,32,35), (32,33), (28 or 73 or 75 or 76), [(63,65) or (67,68)], (78 or 79), (80,82)	p or s	6,5,8
TIV-365	7	1,4,12,15, 17,23,28,119, 120	see TII-365	p	4

318

TIV-379	10	1,3,4,11,13 15,17,27,28, 60	[(21,22) or (55,56,57)], (28 or 73 or 75 or largely p 76), (32,33), [(63,65) or (67,68)], (78 or 79), (80,82)	4
TIV-393	11	1,2,3,4,15, 17,18,27,28, 56,58,65,101, 113,120	30,47,83, (32,36), [(38 or 73 or 74) or 43 or p 75)], (39 or 42 or 70 or 71), (62,63), (78 or 79), (80,82)	4,3
TIV-394	11	1,3,4,15,17 23,27,28,30, 44,75,76,120	(3,8), [(21,22,28) or (55,56,57,73)], (29,32, p 33), (78 or 79), (80,82)	4,9?
TIV-411	11	1,3,13,15, 17,27,28,31, 41,76,83,84, 120	45,61, [(21,22,23) or (3,8,55,56,57)], (28 or p 43 or 73), (32,33), (78 or 79), (80,82)	4,3
TIV-410	7	9,16,17,27 53,56,57,83, 118,120	[(17 or 18a or 18b), (19a or 19b)], [28 or p (38 or 73 or 74) or 43], (39 or 42 or 70 or 71), (45,47), (78 or 79), (80,82), 83	4,3
TIV-427	4	1,3,17,24, 27,121	49,77, (80,82)	p	4
TIV-443	1	32,118	34,83	p	4
TIV-444	1	12	31	p	4
TIV-461	3	31,45,49	10,33, [40 or (46 or 47)]	p	1,7

TABLE 44 (cont.)

Unit Number	Number of Artifact Classes Present*	Artifact Classes Present*	Activities Indicated by Artifact Classes**	Primary or Secondary Deposit	Criterion Used***
TIV-460	1	100	23	P	1,7
TIV-478	2	14,24	31,49	P	1,7
TIV-496	3	17,29,92,117, 121	72,86,87, (5,7), (80,82)	P	1,7
TIV-495	0	121	87	P	1,7
TIV-514	3	10,71,99,120	8,35, (83 or 84)	P	1,7
TIV-515	2	43,73,113, 120	8,30, (44,47)	P	1,7
TV-496	3	17,29,92, 117 121	See TIV-496		
TV-497	5	13,15,17,53 81,117,118, 120	83, [(17 or 18a or 18b)or (3 or 22)], (28 or 72), (32,33), (80,82)	P	1,7
TV-498	4	21,62,67,95, 120	(5,7,14),[(67,68,76) or (70,74)]	P	1,7
TV-479	2	52,58	[(12 or 24), 36]	P	1,7
TV-480	2	17,82,118, 119	83, (43,47), (80,82)	P	1,7
TV-481	3	18,45,60,120	10, (63,67,75)	P	1,7

TV-462	6	29,50,61,62 65,99,117	86, [2 or (8 or (23,28))], [38 or 43 or (75,76)], [40 or (46,47)], (64,67,68)	p	1,7
TV-463	3	17,33,113, 114,118	(29,30,32) (78 or 79), (80,82)	p	1,7
TV-464	6	11,13,16,17 29,53	[3 or (17 or 18a) or (18b,22)], 33, (78 or 79), (80,82), 86	p	1,7
TV-446	7	12,15,17,24 27,45,60,62 117,118	10,49,83, (30,31,32), (67,68,76), (78 or 79), (80,82)	p	4
TV-447	7	1,16,30,31, 56,58,84	[((3,12,72) or (22,24,28)),36],(29,33), [(80,82) or 81]	p	4
TV-448	6	1,15,17,45, 77,92,118	See TII-448	p	4
TV-430	11	1,3,4,9,11, 15,17,18,27, 33, 62 ,113, 115,117	85, (3 or 22 or 57 or 58 or 59 or 60), [(75,76) or 28], (29,30,32), (62,63,67,68), (78 or 79), (80,82)	p	4
TV-431	13	1,3,4,15,16, 17,23,25,27, 28,62,76,100, 113,117,118	See TII-431	s	6,5,8

TABLE 44 (cont.)

Unit Number	Number of Artifact Classes Present*	Artifact Classes Present*	Activities Indicated by Artifact Classes**	Primary or Secondary Deposit	Criterion Used***
TV-432	15	1,3,4,10,15, 17,24,25,27, 61,62,65,94, 101,108,119	47,49, (20 or 27), (32,35), (43 or 75 or 76), (48 or 69), (64,67,68), (78 or 79), (80,82)	s	6,5,8
TV-414	11	1,3,4,11,15, 17,23,28,78, 81,82,120	See TII-414	largely s	4,3
TV-415	10	1,3,4,13,14, 17,23,24,27, 28,35,113,119, 120	23,49, [(21,22,28) or (55,56,57,73)], (30 or 31), (78 or 79), (80,82)	p	4,3
TV-416	10	2,3,4,15,17, 25,27,28,73, 81,100,120	8,23,32, [(43,55,73) or(28,43)], (78 or 79), (80,82), 83	p	1,7
TV-398	7	3,4,14,17, 24,25,27,28	31,49, (53 or 86), (55 or 73), (78 or 79), (80,82)	p	1,7
TV-399	6	1,17,23,27, 28,81	47, [(43,73) or (43,55)],(78 or 79), (80,82)	p	1,7
TV-400	6	6,17,27,53, 56,60,113, 120	30, 87,[(17 or 18a or 18b), [(63,65) or (67, 68)], (78 or 79), (80,82), [28 or 38 or 73 or 74 or 75 or 76], (39 or 42 or 70 or 71)	p	1,7
TV-384	5	4,17,23,24, 83,120	45,49,72, (80,82)	p	1,7

TV-385	4	17,23,25,103, (43,47,48),72,(80,82) 120	p	1,7
TV-386	0	118	p	1,7
TV-370	2	47,77 (4,16)	p	1,7
TV-371	0	none	p	1,7
TV-372	1	36 47	p	1,7
TV-357	3	60,78,89c 47, (8 or 23), [(64,65) or (67,68)]	p	1,7
TV-358	0	none	p	1,7
B-322	8	1,4,11,17, See TIV-322 28,30,52, 108,118	p	1,7
B-334	12	1,4,7,12,17, See TII-334 25,27,28,30, 31,70,86,117	p or s	6,5,8
B-335	9	1,4,11,13, 33,47,83, [(21,22) or (55,56,57)], (28 or 43 17,18,23,27, or 73 or 75), (62,63), (78 or 79), (80,82) 28,81,117,118	p	4
B-336	8	1,4,11,17, See TIV-336 28,33,61,62, 117	s	2

323

TABLE 44 (cont.)

Unit Number	Number of Artifact Classes Present*	Artifact Classes Present*	Activities Indicated by Artifact Classes**	Primary or Secondary Deposit	Criterion Used***
B-337	7	1,4,11,15, 17,27,28,53, 117	32, [(3,17,18a,55,56,57,73) or (21,22,28, 18b)]=(78,79), (80,82)	p	1,7
B-349	12	1,3,4,7,11, 12,14,17,28, 49,54,84,119	See TII-349	p or s	6,5,8
B-350	17	1,3,4,11,15, 16,17,23,25, 26,28,29,30, 61,62,64,70, 117,118,119	See TII-350	p or s	6,5,8
B-351	16	1,2,3,4,10, 11,16,17,23, 28,30,61,62, 64,65,94,117	See TII-351	s	6,5,8
B-364	15	1,3,4,5,10, 11,14,15,16, 17,23,27,28, 60,84,117,118, 119,121	See TIV-364	p or s	6,5,8
B-365	7	1,4,12,15, 17,23,28,119, 120	See TII-365	p	4
B-379	10	1,3,4,11,13, 15,17,27,28,60	See TIV-379	p and s	4

B-380	8	1,3,4,13, 15,17,28,56, 120	See TII-380	p	4,3
B-382	13	1,3,4,10,13, 15,17,18,23, 25,27,28,61, 62,117,120	[(21,22) or (55,56,57)], [28 or 73 or (75, 76)], (32,33,35), (62,63,67,68,69), (78 or 79), (80,82)	s	6,5,8
B-397	10	1,4,15,17, 23,27,28,46, 76,94,99,118	32,83, (8,11,20), (55 or 73), (78 or 79), (80,82)	p and s	6,5,8,4,3
B-412	6	1,15,17,23, 27,28,62,117, 118,119,120	32,83, (55 or 73), [(67,68,76) or (70,74)], (78 or 79), (80,82)	p	4,3
B-413	9	1,3,10,15, 17,27,28,72, 77,89,119	See TII-413	largely s	4,3
B-414	11	1,3,4,11,15, 17,23,28,78, 81,82,120	See TII-414	largely s	4,3,2
B-415	10	1,3,4,13,14, 17,23,24,28, 35,113,119, 120	See TV-415	p	4,3

TABLE 44 (cont.)

Unit Number	Number of Artifact Classes Present*	Artifact Classes Present*	Activities Indicated by Artifact Classes**	Primary or Secondary Deposit	Criterion Used***
B-431	13	1,3,4,15,16, 17,23,25,27, 28,62,76,100, 113,117,118	See TII-431	s	6,5,8
B-432	15	1,3,4,10,15, 17,24,25,61, 62,65,94,101, 108,119	See TV-432	s	6,5,8
B-447	7	1,16,30,31, 56,58,84	See TV-447	p	4

+Collection unit numbers preceded by a "T" indicate those crosscut by the Transect Resistivity Survey. Those preceded by a "B" were encompassed by the Block Resistivity Survey

*Classes present but not included in the counts include 113, 117-121

**Codes are explained in Table 45.

***Codes are explained in Table 46.

TABLE 45
Code numbers for activities which occurred at the Crane site and are listed in Table 44

1. Felling or girdling trees
2. Removing slabs of wood and bark from trees
3. Whittling wood, in general
4. Scraping wood, in general
5. Rough shaping of soft wood forms
6. Scraping soft wood
7. Rough shaping of hard wood forms
8. Scraping hard woods
9. Initial shaping of hard wood shafts, or scraping shafts of bone
10. Smoothing (by scraping) hard wood shafts, ¼" diameter or less
11. Smoothing (by scraping) hard wood shafts ¼ - 11/16" diameter
12. Grooving hard wood, in general
13. Grooving hard wood shafts
14. Sanding wood, other than shafts
15. Sanding wooden shafts
16. Scraping angular surfaces of wood (e.g., edges of bowls, atlatls, shuttles)
17. Drilling soft woods
18a. Drilling hard woods
18b. Drilling bone
19a. Sawing wood
19b. Sawing bone
20. Splitting wood with wedges

21. Generalized manufacturing of bone tools, used in sewing, leather working, weaving, basket making
22. Whittling bone
23. Scraping bone
24. Grooving bone
25. Sanding bone, in general
26. Sharpening pointed bone and possibly wooden implements by sanding

TABLE 45 (cont.)

27. Splitting bone with wedges
28. Boiling bone to soften it prior to working it

29. Manufacturing stone artifacts, in general
30. Manufacturing stone artifacts by percussion, using hammerstones
31. Primary flint knapping: decortication
32. Secondary flint knapping: secondary falking
33. Producing flake or blade knives
34. Manufacturing hoes?
35. Resharpening hoes
36. Manufacturing edges of blades to be used in grooving wood or bone
37. Polishing stone axes or celts

38. Boiling fibrous plant material to break up the parenchymatic tissues, prior to shredding the material and extracting fibers
39. Pounding fibrous plant material prior to shredding it and extracting fibers
40. Shredding and extracting plant fibers
41. Shredding animal sinew
42. Pounding bark to make cloth and basketry
43. Simmering hides with brains, and/or plain warm water
44. Soaking hides in salt water, in soapy water made with roots, in soapy water made with dissolved deer brains, in water with bark, in water with organic dyes, in water with ash (lye), and/or in urine
45. Dehairing hides
46. Defleshing hides
47. Graining and thinning hides
48. Coloring hides

49. Manufacturing pottery
50. Shaping and smoothing the insides of pots
51. Heavy trampling near places where pottery was used
52. Grinding red ochre
53. Manufacturing paint
54. Crushing bone to extract marrow and to prepare it for boiling for tallow and grease

TABLE 45 (cont.)

55. Boiling bone to obtain tallow and grease

56. Butchering of animals, in general
57. Cutting meat
58. Cutting sinew
59. Cutting hide
60. Cutting plant fibers
61. Dismemberment

62. Hulling delicate seeds with grindingstones
63. Grinding delicate seeds with grindingstones
64. Grinding hulled seeds, in general
65. Pounding and hulling seeds in a wooden mortar
66. Pounding nuts, alone
67. Pounding nuts, in general, and/or sunflower seeds
68. Grinding acorn and hazelnut meats
69. Ingredient in bread made from bitter acorns
70. Pulverizing dried fruits, bulbs, roots, and/or rhizomes
71. Pulverizing dried meat

72. Boiling, in general
73. Boiling meat
74. Boiling vegetables
75. Boiling seeds
76. Boiling nuts for oil or to leach out tannic acid (acorns only)
77. Roasting insects or meat in a dry pot, indirectly
78. Roasting or baking meat, directly
79. Fuel for fire
80. Hearth liner: fire
81. Use in construction of earth ovens; fire
82. Cleaning out hearths

83. Serving food?
84. Storing
85. Cultivating the ground
86. Ceremonial
87. Projectile points for hunting

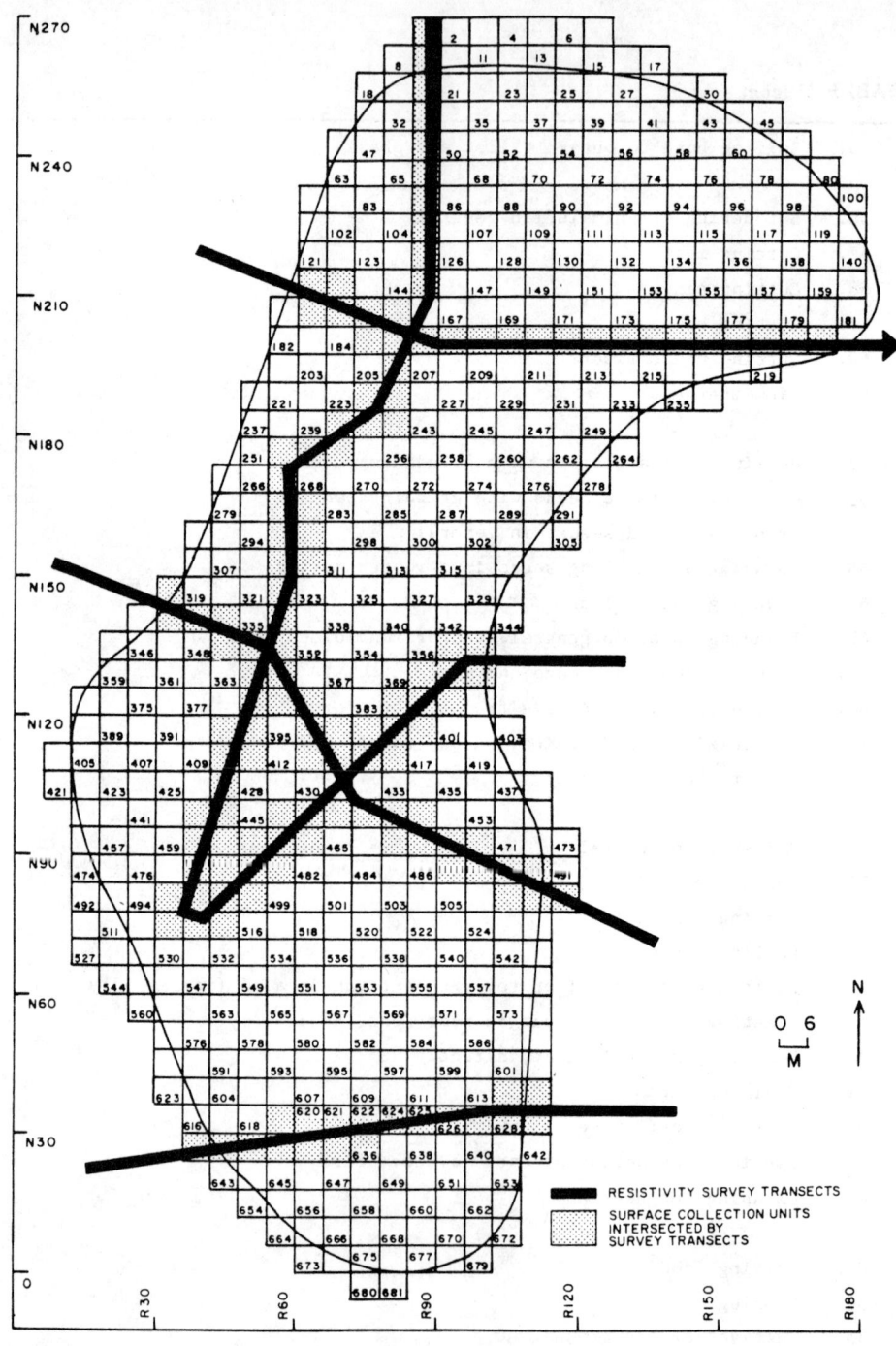

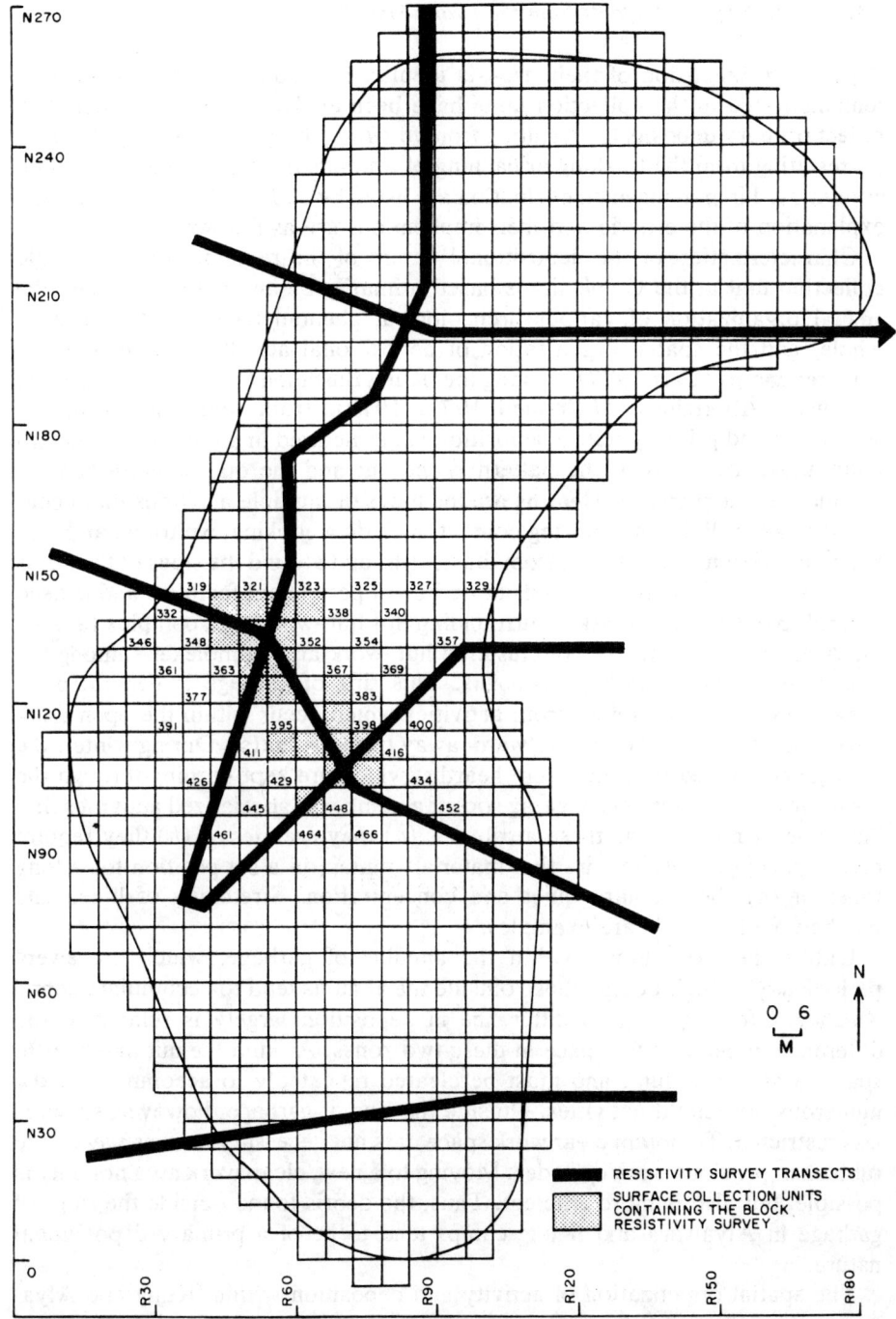

FIGURE 37 *(a, opposite) Spatial distribution of surface collection units in which resistivity and soil surveys were performed in a transect design. (b, above) Spatial distribution of surface collection units in which resistivity and soil surveys were performed in a block design.*

the second dimension of their use—is tabulated in Table 44. The artifactual remains found in the collection units have been evaluated as to whether they reflect primary deposits (p) resulting from *in situ* activity or secondary deposits (s) resulting from the trashing or caching of artifacts used in another area. The criteria used in making each evaluation are listed by code in the same table. An explanation of the criteria and their implications are as follows.

Characterization of the depositional nature of the remains within a single collection unit at the Crane site is based primarily on an analogy between the spatial organization of various archeological phenomena over the site as a whole, and the spatial organization of depositional activities within hunter–gatherer camps (Carr 1977). Among the !Kung Bushmen (Yellen 1974) and the Alyawara Aborigines (O'Connell 1977, 1979), work areas and zones of secondary and primary refuse deposition are structured in an annular (ringlike) pattern. At the focus of the pattern is the hut and the one or more hearths outside the hut entrance. Hearths are the focus of multiple activities that occur in either a familial or socializing context, including cooking, whittling, and flint knapping. Extending outward from the hearths and around the sides of the hut is open work space. This area tends to be free of permanent facilities and is used for multiple, short-term tasks. Surrounding this hut–work-area complex (among the Alyawara) or a number of clustered hut–work-area complexes (among the !Kung) is a ring of garbage deposits. This ring of garbage is generated by persons sweeping the refuse from activities which occur within the open work areas and around the hearth, outward, away from the hut(s). During winter, the garbage ring is fed by ashes from hearths, which are kept burning through the night. Beyond the annulus of garage occur a number of specialized activities that have one or more of the these attributes: (*a*) they are messy; (*b*) they require much space; (*c*) they require that materials remain in a set position for a long time; or (*d*) they require quiet and concentration. Stretching of hides and butchering of animals are examples.

Unlike the work areas within the annulus of garbage, which are swept periodically of their debris, those outside the annulus tend to accumulate debris which is left *in situ*. This difference in deposition largely is related to the difference in demand for space in these two zones. Around the hut and hearth, space is at a premium, and must be cleared repeatedly to accommodate the numerous household activities. Outside the ring of garbage, however, space is less restricted. To obtain clear work space, it is not necessary to sweep away the rubbish of previous work episodes. Moving to a new, clean work area not only is possible, but also is more efficient. Thus, the debris found outside the rings of garbage in Alyawara and !Kung camps tend to be of a primary depositional nature.

The spatial organization of activity and deposition within !Kung and Alyawara camps, then, can be modeled by three concentric zones: (*a*) a central area that contains some but not all of the refuse generated within it; (*b*) a surrounding annulus of secondary refuse derived completely from activities that occurred in the central area; and (c) a yet larger annulus containing debris deposited in the locations in which it was generated.

This model appears to be appropriate to the spatial organization of both activity and deposition of material items at the Crane site. Immediately surrounding the single house at Crane site are clear areas that are devoid of permanent facilities and that might easily have functioned as multipurpose work areas. These clear areas are surrounded by a ring of pit clusters where stationary activities obviously were undertaken. Figure 38 illustrates this annular pattern of space utilization.

If the Alyawara and !Kung model is appropriate to the Crane site, and if the clear areas do represent the sites of multiple tasks, we would expect to find an annulus of secondary trash deposits surrounding the work areas. This appears to be the case, on the basis of two kinds of data. First, surrounding the house and probable work areas are deep midden deposits (Figure 39). These deposits in part might be the result of primary activity that occurred around the clusters of pits outside the central work areas, but also probably represent the sweepings of debris generated in the central work area. The latter translocational process is suggested by the lack of deep middens immediately surrounding the house, where it may reasonably be hypothesized that numerous activities occurred and can be expected that deep midden deposits would otherwise have developed.

A second kind of data that suggests that an annulus of secondary refuse deposits had developed around the house is the spatial distribution of material items. Collection units encompassing a large number of functional artifact types are likely to represent either (a) formalized refuse dumps where the debris from a large number of activities in adjacent locations were swept or dumped, or (b) mixed primary refuse from a number of overlapping activities. If the latter were the major factor determining the spatial arrangement of type-counts among collection units, we would expect the highest counts to occur where work space was at a premium, immediately surrounding the house, and that type-counts per collection unit would decrease steadily away from the house, as work space became more abundant.

We do not find this pattern, however (Figure 40). Grid cells immediately surrounding the house have few types of artifacts and are partially ringed by grid cells having more types of artifacts. This arrangement cannot be explained by reference to an overlap of activities around the house, where space is limited. It can be explained, however, if the archeological deposits in the collection units with many types of artifacts are considered secondary refuse that have been translocated from a number of adjacent work areas close to the house. The low number of artifact types occurring in the collection units immediately surrounding the house and interior to the high count units may be considered the result of sweeping and the removal of debris generated in them.

Thus, on the basis of the spatial arrangement of clear work areas, permanent facilities, deep midden deposits, and artifact type-counts among collection units, the analogy between the spatial organization of activities and depositional

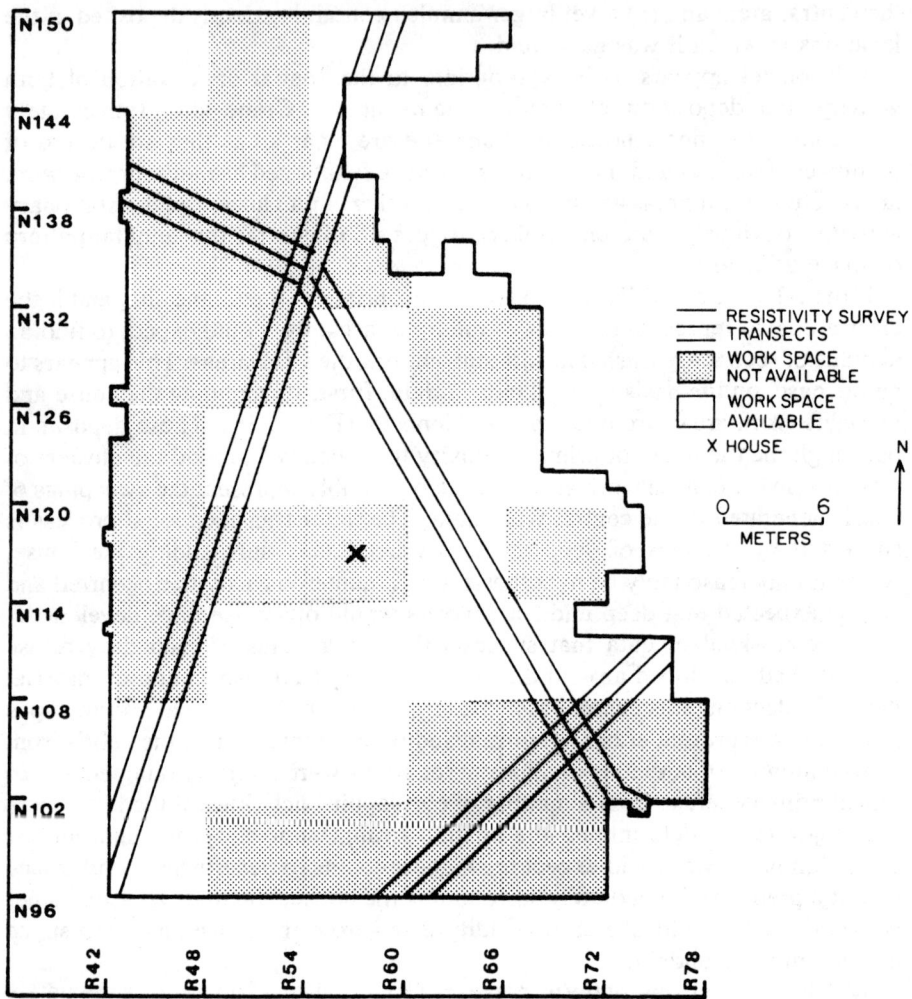

FIGURE 38. *Spatial distribution of areas at the Crane site which are clear of permanent facilities and which might have functioned as multipurpose work areas, as opposed to areas in which permanent facilities occur.*

processes at the Crane site and those in Alyawara and !Kung camps may be accepted—at least with respect to the area close to the house. Beyond the limits of the ring of secondary trash accumulation, however, the analogy has not been tested.

In the Alyawara and !Kung model, debris occurring beyond the annulus of garbage is largely the result of primary deposition from singular, *in situ* activities. This is probably true for much of the area beyond the proposed annulus of garbage at Crane, as well, based on the low number of artifact types

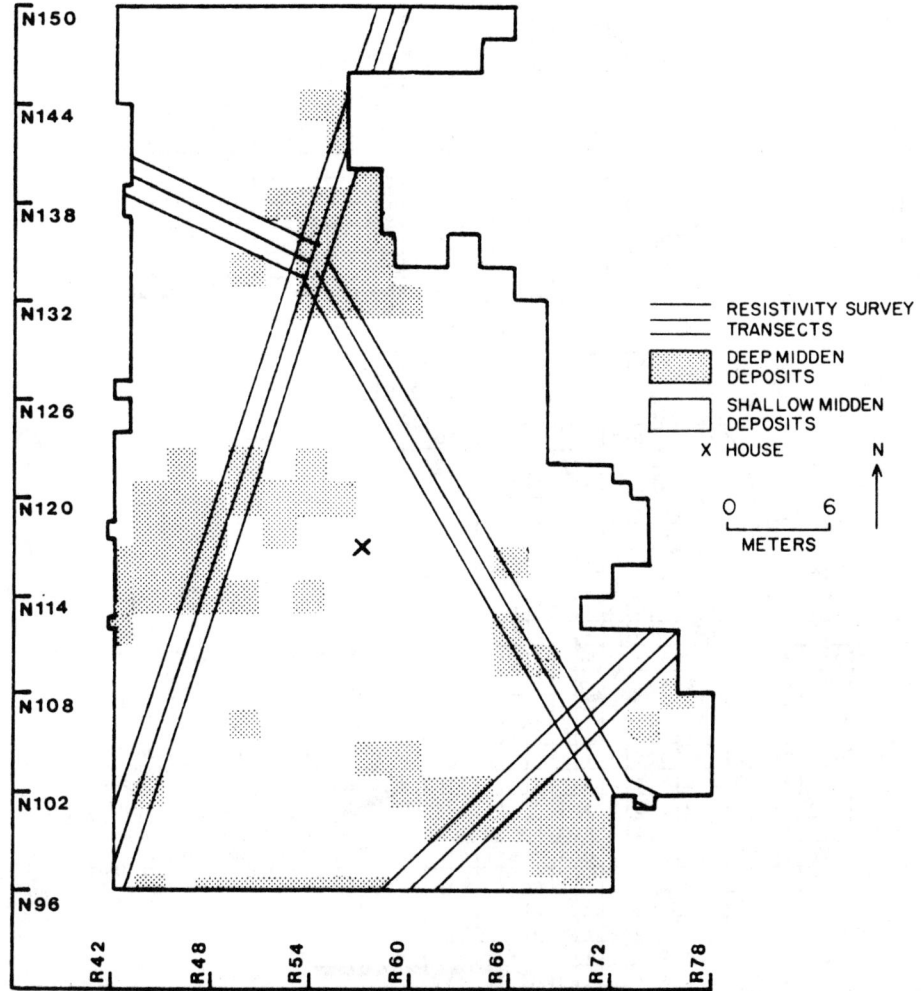

FIGURE 39. *Spatial distribution of areas having deep midden deposits (>35 cm) within the excavated area around the house at the Crane site.*

found in collection units distant from the house (Figure 40). A few collection units, however, have high artifact type-counts, and might represent formalized, secondary trash areas. One such area is known for sure from excavation data. In unit 198R84, which contained a large number of artifact types, a shallow, broad basin ca. 30 cm deep and several meters in diameter, was found. The pit fill contained dense amounts of bone and shell and exhibited lensing as a result of multiple episodes of activity and deposition. It could be interpreted only as a refuse dump associated with a meat and mollusk processing area somewhere nearby. Similar secondary refuse dumps could be located in the other collection

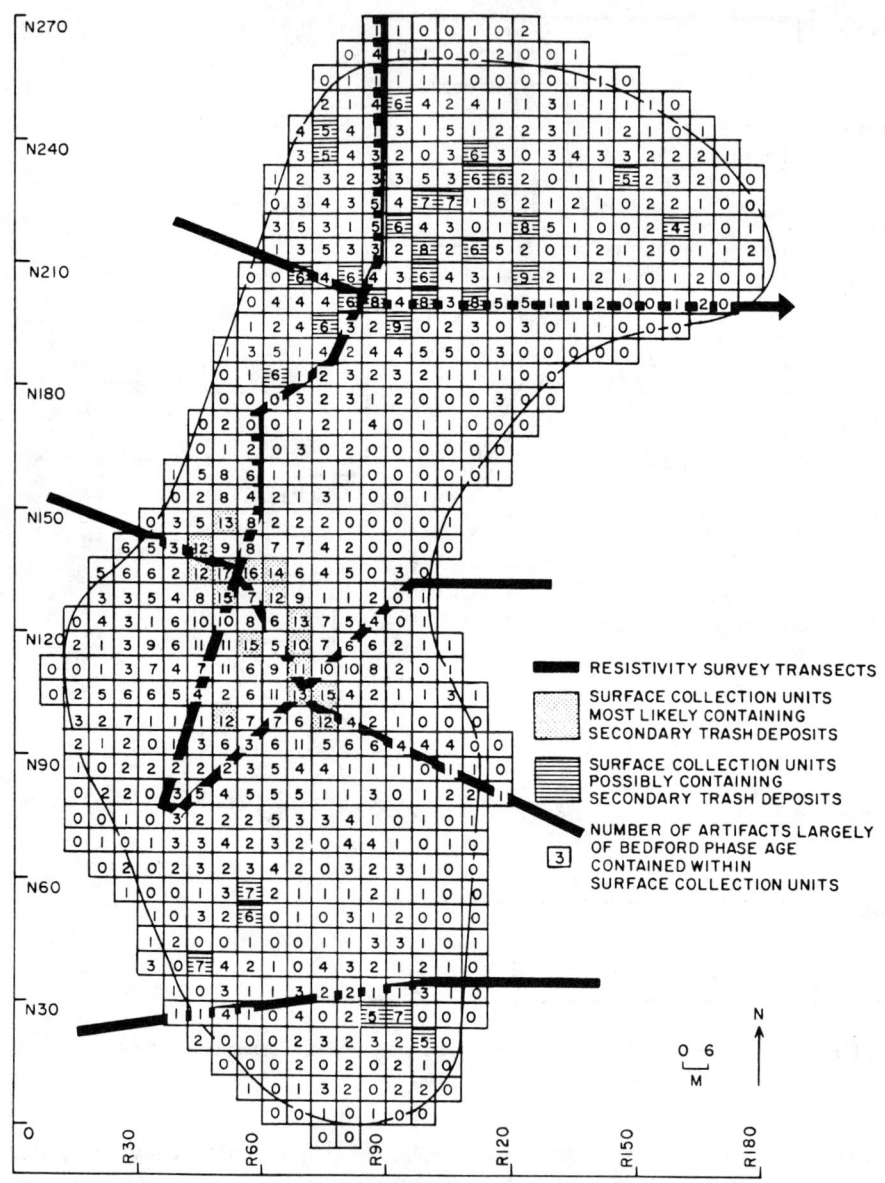

FIGURE 40 *The number of artifacts of Bedford phase age (V1–95, 97–108, 114–116), found in surface collection units at the Crane site. The units shaded with dots most likely represent secondary trash deposits, based upon their high artifact type counts and the annular pattern which they form around the house located at 114R54. The units shaded with lines possibly are secondary trash units, but might easily represent local spatial overlap of multiple activities, as well.*

units that have high type counts and that occur at a distance from the house (Figure 40). In this study, such units with many types of artifacts will be considered areas of secondary trash deposition, on the basis of analogy to the refuse dump in unit 198R84. The remaining units that are away from the house and have few types of artifacts will be assumed to be areas of primary refuse deposition, on the basis of analogy to Alyawara and !Kung camps.

The analysis of the spatial pattern of activity and primary and secondary refuse deposition at the Crane site, which has just been given, refers to the site as a whole. It is concerned with the *overall approximation* of the spatial patterning of particular archeological variables at Crane to the patterns of space use found in the camps of Alyawara Aborigines and !Kung Bushmen. It is not, however, a rigorous analysis of the depositional nature of the archeological remains in each *specific* collection unit. This is to follow.

The depositional nature of the remains in specific collection units was determined by logically concatenating the archeological variables mentioned previously with each other and with several other criteria shown in Table 46. Of particular importance are the additional criteria (5, 8) concerned with locating areas that were reused for multiple activities at different times (Bedford phase or later) and that might be confused for secondary refuse dumps rather than areas having multiple primary refuse deposits. Figure 41 shows those collection units in which pits intrude upon each other, or occur in such high densities that not all could have been used simultaneously (criterion 5). Figure 42 shows collection units in which artifacts diagnostic of other than the Bedford phase are present or occur in "significantly high densities" (see Carr 1977) (criterion 8). In any of these collection units having high pit densities or artifacts diagnostic of other than the Bedford phase, high artifact type-counts cannot be used to indicate secondary refuse accumulations. A high artifact type count might indicate only the reuse of a collection unit for different activities at different times.

The specific ways in which the criteria in Table 46 were concatenated logically to determine the depositional nature of the archeological remains in particular collection units is as follows.

1. Within the area close to the house, those collection units having high artifact type-counts and suspected to be part of a ring of garbage surrounding the house (criterion 6) are characterized as such, providing that there is no evidence of reuse of the area for different purposes at different times. Evidence for reuse includes the occurrence of very high pit density (criterion 5) or artifacts diagnostic of other than the Bedford phase (criterion 8) within the collection unit. If there is evidence that the unit has been reused at several different time periods, no conclusion is drawn as to the depostional nature of the remains within it.
2. Inside the suspected annulus of garbage, the depositional nature of remains within a given collection unit is characterized as *primary* if there is clear work space available within the unit (criterion 4) and if midden deposits are

TABLE 46
Criteria for defining the depositional nature of archeological remains found within the different collection units at Crane site

1. Diversity of artifact types found in the collection unit.
2. Species diversity and density of faunal remains within the collection unit.
3. Thickness of midden deposits within the collection unit.
4. Amount of open, pit-free, potential work space available within the collection unit.
5. Occurrence or absence of pit intrusions or very high densities of pits within the collection unit.
6. Whether the collection unit forms a part of the ring of units which surround the known house at Crane and which have high artifact diversity.
7. Whether the collection unit occurs outside or within the ring of units which surround the known house at Crane and which have high artifact density.
8. Occurrence or absence of several different artifact types diagnostic of different time periods within the same collection unit, i.e., whether multiple archeological components overlap areally within the unit.
9. Whether or not the collection unit occurs within an artifact "sink," such as a house or hut.

not thick (criterion 3), suggesting the area has been swept. The depositional nature of the remains is characterized as *secondary* if there is available work space but midden deposits are thick. The lack of facilities in such areas is probably the result of the area having been used as a dump, and cannot be taken to indicate a multipurpose work area. Units falling interior to the suspected ring of garbage but containing *no* clear work space are considered the locations of primary deposition, regardless of the thickness of the midden deposits within them. The fact that they occur interior to the ring of garbage, have low artifact type counts, and contain pit features that indicate *in situ* activity and deposition suggest the primary nature of their deposits.

3. Outside the suspected ring of garbage (criterion 7), the depositional nature of remains within a collection unit is characterized as primary if there are few artifact types within it (criterion 1). It is characterized as secondary if the artifact type-count is high and if there is no evidence of reuse of the area for different puposes at different times (criterion 8). Artifact type-counts have been defined as high or low in reference to local, average counts rather than any singular threshold value.
4. The deposits in the collection unit containing the house have been

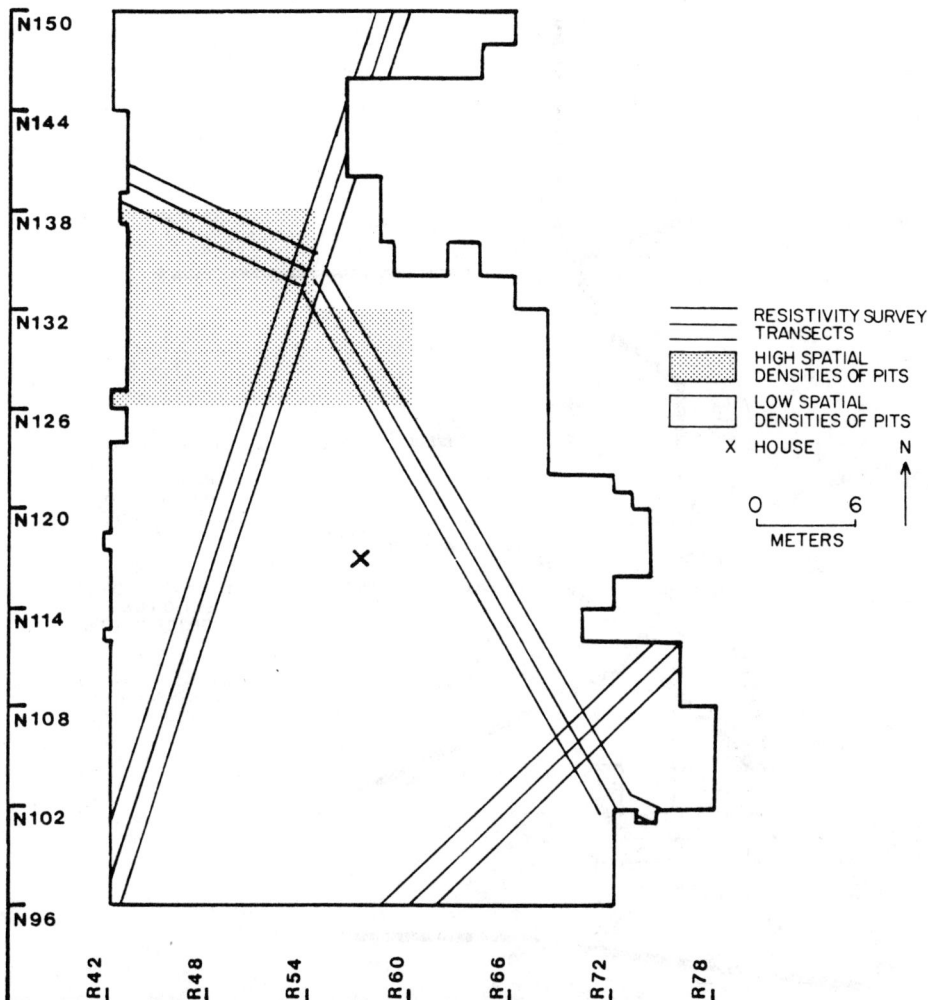

FIGURE 41. *Spatial distribution of areas having or not having pit intrusions or such high densities of pits that not all could have been used simultaneously, within the excavated area around the house at the Crane site.*

characterized as primary, despite the high number of artifact types they possess. The enclosed nature of the house at Crane made it a good artifact "sink" (criterion 9) that could not be thoroughly swept clean of the debris generated within it.

5. As an auxiliary criterion (2), and provided there is no evidence for reuse of an area at different times for different purposes (criteria 5, 8), the secondary

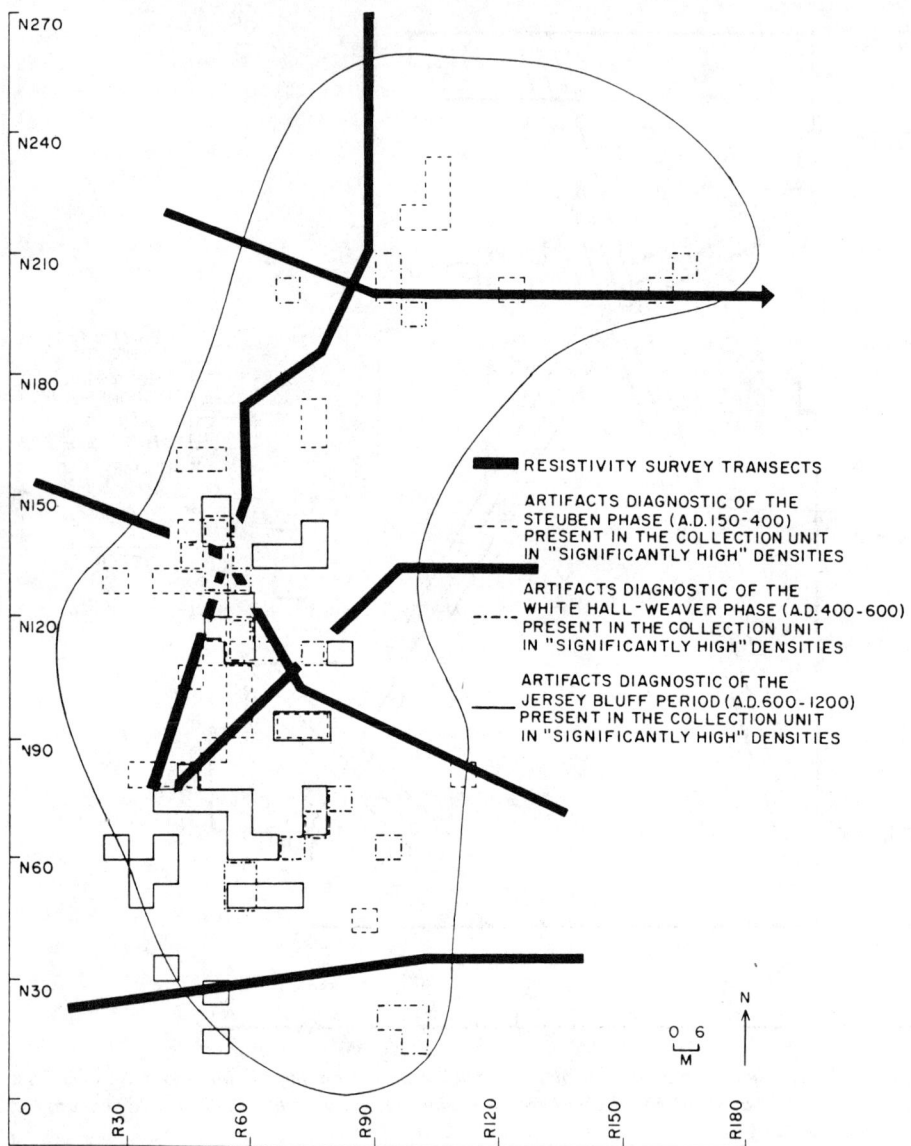

FIGURE 42. *Collection units at the Crane site in which artifacts diagnostic of other than the Bedford phase are either present or occur in "significantly high densities."*

nature of the deposits within a collection unit can be inferred from the presence of high densities of bone from multiple animal species within the unit. Locations having high densities of bone from a *single* species could represent either areas of secondary garbage deposition or specialized animal-processing areas. Locations having high densities from *many* species, however, are more likely to represent secondary refuse areas that received waste products from several different adjacent processing areas. This proposition is based on the reasonable assumption that different animal species are processed in different manners and that some degree of spatial segregation of processing areas is required. For example, among the Indians of the southeastern United States, small mammals were cooked with their guts intact, whereas deer required more extensive butchering. Small mammals were boiled or rolled over and over in embers, as was venison, but venison also was dried on scaffolds and pounded to make pemmican (Swanton 1946:368–375). Locations of butchering and drying deer would likely be different from those used to roast or boil small mammals simply because of the different permanent facilities required for processing them. Thus, a diagnostic of secondary refuse areas would be the joint occurrence of *high* densities of *many* taxa of animals. Those collection units in which 2 × 2-m test excavations yielded high densities of bone from multiple species are shown in Figure 43. Not every collection unit within the vicinity of the house contained test excavation units, so absence of this characteristic from a collection unit is not a significant indicator of the depositional nature of remains within it.

The criteria just described aim at distinguishing collection units having primary archeological deposits from those having secondary *garbage* deposits. They are not, however, capable of distinguishing areas that represent *caches* or stored tools that were put to use elsewhere. Using the methods of classification presented above, caches most likely would be misidentified as primary, *in situ* deposits resulting from a single or a few activities. It is probable, however, that in this inability to distinguish areas of caches from areas of primary refuse is not a major problem. One would not expect a large number of caches to occur on a hunter–gatherer camp such as Crane, compared to the total amount of refuse deposited on the site.

Summary

An attempt has been made to determine those areas of the Crane site where primary archeological deposits occur and to distinguish those areas in regard to the kinds of activities that produced the deposits. This has required (*a*) the assignment of functions to artifacts by means of morphological analysis, ethnographic analogy, and the study of spatial co-occurrences among different artifact types; (*b*) the inventorying of the functional artifact types found in different areas of the site; and (*c*) the determination of whether the artifacts deposited in given areas also were used there, by way of anlogy to the use of space in Aborigine and Bushman camps. The resulting classification of use-space within the Crane site is shown in Tables 44 and 45.

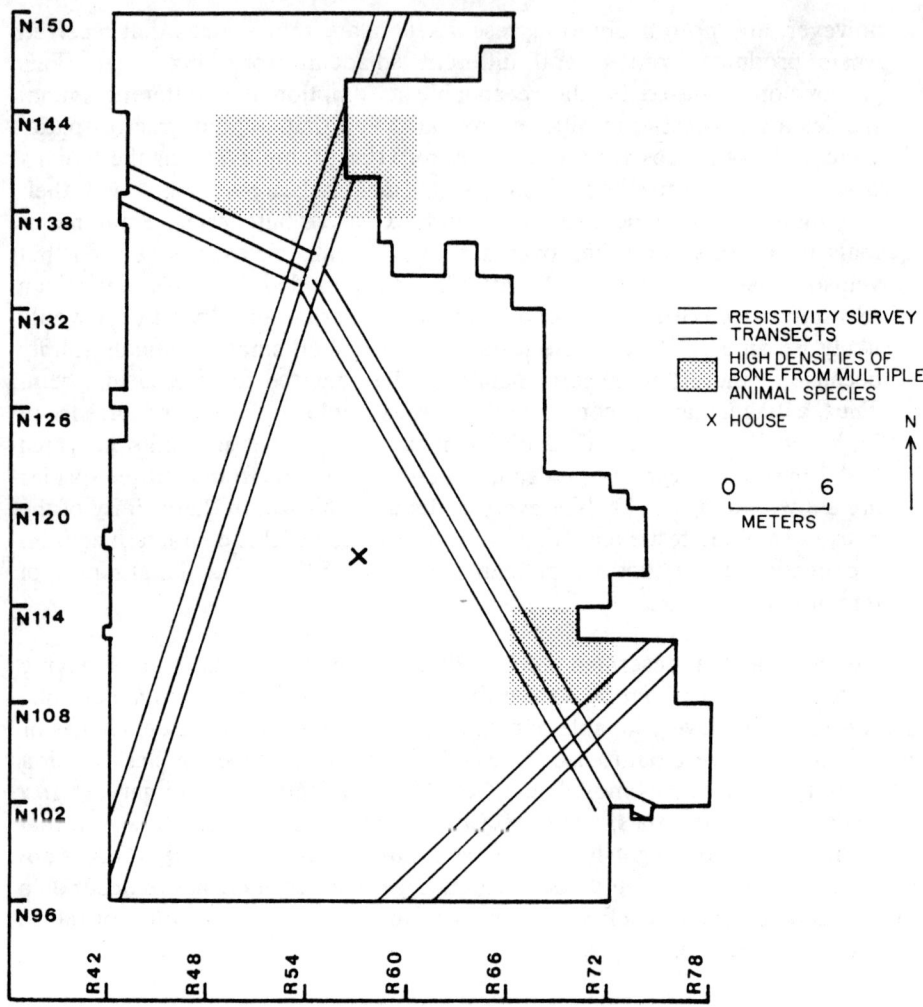

FIGURE 43. *Collection units at the Crane site in which high densities of bone from multiple animal species were found in 2 × 2-m test excavations around the house at the Crane site.*

References

Formation of the Archeological Record; Statistical Methods

Ammerman, Albert J., and Marcus W. Feldman
    1974   On the making of an assemblage of stone tools. *American Antiquity* 39:610–616.

Ascher, Robert
- 1968 Times arrow and the archaeology of a contemporary community. In *Settlement Archaeology*, K.C. Chang (ed.), pp. 43–52. National Press Books, Palo Alto.

Binford, Lewis R.
- 1974 Forty-seven trips. In *Contributions to anthropology: The interior peoples of northern Alaska. Archaeological Survey of Canada, Paper* 49:299–381.

Binford, Lewis R., and Sally R. Binford
- 1966 A preliminary analysis of functional variability in the mousterian of Levallois facies. *American Anthropologist* 68:238–295.

Brown, James, and Leslie Freeman
- 1964 A UNIVAC analysis of sherd frequencies from the Carter Ranch Pueblo, eastern Arizona. *American Antiquity* 31:203–210.

Carr, Christopher
- 1977 The internal structure of a Middle Woodland site and the nature of organization of the archaelogical record. Unpublished preliminary examinations thesis, Department of Anthropology, Univerisity of Michigan.

Clarke, David L.
- 1968 *Analytical archaelogy*. Methuen, London.

Cole, LaMont C.
- 1949 The measurement of interspecific association. *Ecology* 30:411–424.

Cole, A.J., and D. Wishart
- 1970 An improved algorithm for the Jardin–Sibson method of generating overlapping clusters. *Computer Journal* 13:156–163.

Cook, Sherburne F., and Robert F. Heizer
- 1968 Relationships among houses, settlement areas, and population in aboriginal California. In *Settlement archaelogy*, K.C. Chang (ed.), pp. 79–116. National Press Books, Palo Alto.

Cook, Thomas Genn
- 1973 Koster: A lithic analysis of two Archaic phases in west-central Illinois. Chapter 2: Problem orientation and methodology of lithic analysis. Unpublished Ph.D. dissertation, University of Chicago.

Freeman, Leslie A., and Karl Butzer
- 1966 The Acheulean station of Toralba (Spain). A progress report. *Quaternaria* 8:9–21.

Gould, Richard, Dorothy A. Koster, and Ann H. Sontz
- 1971 The lithnic assemblage of the western desert Aborigines of Australia. *American Antiquity* 36:149–169.

Hill, James N.
- 1970 Broken K pueblo: Prehistoric social organization in the American Southwest. *University of Arizona, Anthropological Papers,* 18.

Jardine, N., and R. Sibson
- 1968 The construction of hierarchic and non-hierarchic classifications. *Computer Journal* 11:177–184.

Naroll, Robert
- 1962 *Floor area and settlement population.* *American Antiquity* 27(4):587–589.

O'Connell, James F.
- 1977 Room to move: Contemporary Alyawara settlement patterns and their implications for Aboriginal housing policy. Unpublished manuscript on file, Australian Institute of Aboriginal Studies, Canberra.
- 1979 Site structures and dynamics among modern Alyawara hunters. Unpublished paper presented at the Annual Meetings of the Society for American Archaeology.

Redman, Charles L., and Patty Jo Watson
- 1970 Systematic, intensive surface collection. *American Antiquity* 35:279–291.

Roper, Donna C.
- 1976 Lateral displacement of artifacts due to plowing. *American Antiquity* 41:372–375.

Schiffer, Michael B.
   1972   Archaeological context and systematic context. *American Antiquity* 37:156–165.
   1973   Cultural formation processes of the archaeological record; applications at the Joint site; east central Arizona. Ph.D. dissertation, University of Arizona. University Microfilms, Ann Arbor.
   1975a Behavioral chain analysis: Activities, organization, and analysis in archaeology. *Fieldiana* 65:103–120.
   1975b Factors and "tool kits"; Evaluating multivariate analyses in archaelogy. *Plains Anthropologist* 20:61–70.
   1976   *Behavioral archaeology.* Academic Press, N.Y.
Schiffer, Michael B., and William L. Rathje
   1973   Efficient exploitation of the archaeological record: Penetrating problems. In *Research and theory in current archaeology*, C.L. Redman (ed.), pp. 169–179. John Wiley, New York.
Sneath, Peter H., and Robert R. Sokal
   1973   *Numerical taxonomy.* W.H. Freeman, San Francisco.
Sokal, Robert R., and Peter H. Sneath
   1963   *Principles of numerical taxonomy.* W.H. Freeman, San Francisco.
Speth, John D., and Gregory A. Johnson
   1976   Problems in the use of correlation for investigation of tool kits and activity areas. In *Cultural change and continuity*, C. Cleland (ed.), pp. 35–75. Academic Press, New York.
Struever, Stuart
   1968   Woodland subsistence-settlement systems in the lower Illinois Valley. In *New perspectives in archaeology*, S.R. Binford and L.R. Binford (eds.), pp. 285–312. Aldine, Chicago.
Watson, Patty Jo, Steven A. LeBlanc, and Charles L. Redman
   1971   *Explanation in archaeology.* Columbia University Press, New York.
Whallon, Robert, Jr.
   1973   Spatial analysis of occupation floors I: Application of dimensional analysis of variance. *American Antiquity* 38:266–278.
Yellen, John E.
   1971   Ethnoarchaeology: Bushman settlement patterns. Unpublished manuscript on file at the Smithsonian Institution, Department of Anthropology, Washington, D.C.
   1974   The !Kung settlement pattern: An archaeological perspective. Unpublished Ph.D. Dissertation, Harvard University, Department of Anthropology.

Ethnographic, Experimental and Archeological Documentation of the Functions of Artifacts; Artifact Typology

Ahler, Stanley A.
   1971   Projectile point form and function at Rodgers shelter, Missouri. *Missouri Archaeological Society Research Series,* 8.
Balout, L.
   1967   Procedes d'analyse et questions de terminologie dans L'etude des ensembles industrial du paleolithique inferieur en Afrique du nord. In *Background to evolution in Africa*, W.W. Bishop and J.D. Clark (eds.) University of Chicago Press, Chicago.
Barnes, A.J.
   1932   Modes of prehension of some forms of Upper Paleolithic implements. *Prehistoric Society of East Anglia, Proceedings* 7: 43–56.

Binford, Lewis R., and Sally R. Binford
  1966 A preliminary analysis of functional variability in the Mousterian of Levallois facies. *American Anthropologist* 68:238–295.
Binford, Lewis R., Sally R. Binford, Robert Whallon, and Margaret Ann Hardin
  1970 Archaeology at Hatchery West. *Society for American Archaeology, Memoirs* 24.
Bordaz, Jacques
  1959 *Tools of the old and new Stone Age.* Natural History Press, Garden City.
Bordes, Francois
  1961 *Typologie du Paleolithique ancien et moyen.* Imprimeries Delmas, Bordeaux.
  1969 Reflections on typology and techniques in the Paleolithic. *Arctic Anthropology* 4:1–29.
Brose, David S.
  1968 The archaeology of Summer Island: Changing settlement systems in northern Lake Michigan. Unpublished Ph.D. dissertation, University of Michigan, Department of Anthropology.
  1970 The Summer Island site: A study of prehistoric cultural ecology and social organization in the northern Lake Michigan area. *Case Western Reserve University Studies in Anthropology* 1.
Cook, Thomas Genn
  1973 Koster: A lithic analysis of two Archaic phases in west-central Illinois. Chapter 2: Problem orientation and methodology of lithic analysis. Unpublished Ph.D. dissertation, University of Chicago.
Crabtree, Don E.
  1967 Notes on experiments in flintknapping, 3: The flintknapper's raw materials. *Tebiwa* 10:8–25.
  1968 Mesoamerican polyhedral cores and prismatic blades. *American Antiquity* 33:446–478.
Crabtree, Don (continued)
  1972 An Introduction to flintworking. *Idaho State University Museum, Occasional Papers* 28.
  1973 The obtuse angle as a functional edge. *Tebiwa* 16:46–53.
Crabtree, Don E. and E.L. Davis
  1968 Experimental manufacture of wooden implements with tools of stone. *Science* 159:(3813):426–428.
Crabtree, Don E., and Earl H. Swanson
  1968 Edge-ground cobbles and blade making in the Northwest. *Tebiwa* 11:50–58.
de Heinzelin de Braucourt, Jean
  1962 *Manuel de typologie des industries lithique.* L'Institute Royal des Sciences Naturelles de Belgique, Bruxelles.
Ericson, Jonathon E., and Suzanne P. De Atley
  1976 Reconstructing ceramic assemblages: An experiment to derive the morphology and capacity of parent vessels from sherds. *American Antiquity* 41:484–488.
Ericson, Jonathon E., and E. Gary Stickel
  1973 A proposed classification system for ceramics. *World Archaeology* 4:357–367.
Gould, R.A.
  1971 The archaeologist as ethnographer: A case from the Western Desert of Australia. *World Archaeology* 3:143–178.
Green, F.E.
  1963 The Clovis blades: An important addition to the Llano complex *American Antiquity* 29:145–165.
Hayden, Brian, and Johan Kamminga
  1973 Gould, Koster, and Sontz on 'microwear': A critical review. *Newsletter of Lithic Technology* 2:3–14.
Horne and Aiston
  1924 *Savage life in central Australia.* London.

Keeley, Lawrence H.
　1974　Technique and methodology in microwear studies. *World Archaeology* 5:323–336.
　1977　The functions of Paleolithic flint tools. *Scientific American* 237:108–126.
Kenyon, A.J.
　1927　Stone implements on Aboriginal camping grounds. *Victorian Naturalist* 43.
Kraybill, Nancy
　1977　Pre-agricultural tools for the preparation of foods in the Old World. In *Origins of agriculture,* C.A. Reed (ed.), pp. 485–522. Mouton, The Hague.
Leone, Mark
　1968　Neolithic economic autonomy and social distance. *Science* 162:1150–1151.
Mason, Otis Tufton
　1889　Aboriginal skin-dressing—A study based on material in the U.S. National Museum. *U.S. National Museum, Annual Report*, pp. 553–590.
　1895　*The origins of invention.* Walle Scott, London
　1899　The man's knife among the North American Indians—A study in the collections of the U.S. National Museum. *Smithsonian Institution, Annual Report for the Year 1897:*727–742.
McCarthy, Frederick D.
　1946　The stone implements of Australia. *The Australian Museum, Memoir* 9.
　1967　*Australian Aboriginal stone implements.* V.C.N. Blight, Sidney. Government Printer, New South Wales.
Miles, Charles
　1973　*Indian and Eskimo artifacts of North America.* Bonanza Books, New York.
Mitchell, Stanley Robert
　1949　*Stone Age craftsmen: Stone tools and camping places of the Australian Aborigines.* Tait Book, Melborne.
Nero, Robert W.
　1957　A "Graver" site in Wisconsin. *American Antiquity* 22:300–304.
Odell, G.H.
　1975　Microwear in perspective. *World Archaeology* 7:226–240.
Kidder, Alfred Vincent and Samuel J. Guernsey
　1919　Archeological explorations in northeast Arizona. Bureau of American Ethnology, *Bulletin* 65.
Rau, Charles
　1869　Drilling in stone without metal. Smithsonian Institution, *Annual Report for 1868:*392–400.
　1873　North American stone implements. *Smithsonian Institution, Annual Report for 1872:*395–408.
Riddell, F., and W. Pritchard
　1971　Archaeology of the Rainbow Point site (4-Plu-594), Bucks Lake, Pumas County, California. In *Great Basin Anthropological Conference 1970: Selected papers,* A. Aikens (ed.), pp. 59–102. University of Oregon Anthropological Papers 1.
Semenov, Sergei A.
　1964　*Prehistoric technology: An experimental study of the oldest tools and artifacts from traces of manufacture and wear.* Cory, Adams and MacKay, London.
Shepherd, Walter
　1972　*Flint: Its origins, properties, and uses.* Faber and Faber, London.
Sollberger, J.B.
　1969　The basic tool kit required to make and notch arrow shafts for stone points. *Texas Archaeological Society, Bulletin* 40:231–240.
Sonnenfield, J.
　1962　Interpreting the function of primitive implements. *American Antiquity* 28:56–65.
Spears, Carol S.
　1975　Hammers, nuts and jolts, cobbles, cobbles, cobbles: Experiments in cobble technology in

search of correlates. In *Arkansas Eastman Archaeological Project,* C.M. Baker (ed.), pp. 83–116. Arkansas Archaeological Survey, Research Report 6.

Speth, John D.
1972 Mechanical basis of percussion flaking. *American Antiquity* 37:34–60.

Struever, Stuart
1965 Middle Woodland culture history in the Great Lakes riverine area. *American Antiquity* 31:211–223.

Tixier, Jacques
1963 *Typologie de l'epipaleolithique du Maghref.* Arts et Metiers Graphiques, Paris.

Tringham, R.E.
1972 The function, technology, and typology of the chipped stone industry at Bylany, Czechoslovakia. In *Du Aktuelle Fragen der Bandkeramik,* J. Fitz (ed.), pp. 143–148. Szekesfehervar, Hungary.

White, J.P.
1967 Ethno-archaeology in New Guinea: Two examples. *Mankind* 6:406–414.

Wilmsen, Edwin N.
1968a Functional analyses of flaked stone artifacts. *American Antiquity* 33:156–161.
1968b Lithic analysis in paleoanthropology. *Science* 161:982–987.
1970 Lithic analysis and cultural inference: A Paleo-Indian case. *University of Arizona, Anthropological Papers* 16.

Winters, Howard D.
1967 An archaeological survey of the Wabash Valley in Illinois. Illinois State Museum, *Reports of Investigation* 10.
1969 *The Riverton culture.* Illinois State Museum (Springfield) and Illinois Archaeological Survey (Urbana).

Witthoft, John
1967 Glazed polish on flint tools. *American Antiquity* 32:383–388.

Wobst, Martin H.
1974 The archaeology of band society—Some unanswered questions. Introduction to *A model of band society,* by B.J. Williams. *American Antiquity, Memoir* 29, v–xiii.
1975 Stylistic behavior and information exchange. Unpublished manuscript on file, Department of Anthropology, University of Massachusetts.

Ethnographic Documentation of Material Procurement and Processing Activities in North America and Elsewhere

Battle, Herbert B.
1922 The domestic use of oil among sourthern Aborigines. *American Anthropologist* 24:171–182.

Chestnut, V.K.
1902 Plants used by the Indians of Mendocino County, California. *U.S. National Herbarium, Contribution* 7:295–422.

Driver, Harold E.
1961 *Indians of North America.* University of Chicago Press, Chicago.

Gifford, E.W.
1936 Californian balanophagy. In *Essays in anthropology,* presented to A.L. Kroeber, pp. 87–98. University of California Press, Berkeley.

Gilmore, Melvin R.
1931 Vegetal remains of the Ozark bluff-dweller culture. *Michigan Academy of Science, Arts and Letters, Papers* 14:83–192.

Heider, Karl
  1967 Archaeological assumptions and ethnographic facts: A cautionary tale from New Guinea. *Southwest Journal of Anthropology* 23:52–64.
Leechman, Douglas
  1951 Bone grease. *American Antiquity* 16:355–356.
Mason, Otis Tufton
  1895 *The origins of invention.* Walle Scott, London.
McCarthy, Frederick D.
  1946 The stone implements of Australia. the Australian Museum, *Memoir* 9.
  1967 *Australian Aboriginal stone implements.* V.C.M. Blight, Government Printer, New South Wales.
Merriam, C. Hart
  1918 The acorn, a possible neglected source of food. *National Geographic Magazine* 34:129–137.
Mitchell, Stanley Robert
  1949 *Stone age craftsmen: Stone tools and camping places of the Austrialian Aborigines.* Tait Book, Melbourne.
Osborne, Carolyn M.
  1965 The preparation of Yucca fibers: An experimental study. In *Contributions of the Wetherill Mesa archaeological project,* assembled by Douglas Osborne. Society for American Archaeology, *Memoirs* 19:45–50.
Peale, Titian R.
  1871 On the use of the brain and marrow of animals among the Indians of North America. *Smithsonian Institution, Annual Report for 1870:*390–391.
Swanton, John R.
  1946 Indians of the southeastern United States. *Bureau of American Ethnology, Bulletin* 137.
Waugh, F.W.
  1916 Iroquois foods and food preparation. *Canada Dept. of Mines, Geological Survey, Memoire* 86 *(Anthropological Series,* No. 12).
Wheat, Joe Ben
  1972 The Olsen–Chubbock site: A Paleo-Indian bison kill. *Society for American Archaeology, Memoir* 26.

Illinois Valley Prehistory and History, and Related Archeology

Andreas, Lyter, and Co.
  1873 Atlas map of Greene County, Illinois. Davenport, Iowa.
Anonymous
  1909 Atlas: Greene County, Illinois. On file, City Library, Carrollton, Illinois.
Asch, Nancy B., and David L. Asch
  1977 The economic potential of *Iva Annua* and its prehistoric importance in the lower Illinois Valley. In *The nature and status of ethnobotany*, R.I. Ford (ed.), University of Michigan Museum of Anthropology, *Anthropological Papers* 67:301–341.
Asch, Nancy, Richard I. Ford, and David L. Asch
  1972 Paleoethnobotany of the Koster site: the Archaic horizons. *Illinois State Museum, Reports of Investigation* 24.
Braun, David
  1977 Middle Woodland-(Early) Late Woodland social change in the prehistoric central Midwest. Unpublished Ph.D. dissertation, University of Michigan, Department of Anthropology.

Carr, Christopher
1977 The internal structure of a Middle Woodland site and the nature of organization of the archaeological record. Unpublished preliminary examination thesis, Department of Anthropology, University of Michigan.
Cole, Fay-Cooper, and Thorne Deuel
1937 *Rediscovering Illinois: Archaeological explorations in and around Fulton County.* University of Chicago Press.
Continental Historical Co.
1885 *History of Greene and Jersey Counties, Illinois.* Springfield, Illinois.
Deuel, Thorne, ed.
1952 Hopewellian communities in Illnois. *Illinois State Museum, Scientific Papers* 5. Springfield.
Donnelley, Gassette, and Loyd, Publishers
1879 *History of Greene County, Illinois.* Chicago.
Farnsworth, Kenneth B.
1973 An archaeological survey of Macoupin Valley. *Illinois State Museum, Reports of Investigations* 26.
Griffin, James B.
1952a Some Early and Middle Woodland pottery types in Illinois. In *Hopewellian Communities in Illinois,* T. Deuel (ed.), pp. 93–129. *Illinois State Museum, Scientific Papers* 5.
1955 Observations on the grooved axe in North America. *Pennsylvania Archaeologist* 25:32–44.
1960 Climatic change: A contributory cause of the growth and decline of Northern Hopewellian culture. *Wisconsin Archaeologist* 41:21–33.
1964 The northeast Woodlands Area. In *Prehistoric man in the New World*, J.D. Jennings and E. Norbeck (eds.), pp. 223–258.
1967 Eastern North American archaeology: A summary. *Science* 156(3772):175–191.
Griffin, James B., ed.
1952b *Archaeology of the eastern United States.* University of Chicago Press, Chicago.
Griffin, James B., Richard E. Flanders, and Paul F. Titterington
1970 The burial complexes of the Knight and Norton mounds in Illinois and Michigan. *University of Michigan Museum of Anthropology, Memoir* 2.
Harn, Alan D.
1971 An archaeological survey of the American bottoms in Madison and St. Clair counties, Illinois. *Illinois State Museum, Reports of Investigation* 24, Part II.
Houart, Gail
1971 Koster: A stratified archaic site in the Illnois Valley. *Illnois State Museum, Reports of Investigations* 22.
Johnson, Alfred E. (ed.)
1976 Hopewellian archaeology in the lower Missouri Valley. *University of Kansas, Publications in Anthropology* 8.
McGregor, John C.
1952 The Havana site. In *Hopewellian Communities in Illinois,* T. Deuel (ed.), pp. 45–91. Illinois State Museum, Springfield.
1958 *The Pool and Irving villages. University of Illinois Press, Urbana.*
1959 The Middle Woodland period. In *Illinois archaeology. Illinois Archaeological Survey, Bullentin* 1:21–26.
McKern, William C., Paul F. Titterington, and James B. Griffin
1945 Painted pottery figurines from Illinois. *American Antiquity* 10:295–302.
McPherron, Alan L.
1967 The Juntunen site and the Late Woodland prehistory of the upper Great Lakes area. *University of Michigan Museum of Anthropology, Anthropological Papers* 30.

Maxwell, Moreau S.
   1951   The Woodland cultures in southern Illinois. *Logan Museum Publications in Anthropology, Bulletin* 7.

Meyers, J. Thomas
   1970   Chert resources of the lower Illinois valley. Illinois State Museum, *Reports of Investigation* 18.

Montet-White, Anta
   1963   Analytic description of the chipped stone industry from Snyders site, Calhoun County, Illinois. In *Miscellaneous studies in typology and classification*, A. Montet-White, L.R. Binford, and M.L. Papworth (eds.), *University of Michigan Museum of Anthropology, Anthropological Papers* 19.
   1968   The lithic industries of the Illinois Valley in the Early and Middle Woodland period. *University of Michigan Museum of Anthropology, Anthropological Papers 35.*

Moorehead, Warren K.
   1912   Hematite implements of the United States together with chemical analyses of various hematites. Dept. of Anthropology, *Phillips Academy, Bulletin* 6. Andover, Mass.

Munson, Patrick J.
   1971   An archaeological survey of the Wood River terrace and adjacent bottoms and bluffs in Madison County, Illinois. *Illinois State Museum, Reports of Investigations* 21, Part I.

Munson, Patrick J., and James P. Anderson
   1973   A preliminary report on Kane Village: A Late Woodland site in Madison County, Illinois. *Illinois Archaeological Survey, Bulletin* 9:34–57.

O'Brien, Patricia
   1972   A formal analysis of Cahokia ceramics from the Powell tract. *Illinois Archaeological Survey, Monograph* 3.

Rackerby, Frank
   1973   A statistical determination of the Black Sand occupation at the Macoupin site, Jersey County, Illinois. *American Antiquity* 38:96–101.

Shalkop, Robert L.
   1949   The Jersey Bluff archaeological focus. Unpublished masters thesis, University of Chicago, Chicago.

Struever, Stuart
   1964   The Hopewell interaction sphere in riverine–western Great Lakes culture history. In *Hopewellian Studies*, J.R. Caldwell and R.L. Hall (eds.), pp. 85–106. *Illinois State Museum, Scientific Papers* 12.
   1965   Middle Woodland culture history in the Great Lakes riverine area. *American Antiquity* 31:211–223.
   1968a Woodland subsistence-settlement systems in the lower Illinois Valley. In New perspectives in archaeology, S.R. Binford and L.R. Binford (eds.), pp. 285–312. Aldine, Chicago.
   1968b   A re-examination of Hopewell in eastern North America. Unpublished Ph.D. dissertation, University of Chicago, Chicago.

Struever, Stuart, and Gail Houart
   1972   An analysis of the Hopewell interaction sphere. In *Social exchange and interaction*, E. Wilmsen (ed.), pp. 47–79. *University of Michigan Museum of Anthropology, Anthropological Paper* 46.

Titterington, Paul F.
   1935   Certain bluff mounds of western Jersey County, Illinois. *American Antiquity* 1:6–46.
   1942   The Jersey County, Illinois, bluff focus. *American Antiquity* 9:240–245.

Vogel, Joseph O.
   1964   A preliminary report on the analysis of ceramics from the Cahokia area at the Illinois State Museum. Unpublished report on file at the Illinois State Museum.

Walthal, John
   1978   Galena exchange in the prehistoric eastern United States. Paper presented at the Hopewell Conference, Chillicothe, Ohio.
Whatley, Bonnie L., and Nancy B. Asch
   1975   Woodland subsistence: Implications for demographic and nutritional studies. Unpublished paper presented at the Annual Meetings of the American Anthropological Association.
Witthoft, John
   1955   Worn stone tools from southeastern Pennsylvania. *Pennsylvania Archaeologist* 25(1):16–31.
Witthoft, John, and James Miller
   1952   Grooved axes of eastern Pennsylvania. *Pennsylvania Archaeologist* 12:81–94.
Wray, Donald, and Richard S. MacNeish
   1961   The Hopewellian and Weaver occupations of the Weaver site, Fulton County, Illinois. *Illinois State Museum, Scientific Papers* 7.
Zawacki, April Allison, and Glenn Hausfater
   1969   Early vegetation of the lower Illinois Valley. *Illinois State Museum, Reports of Investigations* 17.

6

Research Designs for Collecting Pedological and Resistivity Data from the Crane Site

As a first step toward determining the relationships holding between prehistoric activities at the Crane site and the alterations they produced in the physical, chemical, and conductive properties of the soils there, use-areas within the site have been reconstructed (Chapter 5). It next is necessary to compare these different use-areas with respect to their anomalous soil attributes and resistivities. In preparation for this comparison, the design by which soil data and resistivity data were collected will be described (this chapter), and non-archeological soil variations—which must not be included in the comparison—will be documented and segregated (next chapter).

Many aspects of the design by which resistivity data were collected at Crane will be found appropriate in *surveying and sampling* other sites. The collection strategy therefore is described in detail as a model for future use. Some aspects, however, are pertinent to only this study and its particular *experimental* goals, and should not be used as part of the sampling programs of routine resistivity surveys of other researchers. Reasons for all aspects of the collection strategy are documented so that the reader can distinguish specific experimental aspects of the collection design from more widely applicable aspects of sampling.

Soil and Resistivity along Transects

Transect Orientation and Time of Survey

Soil and resistivity surveys were performed according to several different experimental designs at the Crane site, in order to answer different sets of questions concerning the relationships between prehistoric activity at the site, soil conditions, and soil resistivity. The first, and largest-scale resistivity and soil survey program was designed to sample as much as possible of the natural and human-related pedological variation on the Crane site under uniform, weather-imposed moisture regimes. A sample including a wide range of soil conditions was desired so that functional or associational relationships between operational variables might be modeled over as wide a range of variation as possible. The constructed models might then be applicable to future resistivity surveys on a large set of sites. Toward this end, five transects of resistivity measurement and soil sampling were made across the site (Figure 44), each oriented so as to crosscut the greatest gradients in topography and artifact concentration on the site in a perpendicular manner. The use of topography and artifact concentration to place the sampling transects was based on two assumptions: first, that any natural soil trends over space would be broadly related to topographic trends, and thus, well encompassed by the surveys; and second, that by sampling areas of overall high, moderate, and low artifact density, a great variety of prehistoric use-areas would be encompassed by the survey program. Although these two assumptions are simplistic, they had to be made when the transect survey samples were chosen. During the initial investigations of the Crane site in 1974, no detailed information was available on the spatial distributions of either different soil parameters, artifact classes, or subsurface features.

Each of the five Transect Surveys was designed so as to *minimize* the effects of specific weather variations that would influence soil conditions and/or resistivity conditions but that were not of interest for investigating *general* relationships between prehistoric space-use, soil properties, and soil resistivity. Each Transect Survey was performed as a unit, in the shortest time possible (one day for Transects I–III, three days for Transects IV and V as a unit) under constant moisture conditions. Not once did rainfall interrupt the surveying of a transect. Survey times also were scheduled at times at least 2 days after a rain, and more when possible. This was done in order to allow enough time for gravitational soil water to drain through the profile and time for the soil profile to reach an equilibrium in the distribution of moisture within it which would not change significantly during the period of survey.

While the meteorological conditions affecting soil moisture and soil resistivity values were approximately constant during the survey of individual transects, the kind of surface vegetation along the transects, the rates of transpiration of those plants, and their effect on soil moisture content could not be controlled. Most of the site at the time of the Transect Survey was overgrown with weedy annuals that had invaded the freshly disked but uncropped soil. The species and density of weeds varied from place to place (20–90% cover) partially in response to the different soil nutrient and soil moisture optima that midden deposits of different organic matter contents presented in different portions of

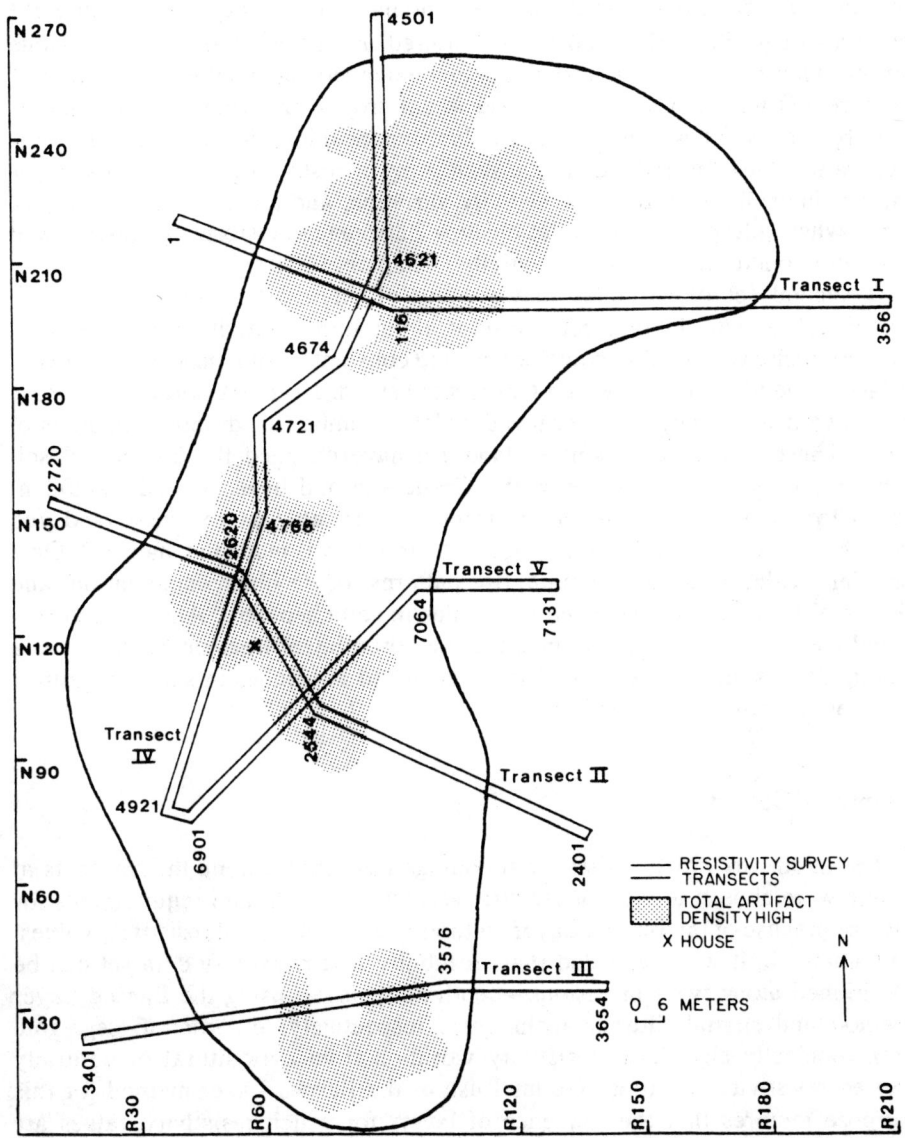

FIGURE 44. *Transects along which soil sampling surveys and resistivity surveys were performed at the Crane site. Roman numerals designate transect numbers; arabic numerals designate the provenience codes of selected benchmarks along the central subtransect of each transect. Shaded areas reflect portions of the site thought to have overall high artifact and debris density prior to analysis of surface artifact collection.*

the site, but also in repsonse to localized anomalies in drainage and the different amounts of fertilizer that had been scattered in particular areas the previous spring. Those portions of the transects that extended beyond the site boundaries (Figure 44) were covered by a dense wheat crop at the time of survey rather than by weeds. These differences in plant composition, density and transpiration rates along the individual transects were a disturbing, extraneous factor influencing soil moisture and resistivity variation, and are a possible source of error when interpreting the relationships between prehistoric space-use, soil moisture conditions, and soil resistivity (Chapters 9, 10).

A second set of variables that are not of interest for modeling general, functional, or associational relationships along human activity, soil properties, and soil resistivity but that could not be held constant within individual transects relate to the historic uses which Crane site served. Different portions of Crane site have been cropped and fertilized or left in timber for different amounts of time. These spatial variations in land-use have changed the balance of soil nutrients in whole subsectors of the Crane site and have created significant disconformities along some of the transects. An attempt to understand and compensate for this undesirable source of variability is made in Chapter 7. On a smaller scale, spatially nonuniform patterns of fertilizer application and dispersal have created localized anomalies in soil chemistry. Also, bulldozing, small amounts of land contouring, and the uprooting of trees on the Crane site during the historic period have formed localized anomalies in soil stratigraphy and soil structure (Chapter 7).

Depth of Survey

The depths to which resistivity soundings were made along the transects at Crane were chosen in light of the stratigraphy of the site and requirements for the optimal use of the Barnes Layer method of calculating soil resistivity values. In Chapter 2, it was suggested that variability in a resistivity data set can be partitioned along two dimensions—depth and space—using the Barnes Layer method and spatial filtering techniques, respectively, in order to segregate archeologically significant resistivity variation from agricultural or naturally caused resistivity variation. Optimal use of the Barnes Layer method for this purpose requires that the sequence of layers for which resistivity values are calculated be as congruent as possible with the actual stratigraphic sequence of agricultural, archeological, and natural soil horizons. Thus, the depths to which resistivity soundings were made at Crane were designed so as to allow the generation of Barnes Layers with depths and thicknesses that corresponded to the depths and thicknesses of the various strata there.

At the time the Transect Survey was designed, very little was known about the stratigraphic and soil horizon sequence. This information was based on a limited soil survey of the site, during which a series of 32 cores were removed in various areas with a 1.9-cm diameter push auger, each core to a depth of 100 cm. The soil in each core was examined for changes or discontinuities in its color and compaction over depth, in order to define major agricultural and cultural stratigraphic units and natural soil horizons. Discontinuities noted in the soil cores suggested the following optimal arrangement of layers for investigation by the Barnes Layer method. A first, surface layer (0–26.47 cm below surface) was chosen to include the plowzone, so that its masking effect might be subtracted out when calculating the resistivity of deeper soil layers with the Barnes Layer method. The second layer (26.47–33.66 cm) was selected to include only midden deposits where they occurred, and neither the plowzone above nor the natural soils below. Determination of the resistivity of this stratum by the Barnes Layer method would allow the segregation of archeological variability within the soils of Crane site. The third layer (33.66–42.54 cm) was designed to encompass the interface between midden deposits, where they occurred, and the natural subsoil of the site. Measurement of the composite resistance of this layer and the two above would allow the effects of these layers to be subtracted from the resistivity soundings made to greater depths, and allow the resistivity of the natural subsoil to be monitored. Also, resistivity variation within this third layer might be used to plot spatial variation in the depth of the base of midden deposits.

The subsoil of Crane site was surveyed with two Barnes Layers: 42.54–64.14 cm, and 64.14–75.56 cm. The boundary between these two layers was not designed to separate different horizons of the subsoil, which in general is uniform in texture (silt loam), structure (weakly prismatic breaking into subangular blocks), and color. Rather, the division was based on the depths below the midden that pit features of different morphological (and possibly functional) types might be expected to penetrate the subsoil. At the Loy site, a Middle Woodland Village that is several miles further up the Macoupin Valley and that was excavated by Farnsworth, pits could be segregated into two distinctly modal types based on the depths to which they extended below the midden (Figure 45; Farnsworth, personal communication). It was hoped that these same classes of pits might occur at the Crane site and that by monitoring separately the resistivity of the layer into which both kinds of pits intruded and the resistivity of the layer into which only the deeper kinds of pits intruded, the relative densities of the two kinds of pits in given areas of the site might be estimated. Additionally, the resistivity of these layers can be used to monitor natural subsoil variation where pit density is low.

Thus, five distinct layers were chosen for sampling the different soil horizons at Crane site for their resistivity. These are named according to a notation which is symbolic of the Barnes Layer subtraction process: Barnes Layer 1–0 (BL1–0), Barnes Layer 2–1 (BL2–1), Barnes Layer 3–2 (BL3–2), Barnes Layer 4–3 (BL4–3), and Barnes Layer 5–4 (BL5–4). The resistivity soundings that extend from the surface to the base of each of these layers and that are used to calculate the resistivity of these layers are denoted Whole Volumes 1 through 5 (WV1 through WV5).

Although this scheme for sampling the several different strata of the soils of

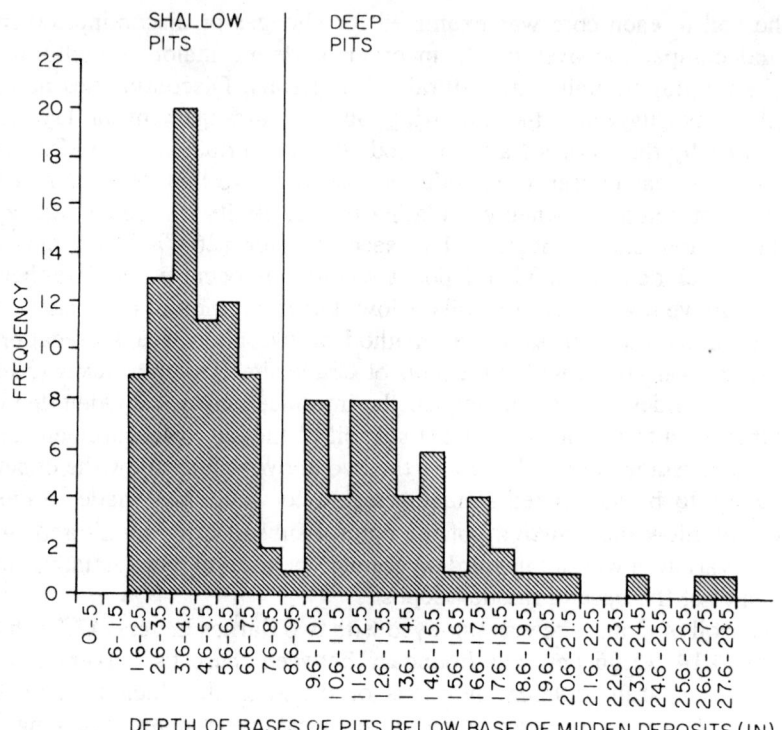

FIGURE 45. *The frequency distribution of depths of the bases of Middle Woodland pits below the base of midden deposits at the Loy site, Illinois. Two modal types of pits can be defined.*

the stratigraphic model upon which it was based served its purpose in allowing the collection of resistivity data (and soil samples) within Barnes Layers that correspond reasonably to stratigraphic units, more recent and broader-scale information on the stratigraphy of the Crane site (Chapter 7) has shown that some of the layers are of a different nature than originally interpreted. Resistivity and soil data from them cannot be used for purposes originally intended. First, the plowzone actually is composed of two distinct stratigraphic units and is deeper than originally estimated. From the surface to 20.32 cm below surface occurs soil that annually has been upturned and churned by moldboard plowing. From 20.32 to 33.02 cm below surface, the Crane soils have been chisel-plowed twice, but not upturned by moldboard plowing. The midden deposits (where present) and/or natural soils within this zone have been altered structurally with respect to the frequency of very large pores within them by the knifelike cutting action of the chisel plow, but chemically and texturally, and with respect to the frequency of large to very small pores within them, they are the same soils which occurred in this zone prior to the historic use of Crane for farming. Thus, the soils encompassed within Barnes Layer BL2–1 may be identified as archeological rather than agricultural with respect to their chemical and textural properties, but not all aspects of their structural organization. Relationships between prehistoric activity, soil alternations, and soil resistivity can be studied with data collected from BL2–1 only in regard to certain soil attributes. The fine points of this qualification will be discussed as individual analyses are presented (Chapters 8, 9).

The second change in the interpretation of the Crane stratigraphy relates to the depth of midden deposits. As originally modeled, midden deposits were thought to be fairly uniform in depth over the site and rarely to extend below the base of BL3–2. This is not true. In a number of locations, midden deposits extend significantly into BL4–3. Spatial variations in the resistivity and other properties of the soils in BL4–3 cannot be considered a result of variations in only the areal densities of pits or the nature of the subsoil. Spatial changes in the depth of midden deposits and changes in the relative amounts of midden deposits and natural subsoil incorporated within BL4–3 also must be considered when interpreting the resistivity and other properties of the soils within this layer. Thus, resistivity data from BL3–2 and BL4–3 may be used to monitor variation in the depth of midden deposits over Crane. In areas where midden deposits do not extend into them, their resistivity can be used to monitor the areal density of pits.

No midden extends into BL5–4. This sampling unit can be used as it was originally designed—to monitor the areal density of deep pits at Crane and natural variations in the resistivity and other properties of the natural subsoil.

Although the interpretation of the various Barnes Layers designed for the Crane resistivity survey changed over the course of the project, the initial reasons for making them given depths and thickness should serve as an example to other suveyors of how Barnes Layers may be generated with optimal metrics in order to serve explicit purposes.

Density of Sampling Locations Where Resistance Measurements Were Made along Transects

Resistance measurements along Transects I–V were made every .5 m. This spacing was designed with respect to the areal expanses of the volumes of soil measured in each of the five Barnes Layers previously defined. The chosen sampling interval was a compromise between two desirable conditions. First, it was desired that adjacent stations be close enough together that the volumes of soil measured at them would touch one another and form a total, *continuous* sample along the direction of the transect for *each* Barnes Layer investigated. This would allow one to monitor a continuous resistivity trace along the transects, and also provide a large as possible a sample of resistivity measurements along them. Second, it was desired that the Barnes Layers of adjacent proveniences be nonoverlapping, and thus, *statistically independent* at *each* level of investigation. This would be a requirement later for performing

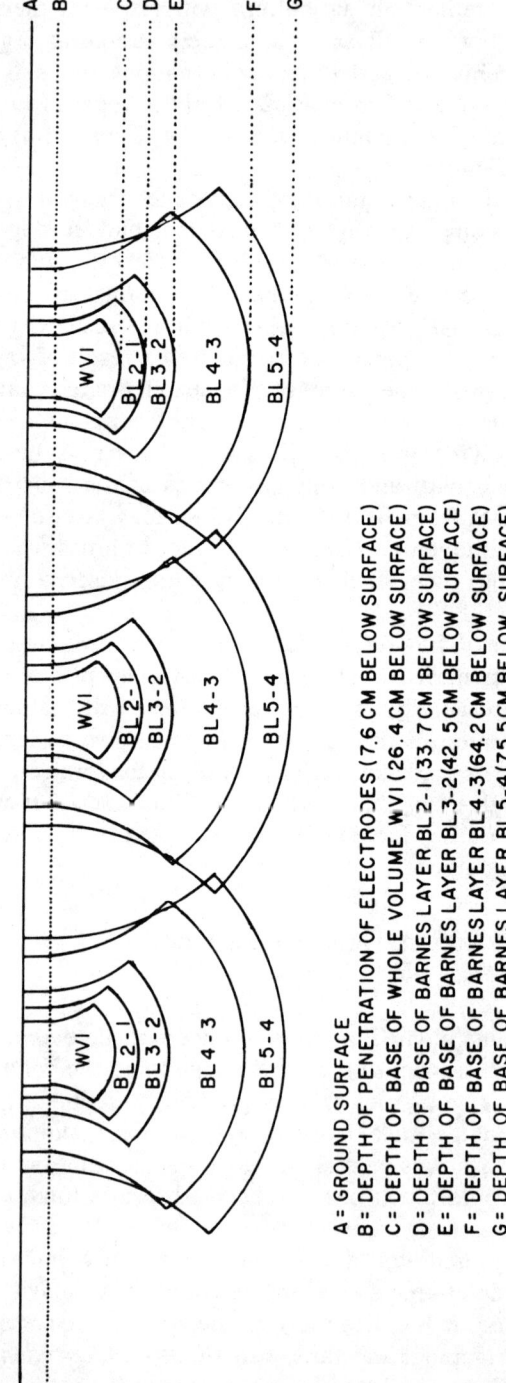

FIGURE 46. *Areal overlap of the discontinuous spacing of Barnes layers in three parallel subtransects, one meter apart. The volumes shown include that soil most affecting resistivity determinations, but not all soil having any significant effect upon the determinations (see Table 47).*

TABLE 47
Approximate diameters of the bases and vertical midpoints of the Barnes Layers used in the transect survey at the Crane site (homogeneous medium assumed)

Barnes Layer	Diameter of the Base of that Volume of Soil Most Effecting the Resistivity Determination	Diameter of the Base of that Volume Including all Soil Having Any Significant Effect Upon the Resistivity Determination
BL1-0	28.5 cm	79.4 cm
BL2-1	39.0 cm	101.0 cm
BL3-2	52.0 cm	127.6 cm
BL4-3	85.0 cm	190.1 cm
BL5-4	105.0 cm	203.8 cm

statistical tests of difference in the mean resistivity values and variances in resistivity values of different subportions of the transects.

Clearly, both these conditions cannot be achieved for all Barnes Layers: the different Barnes Layers have different diameters (Figure 46, Table 47). For example, a sampling interval giving total areal coverage at the level of BL1–0 would imply overlap of the layers in adjacent proveniences at all lower depths. A sampling interval giving complete areal coverage without overlap of layers in adjacent proveniences at the level of BL5–4 would imply incomplete areal coverage at all higher levels. The optimal compromise was found in a 0.5-m sampling interval. If every station along a transect is analyzed, adjacent Barnes Layers at all levels overlap, allowing one to examine a continuous resistivity trace in the direction of the transect. If only every other station is considered, adjacent Barnes Layers BL1–0, BL2–1, and BL3–2 barely overlap or nearly overlap. When every fourth station is used, adjacent Barnes Layers BL4–3 and BL5–4 nearly overlap or overlap slightly. Thus, a sampling interval was designed for taking resistivity measurements along transects: For each level of investigation, a subset of all observations along a transect can be chosen, having observations that (*a*) are approximately nonoverlapping and statistically independent, yet (*b*) are close enough together to allow one to examine a continuous resistivity trace in the direction of the transect.

Density and Arrangement of Locations Where Soil Samples Were Removed along Transects

Soil samples were removed from each Barnes Layer at a subset of all locations surveyed for their resistivity along the transects. Although it would have been advantageous to have sampled each provenience where resistivity measurements were made, this was not possible. Soil sampling procedures were laborious and time consuming. To sample all locations along the transects would have required a longer period of time than that during which soil conditions would have been similar to those conditions when resistivity measurements were made. As the purpose of soil sampling was to compare the properties of the soils at the time of survey to their resistivity, sampling time and the number of samples that were collected had to be limited. Moreover, physical and chemical analyses of the soil samples are expensive, and the total number of soil samples had to be restricted to a minimum. Thus, a design was created for sampling only some of the stations along a transect.

The chosen sampling procedures aimed at obtaining a sample population along each transect characterized by two attributes: (a) a fairly uniform, systematic spatial distribution of sample locations along the transect, documenting gentle, natural soil changes; and (b) a high proportion of sample locations having archeologically significant anomalies relative to the actual low density of such anomalies on the site as a whole. The latter attribute could not be achieved with a uniform sampling scheme. To design a sample having both attributes, sampling locations were chosen in reference to soil resistance values along the transects. As resistivity data were collected along the transects, the resistance values of WV3 and WV5 were examined. It was assumed that low frequency spatial trends in WV5 resistance values indicated natural trends in the properties of subsoils along the transect, whereas deviations from similar trends in WV3 resistance values indicated archeologically significant anomalies. Soil samples were taken along transects so as to yield an even scatter of data points among broad, high and low resistance zones presumed to indicate natural subsoil proveniences; extra samples were taken where anomalies from subsoil trends occurred in WV3 resistance values and might indicate archeological disturbances. It would have been preferable if Barnes Layer resistivity values could have been used rather than whole-volume resistance values for determining the subsoil trends and particularly for determining the archeologically significant anomalies to be sampled, in order to avoid the survey problems discussed in Chapter 1. It was not possible with the equipment available, however, to calculate Barnes Layer values quickly enough in the field to use them to pick the locations that would be sampled for their soil attributes.

The design that was used to pick those locations along transects to be sampled for their soil attributes is considered more appropriate than a simple, systematic sampling design. In testing the ability of resistivity methods to detect archeologically significant anomalies, four situations must be considered:

1. Those in which resistivity data suggest an archeological anomaly occurs and where the anomaly actually does occur
2. Those in which resistivity data suggest an archeological anomaly occurs but it does not exist

3. Those in which resistivity data do not suggest an archeological anomaly occurs but it does
4. Those in which resistivity data suggests an archeological anomaly does not occur and it actually does not.

A simple systematic design would have yielded an undesirable sample population having a high proportion of circumstances 2 and 4 and a low proportion of circumstances 1 and 3, given the low density of subsurface archeological features on the site. Such a sample might have included too few observations from archeologically disturbed areas to estimate the success rates of resistivity methods in detecting archeological anomalies. Thus, a nonsystematic sampling design biased in favor of sampling circumstances 1 and 3 was used.

At a smaller spatial scale, the design used for sampling the soil within the several Barnes Layers at each individual location along the transects must be considered. From each Barnes Layer at each sampled station, enough soil had to be obtained to run all the physical and chemical soil tests that would adequately characterize the nature of the soil in those layers. Given that soil was to be extracted from the several layers with 1.9-cm diameter push augers and on the basis of preliminary studies of the physical and chemical nature of the Crane soils, it was determined that three cores through any layer—even the thinnest (BL2–1)—would provide an adequate volume of soil to run all the desired soil tests. Next, it was necessary to determine an appropriate spatial arrangement for extracting the three cores from each station. The spatial arrangement would have to be such that the soil cores to be drawn would be (*a*) representative of the average soil conditions and range of soil variation within the layer being sampled, and (*b*) composed of soil from the different sectors of the layer in proportion to the relative weight that these sectors would carry in determining the total layer resistivity value. The latter requisite arises from the fact that different equal volumes of soil within a Barnes Layer contribute more or less to the total resistivity of that layer according to their position within the layer. Soil closer to that line described by the electrodes in a Wenner array has more effect on the total resistivity of a Barnes Layer than does soil further from that line. These two demands on the sampling design were met, approximately, by positioning one of the three soil cores at the center of the array and two along the line of the Wenner array at the greatest possible distance from its center but still within the circumference of the Barnes Layers. Placing the extreme cores at the furthest possible distance from each other and the central core increased the probability that maximal soil variation would be encompassed by the soil cores—assuming that variability is a function of distance. Placing all three cores along the line of Wenner array ensured that soil having the greatest effect on the total resistivity of a Barnes Layer would be sampled.

The exact distance chosen for separating the outer two cores from each other and the central core was a compromise between two objectives. On the one hand, it was desired that the two soil samples removed from noncentral positions of each Barnes Layer be separated as widely as possible from each other and the central soil sample, and be situated near the periphery of those layers. Since the different layers have different diameters, the distance between peripheral soil samples could have been varied from layer to layer to maximize it for each layer. This sampling scheme would have required a large number of cores to be taken at different distances from the center of the Wenner array, in order to extract the peripheral soil samples from each Barnes Layer. It also was desired, however, that the amount of time spent in coring out soil samples at individual stations be minimized so that more stations could be sampled. Thus, it was decided to take the peripheral soil samples from each Barnes Layer at a constant distance from each other so that only two cores had to be drawn to obtain the peripheral soil samples for all five Barnes Layers. The maximum distance used to separate the two peripheral cores was restricted by the minimum diameter of the smallest Barnes Layer for which correlations between soil properties and soil resistivity would be sought. This layer was BL2–1,[1] having a diameter at its vertical midpoint equal to 72.7 cm. Peripheral cores were taken at a distance of 50.9 cm from each other, or 10.9 cm interior to the periphery of BL2–1. They were taken slightly interior to, rather than exactly at, the layer's periphery in order to allow a margin of error in placement of soil cores with respect to the placement of resistivity equipment and the Barnes Layers actually measured.

Repetition of Resistivity Measurements along Transects

The transects surveyed at Crane site were positioned and oriented so as to sample as large a number of use-areas as possible. This aspect of the design did not ensure, however, that the number of observations which would be made within each use-area would be sufficient to test the differences between them in a statistical manner. To increase the number of independent observations made within each use-area that occurred along transect lines, each transect across the Crane site was surveyed with resistivity equipment not once, but three times, along three separate but parallel subtransects. Two of the three subtransects in each transect were sampled for their soil attributes according to the design described.

The distance between subtransects, as that between survey stations, represents a compromise between contradictory desires. On the one hand, it was desired that subtransects be in close enough proximity that they would pass

---

[1] Barnes Layer 1–0 (plowzone) was measured for its resistivity simply to allow the removal of its masking effect from deeper layers of interest. It was soil-sampled to monitor agricultural sources of variation that might extend to deeper layers but that could not be removed by the Barnes Layer method (e.g., localized anomalies in fertilizer concentration that could change the chemical balance of the soil to some depth as a result of leaching of the fertilizer from plowzone into lower horizons).

through the same use areas. On the other hand, they had to be far enough apart so that the Barnes Layers of adjacent subtransects at the levels of archeological deposits (BL2–1, BL3–2) would not overlap and so that the resistivity data from adjacent subtransects at these levels would be statistically independent. A one meter distance between subtransects was chosen, this spacing representing the smallest possible under the requirement of nonoverlap of Barnes Layers within the archeological horizons (Table 47, Figure 46).

Measurement of Soil Temperature along Transects

The soil samples collected along the transects allowed a number of variables that determine soil resistivity to be measured at leisure in laboratories distant from the Crane site, after the surveys had been performed. One variable determining soil resistivity that had to be measured directly in the field at the time of survey, however, was soil temperature. At each station from which soil samples were collected, the soil temperature at the base of plowzone was taken. Although it might be thought that it would have been preferable to have taken the temperature of each Barnes Layer at each surveyed station, this operation was unnecessary. Under the conditions of survey, soil temperature variations over depth and at different locations surveyed at different times of the day were minimal with respect to the variations they would cause in soil resistivity values—within each individual transect. For each transect, a single average value of the soil temperature of all Barnes Layers at all times of survey provided an adequate description of temperature as a variable determining soil resistivity. This average was approximated by the average soil temperature found at all stations at the base of plowzone. Different average values were determined for each transect, however, for they were surveyed at different times of the spring and summer in more widely varying temperature regimes.

Soil Analyses Performed on the Soil Samples Collected from Transects I–V

The major properties of soils that can be used to describe and analyze the functional relationships that hold between particular kinds of human activities, the soil alterations they may yield, and the patterns of soil resistivity values determined by those alterations include: (a) the agronomic "availability" of different species of ions and radicals within the soil (on a volume basis); (b) organic matter content; (c) the porosity and pore-size distribution of the soil; (d) soil moisture content (on a volume basis); (e) soil temperature; and (f) soil texture. Had it been feasible economically, values for each of these variables would have been determined for the soil encompassed by each Barnes Layer at each sampled station along each transect. The expense of determining soil texture and pore-size distribution, however, prohibited the use of this experimental design, and these two variables were not assessed for the soil collected along Transects I–V. Texture and pore-size distribution were measured on more limited sets of soil samples collected with greater specificity in relation to the use-areas and expected natural soil trends at the Crane site.

The procedures used to determine each of these variables are as follows. Each sampled station along the transects was represented by 15 soil samples, 3 from each of the 5 layers which were surveyed for their resistivity. The 3 cores from each Barnes Layer were compared visually to determine whether they came from soils that are spatially homogeneous with respect to color, structure, and inclusions, or whether they represented markedly different kinds of soil that might reflect archeological variability, for example, one core may have penetrated a pit intruding into the subsoil whereas the other two cores may not have. If homogeneous, the soil from all three cores was combined into a single, well-mixed sample, later to be subsampled for each soil test. If nonhomogeneous, similar cores (if any) were combined into a single sample whereas dissimilar cores were left separate. The two or three resultant samples from each such inhomogeneous Barnes Layer were handled separately for each soil test made.

Soil moisture content of each sample (combined or single cores) was determined by gravimetric methods, as described by the American Society for Testing Materials (ASTM 1955). The weights of the samples at field moisture were measured. An approximately 20-g subsample then was oven-dried at 100°C and reweighed to determine its moisture loss (moisture content). The moisture content of the total sample then was calculated on a weight basis and volume basis. Moisture content on a volume basis could be calculated because the core samples removed from each Barnes Layer were of a standard diameter (1.9 cm), length (the thickness of the Barnes Layer being sampled), and volume. The error involved in estimating volume-based soil moisture contents is greater than that involved in weight-based moisture contents, however, for core volume was not as accurately controlled in the field as total sample wet weight was measured in the laboratory.

Using the oven-dry weights of the 20-g subsamples and the subsampling proportion, bulk density values (dry weight of soil per unit volume) were calculated for each soil sample (Black 1965:374–399). The porosity (percent pore space on a volume basis) of each sample was calculated using the bulk density estimates and the assumption that particle density for the loessic soils at Crane site is approximately 2.65 g/cm^3 (Brady 1974:50, 53–54; Vomocil 1965:300–301). As with the volume-based moisture contents, estimates of these two volume-based variables include higher percentages of error than variables that have values dependent on laboratory measurements alone.

Chemical analyses were made on the soil remaining in the sample after the 20-g subsamples for moisture content determination were removed. The oven-dried subsamples could not be recombined with the rest of the sample for chemical analysis, for oven drying irreversibly alters the natural chemical availability of ions and radicals within a soil. Clay particles bond together such that some ions on their exchange sites become immobilized and unexchangeable.

Chemical analyses of the soils from the Crane site were performed at the North Carolina Department of Agriculture Agronomic Laboratory, located in Raleigh, N.C. The soil determinations made there include organic matter content, pH, and the availability of nine species of ions and radicals which: (a) usually are found in soil water in concentrations significant enough to be effective in determining soil resistivity values, and (b) are likely to be derived from anthropic as well as natural or agricultural sources (see Chapter 3; Table 11). These nine kinds of ions or radicals in decreasing order of significance include Ca^{2+}, PO_4^{3-}, NH_4^+, K^+, Mg^{2+}, Na^+, SO_3^{2-}, NO_3^{2-}, and Cl^-. The availability values of nitrates and the chloride ion were determined for only a small proportion of the soil samples, for these ions and radicals were found part way through the analysis to occur in negligible concentrations in the soil horizons below plowzone and were not measured thereafter. These ions are subject to an excessive amount of leaching compared to cations, phosphates, and sulfates. Aluminum and silicon, which occur in soil water in concentrations significant enough to be effective in determining soil resistivity values, but which would more likely be of natural origin than anthropic origin, were not assessed for their availability. These analyses were not performed by the soils laboratory that conducted the chemical tests of the Crane soil samples.

The laboratory procedures used in determining pH, nutrient availability, and organic matter content are variations (N.C.D.A. 1972, 1976) of standard methods which are used often in agronomy and soil science (Black 1965). Soil pH was determined by electrometric methods, using a glass–calomel electrode system (Small 1954). Soils were suspended in water rather than a salt solution (Peech et al. 1953; Raupach 1954; Schofield and Taylor 1955), and a soil : water weight ratio of 1 : 1. Soil suspensions were allowed to stand 1 hour before pH determinations were made. The availabilities of the nine species of ions and radicals mentioned above were assessed by leaching a set volume of soil, subsampled from each soil sample, with various extracting agents in constant quantity, and then measuring the concentrations of the ionic species of radicals within the leachate. Phosphorus, potassium, calcium, magnesium, sodium, and ammonium were extracted by the Mehlich double acid method (Mehlich 1953; Nelson et al. 1953; N.C.D.A. 1976), using the weak acids .05 N HCl and .025 N H_2SO_4. Nitrates were extracted with phenoldisulfonic acid, sulfates with .5 N NH_4Cl, and chlorine with distilled water (N.C.D.A. 1972). The concentrations of nutrient ions and radicals within the leachates were assessed as follows: Ca and Mg by atomic absorption spectrophotometry (Prince 1965); K and Na by flame photometer (Rich 1965); PO_4, NH_4, and NO_3 by colorimetric techniques (Bremner 1965; N.C.D.A. 1972; Olsen and Dean 1965); SO_4 by turbidimetric methods using barium acetate hydrate and acetic acid glacial as the precipitating agents (N.C.D.A. 1972); and Cl by turbidimetric methods using silver nitrate and nitric acid to precipitate the chlorine.

The availability of each species of ion and radical for which tests were run was calculated on a weight basis from their concentrations and weights within the leachate. No assumption was made in this calculation that the set volume of leached soil represented any set weight (Mehlich 1973). The actual weights of the leached subsamples were taken into consideration using volume weight determinations. Ion availability on a volume basis then was calculated under the assumption that the bulk density of all sampled soils is constant—1.32 g/cm^3.

The appropriatness of the assumption of a constant field bulk density of all the soils at Crane site in determining the availability of different nutrients on a volumetric basis rightfully may be questioned. The bulk density of the soils at Crane site do vary—from horizon to horizon, and between the natural soils and the archeological deposits (see Chapter 7). They range in bulk density from ca. 1.31 to 1.61 g/cm^3. The bulk densities of most soils at the Crane site (excluding plowzone) are above the assumed value of 1.32 g/cm^3, and consequently, most nutrient availability values are deflated. Actual field availabilities may be as much as 21.97% higher than the values reported by the North Carolina soils lab.

To fully correct for this error, it would have been necessary to have taken special bulk density soil samples for each Barnes Layer at each station sampled, and to have used the true field bulk densities of these soil units in calculating nutrient availability values. Economically, this was not feasible. It was possible, however, to use the volume weights of the ground and sieved soil samples from the Crane site in the laboratory to predict the bulk densities of the undisturbed soils from which they were removed, and to correct the reported nutrient availability values with the predicted bulk density values.

A linear regression model of the relationship between laboratory volume weight and field bulk density was calculated, based on the values of these variables in a random sample of the natural soil horizons and the archeological deposits coming from all sectors of the Crane site. The data used in calculating the regression model are shown in Table 48 and are plotted in Figure 47 along with the regression line. The model itself is as follows:

$$\text{Bulk density}_i = 1.3251 \,(\text{volume weight}_i) - .041540$$

Statistics for the evaluation of the regression analysis are shown in Table 49. The assumptions behind these statistics—that residuals are approximately normally distributed about the regression line, and that the variance of the residuals about the regression line along its entire length is approximately constant—were met, so the statistics may validly be used in assessing the regression.

Although the scatterplot of bulk density against volume weight (Figure 47) would suggest that a linear model is the most appropriate form of functional

TABLE 48
Field bulk density and laboratory volume weight of a random sample of natural soil horizons and archeological deposits at Crane site

Volume Weight (gm/cc)	Bulk Density (gm/cc)	Volume Weight (gm/cc)	Bulk Density (gm/cc)
1.0900	1.5700*	1.1900	1.5545
1.1150	1.5050	1.1270	1.4087
1.1400	1.5750*	1.1500	1.3470
1.1650	1.2545*	1.1470	1.3777
1.1500	1.5300	1.0900	1.3435
1.1330	1.4270	1.1667	1.5045
1.1130	1.3800	1.1600	1.4397
1.1250	1.4650	1.1450	1.4540
1.1600	1.5530	1.1300	1.4860
1.1140	1.5000	1.1300	1.4685
1.1600	1.5030	1.1230	1.3897
1.0900	1.4440	1.1000	1.4110
1.1500	1.5210	1.1100	1.5505*
1.2000	1.5570	1.1150	1.4750
1.1370	1.4760	1.1200	1.4047
1.1533	1.5320	1.1100	1.3360
1.1470	1.5103	1.2150	1.5070*
1.1250	1.5113	1.2030	1.4730*
1.1700	1.5215	1.2000	1.4440*
1.1250	1.5240	1.2200	1.4965*
1.1400	1.4530	1.1900	1.5290
1.1667	1.4843	1.2230	1.3950*
1.1100	1.4760	1.2500	1.6065*

*not used in regression analysis

relationship for predicting bulk density from volume weight, the best-fit linear model still is not a good one. The model explains only 33% of the variance in bulk density values ($R^2 = .328$), and there is a significant partial correlation (.573) between the independent variable and the unexplained residual variability. Thus, it is possible to make some adjustments in the reported nutrient availability values for the incorrect assumption that the field bulk densities of all

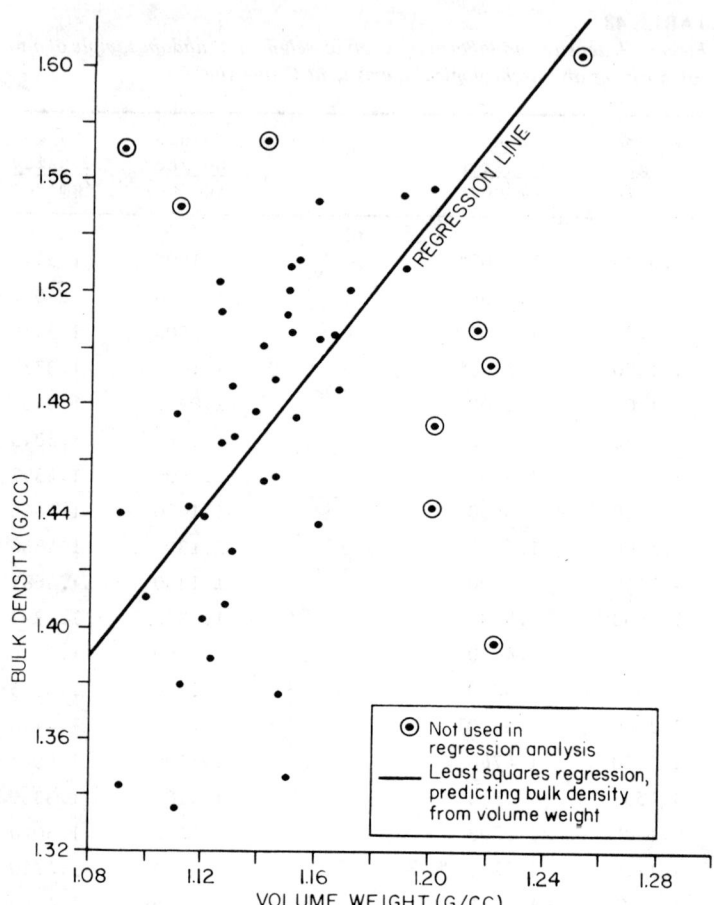

FIGURE 47. *Scatterplot of the volume weights and bulk densities of natural soil horizons and archeological deposits randomly selected from the Crane site. The line through the data swarm is that defined by a least-squares regression model for the data, using volume weight as the independent variable and bulk density as the predicted variable.*

soil samples were 1.32 g/cm³, but the model to be used in this correction procedure still leaves a substantial range of errors within the data. Use of the model allows a reduction in the *average* error from 10.28%[2] within the uncorrected data set to 4.39% within the transformed data set, but the *maximum* error within the transformed data set could still be at least as high as 13.11%.[3] These error values reflect error attributable to only the problem of estimating

---

[2] (Average bulk density in Table 48 − 1.32)/(Average bulk density).

[3] Maximum departure of actual bulk density values in Table 48 from bulk density as predicted by volume weight.

TABLE 49
Statistics for regression analysis of bulk density and volume weight values in Table 48

Source of Variance	Degrees of Freedom	Sum of Squares	Mean Square	F-Statistic	Level of Significance
Regression	1	.044710	.044710	16.601	.0003
Error	34	.091566	.002693		
Total	35	.136276			

Constants in Model	Coefficient	Standard Error	T-Statistic	Significance Level of Coefficient
a	-.041540	.37028	-.11219	.9113
b	1.3251	.32521	4.0745	.0003

$R = .57279$

$R^2 = .32808$

Partial correlation between independent variable and unexplained variance = .57279

bulk density and do not include additional laboratory errors in measuring ion availability.

The final test performed by the North Carolina laboratory on the soil samples was the determination of their percent organic matter content. This was done by sulfuric acid digestion and colorimetric procedures similar to those designed by Walkley and Black (N.C.D.A. 1972). The percent organic matter contents calculated from the results of these procedures assumes that the volume weight of the tested sample is 1.25 g/cm^3. For those soil samples having volume weights of other than 1.25 g/cm^3, a correction of the determined values was made by multiplying them by the factor 1.25 over volume weight of sample.

Coding and Presentation of the Data Collected for the Transect Survey

The resistivity and soil data that were collected from Transects I–V are presented in the matrices in Appendix 5. The transects made across the Crane site have been labeled I–V. Their placements and orientations are shown in

Figure 47. Each station along Transects I–V that was surveyed for its soil resistivity was sequentially given a provenience number, ranging from 1 through 7531. Subtransect provenience numbers begin on even 100s, and the adjacent three stations on different parallel subtransects of the same transect have provenience numbers that correspond in their last two digits. The grid locus and provenience number of each survey station in each transect are given in Table 1 of Appendix 5.

From the resistance data collected at each station were calculated 5 whole volume resistivity values pertaining to soil extending from the surface to given depths of investigation, and 15 Barnes Layer values corresponding to various subsurface horizons or combinations of horizons. The codes given to these whole volumes and Barnes Layers are shown in Table 50, along with the constants used in Wenner's equation, to calculate resistivity values from

TABLE 50
Constants used in calculating resistivity values from resistance values gathered along Transects I-V

Barnes Layer	Electrode Spacing (cm) A_1	A_2	Layer Thickness (cm) A^{**}	Average Electrode Separation (cm) A^{***}	Electrode Penetration (cm) B^{****}	Parameter N
BL1-0*	18.838	0.	18.838	9.417	7.35	1.627
BL2-1	26.035	18.838	7.197	22.437	7.35	1.708
BL3-2	34.925	26.035	8.890	30.480	7.35	1.819
BL4-3	56.515	34.925	21.590	45.720	7.35	1.911
BL5-4	67.945	56.515	11.430	62.230	7.35	1.950
BL2-0*	26.035	0.	26.035	13.017	7.35	1.766
BL3-1	34.925	18.383	16.087	26.882	7.35	1.778
BL4-2	56.515	26.035	30.480	41.275	7.35	1.893
BL5-3	67.945	34.925	33.020	51.440	7.35	1.929
BL3-0*	34.925	0.	34.925	17.462	7.35	1.856
BL4-1	56.515	18.383	37.677	37.677	7.35	1.874
BL5-2	67.945	26.035	41.910	46.990	7.35	1.915
BL4-0*	56.515	0.	56.515	28.257	7.35	1.940
BL5-1	67.945	18.383	49.107	43.392	7.35	1.902
BL5-0*	67.945	0.	67.945	33.972	7.35	1.958

* Barnes Layers BL1-0, BL2-0, BL3-0, BL4-0, BL5-0 are equivalent, respectively, to Whole Volumes WV1, WV2, WV3, WV4, and WV5
** A in the numerator of Wenner's equation
*** A in the denominator of Wenner's equation
**** B in the denominator of Wenner's equation

resistance values. The depths of the basal boundaries of the whole volumes and Barnes Layers are 7.62 cm greater than the electrode spacings used in calculating their resistivity. This circumstance arises from the fact that the source and sink for the current drawn through the soil while making resistivity measurements were not located at the surface, but at the tip of electrodes, which were inserted 7.62 cm into the ground (Figure 46). The resistivities of all 5 whole volumes and all 15 Barnes Layers for each surveyed station are recorded in Table 2 of Appendix 5.

The physical and chemical properties of the soil samples collected along Transects I–V are summarized in Tables 3 and 4 of Appendix 5. In the table of chemical properties, a three-place alphabetic code used by the North Carolina Department of Agriculture soils laboratory to identify the samples is listed along with the provenience designation of each soil sample. The first two places of the code specify provenience. The last place specifies the Barnes Layer from which the sample was removed, using the five letters A through E for the five Barnes Layers, BL1–0 through BL5–4.

The nutrient availability levels recorded in Table 4 for the nine different ions and radicals are given as they were reported by the North Carolina soils laboratory, on a pseudovolumetric basis (kilograms per hectare to a depth of 20 cm) and assuming a field bulk density of 1.32 g/cm^3. Correction of these values for the variable bulk density of the soils from which the samples were derived can be made using the volume weight of the samples and the regression model between volume weight and bulk density that was described in the previous section of this chapter. In the body of this work, however, references to nutrient availabilities always take into consideration the variation of their bulk densities from 1.32 g/cm^3. The nutrient availabilities discussed have been corrected for this variation using the volume weight of the samples and the regression model between volume weight and bulk density.

The Block Survey: A More Selective Soil and Resistivity Survey

The Transect Survey was designed to sample the soil resistivity and other selected soil properties of as many different prehistoric use-areas on the Crane site as possible. This project was not adequate, however, for gathering *all* the information necessary for defining the relationships between prehistoric space-use, soil alteration, and soil resistivity at Crane. First, although the survey was adequate in sampling a large variety of use-areas having different kinds of *midden deposits* generated by different activities, it did not provide a large sample of use-areas having different *pit densities*. Most of the data collected with the Transect Survey design pertain to areas of the site that had very low areal densities of pits or that were totally devoid of pits. Also, very few of the surveyed areas were excavated and documented for their pit densities. Since one aspect of the relationship between prehistoric space-use and resistivity patterns, which we wish to investigate, concerns the effect of different areal densities of subsurface features on the variability of a soil resistivity signature, it was necessary to design and carry out a second soil and resistivity survey that would increase the sample of zones having moderate or high pit densities.

A second inadequacy of the Transect Survey is that it did not include the collection of information on soil texture and soil pore-size distribution—two variables that determine soil resistivity and that should be monitored when investigating relationships between prehistoric space-use, soil alterations, and soil resistivity. These data also had to be collected by further survey. Third, although the Transect Survey did sample a wide range of *natural* soil conditions at the Crane site, it was not designed explictly to reconstruct the spatial patterning of natural soil variations or to understand the origin and natural processes of formation of that variability separate from the effects of human occupation upon the Crane soils. An understanding of the natural soils of an archeological site is vital to defining and interpreting pedological and resistivity anomalies of prehistoric human origin within it; anthropic anomalies must be defined in relation to *local, natural norms*, which may vary over the site. Thus, additional data had to be collected on natural soil variation at Crane.

To meet these needs, four survey programs additional to the Transect Survey were designed and carried out at Crane. The first of these was designed to increase the sample of surveyed zones having moderate to high areal densities of pits, and was similar to many respects to the Transect Survey. In 1975, Kenneth Farnsworth decided to excavate a large, continuous block on a portion of Crane known from previous test excavations to have a high density of pits. This block centers on the one known house (grid locus 116R58) and includes a zone of open work space that immediately surrounds the house as well as a more peripheral annulus containing high densities of pit features (Figure 41, Chapter 5). The decision to open this wide area and to document the location and nature of a large number of subsurface pits gave an excellent opportunity to investigate the relationship between variability within resistivity survey data sets and the areal density of pits. Consequently, it was decided to survey a portion of the block prior to excavation, for both resistivity and soil attributes.

The areas that were surveyed within the block and the relative densities of pits within them are shown in Figure 48. Choice of the total amount of area to be surveyed was based on the known rates at which surveying could be done, and on the desire to complete all surveying within one work day of uniform weather conditions. Choice of the locations to be surveyed was based on the known distribution of pits within random test excavations that had been dug within the block during the previous field season. It was hoped that the areas chosen for survey would include zones of low, moderate, and high areal densities of pits. The areas to be surveyed also had to be planned around and be distant from the infilled test squares that previously had been dug within the block, for the test

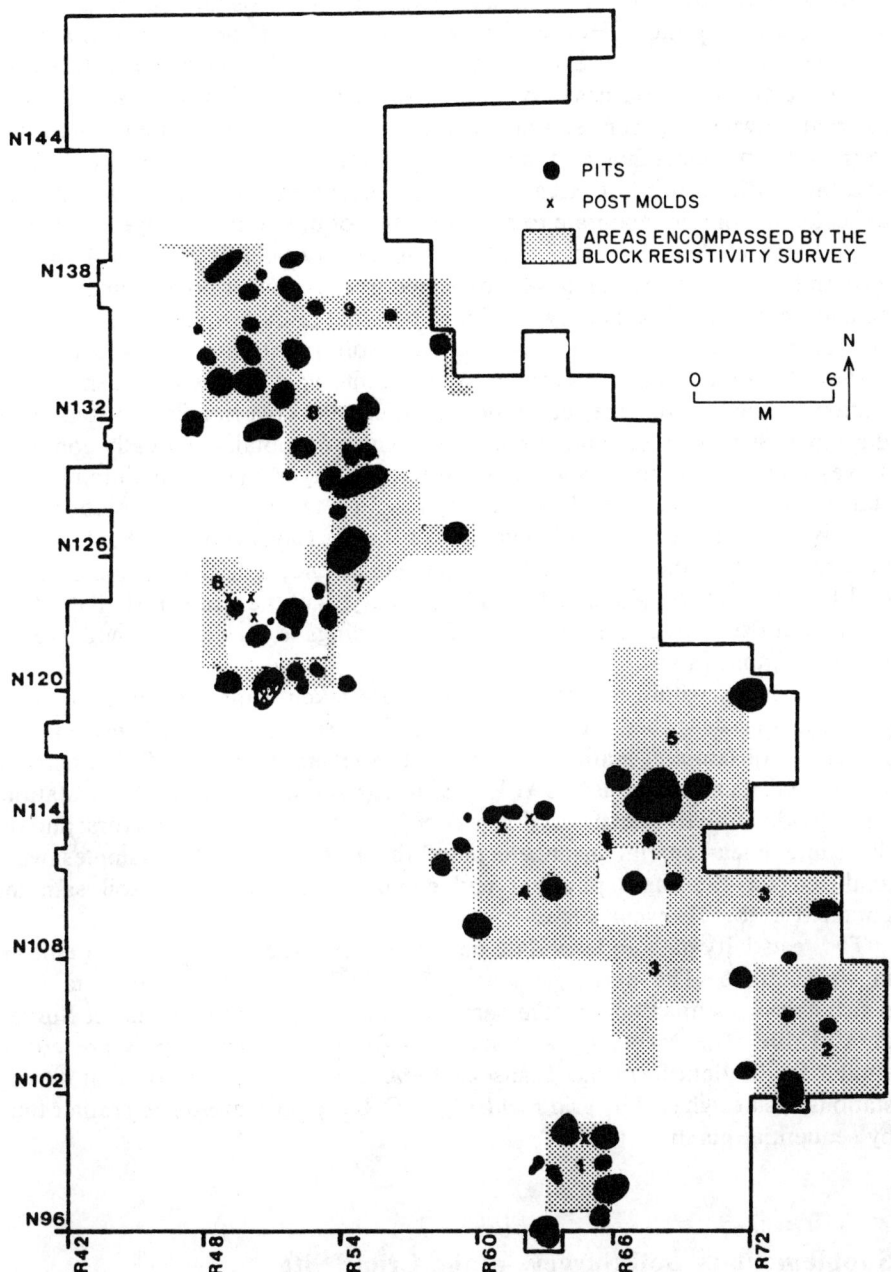

FIGURE 48. *Submidden floor plan of pits and post molds within the block excavation at the Crane site. Areas surveyed for their resistivity during the block survey are shaded.*

squares served as major disturbances to the natural moisture equilibria of the soils surrounding them. Hence, the irregular outline of some surveyed areas.

The stratigraphic layers chosen for investigation within the block at the Crane site were the same Barnes Layers investigated in the Transect Survey. The interval between adjacent stations at which soil resistance was measured was .5 m or 1.0 m, depending on the area of the block sampled. Areas 3, 6, 7, 8, 9, and the northern third of Area 4 and the western half of Area 5, which were expected to contain moderate to high densities of pits, were surveyed with a .5-m grid lattice. Areas 1 and 2, and the southern two-thirds of Area 4 and the eastern half of Area 5, which were expected to have few or no pits, and where a smaller number of resistivity readings per unit area would more likely be adequate for describing the less variable soil resistivity over space, were surveyed with a 1-m grid lattice. As with the intervals between adjacent survey stations along the transects at Crane, the intervals between adjacent stations in the Block Survey reflect an optimal compromise for obtaining areally complete survey coverage of the soils within each Barnes Layer, yet also maintaining the statistical independence of the resistivity readings taken at adjacent stations, for each Barnes Layer. The list of diameters of Barnes Layers given in Table 47 can be used to determine whether every adjacent, second, or fourth station of the .5 or 1.0-m grid lattices should be used in an analysis of the Crane resistivity data, to maintain statistical independence of all readings at a given Barnes Layer of interest (Table 51).

Soil samples within the Block Survey were taken using the same sampling design as that described previously for the Transect Survey, in all respects: (*a*) placement of sampling stations with respect to general trends in WV5 resistance values and local anomalies in WV3 values; (*b*) three cores taken per station, each divided into five segments corresponding to the five Barnes Layers; and (*c*) the same linear spatial arrangement of the three cores. Soil samples were analyzed for the same physical and chemical tests as those soil samples obtained in the Transect Survey.

The resistivity data and soil data obtained during the Block Survey are listed in Appendix 6. The parameters used to calculate the resistivity data from resistance measurements are the same as those used for the Transect Survey data (Table 50). Resistivity and soil data from the Block Survey are coded similar to the data from the Transect Survey, with the exception that survey stations are designated by grid loci (origin, ORO, placed at 96R36) rather than by sequential numbers.

Supplementary Soil Surveys at the Crane Site

In addition to the Block Survey, three other survey programs were carried out at Crane site to supplement information obtained by the Transect Survey. All

TABLE 51
Points within the block resistivity survey grid lattices which may be used if their statistical independence is required

Barnes Layer	Points Used	
	.5 m Grid Lattice	1m Grid Lattice
BL1-0	every second station	every station
BL2-1	every second station	every station
BL3-2	every second station	every station
BL4-3	every fourth station	every second station
BL5-4	every fourth station	every second station

three involved only soil sampling and soil analysis; no resistivity surveying was involved.

Survey of Soil Pore-Size Distribution and Texture along the Transects and within the Block at Crane

One of the supplementary soil surveys that was carried out at Crane was aimed at obtaining information along Transects I–V and within the excavated block on two variables: soil texture and soil pore-size distribution. It was not possible economically to monitor these variables at each station in the Transect Survey and Block Survey that was sampled for other soil attributes. Information of soil bulk density, which was monitored but not very accurately in the Transect and Block Surveys, also was collected. Eighteen locations for sampling pore-size distributions, texture, and bulk density were selected, as shown in Figure 49. Their positions along the transects and within the block were not chosen with respect to resistance values at those locations, but rather, were chosen so as to fall with a large number of different kinds of prehistoric use-areas, as evidenced by surface materials.

At each chosen sampling location, a test excavation unit was dug through the archeological deposits and extended into the underlying B and C soil horizons. Excavation units situated along transects were 3×2 m, whereas those within the block were restricted to the 2×2-m size of all excavation units there. Excavation units along transects were dug with both their 3-m-long faces centered over the middle subtransect, crossing the subtransect perpendicularly, and positioned along the transect such that the faces coincided with resistivity survey stations (Figure 50). Excavation units within the block were dug in an analogous manner, but with only 2-m-long faces (Figure 50). They consequently encompassed slightly less soil variability.

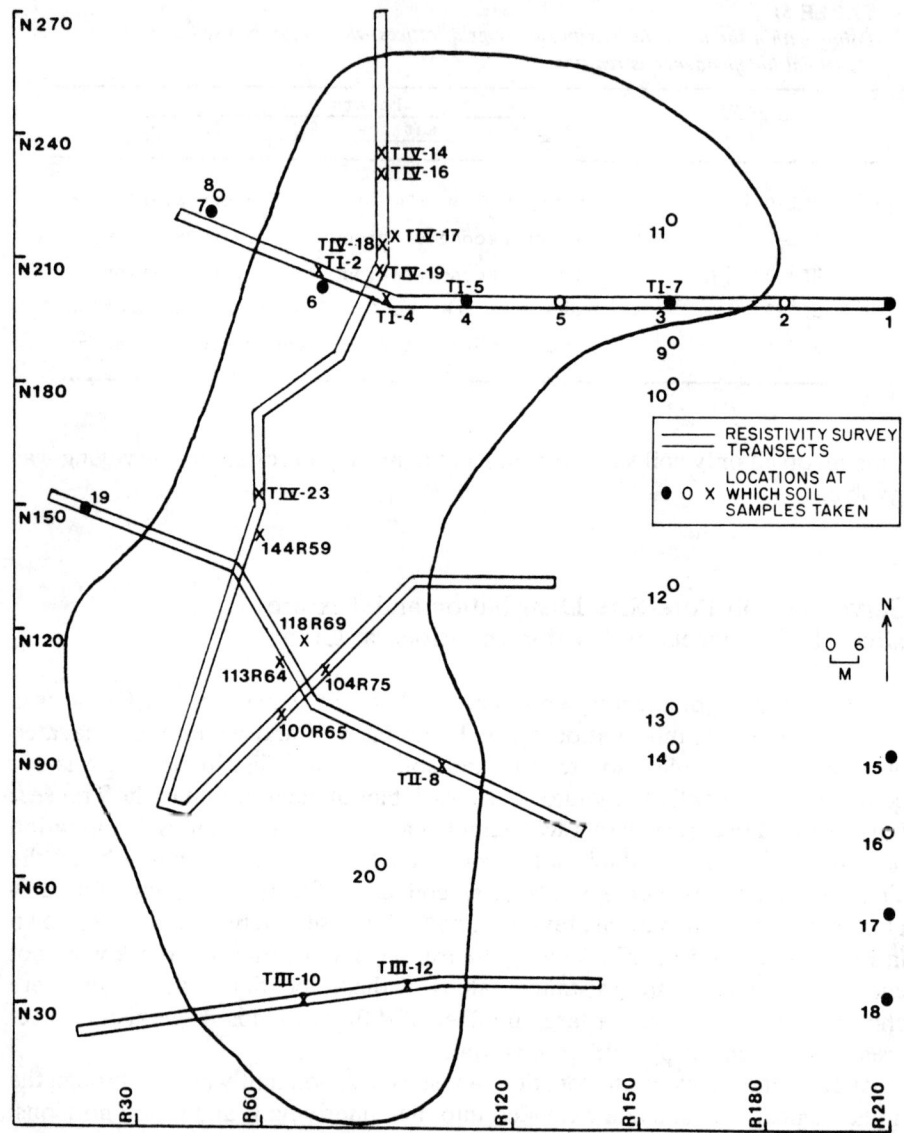

FIGURE 49. *Locations sampled for soil texture, pore size distribution, and natural soil variation. Circles and dots indicate locations at which field observations and soil descriptions were made for each soil horizon, as given in Appendix 7, Table 5. Dots indicate the subset of such locations, where soil samples were removed from each soil horizon for laboratory analysis of their pH, nutrient availabilities, organic matter content, soil texture, pore size distribution, and bulk density. Xs indicate locations where incomplete field observations of soil profiles were made, and where samples for laboratory analysis were removed from only midden deposits and B1 or B2 horizons.*

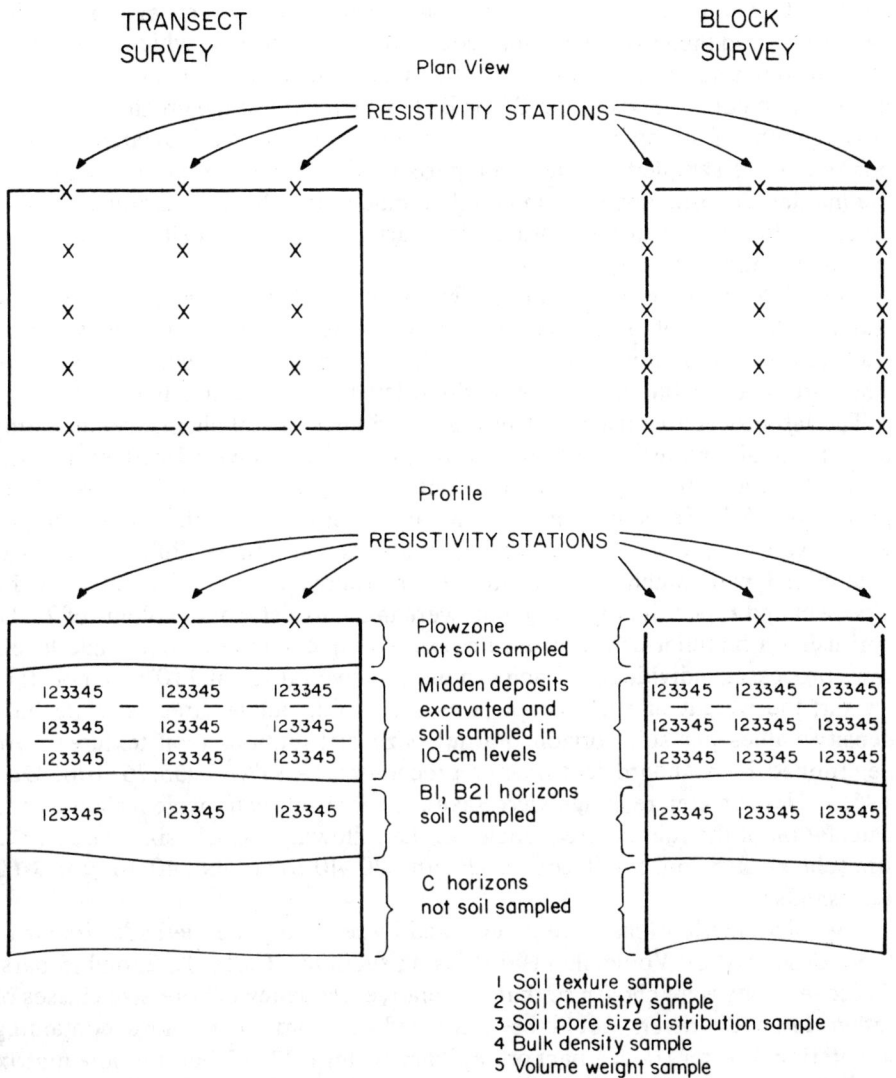

FIGURE 50. *Design for sampling selected soil profiles along Transects I–V and within the block at the Crane site for soil texture, soil pore size distribution, and bulk density, as well as soil chemistry.*

From one wall of each excavation unit, soil samples at each of several different depths were removed, for determining soil texture, soil pore-size distribution, bulk density, and soil chemistry. Samples for determining volume weight and correlations between bulk density and volume weight also were taken. The particular design used for collecting soil for each soil test is shown in

Figure 50. Samples for all tests were taken from each 10-cm level of midden excavated, and then from that submidden natural soil horizon which structurally was presumed to be most similar to the midden deposits. For the Crane soils, this horizon corresponds to the B1 or B21, which exhibit the greatest degree of development of structural joints not cemented together by illuviated clays. B horizons were sampled so that a comparison of their soil structures to those of the midden deposits above them could be made, in order to determine whether buried B horizons can be confused for buried midden deposits by resistivity survey methods (see Chapter 8).

At each level of the soil profile that was sampled, for each test, three separate soil samples were obtained. Triads were dispersed spatially over the sampled wall, in order to obtain a more representative sample of the soil at each level. The particular spatial arrangement within levels is shown in Figure 50.

The lab procedures used to determine attributes of soil chemistry, including pH, the availabilities of the nine ions and radicals investigated during the transect survey, and organic matter content, are the same as those described previously. All three soil samples from each level were combined into a single, well-mixed sample on which chemical analyses were made. Bulk density was determined with much more control over volume than was possible in the Transect and Block Surveys. Samples were taken to a standard volume of 70.18 cm^3 using a hammer-driven ring sampler. Multiple samples from single levels were processes individually. Each sample was oven-dried at 110°C to constant weight; the weight of each sample was used to calculate three separate bulk density values per soil horizon. Particle-size distributions (soil texture) were determined by standard hydrometer procedures (ASTM 1955:76–106; Day 1965). Hydrometer readings were taken at time intervals which allowed the calculation of the relative frequencies of the following particle size classes: <2 μm (clays); 2–5 μm, 5–10 μm, 10–20 μm, 20–40 μm (silts), 40–62 μm, >62 μm (sands).

Pore-size distributions were determined by tension plate methods similar to those described by Vomocile (1965), using suctions of .05, .1, .2, and .5 bars. These tensions were chosen so that the relative frequency of pore-size classes of rather large diameters could be examined—in particular, those containing gravitational water (0–.33 bar) or capillary water (.33–15 bars) at low matrix potentials. The relative frequencies of pores of these larger size classes are of interest because the primary effect of organic enrichment to a soil (by human or natural cause) on its pore size frequency distribution occurs within the large pore-size range (see Chapter 3). Congruently, in a preliminary study of the Crane soils, it was found that midden deposits on site could be distinguished from natural soil horizons at similar depths off site most easily at lower tensions within their moisture characteristic curve, which pertain to pores of larger size. Midden deposits in different areas of Crane site also were distinguished from each other more easily at lower tensions. Thus, it was decided to concentrate the study of the structure of the Crane soils on pores of larger size classes— specifically those holding water at tensions of .5 bar or less.

The sampling design (Figure 50) that was used for collecting soil samples from each unit to be characterized for its pore-size distribution depended on the number of tensions and the number of independent observations per tension that were expected to be required in order to distinguish statistically the structure of different kinds of deposits. Based on a preliminary study of within-deposit and between-deposit variability at Crane, three observations at each of the four tensions just mentioned—or a total of 12 samples per unit being characterized—were determined to be adequate. These samples were obtained with a 1.9-cm-diameter push auger from the profile wall of the excavation pits in which the soil units and midden deposits were exposed (Figure 50). Three pairs of cores, 5 cm long, were removed from each unit at three dispersed sampling locations within it. Cores were halved, providing four samples from each of the three different sampling locations. The four halves, having come from a limited area, were presumed to be homogeneous, were subject to the four different suctions mentioned, and represented one characterization of the structure of the soil unit at four different suctions. The three different sets of four halved cores, having come from separate areas of the soil unit, represented three independent characterizations of the soil's structure.

Soil units within the different locations of the Crane site sampled in this survey are coded in the following manner. Location is designated first, indicating the transect and the particular excavation unit along the transect (e.g., TIV–23), or in the case of those locations within the block, the grid locus of the center of the profile wall (e.g., 100R64). The 10-cm-thick midden units within the excavated pits are numbered sequentially 01, 02, 03, etc., and have been appended to the location indicators (e.g., TIV–2301; 100R64–01). Samples from the B1 or B21 horizons below middens are indicated by the code "sub," which similarly is appended to the location indicators (e.g., TIV–23 sub; 100R64 sub). The physical and chemical soil data collected from each location are given in Appendix 7, Tables 1 and 2.

Survey of the Soil Attributes of Pit Fills and Their Natural Soil Matrices

A second supplementary soil survey that was carried out at Crane was aimed at investigating the physical and chemical differences between the fills of pits and their natural surrounding matrices (A3 soil horizons). Where pits were revealed by excavation within the 18 sampling locations just discussed, soil samples were removed from both their fills and their matrices. Two samples for soil chemical analysis (combined), two for determination of bulk density, two for determination of soil textural distribution, and four for pore-size analysis were removed from the pit fills near the centers of the pits. The same total number of samples were removed from the natural soil matrices, half of each type on either side of the pit and ca. 20 cm away from any visible signs of anthropic soil disturbance. Samples within the pit fills and within their matrices were removed from the same depths below surface, in order to keep constant natural sources of variation such as clay eluviation–illuviation and chemical leaching.

In overall design, the sampling program is of the "randomized block" type, and the data obtained can be analyzed in only certain, restricted ways. To determine the effect of human refuse deposition on the soil fills of pits, comparisons can be made only between fills and their corresponding matrices as individual *paired* units. It is not possible to consider the difference between fill materials as a whole population and matrix materials as a whole population, for neither the fills nor their matrices comprise homogeneous populations with respect to the determinants of their physical and chemical natures. The pairs come from different portions of Crane where: (a) natural soil texture varies, (b) different patterns of historic land-use have altered the chemical attributes of the soils in different ways or to different extents (see Chapter 7), and (c) prehistoric human activity and its effect on the soil varied in kind and extent.

Particle-size distribution, bulk density, pore-size distribution, pH, the availability of the same nine ions and radicals mentioned previously, and organic matter content were determined on the samples by the same procedures already described. The data obtained from the soil tests are given in Appendix 7, Tables 3 and 4. Samples from different pits and matrices are coded according to the location of the pit or matrix, in a manner similar to the way in which midden deposits were coded (e.g., Pit TIV–16F1 is a sample from the first archeological feature found in excavation unit TIV-16).

Survey of Natural Soil Variation within and around the Crane Site

A final soil survey program that was performed to supplement the information collected by the Transect Survey was designed specifically to monitor natural soil variation. In particular, it was aimed at determining if the properties of parent materials or the degree of soil development at Crane varied systematically with distance from Macoupin Creek or with respect to topography (see Chapter 7). To investigate these possibilities, 19 locations (Figure 49) within and around the Crane site were described and/or sampled for the nature of their soil profiles and parent materials. Most of the locations are positioned along two transects: one oriented east–west and with respect to distance from precanalized Macoupin Creek; the second running north–south, and perpendicular to the direction of the two major ridges on which the Crane site is situated (Figure 49). Two sample locations also were situated in the precanalized bed of Macoupin Creek, which was included within the Transect Survey.

At each of the 19 different locations, soil profiles were described for those of their attributes that were observable in the field and that related to soil profile development. These include: textural and structural changes over depth, development of clay and humus coatings on soil peds, and attributes related to drainage conditions such as the occurrence of iron and manganese concretions, salt accumulation, and mottling. These observations are summarized in Appendix 7, Table 5.

Those locations where soil profiles showed the greatest differences in their observable attributes and that might serve as benchmarks for understanding sitewide trends in soil profile development and parent material were sampled for soil for laboratory analyses that would allow detailed characterization. At each such location, samples were removed from each natural soil horizon that could be identified in the field. The laboratory tests run on these samples include: pH, the availability of the same nine ions and radicals described, organic matter content, texture, pore-size distribution, and bulk density. The laboratory procedures used to make these determinations are the same as those described before. The resultant data are summarized in Appendix 7, Tables 6 and 7.

Summary

Resistivity measurements and soil samples were taken at the Crane site as a part of several different surveys having different designs and purposes. Resistivity and soil determinations were made along transects in order to sample a wide range of use-areas and naturally varying zones within the site for the physical, chemical, and electrical attributes of their soils. The data from this survey will be used to determine general correlations between these attributes, and their degree of association with use-areas of particular kinds. Resistivity and soil surveys within a later excavated block were carried out in order to gather data for comparing the variability of resistivity values within use-areas to the spatial density of features within them. Several supplemental soil surveys were made in order to (a) gather information from a few, select locations on soil attributes that were expensive to measure and that could not be determined routinely as were those attributes measured during the Transect and Block Surveys; (b) collect soil from pits and their surrounding matrices for comparison; (c) determine how the properties of parent materials and the degree of soil development varied naturally across Crane.

The physical and chemical determinations made on the Crane soils include: pH, the availability of nine ions and radicals, organic matter content, texture, pore-size distribution, and bulk density. The resistivity and soil data collected by the several surveys are summarized in Appendices 5, 6, and 7.

References

Methods of Soil Analysis

American Society for Testing Materials
    1955   *ASTM Standards,* Part 3, ASTM, Philadelphia.
Black, C. A.
    1965   *Methods of soil analysis.* American Society of Agronomy, Madison, Wisconsin.
Black, G. R.
    1965   Bulk density. In *Methods of soil analysis,* C. A. Black (ed.), pp. 374–399. American Society of Agronomy, Madison, Wisconsin.
Brady, Nyle C.
    1974   *The nature and property of soils.* Macmillan, New York.
Bremner, J. M.
    1965   Inorganic forms of nitrogen. In *Methods of soil analysis,* C. A. Black (ed.), pp. 1179–1237. American Society of Agronomy, Madison, Wisconsin.
Day, Paul R.
    1965   Particle fractionation and particle size analysis. In *Methods of soil analysis,* C. A. Black (ed.), pp. 545–567. American Society of Agronomy, Madison, Wisconsin.
Mehlich, A.
    1953   Determination of P, Ca, Mg, K, Na, and NH_4 by the North Carolina Soil Testing Laboratory. Mimeograph on file at the North Carolina Department of Agriculture, Agronomic Division Soils Lab, Raleigh.
    1973   Uniformity of soil test results as influenced by volume weight. *Communications in Soil Science and Plant Analysis* 4(6):475–486.
Nelson, W. L., A. Mehlich, and E. Winters
    1953   The development, evaluation, and use of soil tests for phosphorus availability. In *Soil and fertilizer phosphorus, Vol. 4, Agronomy,* W. H. Pierre and A. G. Norman (eds.), pp. 153–188. Academic Press, New York.
North Carolina Department of Agriculture
    1972   Analytical methods used by the Soils Testing Division, North Carolina Department of Agriculture, Raleigh, North Carolina. Unpublished manuscript on file, N.C.D.A. Soil Testing Division Laboratory, Raleigh.
    1976   Determination of phosphorus, potassium, calcium, magnesium, and sodium in soils by the double acid method. Unpublished manuscript on file, N.C.D.A. Soils Testing Division Laboratory, Raleigh.
Olsen, S. R., and L. A. Dean
    1965   Phosphorus, In *Methods of soil analysis,* C. A. Black (ed.), pp. 1035–1049. American Society of Agronomy, Madison, Wisconsin.
Peech, M., R. A. Olsen, and G. H. Bolt
    1953   The significance of potentiometric measurements involving liquid junction in clay and soil suspensions. *Soil Science Society of America, Proceedings* 17:214–218.
Prince, G. B.
    1965   Absorption spectrophotometry. In *Methods of soil analysis,* C. A. Black (ed.), pp. 866–878. American Society of Agronomy, Madison, Wisconsin.
Raupach, M.
    1954   The errors involved in pH determination in soils. *Journal of Agronomic Research* 5:716–729.
Rich, C. I.
    1965   Elemental analysis by flame photometry. In *Methods of soil analysis,* C. A. Black (ed.), pp. 849–865. American Society of Agronomy, Madison, Wisconsin.

Schofield, R. K., and A. W. Taylor
   1955   The measurement of soil pH. *Soil Science Society of America, Proceedings* 19:164–167.

Small, James
   1954   *Modern aspects of pH*. Ballière, London.

Vomocil, James A.
   1965   Porosity. In *Methods of soil analysis,* C. A. Black (ed.), pp. 299–314. American Society of Agronomy, Madison, Wisconsin.

7
Natural Soil Variations and Soil Alterations Produced by Historic Human Activity at the Crane Site

In analyzing the nature of the relationships holding between particular prehistoric human activities on archeological sites and their effects on soils, it is the alterations, or *deviations* in soil properties from *local, natural norms,* rather than the absolute soil properties, themselves, which are of interest. Consequently, this study of prehistoric human effects on the soils of the Crane site must be made with reference to the natural soil variations that occur there. Also, those deviations in soil properties from local natural norms that are the result of historic or contemporary land-use patterns must be sorted out from those that are more likely a result of prehistoric activity. This chapter is concerned with providing the understanding of natural, historic, and contemporary soil variations on Crane, which later will be necessary for interpreting the effects of prehistoric human activity on the Crane soils (Chapters 8, 10).

The information presented in this chapter is as detailed a record of the natural soil variation and historic and contemporary land-use patterns at Crane as was possible to obtain and reconstruct from available soil data, historic records, and personal interviews with Donald Crane. Although not all of the information is used directly in the analyses to follow in later chapters, it is presented fully for those researchers who might wish to rework the analyses or take them further.

Overview of the Parent Materials and Soils at Crane Site

Geomorphologically, the Crane site is located on an alluvially redeposited, stratified, loessic parent material. Two different strata are identifiable within the deposits. The upper member of the profile, extending from the surface to ca. 86–94 cm below surface, ranges from a silt loam to a clayey silt loam in texture and is unstratified. The lower member, extending from 86–94 cm below surface to a depth of at least 203 cm below surface, ranges in texture from a silty clay loam to a sandy loam, and is stratified (U.S.D.A. 1974:32, 35). Both members of the parent material are composed largely of Wisconsinan-age Peorian loess (Wascher et al. 1971), which has been transported by Macoupin Creek from deposits further upstream. They also may be composed, to a limited extent, of eroded and reworked Illinoisan-age tills, which underlay the Peorian loess. These materials were deposited at the Crane site during the Wisconsinan or Holocene.

This interpretion of the age of the loess from which the Crane deposits were derived and the time at which they were eroded, transported, and redeposited is based on the elevation of the site and the elevation at which Wisconsinan and Illinoisan alluvial deposits in Macoupin Valley are know to occur. Illinoisan-age alluvial features within secondary river valleys tributary to the lower Illinois Valley belong to the Deerplain slackwater terrace system, and are found at elevations between 500 and 540 ft. (152 and 165 m) (Butzer 1977, Ruby 1952). This terrace system either did not form, or has not been eroded away, at lower elevations, where Wisconsinan age or Recent alluvial features predominate. Crane site occurs at an elevation of 400–420 ft. (122–127 m), well below those elevations at which remnants of the Deerplain terrace are found. Consequently, the age of alluvial deposition of the parent materials upon which the Crane soils developed probably is Wisconsinan or Recent.

It also is most likely that the *loessic materials*, themselves, which form the Crane deposits and which were transported from upstream are of Wisconsinan-age rather than Illinoisan-age. Most of the Wisconsinan and Recent landscapes beyond the Illinois Valley trench are dominated by a thick mantle of Peorian loess, from 1.9–3.8 m in depth (Frye and Willman 1970). Exposed Illinoisan-age features—till plain or remnants of the Deerplain terrace system—formed only a small percentage of the drainage basin of Macoupin Creek at these times. If the Crane site parent materials were deposited during the Wisconsin or Recent periods, as they seem to have been, after much of the Illinoisan landscape (Deerplain terrace) had been buried by Peorian loess or eroded away, then it is most likely that the alluvial deposits at the Crane site were derived from Wisconsinan sources.

The general nature of the soils that developed upon the alluvial parent materials at Crane and the specific processes by which they were formed have been described by the United States Soil Conservation Service (U.S.D.A. 1974:32). According to the SCS, the soils at Crane site belong to a class of silt loams—the Proctor series—that developed under prairie vegetation, in conditions of moderate to good drainage. This classification, and the soil formation processes and local vegetational communities implied by it, however, are not in accord with present knowledge of and interpretation of the Crane soils.

Since soil formation processes are critical to this study, it is necessary to consider the differences between the view taken here and that of the SCS.

In particular, it does not appear that the Crane soils are a product of either prairie vegetation or moderate to good drainage conditions. A number of independent data suggest that the soils around Crane developed under a forest community rather than prairie vegetation. First, historical documents specify that when the area around Crane first was settled in 1819 by Christopher J. Gardiner and his son, James B. Gardiner, it was in virgin timber and had to be cleared in order to cultivate the soil (Andreas et al. 1873). This historical account can be verified by independent data provided by the 1820 Land Survey Records of the area around Crane site (Illinois State Museum 1820) that make some references to local vegetation along survey transects. A map of the distribution of forest and prairie communities within the vicinity of the Crane site, as reconstructed from the 1820 Land Records by the Botany Laboratory of the Center for American Archeology, is shown in Figure 51.

A second line of evidence suggesting that the soils of Crane site formed under forest vegetation is the documented patterns of land use and natural succession involving clear cutting and reforestation on Crane. Aerial photographs taken by the United States Department of Agriculture for the region around the site (U.S.D.A. 1937) show that in 1937, the northern portion of Crane was covered with dense stands of forest, as well as some uncultivated ground apparently used for a bluegrass horse pasture (Donald Crane, personal communication). Aerial photos taken in 1950 (U.S.D.A. 1950) show the uncultivated ground in secondary forest and shrubbery. This process of reforestation was not the product of conscious efforts by the owners of the Crane site to convert the pasture into forest land, but a natural successional sequence (Crane, personal communication). That a forest community rather than a prairie community began to develop on the untended property suggests that local environmental conditions were suited more to the requirements of forest than prairie communities. If these conditions can be extended back in time, they would imply that the prehistoric vegetation within the vicinity of Crane more probably was forest than prairie.

Finally, several properties of the Crane soils, themselves, suggest that they developed under forest cover. Most importantly, over much of the Crane site and its immediate vicinity, soil profiles indicate eluviation and illuviation processes and B-horizon development that typify the braunerde (Kubiena 1953) and gray–brown podsolic soils (Afisols) of temperate deciduous forests rather than the chernozems (Mollisols) of midlatitude prairies. In particular, the Crane soils exhibit B horizons that have illuviated clay and humus coatings on ped surfaces and that contain small (1–2 mm) iron concretions. In a few areas of the site, podsolization has proceeded to the point where fine silt particles as well as iron sesquioxides, clays, and humus have been eluviated into the B horizon and coat peds, there. The prismatic structure (weak) of the B horizons of some soils on the Crane site also evidences the vertical movement of organics and

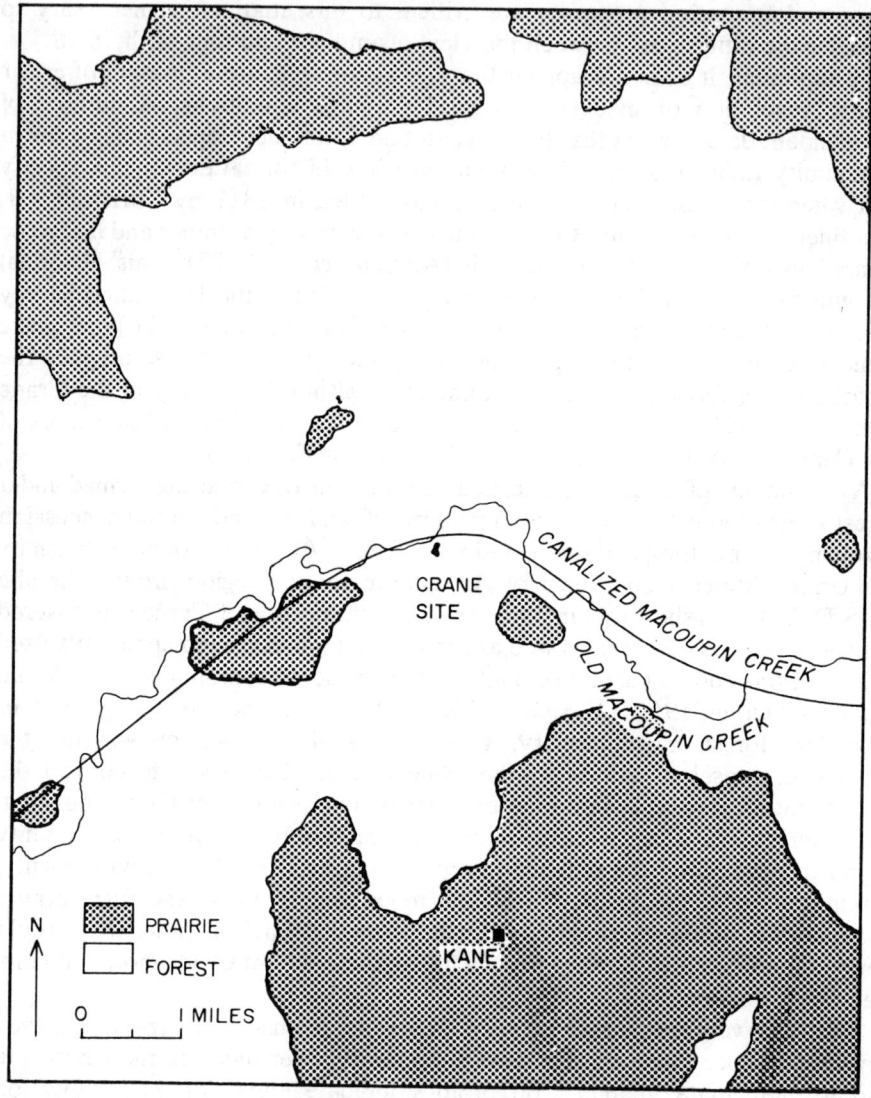

FIGURE 51. *Distribution of forest and prairie communities within the vicinity of the Crane site, as reconstructed from 1820 Land Survey records (Illnois State Museum 1820) for the period prior to white settlement (1819). The reconstruction was made by the Botany Laboratory of the Center for American Archeology.*

inorganics. These soil characteristics are not found in prairie soils, which are typified by AC profiles (Brady 1974:326–330; Butzer 1971).

At the same time, however, profiles of the soils on Crane do not have an observable A2 horizon from which the iron, clays, and humus in their B horizons presumably were derived. This is true even in locations where B horizon development is strongest, where peds in the B horizon have silty coatings and where prismatic structure is moderately developed. This situation does not contradict the interpretation that the Crane soils developed under forest vegetation (the B horizons still exist), but does suggest that they have been degraded recently. It is possible that the historic use of some portions of Crane site and its surroundings for pasture (Crane, personal communication) has contributed to the loss of the A2 horizon, there. The aboriginal occupation of Crane and the enrichment of the upper horizons of the Crane soils with organic matter and nutrients, however, is a circumstance that more likely explains the loss of the A2 horizon from the Crane soil profiles. Thus, the soils in the vicinity of Crane may be interpreted as degraded forest soils. In the framework of the soil classification system of the SCS (U.S.D.A. 1974), they belong to the same catena as Cambden, Starks, and Sexton forest soils but degradation of their A2 horizons has given them some characteristics transitional to Proctor and Worthen soils, which developed under prairie vegetation.

The second genetic feature of the Crane soils that is not in accord with their classfication as a Proctor silt loam is their drainage. Soils of the Proctor series are found in moderately well- to well-drained regions, whereas those in the immediate vicinity of Crane site are poorly drained. Several data suggest the poor drainage of the Crane soils. First, while collecting soil samples from the east end of Transect I (Figure 44), a perched water table was found at a depth of 64 cm below surface (BL4–3). Second, from discussions with Donald Crane, who farms the field in which Crane site occurs, it is known that other areas immediately east of Crane site also do not drain easily. Mr. Crane had to lay drainage tiles in these areas to farm the land. Finally, several attributes of the soils, themselves, in some areas of the site and its immediate vicinity suggest imperfect drainage conditions. These include: (a) the accumulation of salts (Ca or Na) and Mn (in the form of concretions 1–2 mm in diameter) within B horizons, rather than a flushing of these elements from the soil; and (b) the occurrence of gray mottles within generally orange–brown B horizons. Gray mottles indicate zones of poor aeration and reduction of iron (Brady 1974:265; U.S.D.A. 1960:249). Thus, the soils of Crane site are best characterized as somewhat poorly drained. They fall within the range of drainage conditions under which Starks soils, of the Cambden–Starks–Sexton catena of forest soils, develop (U.S.D.A. 1974:35).

In summary, within the framework of the soil classification system offered by the SCS (U.S.D.A. 1974), the soils in the area of Crane site may be characterized as a member of the Starks series, but transitional to a Proctor silt loam with respect to their degraded eluvial A2 horizon. This classification properly reflects the development of the Crane soils under forested vegetation and within poorly drained parent materials, and the later disturbance of soil development by the prehistoric occupation at Crane and/or the use of the site for agricultural purposes.

Natural Spatial Variation of Parent Materials at the Crane Site

Given the alluvial nature of the parent materials and the variation of topograghy at the Crane site, it would be naive to assume that parent materials and degree of soil development are uniform over the whole site. In this section, an attempt will be made to define subareas of Crane that are more uniform with respect to the physical and chemical properties of their parent materials, either by identifying truly localized bounded areas of similarity, or by defining gradients of change that can be artifically subdivided into reasonably homogeneous strata. Within such subareas, more meaningful associations between different kinds of use-areas and their anomalous soil properties may be made.

Spatial variation in parent materials could imply variation in a number of different physical and chemical properties, including the availability of different ions and radicals, texture, fabric, and total porosity. Of these variables, only soil texture cannot be significantly altered by human activity; consequently, this variable can be used in an unambiguous way to investigate natural variation in the parent materials of Crane. Definition of subareas of Crane that are supposed to be uniform in general in the physical and chemical properties of their parent materials, thus, will be based only on a study of soil texture and its spatial variation. Spatial strata homogeneous in texture will be assumed to be homogeneous in regard to other physical and chemical properties of their parent materials.

In examining spatial variation in the parent materials at Crane, it is necessary to consider the geomorphological processes by which they formed and the manners of their patterning that are most likely. The parent materials at Crane are alluvial in origin, and could reflect several different processes of floodplain development and spatial patterns of variation. On the one hand, they could exhibit a variegated, mosaic pattern reflecting a complex alluvial history related to the meandering of Macoupin Creek and the infilling of cutoff meanders. Macoupin is a fully mature stream, and prior to canalization, exhibited an intricate system of meanders. Within the vicinity of Crane, infilling cutoff meanders evidence the changes that Macoupin Creek has made in its course over time. Crane site, itself, is surrounded by a meander of precanalized Macoupin Creek and at least one infilling cutoff meander (U.S.D.A. 1937).

On the other hand, it might be assumed that the parent materials within depths of interest to us were all deposited during the period when Macoupin Creek followed its most modern precanalized course. In this case, they might exhibit a less complex pattern of variation related to distance from a single source location. In particular, the percentage of clay-sized particles in the parent material might increase with distance from precanalized Macoupin Creek, whereas the percentage of sand and coarser silt-sized particles might decrease. Such a pattern is common in floodplains, and arises from the different rates at which sand, silt, and clay-sized particles settle from floodwaters as they cover a landscape and loose their velocity and kinetic energy for suspending particles within them. Finer particles can be suspended by slower moving flood waters more distant from a river channel whereas coarser particles cannot, and settle out closer to the river channel (Strahler 1970:310).

The soil texture data that will be used to investigate the Crane parent materials come from two of the supplementary soil surveys described in the latter portion of Chapter 6. The locations at which soil samples have been analyzed for texture are shown in Figure 49. The relevant textural data are given in Appendix 7, Tables 1 and 6.

As a first step in the study of spatial variations in texture, an attempt was made to determine whether any simple, unidirectional spatial trends occur in the relative frequencies of different particle-size classes—either with respect to distance from Macoupin Creek or with respect to topography. The ratios of percentage clay/percentage sand (>40 μm)[1] and percentage clay/percentage coarse sand (>62 μm) were plotted over depth at each of several sampling stations along a west-to-east (W–E) transect (Sites 6, 4, 3, 1; Figure 49) and a north-to-south (N–S) transect (Sites 1, 15, 17, 18; Figure 49) in the greater Crane area. These plots are shown in Figures 52 through 55. The W–E transect runs more perpendicular to Macoupin Creek, whereas the N–S transect runs less perpendicular to it and crosscuts the two ridges at Crane that form its major topographic relief. In plotting texture over depth at any single location, textures are not presented as point determinations at unique depths, but rather as bands, each encompassing a range of depths. The range is that over which soil was removed for making a single determination of the texture of a single soil horizon as a whole.

To interpret Figures 52 through 55, spatial variation in the degree of soil profile development and eluviation/illuviation of clays must be taken into consideration as well as the possibility of spatial variation in the texture of the parent materials prior to soil profile development. Variation within locations due to *profile development* is manifested as an increase over depth in the clay/sand ratios until maximal B-horizon development is reached, at which point the ratio begins to decrease. Spatial variation in *parent material* is manifested by the shifting of such profiles to the left or right, either as wholes, or possibly in segments if the parent material is stratified. The expected pattern, if clay and sand percentages are related to variation in parent material composition and are a function of distance from Macoupin Creek, is that profiles further from the Creek will be shifted further to the right (i.e., have greater amounts of clay and less sand).

Examining the plots for the W–E transect first (Figures 52, 53), no

---

1. Sand particles have diameters of 50–1000 μm, as defined by the United States Department of Agriculture (1960). Particles classified as sand here include some that standardly would be classified as silt.

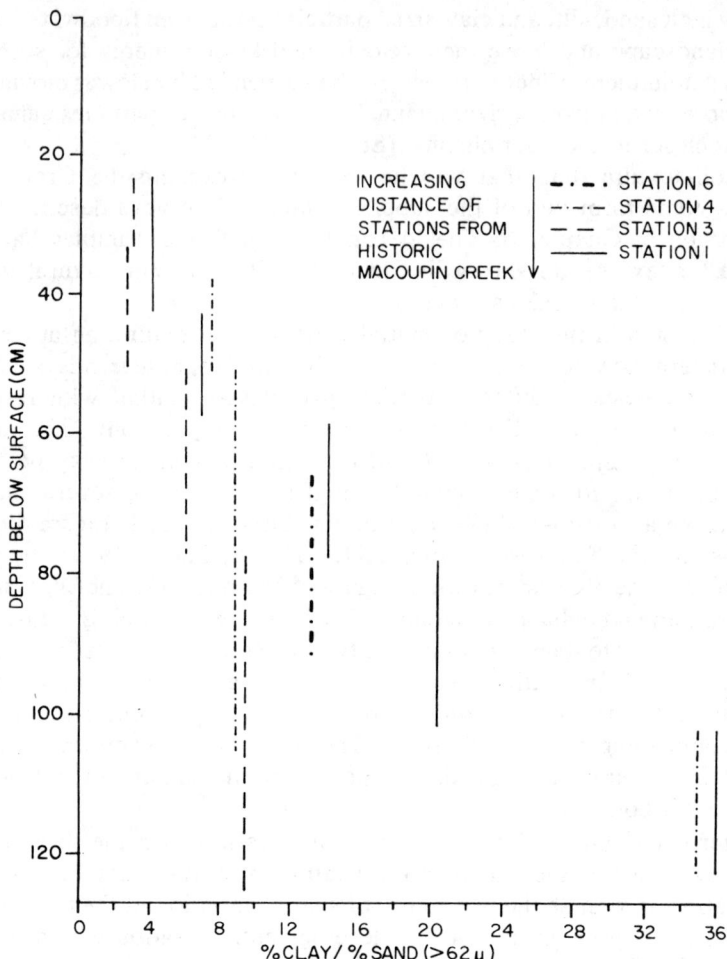

FIGURE 52. Spatial variation in % clay/% sand (>62 μm) ratios in soil profiles along a west-to-east transect across the Crane site, including the sampling stations 6, 4, 3, and 1 shown in Figure 49.

monotonically increasing trend in the clay/sand ratios can be found with respect to distance from precanalized Macoupin Creek. The order of station numbers from closest for furthest from Macoupin Creek is 6, 4, 3, 1. The order of station numbers for increasing clay/coarse-sand ratios (Figure 52) is 3, 4, 6, 1 below ca. 54 cm and is 3, 1, 4 above that. The order of station numbers for increasing clay/sand ratios (Figure 53) is 3, 6/4, 1 between ca. 54 and 107 cm below surface, and 3, 1, 4 above and below that. At no single depth do the observed ordering of stations by their relative percentages of clay and/or coarse sand approach their spatial ordering with respect to distance from precanalized Macoupin Creek.

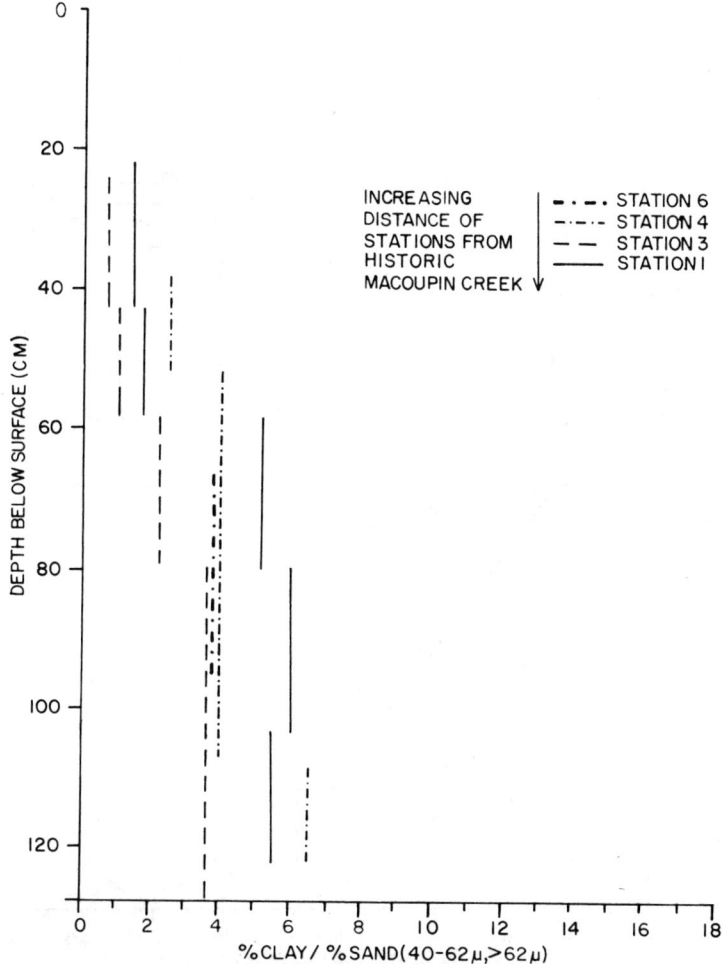

FIGURE 53. *Spatial variation in % clay/% sand (>40 μm) ratios in soil profiles along a west-to-east transect across the Crane site, including the sampling stations 6, 4, 3, and 1 shown in Figure 49.*

Along the N–S transect, clay/sand ratios at some depths do increase in a monotonic fashion over space, but *opposite* the expected direction. The order of station numbers from closest to furthest from precanalized Macoupin Creek is 1, 15, 17, 18. Below ca 73 cm, the order of station numbers for increasing clay/coarse-sand ratios (Figure 54) is 18, 17, 15, 1. The same ordering of station numbers is found at similar depths (below 79 cm) for increasing clay/sand ratios (Figure 55). Above 73 cm and 79 cm in Figures 54 and 55 there is no single ordering of stations by particle-size and class ratios.

It is possible to integrate these patterns and lack of patterns found in particle-

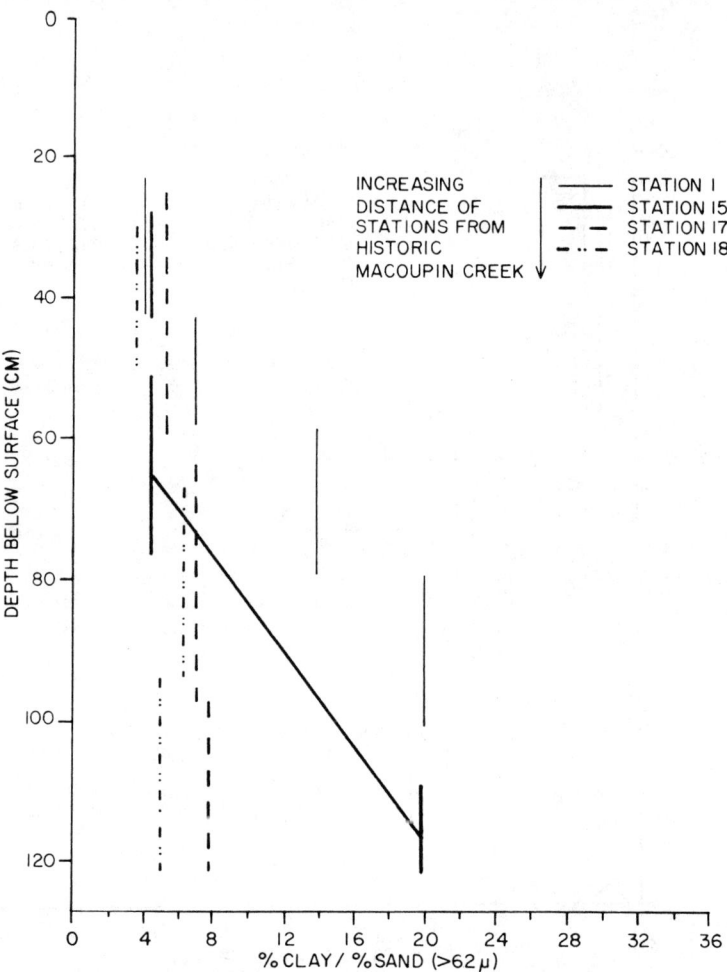

FIGURE 54. *Spatial variation in % clay/% sand (>62 μm) ratios in soil profiles along a north-to-south transect east of the Crane site, including the sampling stations 1, 15, 17, 18 shown in Figure 49.*

size–class ratios along both transects and over depth within a single model of the floodplain development in the vicinity of Crane site. The parent material at Crane site is composed of at least two members: an unstratified upper stratum from the surface to ca 86–94 cm below surface; and a stratified, lower stratum, from ca. 86–94 cm below surface to at least 203 cm below surface. Spatial trends in soil texture have been documented for only the lower member, and in a N–S direction. It is possible that the sediments composing the lower member were deposited by Macoupin Creek when it flowed south rather than east and north of Crane. This would explain the increase in clay/sand ratios from south

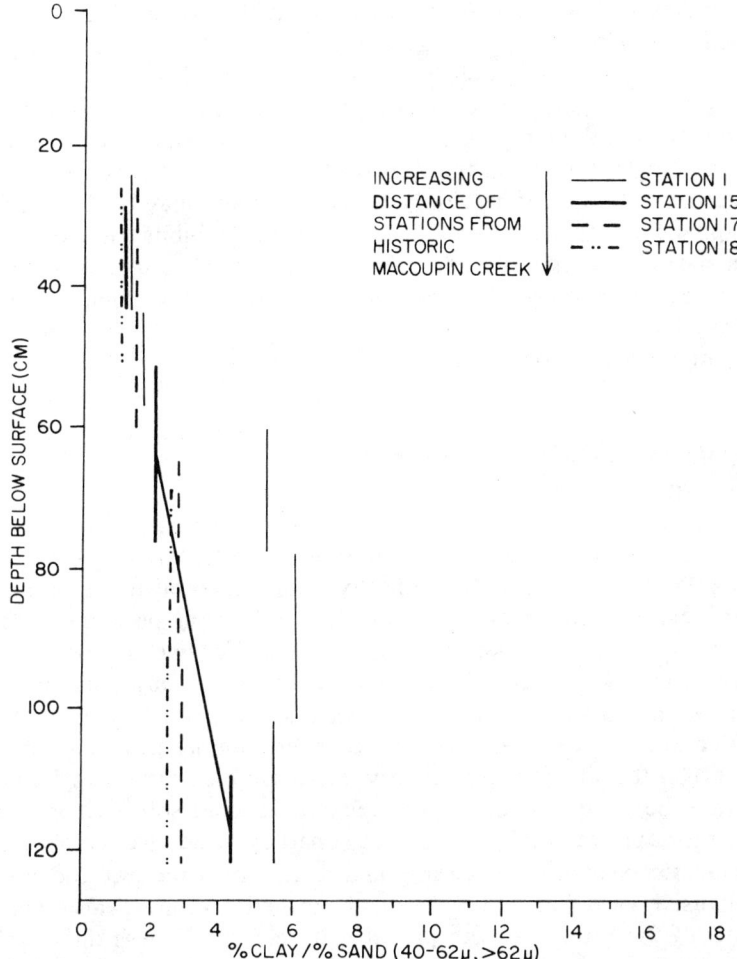

FIGURE 55. *Spatial variation in % clay/% sand (>40 μm) ratios in soil profiles along a north-to-south transect east of the Crane site, including the sampling stations 1, 15, 17, 18 shown in Figure 49.*

to north rather than north to south in the vicinity of Crane. It also would explain the lack at these depths of any spatial trends in clay/sand ratios along the W–E transect, which parallels the old course of Macoupin. The sediments in the lower member of the parent materials probably were deposited over a long period of relatively stable stream volume and flow, without violent floods. This is indicated by the stratifed nature of the materials, which is observable.

The upper member of the Crane parent material exhibits neither stratification nor monotonic spatial trends in clay/sand ratios. It is hypothesized that the sediments within this member were deposited in a relatively short period of active flooding and adjustment of stream parameters (including stream course and length) to increasing discharge. Greater amounts of deposition per flood over a shorter time span would explain the "unstratified" (or less obviously stratified) nature of the sediments within the upper member. Adjustments in stream course and the location of the source of flood waters would explain the lack of any single directional trend in clay/sand ratios over space.

In summary, the parent materials of the Crane site may be divided into two stratigraphic units, only the lower member of which exhibits a monotonic spatial trend in particle-size distribution. This circumstance is understandable with respect to geomorphological processes by which the floodplain of Macoupin Creek may have developed. The subsuming interpretive framework, however, remains an untested hypothesis.

Spatial Variations in the Upper Member Parent Material

The pattern found in the spatial variation of the texture of parent materials below ca. 90 cm at Crane allow one to predict textural distributions at that depth in locations other than those sampled. No simple pattern was found for the upper-member parent material, however, which allow such preditions to be made for it, too. A next step in the analysis of spatial variations in the Crane parent materials, therefore, must be to document the more complex pattern of textural variations over space within the upper member using data from more locations over the site. It is particularly important that natural variation within this stratum be documented, for it encompasses those soil horizons that were examined for their resistivity and other properties, which were most altered by the prehistoric occupants of Crane, and from which natural and prehistoric human-caused variations must be distinguished. Natural variations in the sediments of the lower-member parent materials are of less direct relevance to this study, but are critical with respect to drainage.

When surveying the Crane site for its natural soil variation, it was not thought that spatial variability was as complex as it presently is known to be in the upper horizons. The number of locations sampled for their soil texture is inadequate for constructing a satisfactorily detailed model of textural variation within the upper-stratum parent material at Crane. Some very general trends in soil texture, however, can be documented.

The textural data to be analyzed come from soil samples that were removed at the ca. 60-cm depth from those locations shown in Figure 49. These data are compiled in Table 52. In some cases, it has been necessary to interpolate clay, silt, and sand percentages from samples taken above and below the 60-cm depth, when no samples were removed at or close to that depth.

Variation in the percentages of clay, silt, and sand among the different sampling locations and over space may be a result of either spatial variation in

TABLE 52
Particle distributions of soils sampled from locations shown in Figure 49, 60-cm depth

Location	Particle Size Distribution		
	% Clay <2μ	% Coarse Silts 10–40μ	% Sands >40μ
TIV-14	.259	.453	.101
TIV-16	.205*	.470*	.105*
TIV-19	.306	.433	.130
TI-2 and 6, averaged	.308*	.444*	.104*
TI-5 and 4, averaged	.260*	.439*	.117*
TI-7 and 3, averaged	.242*	.452*	.126*
1	.251*	.470*	.082*
TIV-23	.244	.417	.116
144R59	.288	.413	.125
118R69	.349	.367	.075
104R75	.353	.305	.073
100R65	.389	.348	.072
TII-8	.250*	.530*	.070*
TIII-10	.292	.423	.085
TIII-12	.334	.357	.110
15	.255*	.450*	.123*
17	.241*	.467*	.115*
18	.340*	.375*	.140*

*Value interpolated from soil samples taken above and below 60cm depth

the upper-member parent material, or spatial variation in soil profile development and clay eluviation/illuviation, or both. For the moment, for analytical purposes, it will be assumed that all spatial variation in texture is contributed by differences in parent material, with no variation attributable to soil profile development.

Plotting the percentages of clay, coarse silts (10–40 μm) and sands (>40 μm) by location revealed some patterning in the spatial arrangement of anomalies for each particle-size class. This patterning is summarized in Figure 56. Areas having anomalously high percentages of coarse silts with respect to the sitewide mean for this particle-size class occur along the entire length of the two ridge crests that crosscut Crane in an east–west direction. Areas with anomalously high percentages of sands also occur along these ridge crests but in isolated pockets. In at least one instance (Station 18, Figure 49), however, anomalously high percentages of sands occur on lower, sloping ground. Locations having anomalously high percentages of clays are restricted to south of the gulley that crosscuts Crane, but are not associated with any particular topographic feature.

These results are very general and must be taken in light of the paucity of locations for which textural data are available. The boundaries shown in Figure 56 have been sketched in, not plotted, and should not be considered accurate with respect to the precise locations where resistivity survey transects I–V or particular artifact collection units pass over different parent materials. Moreover, the lines drawn in Figure 56 probably should be interpreted as contours for gentle gradients in textural change rather than sharp boundaries of spatially well-defined parent material units. Nonetheless, the patterns shown in the figure do give some indication of areas of the Crane site that are or are not of similar parent material, and where soil differences may or may not be related to prehistoric human activity, alone.

The model summarized in Figure 56 pertains to spatial variations in the texture of parent materials at the ca. 60-cm depth. Can it be used, however, to characterize the upper-member parent materials in general?

To answer this question, consider a plot of the variation of the ratio, percentage coarse silt (10–40 μm)/percentage sand (>40 μm), over the depth of the upper-member parent material at each of several different locations that are exterior to the Crane site and that have completely undisturbed, natural profiles from the base of plowzone to the base of the upper member. Changes in this ratio over depth largely will reflect variations in parent material rather than degree of profile development, for coarse silts and sands are not as likely to be mobile within a soil profile as are clays.[2] If the model of spatial variation in the texture of parent materials at the 60-cm depth holds for shallower levels, a plot of the ratio percentage silt/percentage sand for each sampled location and over depth should have the following characteristics:

1. Changes in the ratio over depth within stations should be minimal from the surface to the base of the upper member parent material.
2. The ordering of stations should be the same at each depth with respect to the ratio.

Ideally, such a graph with texture on the horizontal axis and depth on the vertical axis should look like parallel, vertical lines.

---

2. A ratio of silt and sand percentages must be used instead of the individual percentages of sand and silt because the latter two are not statistically independent from the variable percentage clay and actually do reflect information on clay movement. The ratio percentage silt/percentage sand is statistically independent from the variable percentage clay, and given a lack of silt and sand eluviation, should reflect information on only the nature of the parent material.

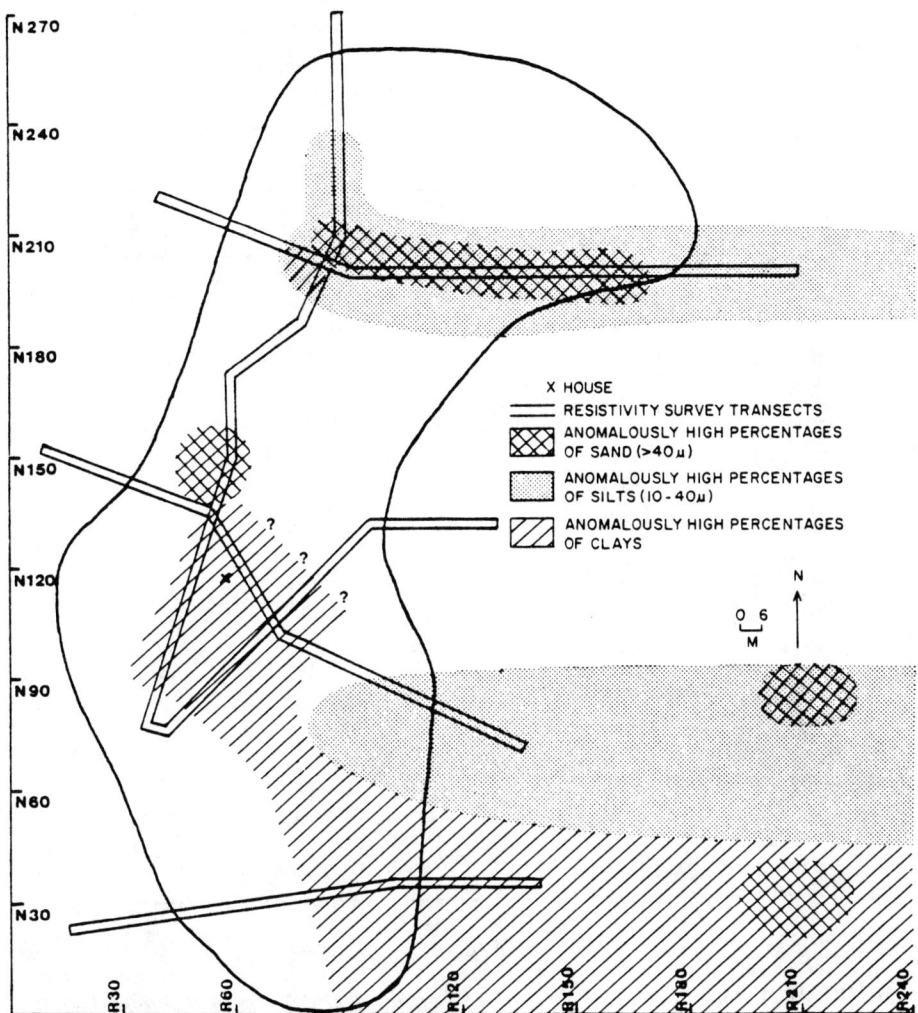

FIGURE 56. *General zones at the Crane site having anomalously high percentages of clays, coarse silts (10–40 μm) and sands (>40 μm) relative to sitewide mean percentages of these same particle size classes. The depth represented is ca. 60 cm below surface.*

The plot of the relevent data from Crane site is shown in Figure 57. In general, it suggests that the model of spatial variation in the texture of parent materials at the 60-cm depth holds for shallower levels, as well. Each of the four sampled stations, 3, 15, 17, and 4, show a nearly constant silt/sand ratio from the base of the plowzone to the base of the upper-member parent material, suggesting the homogeneity of the parent materials within each of these stations. The stations also maintain the same ordering, with respect to the value of this

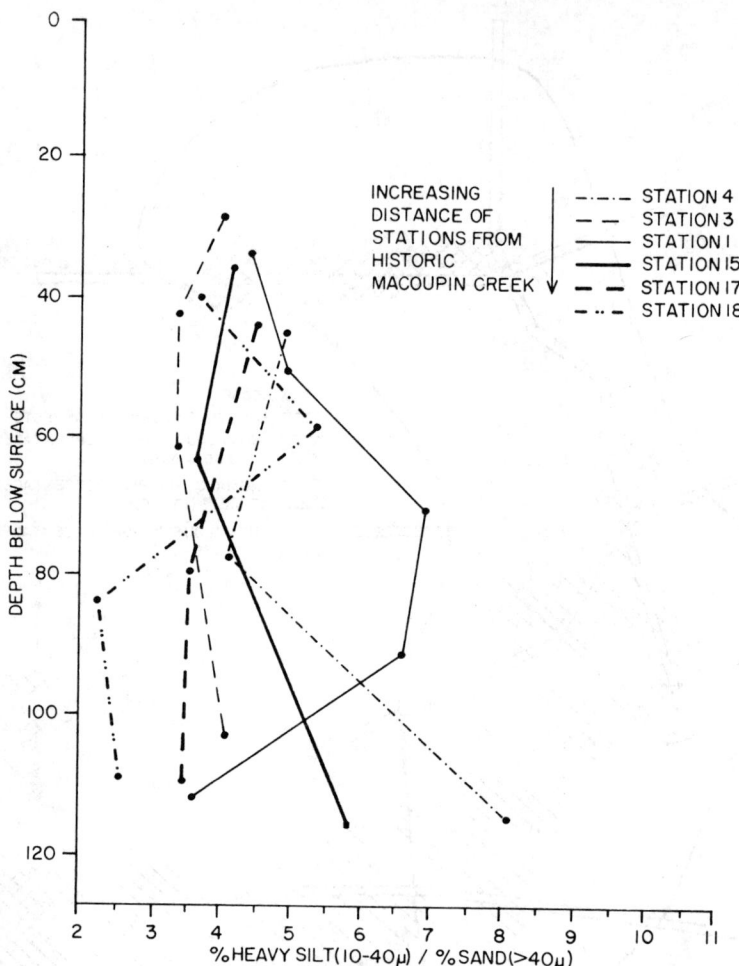

FIGURE 57. *Variation in percentage heavy silt (10–40 μm)/percentage sand (>40μm) ratios over space and depth within undisturbed natural soils east of the Crane site.*

ratio, at almost every depth above the base of the upper member, suggesting the similar spatial patterning of textural variation at all depths within the upper-member parent material. Locations exceptional to these spatial and stratigraphic patterns include Sites 1 and 18, which have convex profiles over depth and which change order with other sampling stations over depth in relation to the value of the silt/sand ratio. The patterns found at these stations apparently reflect more intensive profile development and the downward movement of silts, rather than variation in the nature of the parent material. Silt coatings were observed on the peds within the B horizon at Station 1, indicating the mobility of silt there. Although silt coatings were not observed within the soil profile at Station 18, it is probable that silt is mobile there, too. The degree of profile development at Station 18 is similar to that at Sation 1 (see following section and Table 53). Thus, it would appear that, excluding the effects of soil profile development, the model of spatial variation in the texture of parent materials at the 60-cm depth holds for the upper-member parent material as a whole.

Spatial Variations in the Degree of Soil Development at the Crane Site

The spatial variations in texture just described for the upper-member parent material was assumed to be primarily a result of differences in parent material composition rather than the intensity of soil profile development. In this and the next section, spatial variation in soil profile development at Crane will be examined and compared to the preceding model of textural variation in order to qualify this assumption and to understand variation within the upper-member parent material as a *product* of both differences in parent material composition and intensity of soil profile development.

The data that can be used to examine profile development are limited to those collected from locations *beyond* the site border and within those portions of the site where prehistoric human activity has resulted in little addition of organic matter. Locations where there has been much organic enrichment by prehistoric activity must be excluded from the analysis because organic matter encourages soil profile development[3] and because variations in profile development resulting from human activity are not of interest here. The locations to be used are a subset of those shown in Figure 49, including Stations 1–6 and 9–18. The latter set of stations are located east of Crane along a north–south transect, and generally include no midden horizons. The first set of stations crosscut the northern portion of Crane in an east–west direction, through areas where prehistoric human activity was not intense and where organic enrichment is minimal.

At each location, field observations were made as to the degree of soil profile development, including attributes such as textural and structural changes over depth, and development of clay and humus coatings on soil peds. These observations are summarized in Appendix 7, Table 5. Soil samples for textural determinations were taken from each horizon indentified in the field at those

---

3. Organic matter encourages granulation and improved drainage in the upper horizons, which facilitates the eluviation of clays and humus and the leaching of sesquioxides of Fe and Al into lower horizons. The degree of accumulation of these materials at lower levels within the soil is one indication of intensity of soil profile development. Additionally, organic matter eluviated into a B horizon facilitates the development of soil structure, there—a second indicator of intensity of soil development.

TABLE 53
Soil attributes indicating the relative degree of soil profile development at several locations on Crane site

Attribute of B horizon or Profile	Along the West-East Transect						Sampling Location Numbers				Along the North-South Transect						
	6	4	5	3	2	1	11	3	9	10	12	13	14	15	16	17	18
Basal depth of B_{max} (cm)	130	107	107	127	112	102	79	127	>122	94	20	no B	102	109	97	97	94
% clay, B_{max}	n.d.	n.d.	n.d.	2.950	n.d.	2.135	n.d.	n.d.	n.d.	n.d.	n.d.	n.d.	n.d.	1.589	n.d.	1.621	2.387
% clay, A_3																	
Degree of development of clay coatings	2	2	2	3	2	2	0	3	0	0	0	0	1	0	1	2	2
Degree of development of prismatic structure	1	1	1	3	1	0	0	3	1	0	0	0	3	0	1	1	3
Degree of development of subangular-blocky structure	3	2	2	3	3	1	1	3	2	1-2	0	0	n.d.	1	2	2	3
Midden deposits present in profile?	no	yes	yes	no	no	no	no	no	no	no	no	no	yes	no	no	no	no

Gulley and erosion

n.d. = no data

locations showing the greatest differences in soil profile development. These data are given in Appendix 7, Tables 1 and 6.

The relative intensity of soil profile development at each station along the two transects can be judged on the basis of a number of their profile attributes (Table 53). The depth at which maximum B-horizon development (B22, B2T) occurs can be used as an indicator of the relative intensity with which the parent material at each location has been weathered and the degree to which soil genesis has proceeded. The ratio of the percentage of clays in the horizon of maximum clay accumulation (B22, B2T) to the percentage of clays in the horizon of maximum eluviation (A3 at Crane) provides an indication of the relative amount of clay that has been mobile within the profiles. Variations among different locations in parent materials and their percentage clay composition as a whole do not affect this variable as a measure of profile development. The measure does assume, however, that if the percentage of clays varied from level to level in the unweathered parent materials at Crane, the relative differences between levels is the same at all stations examined. Two other soil attributes that will be used to indicate the degree of soil horizonization are the intensity of development of clay skins on ped surfaces in B horizons and the degree of development of soil structure within B horizons. Clay skins evidence the mobility of clays within a soil profile and illuviation of clays within a B horizon. Development of structural aggregates of soil within a B horizon reflects the downward movement and accumulation of several materials on which the development of soil structure depends: humus, gums, and polysaccharides; clays; and polyvalent cations (Ca^{2+}, Mg^{2+}, Fe^{2+}, Al^{3+}) (Brady 1974:58–60).

Values for each of the five variables just described for each sampling location are summarized in Table 53. Locations where midden deposits occur within the soil profile and where natural profile development could have been hastened by organic enrichment of the soil from human occupation are shown. Degree of development of clay coatings and soil structure have been assessed on an arbitrary scale from 1 to 3, ranging from incipient development to aproximately "weak" to "moderate" development in the terminology of the U.S. Soil Conservation Service (U.S.D.A. 1960). Development of prismatic soil structure within the B horizon of soils at Crane reflects greater profile differentiation than does development of subangular blocky structure. Prisms apparently form only after blocky structure is relatively well developed within the Crane soils. Prisms within the Crane soils are larger aggregates than the blocky peds, and therefore presumably reflect more intense formation processes (Brady 1974:60). Prisms also are composed of subangular blocky peds, but not vice versa.

From Table 53, some generalities about spatial variation in the degree of soil profile development over the Crane site are apparent. First, of all observed locations, Station 3 exhibits the greatest degree of profile development—with respect to *all* attributes. It has the deepest B horizon, the highest ratio expressing clay mobility, and the strongest development of structural aggregates and clay coatings on those aggregates. South of Station 3 and approaching the gully (Figure 49), profile development decreases as the slope of the landscape increases, as the amount of infiltration of precipitation compared to runoff decreases, and as soil erosion increases. South of the gully, where topography begins to level toward the crest of the southern ridge at Crane, and between Stations 15 and 18, profile development again increases, but not to the intensity expressed at Station 3. Station 14 represents an anomalous high in the sequence of profile developments from the gully through Station 18, which otherwise would define a monotonically increasing series of changes. This anomaly is apparently a result of the presence of midden deposits there and the effects of anthropic organic enrichment to the soil profile rather than a result of natural soil variation. Thus, in general, changes in soil profile developments south of Station 3 appear to be closely associated with topographic trends and developmental processes related to slope.

North of the location of maximal soil profile development (Station 3), there is a marked decrease in the degree of horizon differentiation (Station 11). This decrease does not appear to be related to topography, for both Station 11 and Station 3 occur in ridge–crest positions.

Along an east–west axis (Stations 1–6) and running along the crest of the northern ridge at Crane, soil profile development decreases in a smooth, monotonic manner in both directions away from the maximum at Station 3. There is a significant increase in the degree of profile development, however, at the far west end of the northern ridge (Site 6), where slope increases. All of these spatial trends are independent of or contrary to what one would expect from the lay of the landscape, if infiltration of precipitation and erosion were the primary factors governing soil profile development.

In summary, natural variations in soil profile development along the two sampled transects at Crane site follow a few regular trends that can probably be generalized to Crane site as a whole. North of the gully, the strength of soil profile development falls with increasing distance from ca. 200R160 (Station 3), more rapidly so in the northerly direction than any other. A secondary maximum of profile differentiation occurs around ca. 204R76 (Station 6). South of the gully, there is a uniform trend in natural profile development from north to south, and possibly from northwest to southeast along the southern ridge crest, if one can extrapolate from observations made for the northern ridge crest.

Spatial Variations in Soil Profile Development Compared to Spatial Variations in Parent Material Composition

The patterns of soil profile development already described have been reconstructed with data independent of those used to model spatial variation in the composition of the upper-member parent material. They may be compared to that model to determine whether the textural variations found in the upper member are primarily a result of differences in parent materials, as assumed, rather than degree of profile development. If textural variations at the 60-cm depth were the result of differential profile development and clay mobility rather than parent material composition, one would expect that those stations showing strongest profile development also would be locations having anomalously high percentages of clay at the 60-cm depth (the depth of B1, B21 horizons). One also would expect those stations having weakly differentiated soil horizons to be the location of anomalously high percentages of sand and silt compared to elsewhere on the site. An examination of the locations of clay, silt, and sand anomalies in Figure 49 clearly shows these expectations not be the case. Locations of relatively strong profile development (Stations 13, 6, 18) exhibited anomalously high percentages of sand and silt rather than clay at the 60-cm depth (i.e., in their horizons of clay accumulation). Location of weak profile development (Stations 1, 11, 10, 12, 13) do not have anomalously high percentages of sand and silt relative to other areas of the site. Thus, it would appear that the pattern of textural variation over space at the 60-cm depth *is* primarily a result of differences in the composition of parent materials, as assumed, and not simply variation in profile development. Variation in the composition of parent materials and the degree of soil profile development in the upper member at Crane follow independent spatial patterns.

In summary, three independent spatial patterns of natural soil variation at the Crane site can be reconstructed: one concerned with the composition of parent materials below 90 cm (lower member); a second concerned with the composition of parent materials above 90 cm (upper member); and a third concerned with the alteration of the upper-member parent materials by those soil formation processes responsible for profile development. None of these spatial patterns is uniform. Consequently, when determining associations between prehistoric use-areas and soil properties for the purpose of defining the kinds of soil changes produced by particular kinds of activities (Chapter 8), it will be necessarily to consider the several different kinds of natural variations documented previously. Not all soil variations over space at Crane are a result of human activity.

Chemical Soil Vairation at Crane Site Related to Historic Land-Use Patterns: Deforestation and Cultivation

Natural soil variability is one kind of soil variability that must be held constant when determining the soil alterations produced by different kinds of prehistoric activities at Crane site. Variability resulting from *modern agricultural practices* and from the different ways in which portions of the site were *used during the historic period* is another. In this section, several different land-use patterns at Crane will be documented and the spatial variation in soil chemistry that resulted from them will be described. Spatial strata that are homogeneous with respect to historic land-use will be defined. In addition to documenting *relative* differences between areas in the chemical changes produced in their soils, the *absolute* changes in soil chemistry that have occurred over time in them will be described. Both spatially uniform and spatially disuniform land-use practices will be documented for these two purposes.

Overview of Historic and Contemporary Land-Use Patterns at Crane

In a relativistic framework, a primary historic source of soil variation over Crane is the length of time that has passed since different sectors of the site were cleared of their natural forest vegetation and have been cropped and fertilized. The Crane site can be divided into three sections that have been cropped or used as pastureland for different amounts of time (Figure 58; Table 54). Land comprising Areas 1 and 2 in Figure 58 were initially cleared of their primary forest sometime between 1819 and 1873. The year 1819, marks the purchase of the land on which the Crane site is located by Christopher Gardiner, and the beginning of his efforts to clear his property of virgin timber for agricultural fields (Andreas *et al.* 1873:27, 30). It is probable that the land on which Crane is located was not cleared immediately, for Crane and the Gardiner homestead lie at opposite ends of the Gardiner estate, separated by ca. 1.1 miles. Timber in Areas 1 and 2 had definitely been removed by 1873, however, when the first published map of this area of Green County shows timber in only Area 3 of Crane site (Andreas *et al.* 1873).

It is not known whether all of Areas 1 and 2 were used for the same or different purposes initially after they were cleared of woods. By the turn of the century, however, Area 2 had been sown in bluegrass and lesperdeezer for a horse pasture whereas the southern portion of Crane (Area 1) was cropped (Crane, personal communication; U.S.D.A. 1937). Sometime after the early 1900's, Area 2 fell into disuse and was allowed to revert to timber. Reforestation definitely was in progress by 1937 and complete coverage of ground by a secondary growth of trees and shrubs was achieved by 1950, as evidenced by aerial photographs of Crane site at these times (U.S.D.A. 1937, 1950).

Between 1950 and 1968, Areas 2 and 3 remained in forest while Area 1 was cultivated. In fall of 1968, farmer Crane deforested all of Areas 2 and 3 using bulldozer and chaining methods and claimed the land in this northern portion of Crane site for cultivation. Areas 2 and 3 were allowed to lie fallow one year, and then were planted in winter wheat in fall of 1969. All of the Crane site has been

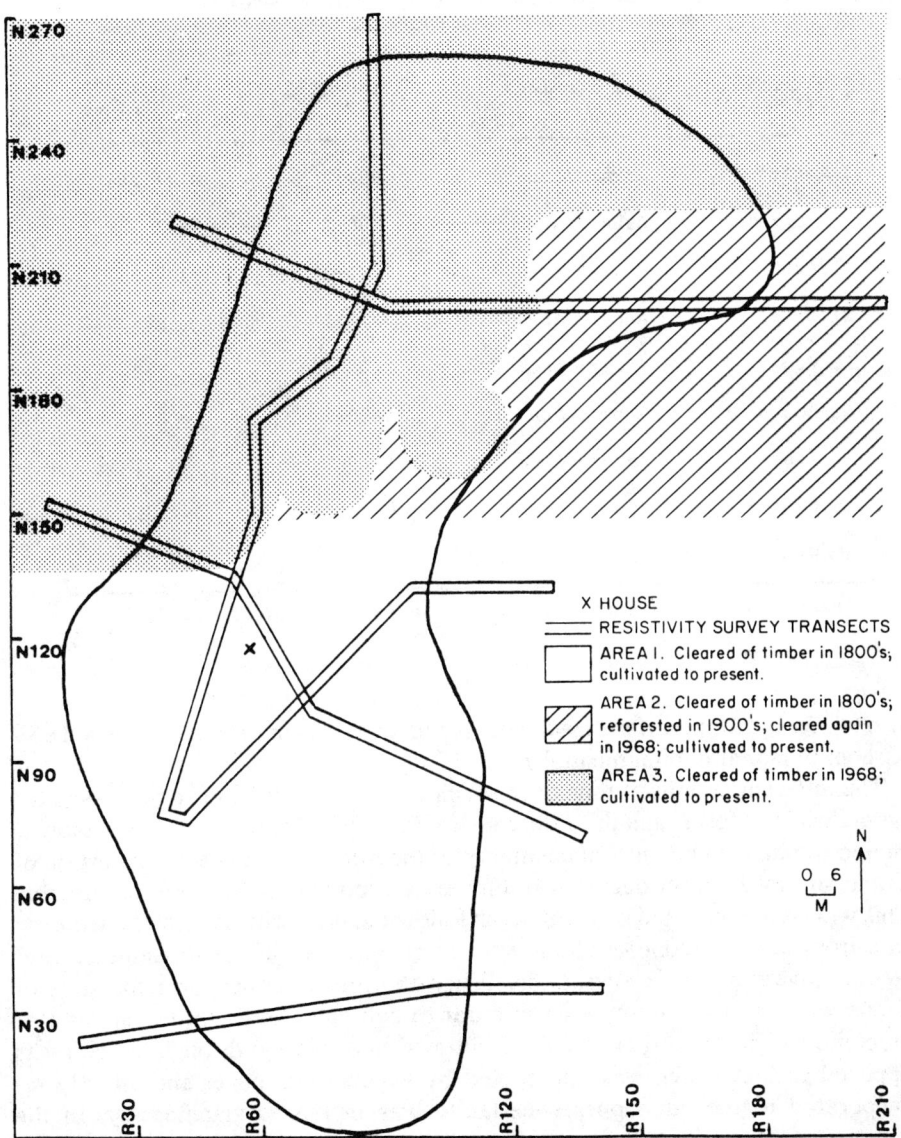

FIGURE 58. *Three portions of the Crane site which were deforested of virgin timber at different times in recent history and which have been used for cropping or pastureland different lengths of time.*

TABLE 54
*Approximate locations of intersection of Transects I-V with boundaries between areas of different historic land-use at Crane site**

Transect Subtransect	Boundary	Provenience	Approximate Grid Lines
I North	Area 3/Area 2	990±2	R127.25
I Central	Area 3/Area 2	590±2	R127.25
I South	Area 3/Area 2	190±2	R127.25
II North	Area 3/Area 1	2222±2	R53
II Central	Area 3/Area 1	2625±2	R52
II South	Area 3/Area 1	3026±2	R51
IV East	Aera 3/Area 1	6176±2	144.72R58
IV Central	Area 3/Area 1	4781±2	142.72R58
IV West	Area 3/Area 1	5407±2	140.72R54

*Based on Soil Conservation Service aerial photos for 1937, 1968 (USDA 1937, 1968)

in one field, cropped, fertilized, and limed in the same manner, since 1969 (Crane, personal communication).

The different lengths of time over which Areas 1, 2, and 3 have been used for agricultural purposes, and the different specific ways in which each was exploited altered the natural balance of chemistry of the forest soils in each subportion of Crane site to different degrees, in different directions, and to different depths. This will be documented in detail in the following, primarily using soil chemistry data from the plowzone of Crane site. First, however, let us examine several factors governing these alterations. When the forest canopy over the soils of Crane was removed in any area in order to cultivate the earth, the amount of precipitation per year that actually infiltrated and leached through the soil—as opposed to that which was intercepted by vegetational cover and directly re-evaporated to the atmosphere—certainly was increased significantly. In the temperate eastern United States, interception and re-evaporation may account for 10–20% of the rain that falls on corn, soybean, and wheat fields during their growing seasons. In contrast, it may account for 30–50% of the total annual precipitation that falls on a deciduous forest (Brady 1974:200–201) and even greater proportions of the rain that falls during the growing season. Part of the increased percentage of annual precipitation that reached the soils of Crane after deforestation probably was lost in runoff, but the residual would have infiltrated the soil and increased the total amount of nutrients that were leached from the soils per annum. This increase in leaching would not have been uniform with respect to all species of ions and radicals, but in inverse proportion to their strength of fixation within the soil (see Chapter 4). It would have altered the *relative proportions* of different nutrients as well as their *total availability* in the soil. Thus, the different lengths of time over which the soils in Areas 1, 2, and 3 have been free of a protective forest canopy has led to their differentiation in the magnitudes and spectra of their nutrients.

Another factor that has differentiated the soils in Areas 1, 2, and 3 is the different lengths of time during which they have been a part of artificial, nutrient-*replacing* agricultural systems rather than natural nutrient-*recycling* forest ecosystems. When the Crane site was covered with forest vegetation, nutrients which were taken from the soils were, *in the long run*, recycled to the soil in the same amounts and proportions that they were taken from it. With respect to vegetational uptake and replacement of nutrients, the chemical composition of the soil was maintained. At any *one* time, the relative proportions of different nutrients required by the forest cover and withdrawn from the soil might have been different from the relative proportions of different kinds of nutrients available in the soil. The relative proportions of different kinds of nutrients returned to the soil via decomposition also might have been different from both the relative proportions of different kinds of nutrients withdrawn from the soil and available within the soil. But *over time,* the soil in the forested areas may be seen as having been in equilibrium with respect to those vegetational nutrient requirements withdrawn from it and those nutrients returned to it.

This state of equilibrium and soil nutrient balance was disrupted in each area when it was deforested and began to be used for cropping. Nutrients uptaken by harvested crops were not returned to the soil. Nutrients returned to the soil through fertilization were not necessarily added in the same amounts and proportions with which they were removed. At the time of this study, the degree of alteration of soil nutrient balance in each subarea of Crane from that which existed under forested conditions was related to (a) the length of time each had been cultivated, (b) the different kinds of crops grown in each area, and (c) the different kinds of fertilizers used in each area. With respect to factors b and c, it is known specifically that Area 1 was cropped while Area 2 was used as pastureland for horses around the turn of the century.

This contrast between nutrient-recycling and nutrient-replacing systems must be qualified slightly. The forest that stood on the Crane site apparently was harvested of its wood to some degree for lumber and firewood during the historic period. The timber that was clear-cut from the northern portion of Crane during 1968 was a *secondary* stand of oak, cottonwood, walnut, and maple (Crane, personal communication) rather than primary forest. Thus, during the historic period, in the forested areas of Crane, nutrients taken up from the soils by the forest were not completely recycled to the soils. Soils in the forested areas were subject to some degree of nutrient loss and alteration in their nutrient balance through harvesting as were the cultivated areas. This circumstance will be considered in greater detail later in this chapter.

The major historic factors which have just been discussed as causing alterations in the magnitudes and proportions of nutrients within the soils at Crane site have been viewed largely from the perspective of the *spatial variability* they developed in the Crane soils. *Relative change* in nutrient status and nutrient balance from area to area of Crane site has been stressed. It must also be remembered, however, that these same factors also were sources of *absolute* changes in nutrient status and nutrient balance. As each subarea of Crane site came under cultivation, differences in the magnitudes and proportions of nutrients between distinct prehistoric use-areas of Crane site were lessened and masked over by historic, agricultural sources of soil variation. Even after 1968, when all of Crane site became incorporated within one agricultural field and was cropped, fertilized, and rained on in a more uniform manner, the magnitudes and proportions of soil nutrients that had, in part, a prehistoric cause continued to be altered. Both spatially uniform and spatially disuniform alteration of the soil by modern agricultural practices is of importance to this study.

Recent Cropping and Fertilizing Practices at Crane, Documented in Detail

The most recent cropping and fertilizing practices at Crane and the soil alterations they produced are those most likely to have been apparent in 1974 and 1975 when soil and resistivity surveys were undertaken at the site. Detailed records (Crane, personal communication) of these practices extend back as far as 1959 or 15 years prior to the resistivity survey at the Crane site. During this period, those portions of the site that were farmed generally were planted in either corn (Pioneer #3369), soy beans (Wayne), or winter wheat (Arthur) with clover, in a 3-year rotation. The rotation sequence was interrupted between 1969 and 1971, immediately after Area 3 of Crane had been cleared of forest, and when wheat was grown on the site for 3 consecutive years (Table 55). Prior to 1959, those portions of the site that were farmed probably were planted in corn, beans, and wheat, but there are no records to indicate how far back this system was used.

Since 1959, fertilizer applications have followed a regular pattern, coordinated with crop rotation. In general, when a corn crop was planted, the following nutrients were added to the soil: 100 lb/acre (112 kg/ha) NH_3 in the form of 100% liquid ammonia; 9 lb/acre (10 kg/ha) N in an undetermined form, 15.7 lb/acre (17.6 kg/ha) P as a phosphate, and 29.8 lb/acre (33.4 kg/ha) K in elemental form. The latter three nutrient increments were applied as 150 lb of 6–24–24 superphosphate. When winter wheat and clover were planted in fall, nitrogen was applied to the field in the form of ammonium nitrate, at the rate of 7.4 lb/acre (8.3 kg/ha) NH_4 and 25.6 lb/acre (28.7 kg/ha) NO_3. No fertilizers were applied to Crane when it was planted in soy beans.

TABLE 55
Sequence of crops harvested from and fertilizers applied to Crane site, Areas 1, 2, 3

Year Crop Harvested	Crop in Area 1	Crop in Areas 2,3	NH_4*	NO_3*	N?	P	K	Other Activities
			Fertilizers Applied in Cultivated Areas (lbs/A)					
1959	beans	in forest						
1960	wheat	in forest	7.44	25.56				
1961	corn	in forest			9.00	15.71	29.89	
1962	beans	in forest						
1963	wheat	in forest	7.44	25.56				
1964	corn	in forest			9.00	15.71	29.89	
1965	beans	in forest						
1966	wheat	in forest	7.44	25.56				
1967	corn	in forest			9.00	15.71	29.89	
1968	beans	no crop						deforestation of Areas 2 and 3 in Fall
Fall 1969	wheat	wheat				200.75		1832 lbs/A Ca; 1095 lbs./A Mg
1970	wheat	wheat			18.00	31.42	59.78	
1971	wheat	wheat						
1972	corn	corn			9.00	15.71	29.89	
1973	beans	beans						
1974	wheat	wheat	7.44	25.56				transect survey
1975	corn	corn			--	--	--	block survey

*Applications in fall of year preceding the harvest of wheat

Exceptions to this sequence of fertilizer applications are few, and largely relate to the time when area 3 was deforested and the soils had to be "primed" for cropping (Table 55). In 1969, when all of Crane was planted in wheat, 200.8 lb/acre (225 kg/ha) of P were applied to the whole site, in phosphate form. Additionally, 3 tons/acre (6720 kg/ha) of crushed dolomitic limestone were sown on the whole site. This limestone was obtained from Valstead quarry, near Kane, Illinois, and has been determined by the Illinois State Geological Survey (James Bradbury, personal communication) to be approximately 76% calcite and 6% dolomite, the remainder consisting of inert chert. The elemental calcium and magnesium added to the Crane soils by the 1969 liming would be approximately 1832 lb/acre (2052 kg/ha) and 109 lb/acre (122 kg/ha), respectively. In 1971, when the Crane site still was planted in wheat and the northern portion was thought unsuited for growing corn or beans, 300 lb/acre (336 kg/ha) of 6–24–24 superphosphate (18 lb/acre or 20 kg/ha of N; 31.4 lb/acre of P; 59.8 lb/acre or 67 kg/ha of K) were applied to the whole site rather than the usual quantity of ammonium nitrate. Finally, in 1975, no fertilizers were applied to the block area of the site, which was surveyed then for its soil resistivity and other soil attributes.

With respect to the rates at which the nutrients in the several fertilizers applied to Crane site become available after broadcasts, none of the fertilizers was designed for slow nutrient release. Of the granular fertilizers, neither the 6–24–24 superphosphate nor the ammonium nitrate were coated with biodegradable impermeable membranes or semipermeable membranes (Brady 1974:509) to slow their rates of solubility. Ammonium was applied to the Crane soils as a liquid. For all practical purposes, all the NH_4, NO_3, PO_4, and K applied to the Crane soils in a given growing season may be considered to have been released to the soil water within the period of that growing season. This does not, of course, imply that the total quantity of broadcast nutrients remained in *available* form after their release to the soil water. In particular, most of the phosphate that was released to the soil water at the Crane site during any one growing season and that was not withdrawn by crops probably was fixed within the growing season of its application. Calcium; hydrous oxides of iron, aluminum, and magnesium; and silicate clays would have been the major agents of fixation (Brady 1974:462–465), precipitation by calcium being particularly important. Derived from a loessic parent material, the soils of Crane site are rich in available calcium capable of fixing phosphates. Likewise, the nitrate applied to the soils probably did not remain available past their year of application. As an anion, nitrate would easily have been leached from the Crane soils.

Absolute Changes in the Chemistry of the Soils

The most recent cropping and fertilizing practices at the Crane Site just inventoried were, until 1968, restricted to Area 1 (Figure 58) in the southern portion of the site, but have been applied uniformly to the whole site since then. If it is assumed that the most recent agricultural disturbances to the soils of Crane are those which would have been most apparent during the 1974 and 1975 soil surveys, it can be asked whether there is any pedological evidence of these sitewide disturbances. The approach that will be taken in answering this question is an *absolute* one, concerned with the *absolute magnitudes and ratios* of nutrients over the *whole site*, rather than a relativistic one concerned with differences in nutrient status and balance in different portions of Crane.

Of the several agricultural factors influencing the nutrient status and nutrient balance of the Crane soils—including leaching, cropping, and fertilizing—the only one for which data are available for analysis within an absolute rather than relative framework is fertilization. The absolute amounts of nutrients applied to the Crane site each year as fertilizer are known, whereas those lost through leaching and cropping are not known. Two questions may be asked about fertilizer applications: (a) Have rates of application of fertilizer been large enough that the ratios of different kinds of nutrients in the soil have been altered toward the constant ratios of nutrients found in the fertilizers?; (b) to what depths have fertilizer applications been effective in changing soil nutrient status and balance?

The largest single input of nutrients to the Crane soils that has occurred most recently came in fall of 1969, when 3 tons of agricultural lime were applied to each acre of the site. The lime that was used was dolomitic in quality, and had a Ca:Mg ratio somewhere within the range of 16.72 and 23.42. In comparison, the mean ratio of available calcium to available magnesium in the plowzone of Crane is estimated at 18.60.[4] The figure falls well within the range of possible Ca:Mg ratios found within the agricultural lime and suggests that the ratio of Ca to Mg in the Crane soils have been altered significantly by the liming application toward the constant ratio of Ca to Mg found in the lime. It also suggests that the total quantity of available Ca and Mg in the Crane soils may have been significantly altered as well. To determine the actual extent to which the balance and status of Ca and Mg in the soils of Crane site have been altered, it is possible to compare the absolute availabilities of Ca and Mg and the ratio of available Ca to Mg found in the Crane soils to those same attributes found in uncultivated, undisturbed loessic soils elsewhere in the Illinois Valley.

In choosing undisturbed loessic soils to which the soils at Crane site might be compared, it is important that the comparative soils be as similar as possible to those at Crane site in their natural properties. In particular, the sources of the windblown loess, the distance between the sources and the sites being examined for their soil attributes, and the vegetation under which the soils developed must be held constant. The source of the loessic parent materials of the comparative soils and the Crane soils must be similar in order to ensure similarity in their overall minerology and their natural Ca and Mg content. The mineral composition of Peorian loess along the Illinois River Valley differs north and south of Hennepin, Illinois. North of Hennepin, Peorian loess deposits were derived from the valley train of the Ancient Illinois River alone, and reflect the

---

4. This figure is based on the ratio of available Ca to available Mg found in soil samples removed during the transect survey from BL1–0 and over most of Crane site, excluding only the gully and the old bed of Macoupin Creek. The location codes of the samples used in this estimation are those that fall within the following set of proveniences along Transects I–V: 74–356, 474–756, 2001–2230, 2401–2631, 3401–3654, 4201–4454, 4501–4658, 4766–4935, 5901–6057, 6166–6335, 6901–7131, 7201–7531.

minerology of the Wisconsinan tills found within the drainage of the Illinois River in the northeastern portion of the state. South of Hennepin, Peorian loess deposits were derived from the valley train of the Ancient Illinois River, but also that of the Ancient Mississippi, which conjoined with the Illinois and flowed through its valley south of Hennepin during the early Woodfordian. The composition of Peorian loess deposits south of Hennepin thus reflects the minerology of Wisconsinan tills found in a broader area drained by both the Ancient Illinois and Ancient Mississippi (Frye et al. 1968). Crane site is located south of Hennepin, Illinois, and consequently, the undisturbed loessic soils chosen for comparison with the Crane soils must be limited to those within the lower portion of the Illinois Valley.

Just as important, the distances of the comparative soils from the source of their parent materials—the Illinois Valley—must be similar to or greater than the distance of Crane site from the Illinois Valley. This condition is necessary in order to hold constant as best as possible the degree of weathering and loss of Ca and Mg that the Crane soils and comparative soils have undergone. In general, as the distance between the Illinois Valley and loess deposits in the surrounding uplands increase, average loess particle-size decreases, and consequently, the degree to which weathering has proceeded increases (Beavers et al. 1963; Jones et al. 1967; Jones and Beavers 1966; Smith 1942; Ulrich 1949, 1950). Degree of weathering affects both the absolute quantities of available Ca and Mg within a loess soil and their relative proportions (Jones et al. 1967; Ulrich 1950). Calcium is fixed less strongly and is leached more quickly from soils in general than is magnesium (Kardos 1964:191; Wilklander 1964:136), and the ratio of available Ca to Mg in loessic soils decrease over time with their greater weathering.

Finally, the vegetative cover under which the comparative soils should develop be as similar as possible to the forested conditions of the Crane soils. Surface vegetation plays a major role in determining the pH of a soil profile and the amount of organic matter in a soil profile, which in turn affect the mobility of different nutrients within it, their loss, and total nutrient status.

Taking these factors into consideration, three comparative soils profiles within loessic deposits along the Illinois River were found in the literature (Wascher et al. 1971) and compared to soil profiles east of the Crane site and free of prehistoric human disturbances. The comparative soils are similar in the minerology of their parent material to that of the Crane soils and developed under forests of composition (oak–hickory) similar to that at Crane. The relevant data are summarized in Tables 56 and 57.

As can be seen, horizon for horizon, the natural, but limed soils away from Crane have higher ratios of available Ca to Mg than do the unlimed comparative soils. This is true even for those comparisons in which the Crane soils are much more distant from the Illinois Valley than the comparative soils, and should be more weathered and have lower Ca:Mg ratios than those of the comparative

TABLE 56
Status and balance of available calcium and magnesium in Peorian loessic soils not under cultivation*

Horizon	Available Ca meq/100g Soil			Available Mg meq/100g Soil			Available Ca/Mg meq/100g Soil		
Station	13	16	17	13	16	17	13	16	17
A1	10.3	12.7	8.3	2.2	3.8	4.4	4.681	3.342	1.886
A2	6.0	5.9	2.3	.9	2.5	1.6	6.667	2.360	1.438
A3	8.4	--	3.2	1.3	--	3.6	6.462	--	.889
B1	10.4	7.8	4.4	2.7	5.0	6.3	3.852	1.560	.698
B21	11.7	11.3	5.5	5.1	7.8	11.4	2.290	1.449	.482
B22	10.0	14.8	7.0	7.5	10.4	12.7	1.333	1.423	.551
B23	7.6	13.0	8.1	7.6	10.4	13.9	1.000	1.250	.583
B24	6.4	--	--	7.4	--	--	.865	--	--
B3	6.6	12.1	8.4	7.0	10.3	11.7	.943	1.174	.718
C1	7.8	--	--	7.2	--	--	1.083	--	--

Site 13: Grazed woodland pasture, 1.5 mi east of Illinois River bluffs, Woodford County T27N, R3W, Sec. 8, SE160 SW40 SW10.

Site 16: Grazed woodland pasture 3.25 mi southeast of Illinois River bluffs, Woodford County T27N, R3W, Sec. 10, SW160, NE40, NE10.

Site 17: Grazed woodland pasture, 19 mi southeast of Illinois River bluffs, McLean County T24N R1W, Sec. 10, SW160, NW40, NE10.

* Wascher et al. (1971)

soils. Thus, it may be concluded that liming has significantly increased the ratio of available Ca to Mg within the soils at Crane.

Although the *ratio* of Ca to Mg within the Crane soils does appear to have been altered by liming, the *total quantity* of Ca and Mg does not appear to have been increased significantly. In comparing the availabilities of Ca and Mg in the limed soils located east of Crane (Table 57) to the availabilities of Ca and Mg in the unlimed comparative soils shown in Table 56, horizon by horizon, the Crane soils show *less* or similar amounts of Ca and Mg than are found in the comparative soils. They do not show an increase in Ca and Mg. This result is predictable, if it is considered that (*a*) most of the exchange sites on clay and humus colloids within the loessic soils of Crane would have been occupied already by Ca and Mg prior to liming (Wascher et al. 1971); and (*b*) the cation exchange capacity and total number of exchange sites potentially available for

TABLE 57
Status and balance of available calcium and magnesium in four locations east of Crane site (Figure 49)

Horizon	Available Ca meq/100g Soil				Available Mg meq/100g Soil				Available Ca/Mg meq/100g Soil			
Station	1	15	17	18	1	15	17	18	1	15	17	18
AP	6.0748	7.7273	5.4348	5.8962	.869	.905	.594	.660	11.504	14.050	15.060	14.706
A3	4.6875	6.3679	5.0885	5.0000	1.117	.846	.779	.778	6.908	12.385	10.748	10.577
B1	2.9204				1.413				3.402			
B21	4.7727	6.1905	6.1926	6.4286	2.866	2.242	2.246	2.238	2.742	4.546	4.538	4.728
B22	5.9524			6.1364	3.574			2.364	2.741			4.272
IIB3	6.1881	6.3636	6.8396	n.d.	3.765	2.866	3.067	n.d.	2.706	3.6554	3.671	n.d.

n.d. = no data

occupation by Ca and Mg probably was not increased much, if any, by liming procedures. The Ca and Mg ions that were freed to the Crane soils when they were limed would have been able to alter the ratio of available Ca to Mg by replacing these same nutrients on colloid exchange sites previously occupied by them, and also to a limited extent by replacing other nutrients on the few exchange sites not already occupied by Ca or Mg. There would not, however, have been much opportunity for the added Ca and Mg ions to significantly increase the total amount of exchangeable Ca and Mg within the Crane soils, given that most of the exchange sites were occupied already by Ca or Mg prior to liming, and given that the total number of exchange sites probably was not increased much by liming.

To what depths within the Crane soils was liming effective in altering the ratio of available Ca and Mg? It is apparent from Tables 56 and 57 that, with greater depth, the deviation of the ratio of available Ca to Mg within the Crane soils from that ratio within the comparative soil profiles becomes less and less significant, perhaps asymptotically approaching differences that are due solely to minerology. To clarify this situation, a graph was made of variation in the Ca:Mg ratio over depth for several soil profiles located east of the site (Figure 59). This graph shows that the ratio of available Ca to Mg decreases steadily to approximately 50.8–63.5 cm below surface, at which point it essentially becomes constant. This would seem to suggest that the lime which was applied to the Crane soils was effective in altering the ratio of available Ca to Mg within those soils to only that depth, the ratio at depths below that reflecting the minerology of the parent material. The results obtained here are in agreement with those obtained by Beavers *et al.* (1963:410) who found that differences in the amount of weathering and loss of total Ca from loessic soil profiles located at different distances from the source of their parent material become insignificant below about 61 cm.

Just as the ratio of available Ca to Mg within the soils at the Crane site has been examined to determine whether lime applications were large enough to have altered this ratio toward the constant ratio of these nutrients within the lime, so it is possible to examine the ratios of other nutrients within the Crane soils for their possible alteration by fertilization practices. Of those nutrients added to the Crane soils by fertilization, the ionic species and radicals most suited to this kind of analysis include NH_4 and K. Both of these tend to remain in exchangeable forms that are not easily replaced by other nutrients (Kardos 1964; Wilklander 1964) on clay–humus micelles. They stand in contrast to the nitrate radical, which as an anion, is quickly washed from the soil, and the phosphate radical, which is quickly fixed in the soil in unavailable forms.

Ammonium and potassium were applied to the soils of the Crane site in a constant ratio over each crop rotation cycle of 3 years, which was completed in the period 1959–1969 and 1972–1974 (Table 55). This ratio of NH_4 to K lies somewhere within the range of 3.89 and 3.60, depending on whether ammonium nitrate was the only source of NH_4, or whether the N in the 6–24–24

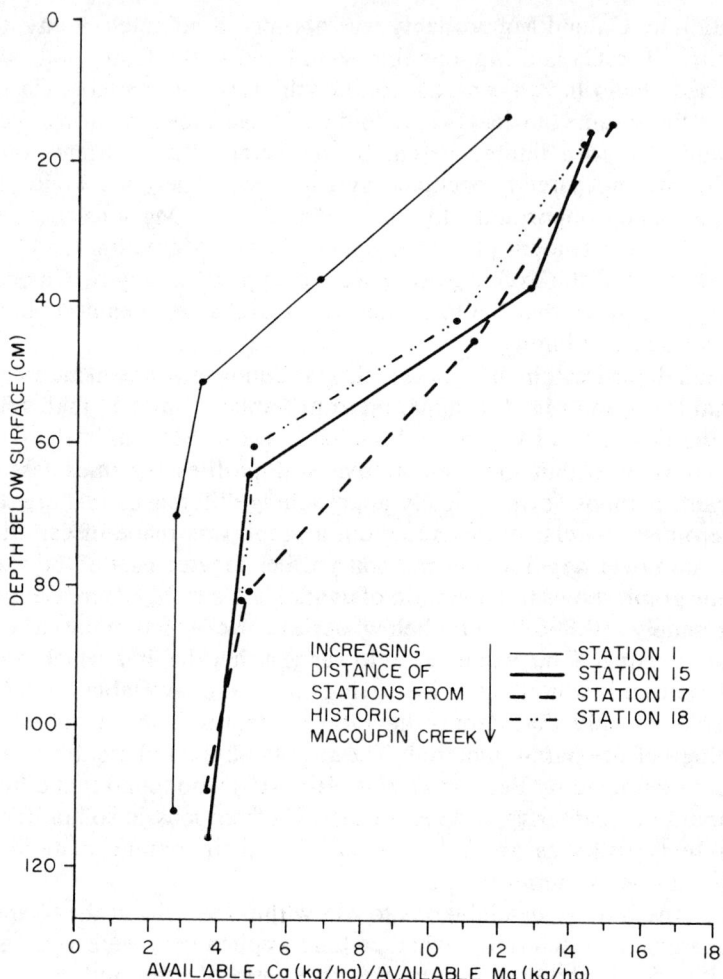

FIGURE 59. *Variation in the ratio of available Ca to Mg over depth at locations east of the Crane site. Locations are shown in Figure 49.*

superphosphate that was applied to the Crane soils also was in the form of ammonia. In contrast, the ratio of available NH_4 to K in the plowzone of Crane as a whole[5] was only .28 and the same ratio in the plowzone of off-site stations 1, 15, 17, and 18 (Figure 49) ranged from .22 to 1.10. Thus, on the basis of the ratio of available NH_4 to K within the plowzone at Crane, there is no evidence that NH_4 and K were applied in such huge quantities as to alter their ratio in the

---

5. The same soil samples were used as those considered in the analysis of Ca:Mg ratios, immediately prior.

soil toward their ratio in the superphosphate and ammonium nitrate fertilizers applied to Crane.

Comparative soil data from loessic Alfisols similar to those at Crane site but not fertilized could not be found in the literature to further investigate the degree to which fertilization of the soils at Crane has altered the ratio of available NH_4 to K in them. Nor could data be found on the absolute availability of NH_4 in loessic Alfisols. Data on the absolute availability of K were obtainable (Wascher et al. 1971), however, as was information on the proportional availability of K to Na. Both of these kinds of data—*absolute* and *proportional*—can be used to determine whether K was applied to the soils at Crane in quantities sufficient enough to have significantly altered its *absolute* availability.

The appropriateness of using comparative data on the proportional availability of K to Na requires some explanation. Among the univalent nutrient cations, K has a relatively strong replacing power (Kardos 1964). Had excessive concentrations of K been freed to the Crane soils by fertilization, they should have been able to have replaced Na ions on the exchange sites of clay–humus micelles, which would have increased the proportional availability of K relative to Na as well as the total availability of K.

A comparison of the soils that lay east of the Crane site and that are free of prehistoric human disturbances to loessic Alfisols similar to Crane's but that have not been farmed is presented in Table 58 with respect to their absolute and proportional availabilities of K. Neither the absolute availability of K nor its availability relative to Na in the Crane soils is greater than that in the comparative, unfertilized soils, horizon for horizon. Thus, K recently added to the Crane soils in the form of fertilizer does not appear to have altered the absolute availability of this nutrient. In fact, it most likely was *not* added in quantities sufficient enough to balance the loss of K through cropping, the increased rates of leaching brought on by deforestation (see the following pages), and possibly its replacement by Ca when the Crane soils were limed.

In summary, the applications of fertilizer which have been made to Crane site as a whole in recent years have had some effect on the availabilities of some nutrients in the soils there. The proportional availabilities of Ca and Mg have been altered with respect to each other, and probably with respect to the other nutrients that might have been replaced by Ca and Mg on clay and humus colloid exchange sites—particularly Na (Kardos 1964; Wilklander 1964). The absolute amounts of available Ca, Mg, and K within the Crane soils and the proportional amounts of available K and Na with respect to each other, however, do no appear to have been significantly changed by recent fertilizer and lime applications. These conclusions have implications on the significance of the associations found between chemical properties of the soils at Crane site and prehistoric activity there. First, should anomalously high or low ratios of available Ca to Mg or ratios of available Ca or Mg to other nutrients be found to associate with particular kinds of use-areas, the significance of the association

TABLE 58
Status of potassium and balance of potassium and sodium in Peorian loessic soils not fertilized and in the Crane soils

Horizon	Unfertilized Soils*						Horizon	Soils at Crane Site**					
	Available K (meq/100g Soil)			Available K/Na (meq/100g Soil)				Available K (meq/100g Soil)			Available K/Na (meq/100g Soil)		
Station	13	16	17	13	16	17	Station	1	15	17	18		
												1	15	17	18
A1	.4	.2	.2	4.	2.	2.	AP	.0968	.2720	.0979	.1399	3.173	6.886	3.694	3.102
A2	.1	.2	.1	1.	2.	1.									
A3	.2	n.d.	.1	2.	n.d.	1.	A3	.0891	.0941	.0883	.0791	2.296	2.295	1.911	1.669
B1	.3	.2	.3	3.	2.	3.	B1	.1064				2.128			
B21	.3	.3	.3	3.	3.	3.	B21	.1628	.1218	.0880	.1291	2.058	2.100	1.103	2.011
B22	.4	.4	.4	4.	4.	4.	B22	.1802			.1628	1.673			1.871
B23	.4	.4	.3	4.	4.	3.									
B24	.3	n.d.	n.d.	3.	n.d.	n.d.									
B3	.3	.3	.3	3.	3.	3.	IIB3	.1823	.1558	.1882	n.d.	1.367	1.791	1.639	n.d.
C1	.3	n.d.	n.d.	3.	n.d.	n.d.									

*Same stations shown in Table 56, as documented by Wascher et al. (1971)
**Station locations shown in Fig 49.
n.d. = no data

need not be questioned (liming was uniform over all of Crane) but the particular values taken by the ratio will not be of consequence for soil and resistivity studies made at other sites—even those sites located on Peorian loess. The values taken by such ratios at Crane site are a product of recent agricultural altertions as well as possibly prehistoric human disturbances. On the other hand, the anomalous values taken by ratios of nutrients other than Ca and Mg may be of importance at other archeological sites located on Peorian loess. This conclusion will be qualified in the next section, when withdrawal of nutrients from the Crane soils by cropping as well as increment of nutrients by fertilization and liming are considered.

Relative Differences between Portions of Crane in the Chemical Alterations Which Were Produced Historically in Their Soils

The stipulations drawn thus far for the determination of significant associations between soil chemistry and prehistoric activity at the Crane site concern the most recent agricultural soil disturbances that have occurred over the site *as a whole*. Further stipulations are required if historic land-use patterns having greater time depth and having produced different soil changes in different portions of the site are considered. When seeking associations between those areas of Crane where particular kinds of activities occurred and those areas exhibiting soil anomalies in the magnitude of availability of particular kinds of nutrients or ratios of availability of nutrients, cases from all historic land-use strata can be used only if historic disturbances to the magnitudes and ratios of availability of the nutrients in question are similar over all strata. Accordingly, this section will document the similarities and differences between the three historic land-use areas of Crane (Figure 58) in both the *magnitudes of availabilities* of given nutrients and the *proportional availabilities* of pairs of nutrients within their soils.

Means and variances of the magnitudes of nutrient availabilities and the ratios of nutrient availabilities were calculated for different populations of soil samples taken within the plowzone of the different historical land-use strata, in off-site locales. Differences in the means and median of the populations were tested for their statistical significance using a Student's t-test and a nonparametric median test (Dixon and Massey 1969:351). Only data from Areas 1 and 3—those areas cultivated for the longest and shortest times, respectively—were compared. Not enough soil samples were available from Area 2 within areas of little or no prehistoric activity and within areas exhibiting nonextreme degrees of soil profile development to make a meaningful comparison. The soil samples used to characterize the plowzone of Area 1 come from those proveniences along Transects II, III, and V, which lie east of the border of Crane site.[6] Those used to characterize the plowzone of Area 3 occur along Transect IV,[7] in areas of apparently little prehistoric activity. No samples within Area 3 but off-site were available for study. Paucity of prehistoric activity within the areas along Transect IV that were analyzed is suggested by the low number of artifacts found on their surface per 6 × 6-m-square (Figure 30), by their thinner midden deposits, and by the low to moderate organic matter content, pH, and nutrient availabilities of their soils to the depth of plowzone. The latter soil attributes used for choosing the soil samples to represent Area 3 are not considered to have biased the results of the comparison of land-use areas for their soil properties.

The results of the tests of significance and ancillary statistics are shown in Tables 59 and 60. Characterization of geographic strata as comparable or not comparable with respect to either the magnitudes or ratios of availabilities of particular nutrients is based upon Student's t-tests, when the assumptions of the test have been met. Otherwise, the results of the median test have been used in making the determination.

Magnitudes of Nutrient Availabilities

With respect to the magnitudes of availabilities of the nutrients in Area 1 versus Area 3, some very consistent patterns exist. First, in general, the availabilities of *cation* nutrients are significantly less in Area 1 than in Area 3. This is an expectable circumstance, for at least four possible reasons. First, the soils in Area 1 have been free of their natural forest canopy longer than have the soils in Area 3, and subject to greater amounts of precipitation and leaching than have the soils in Area 3.

Second, the soils in Area 1 have been cropped for a longer period of time. The depressed levels of nutrients within the soils of Area 1 compared to Area 3 reflect the longer period of time over which nutrients have been withdrawn from them by crops and insufficiently replaced by fertilizer applications. Nutrients in the soils of Area 3, on the other hand, have been subject to natural and more complete recycling within a forest ecosystem until very recently.

Both of these interpretations of the observed pattern—greater leaching and less complete nutrient replacement within the soils of Area 1—assume that the cation exchange capacities of the soils in Areas 1 and 3 are similar. The pattern also could be explained, however, by the alternative but not mutually exclusive hypothesis that the cation exchange capacity of the soils of Area 1 is less than

---

6. All soil sampling stations with the lengths of transect encompassed by the following provenience numbers are included: 2001–2003, 2401–2429, 3613–3649, 4409–4437, 7103–7131, 7506–7527.

7. All soil sampling stations within the lengths of transect encompassed by the following provenience numbers are included: 4682–4729, 6097–6156.

TABLE 59
Differences in nutrient availability between Area 1 and Area 3 (Figure 58) at Crane site horizon: Plowzone

Soil Attribute	Area 1: Cultivated μ±σ	Area 3: Forested μ±σ	N_1	N_2	T-Statistic	F-Statistic	Significance: One-Sided T-Test on μ	Significance: F-Test on σ²	Distribution Approximately Normal? X_1	X_2	Significance: Median Test, Two-Sided	Nutrient Availabilities Comparable?	Higher Value
pH	6.14±.1547	6.43±.2160	29	6	-3.934	1.949	.0002	.0588	yes	yes	.0076	no	forested
available P	49.7±16.99	169.±79.14	29	6	-7.736	21.70	.0000	.0000	yes	yes	.0167	no	forested
available Mg	216.±42.97	314.±34.19	29	6	-5.238	1.580	.0000	.3238	+S	yes	.0167	no	forested
available Ca	3421±771.2	4690.±624.0	29	6	-3.768	1.527	.0003	.3396	+K −S	yes	.0167	no	forested
available Na	18.8±3.522	16.8±3.125	24	6	1.268	1.270	.1075	.4305	yes	yes	.5310	≈	≈
available K	129.±23.28	269±161.7	29	6	-4.698	48.22	.0000	.0000	+K	+S	.0167	no	forested
available NH_4	53.0±16.81	82.2±22.19	28	4	-3.141	1.743	.0019	.1819	yes	−K	.0506	no	forested
available NO_3	17.2±11.18	23.07±18.30	12	3	-.7248	2.680	.2407	.1127	−K	−K	.5538	≈	≈
available SO_4	12.83±7.531	13.5±6.807	29	4	-.1689	1.224	.4335	.5046	+S	−K	.5738	≈	≈
available Cl	39.4±49.23	3.15±1.485	15	2	1.013	1099.	.1635	.0236	−K	yes	.5147	no	?
% organic matter	1.43±.3276	1.85±.8167	29	6	-2.114	6.214	.1211	.0005	yes	yes	.3576	no	forested

±K: positively or negatively kurtotic distribution
±S: positively or negatively skewed distribution

The soil samples used to characterize Area 1 come from those stations along Transects II, III, and V which lie east of the border of Crane Site and which are encompassed by the following provenience numbers: 2001-2033, 2401-2429, 3613-3649, 4409-4437, 7103-7131, 7506-7527. Those soil samples used to characterize Area 3 come from stations along TIV which fall in areas of apparently little prehistoric activity, and which are encompassed by the following provenience numbers: 4682-4729, 6097-6156.

TABLE 60

Differences in ratios of nutrient availability between Area 1 and Area 3 (Figure 58) at Crane site; horizon: Plowzone

Ratio	Area 1: Cultivated $\mu \pm \sigma$	Area 3: Forested $\mu \pm \sigma$	N_1 N_2	T-Statistic	F-Statistic	Significance: One-Sided T-Test on μ	Significance: F-Test on σ	Distribution Approximately Normal X_1	X_2	Significance: Median Test: Two-Sided	Nutrient Ratios Comparable?
V2/V3	.1385±.04975	.4351±.0149	29 6	4.5860	11.093	.0000	.0068	yes	yes	.0049	no
V2/V4	.02930±.004844	.02068±.002478	29 6	4.2136	3.8204	.0001	.0699	yes	yes	.0049	no
V2/V5	.002169±.002145	.001397±.0002361	29 6	.8709	82.56	.1950	.0001	yes	yes	.0762	no
V2/V6	.3363±.05936	.3959±.09124	24 6	-1.974	2.362	.0291	.0720	+K	yes	.3257	
V2/V7	.04883±.00944	.02835±.009486	29 6	5.011	1.100	.0000	.3824			.0076	no
V2/V8	.1266±.03772	.8367±.02457	28 4	2.191	2.357	.0182	.2628			.0506	no
V2/V9	.6970±.8253	.3882±.2260	12 3	.6260	13.34	.2711	.0718	+S +K	-K	.5538	
V2/V10	.7224±.6667	.5653±.2670	29 4	.4608	6.2344	.3241	.0779	+S +K	-K	.6760	
V2/V11	1.223±1.194	2.262±1.111	15 2	-1.160	1.155	.1320	.3160			.2647	
V3/V4	.2432±.1009	.5500±.2696	29 6	-4.879	7.140	.0000	.0002	-K	yes	.0114	no
V3/V5	.1878±.02456	.3705±.01804	29 6	-1.7201	-1.853	.0474	.2555			.0114	
V3/V6	2.612±.9423	10.36±4.634	24 6	-7.949	24.19	.0000	.0000	yes	yes	.0084	no
V3/V7	.3915±.1368	.7451±.4166	29 6	-3.840	9.283	.0002	.0000	-K	yes	.1012	no
V3/V8	1.003±.3781	1.793±.3074	28 4	-3.979	1.512	.0002	.4169			.0506	
V3/V9	6.4752±8.802	13.120±14.04	12 3	-1.051	2.546	.1561	.1234	+S +K	yes	.5538	
V3/V10	5.941±5.808	16.86±15.61	29 4	-2.786	7.222	.0045	.0005	+S +K	-K	.0748	no
V3/V11	11.02±12.01	44.64±17.68	15 2	-3.582	2.165	.0013	.1633			.2647	
V4/V5	.07412±.0675	.06735±.004421	29 6	.2427	233.1	.4048	.0000	+S +K	yes	.0114	no
V4/V6	12.11±.3270	19.44±5.653	24 6	-4.217	2.988	.0001	.0320		yes	+S	.0084	no
V4/V7	1.710±.4258	1.382±.5164	29 6	1.658	1.471	.0534	.2308			.3006	
V4/V8	4.497±1.894	4.218±1.660	28 4	.2784	1.302	.3913	.4780			.7002	
V4/V9	22.45±26.95	19.99±11.63	12 3	.1508	5.370	.4412	.1674			.5538	
V4/V10	24.59±20.65	27.64±12.11	29 4	.2867	2.905	.3881	.2064			.6760	
V4/V11	42.33±42.94	106.0±46.80	15 2	-1.956	1.188	.0347	.2942			.2647	
V5/V6	196.3±46.84	289.0±83.22	24 6	-3.682	3.156	.0005	.0259	yes	+S	.0084	no
V5/V7	27.12±7.868	20.62±7.76	29 6	1.848	1.028	.0368	.5484			.0762	no
V5/V8	71.88±29.88	61.19±25.72	28 4	.6781	1.350	.2515	.4630			.7002	
V5/V9	269.57±219.5	304.4±185.3	12 3	-.2517	1.402	.4026	.4885			.5538	
V5/V10	401.3±393.6	429.3±205.9	29 4	-.1385	3.654	.4553	.561			.6760	
V5/V11	627.2±697.5	1648.±835.06	15 2	-1.916	1.433	.0373	.2511			.2647	
V6/V7	.1451±.03106	.7764±.03411	24 6	4.671	1.206	.0001	.3377			.0084	no
V6/V8	.3848±.1273	.2095±.1047	23 4	2.592	1.478	.0078	.4243			.2689	
V6/V9	1.602±1.148	1.135±.7014	9 3	.6527	2.679	.2643	.3001			.5000	
V6/V10	1.931±1.621	1.649±.8444	24 4	.3369	3.685	.3694	.1546			.7022	
V6/V11	3.178±4.021	6.548±4.209	11 2	-1.086	1.096	.1504	.3198			.2692	
V7/V8	2.691±1.054	3.790±2.862	28 4	-1.525	7.377	.0689	.0009	+S +K	yes	.7002	
V7/V9	14.82±16.25	11.67±7.820	12 3	.3193	4.320	.3772	.2029			.5538	
V7/V10	14.04±9.005	18.52±10.82	29 4	-.9143	1.444	.1838	.2510			.6760	
V7/V11	26.12±25.73	76.54±47.31	15 2	-2.419	3.382	.0143	.0436	-K	-K	.2647	

TABLE 60 (cont.)

V8/V9	6.349±7.578	3.047±1.090	12	2	.5953	48.33	.2813	.1118			.7692
V8/V10	6.143±5.518	43.78±2.682	28	2	.4430	4.232	.3306	.3692			.7586
V8/V11	12.64±12.85	23.91	15	1	--	--	--	--			?
V9/V10	2.719±4.113	1.585±1.073	12	3	.4612	14.70	.3261	.0654	+S +K	−K	.5538
V9/V11	4.058±5.135	10.50	7	1	--	--	--	--			?
V10/V11	2.405±2.404	$3.810± 2.2 \times 10^{-16}$	15	2	−.8032	1×10^{32}	.2672	.0000	−K	−K	.2647

V2: pH
V3: available P
V4: available Mg
V5: available Ca
V6: available Na
V7: available K
V8: available NH_4
V9: available NO_3
V10: available SO_4
V11: available Cl

∗K: positively or negatively kurtotic distribution
∗S: positively or negatively skewed distribution

The soil samples used to characterize Area 1 come from those stations along Transects II, III, and V which lie east of the border of Crane Site and which are encompassed by the following provenience numbers: 2001-2033, 2401-2429, 3613-3649, 4409-4437, 7103-7131, 7506-7527. Those soil samples used to characterize Area 3 come from stations along TIV which fall in areas of apparently little prehistoric activity, and which are encompassed by the following provenience numbers: 4682-4729, 6097-6156.

that of the soils in Area 3. Lower availabilities of cations in Area 1 would be a function of the fewer number of exchange sites available for cations to bond to and prevent themselves from being leached from the soil. This hypothesis is not unreasonable. One possible source for a difference between Areas 1 and 3 in the cation exchange capacities of their soils would be the different percentage of organic matter contained within them. The soils in Area 1 have been plowed and aerated, and have had their organic residues broken up and brought into greater contact with soil organisms. They consequently have been made susceptible to rapid decomposition and loss of their organic matter (Brady 1974:158–161) for a longer period of time.

Finally, the difference in cation nutrient levels between Areas 1 and 3 could reflect the smaller reserve of nutrients in the form of organic matter within the soils of Area 1, and consequently, the release of less nutrients annually through organic matter decomposition.

At least two, and possibly all of these explanations of the observed patterns are true. This can be shown by restating and grouping the hypotheses in a more operational manner. On the one hand, the lower availabilities of cations in Area 1 and in Area 3 could reflect that a lower *proportion* of exchange sites are occupied by nutrient cations as opposed to hydrogen ions in Area 1. This could have resulted from either the longer time Area 1 has been subject to leaching, or the longer time Area 1 has been subject to cropping and insufficient nutrient replacement by fertilization, or the loss of nonexchangeable nutrient reserves in the form of organic matter. On the other hand, the lower availabilities of cations in Area 1 could reflect that a lower *number* of exchange sites are available for occupation by cations in Area 1 as a result of the longer period over which the soils there have been subject to rapid decomposition and loss of their organic micelles. One or more of the first set of explanations is evidenced by the lower pH (higher proportion of exchange sites occupied by H rather than nutrient cations) within the soils in Area 1 than in Area 3 (Table 59). The second explanation is evidenced by the lower organic matter content (number of exchange sites) in the soils of Area 1 than Area 3 (Table 59).

A second pattern shown in Table 59—one in contrast to that found for cation nutrients—is the similar mean levels of availability of anions between Areas 1 and 3 relative to within-strata variances. This could reflect the smaller population sizes that were tested for differences in mean or median for the anions. It also, however, may reflect the fact that anions are very easily leached from soils, and that their availabilities in 1974 may largely have been a product of only the most recent, spatially uniform inputs of NO_3, SO_4, and Cl from precipitation, mineral weathering, and fertilizer applications.

On a more subtle and statistically less significant level, though, the anions Cl and NO_3 do show logical and predictable differences in their availabilities. The slightly greater amounts of nitrate in the soils in Area 3 than in Area 1 possibly reflect the greater amounts of organic matter, there, and consequently, the greater annual release of ammonium and nitrification of ammonium to nitrate within Area 3. The greater quantities of chlorine in the soils of Area 1 than those in Area 3 might be the residual effect of fertilization of Area 1 with muriate of potash (KCl) prior to 1959. Although there are no records of this fertilizer having been used at Crane site, muriate of potash was used very extensively in the United States during the past, and still is today, as a carrier of potassium (Brady 1974:513–514; Tisdale 1956). One datum that gives some strength to this interpretation is that the ratio of available Cl to K found within the soils in Area 1 is somewhat closer to the ratio of Cl to K found in muriate of potash (1.0) than is the ratio of available Cl to K within the soils of Area 3. The ratios for Areas 1 and 3 are, respectively, .0383 ± .0389 and .0131 ± .0211 (Table 60).

Whether the differences in the availabilities of Cl and NO_3 between Areas 1 and 3 actually reflect differences in land-use, or whether they are truly insignificant differences is equivocal and would require larger sample populations to determine. The differences probably reflect differential land-use. If all soil samples taken from plowzone in Areas 1 and 3 are used to test for the differences of interest, regardless of whether the samples occur in locations of little prehistoric activity or intense prehistoric activity, the pattern remains the same—with higher chlorine availabilities in Area 1 and higher nitrate availabilities in Area 3—but the differences in mean values between areas are statistically significant at high levels of confidence (Table 61). Unfortunately,

TABLE 61
Differences in the availabilities of nitrates and chlorides between Areas 1 and 3 at Crane site using on-site and off-site soil samples; horizon: Plowzone

Variable	Area 1 $\mu \pm \sigma$ (cultivated)	Area 3 $\mu \pm \sigma$ (forested)	N_1	N_2	T-Statistic	F-Statistic	Significance: One-Sided T-Test on μ	Significance: F-Test on σ	Higher Value
NO_3^-	13.2±13.58	31.3±14.94	114	23	-5.717	1.210	.0000	.1270	forested
Cl^-	19.8±40.09	2.42±2.952	112	23	2.071	184.4	.0201	.0000	cultivated

these statistics can not be used in a definitive manner to test the hypotheses concerning recent differences in land-use, for the data include soil variability resulting from prehistoric differences in space-use, as well.

In summary, the nutrients P, Mg, Ca, K, NH_4, and possibly NO_3 and Cl have different availabilities in the soils of historic land-use Areas 1 and 3, at Crane—the cations and NO_3 occurring in greater quantities in Area 3, Cl occurring in greater quantities in Area 1. The percentage of organic matter exhibits spatial variation following the pattern of cations and NO_3. When seeking associations between prehistoric use-areas and soil properties at the Crane site, cases from both Areas 1 and 3 can not be used simultaneously if the soil properties concerned are the absolute availabilities of P, Mg, Ca, K, NH_4, NO_3, or Cl, or percentage of organic matter content. The availabilities of sodium and sulfate, however, can be compared between different prehistoric use-areas falling within different historic land-use strata. These conclusions probably hold with respect to Area 2, as well. To be on the conservative side, they will be assumed to be true.

Ratios of Nutrient Availabilities

The soils in historic land-use Areas 1 and 3 differ not only in the magnitudes of availability of some nutrients, but also the ratios of their availabilities. Table 60 lists all possible ratios of nutrient availabilities for the soils within Areas 1 and 3 and suggests which ratios are unequal between those areas. The following patterns can be found within the table. First, in general, those cation nutrients that were found to have absolute availabilities that were not similar between land use strata are the same as those that form ratios of availabilities that are not similar between strata. Included among the latter set are H (pH), P, Mg, Ca, Na, and K. One exception to the pattern is Na, which was found to be comparable between strata in the magnitude of its availability but not in its proportional availability. The dissimilar proportional availability of sodium, however, occurs simply by default of its forming ratios with cations that are not similar between strata in their absolute nutrient availabilities. A true exception to the pattern is ammonium, which was not found to be similar between strata in the magnitude of its availability, but was found to be comparable in its proportional availability with K, Ca, Mg, Na, and SO_4. Also, the ratio of the availabilities of K to Mg are similar for Areas 1 and 3, despite the fact that the magnitudes of availability of each differ between these areas.

A second pattern apparent in Table 60 is that ratios of availabilities of nearly any nutrients to any anions are similar between Areas 1 and 2. An exception is the ratios of the availabilities of PO_4 to SO_4. Whether this general pattern reflects a similarity of soil formation processes in Areas 1 and 3 with respect to the proportional availabilities of anions, or whether it reflects only the statistical inadequacies (small population sizes, high σ^2/μ ratios) of the populations being examined is equivocal.

The results obtained in Table 60 do not completely clarify which pairs of nutrients have availabilities that form ratios that are similar for Areas 1 and 3. Ammonium can not be available in Areas 1 and 3 in equivalent proportions relative to all of the nutrients, K, Ca, Mg, Na, and SO_4. If this were true, by default, the ratios of available nutrients K:Ca, K:Mg, K:Na, K:SO_4, Ca:Mg, Ca:Na, Ca:SO_4, Mg:Na, Mg:SO_4, and Na:SO_4 *also* would have to be equivalent between strata. This is not so; there obviously are some internal inconsistencies in the results shown in the table. The only proportional nutrient ratios found to be equivalent between Areas 1 and 3 and which form an internally consistent set are NH_4:K, NH_4:Mg, and K:Mg. Thus, an attempt must be made to clarify the pattern found in Table 60. To do so, the difference between the nutrient budgets of oak–hickory forest ecosystems and the nutrient budgets of 3-year rotational cropping systems similar to the ones at Crane and documented in the literature will be found, in order to *predict* which nutrient ratios should be similar or dissimilar between forested Area 3 and cultivated Area 1. The predicted similarities and differences then will be compared to the data shown in Table 60 in order to locate the sources of inconsistency within them.

The consideration given to the nutrient requirements of the forest vegetation in Area 3 in addition to the nutrient requirements of the crops grown in Area 1 requires explanation. In the preceding discussions, the soils in Area 3 have been characterized as more similar to natural soil conditions than the soils in Area 1. The soils in Area 3 have been a part of a natural nutrient-recycling forest system until very recently, whereas those in Area 1 have been subject to net losses of nutrients through cropping for a long time. One might think that in order to predict which ratios of nutrient availabilities should differ between Areas 1 and 3, it would be necessary to examine the nutrient requirements of only the crops grown in Area 1. These requirements would represent the annual losses to which the soils in Area 1 have been subjected and which have altered those soils from natural conditions approximated by the soils in Area 3. This viewpoint is not correct, however, the difference between the requirements of crops grown in Area 1 and the requirements of the forest vegetation that grew in Area 3 must be examined in order to predict which nutrient ratios should be similar or dissimilar between Areas 1 and 3, *as if both vegetational covers were being harvested and as if the soils in both areas were loosing the nutrients extracted by their vegetational covers.*

The forest–soil system within Area 3 was *not* closed with respect to the nutrients taken from and returned to the soil; it did include "harvesting" as a pathway of nutrient loss. The forest present in Area 3 of Crane site in 1968 was a secondary forest, not virgin timber, and apparently had been harvested of wood for lumber and firewood to some extent since the early 1800s. It is possible that some logging was done on Crane during the early historic period, but this is only hinted at by the location of an old logging road nearby the site (Crane, personal communication). Also when Area 3 was deforested in 1968 for cultivation, very little of the nutrients tied up in the timber grown on the land were returned to the soil. Trees were uprooted with a bulldozer and pushed outside the borders of Crane site for burning. If the average age of these trees was, say, 30 years, the act of deforestation would have been equivalent to having farmed the soil in Area 3 annually for 30 years with crops demanding nutrients in those proportions demanded by the oak-hickory forest in Area 3. Thus, nutrients have been lost from the soils in Area 3 through tree harvesting. Consequently, the difference between the nutrient requirements of the forests in Area 3 and those of the crops grown in Area 1, rather than simply the nutrient requirements of the crops grown in Area 1, must be considered in the analysis.

Ideally, a complete input–output analysis of all gains and losses of nutrients should be computed for both systems, but this can not be done, for lack of data. Instead, only the balance of nutrients required by the vegetational components of the two systems will be considered when predicting which nutrient availability ratios should be similar or different between Areas 1 and 2. This alternative approach is not unreasonable, if the major soil chemistry changes within Area 1 have resulted from nutrient losses produced by cropping rather than by the increases in effective precipitation and nutrient leaching that accompanied deforestation.

The comparative data on the proportional nutrient requirements of oak–hickory forests and corn–bean–wheat cropping systems that will be used to evaluate the statistics on the nutrient availability ratios of the soils in Area 1 and 3 are summarized in Table 62. The requirements shown for each vegetational system are averages of the requirements reported by multiple studies (see Table 62 for references). When possible, studies have been chosen where the forests or crops grew under soil and climate conditions as similar to those at the Crane site as possible.

Proportional annual nutrient requirements estimates for oak forests are based on detailed ecological studies that document the total amounts of different nutrients uptaken by stands annually versus that returned annually as litter. The data pertain to oak forest communities at large and include the understory groundcover as well as the trees, themselves. Forest compositions range from mixed oak–beech associations to nearly pure oak stands. All stands were of species of oak different from those that comprised the forest at Crane site, including *Quercus ruba* and *Q. petraea* rather than *Quercus velutina* and *Q. alba*.

Proportional nutrient requirement estimates for corn, soy beans, and winter wheat are based on studies of elemental plant composition. For corn, composition generally was determined from sample of leaves, whereas the compositions of soy bean and wheat usually were determined using whole tops. Most of the studies of corn, beans, and wheat, which were referenced previously, present the results of experiments documenting changes in plant composition with variation in one or more parameters, including (a) the amount of fertilizer containing one or more nutrients applied to the crop, (b) transpiration rates, (c) the variety of the plant, (d) form of root development—adventurous or primary tap roots, (e) the particular stage of growth, and (f) the portion of the plant sampled. Proportional nutrient requirements have been estimated using the

TABLE 62
Proportional nutrient requirements of oak forests and corn-bean-wheat cropping systems similar to those at Crane site compared to proportional nutrient availabilities within Area 1 and Area 3 at Crane site

		\	\	\	Nutrient in Numerator of Ratio	\	\	\	\
		N	P	K	Ca	Mg	S	Key	
Nutrient in Denominator of Ratio	P	.5577 no .9970 12.168±1.267 10.27±1.459 14.21±3.359 5.771±.8775						Proportional nutrient availability within Area 3 (forested) { .5577 no .9970 } Are proportional nutrient availabilities stastically similar between Areas 1 and 3?	
								Proportional nutrient availability within Area 1 (cultivated) { 12.168±1.267 10.27±1.459 14.21±3.359 5.771±.8775 } μ±σ for proportional nutrient requirements for: oak forests, corn, soy beans, and winter wheat, top to bottom	
	K	.2639 yes .3716 no 1.702±.2906 1.464±3.891 3.502±.3182 4.938±.2263	.7451 .3915 no .1410±.0287 .1445±.0402 .2473±.0415 .8627±.0921					Proportional nutrient requirements are based upon the following references:	
	Ca	.01634 yes .01391 1.379±.4588 7.742 2.826 7.843	.3705 .1878 no .1173±.03764 .7422 .1550±.01117 1.340	.04850 no .03687 .8609±.3332 7.351 1.238±.5318 3.968				Oak forests: Duvigneaud and Denaeyer 1970; Ovington 1962; Remezov and Pogrebnyak 1969. Corn: Bear 1951; Bennett et al. 1953; Robertson et al. 1965; Viets et al. 1954.	
	Mg	.2371 yes .2224 8.059±.5321 12.426±.5445 6.890 2.435	.5500 no .2432 .6897±.1007 1.778±.8902 .3236±.1043 4.161	.7236 yes .5848 4.956±.7608 12.24±.7283 2.656±.7739 12.32	14.85 no 13.49 6.356±2.010 1.415±.2828 2.144±.5198 3.104			Soy beans: Bear 1951; F.A.O.U.N. 1970; Fletcher and Kurtz 1964; Hanway and Weber 1971; Hymowitz and Walker 1970; Oliver and Barber 1966; Raper and Barber 1970. Winter wheat: Bear 1951; F.A.O.U.N. 1970; Macy 1936.	
	Na	4.773 yes 2.599 no 62±26.16 -- -- --	10.36 no 2.612 5.5±2.828 -- -- --	1.288 no 6.892 42.88±23.51 37.44 -- --	289.0 no 196.3 38.62±23.86 22.69 -- --	19.44 no 12.11 7.875±3.005 13.22 -- --			
	S	4.378 yes 6.143 6.818 9.476 15.54 11.17±4.007	16.86 yes 5.941 .5000 1.769 1.458 2.139±.3599	18.52 yes 14.04 3.636 9.00 4.712 4.854±3.152	429.3 yes 401.3 .8580 -- -- 1.785	27.64 yes 24.59 1.273 .7217±.0222 -- .575	1.649 yes 1.931 -- -- -- --		

results of only those experiments in which (*a*) the values taken by the several variables were consistent with growing conditions at the Crane site; and (*b*) the sampling procedures used are relevant for estimating *total* above-ground plant requirements for an entire growing season.

The data summarized in Table 62 have several inadequacies that should be recognized. First, proportional nutrient requirements for forest and cropping systems were not available for all nutrient pairs. Relevant data for sodium and the anions, SO_4, NO_3, and Cl, are lacking in the literature of agronomy and forest ecology. Second, the data cannot be used in a statistical manner, testing for significant differences between forest and crop requirements. In many cases, estimates of standard error are missing. Finally, the proportional annual nutrient requirements for corn, beans, and wheat can not be combined to calculate average proportional annual requirements pertaining to a whole 3-year rotation, to be compared with the proportional annual requirements of oak forests. The appropriate weighting factors for averaging the proportional requirements of corn, beans, and wheat are not known. Thus, their proportional nutrient requirements must be compared separately to those of oak forests.

Accepting these inadequacies, if a comparison is made of the proportional nutrient requirements of oak forests and those for corn, beans, and wheat, where data are available, they substantiate the data on proportional soil nutrient availabilities shown in Table 60. In general, those nutrient pairs having proportional availabilities that differ within the soils of Area 1 and Area 3 at Crane are required by oak forests and corn–bean–wheat crops in different proportions (P:K, P:Ca, P:Mg, P:SO_4, K:Ca, K:Na, Mg:Na, N:P?) The directions of these inequalities within the nutrient requirement data and within the nutrient availability data are consistent, as well. Where proportional nutrient requirements are higher for forest ecosystems than cropping systems and more demanding on the soil, the proportional availabilities of those nutrients are less in Area 3 than in Area 1 (P:K, P:Ca, P:Mg, P:SO_4, K:Ca, Mg:Na). Where proportional nutrient requirements are lower for forest ecosystems than cropping systems and less demanding on the soil, the proportional availabilities of those nutrients are greater in Area 3 than in Area 1 (K:Na, N:P?). Thus, there is good agreement between the differences in proportional nutrient availabilities that would be expected between Areas 1 and 3 on the basis of nutrient requirement data for forest and cropping systems, and those that are actually found.

With respect to those nutrient pairs found to have similar proportional availabilities between Areas 1 and 3 at Crane site—NH_4:K, NH_4:Ca, NH_4:Mg, NH_4:Na, and K:Mg—and that are sources of inconsistency within Table 60, the nutrient requirement data can be used to locate the problematical ratios. Oak forests and corn–bean–wheat agricultural systems similar to those at Crane have similar proportional nutrient requirements with respect to the nutrient ratios N:K and N:Mg. The similarity of forest and agricultural systems in their proportional nutrient requirement, N:Na, can not be assessed, but these systems appear to differ in their proportional nutrient requirements N:Ca and K:Mg.

Thus, nutrient requirement data would suggest that the set of nutrient availability ratios that may be considered similar between Areas 1 and 3 at Crane site should be restricted to NH_4:K and NH_4:Mg, and does not include NH_4:Ca, K:Mg, or the large group of nutrient availability ratios that are defined by all combinations of NH_4, K, Ca, Mg, and Na and that would also have to be included in the set. The nutrient availability ratio K:Mg also should probably be considered equivalent in Areas 1 and 3—even though this is not substantiated by the nutrient requirement data—in order to achieve optimal consistency within both the nutrient requirement and nutrient availability data. The nutrient requirement ratios N:K and N:Mg are similar for the forest and cropping systems, which would imply that the ratio K:Mg also should be similar for the two systems.

In summary, with respect to the primary purpose of the analysis, it would seem that when determining associations between prehistoric use-areas at Crane site and the spectra of nutrients within the soils of those areas, cases from both the historic land-use strata at Crane—Areas 1 and 3—may be used only if the nutrient ratios of interest are NH_4:K and NH_4:Mg, and probably K:Mg, as well. Observations from both Areas 1 and 3 might be usable when investigating associations between use-areas and nutrient availability ratios involving the anions NO_3, SO_4, and Cl, but to be conservative, analyses examining these ratios, as well, will be restricted to observations within single land-use strata. Additionally, ratios of nutrient availabilities that are thought not to be comparable between Areas 1 and 3 will be considered uncomparable between areas 2 and 1 and Areas 2 and 3.

Physical Soil Disturbances at Crane Site Related to Historic Land Use: Stratigraphy

The soil alterations that occurred at the Crane site as a result of its historic use for farmland and pasture are not restricted to changes in nutrient status and balance. The stratigraphy of the site also has been disturbed in various locations by a number of modern agricultural activities. Aside from the obvious incorporation and reworking of the upper portions of archeological deposits within plowzone, the most widespread disturbance of the Crane stratigraphy relates to the process by which the forest in Area 3 of the site was removed. Whole trees were uprooted and displaced by bulldozing and chaining techniques. Large masses of soil were uprooted along with the trees, creating pot holes up to several meters in diameter (Crane, personal communication). Holes were filled in with surface soils, creating features essentially similar to archeological pits with respect to their morphology as well as the high organic matter content and high porosity of their fills. The remnants of these features still were apparent in 1975, 7 years after they had been created, as local depressions within the landscape that collected water. Their locations also were apparently observable from the air in 1974 as vegetational anomalies in the weeds (largely pioneer annuals), which grew up on Crane site after it was disked and surface-collected in 1974. The anomalies had vegetation of a darker green coloration and denser growth—apparently as a result of the greater depth to which organically enriched soils with more moisture and nutrients extended in these areas.

A second source of disturbance of the stratigraphy at Crane site was a minor amount of landscaping that Donald Crane did along the bank of the old bed of Macoupin Creek, south of the gully (Figure 29). Mr. Crane lowered the crest of the bank between ca. 96R18 and 150R46 with a bulldozer, pushing the material to the northwest into the old bed of Macoupin Creek. This was done in order to decrease the slope of the land at this point. The amount of soil removed ranged from ca. 1.2 m at the southern end of the work area to nothing at the northern end. The effects of bulldozing were apparent within the block that was excavated at Crane site in collection units 320 and 334 (Figure 37) where midden deposits suddenly became thinner compared to those to the east. Those proveniences along Transect II occurring within the area which was lowered or infilled include those west of the R44 line (i.e., those west of proveniences 2242, 2642, and 3042) to an unknown distance.

Associated with either the removal of trees from Area 3 of Crane or the landscaping along the old bed of Macoupin Creek was the disturbance of the stratigraphy along Transect IV, between the North 146 and 158 lines, proveniences 6172–6150, 4772–4750, and 5472–5450, and in an east–west direction between R46 and at least R74. Organically enriched deposits in this area extended to unusual depths, as much as 57 cm below surface, and contained large clods of sterile soil. The deposits could represent midden and subsoil that have been churned, locally, or modern fill from elsewhere on top of intact midden deposits. In either case, the area will not be considered in the analysis of soil and resistivity data.

A final source of stratigraphic disturbance at Crane is Mr. Crane's bulldozing near the base of the gully along one side (which side is unknown) in 1968. The amount of soil removed from the gully side was probably around 1 ft—just enough to remove drainage tiles that were beginning to erode out of the gully side and that Mr. Crane's plow would strike 8 in. or less below surface. Bulldozing was restricted to the lower sides of the gully, and did not extend onto the crests surrounding the gully banks.

Summary

In this chapter, several descriptive models of those dimensions of soil variability at the Crane site that are *not* a result of prehistoric human activity have been presented. These include:

1. A model of spatial variation in the texture and presumably the composition of the lower-member parent material (90 cm or more below surface)

2. A model (Figure 56) of spatial variation in the texture and presumably the composition of the upper-member parent material (surface to 90 cm below surface)
3. A model (Table 53; Appendix 7, Table 5) concerned with spatial variation in natural soil profile development
4. A model of spatially uniform alterations of the proportional availabilities of Ca and Mg that have been made within plowzone and over depth by recent liming practices
5. A model of spatially nonuniform alterations of absolute nutrient availabilities (Table 59), which have occurred in Areas 1 and 3 of Crane site as a result of historic land-use patterns
6. A model of spatially nonuniform alterations of proportional nutrient availabilities (Table 60) that have occurred in Areas 1 and 3 as a result of historic land-use patterns
7. A list of of locations within Crane site that have been stratigraphically disturbed by recent deforestation and landscaping.

All of these models describe dimensions of soil variability that are not pertinent to an understanding of the different kinds of soil alterations produced by different kinds of prehistoric activity, and that must be segregated from the analysis of these relationships. Each model allows the definition of spatial strata across the Crane site, which are relatively homogeneous with respect to natural and historic sources of soil variability, and within which more meaningful associations between different kinds of prehistoric use-areas and their soil properties may be sought. Cross-site strata can be and will be defined in multiple ways using these models, depending on the particular soil attribute being investigated and which sources of natural or historic soil varability must be held constant in the analysis.

References

Natural Soil Variation at the Crane Site and Related Topics

Beavers, A.H., J.B. Fehrenbacher, P.R. Johnson, and Robert L. Jones
   1963   CaO–ZnO$_2$ molar ratios as an index of weathering. *Soil Science Society of America, Proceedings* 27:408–412.
Brady, Nyle C.
   1974   *The nature and property of soils*. Macmillan, New York.
Butzer, Karl W.
   1971   *Environment and archaeology*. Aldine, Chicago.
   1977   Geomorphology of the lower Illinois Valley as a spatial–temporal context for the Koster Archaic site. *Illinois State Museum, Reports of Investigations* 34.
Frye, John C., H.D. Glass, and H.B. Willman

1968 Mineral zonation of Woodfordian loesses of Illinois. *Illinois State Geological Survey, Circular* 427.
Frye, John C., and H.B. Willman
1970 Loess thickness in Illinois. Map distributed by the Illinois State Geological Survey, Urbana.
Jones, Robert, and A.H. Beavers
1966 Weathering in surface horizons of Illinois soils. *Soil Science Society of America, Proceedings* 30:621–624.
Jones, Robert L., B.W. Ray, J.B. Fehrenbacher, and A.H. Beavers
1967 Mineralogical and chemical characteristics of soils in loess overlying shale in northeastern Illinois. *Soil Science Society of America, Proceedings* 31:800–804.
Kardos, Louis T.
1964 Soil fixation of plant nutrients. In *Chemistry of the soil*, F.E. Bear (ed.), pp. 177–199. Reinhold, New York.
Kubiena, W.L.
1953 *The soils of Europe*. T. Murby, London.
Ruby, William W.
1952 Geology and mineral resources of the Hardin and Brussels quadrangles (in Illinois). *United States Geological Survey, Professional Papers*, 218.
Smith, G.D.
1942 Illinois loess, variations in its properties and distribution, a pedologic interpretation. *Illinois Agricultural Experimental Station, Bulletin* 490.
Strahler, Arthur N.
1970 *Introduction to physical geography*. John Wiley, New York.
Tisdale, Samuel L. and Werner L. Nelson
1956 Soil fertility and fertilizers. Macmillan, New York.
Ulrich, Rudolf
1949 Some physical changes accompanying Wiesenboden and Planosol soil profile development from Peorian loess in southwestern Iowa. *Soil Science Society of America, Proceedings* 14:287–295.
1950 Some chemical changes accompanying profile formation of the nearly level soils developed from Peorian loess in southwestern Iowa. *Soil Science Society of America, Proceedings* 15:324–329.
United States Department of Agriculture
1960 Soil classification, a comprehensive system: 7th approximation. U.S.D.A. Soil Conservation Service, Washington, D.C.
1974 Soil survey of Greene County, Illinois.
Wascher, H.L., B.W. Ray, J.D. Alexander, J.B. Fehrenbacher, A.H. Beavers, and R.L. Jones
1971 Loess soils of northwest Illinois. *University of Illinois College of Agriculture, Bulletin* 739.
Wilklander, Lambert
1964 Cation and anion exchange phenomena. In *Chemistry of the soil*, F.E. Bear (ed.), pp. 107–148. Reinhold, New York.

History of the Crane Site

Andreas, Lyter, & Co.
1873 Atlas map of Greene County, Illinois. Davenport, Iowa.
Anonymous
1909 Atlas: Greene County, Illinois. On file, City Library, Carrollton, Illinois.
Continental Historical Company
1885 *History of Greene and Jersey Counties, Illinois*. Springfield, Illinois.

Donnelley, Gassette & Loyd, Publishers
 1879   *History of Greene County, Illinois*. Chicago.
Illinois State Museum
 1820   Illinois land records. Original field notes, books 94, 97. On file in the Illinois State Museum.
United States Department of Agriculture
 1937   Aerial photographs SH-12-954, SH-12-955. On file in the U.S. National Archives.
 1950   Aerial photographs SH4G33BC, SH4G34TC. On file in the U.S. National Archives.
 1956   Aerial photographs SH4R58D, SH4R57L. On file in the U.S. National Archives.
 1962   Aerial photographs SH2CC95D, SH2CC94L. On file in the U.S. National Archives.
 1968   Aerial photographs SH1JJ201, SH1JJ200. On file in the U.S. National Archives.

Crop and Forest Nutrient Requirements

Bear, Firman E.
 1924   *Soil management*. John Wiley, New York
 1951   *Soils and fertilizers*. John Wiley, New York.
Bennett, W.F., G. Stanford, and L. Durneil
 1953   Nitrogen, phosphorus and potassium content of corn leaf and grain related to nitrogen fertilization and yield. *Soil Science Society of America, Proceedings* 17:252–258.
Boatwright, G.O., and Hayden Ferguson
 1967   Influence of primary and/or adventitious root systems on wheat production and nutrient uptakes. *Agronomy Journal* 59:299–302.
Dickson, C.H., and G.J.F. Pugh
 1974   *Biology of plant litter decomposition*. Vol. I, II. Academic Press, New York.
Dumenil, Lloyd
 1961   Nitrogen and phosphorus composition of corn leaves and corn yields in relation to critical levels and nutrient balance. *Soil Science Society of America, Proceedings* 25:295–298.
Duvigneaud, P., and S. Denaeyer de Smet
 1970   Biological cycling of minerals in temperate deciduous forests. In *Analysis of Temperate Forest Ecosystems*, D.E. Reichle (ed.), pp. 299–325. Springer-Verlag, New York.
Ellis, B.G., C.J. Knauss, and F.W. Smith
 1956   Nutrient content of corn as related to fertilizer application and soil fertility. *Agronomy Journal* 48:455–459.
Fletcher, H.F., and L.T. Kurtz
 1964   Differential effects of phosphorus fertility on soybean varieties. *Soil Science Society of America, Proceedings* 28:225–228.
Food and Agriculture Organizations of the United Nations
 1970   *Fertilizers and their use*. FAOUN, Rome.
Freeland, R.O.
 1937   Effect of transpiration upon the absorption of mineral salts. *American Journal of Botany* 24:373–374.
Hamilton, Herman A.
 1966   Effect of nitrogen and potassium salts with phosphate on the yield and nitrogen, potassium, and manganese contents of oats. *Soil Science Society of America, Proceedings* 30:239–241.
Hanway, J.J., and C.R. Weber
 1971a   N, P, and K percentages in soybean plant parts. *Agronomy Journal* 63:286–290.
 1971b   Accumulation of N, P, and K by sobean plants. *Agronomy Journal* 63:406–408.
Hymowitz, T., and W.M. Walker
 1970   Leaf analyses as a selection index for soybean seed oil and protein. *Agronomy Journal* 62:631–632.

Macy, P.
 1936 The quantitative mineral nutrient requirement of plants. *Plant Physiology* 11:749–764.
Oliver, S., and S.A. Barber
 1966 An evaluation of the mechanisms governing the supply of Ca, Mg, K and Na to soybean roots. *Soil Science Society of America, Proceedings* 30:82–86.
Ovington, J.D.
 1962 Quantitative ecology and the woodland ecosystem concept. *Advances in Ecological Research* 1:103–192.
Raper, C.D., Jr., and S.A. Barber
 1970 Rooting systems of soybeans, II: Physiological effectiveness as nutrient absorption surfaces. *Agronomy Journal* 62:585–588.
Remezov, N.P., and P.S. Pogrebnyak
 1969 *Forest soil science*. Translated from Russian. Israel Program for Scientific Translations, Ltd., Jerusalem.
Rennie, J. Peter
 1955 The uptake of nutrients by mature forest growth. *Plant and Soil* 7:49–95.
Robertson, W.K., L.C. Hammond, and L.G. Thompson
 1965 Yield and nutrient uptake by corn for silage on two soil types as influenced by fertilizer, plant population and hybrids. *Soil Science Society of America, Proceedings* 29:551–554.
Rodin, L.E., and N.I. Bazilevich
 1967 *Production and mineral cycling in terrestrial vegetation*, G.E. Fogg (trans.), Oliver and Boyd, London.
Rolfe, G.L., M.A. Akhtar, and L.E. Arnold
 1978 Nutrient distribution and flux in a mature oak–hickory forest. *Forest Science* 24:122–130.
Sayre, J.D.
 1948 Mineral accumulation in corn. *Plant Physiology* 23:267.
Tyner, E.H.
 1946 The relation of corn yields to leaf nitrogen, phosphorus and potassium content. *Soil Science Society of America, Proceedings* 11:317–323.
Viets, F.G., Jr., C.E. Nelson, and C.L. Crawford
 1954 The relationships among corn yields, leaf composition, and fertilizers applied. *Soil Science Society of America, Proceedings* 18:297–302.

Statistical References

Dixon, Wilfrid J. and Frank J. Massey
 1969 *Introduction to statistical analysis*. McGraw-Hill, New York.

8
Soil Alterations Produced by Prehistoric Human Activities at the Crane Site

In Chapter 4, the literature documenting some of the effects that human occupation can have on the soils of archeological sites was reviewed. These studies suggest that the soils in different use-areas of archeological sites may be differentiable in their chemical and physical characteristics, but this idea has not been investigated systematically. Only limited attention (Cook and Heizer 1965; Hurley and Heidenreich 1971) has been given to the different spectra of nutrients found in different use-areas; alterations of soils structure and moisture equilibria have not been investigated at all.

In this chapter, several kinds of data will be assembled in order to model or exemplify systematically the kinds of soil alterations that different prehistoric activities can produce on earthen archeological sites. On a large geographic scale, the use-areas at the Crane site defined in Chapter 5 will be examined and compared for those of their chemical soil characteristics that were monitored by the Transect Survey and Block Survey described in Chapter 6. On a more local scale, pit fills and midden deposits within different use-areas will be compared to each other or to natural soil horizons surrounding or underlying them with respect to those physical soil attributes measured during the supplementary soil surveys described in Chapter 6.

The analyses to follow vary in their purpose and level of generality. Some are meant only to illustrate the general relationships between prehistoric activities and the soil alterations they produce, which have been modeled in Chapters 3 and 4. The results of some other analyses will constitute either general models, which have not been presented earlier and which are applicable to all earthen sites, or more specific models, which are restricted in their applicability to sites located in climatic and pedological conditions similar to Crane. The purpose and level of generality of each analysis will be pointed out as it is discussed.

Relationships between Use-Areas and Soil Chemistry at the Crane Site: Differentiation of Use-Areas by the Spectra of Their Natural Anomalies

Anomalous Magnitudes of Nutrients versus Anomalous Spectra of Nutrients

The first analysis aims at defining the different spectra of nutrients that have anomalously high availabilities within the soils of different kinds of use-areas at Crane, in order to illustrate the principles described in Chapters 3 and 4. By "spectrum" is meant the magnitudes of nutrient availability anomalies *relative* to each other, as opposed to their *absolute* magnitudes. The spectrum of anomalous nutrient availabilities within the soils of a use-area may be quantified either as a set of *ratios* of anomalous nutrient availabilities, or as a set of presence–absence states of anomalous nutrient availabilities defined in reference to some threshold.

It is important to make the distinction between the absolute magnitudes and the spectra of nutrient anomalies found within use-areas. The absolute magnitudes of nutrient anomalies found within various use-areas are dependent on a site-specific factor—the different lengths of time that the different use-areas within the site were used and the different amounts of refuse deposited within them. Absolute magnitudes of nutrient anomalies consequently are not useful for illustrating or modeling general relationships linking activities to soil change. The spectra of anomalous nutrient availabilities found within different use-areas of a site, however, are not dependent on the length of time the areas were used. They are related to more general phenomena of interest, such as the relative proportions of nutrients found in the refuse deposited within them and the general soil processes of nutrient maintenance and loss described in Chapter 4. Thus, in the analysis to follow, the *spectra* of anomalous nutrient availabilities within use-areas will be examined.

Spectra will be quantified as presence–absence states of anomalous nutrient availabilities rather than ratios of these availabilities. This is necessary because the nutrient enrichments that will be reconstructed as having occurred in different use-areas and that will be compared to the anomalous nutrient availabilities are quantified on only a presence–absence scale.

Definition of Activity Groups and the Residues Associated with Them

To illustrate the relationships holding between prehistoric activities and the soil alterations that they produce, three independent Crane site data sets must be considered, documenting (*a*) the activities that occurred in different use-

Use Areas and Soil Chemistry: Spectra of Natural Anomalies 443

areas of Crane and the refuse materials they deposited within those areas; (*b*) the chemical nature of the refuse materials (i.e., which nutrients are most augmented to the soils of use-areas); and (*c*) the actual soil anomalies that occur in the different areas. The next three subsections describe the nature and organization of these three data sets, in preparation for the analysis of the interrelationships that will follow.

As a first step in the analysis, data on the activities that were reconstructed to have occurred in different portions of the Crane site (Chapter 5) had to be reorganized into a form that would allow them to be compared with soil data in a meaningful way, that is, one emphasizing the *nutrient contributions* made by the activities to the soil rather than one considering the *type* of activity, per se. It will be recalled that in Chapter 5, each 6 × 6-m surface collection unit on Crane site that falls within the areas surveyed for their soil and resistivity attributes was characterized by: (*a*) the artifact types found within it, (*b*) the activities implied by those artifact types, and (*c*) whether or not the implied activities actually occurred in the unit or elsewhere, that is, whether artifact deposition was of a primary or secondary nature (Table 44). An attempt was made to be as *specific* as possible in describing the functions of artifact classes and the activities associated with particular units, making fine distinctions such as "scraping soft wood" and "scraping hardwood" rather than broader classes such as woodworking in general. Detailed functions, when known, were specified in order to facilitate the process of assigning functions to artifacts of a more ambiguous nature by way of spatial association between the known and unknown types. Also, when considering the debris categories (as opposed to tool types) present in collection units, the analysis placed emphasis on describing the *activities* that generated the debris and that are implied by those debris rather than the *act of deposition* of specific types of debris as an activity in its own right. For example, the presence of unburnt bone in a surface collection unit was taken to indicate activities that generate bone refuse—bone working, boiling bone, etc.—rather than the act of deposition of bone refuse, itself. Again, this was necessary in order to evaluate more clearly the activities that might have occurred in specific collection units where the tool types present in them had multiple or ambiguous functions.

In examining the relationship between particular activities and the soil changes produced by them, however, such detail in the definition of activities is unnecessary, and emphasis on the activities generating the refuse rather than refuse deposition, itself, is inappropriate. For our purposes, finely differentiated activities leaving residues that have similar effects on the chemical nature of soils should be grouped together analytically. Also, the debris found within collection units should be taken to indicate simply the deposition of those kinds of debris, and not the multiple activities which may have produced them. Thus, it was necessary to reconsider the activities which had been associated with given collection units in Chapter 5 within a new framework and for new purposes, stressing the *general* categories of *refuse materials*, which most probably were deposited within the collection units, rather than the *specific, primary activities,* themselves.

To accomplish this, first a list was compiled of 24 sets of activities, or "activity groups," each of which (*a*) encompass one or more of the finely differentiated activities reconstructed in Chapter 5 (Table 63); and (*b*) generate a different kind of refuse (Table 64) having different chemical attributes. Particular activity groups and the deposited refuse materials implied by them then were associated with particular collection units according to the inventories of finely differentiated activities which were found to characterize the units in Chapter 5.

Only the *presence or absence* of activity groups and of deposition of particular kinds of refuse within collection units was reconstructed; the amount of refuse deposited within collection units was not assessed. Although the latter characterization of collection units would have been preferable and might be thought possible through the use of counts of artifact types within collection units, it was not attainable. There are two reasons for this circumstance. First, different activity groups (and the finely differentiated activities they encompass) certainly are characterized by the use and disposal of different numbers of artifacts per unit of work and per unit of refuse generated. This prohibits the uniform use of artifact counts as a measure of quantity of refuse deposition when making comparisons *between* collection units having *different* activity groups. Second, when making comparisons between collection units having the *same* activity groups present within them, the frequency of a specific artifact type within a unit can be taken as a measure of the quantity of refuse deposited within the unit only on the roughest of scales. The numerous formation processes discussed in Chapter 5 as producing depositional sets of polythetic organization (e.g., curation, rates of breakage) distort the frequencies with which artifacts are discarded in a work area relative to the amount of work done and the amounts of refuse generated.

It was possible, however, to use data on the quantity and variety of artifacts found within surface collection units to structure the presence and absence states of activity groups within collection units in an optimal manner. Activity groups represented by debris classes (e.g., burnt bone) were coded present only in those collection units where the debris occurred in high densities, evidenced significant nutrient input, and where the *likelihood of soil alterations having occurred* was greatest. "High density" was defined with respect to local density norms for the debris class, that is, only *anomalously* high debris concentrations within general debris scatters were coded with presence states for the analysis. Ignoring the presence of activity groups in collection units where their diagnostic debris classes were found in lower densities as part of a general scatter is acceptable, for the nutrient anomalies to be correlated with the data are defined in an analogous manner (see page 468).

Optimal coding of the presence or absence of activity groups represented by tool types as opposed to debris classes was done in a slightly different, but

TABLE 63
Definition of activity groups

Activity Group Number	Descriptive Title of Activity Groups	Activities (Table 45) Included in Activity Group or Raw Materials Used to Indicate the Activity Group
1	wood working	3-20*
2	pounding bark	42
3	hearth dumpings	sandstone, fire-cracked igneous cobbles, clay blobs
4	bone working	18b, 19b, 22-27*
5	crush bone	54
6	burnt bone, deposition	burnt bone
7	unburnt bone, deposition	unburnt bone
8	mollusk processing	mollusk shell concentrations found in test excavations
9	dehair hides, alone	45
10	dehair and soak hides	45,44, both required
11	dehair and simmer hides	45,43, both required
12	grain hides, alone	47
13	grain and soak hides	47,44, both required
14	grain and simmer hides	47,43, both required
15	uncooked animal tissue, deposition	41,46,56,57,58,59,61,71
16	uncooked plant tissue, deposition	39,40,60
17	pulverized roots, tubers, bulbs, rhizomes, fruits, dried meat	70,71
18	uncooked seeds and seed coats, deposition	62,63,64,65

TABLE 63 (cont.)

Activity Group Number	Descriptive Title of Activity Groups	Activities (Table 45) Included in Activity Group or Raw Materials Used to Indicate the Activity Group
19	uncooked nuts, deposition	66,67,68
20	uncooked seeds or nuts, deposition	62-68
21	boil bone, meat	28,55,73
22	boil nuts, seeds	75,76
23	boil greens	38,74
24	red ochre, deposition	red ochre

*Activities considered discrete in totalling the number of woodworking activities found in a surface survey unit are: whittling (3), scraping (4,5,6,7,8,16), grooving (12), sanding (14), drilling (17,18a), sawing (19a), splitting (20), scraping shafts (9,10,11), grooving shafts (13), sanding shafts (15).

**Activities considered discrete in totalling the number of bone working activities found in a surface collection unit are whittling (23), scraping (23), grooving (24), sanding (25), drilling (18b), sawing (19b), splitting (27), sharpening pointed bone implements by sanding (26).

analogous manner. In some cases, activity groups are represented by a number of tool types that function in specific subtasks belonging to one general kind of activity and producing one general type of debris. For example, the woodworking activity group (#1, Table 63) is represented by artifact types that functioned in scraping, drilling, sanding, grooving, etc. In such cases, it was possible to tabulate for each surface collection unit the number of different kinds of subtasks of the activity group that are represented by the unit's artifact inventory. Representation of a high variety of subtasks within a collection unit was taken to indicate greater amounts of activity and a greater *probability* that a significant amount of refuse was deposited there. Lower variety could reflect either less total activity and deposition, or greater specialization of activity along certain subtask lines. Activity groups were coded present in only those collection units where a high number of their subtasks occurred. "High" was defined with respect to local variety norms. Thus, it was possible to pinpoint particular collection units where soil anomalies would be most expected to occur. In other cases, activity groups are represented by artifact types of only one function, without specification of subtasks. In these cases, activity groups

TABLE 64
Residues left by each activity group

Activity Group Number	Descriptive Title of Activity Groups	Residue
1	wood working	unburnt wood
2	pounding bark	bark
3	hearth dumpings	ash
4	bone working	dry, unboiled bone
5	crushed bone	dry, unboiled bone
6	burnt bone, deposition	dry unboiled bone meat drippings from roasting?
7	unburnt bone, deposition	dry unboiled bone (bone artifact manufacture) dry boiled bone (boiling meat foods, bone artifact manufacture)
8	mollusk processing	mollusk shell, mollusk meat
9	dehair hides, alone	deer meat (putrefaction)?
10	dehair and soak hides	urine? ash? (most probable)
11	dehair and simmer hides	urine? ash?
12	grain hides, alone	urine? ash? (most probable) brains? meat drippings? broth of meat? broth of bone? bone dust?
13	grain and soak hides	urine? (possible) ash? brains?
14	grain and simmer hides	brains? ash? (most probable)

TABLE 64 (cont.)

Activity Group Number	Descriptive Title of Activity Groups	Residue
15	uncooked animal tissue	deer meat small mammal meat turtle meat
16	uncooked plant tissue, deposition	greens
17	pulverize roots, tubers, bulbs, rhizomes, fruits, dried meats	roots, tubers fruits powdered deer meat (dried)
18	uncooked seeds and seed coat, deposition	seeds in general chenopods
19	uncooked nuts, deposition	nutmeats in general hickory nut meats nutshells
20	uncooked seeds or nuts, deposition	seeds in general chenopods nutmeats in general hickory nut meats nutshells
21	boil bone, meat	bone broth meat broth
22	boil seeds or nuts	seeds in general chenopods nutmeats in general hickory nutmeats nutshells nut oil
23	boil greens	greens
24	red ochre, deposition	red ochre

were coded present in only those collection units where the tools occurred in high densities, "high" being defined here with respect to local density norms.

For most activity groups, there are some collection units in which it is questionable whether the activity group of interest is represented by the artifact inventory. These cases occur where the function of one or more of the artifact types is ambiguous or where several activities could be indicated by the joint occurrence of several artifact types. In most cases, activity groups were not coded present in those units where their representation is questionable. Exeptions include Activity Groups 2, 5, 16, and 17, where the majority or all of the cases of occurrence are questionable. These activity groups were coded present in any collection unit where they possibly are represented.

The specific criteria used for each activity group in assessing whether it should be considered present or absent from collection units are shown in Table 65. The only activity group for which criteria more complex than those just described were used is activity group 3, hearth dumpings. Here, the effective residue is ash, and the diagnostic artifact types are several debris classes, all of which could have been used in making hearth retaining walls or in the construction of eath ovens: fire-cracked igneous rock, sandstone, and clay blobs. Since the multiple diagnostic artifact classes are functional *alternatives* and probably were used mutually exclusively from each other, they are not analogous to multiple artifact classes that indicate different subtasks of a general activity group and that are likely to co-occur. Consequently, tabulations were not made of the number of different kinds of hearth dumping materials found within collection units or the total number of different kinds of hearth dumping materials found within collection units as an indicator of general hearth activity and of the likelihood of whether ash was deposited in them in significant quantities. Rather, likelihood of significant ash deposition within a collection unit was evaluated by whether or not *one* of the diagnostic debris classes is present in the unit in high densities—"high" being defined in reference to local density norms.

Finally, it is necessary to consider the sample of collection units used in the analysis. The collection units appropriate for the analysis include (*a*) those within which soil chemistry samples were removed and analyzed for nutrient availabilities; (*b*) those where artifacts were apparently deposited in primary rather than secondary contexts; (*c*) those where no gully erosion was evident; and (*d*) those where no bulldozer disturbances are suspected. Primary deposition of the artifacts within a collection unit was required in order to ensure that the refuse generated by the activities indicated for the unit actually were deposited there, in contrast to locations of secondary deposition where only the tools used in the activities, and not the refuse generated by the activities, were deposited. Collection units where deposition was of a secondary nature *were* considered, however, when the activity groups under examination were those indicated by debris types that have direct effect, themselves, on the soil.

The final recompilation of the activity data from Crane site, for comparison with soil data, is shown in Table 66.

Chemical Nature of the Residues Deposited by Activity Groups

An analysis of the changes in the spectra of soil nutrient availabilities caused by different prehistoric activities at the Crane site and maintained over time could be done inductively, simply by defining associations between particular classes of use-areas and the particular kinds of nutrient found to be anomalous

TABLE 65
Attributes of the artifact inventories of surface collection units which were used in determining the most probable units in which activity groups occurred

Activity Group Number	Attribute
1	<u>two</u> or more "discrete" woodworking activities (see Table 63) which <u>definitely</u> occur in the collection unit
2	all cases, possible or definite occurrences of the activity
3	high densities of one of the following -- fire-cracked igneous rock, igneous cobbles, sandstone or clay blobs -- plus low, moderate, or high densities of another of them. Densities considered low, moderate, and high, respectively, are: 34-38, 39-44, ≥45 (fire-cracked igneous cobbles); 1, ≥2 (igneous cobbles); 5, 6-7, ≥8 (sandstone lumps); ≤300, 301-511, ≥512 (clay blobs).
4	<u>one</u> or more discrete bone working activities which <u>definitely</u> occur in the collection unit
5	all cases, possible or definite occurrences of the activity
6	high densities of burnt bone, ≥63 grams
7	high densities of unburnt bone ≥ 20 grams
8	occurrence in the surface collection unit of test excavations which revealed mollusk shell concentrations
9	all definite cases of occurrence of the activity
10	all definite cases of occurrence of both of the activities
11	all definite cases of occurrence of both of the activities
12	all definite cases of occurrence of the activity
13	all definite cases of occurrence of both of the activities
14	all definite cases of occurrence of both of the activities
15	all definite cases of occurrence of one or more of the activities

TABLE 65 (cont.)

Activity Group Number	Attribute
16	all possible or definite cases of occurrence of one or more of the activities which cluster separately
17	all possible or definite cases of occurrence of the activities
18	all definite cases of occurrence of one or more of the activities
19	all definite cases of occurrence of one or more of the activities
20	all definite cases of occurrence of one or more of the activities
21	all definite cases of occurrence of one or more of the activities
22	all definite cases of occurrence of one or more of the activities
23	all definite cases of occurrence of one or more of the activities
24	higher densities of red ochre, ≥ 2 lumps per collection unit

within their soils. The results would be site-specific, however, dependent on the particular soil conditions found at the Crane site. A more reasonable approach yielding results of more general applicability is a deductive one, involving:

1. The development of a model of the relative proportions of nutrient found in different residues left by different kinds of activities
2. The prediction of the particular nutrients that should be found in anomalously high concentrations within particular use-areas at Crane, based on the model
3. The comparison of empirical soil data from Crane with the predicted patterns of association
4. The attempt to explain deviations in the soil data from expectation by way of reference to both (a) general soil processes that can alter anthropic nutrient enrichments and their relative proportions over time (Chapter 4), and (b) site-specific soil conditions.

TABLE 66

Activity groups present or absent in various surface collection units at the Crane site*

Surface Collection Units	1	2	3	4	5	6	7	8	9	10	11	12	13	14	15	16	17	18	19	20	21	22	23	24
9	1	0	0	0	0	0	0	1	0	0	1	0	0	0	0	1	1	1	0	0	0	0	0	0
33	1	0	0	0	0	0	0	0	0	0	1	0	0	0	0	0	0	0	0	0	0	0	0	1
49	1	0	0	0	0	0	0	0	0	0	0	0	0	0	0	0	0	0	0	0	0	0	0	0
66	1	0	0	0	0	1	0	0	0	0	0	0	0	0	0	0	0	0	0	0	0	0	0	0
85	0	0	0	0	0	1	0	0	0	0	0	0	0	0	0	0	0	0	0	0	0	0	0	0
105	1	0	0	0	0	1	1	0	0	0	0	0	0	0	0	0	0	0	0	1	0	0	0	0
125	1	0	0	0	0	1	0	0	0	0	0	1	0	0	0	0	0	0	0	0	0	0	0	1
141	1	0	0	1	0	0	0	0	0	0	0	0	0	0	0	0	0	0	0	0	0	0	0	0
145	1	0	0	0	0	1	0	0	0	0	0	0	0	0	0	0	0	0	0	0	0	0	0	0
163	1	0	0	1	0	0	0	0	0	0	0	0	0	0	1	0	0	1	1	1	0	0	0	0
164	1	1	0	1	0	0	0	0	0	0	0	1	0	0	1	1	1	0	0	0	0	0	0	0
165	1	1	0	0	0	0	0	0	0	0	0	0	0	0	0	1	1	1	0	0	0	0	0	0
166	0	0	0	1	0	1	1	0	0	1	0	0	0	0	0	0	0	0	0	1	0	0	0	1
186	1	0	0	0	0	1	1	0	0	0	0	0	0	0	1	0	0	0	0	1	0	0	0	0
187	1	0	0	1	0	0	0	1	0	0	0	0	1	0	1	0	0	1	1	1	1	0	0	0
188	0	0	0	1	0	0	0	0	0	0	0	0	0	0	0	0	0	1	1	1	1	1	0	0
189	1	0	0	0	0	0	0	0	0	0	0	0	0	0	0	0	0	1	1	1	0	1	0	1
190	1	0	0	0	0	0	0	0	0	0	0	0	0	0	0	0	0	0	0	0	0	0	0	0
191	1	0	0	0	0	0	0	0	0	0	0	0	0	0	0	0	0	0	0	0	0	0	0	0
193	0	0	0	0	0	0	0	0	0	0	0	1	0	1	0	0	0	0	0	0	0	0	0	0
200	0	0	0	0	0	0	0	1	0	0	1	0	0	0	0	0	0	0	0	0	0	0	0	0
206	0	0	0	0	0	0	0	0	0	0	1	0	0	1	0	0	0	0	0	0	0	0	0	0
225	0	0	0	0	0	0	0	0	0	0	1	0	0	1	0	0	0	0	0	0	0	0	0	0
336	0	0	0	0	0	1	1	0	0	0	0	0	0	1	0	0	0	1	1	1	0	1	0	0
349	1	0	1	1	0	1	1	0	0	0	0	0	0	1	1	1	0	0	0	0	1	0	0	0

*Only those collection units from which one or more soil samples were removed are listed. Activity Groups are listed by their number, as shown in Table 63. 1's indicate presence of the Activity Group in the collection unit. 0's indicate absence.

TABLE 66 (cont.)

Surface Collection Units	1	2	3	4	5	6	7	8	9	10	11	12	13	14	15	16	17	18	19	20	21	22	23	24
350	0	0	1	1	0	1	1	0	0	0	0	0	0	0	0	0	1	1	1	1	1	0	0	1
351	0	0	1	1	0	1	1	0	0	0	0	0	0	0	0	0	1	1	1	1	1	0	0	1
364	1	0	1	1	0	1	1	0	0	0	0	0	0	0	0	0	1	1	1	1	0	1	0	0
365	0	0	1	0	0	1	1	0	0	0	0	0	0	0	0	0	0	0	0	1	0	0	0	1
372	0	0	0	0	0	0	0	0	0	1	0	0	0	0	0	0	0	0	0	0	0	0	0	0
379	0	0	0	0	0	1	1	0	0	0	0	0	0	1	0	0	1	1	1	1	0	1	1	0
381	1	0	1	0	0	1	1	0	0	0	0	0	0	0	0	0	0	0	0	1	0	0	0	0
393	1	1	0	0	0	1	0	0	0	0	0	0	1	1	1	1	1	0	1	1	1	1	1	0
394	1	0	1	1	0	1	1	0	0	0	0	0	0	1	0	0	0	0	0	1	0	0	0	0
396	1	0	1	1	0	1	1	0	0	0	0	0	0	0	0	0	0	0	0	1	0	0	0	0
399	0	0	1	0	0	1	1	0	0	0	0	0	1	0	0	0	0	0	0	1	0	0	0	0
410	1	1	0	1	0	1	0	0	0	1	0	0	1	1	1	0	0	0	0	1	0	1	0	0
413	1	0	1	0	0	1	1	0	0	0	0	0	0	0	0	0	0	0	0	1	0	0	0	1
414	1	0	0	1	0	1	1	1	0	0	0	0	1	0	0	0	0	0	0	1	0	0	0	0
415	0	0	1	1	0	1	1	0	0	0	0	0	0	0	0	0	0	0	0	1	0	0	0	1
427	0	0	0	0	0	1	0	0	0	0	0	0	0	0	0	0	0	0	0	0	0	0	0	0
431	0	0	1	1	0	1	1	0	0	0	0	0	0	0	0	0	0	1	1	1	1	0	0	1
444	0	0	0	0	0	0	0	0	0	0	0	0	0	0	0	0	0	0	0	0	0	0	0	1
447	1	0	0	1	0	0	0	0	0	0	0	0	0	0	0	0	0	0	0	1	0	0	0	0
448	1	0	0	0	0	0	0	0	0	0	0	0	0	0	0	0	0	0	0	0	0	0	0	0
449	1	0	0	0	0	1	0	0	0	0	0	0	0	0	0	0	0	0	0	0	0	0	0	0
450	1	0	0	1	0	1	0	0	0	0	0	0	0	0	0	0	0	0	0	0	0	0	0	0
463	0	0	0	0	0	1	0	0	0	0	0	0	0	0	0	0	0	0	0	0	0	0	0	0
467	1	0	0	0	0	1	0	0	0	0	0	0	0	0	0	0	0	0	0	0	0	0	0	0
468	0	0	0	0	0	0	0	0	0	0	0	0	1	1	0	0	0	0	0	0	0	0	0	0
469	1	0	0	0	0	0	0	0	0	0	0	0	0	0	0	0	0	1	1	0	0	0	0	0
480	0	0	0	0	0	0	0	0	0	0	0	0	1	0	0	0	0	0	0	0	0	0	0	0
481	1	0	0	0	0	0	0	0	0	0	0	0	0	0	0	0	1	0	1	1	0	0	0	1
489	0	0	0	0	0	0	0	0	0	0	0	0	0	0	0	0	0	0	0	0	0	0	0	1
490	0	0	0	0	0	0	0	0	0	0	0	0	0	0	0	0	0	0	0	0	0	0	0	1

TABLE 66 (cont.)

Surface Collection Units	1	2	3	4	5	6	7	8	9	10	11	12	13	14	15	16	17	18	19	20	21	22	23	24
496	1	0	0	0	0	0	0	0	0	0	0	0	0	0	0	0	0	0	0	0	0	0	0	0
497	1	0	0	1	0	0	0	0	0	0	0	0	0	0	0	0	0	0	0	0	1	0	0	0
498	1	0	0	0	0	0	0	0	0	0	0	0	0	0	0	0	1	0	1	1	0	1	1	0
508	1	0	0	1	0	0	0	0	0	0	0	0	0	0	0	0	0	0	0	0	1	0	0	1
515	1	0	0	0	0	0	0	0	0	0	0	0	1	0	0	0	0	0	0	0	0	0	0	1
621	0	0	0	1	0	0	0	0	0	0	0	0	0	0	0	0	0	0	0	0	0	0	0	0
622	0	0	0	0	0	0	0	0	0	0	0	1	0	0	0	0	0	0	0	0	0	0	0	0
624	0	0	0	0	0	0	0	0	0	0	0	0	0	0	1	0	1	0	1	1	0	0	0	0
625	1	0	0	1	0	0	0	0	0	0	0	0	0	0	0	0	0	0	0	0	0	0	0	0
627	0	0	0	0	0	0	0	0	0	0	0	1	0	0	0	0	0	0	0	0	0	0	0	0
628	1	0	0	0	0	0	0	0	0	0	0	0	0	0	0	0	0	0	0	0	0	0	0	0
631	0	0	0	0	0	0	0	0	0	0	0	1	0	0	1	1	0	0	0	0	0	0	0	0
632	1	0	0	0	0	0	0	0	0	0	0	0	0	0	0	0	0	1	0	1	0	1	0	0
633	0	0	0	0	0	0	0	0	0	0	0	1	0	0	0	0	0	0	0	0	0	0	0	0
635	0	0	0	1	0	0	0	0	0	0	0	0	0	0	0	0	0	1	1	1	0	0	0	0

By approaching the problem of the relationship between activity and anthropic soil anomalies in this manner, the effects of *general* phenomena, such as residue nutrient composition and soil processes, on soil characteristics can be made explicit and illustrated free of the effects of site-specific circumstances. In the inductive approach, both general and site-specific relationships would be compounded and confused. Thus, the deductive approach was chosen, and it became necessary to model the approximate relative proportions of nutrients found in different refuse materials left by different kinds of activities on the Crane site.

Data on the nutrient compositions of different refuse materials believed to have been deposited on the Crane site (Table 64) are summarized in Table 67. In Table 68, the data are reorganized, listing for each nutrient the refuse materials containing the nutrient, in decreasing order of concentration. This list of refuse materials according to their chemical attributes, together with the list of residues left by different activity groups (Table 64) constitute a predictive model specifying which kinds of use-areas should be expected to have more or less of a given nutrient—not taking into consideration soil processes or the intensity of activity in the different areas. They form a general model, applicable to all sites where the activities listed in Table 64 have occurred.

The data summarized in Table 67 are considered accurate enough to define on a *comparative, qualitative* scale (high, medium, low; Table 68) the approximate nutrient contents of the particular refuse materials deposited on Crane; they serve our purposes well. Readers wishing to use the data in an *absolute, quantitative* manner, requiring the precise compositions of the materials, in order to analyze the Crane data further than has been done here or to analyze data from other archeological sites, should be aware of certain limitations in the material compositions listed.

A first concern is the quantitative error involved in estimating the nutrient contents of the various raw materials. The concentration of Na within food classes may be true or somewhat overestimated, depending on the laboratory techniques used in making the assays. Determinations dating to the 1940s and 1950s were made with gravimetric methods and often included K in their estimation of the Na content of foods. These values for Na in Table 67 are most likely inflated. Statistics in Table 67, which are referenced to more recent literature, are probably more accurate for Na content. Determinations in most cases have been made with either the magnesium uranyl acetate method or colorimetric and flame photometric methods (Watt *et al.* 1963:165). Estimates of the magnesium contents of the food materials in Table 67 should be considered as only rough approximations. The content of Mg in foods has been studied only a short time in comparison to the contents of other nutrients, and most figures are based on only single or a few determinations (Watt *et al.* 1963:166).

A second concern with the quantitative accuracy of the data presented in Table 67 is the specific raw materials, their conditions, and their portions used in representing the average nutrient contents of *broad* categories of materials, such as greens, or nuts in general. In all cases, I have tried to use the very same species of plants or animals and the same kind of raw materials that predominate at Crane as representatives of the broader classes of resources. For example hickory nuts, white-oak acorns, hazelnuts, butternuts, and black walnuts were probably the main kinds of nuts utilized at Crane site. The nutrient contents of these nut species have been averaged to estimate the nutrient contents of nuts in general. In some cases, however, it was not possible to find data on the concentration of nutrients within the specific kinds of raw materials utilized at Crane, and values for similar materials had to be substituted (e.g., the use of domestic grapes as a substitute for winter grape). Raw materials for which others have had to be substituted to obtain estimates of nutrient concentration are pointed out in the footnotes of Table 67. Thus, it is suggested that the nutrient data in Table 67 are not amenable to extremely fine, quantitative analyses, should the reader wish to analyze the Crane data in greater detail than has been done here.

Similarly, the statistics reported in Table 67 for general food categories and raw materials, such as roots, fruits, greens, unburnt wood, mollusk shell, consider the nutrient contents found in *only those species represented at Crane*, and not necessarily those species that are found at other sites. Consequently, the

TABLE 67
Proportions of nutrients, ash, and fiber in raw materials used at the Crane site and left as residues

Raw Materials	N	P	K	Ca	Mg	Na	S	Cl	Total Ash	Fiber Carbohydrate
hickory meat	2.49	.36	?	Tr	.16	?	?	?	2.0	1.9
nutmeats in general[1]	3.30±1.58	.422±.128	.555±.211	.069±.121	.178±.016	.0025±.0005	.104	.023	2.6±.442	1.87±.35
nut oil	0	0	0	0	0	0	0	0	0	0
chenopod seeds[2]	2.01	.442	?	.128	?	?	?	?	3.8	7.8
seeds in general[3]	1.99±.58	.371±.064	.402±.0597	.0586±.0469	.138±.0318	.0025±.0005	.0865	?	2.25±1.03	2.00±.3
roots, tubers[4]	2.44±.075	.0432±.0161	.340±.110	.0322±.0156	.0224±.0076	.0255±.0227	.0376±.0180	.0518±.0242	.875±.198	.975±.483
fruits powdered[5]	.16±.045	.022±.005	.24±.099	.030±.003	.019±.015	.001±0	.005	.011	.7±.2	2.8±1.838
fruits not powdered[6]	.169±.054	.019±.005	.161±.019	.019±.005	.0114±.002	.002±.0008	.009±.0044	.006±.008	.508±.102	1.0±.675
greens[7]	.432±.081	.054±.016	.397±.158	.121±.088	.036	.021±.020	?	?	1.68±.60	1.125±.236
deer[8]	3.36	.249	.364	.010	.033	.086	.321	.089	1.0	0
small mammal meat[9]	3.35	.352	.385	.020	.030	.043	.347	.108	0	0

turtle meat	3.17	?	?	?	?	?	?	?	1.2	0
mollusk meat[10]	1.755±.375	.162±.042	.114±.094	.116±.067	.00006	.13±.01	?	?	1.6±.5	0
unburnt wood[11]	.368±.0516	.0415±.0247	.18±.0354	.165±.099	.031±.030	.021±.0169	.014±.004	.005±.001	.4	>50%
ash of wood[12]	Tr	9.36	40.59	37.20	6.99	4.74	Tr	1.13	1.00	?
bark[13]	.46±.094	.03±.006	.23±.03	3.1±.252	.327±.038	?	.064	?	?	>50%
dry, unboiled bone without marrow[14]	4.76±.048	3.933±.004	.053±.001	25.83±.014	.422±.009	.707±.014	.0008	.0745±.004	--	?
dry, boiled bone without marrow[15]	0.	32.34±.665	.291±.100	56.12±.181	.954±.232	1.229±.041	?	.0563±.038	--	?
broth of bone[16]	.940	.009	.048	.008	.002	.122	?	.122	?	0
broth of meat[17]	Tr	.047	.181	.002	.001	.031	Tr	.031	?	0
meat drippings[18]	Tr	.013	.004	.0008	Tr	.005	.009	.002	?	0
brains[19]	1.66	.312	.219	.010	.013	.125	.132	.167	1.4	0
urine[20]	2.11	.044	.15	.015	.008	.48	.0599	.6	--	--
nut shells[21]	.15	.002	.08	.09	.009	?	.028	?	?	--
mollusk shells[22]	Tr	.0039	.00256±.00307	35.0	.00423±.00295	.184±.0122	Tr	Tr	--	--
red ochre[23]	--	maintains PO_4	--	--	--	--	maintains SO_4	--	--	--

TABLE 67 (cont.)

[1] N = hickory, white oak, hazelnut, butternut, black walnut; Asch et al. (1972)
Fiber = ibid.   Ash = ibid.
P,K,Ca,Mg,Na = hickory, hazelnut, black walnut; Watt et al. (1962)
S,Cl- = black walnut; McCance and Widdowson (1947)

[2] McCance and Widdowson (1947)

[3] fiber, ash, N,P,Ca = chenopods, wheat, rye, sorghum; White et al. (1955), Watt et al. (1963)
K,Mg, Na = wheat, rye, sorghum; Watt et al. (1963)
S = shredded wheat; McCance and Widdowson (1947)

[4] beets, carrots, leeks, onions, parsnips, potatoes, radishes, turnips N,P,K,Ca,Mg,Na, ash, fiber; Watt et al. (1963)
S,Cl = McCance and Widdowson (1947)

[5] persimmon and blackberry (as a substitute for hackberry) (Watt et al. 1965). These fruits have high fiber contents and were dried and powdered in food preparations as well as used fresh aboriginally (Zawacki and Hausfater 1969).

[6] value for sour red cherry (as a substitute for black cherry, choke cherry), domestic grapes (as a substitute for winter grape), prune-type plums (as a subsistute for wild plums), gooseberries, strawberries are all given in Watt et al. (1963). Values for black and white mulberries (as a substitute for red mulberries) are given in Leung et al. (1972). Nutrient values for all 6 kinds of fruits were averaged. None of these fruits were powdered aboriginally in food preparation, although some were dried.

[7] amaranth, dock, dandelion, cress, poke; Watt et al. (1963)

[8] K,Na,S,Cl from McCance and Widdowson (1947)

[9] rabbit; Mg,S,Cl from roasted specimens and inflated; McCance and Widdowson (1947).

[10] All nutrient values except Mg are for clams, hard and soft, of unspecified variety, edible meat parts only, and without liquid, as given by Watt et al. (1963); and for freshwater mussels, as given by Leung et al. (1972). Mg values are for the entire soft parts of Anodanta sp. (Segar et al. 1972), which is found in the lower Illinois River and was exploited by the Middle Woodland occupants of the Apple Creek Site (Parmalee et al. 1972).

[11] N is for oak; Schorger (1926:63)
P,K,Ca,Mg,Na,S,Cl are for white oak, hickory (Schorger 1926:51).
Ash content is for oak; Browning (1963:74), Buchanan (1963:355), Wise (1952:657).
The cellulose content of wood is taken as a minimal estimate of fiber content; Browning (1963).

[12] Ash elements in their relative proportions found in unburnt oak and hickory.

[13] bark of red oak; Duvigneaud and Denaeyer-de Smet (1970).

[14] Values for all elements except S are taken from Dallemagne and Richelle (1973:26), adjusted for the fat content of undemineralized bone. A fat content of 3.27% was used based upon the total percentage of lipids in demineralized bone (4.7%, Shapiro 1973:125) and the weight ratio of demineralized bone to undemineralized bone (69.66%, Eastoe 1956:88). Sulfur content is that found in the micropolysaccharide protein complexes (1.07% SO_4; Eastoe 1973:98) which comprise 0.24% of air dry undemineralized bone (Eastoe 1973:88).

[15] Values for mineral bone leached of all organic matter as reported by Dallegmagne and Richelle (1973:25).

[16] Nutrients extracted when simmering crushed veal bone 7 hours, as a percentage of the original weight of the bone.

[17] Difference in nutrient values of lean raw beef and lean boiled beef; McCance and Widdowson (1947).

[18] drippings from beef; McCance and Widdowson (1947).

[19] N,P,K,Ca,Na, cow, calf, hog, sheep brain; Watts et al. (1963). Mg,S,Cl; beef brain only; McCance and Widdowson (1947).

[20] Nutrient values are the average of those reported by Bard (1961:333) and Bell et al. (1965). Proportions for elemental sulfur and phosphorus have been adjusted from those for sulfate and phosphate.

[21] Values for nutrient concentrations in nut shells are taken to be approximated by nutrient concentrations in the heartwood of the trunks of red oak (Duvigneaud and Denaeyer-de Smet 1970:212). The genesis of both heartwood and nut shells involves the accumulation of cellulose and lignin and a porportional decrease in the concentration of ash elements.

[22] All ash element concentrations are based on those found in Anodanta sp. (Segar et al. 1971); Mg,Na, and K concentrations are additionally based on those found in Elliptio crassidens and Quadrula pustulosa (Nelson et al. 1966). Nitrogen occurs in only trace concentrations as part of the organic framework of shell (Wilbur 1960:17). Nutrient concentrations in the samples examined above apparently are not biased by the ages or sex of the clams, at least for Ca and Na (Saville and Sturett 1974). All three species are found in the Illinois River and were exploited by Middle Woodland Indians at the Apple Creek Site, lower Illinois Valley (Parmalee et al. 1972:6).

[23] Red ochre lumps are yellow limonite concretions which have had their iron contents oxidized (by either natural weathering or by purposeful firing of the limonite nodules) giving them their characteristic red color and friable nature. Chemically, the oxidized portions of such nodules are iron sesquioxide (Fe_2O_3) while any reduced portions are a hydrous oxide of iron (goethite, $HFeO_2$; Winters 1969:26). The iron hydrous oxide content is insoluble to the soil water solution but nevertheless is capable of precipitating and fixing phosphate and sulfate ions (Brady 1974:452-453; 463-465). Solecki (1950) noted that red ochre nodules found in archaeological contexts of high phosphate concentration and under acidic soil conditions had very high phsophate contents.

TABLE 68
Ordering of refuse materials by their relative concentrations of given nutrients

Proportion of Nutrient in Refuse Material*	Refuse Material	Subjective Level of Nutrient in Refuse Material
Nitrogen		
4.76	dry unboiled bone	H
3.36	deer meat	H
3.35	small mammal meat	H
3.30	nut meats in general	H
3.17	turtle meat	H
2.49	hickory meats	M
2.11	urine	M
2.01	chenopods	M
1.99	seeds	M
1.75	mollusk meat	M
1.66	brains	M
.940	broth of bone	M
.460	bark	L
.432	greens	L
.368	unburnt wood	L
.244	roots, tubers	L
.169	fruits not powdered	L
.16	fruits powdered	L
.15	nut shells	L
Tr	meat drippings	L
Tr	broth of meat	L
Tr	ash of wood	L
Tr	mollusk shell	L
0	nut oil	L
0	dry boiled bone	L
Phosphorus		
32.34	dry boiled bone	H

TABLE 68 (cont.)

Proportion of Nutrient in Refuse Material*	Refuse Material	Subjective Level of Nutrient in Refuse Material
Phosphorus (cont.)		
9.36	ash of wood	H
3.93	dry unboiled bone	H
.442,	chenopods	M
.422	nuts in general	M
.371	seeds in general	M
.36	hickory meat	M
.352	small mammal meat	M
.312	brains	M
.249	deer meat	M
.162	mollusk meat	M
.054	greens	L
.047	broth of meat	L
.044	urine	L
.043	roots, tubers	L
.042	unburnt wood	L
.03	bark	L
.022	fruits powdered	L
.019	fruits not powdered	L
.013	meat drippings	L
.009	broth of bone	L
.0039	mollusk shell	L
.002	nut shells	L
0.	nut oil	L
?	turtle meat	?
Potassium		
40.59	ash of wood	H
.555	nut meats in general	H
.402	seeds in general	H
.397	greens	H
.385	small mammal meat	H

TABLE 68 (cont.)

Proportion of Nutrient in Refuse Material*	Refuse Material	Subjective Level of Nutrient in Refuse Material
Potassium (cont.)		
.364	deer	H
.340	roots, tubers	H
.291	dry, boiled bone	H
.24	fruits powdered	M
.23	bark	M
.219	brains	M
.181	broth of meat	M
.181	unburnt wood	M
.161	fruits not powdered	M
.15	urine	M
.114	mollusk meat	M
.08	nut shells	L
.053	dry unboiled bone	L
.048	broth of bone	L
.004	meat drippings	L
.0025	mollusk shell	L
0.	nut oil	L
?	hickory meats	?
?	chenopods, seeds	?
?	turtle meat	?
Calcium		
56.12	dry boiled bone	H
37.20	ash of wood	H
35.0	mollusk shell	H
25.83	dry unboiled bone	H
3.1	bark	H
.165	unburnt wood	M
.128	chenopod seeds	M
.121	greens	M
.116	mollusk meat	M

TABLE 68 (cont.)

Proportion of Nutrient in Refuse Material*	Refuse Material	Subjective Level of Nutrient in Refuse Material
Calcium (cont.)		
.09	nut shells	M
.069	nut meats in general	L
.0586	seeds in general	L
.0322	roots, tubers	L
.030	fruits powdered	L
.02	small mammal meat	L
.019	fruits not powdered	L
.015	urine	L
.01	deer meat	L
.01	brains	L
.008	broth of bone	L
.002	broth of meat	L
.0008	meat drippings	L
Tr	hickory meat	L
0	nut oil	L
?	turtle meat	?
Magnesium		
6.99	ash of wood	H
.954	dry boiled bone	H
.422	dry unboiled bone	H
.327	bark	H
.178	nut meats in general	M
.160	hickory meat	M
.138	seeds in general	M
.12-.48?	mollusk shell	M
.036	greens	L
.033	deer meat	L
.031	unburnt wood	L
.030	small mammal meat	L
.0224	roots, tubers	L

TABLE 68 (cont.)

Proportion of Nutrient in Refuse Material*	Refuse Material	Subjective Level of Nutrient in Refuse Material
Sodium (cont.)		
.0025	nut meats in general	L
.0025	seeds in general	L
.002	fruits not powdered	L
.001	fruits powdered	L
0.	nut oil	L
?	hickory nut meat	?
?	chenopod seeds	?
?	turtle meat	?
?	bark	?
?	nutshells	?
Sulfur		
.347	small mammal meat	H
.321	deer meat	H
.132	brains	H
.104	nut meats in general	H
.0865	seeds in general	M
.064	bark	M
.0599	urine	M
.0376	roots, tubers	M
.028	nut shells	M
.014	unburnt wood	L
.009	meat drippings	L
.009	fruits not powdered	L
.005	fruits powdered	L
.0008	dry unboiled bone	L
Tr	ash of wood	L
Tr	broth of meat	L
Tr	mollusk shell	L
0.	nut oil	L

TABLE 68 (cont.)

Proportion of Nutrient in Refuse Material*	Refuse Material	Subjective Level of Nutrient in Refuse Material
Magnesium (cont.)		
.019	fruits powdered	L
.013	brains	L
.0114	fruits not powdered	L
.009	nut shells	L
.008	urine	L
.004	mollusk shell	L
.002	broth of bone	L
.001	broth of meat	L
.00086	mollusk meat	L
Tr	meat drippings	L
0.	nut oil	L
?	chenopod seeds	?
?	turtle meat	?
?	mollusk meat	?
Sodium		
4.74	ash of wood	H
1.229	dry boiled bone	H
.707	dry unboiled bone	H
.480	urine	H
.184	mollusk shell	M
.13	mollusk meat	M
.125	brains	M
.122	broth of bone	M
.086	deer meat	M
.043	small mammal meat	L
.031	broth of meat	L
.0255	roots, tubers	L
.021	greens	L
.021	unburnt wood	L
.005	meat drippings	L

TABLE 68 (cont.)

Proportion of Nutrient in Refuse Material*	Refuse Material	Subjective Level of Nutrient in Refuse Material
Sulfur (cont.)		
?	hickory meat	?
?	chenopod seeds	?
?	greens	?
?	turtle meat	?
?	mollusk meat	?
?	dry boiled bone	?
?	broth of bone	?

*Likelihood of refuse material causing soil anomaly.

data should not be subjected to extremely fine, quantitative analysis when analyzing pedological and archeological data from other sites either.

The portion and condition of the food materials for which nutrient data are reported in Table 67 also influences the accuracy of the data and the manner in which it can be handled. For food items, the nutrient contents of only the *edible* portions have been reported. The refuse left on Crane or other sites during the processing of such foods, however, may have included both edible and nonedible portions, and the nutrient spectra given in Table 67 may be somewhat misleading in its partial nature. Similarly, the conditions of the food items for which nutrient contents are reported may be somewhat different than the conditions in which food refuse were deposited on Crane or other sites. For example, dried fish and meat have fewer nutrients than fresh raw fish and meat even when placed on a comparable scale of moisture content. The nutrient values reported in Table 67 are for fresh, raw meat, alone. Also, the nutrient composition of game meat depends on the season of kill and the amount of fat on the animal at that time. Such variability in the nutrient content of meat refuse is not taken into consideration in Table 67.

In summary, the nutrient data shown in Table 67 and reorganized in Table 68, along with Table 64, provide a model for assessing in a qualitative manner which of the refuse materials deposited at the Crane site or other archeological sites have more or less of specific nutrients, and which kinds of use-areas should have greater or lesser enrichments of these nutrients. Any quantitative use of the nutrient data to analyze the Crane site further than will be done here, or to analyze the refuse and soils of other archeological sites should be undertaken with great care for the qualifications discussed previously.

Definition of Soil Anomalies within the Crane Site

The third and final set of data that had to be compiled and organized for the analysis of the spectra of nutrients found in different use-areas is the soil chemistry data, itself. The particular data that were used are the availabilities of nutrients found in different locations along Transects I–V and within different levels (Chapter 6). These data had to be transformed in several ways to be made comparable with activity groups data.

First, it had to be decided how to define archeologically significant soil variation apart from natural and historically caused soil variability. The soils of Crane site can be subdivided into a large number of spatial strata which differ in: (a) the texture and presumably the chemical composition of their parent materials, (b) their degree of profile development, and (c) the degree to which their chemistry has been altered by historic land-use patterns (Chapter 7). A number of analytical approaches could have been used to sort out these sources of spatial variability from those due to prehistoric human activity, but all but one of them were found to have major drawbacks, either on general principles or with respect to the specific nature of the Crane data. I shall briefly describe these several alternative approaches and their drawbacks to help explain why archeologically significant soil anomalies have been defined in the particular manner they have.

First, the effects of parent material, profile development, and historic land-use on soil nutrient availability could have been held constant by performing separate analyses within each area of Crane homogeneous with respect to these variables. Archeologically significant anomalies could have been defined within such strata simply by the comparison of nutrient concentrations in the soils of intensely used areas with those of unused or slightly used areas, subtracting the latter from the former. Association analyses of particular kinds of nutrient anomalies and use-areas could have proceeded from there. This approach would have been optimal, but not enough use-areas of different kinds occurred within individual homogeneous strata to make this approach possible.

A second means by which archeologically significant anomalies could have been defined, holding parent material, profile development, and historic land-use constant would have involved two parallel transects of soil samples, one on-site and crosscutting different use-areas, and the other off-site, but both crosscutting natural and historic spatial variation perpendicularly and in *synchrony*. Nutrient values from the off-site locations could have been subtracted from nutrient values of *corresponding* locations onsite, so as to hold natural and historic soil variations more or less constant and to define anomalies largely of archeological significance. This approach is restricted to sites where natural and historic soil variation run largely in one direction. It could have been used with some success at Crane, had a more complete transect of soil samples been obtained in off-site locations to the east of the site.

468 8. *Soil Alterations Produced by Prehistoric Human Activities*

A third, and less satisfactory method which could have been used to define archeologically significant anomalies is spatial filtering. In a manner similar to that described for the manipulation of resistivity data in Chapter 2, low-frequency nutrient trends could have been generated from a raw data series of nutrient values at the depth of archeological interest and subtracted from them in order to define high-frequency anomalies. The low-frequency nutrient trends presumably would have represented broad-scale natural and historic soil variation, whereas the positive anomalies from these trends would have represented archeologically significant soil variability. A major drawback to this approach, however, is that the smoothed series would have included information on not only natural and historically caused soil variation, but also archeologically significant variation (e.g., the gentle rise and fall of nutrient availability around major centers of activity, such as the house at Crane), which would have been lost in the subtraction of the smoothed from the raw data series. This is the same kind of information loss that occurs when spatial filtering techniques are used on resistivity data that have not first been interpreted by the Barnes Layer method and which contain information on both archeological and natural–historic soil variation.

A fourth method for defining archeologically significant anomalies, and that which was used to analyze the Crane data, requires the application of both stratigraphic and spatial filtering techniques, and is analogous to the methods of resistivity data interpretation advocated in Chapter 2. First, at each location where soil samples were taken, and for each nutrient of interest, the availability of the nutrient within submidden, natural soil horizons (Barnes Layer BL5–4) was subtracted from its availability within layers encompassing archeological deposits (BL2–1 or BL3–2). The resulting statistic is presumed to be more or less free of the effects of variation in parent material minerology on nutrient availability and allowed the comparison of nutrient availability data from use-areas in portions of Crane that differ in their parent materials. All three Barnes Layers encompass parent materials of a single type—the upper-member alluvium (see Chapter 7)—making it possible to cancel out its effect on the data by the subtraction procedure. Whether nutrient data for BL3–2 or BL2–1 was used to represent prehistorically disturbed soil in particular locations depended on the depth to which midden deposits extended, there. Data from layer BL3–2 was used whenever possible, where midden deposits extended through it, for two reasons: (*a*) It encompasses more soil, which facilitated more accurate laboratory determinations of nutrient availabilities; and (*b*) it had not been disturbed by chisel plowing as had BL2–1. Chisel plowing presumably affected only the structural nature of the soils within BL2–1, but there is the chance that in some areas, it may have introduced plowzone soils into BL2–1 and affected the nutrient status of the soils in BL2–1. This source of chemical contamination, if it occurred, had to be avoided as much as possible.

The difference statistic created by subtracting nutrient availabilities in BL5–4 from those in BL2–1 or LB3–2 is assumed to be free of not only the effect of parent material chemistry on nutrient availability, but also to contain less information on the effect of historic land-use patterns on nutrient availability than do the original nutrient data. As shown in Chapter 7, the various nutrient enrichments or losses that have resulted from liming, fertilizing, and cropping extend well below plowzone, into the natural soil. Subtraction of nutrient availabilities within the archeological layer from those within the natural layer removed from the data those historical enrichments or losses common to both horizons. Historically caused nutrient availability was not removed completely in those cases where the enrichments or losses are not equivalent in the two layers.

After subtracting the availability of each nutrient within BL5–4 from their availabilities in BL2–1 or BL3–2 at each location along Tansects I–V, the resulting difference statistics then were filtered spatially to separate archeologically significant anomalies from the low-frequency, natural trends reflecting spatial variation in: (*a*) the degree of profile development, (*b*) the amount of clay eluviation and nutrient leaching, and (*c*) that portion of the cation exchange capacities of the Crane soils attributable to clays. Some archeologically significant information certainly was lost in the process of filtering the difference statistics, but not nearly as much as would have been, had the data not been manipulated stratigraphically, first. By initially removing some naturally and historically caused soil variation by stratigraphic filtering, the variability that was left in the data and that was due to nonprehistoric factors was reduced. This made it possible to filter the data with lower-frequency filters and to remove less information on prehistoric anthropic nutrient availabilities occurring within the moderate to low frequency range when defining archeologically significant anomalies.

The methods used to filter the soil nutrient data spatially are not those that were described in Chapter 2, although they certainly could have been applied. Rather than calculating a smoothed, low-frequency data series for each series of difference statistics along each subtransect and then subtracting the difference statistics from the smoothed series, I used a straight-edge to find general trends through the data and tabulated the *presence or absence* of positive anomalies at each location where chemistry data existed. A filter width of approximately 100 stations (50 m) was used. The use of a hand-placed straight-edge rather than a calculated low-frequency series to judge anomalies was felt appropriate because (*a*) the desired filter width was very wide, approximating a linear fit to the data; (*b*) only the presence and absence of positive anomalies had to be noted, not their magnitude (i.e., the deviation of the difference statistic from the smoothed series); (*c*) the method was very quick; and (*d*) the observations to be filtered are not uniformly spaced, which would have made the designing of an accurate filter operator difficult. Only the presence and absence of nutrient anomalies, and not their magnitudes, were recorded because our concern is with the *spectra* of anomalous nutrient availabilities found in *generally high or low* concentrations within use-areas rather than the *absolute magnitudes* of the anomalous availabilities.

The final form of organization of the soil nutrient data to be compared with data on prehistoric activity is presented in Table 69, as a matrix of presence and absence states of anomalous nutrient availabilities for various locations. Several aspects of these data should be mentioned. First, the nutrients, chloride and nitrate, are not included for analysis. Chloride and nitrate concentrations within soils are poor indicators of the functions that areas of earthen archeological sites served prehistorically, even though the kinds of refuse deposited in them may have been distinguished in their chloride and nitrogen contents (Table 67). Chloride and nitrate enrichments to a soil are easily leached from it. This circumstance relates to the fact that anions, in general, are adsorbed only weakly by clay micelles and also to the lack in soils of any mechanism for permanently fixing chloride or nitrate ions. At the Crane site, these ions were found to be present in measurable quantities in less than half of the soil samples removed from the site.

Second, the manner in which the presence or absence of nutrient anomalies within collection units have been defined is congruent to the manner in which the presence or absence of nutrient enrichments to the soils of collection units have been reconstructed. In both cases, presence and absence have been assessed with respect to local norms. Widespread, low-density scatters of artifacts were ignored when coding the presence of artifact types and activity groups within collection units and when reconstructing the residues and nutrients enriched to their soils. Nutrient availabilities of low spatial frequency were filtered from the soil nutrient data when defining archeologically significant nutrient availability anomalies. The analogous manner of coding the presence or absence of these phenomena was necessary if the degree of association between activities and nutrient anomalies was to be measured with as little distortion as possible.

Results I: Maintenance of the Spectra of Anomalous Nutrient Availabilities within Use-Areas

Given the structure of the archeological and pedological data described previously, it is possible to investigate two questions:

1. Have the spectra of nutrients that were enriched to the soils within particular use-areas been maintained in them over time, that is, do those nutrients that we would predict should occur in anomalously high concentrations within particular use-areas, based on the spectra of nutrients in the refuse materials

TABLE 69
Locations of archeologically significant nutrient anomalies within the Crane site*

Soil Sampling Station	Collection unit in which soil sampling station occurs	Nutrient occurring or not occurring in anomalous concentrations in the soils at these stations							
		pH	PO_4	Mg	Ca	Na	K	NH_4	SO_4
5915	9	0	0	0	0	1	1	n.d.	0
4540	33	0	1	0	0	1	n.d.	n.d.	0
5954	49	0	0	0	1	0	1	0	0
4568	66	1	1	1	1	n.d.	1	1	0
5968	66	1	1	1	1	1	1	1	0
4579	85	0	n.d.	n.d.	n.d.	n.d.	n.d.	n.d.	n.d.
5977	85	1	1	1	1	0	1	0	n.d.
5987	105	0	0	0	0	1	0	1	1
4605	125	0	1	0	0	0	0	0	n.d.
455	141	1	1	1	0	1	1	0	1
4614	145	1	0	0	0	1	1	0	0
6012	145	1	1	1	1	0	1	1	0
65	163	0	n.d.	n.d.	n.d.	n.d.	n.d.	n.d.	n.d.
73	163	0	1	0	0	0	0	0	0
468	163	0	1	1	0	1	0	0	0
76	164	1	1	0	1	1	1	1	0
82	164	1	1	0	1	1	0	0	1
479	164	1	1	0	1	1	0	1	1
487	165	1	1	0	1	0	1	1	1
6028	166	1	1	1	1	1	1	n.d.	1
93	186	0	1	1	1	0	0	1	1
99	186	0	1	1	0	0	0	1	0
100	187	0	1	1	0	n.d.	1	1	0
103	187	0	1	1	1	0	1	0	0
499	187	0	1	1	1	0	1	0	1
511	187	0	0	0	0	1	1	1	0

* 1s indicate presence of a nutrient anomaly at the location. 0s indicate absence. n.d. = no data

TABLE 69 (cont.)

Soil Sampling Station	Collection unit in which soil sampling station occurs	Nutrient occurring or not occurring in anomalous concentrations in the soils at these stations							
		pH	PO_4	Mg	Ca	Na	K	NH_4	SO_4
4635	147	1	1	0	1	0	1	1	n.d.
6042	187	0	1	1	1	1	1	1	0
121	188	1	1	1	1	1	1	0	1
524	188	0	1	1	1	0	1	0	1
530	189	1	1	1	1	n.d.	1	1	1
145	190	1	1	1	1	0	1	0	0
542	190	0	1	1	1	n.d.	1	0	0
550	190	1	1	0	1	0	1	n.d.	0
162	191	1	1	0	1	1	1	1	0
563	191	0	1	0	0	0	0	1	1
181	193	0	1	1	1	0	0	1	0
266	200	1	0	1	1	0	1	1	0
669	200	1	1	1	1	n.d.	0	1	0
4651	206	1	1	1	1	1	1	1	0
4659	206	0	1	1	1	0	1	0	1
6051	206	0	0	0	0	0	0	0	1
4668	225	0	0	1	0	0	1	1	0
6068	225	0	0	0	0	1	1	1	1
4783	336	0	1	1	1	1	1	1	1
4790	336	1	1	n.d.	1	1	1	1	1
6181	336	1	1	1	0	0	1	n.d.	0
2240	349	0	1	1	0	0	1	0	1
2223	350	1	1	1	1	1	1	1	1
2231	350	0	1	0	0	n.d.	0	1	0
2633	350	0	1	1	1	1	1	0	0
2619	351	1	1	1	1	1	0	1	1
6191	351	1	1	0	1	1	1	0	n.d.

TABLE 69 (cont.)

Soil Sampling Station	Collection unit in which soil sampling station occurs	Nutrient occurring or not occurring in anomalous concentrations in the soils at these stations							
		pH	PO_4	Mg	Ca	Na	K	NH_4	SO_4
6202	351	0	1	0	0	1	0	0	0
4804	364	1	1	1	1	1	1	1	1
4815	364	1	1	0	1	1	1	1	1
2207	365	1	1	0	1	1	0	0	1
2213	365	1	1	1	1	1	0	1	0
2216	365	1	1	0	1	n.d.	1	n.d.	1
2606	365	0	1	0	0	n.d.	0	0	n.d.
7373	372	0	0	0	0	0	0	0	0
7374	372	1	0	0	0	1	0	0	0
4827	379	1	1	0	1	1	1	1	0
6216	379	1	1	1	1	1	1	n.d.	0
6226	379	1	1	n.d.	1	0	0	1	1
2192	381	0	1	0	1	1	1	1	1
2595	381	1	1	0	1	n.d.	1	1	0
4841	393	0	1	0	0	1	0	1	1
4833	394	0	1	1	1	1	1	n.d.	0
6240	394	0	1	0	0	0	0	1	0
2180	396	1	1	1	1	0	1	1	n.d.
2183	396	0	1	0	1	1	1	1	0
2185	396	0	1	0	0	n.d.	0	1	0
7021	399	1	1	1	1	1	1	1	n.d.
7321	399	0	1	0	1	1	1	1	0
7326	399	0	1	0	1	0	1	1	1
4846	401	0	1	1	1	1	0	1	1
6248	410	0	1	1	1	1	1	1	1
2170	413	1	1	1	1	1	0	1	1
2161	414	0	1	0	1	1	1	n.d.	0
2569	414	1	1	1	1	1	0	0	0
6999	415	1	1	1	1	n.d.	0	1	n.d.

TABLE 69 (cont.)

Soil Sampling Station	Collection unit in which soil sampling station occurs	Nutrient occurring or not occurring in anomalous concentrations in the soils at these stations							
		pH	PO_4	Mg	Ca	Na	K	NH_4	SO_4
7010	415	1	1	0	0	n.d.	1	1	0
7301	415	1	1	0	1	1	1	1	n.d.
7307	415	1	1	1	1	1	1	1	1
4857	427	1	1	1	1	1	0	0	0
4865	427	1	0	1	0	1	1	0	1
6265	427	0	1	0	0	1	0	1	1
2153	431	0	1	0	1	1	1	0	1
2550	431	1	1	0	1	1	1	1	n.d.
6982	431	1	1	1	1	1	0	1	1
6985	431	1	1	1	1	1	1	n.d.	0
7289	431	1	1	1	1	1	1	1	0
4877	444	0	0	0	0	0	0	n.d.	0
6964	447	1	1	0	1	1	1	1	0
6966	447	1	1	1	1	1	1	1	0
6971	447	0	1	0	1	1	0	1	1
7275	448	0	1	0	1	1	1	0	n.d.
2137	449	1	1	1	1	1	1	1	0
2539	449	0	1	1	1	1	0	0	1
2129	450	1	1	0	0	0	1	1	0
2225	450	0	0	0	1	0	1	0	1
2519	450	0	1	1	0	0	0	0	0
2528	450	1	1	0	0	1	1	1	0
7253	463	0	1	0	1	0	0	0	1
2118	467	0	1	1	1	1	0	0	1
2105	468	1	1	0	0	1	1	1	n.d.
2109	468	0	0	0	0	0	1	0	1
2111	468	0	0	0	0	0	0	0	0
2098	469	0	0	0	1	1	0	0	0
2497	469	0	1	0	0	1	0	0	1
2503	469	0	0	0	0	n.d.	0	1	0

TABLE 69 (cont.)

Soil Sampling Station	Collection unit in which soil sampling station occurs	Nutrient occurring or not occurring in anomalous concentrations in the soils at these stations							
		pH	PO_4	Mg	Ca	Na	K	NH_4	SO_4
6931	480	0	1	1	1	0	1	0	1
7241	481	0	1	1	0	0	1	0	1
2067	489	1	1	n.d.	1	1	0	0	0
2071	489	1	0	1	1	0	1	0	1
2073	489	1	1	0	1	0	0	1	0
2473	489	1	0	1	1	0	1	1	0
2461	490	0	0	0	0	1	0	1	1
2462	490	0	0	1	1	0	1	0	0
4910	496	0	0	0	0	0	0	1	n.d.
6309	496	0	1	0	1	1	0	1	1
6914	497	1	0	0	0	0	0	0	1
6224	497	0	0	1	0	1	1	1	0
7212	497	1	1	0	0	0	0	1	0
7224	498	0	0	1	1	1	1	0	1
2053	508	0	0	0	1	1	1	1	1
2056	508	0	1	0	n.d.	n.d.	0	1	0
2059	508	0	1	1	1	1	0	n.d.	n.d.
2060	508	0	1	1	1	1	0	0	1
2453	508	0	1	1	1	0	0	0	0
4928	515	0	0	0	0	0	0	0	1
6318	515	0	0	0	0	n.d.	0	0	0
7204	515	0	0	1	0	1	1	0	0
3507	621	1	1	0	1	0	0	0	1
3501	621	1	1	0	1	0	1	1	n.d.
3515	621	1	1	0	1	0	0	n.d.	0
3520	622	0	0	1	1	1	0	0	1
3524	622	0	1	1	1	1	1	1	1
4327	622	1	1	1	1	1	1	1	1
4330	624	1	1	1	1	0	0	1	1
3542	625	0	1	0	1	1	1	0	n.d.

TABLE 69 (cont.)

Soil Sampling Station	Collection unit in which soil sampling station occurs	Nutrient occurring or not occurring in anomalous concentrations in the soils at these stations							
		pH	PO_4	Mg	Ca	Na	K	NH_4	SO_4
3551	625	1	1	0	1	0	0	1	1
4340	625	0	1	1	1	1	1	1	0
4345	625	0	1	1	1	0	0	0	1
4348	625	1	1	1	1	0	0	1	1
4351	625	1	0	0	0	0	1	1	0
3564	627	0	1	1	1	1	0	1	1
3575	627	0	0	1	1	1	0	0	1
4381	628	0	1	1	0	1	0	0	0
3457	631	0	1	1	1	1	1	0	0
3461	631	1	1	1	0	n.d.	1	1	0
4259	631	1	1	1	1	1	1	1	1
4261	631	1	1	1	0	1	1	n.d.	1
4263	631	1	1	1	n.d.	1	1	1	0
4264	631	1	1	1	1	1	0	1	1
3469	632	1	1	1	0	1	1	n.d.	0
3474	632	0	0	0	0	0	0	0	1
4268	632	1	1	1	1	0	0	0	0
4274	632	0	0	0	0	0	1	1	1
4277	632	1	0	0	1	0	0	1	0
3485	633	0	0	0	0	0	0	1	0
3489	633	0	1	0	0	0	0	1	1
4282	633	0	0	0	0	0	1	1	1
4284	633	1	0	0	0	0	0	0	1
3504	635	1	0	1	1	1	0	0	1
4315	635	1	1	0	1	0	1	1	1

deposited within them, actually occur within them in anomalously high concentrations?
2. Are the soils in different kinds of use-areas distinguishable in the spectra of their nutrient anomalies?

Both these questions are general rather than site-specific in nature, for they are concerned with the spectra rather than the magnitudes of nutrient availability anomalies.

In phrasing Question 1, it should be noted that care has been taken to consider only whether *anomalously high concentrations* occur within the use-areas we would predict they should; it has not been asked whether *lack* of nutrient anomalies occur where we would predict they should. Both kinds of questions can be asked based on knowledge of the relative concentrations of nutrients found in different refuse materials, and the locations of activities yielding these refuse materials; only the former, however, can be answered with the Crane data. As Table 66 suggests, there was a fair amount of spatial overlap of prehistoric activities at the Crane site, many collection units evidencing multiple activity groups. As a result, an area where we might expect to find low concentrations of a given nutrient, based on the occurrence of a given activity group in the area, may in fact have high concentrations of the nutrient due to contributions made by other activity groups also occurring in the area. Activity overlap at Crane consequently prevents us from investigating completely the patterns of association and dissociation between particular kinds of use-areas and nutrients. In terms of a fourfold contingency table that might be used in an association analysis of use-areas and nutrient anomalies, information for only the a and c cells were available for study:

	Soil anomalies expected or not	
	+	−
Soil anomalies found or not found +	available	not available
Soil anomalies found or not found −	available	not available

Activity overlap at Crane site thus prevented the use of a most direct approach—chi-square analysis—that otherwise would have been taken in testing relationships between particular kinds of use-areas and nutrients found in anomalous concentrations. An alternate approach therefore was devised. For each activity group and *only* those that produce refuse with a high concentration of a particular nutrient of interest, a tabulation was made of the proportion of soil samples located in surface collection units associated with the activity group that had anomalously high concentrations of the nutrient (Table 70). It is expected that this proportion should be high relative to off-site conditions, if the

TABLE 70
Proportion of soil samples which are located in surface collector units associated with particular activity groups and which have anomalously high concentrations of given nutrients

												Activity Group												
Nutrient/Soil Attribute		1	2	3	4	5	6	7	8	9	10	11	12	13	14	15	16	17	18	19	20	21	22	23	24
	pH	53.8	33.3	57.1	45.5	nd	57.1	41.1	50.0	33.0	0	nd	48	50	50	52.9	46.2	66.7	75.0	44.4	56	100	75.0	41.7	66.7
	PO_4	53.8	100.0	82.1	63.6	nd	85.7	82.4	87.5	0.	100	nd	52	0	50	52.9	61.5	77.8	83.3	100.0	76	66.7	75.0	91.7	100.0
	Mg	53.8	50.0	64.3	72.7	nd	57.1–60.7	70.6	87.5	66.7	100	nd	64	25	100	70.6–76.5	69.2	66.7	66.7	22.2	48–52	100	58.3	50–58.3	66.7
	Ca	46.1	66.7	71.4–75	90.9	nd	75.0	76.5	87.5	66.7	0	nd	72	75	75	82.4	69.2	66.7	66.7	66.7	64	66.7	58.3	75.0	77.8
	Na	61.5	83.3	71.4–75	54.5–63.6	nd	64.3–67.8	64.7–70.6	50.0	100.0	100	nd	56	0	75	47.1	46.2	55.6	41.7–50	77.8–88.9	68–72	33.3	41.7–50	75.0	77.8–88.9
	K	69.2–76.9	50.0	53.6	63.6	nd	53.6	52.9	37.5	66.7	100	nd	40–44	75	50	35.3	30.8	77.8	58.3	33.3	60	66.7	66.7	66.7	66.7–77.8
	NH_4	61.5	83.3	64.3	90.9	nd	57.1–60.7	76.5	100.0	66.7–100	0	nd	68–72	25	50	88.2	76.9	88.9	83.3	88.9	84–88	100	83.3	83.3	55.6–66.7
	SO_4	38.5–53.8	83.3	32.1–46.4	81.8–100	nd	46.4	29.4	12.5	100	100	nd	48–52	50	25–50	58.8–64.7	76.9	66.7	76.0	66.7	76	66.7	66.7	66.7	55.6
	OM	38.5	66.7	53.6	63.6	nd	57.1	70.6	87.5	66.7	100	nd	44	50	50	52.9	38.5	77.8	75.0	55.6	64	66.7	66.7	50.0	44.4
number of soil samples analyzed		13	6	28	11	0	28	17	8	3	1	0	25	4	4	17	13	9	12	9	25	3	12	12	9

nd = no data

enrichments of the nutrient of interest have been maintained within the soils of those areas. Next, for each nutrient of interest, and using soil samples located off-site, a calculation was made of the proportion of soil samples that had anomalously high concentrations of the nutrient of interest (Table 71). These statistics represent high-frequency background levels of noise—soil variations of natural or historic cause that were *not* removed by the stratigraphic and spatial filtering techniques just described. A comparison then was made (Table 72) between the proportion of soil samples having anomalously high concentrations of the nutrient of interest within the collection units where it was enriched to the soil, and the background noise-level of the nutrient. The conclusion was drawn that the nutrient was maintained in significant quantities within the soil if the proportion of soil samples exhibiting anomalies is greater than the background level of noise. Note that only information pertaining to the a and c cells of the contingency table are considered by this method of analysis.

Most of the activity groups defined in Table 63 could be investigated for their effects upon the Crane soils, but in some cases, spatial overlapping of activities prevented this. Where the majority of the soil samples spatially associated with an activity group of interest are shared by another activity group, the possibility exists that the nutrient anomalies observed are a result of the overlapping activity group rather than the activity group of interest. In this case, it was not possible to draw any conclusion about the relationship between the activity group of interest, the nutrient anomalies found, and maintenance of nutrient anomalies. Such cases are noted in Table 72. The degrees of independence and overlap of sets of soil samples taken from locations of different activity groups are summarized in Table 73 (Analysis I).

The results of the analysis summarized in Table 72 indicate that for nearly all activity groups, in those locations where they are present, significantly high proportions of soil anomalies (relative to background noise levels) were found for those nutrients expected to be anomalous, based on their concentrations within the refuse yielded by the activity groups. The spectra of nutrients characterizing the refuse of particular activity groups at Crane were maintained within the soils of some areas from the time of occupation to the present. The spectra of N, P, K, Ca, Mg, Na, and S all were maintained.

It should be pointed out, however, that not all use-areas at Crane are distinguished by the anomalous nutrient availabilities that were expected. In fact, the proportions of sampled locations of particular use-area types where nutrients expected to occur in anomalous concentrations actually did occur in anomalous concentrations is rather low. If for each nutrient of interest we subtract their background noise levels from the proportions of sampled locations that fall within a given type of use-area and that show anomalous concentrations of these nutrients, the resulting proportions of locations registering archeological significant nutrient anomalies is less than 50% in many case. This circumstance should not be taken to indicate, however, only the degree to which nutrients of prehistoric anthropic origin could be maintained within the Crane soils. It also

TABLE 71
Background level of noise in the relative frequency of nutrient anomalies found in natural soils east of Crane site

Nutrient	Relative Frequency of Anomalies
NH_4	54–57%
PO_4	39%
K	36%
Ca	48%
Mg	36%
Na	33–45%
SO_4	39–54%

Soil Samples Used:

Transect I, East end, off site: proveniences 289, 299, 309, 325, 331, 335, 339, 346, 354, 705, 725, 731, 736, 747, 756

Transect II, East end, off site: proveniences 2003, 2008, 2016, 2024, 2033, 2404, 2405, 2421, 2429

Transect III, East end, off site: proveniences 3603, 3613, 3616, 3620, 3626, 3633, 3636, 3643, 3644, 3649, 4406, 4409, 4413, 4419, 4424, 4437

Transect V, East end, off site: proveniences 7083, 7103, 7122, 7131, 7406, 7414, 7427

reflects a number of site- or analysis-specific factors that have a negative effect: (*a*) Some use-areas may not have been used long enough to have detectably enriched the soils within them; (*b*) soil samples analyzed as if they occurred within particular use-areas may actually have been taken outside the use-areas (the boundaries of use-areas on Crane are defined grossly by surface collection unit rather than in detail); (*c*) some kinds of use-areas were sampled with only a few soil cores, and the proportions calculated may be poor estimates of the actual proportion of the total area of the use-areas of those kinds having anomalous nutrient concentrations.

Because of these site-specific factors, the precise extent to which significant enrichments of nutrients to the soils of Crane have been maintained over time, and the degree to which use-areas are detectable by their soil chemistry and resistivity, cannot be assessed. Nevertheless, it is apparent that certain kinds of use-areas, where certain kinds of nutrients have been added significantly to the

TABLE 72
Comparison of refuse material expected with greater probability to yield soil anomalies to the proportion of anomalies actually found in areas of their deposition

Refuse Material Expected with Greater Probability to Yield Soil Anomalies*	Activity Group Depositing the Raw Material	Proportion of Anomalies Found in Soils From Surface Collection Units Containing the Activity Group	Sample Size	Significance of the Proportion Relative to the Background Level of Noise in Nutrient Anomalies
Nitrogen				
dry unboiled bone	4	90.9%	11	yes
	6	57.1-60.7	28	yes
	7*6	76.5	17	yes
	12	68-72	25	yes
deer meat, small mammal meat	9?*12	66.7-100	3	yes
	15	88.2	17	yes
	17?	88.9	9	yes
nuts in general, hickory nuts	19	88.9	9	yes
	20*18,20	84-88	25	yes
	22	83.3	12	yes
urine	13?	25.	4	no
seeds in general, chenopods	18*22	83.3	12	yes
	20*18,20	84-88	25	yes
	22	83.3	12	yes
mollusk meat	8	100.	8	yes
brains	--	---	--	---
broth of bone	7?**	76.5	17	yes
	21*17	100.	3	yes
Phosphorus				
dry, boiled bone	7*3,6	82.4	17	yes
ash of wood	3	82.1	28	yes
	10	100.	1	yes

TABLE 72 (cont.)

Refuse Material Expected with Greater Probability to Yield Soil Anomalies	Activity Group Depositing the Raw Material	Proportion of Anomalies Found in Soils From Surface Collection Units Containing the Activity Group	Sample Size	Significance of the Proportion Relative to the Background Level of Noise in Nutrient Anomalies
ash of wood (cont.)	12	52.0%	25	yes
	14*[15]	50.0	4	yes
	4	63.6	11	yes
dry unboiled bone	6	85.7	28	yes
	7*[3,6]	82.4	17	yes
seeds in general, chenopods.	18*[22]	83.3	12	yes
	20	76	25	yes
	22*[18,20]	75	12	yes
nuts in general, hickory nuts	19	100.	9	yes
	20	76	25	yes
	22	75	12	yes
deer meat, small mammal meat, turtle meat	9?*[12]	0	3	
	15	52.9	17	yes
	17?	77.8	9	yes
mollusk meat	8	87.5	8	yes
Potassium				
ash of wood	3	53.6	28	yes
	10	100.	1	yes
	12	40–44	25	yes
	14*[15]	50	4	yes
nuts in general	19	33.3	9	no
	20	60	25	yes
	22*[18,20]	66.7	12	yes
seeds in general	18*[22]	58.3	12	yes
	20	60	25	yes
	22*[18,20]	66.7	12	yes
greens	16	30.8	13	no
	23	66.7	12	yes

TABLE 72 (cont.)

Refuse Material Expected with Greater Probability to Yield Soil Anomalies	Activity Group Depositing the Raw Material	Proportion of Anomalies Found in Soils From Surface Collection Units Containing the Activity Group	Sample Size	Significance of the Proportion Relative to the Background Level of Noise in Nutrient Anomalies
small mammal meat	9?*[12]	66.7%	3	yes
	15	35.3	17	no
deer meat	9?*[12]	66.7	3	yes
	15	35.3	17	no
	17	77.8	9	yes
roots, tubers	17?	77.8	9	yes
dry boiled bone	7*[3]	52.9	17	yes
Calcium				
mollusk shell	8	87.5	8	yes
dry boiled bone	7*[3,6]	76.5	17	yes
ash of wood	3	71-75	28	yes
	10	0	1	no
	12	72	25	yes
	14	75	4	yes
dry unboiled bone	4	90.9	11	yes
	6	75	28	yes
	7*[3,6]	76.5	17	yes
bark	2*[16]	66.7	6	yes
unburnt wood	1	46.7	13	yes
chenopod seeds	18*[22]	66.7	12	yes
	20	64.0	25	yes
	22*[18,20]	58.3	12	yes
greens	16	69.2	13	yes
	23	75	12	yes
mollusk meat	8	87.5	8	yes

TABLE 72 (cont.)

Refuse Material Expected with Greater Probability to Yield Soil Anomalies	Activity Group Depositing the Raw Material	Proportion of Anomalies Found in Soils From Surface Collection Units Containing the Activity Group	Sample Size	Significance of the Proportion Relative to the Background Level of Noise in Nutrient Anomalies
nutshell	19	66.7%	9	yes
	20	64	25	yes
	22*[18,20]	58.3	12	yes
Magnesium				
ash of wood	3	64.3	28	yes
	10	100	1	yes
	12	64	25	yes
	14	100	4	yes
dry boiled bone	7*[3,6]	70.6	17	yes
dry unboiled bone	6	57.1–60.7	28	yes
	7*[3,6]	70.6	17	yes
bark	2	50	6	yes
nuts in general, hickory meat	19	22	9	no
	20	48–52	25	yes
	22	58.3	12	yes
seeds in general	18*[22]	66.7	12	yes
	20	48–52	25	yes
	22*[18,20]	58.3	12	yes
Sodium				
ash of wood	3	71.4–75	28	yes
	10	100	1	yes
	12	56	25	yes
	14*[15]	75	4	yes
dry, boiled bone	6	64.3–67.8	28	yes
dry, unboiled bone	4	54.5–63.6	11	yes
	6	64.3–67.8	28	yes
	7*[3,6]	64.7–70.6	17	yes

TABLE 72 (cont.)

Refuse Material Expected with Greater Probability to Yield Soil Anomalies	Activity Group Depositing the Raw Material	Proportion of Anomalies Found in Soils From Surface Collection Units Containing the Activity Group	Sample Size	Significance of the Proportion Relative to the Background Level of Noise in Nutrient Anomalies
mollusk shell	8	50 %	8	yes
mollusk meat	8	50	8	yes
brain	--	--	--	---
broth of bone	7?*[3,6]	64.7-70.6	17	yes
	21*[17]	33.3	3	no
deer meat	9?*[12]	100	3	yes
	15	47.1	17	yes
	17?	55.6	9	yes
Sulfur				
small mammal and deer meat	9?*[12]	100	3	yes
	15	58.8-64.7	17	yes
	17	66.7	9	yes
brains	--	---	--	---
nut meats in general	18*[22]	75.	12	yes
	20	76	25	yes
	22*[18,20]	66.7	12	yes
seeds in general	19	66.7	9	yes
	20	76	25	yes
	22	66.7	12	yes
bark*	2	83.3	6	yes
urine	13?	50	4	yes
roots, tubers	17?	66.7	9	yes
nut shells	18	75	12	yes
	20	76	25	yes
	22*[18,20]	66.7	12	yes

TABLE 72 (cont.)

*Refuse materials are ordered under each nutrient by their relative concentrations of the given nutrient, from greater to lesser.

*[6] Activity group shares the majority of its soil samples with another activity group, which <u>also is expected</u> to <u>yield a high proportion of soil anomalies</u> for the given nutrient. The proportion of anomalies found in the soil samples from the surface collection units containing the activity, therefore, do not necessarily reflect the effect of the activity group upon the soil chemistry there. The activity group with which the activity group of interest overlaps is given to the right of the asterisk.

soils, are more detectable than others because some nutrients have been maintained in available form within the Crane soils more so than others. Of all the nutrients investigated, P, N, and Ca occur in anomalous concentrations in a high proportion of soil samples from most kinds of use-areas (70–90%), whereas Na and SO_4 and Mg and K occur in anomalous concentrations in lower proportions of the soil samples from most kinds of use-areas (60–75% and 50–70%, respectively).

The higher proportions associated with N may be attributed to its high background noise—a site-specific phenomenon—but the higher proportions found for P and Ca are significant in a more general way. Phosphates and calcium of prehistoric anthropic origin presumably have been maintained to a greater degree within the Crane soils than have other nutrients because of the natural soil processes influencing their mobilities and availabilities (Chapter 4). Most importantly, phosphorous and calcium are precipitated in alkaline soil conditions (such as those predominating at Crane) to form highly insoluble, nonleachable compounds that nevertheless are in equilibrium with the soil water solution. Sodium, on the other hand, is not fixed in insoluble forms within soils after it is released from decomposing organic matter and is more readily leached from soils. As a univalent cation having less charge than bivalent Ca and being found in lower concentrations at Crane, Na also is less strongly adsorbed by clay and humus micelles than is Ca. The generally lower degree to which sulfates were found within the soils of Crane compared to P and Ca is expectable considering that sulfates are fixed to only a limited degree within soils (within the structure of humus molecules) and that, as anions, they are only weakly adsorbed to clay and humus micelles.

While the lower proportions of Na and S anomalies than P and Ca anomalies within use-area where they would be expected can probably be attributed to the different degrees to which these nutrients have been leached from the Crane soils, the lower proportions of Mg and K anomalies probably are attributable to their fixation in forms that only slowly come into equilibrium with the soil water solution and which are unavailable. Both K and Mg could have been fixed in considerable quantities by isomorphic substitution within the expandable illitic clays that predominate in the loessic soils in the Crane area (Frye *et al.* 1968).

TABLE 73
Degrees of independence of sets of soil samples taken from locations of different activity groups

Activity Group #1	Activity Group #2	Analysis I			Analysis II		
		Number of Soil Samples in Surface Collection Containing the Activity Groups		Soil Samples Shared by the Two Activity Groups	Number of Soil Samples in Surface Collection Containing the Activity Groups		Soil Samples Shared by the Two Activity Groups
		N_1	N_2		N_1	N_2	
1	3	13	28	3	13	12	2
1	6	13	28	1	13	12	0
1	7	13	17	3	13	17	3
1	12	13	25	2	13	25	2
1	16	13	13	1	13	9	1
1	17	13	9	1	13	9	1
1	18	13	12	5	13	12	5
1	20	13	25	5	13	21	5
1	21	13	3	1			
1	22	13	12	6	13	12	6
1	23	13	12	1	13	8	1
1	24	13	9	3	13	9	3
2	12	6	25	3			
2	16	6	13	6			
2	17	6	9	4			
2	18	6	12	1			
2	20	6	25	1			
2	22	6	12	1			
2	23	6	12	3			
3	4	28	11	4	12	11	4
3	6	28	28	16	12	12	0
3	7	28	17	15	12	17	8

TABLE 73 (cont.)

Activity Group #1	Activity Group #2	Analysis I			Analysis II		
		Number of Soil Samples in Surface Collection Containing the Activity Groups		Soil Samples Shared by the Two Activity Groups	Number of Soil Samples in Surface Collection Containing the Activity Groups		Soil Samples Shared by the Two Activity Groups
		N_1	N_2		N_1	N_2	
3	16	28	13	1	12	9	0
3	18	28	12	3	12	12	0
3	19	28	9	3	12	9	0
3	20	28	25	5	12	21	1
3	22	28	12	3	12	12	0
3	24	28	9	3	12	9	0
4	7	11	17	4	11	17	4
4	17	11	9	2	11	9	2
4	18	11	12	2	11	12	2
4	20	11	25	2	11	21	2
4	21	11	3	2			
4	22	11	12	2	11	12	2
6	7	28	17	14	12	17	7
6	8	28	8	2	12	8	2
6	15	28	17	3	12	13	3
6	16	28	13	1	12	9	1
6	18	28	12	3	12	12	0
6	19	28	9	3	12	9	0
6	20	28	25	8	12	21	4
6	22	28	12	3	12	12	0
6	23	28	12	2	12	8	1
6	24	28	9	8	12	9	5
7	8	17	8	2	17	8	2
7	16	17	13	1	17	9	1
7	18	17	12	3	17	12	3
7	19	17	9	3	17	9	3

TABLE 73 (cont.)

Activity Group #1	Activity Group #2	Analysis I Number of Soil Samples in Surface Collection Containing the Activity Groups		Soil Samples Shared by the Two Activity Groups	Analysis II Number of Soil Samples in Surface Collection Containing the Activity Groups		Soil Samples Shared by the Two Activity Groups
		N_1	N_2		N_1	N_2	
7	20	17	25	3	17	21	3
7	22	17	12	3	17	12	3
7	24	17	9	3	17	9	3
9	12	3	25	3	3	25	3
9	20	3	25	1	3	21	1
12	15	25	17	6	25	13	2
12	16	25	13	6	25	9	2
12	20	25	25	1	25	21	1
13	15	4	17	1			
14	15	4	17	3			
15	16	17	13	6	13	9	2
15	17	17	9	1	13	9	1
15	19	17	9	3	13	9	3
15	20	17	25	7	13	21	1
15	23	17	12	3	13	8	0
16	17	13	9	4	9	9	4
16	18	13	12	1	9	12	1
16	20	13	25	1	9	25	1
16	22	13	12	1	9	12	1
16	23	13	12	3	9	8	3
17	18	9	12	3	9	12	3
17	20	9	25	2	9	21	2
17	21	9	3	2			
17	22	9	12	4	9	12	4
17	23	9	12	2	9	8	2
18	19	12	9	3	12	9	3

TABLE 73 (cont.)

Activity Group #1	Activity Group #2	Analysis I			Analysis II		
		Number of Soil Samples in Surface Collection Containing the Activity Groups		Soil Samples Shared by the Two Activity Groups	Number of Soil Samples in Surface Collection Containing the Activity Groups		Soil Samples Shared by the Two Activity Groups
		N_1	N_2		N_1	N_2	
18	20	12	25	10	12	21	10
18	21	12	3	2			
18	22	12	12	11	12	12	11
18	23	12	12	1	12	8	1
18	24	12	9	3	12	9	3
19	20	9	25	9	9	21	6
19	23	9	12	3	9	8	3
19	24	9	9	3	9	9	3
20	22	25	12	9	21	12	9
20	23	25	12	8	21	8	4
21	22	3	12	2			
22	23	12	12	2	12	12	2
22	24	12	9	3	12	9	3

In summary, the nutrient data given in Table 72 do suggest that prehistoric anthropic nutrient enrichments to the soils at the Crane site were maintained in some areas of the site over extended periods of time since occupation. Those nutrients that one would predict should occur in anomalously high concentrations within particular use-areas, based on the relative proportions of nutrients in refuse materials deposited within them, actually did occur in anomalously high concentrations in a significantly high proportion of the areas investigated (relative to background noise levels). The degree to which nutrient enrichments were maintained within the Crane soils, however, varied with the nutrient of interest and how tightly it was fixed within the Crane soils by natural mechanisms and processes. The particular nutrients maintained to greater or lesser degrees within the Crane soils and the kinds of use-areas often found to manifest nutrients in anomalous concentrations probably reflect general circumstances that can be found at earthen sites on loessic soils within the Midwest United States at large. The extent to which use-areas on other sites are detectable in their soil chemistry and resistivity, however, will depend on site-specific factors as well as the general factors examined here. The length of time a work area was used and the amount of refuse deposited within it are critical factors determining detectability.

More Specific Results

Having overviewed the soil alterations associated with all activity groups defined at Crane, it is instructive to consider the effects of several activity groups in greater detail. The first is activity group 24, red ochre deposition. The prediction was made (Table 67) that the addition of red ochre to soils should increase the potential of a soil for fixing and maintaining phosphates and sulfates. Red ochre is one form of a general class of hydrous oxides capable of maintaining phosphates and sulfates within soils in at least two ways. First, they can precipitate phosphates and sulfates within insoluble yet available compounds, thus protecting them from leaching but keeping them detectable. Second, hydrous oxides are capable of cation and anion exchange. Their enrichment within the soil increases the ion exchange capacity of the soil and its capability for adsorbing phosphates and sulfates.

The effect of red ochre deposition on phosphate and sulfate retentive capacities of soils cannot be questioned, but the significance of this activity is debatable. Could enough red ochre be deposited on a site like Crane—which was probably occupied less than 10 years by one household—to have had a significant effect on phosphate and sulfate retention in some areas? It would appear that this is the case. Phosphate and sulfate anomalies both were found in significantly high proportions (100%, 55.6%, respectively) within soil samples located in areas of red ochre deposition. The high proportion of phosphate anomalies found probably in part reflects the spatial overlap of burnt bone depositional activity with red ochre depositional activity, but the significant proportion of sulfate anomalies can be attributed only to the enrichment of red ochre. It also is significant that phosphate anomalies occur in higher proportion than sulfate anomalies within the use-areas where red ochre was deposited. Phosphates are retained more strongly by hydrous oxides of iron than are sulfates (Brady 1974:453).

A second set of activity groups that are of interest are those involved in the dressing of hides. Various raw materials were used ethnographically in North America to dress hides, among these being wood ash, urine, brains, broth of bone, broth of meat, meat drippings, and bone dust (Mason 1889). Different materials were used at different stages of dressing (dehairing, graining) and depending on whether soaking and/or simmering of the hide was involved. It was necessary to determine which of these raw materials, if any, were used in hide-dressing activities at Crane, in order to make the comparison presented above, between locations where soil nutrient anomalies are expected and those where they were found at Crane. This was done using data on the full spectra of soil anomalies found in locations where different kinds of hide-working activities

8. Soil Alterations Produced by Prehistoric Human Activities

are reconstructed to have occurred at Crane. It should be pointed out that this analysis not only is relevent to this work, but also is of general archeological interest in itself, with respect to reconstruction of prehistoric lifeways in the Eastern Woodlands.

The data used in the analysis are shown in Table 74. For each activity group involved in hide working (Groups 9, 10, 11, 12, 13, 14), a tabulation was made of the proportion of soil samples that had anomalously high concentrations of specific nutrients within collection units where the activity group is present, and also whether or not the proportions are significant compared to background noise levels for the nutrients. The spectra of nutrients found to be anomalous in significant proportions then were compared with the spectra of nutrients in the various raw materials that might have been used in hide dressing to determine which of the raw materials, if any, were deposited in hide-dressing areas. The results are as follows. For activity groups 10, 12, and 14—dehairing and soaking hides, graining hides, and graining and simmering hides— correspondence between the spectra of soil nutrient anomalies and the proportions of nutrients within the raw materials that might have been used was greatest for wood ash, compared to other materials. The probable use of wood ash in *multiple* stages of hide dressing is a significant pattern in itself and lends some support to the individual correspondences found. It is unclear whether the nutrient spectra found within the soils of areas where Activity Group 13 (graining and soaking hides) is present corresponds more closely to the nutrient compositions of ash or urine. The correspondence is not good in either case and neither material may have been used prehistorically in these use-areas.

The spectra of nutrient anomalies found in the collection units where activity groups 10, 12, or 14 are present cannot be explained by nutrient enrichments from those activities other than hide dressing that also occurred in these units. Only activity group 16 (uncooked plant tissue deposition) overlaps spatially with activity groups 10, 12, or 14 to any significant extent (Table 73); the refuse materials produced by it do not have compositions that could produce the spectra of anomalous nutrient availabilities observed in the collection units under consideration. The creation of these anomalies by ash deposition related to hide dressing seems likely.

Some lack of correspondence between the relative proportions of nutrients within ash and the proportions of nutrient anomalies found within the areas encompassing activity groups 10, 12, and 14 did occur, but most of the disagreements appear to be explainable. Disagreements in regard to activity group 10 may simply be a function of the small number of soil samples that were used to evaluate soil nutrient spectra. The unexpectedly high proportion of soil samples found to have anomalous concentrations of NH_4 in use-areas where activity groups 12 and 14 occurred may reflect the deposition of animal tissue from defleshing operations that occurred in the same locales. Animal meat has a relatively high proportion of N compared to other raw materials (Table 68).

TABLE 74
Statistics used in determining the raw materials used in working hides

Activity Group 9: Dehair hides   Sample Size = 3

Complete overlap of surface survey units having this Activity Group with surface survey units having Activity 12. No conclusions possible.

Activity Group 10: Dehair and soak hides.   Sample Size = 1

	PO_4	K	Mg	Na	SO_4	NH_4	Ca
% anomalies found	100	100	100	100	100	0	0
Background noise level	39%	36%	36%	33-45%	39-54%	54-57%	48%
% anomalies significant?	yes	yes	yes	yes	yes	no	no
Expected anomaly level:							
wood ash	H	H	H	H	L	L	H
urine	L	M	L	H	M	M	L

Activity Group 11: Dehair and simmer hides   Sample Size = 0

Activity Group 12: Grain hides, alone   Sample Size = 25

	Ca	NH_4	Mg	Na	PO_4	SO_4	K
% anomalies found	72	68-72	64	56	52	48-52	40-44
Background noise level	48%	54-57%	36%	33-45%	39%	39-54%	36%
% anomalies significant?	yes	yes	yes	yes	yes	no	yes?
Expected anomaly level:							
wood ash	H	L	H	H	H	L	H
urine	L	M	L	H	L	M	M
brains	L	M	L	M	M	H	M
broth of bone	L	M	L	M	L	?	L
broth of meat	L	L	L	L	L	L	M
meat drippings	L	L	L	L	L	L	L
boiled bone, dust	H	L	H	H	H	?	L

In summary, soil chemistry data from the Crane site would suggest that wood ash was used in many stages of hide dressing there, and that the soil anomalies resulting from its use have been maintained since the time of occupation. Ash was among the most common processing materials used in dressing hides in North America, and the suggestion of its use at Crane based on soil data is not unreasonable. It therefore was used as the residue material in the comparisons made above between locations where soil nutrient anomalies are expected and

TABLE 74 (cont.)

Activity Group 13: Grain and soak hides   Sample Size = 4

	Ca	K	SO$_4$	NH$_4$	Mg	PO$_4$	Na
% anomalies found	75	75	50	25	25	0	0
Background noise level	48%	36%	39–54%	54–57%	36%	39%	33–45%
% anomalies significant?	yes	yes	no?	no	no	no	no
Expected anomaly level:							
wood ash	H	H	L	L	H̲	H̲	H̲
urine	L̲	M	M̲	M	L	L	H̲
brains	L̲	M	H̲	M̲	L	M̲	M̲

Activity Group 14: Grain and simmer hides   Sample Size = 4

	Mg	Ca	Na	NH$_4$	PO$_4$	K	SO$_4$
% anomalies found	100	75	75	50	50	50	25–50
Background Noise level	36%	48%	33–45%	54–57%	39%	36%	39–54%
% anomalies significant?	yes	yes	yes	no	yes	yes	no
Expected anomaly level:							
wood ash	H	H	H	L	H	H	L
brains	L̲	L̲	M	M̲	M	M	L

L = low proportion of anomalies expected
M = medium proportion of anomalies expected
H = high proportion of anomalies expected
underlined letters designate locations of disagreement between the expected and actual proportions of anomalies found in the given nutrient.

locations where they actually were found at Crane for the hide-dressing activity groups 10, 12, and 14.

Results II: Differentiation of Use-Areas by the Spectra of Their Anomalous Nutrient Availabilities

It has been shown that prehistoric nutrient enrichments to soils at Crane have been maintained in some areas since occupation and that anomalously high nutrient availabilities occur within the soils of use-areas where they are expected. It has yet to be determined, however, whether the different kinds of use-areas at Crane are *distinguishable* from each other in the spectra of nutrient anomalies within their soils.

To investigate the distinguishability of soils in different use-areas at Crane, a three-step analysis was performed. First, all of the refuse materials for which nutrient composition data had been collected (Table 67) were subjected to cluster analysis to determine which materials were most similar and dissimilar in their nutrient contents.[1] The results of this analysis suggest which use-areas might be distinguishable under *ideal* soil conditions, where all aspects of the nutrient spectra distinguishing the different kinds of *refuse* produced by the different kinds of activities have been preserved in the soil. Second, the several types of activity areas defined at Crane site were subjected to cluster analysis using data on the proportions of soil samples within them having nutrient anomalies in order to determine which types of use-areas at Crane were most similar in their soil attributes. The two cluster analyses then were compared to determine how natural soil processes had altered the nutrient contents of refuse materials incorporated within the soils and altered the distinguishability of the soils within various types of use-areas from that which might be expected under ideal conditions.

In both cluster analyses, the several nutrient variables that were used to determine the degrees of similarity between refuse materials and between the soils within use-areas were weighted equally in importance rather than in proportion to their absolute magnitudes. All variables were standardized such that their case values had a mean of 0 and a variance of 1. This was necessary in order to prevent clusters from being defined largely with respect to the one or few nutrients occurring in greatest concentrations within the refuse materials (N, K, P) and soils (P, N, Ca). It was also necessary in order to maintain the criteria used to define clusters constant between the two cluster analyses.

Cluster Analysis of Refuse Materials

Keeping this overview in mind, let us now consider the details of the first cluster analysis—that of the different refuse materials presumed to have been produced by the different activities at Crane. The proportions of nutrients that occur in the different refuse materials and that were used in defining their degrees of similarity are shown in Table 67. Only the variables N, P, K, Ca, Mg, Na, and S were used in the analysis—those variables for which information on their availabilities within the Crane soils also is available for clustering. The refuse materials, chenopod seeds, turtle meat, and red ochre were eliminated from the analyses because too little was known about their nutrient contents.

---

1. The nutrient data in Table 67 are considered accurate enough to perform such a cluster analysis since the conclusions drawn from the analysis are only *qualitative* statements of which refuse materials are similar or dissimilar, rather than absolute estimates of their degrees of similarity or dissimilarity.

Trace concentrations of nutrients, where found, were considered 0. A number of missing nutrient concentrations were estimated from their values in similar materials, and are summarized in Table 75. From the completed data matrix, a matrix of dissimilarity coefficients was calculated (Table 76), standardizing within variables and using a Euclidean distance coefficient as the measure of dissimilarity. A dendrogram (Figure 60) was constructed from the dissimilarity matrix using an unweighted average linkage clustering algorithm. The dendrogram is a reasonable one-dimensional representation of the multidimensional relationships defined by the dissimilarity matrix, the cophenetic correlation coefficient being .9888. A number of clusters of refuse materials at various levels of similarity are readily definable in the dendrogram, and are summarized in Table 77.

The clusters of refuse materials shown in Table 77 suggest that certain types of use-areas, even under soil conditions ideal for the maintenance of anthropic nutrient anomalies, should be more distinct than are others. Most distinctive are those use-areas where wood ash, dry boiled bone, dry unboiled bone, mammal meat, mollusk shell, or nut meats in general have been deposited. Most of these distinctive refuse materials are characterized either by their inorganic nature and/or by the occurrence of a few nutrients in high concentrations within them (Ca and P in bone; Ca in shell; and K and Ca in ash). At the other extreme, use-areas where roots and tubers, nut shells, greens, or unburnt wood were deposited would be more difficult to distinguish with soil chemistry data. Likewise, use-areas where meat broth or drippings, or fruit were deposited would be difficult to distinguish. These results are general in nature rather than site-specific, with the exception of the particular species used to characterize the broad material categories such as greens, wood, and nut meats.

Cluster Analysis of Use-Areas

Data on the proportion of soil samples having anomalously high concentrations of particular nutrients within different kinds of use-areas at Crane site were subjected to a cluster analysis congruent with that performed for refuse materials. The data used are shown in Table 78. These data differ from those given in Table 70 in several respects. First, activity groups 2, 5, 9, 10, 11, 13, 14, and 21 were eliminated from the analysis. Use-areas associated with these activity groups were sampled with only a few soil samples. The nutrient anomaly proportions calculated for these activity groups consequently are restricted to specific values (0–100%, 0–50–100%, or 0–33–66–100%, etc.) rather than capable of taking almost any value between 0 and 100%, and are poor estimates of the actual proportion of soil cores that would be expected to have anomalous nutrient concentrations within the use–areas of interest, given an adequate sample. Only activity groups represented by 8 or more soil samples and allowing the calculation of nutrient anomaly proportions within 12.5%

TABLE 75
Estimation of the nutrient contents of refuse materials for which nutrient data in Table 67 are missing

---

1. K content of hickory nuts estimated by K content of nuts in general
2. Na content of hickory nuts estimated by Na content of nuts in general
3. S content of hickory nuts estimated by S content of nuts in general
4. Na content of bark estimated by Na content of wood
5. Na content of nutshells estimated by ½ Na content of wood (most nutrients in nutshells occur in ca. ½ their concentration found in wood)
6. S content of greens estimated by S content of roots and tubers
7. S content of mollusk meat estimated by ½ S content of deer meat
8. S content of dry boiled bone estimated by S content of dry unboiled bone
9. S content of broth of bone estimated by S content in dry unboiled bone (ca. 0)

---

intervals[2] were admitted to the analysis. Second, the nutrient anomaly proportions in Table 78 differ from those in Table 70 in that the different background levels of noise for each nutrient have been subtracted from them. This was necessary in order to prevent natural and historically caused soil variation from entering into the similarities calculated between different activity groups. Finally, for some activity groups, the proportions of soil samples with anomalous nutrient concentrations have been altered from those given in Table 70 by eliminating cases (use-areas) where activity groups overlapped. For each activity group analyzed, an attempt was made to keep the percentage of soil samples shared with any one other activity group below 50%. For most activity groups, the amount of overlap is much less than this (Table 73, Analysis II). In a few cases, however, it was not possible to eliminate soil samples without reducing the sample size below the required number of 8, and significant activity overlap was retained within the analysis.

As in the cluster analysis of refuse materials, the nutrient variables used to measure similarity were standarized, a matrix of dissimilarity coefficients was calculated using a Euclidean measure of distance (Table 79), and a dendrogram of activity group clusters was constructed with an average linkage clustering algorithm. Unfortunately, the dendrogram was characterized by a low (.6751) cophenetic correlation coefficient, and could not be used to examine visually the

---

2. An arbitrarily set degree of accuracy.

TABLE 76
Dissimilarity of various refuse materials deposited at the Crane site

Refuse Material				
Seeds in general	.22523 -1*			
Brains	.64728 -1	.43824 -1		
Mollusk meat	.97833 -1	.35138 -1	.12098	
Urine	.75779 -1	.46384 -1	.11503	.29920 -1
Nutmeats in general	.46161 -1	.12567	.20579	.22815
Nut oil	.61044	.39930	.47315	.25122
Meat broth	.61010	.39907	.47200	.25015
Meat drippings	.58196	.37581	.43649	.23945
Fruits powdered	.53925	.34233	.41685	.20676
Fruits not powdered	.52425	.33011	.39904	.19984
Roots, tubers	.42643	.25361	.28419	.16312
Nut shells	.47910	.29413	.33432	.18781
Greens	.36922	.20969	.24943	.12590
Unburnt wood	.44702	.26971	.34035	.15232
Bone broth	.34534	.20031	.31358	.78960 -1
Bark	.32325	.18093	.18877	.13786
Mollusk shells	1.3347	1.1212	1.1907	.96414
Dry, unboiled bone	1.0377	1.1608	1.4397	1.1565
Deer meat	.80345	1.0073	.77092	1.3917
Small mammal meat	.99147	1.2091	.93615	1.6291
Dry, boiled bone	5.7778	5.5641	5.6249	5.4272
Ash of wood	11.263	.11.095	11.157	10.961
	Hickory meats	Seeds in general	Brains	Mollusk meat
Nut shells	.22728 -2			
Greens	.25117 -2	.73311 -2		
Unburnt wood	.99986 -2	.65313 -2	.92339 -2	
Bone broth	.58047 -1	.58113 -1	.42394 -1	.27681 -1
Bark	.26206 -1	.39594 -1	.22146 -1	.51323 -1
Mollusk shells	.74885	.73434	.75448	.72946
Dry, unboiled bone	1.9689	2.0222	1.8493	1.8688
Deer meat	1.9591	2.0894	1.8793	2.1274
Small mammal meat	2.1993	2.3372	2.1197	2.3872
Dry, boiled bone	5.2653	5.2641	5.2662	5.2440
Ash of wood	10.831	10.893	10.819	10.835
	Roots, tubers	Nut shells	Greens	Unburnt wood

*Exponent of a base 10 multiplier factor

.16571						
.40292	.94071					
.39876	.94035	.20986 −3				
.38639	.91222	.12909 −2	.14499 −2			
.34797	.85124	.23396 −2	.23566 −2	.21937 −2		
.33862	.83523	.33576 −2	.34253 −2	.20680 −2	.27661 −3	
.28225	.72884	.27010 −1	.26731 −1	.17504 −1	.17487 −1	.13543 −1
.31771	.79225	.14075 −1	.14132 −1	.73430 −2	.84852 −2	.57879 −2
.23609	.65017	.36064 −1	.35778 −1	.26563 −1	.22218 −1	.18071 −1
.27692	.73527	.12849 −1	.12725 −1	.10100 −1	.44142 −2	.32735 −2
.17155	.56854	.64247 −1	.63340 −1	.65363 −1	.45352 −1	.45164 −1
.23424	.60007	.93020 −1	.92816 −1	.75971 −1	.73659 −1	.66406 −1
1.1016	1.6621	.72452	.72302	.72551	.72556	.72698
1.0050	.82124	2.1146	2.1073	2.1143	2.0070	2.0025
1.2152	.75088	2.4321	2.4312	2.3415	2.3074	2.2635
1.4452	.94001	2.7027	2.7022	2.6047	2.5741	2.5269
5.4843	6.0892	5.2620	5.2418	5.2587	5.2549	5.2578
10.637	11.570	10.919	10.844	10.912	10.860	10.881
Urine	Nut meats in general	Nut oil	Meat broth	Meat drippings	Fruits powdered	Fruits not powdered

.95575 −1						
.78203	.68876					
1.5246	1.7803	1.7394				
2.0492	1.6527	3.1517	2.2644			
2.3226	1.8714	3.4241	2.5471	.11043 −1		
5.2809	5.0932	3.6128	4.6083	7.6022	7.8670	
10.808	10.499	9.8602	10.319	13.130	13.452	9.2985
Bone broth	Bark	Mollusk shells	Dry, unboiled bone	Deer meat	Small mammal meat	Dry, boiled bone

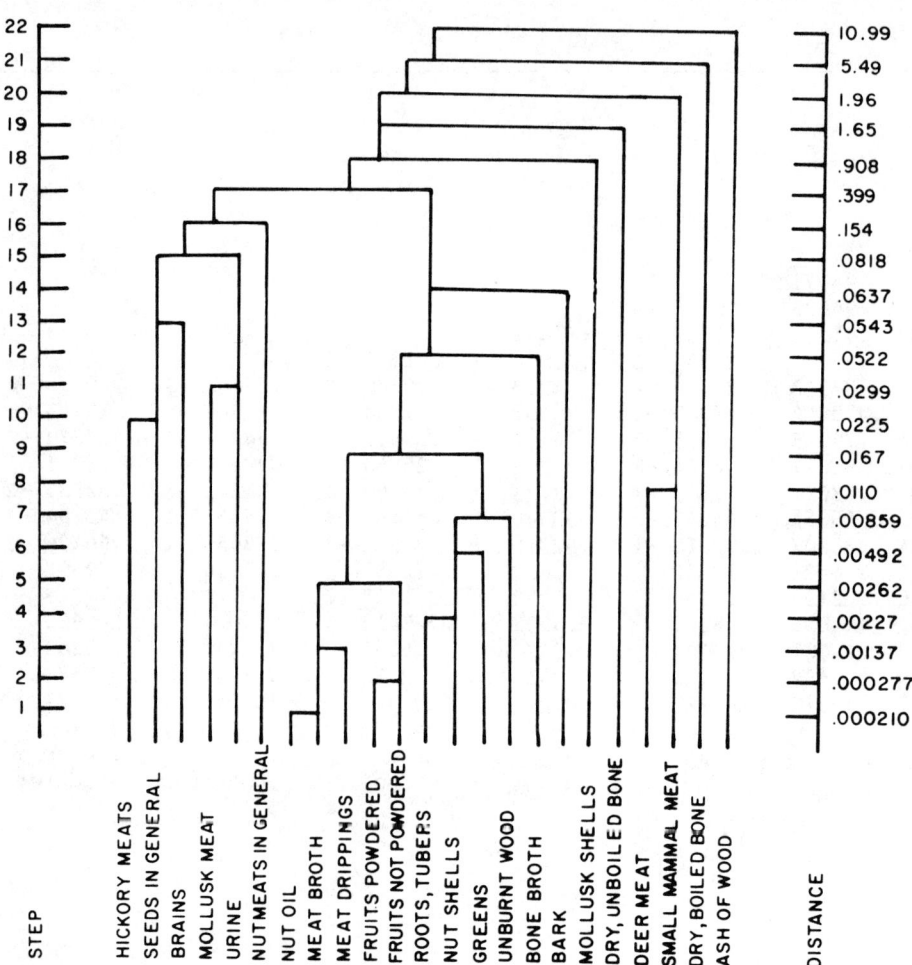

FIGURE 60. Dendrogram showing the similarity of various refuse materials with respect to their nutrient compositions.

relationships held within the dissimilarity matrix. As a partial solution to this problem, a single-linkage algorithm was used to construct a dendrogram that would indicate at least those activity groups with use-areas most distinct in their soil attributes. Distinctiveness in this case was defined on the basis of whether or not the activity group chained individually to the trunk of the dendrogram. The activity groups found to be distinctive by this mode of analysis include: 3, hearth dumpings; 8, mollusk processing; 19, uncooked nut deposition; and 1, woodworking.

To define activity groups with use-areas that were most similar in their soil attributes at Crane, it was necessary to examine the dissimilarity matrix

TABLE 77
Clusters of refuse materials defined on the basis of their nutrient proportions using standardized nutrient variables, the Euclidean distance measure of dissimilarity, and an average linkage routine

Cluster 1

hickory meats
seeds in general
brains

Cluster 2

urine
mollusk meat

Cluster 3

a. nut oil
   broth of meat
   meat drippings
   fruits powdered
   fruits not powdered

b. roots, tubers, bulbs, rhizomes
   nut shells
   greens
   unburnt wood

c. broth of bone

d. bark

Refuse Materials Not Forming Clusters, and Chaining at the end of the Analysis

mollusk shell
dry, unboiled bone
meat of deer, small mammals
dry boiled bone
ash of wood
nut meats in general

---

directly. For each activity group, an ordered list was made of those activity groups more or less similar to the activity groups of interest, based on their dissimilarity coefficients (Table 80). These lists could have been carried out until all activity groups were enumerated. Instead, they were truncated at points where ordering no longer was meaningful. Two critical truncation points are (*a*) when the magnitude of the distance coefficient surpassed the level of accuracy of the coefficients for the activity group of interest (Table 81), level of accuracy being based on the number of soil samples used in defining the nutrient anomaly

TABLE 78

*Proportions of soil samples which are located in surface collection units associated with particular activity groups and which have anomalously high concentrations of given nutrients**

Nutrient	Activity Groups															
	1	3	4	6	7	8	12	15	16	17	18	20	22	23	24
N	6.0	27.8	35.4	11.2	21.0	44.5	14.5	29.1	11.2	33.4	27.8	25.5–30.3	27.8	19.5	5.6
P	14.8	44.3	24.6	52.7	43.4	48.5	13.0	14.8	27.7	38.8	44.3	32.4	36.0	48.5	61.0
K	37.1	39.0	27.6	39.0	10.9	1.5	6.0	10.2	8.4	41.8	22.3	25.9	30.7	39.0	36.2
Ca	0.	35.3	42.9	43.7	28.5	39.5	24.0	36.6	18.7	18.9	18.7	9.1	10.3	14.5	29.8
Mg	17.8	55.7	36.7	39.0–47.3	36.4	51.5	28.0	33.2–40.9	30.7	30.7	30.7	16.4–21.1	0.	26.5–39.0	30.7
Na	22.5	44.3–52.7	20.1	27.7–36.0	28.6	11.	17.0	22.5	27.7	16.6	6.8	27.7–32.4	6.8	36.0	44.4
S	0.	0.	44.4	20.2	0	0	3.5	15.0–22.7	42.4	20.2	28.5	29.7	20.2	16.0	9.1

*Background noise levels have been removed and the number of soil samples shared between activity groups has been minimized

TABLE 79
Dissimilarity of activity groups with respect to the soil chemistry of the collection units in which they occur

Activity Group

	16	7	12	15	17	18	20	23
7	1.6210							
12	1.3033	.83865						
15	1.1586	.92071	.62398					
17	1.7965	1.0841	1.8522	1.3445				
18	1.0981	1.1205	1.4938	1.1814	.42994			
20	.80471	1.2564	1.5365	1.3252	.60546	.71546		
23	1.4990	.72815	1.9904	1.9060	.62545	1.1150	.64680	
22	2.0173	1.9061	1.8284	2.1822	.83823	.77123	.75717	1.6186
4	1.4759	2.1053	2.3759	.78481	1.1877	1.1543	1.5056	2.1697
1	2.3204	1.9000	1.3058	2.6831	1.9026	2.3520	1.5214	1.5229
3	3.2979	1.0184	2.9189	2.0903	1.8172	2.8261	2.5089	1.0863
6	1.9627	1.0081	2.5067	1.8921	1.4953	1.7374	2.1159	.90747
24	2.2969	.96999	2.7389	2.8575	2.0175	2.3977	2.0233	.60714
19	2.4014	1.8449	3.1132	2.8699	2.6667	2.3421	1.4865	2.0485
8	3.6254	1.3304	2.3460	1.4759	2.1976	1.8094	3.2087	3.0738

	22	4	1	3	6	24	19
1	2.4265						
3	1.5947	4.2228					
6	4.2676	2.6828	3.5631				
24	3.1391	1.7508	3.2565	1.0202			
19	3.1181	3.3454	2.5826	1.2003	.55706		
8	2.2747	3.5118	4.1790	3.6250	3.3943	2.4919	
	3.6075	2.4375	5.2972	2.4622	2.8773	3.9203	3.4459

Activity Group

TABLE 80
Activity groups most similar to other activity groups in their proportion of nutrient anomalies

Activity Group of Interest	Empirically Similar Activity Groups	Distance Coefficient	Similar Activity Group Expected To Be Similar?	Unexpectedly Similar Activity Group Explainable? How?
1 woodworking	12 grain hides, alone	1.305	no	unexplainable
	20 uncooked seeds or nuts, deposition	1.521	no	yes, overlap
	23 boil greens	1.523	yes	
3 hearth dumpings	7 unburnt bone, deposition	1.018	yes	
	6 burnt bone, deposition	1.020	yes	
4 bone working	15 uncooked animal tissue	.785	yes	
	18 uncooked seeds and seed coats, deposition	1.154	no	yes, scales of 4,18
	17 pulverize roots, fruits or dried meat	1.188	yes	
	16 uncooked plant tissue, deposition	1.476	no	yes, scales of 4,16
	20 uncooked seeds or nuts, deposition	1.506	no	yes, scale of 4
	6 burnt bone, deposition	1.751	yes	
6 burnt bone deposition	24 red ochre deposition	.557	yes	
	23 boil greens	.907	no	yes, scale of 23
	7 unburnt bone deposition	1.01	yes	
7 unburnt bone deposition	23 boil greens	.728	no	yes, scales of 7,23
	12 grain hides, alone	.839	yes	
	15 uncooked animal tissue	.921	yes	
	24 red ochre deposition	.970	yes	
	6 burnt bone deposition	1.008	yes	

Category	Item	Value	Y/N	Scale
7 unburnt bone deposition (cont.)	18 uncooked seeds and seed coats, deposition	1.121	no	yes, scales of 7,18
	17 pulverize roots, fruits or dried meat	1.084	yes	
	3 hearth dumpings	1.018	yes	
8 mollusk processing	7 unburnt bone deposition	1.330	yes	
	15 uncooked animal tissue	1.476	yes	
	18 uncooked seed and seed coat deposition	1.809	no	yes, scale of 8
	17 pulverize roots, fruits or dried meat	2.198	yes	
	12 grain hides, alone	2.346	yes	
	6 burnt bone deposition	2.877	yes	
12 grain hides, alone	15 uncooked animal tissue	.624	yes	
	7 unburnt bone deposition	.838	yes	
15 uncooked animal tissue	12 grain hides, alone	.624	yes	
	4 bone working	.785	yes	
	7 unburnt bone deposition	.921	yes	
	24 red ochre deposition	.970	yes	
	6 burnt bone deposition	1.008	yes	
	3 hearth dumpings	1.018	yes	
	17 pulverize roots, fruit or dried meat	1.084	yes	
16 uncooked plant tissue	20 uncooked seeds or nuts; deposition	.805	no	yes, scale of 16
	18 uncooked seeds and seed coat deposition	1.098	no	yes, scales of 16,8
	15 uncooked animal tissue	1.159	no	yes, scale of 16
	12 grain hides, alone	1.303	no	yes, scale of 16
	4 bone working	1.476	no	yes, scales of 4,16

TABLE 80 (cont.)

Activity Group of Interest	Empirically Similar Activity Groups	Distance Coefficient	Similar Activity Group Expected To Be Similar?	Unexpectedly Similar Activity Group Explainable? How?
17 pulverize roots, fruits, meat	Level of error in scale so large that nearly all Activity Groups have Distant Coefficients less than the error level.			
18 uncooked seeds and seed coats	17 pulverize roots, fruits, dried meat	.430	no	yes, scales of 17,18
	20 uncooked seeds or nuts, deposition	.715	yes	
	22 boil nuts, seeds	.771	yes	
20 uncooked seeds or nuts	17 pulverize roots, fruits, dried meat	.430	no	yes, scales of 17,20
	18 uncooked seeds and seed coats, deposition	.715	yes	
	22 boil nuts and seeds	.771	yes	
22 boil nuts, seeds	20 uncooked seeds or nuts, deposition	.757	yes	
	18 uncooked seeds and seed coats, deposition	.771	yes	
23 boil greens	24 red ochre deposition	.607	no	yes, scales of 23,24
	17 pulverize roots, fruits or dried meat	.625	yes	
24 red ochre deposition	6 burnt bone deposition	.557	yes	yes, scales of 23,24
	23 boil greens	.607	no	
	7 unburnt bone deposition	.970	yes	
	3 hearth dumpings	1.200	yes	

TABLE 81
Accuracy of estimates of proportions of nutrient anomalies corresponding with given activity groups

Activity Group Number	Number of Cases	Smallest Increment in Proportions Observable, Non-Standardized Data	Smallest Increment in Proportions Observable, Standardized Data
1	13	7.692%	1.491%
3	28	3.570	.6919
4	11	9.091	1.762
6	28	3.570	.6919
7	17	5.882	1.140
8	8	12.500	2.422
12	25	4.000	.7753
15	17	5.882	1.140
16	13	7.692	1.491
17	9	11.111	2.154
18	12	8.330	1.615
19	9	11.111	2.154
20	25	4.000	.7753
22	12	8.330	1.615
23	12	8.330	1.615
24	9	11.111	2.154

proportions; and (*b*) when the list began to include activity groups that we would *not* expect to be similar to the activity group of interest, based on the dissimilarity of the refuse materials they produced from the refuse materials produced by the activity group of interest.

Comparison of the Cluster Analyses of Refuse Materials and Use-Areas

The data summarized in Tables 77, 79, and 80 describe those refuse materials most similar or distinct in their nutrient contents and those activity groups most similar or distinct in the nutrient anomalies they produced in the soils of the Crane site. These data may be compared to examine how natural

508   8. Soil Alterations Produced by Prehistoric Human Activities

soil processes have altered the nutrient contents of the refuse materials incorporated within the soils at Crane and have altered the distinguishability of the soils within various types of use-areas over time.

First, let us compare the activity groups found to be *most distinctive* in the nutrient anomalies within their corresponding use-areas to those activity groups that would be expected to be most distinctive based on the nutrient spectra of the refuse they produce. Table 82 shows that when the data allow comparison and spatial overlap among activity groups is not a confusing factor, the correspondence between expectation and fact is fairly good. Of the four activity groups found to be most distinctive in the spectra of nutrients within the soils of their use-areas, three were expected, based on the distinctive nutrient compositions of the residues they produced. Of the six refuse materials most distinctive in their nutrient contents and having corresponding activity groups that did not overlap significantly with other activity groups, three are associated with activity groups that were found to be most distinctive in their effects on the soils within their use-areas. Based on the nutrient contents of refuse materials that they produced, bone working and meat processing should have been found to be distinctive activity groups but were not when characterized by the soil data. Woodworking was found to be distinctive when characterized by the soil data but was not expected to be.

In general, the analysis suggests that the nutrient spectra of the most distinctive refuse materials deposited on Crane were maintained within the soils where they were deposited. The natural processes of incorporation of refuse materials into soils, including decomposition and resynthesis, and subsequent processes of alteration did not obscure the distinctive nutrient spectra. This circumstance probably is of a general rather than a site-specific nature. Where the distinctive refuse materials analyzed here have been deposited on other earthen sites located on loessic soils within the Midwest United States, they probably can be expected to have produced distinctive soil alterations (i.e., chemically distinct use-areas). Where lacks of correspondence were found between activity groups expected to be distinctive and those actually found to be distinctive at Crane, it is suggested that site-specific processes may be reflected. Bone working, meat processing, woodworking, and graining hides (a possible case of lack of correspondence) all are activities for which smaller amounts of refuse deposition and the creation of smaller soil anomalies within use-areas are predictable at Crane compared to, for example, the deposition of shell within mollusk processing areas or ash in hearth dumping locations. The smaller soil anomalies possibly produced by these activity groups would have allowed greater degrees of error in estimating nutrient anomaly proportions and could account for the lack of correspondence found in the analysis. The activity groups and refuse materials for which lack of correspondence were found do not form other patterns along more general, non-site-specific dimensions, such as the particular nutrients occurring in high or low concentrations within the refuse materials.

TABLE 82
A comparison of refuse materials most distinctive in their nutrient composition and activity groups at Crane site found to be most distinctive in the soils within their use-areas

Most Distinctive Refuse Materials	Corresponding Activity Groups Expected to be Most Distinctive	Corresponding Activity Groups Found to be Most Distinctive	Spatial Overlap in Activity Groups Explain Their Lack of Distinctiveness?
ash of wood	hearth dumping; graining hides, alone	hearth dumping	
dry boiled bone	unburnt and burnt bone deposition, bone working	corresponding activity groups not found to be distinctive	yes, except for bone working
dry unboiled bone	unburnt and burnt bone deposition, bone working	corresponding activity groups not found to be distinctive	yes, except for bone working
mammal meat	uncooked animal tissue deposition	corresponding activity groups not found to be distinctive	no
mollusk shell	mollusk processing	mollusk processing	
nut meats in general	uncooked nuts, deposition	uncooked nuts, deposition	
corresponding refuse not found to be distinctive	----	woodworking	

Next, let us consider the activity groups at Crane that were found to be *most similar* in their effects on soil chemistry compared to those we would expect to be similar based on the nutrient contents of the refuse they yield. Table 80 shows the relevant data. Again, the data suggest that the relative degrees of distinctiveness of different refuse materials have been maintained over time after their deposition and incorporation within the Crane soils. In all but one case, where the degree of activity overlap was low enough and the degree of accuracy of nutrient anomaly proportions was high enough to allow comparison, those activity groups expected to be most similar in their effects on soil chemistry were found to be most similar. The exception is activity group 16 (uncooked plant tissue deposition). None of the activity groups found to be similar to activity group 16 were expected to be similar. This might be a result of the inaccuracy of the nutrient anomaly proportions calculated for activity group 16, but the *total* lack of correspondence would suggest a systematic source of error rather than random measurement errors. It is possible, as stated in Chapter 5, that the lithic tools assigned the function of shredding plant tissue were in fact used for other unknown purposes.

As a check on the analysis just described, and to ensure that its outcome is not an artifact of the particular point at which the list of activity groups "most similar" to each activity group of interest was truncated, another analysis of the relative distinctiveness of activity groups was performed. For each activity group, all other activity groups were divided into two sets—those that would be expected to have more similar effects on the Crane soil chemistry and those that would be expected to have less similar effects (Table 83), based on the nutrient compositions of the refuse they produce. For each of these sets, the distance coefficients relating the members of the set to the activity group of interest were averaged. If the relative degrees of distinctiveness of different refuse materials have been maintained over time at the Crane site within the soils where they were deposited, then for each activity group of interest, the average distance coefficients of those activity groups expected to be most similar to it should be less than the average distance coefficient of those activity groups expected to be less similar to it. These comparisons were made, and are shown in Table 84. In nearly all comparisons, the expected ordered relation of means was found. Where it was not, the unexpected results usually can be explained by the low level of accuracy of the distance coefficients, which is related to the low level of accuracy of the calculated nutrient anomaly proportions (Table 81).

Two activity groups for which measurement inaccuracies cannot explain the unexpected order relation of means are 16^3 (uncooked plant tissue deposition)

---

3. Activity group 16 was not used within any of the comparisons made for other activity groups, in consideration of the results obtained in the previous analysis and the possibility of misidentification of the function of artifacts representing this group. It was, however, analyzed as an activity group of interest.

In most comparisons, activity group 20, a composite of activity groups 18 and 19, was used instead of 18 and 19, individually, in order to increase the number of soil samples from which

TABLE 83
Activity groups expected to be similar and dissimilar to each other, based upon the similarity of the nutrient compositions of the refuses they yield (Figure 60)

Activity Group of Interest	Activity Groups Expected to be Similar	Activity Groups Expected to be Dissimilar
1	16,17,19,22,23	3,4,6,7,8,12,15,18,24
3	4,6,7,8,12,15,24	1,16,17,20,22,23
4	3,6,7,8,12,15,24	1,16,17,20,22,23
6	3,4,7,8,12,15,24	1,16,17,20,22,23
7	3,4,6,8,12,15,24	1,16,17,20,22,23
8	3,4,6,7,12,15,24	1,16,17,20,22,23
12	3,4,6,7,8,15,24	1,16,17,20,22,23
15	3,4,6,7,8,12,24	1,16,17,20,22,23
16	1,17,19,22,23	3,4,6,7,8,12,15,18,24
17 (roots)	1,16,19,22,23	3,4,6,7,8,12,15,18,24
17 (meat)	3,4,6,7,8,12,15,24	1,16,20,22,23
18	22	1,3,4,6,7,8,12,15,16,17,23,24
19	22	1,3,4,6,7,8,12,15,16,17,23,24
20	22	1,3,4,6,7,8,12,15,16,17,23,24
22	20	1,3,4,6,7,8,12,15,16,17,23,24
23	1,17,19,22	3,4,6,7,8,12,15,18,24
24	3,4,6,7,12	1,8,15,16,17,20,22,23

and 12 (graining hides, alone). As in the previous analysis, the results obtained for Activity Group 16 may reflect the misidentification of the artifacts representing the activity group. The unexpected relation found for Activity Group 12 may result from the smaller soil anomalies produced by it and the greater degrees of error this allowed in estimating nutrient anomaly proportions and distance coefficients. This is a very uncertain explanation, however.

---

nutrient anomaly proportions were calculated and to increase the level of accuracy of the distance coefficients (Table 81). Activity groups 18 and 19 had to be used when making comparisons for activity groups 16 and 17, however, in order to distinguish between activities yielding nut refuse (group 19) and those yielding seed refuse (group 18).

TABLE 84
Comparison of average distance coefficients of activity groups expected to be similar versus dissimilar to a given activity group

Activity Group of Interest	Distance Coefficients of Activity Groups Expected To Be Similar $\mu_1 \pm \sigma$	N	Distance Coefficients of Activity Groups Expected To Be Dissimilar $\mu_2 \pm \sigma$	N	Mann-Whitney U Statistic	Significance Level One-Sided Test	Distance Coefficients of Activity Groups Expected to be Similar < Distance Coefficients of Activity Groups Expected to be Dissimilar?	Unexpected Results Explainable by High Level of Error in the Distance Coefficients of the Activity Group of Interest?
1	2.3039±1.0943	5	3.01812±1.2210	9	13.0	.0664	yes	
3	1.9130±.8209	7	2.7568±1.1798	6	11.0	.1917	yes	
4	2.2118±.8004	7	2.1647±1.1106	6	----	----	no	yes
6	1.6588±.8476	7	1.9241±.8264	6	14.0	.1917	yes	
7	1.2525±.5212	7	1.4159±.4744	6	16.5	.0387	yes	
8	2.4070±.8704	7	3.5017±1.0218	6	9.0	.0387	yes	

12	2.0498±.9251	7	1.6361±.2963	6	---	---	no
15	.9717±.2633	7	1.6361±.2963	6	3.0	.0041	yes
16	2.007±.3729	5	1.9821±.9246	9	---	---	no
17, roots	1.5659±.8355	5	1.4918±.5498	9	---	---	no yes
17, meat	1.6245±.4053	8	1.1536±.6429	6	---	---	no yes
18	.771	1	1.5596±.6888	12	---	---	yes
19	2.2747	1	1.5596±.6888	12	---	---	no yes
20	.771	1	1.5596±.6888	12	---	---	yes
22	.757	1	1.5596±.6888	12	---	---	yes
23	1.4717±.6017	4	1.5578±.8057	9	18.0	.1713	yes
24	1.7623±1.210	5	2.4279±.9693	8	14.0	.2063	yes

In summary, the analyses presented in this section suggest that at the Crane site and other midwestern sites located on loessic soils, different kinds of use-areas can be distinct in the spectra of nutrient anomalies within their soils. The degree to which they are differentiated depends on the degree to which the refuse materials deposited within them differ in their nutrient compositions. The distinctive spectra of nutrients within refuse materials is maintained over time in the soils of the area where they are deposited, in most cases.

Conclusions

Analyses of the soil chemistry data from the Crane site would suggest several generalities.

1. Enrichments of N, P, K, Ca, Mg, Na, and S made to the soils of Crane site during its occupation were maintained in some areas of the site over extended periods of time.
2. The degree to which nutrient enrichments were maintained within the Crane soils depended on the nutrient of interest and how tightly it was fixed within the Crane soils by natural mechanisms and processes. The occurrence of detectable nutrient anomalies also probably depended on the amount of the nutrients enriched to the soil.
3. The soils within different kinds of use-areas at Crane were distinguishable in the spectra of their nutrient anomalies.
4. The spectra of nutrient anomalies characterizing the soils of the different use-areas are predictable, based on the nutrient contents of the refuse materials deposited within them.
5. Natural soil processes, over time, certainly altered the ratios of the magnitudes of anomalies of different nutrients within the soils of all use-areas at Crane site. The more general spectra of presence and absence states of high nutrient concentrations found within the refuse materials deposited in different use-areas, however, were maintained within the soils of those use-areas by such processes.

Each of these conclusions in based on activities and soil conditions specific to the Crane site. The conclusions drawn from the Crane data are, nevertheless, probably applicable to a wide variety of sites located on loessic soils within the Midwest, where natural soil processes and climatic conditions are similar to those at Crane. Their applicability to sites located on coarse-textured acidic soils where loss of nutrients by leaching is more significant is questionable.

It is important that these general conclusions are based on chemistry data pertaining to *exchangeable* nutrient availabilities rather than *total* nutrient concentrations—fixed and available. It is the exchangeable nutrients, rather than the fixed, which are in dynamic equilibrium with ions in the soil water solution and which have a direct effect on soil conductivity values. The results obtained allow one to argue that different kinds of use-areas should be distinguishable in their soil conductivity values as well as their soil spectra, all else held constant, including the magnitude of anomalous nutrient availabilities found within them.

This cannot be concluded from most previous studies of the soils chemistry of archeological sites, which examine total nutrient concentrations rather than exchangeable nutrient availabilities.

Relationships between Use-Areas and Soil Chemistry at the Crane Site: Differentiation of Use-Areas by the Magnitudes of Their Nutrient Anomalies

In the previous section, concern has been with particular *nutrients* and *nutrient spectra*—whether they were maintained within the Crane soils and whether they distinguished different kinds of use-areas. The *magnitudes* of nutrient anomalies were ignored in the previous analyses and were considered site-specific phenomena rather than conditions shedding light on more general archeological formation processes and soil processes. It is obvious from former studies of the chemistry of soils in archeological sites (Chapter 4), however, that in many circumstances, the magnitude of nutrient anomalies within soils, as well as the spectra of nutrient anomalies within soils, carries significant information about the nature of activities and use-areas within sites.

The amounts of nutrients enriched and maintained within the soils of a use-area minimally depends on the following factors: (*a, b, c*) the amount of refuse deposited within it as a function of the *length of time* it was used, the *kinds of activities* that occurred in it, and the *number of activities* which occurred within it; (*d*) the *concentrations of nutrients* within the refuse materials deposited; and (*e*) local soil, topographic, vegetational, and other *natural conditions*. The length of time an area is used and the number of activities for which it is used may depend upon site-specific factors such as locations of shade or of pleasant working conditions. They also may depend, however, on more general factors. For example, house locations, multipurpose work areas around them, and areas for dumping domestic refuse can be expected to receive additions of refuse more often and from more activities than can some more specialized work areas located away from such foci of activity. The particular amounts of debris produced by given kinds of activities per unit activity and the concentration of nutrients within those debris also are general factors linking the magnitude of soil nutrient enrichments to the nature of activities and use-areas. For example, hide graining, bone working, meat processing, and possibly woodworking can be expected to produce less debris per unit activity than mollusk processing or hearth cleaning. Also, the debris produced by the former are less nutrient-rich and less capable of yielding significant soil anomalies than those produced by the latter.

To illustrate the information on prehistoric activity that the magnitudes of nutrient anomalies within use-areas may carry, let us consider the use-areas at Crane immediately surrounding the Middle Woodland house within the block excavation (Figure 48). In Chapter 5, it was shown that a number of archeological variables, including the spatial distribution of pit features and open work areas (Figure 38), the depth of midden deposits (Figure 39), and the variety of artifact types found within surface collection units (Figure 40), had an annular pattern, centering around the house. Open work areas tend to occur immediately adjacent to the house, with clusters of pits surrounding them. It was suggested that the spatial distribution of midden deposits and artifact types was in a part a result of the Crane occupants having periodically swept multipurpose open work areas free of debris, and having deposited the refuse in localized garbage dumps exterior to the work areas.[4] If this interpretation is correct, it is expectable that the enrichment of nutrients to soils within the dumping areas was greater than that within the swept areas, and that these two generalized use-zones might be differentiated, on the average, by the magnitudes of nutrient anomalies within them. We are interested here in the *total* nutrient enrichments to the soils of these two zones rather than the *spectra* of their nutrient enrichments since the two zones contain debris from the *same* set of activities, only in differing amounts. Let us compare these two kinds of use-zones with respect to their average total nutrient enrichments, to see whether expectations are met.

As a first step in the analysis, it was necessary to classify portions of the block as either swept work areas or dumping locations. This was not done in Chapter 5, where only the general annular pattern of archeological variables was noted. Using the spatial distribution of pit features within the block, it is possible to divide this area into 11 spatial strata believed to be homogeneous in their use as working or dumping locations (Figure 61). These 11 areas include: (*a*) rings of pits spatially isolated by themselves (Strata 1–3), (*b*) dense clusters of pits, sometimes intrusive upon each other (Strata 4–5), indicating repeated reuse of the area, and (*c*) blank areas within a matrix of pitted areas (Strata 6–11). Rings of pits isolated by themselves obviously were foci of tasks, and presumably were swept clean periodically of accumulating debris. They therefore were used to indicate swept areas.[5] Dense clusters of pits were used to indicate dumping areas. In these areas, not all of the pits could have been used simultaneously, for lack of work space, and several episodes of pit digging, working, and pit infilling are certain. Garbage was used to fill many of the pits in these areas. Although

---

[4]. Sweeping was incomplete, however; swept areas, contain some midden accumulation, although less than that within the dumping areas.

[5]. An exception is Stratum 3, which had extremely deep midden deposits and a high variety of artifact types.

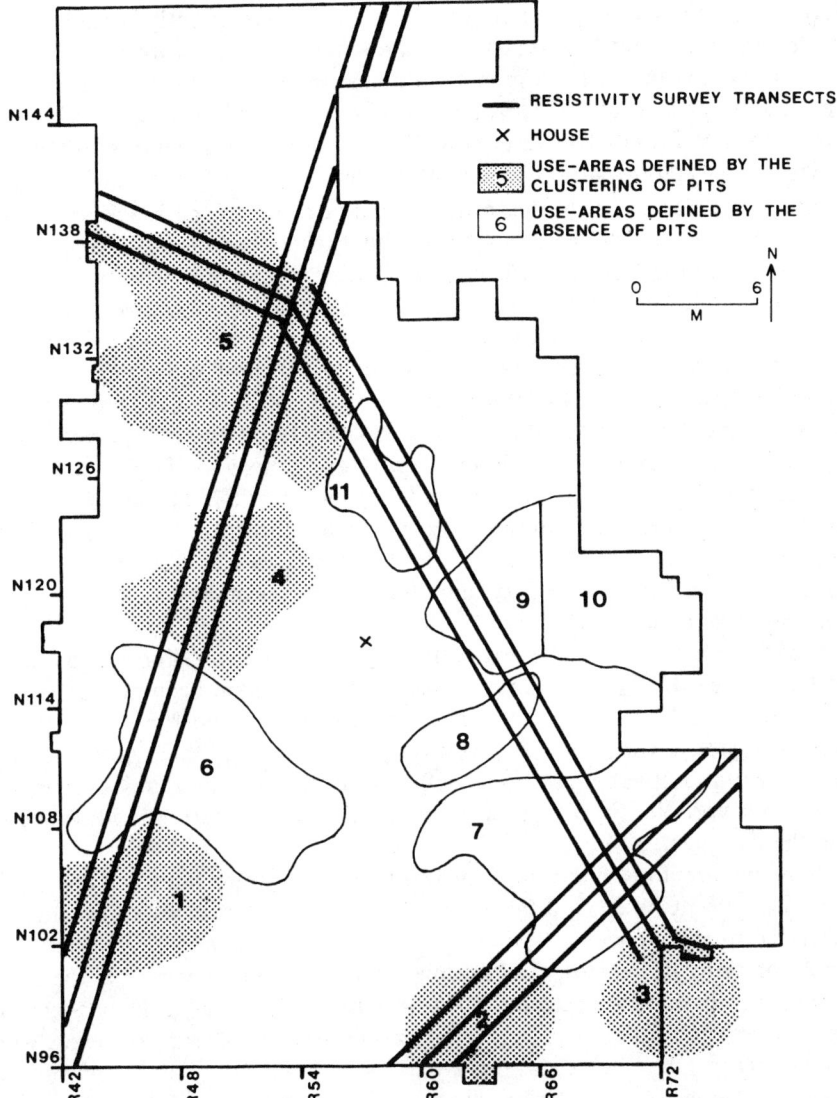

FIGURE 61. *Locations of several different use-areas around the house at the Crane site.*

such areas were locations of work, the *ultimate* result of their repeated reuse was garbage accumulation. Areas that lack pits could have been used as either multipurpose work areas or locations of refuse dumps. To decide which of these functions should be assigned to particular spatial strata, two variables were examined: the variety of artifact types found within surface collection units encompassing them, and midden depth. It was assumed that garbage dumps received refuse from multiple open work areas, and consequently should contain a greater variety of artifact types than the singular, open work areas around them. They also should have greater midden accumulations. On the basis of these assumptions, Strata 6, 8, 9, and 11 were classified as swept work areas, whereas Strata 7 and 10 were classified as dumping locations. Table 85 summarizes the nature, archeological attributes, and function of all 11 strata.

Next, it was necessary to determine which portions of the block and which soil samples taken within the block during the transect and block surveys should be considered in the analysis. Since the analysis is concerned with the absolute magnitudes of nutrient anomalies and not just their presence or absence, it was necessary to include only those soil samples that were taken from portions of the block homogeneous in parent material, degree of soil profile development, and historic land-use. Divisions of the block along these dimensions are shown in Figures 56 and 58. A discontinuity in parent material occurs approximately along the N138 line, soils at the 60-cm depth having more sand north of this line and more clay south of it. The boundary between historic land-use Areas 1 and 3 crosscuts the block in a southwest to northeast direction between approximately the N136 and N145 lines. To keep the analysis as free as possible of natural and historic soil variability, only those portions of the block south of the N138 line were considered.

The horizon for which archeologically significant soil variation was assessed is Barnes Layer BL2–1. This layer contains only midden accumulations, and does not include any of the submidden, natural soil horizons anywhere within the areas of study. Consequently, any variations found in the availabilities of nutrients within soil samples should reflect only differences in midden chemistry, and not variation in the proportions of midden and natural soil occurring within the samples.

Unlike the previous analysis, it was not necessary to subtract the availabilities of nutrients in BL5–4 from those in BL2–1 to control for soil variations resulting from differences in parent material. This source of variation was removed spatially by selecting soil samples from only those portions of the block homogeneous in parent material. *Constant* natural and historic background levels of nutrient availability, pH, and organic matter content, however, were subtracted from the soil data for BL2–1, in order to give a measure of archeologically significant, *anomalous* nutrient availability. The appropriate constant (Table 86) subtracted from each characteristic was determined using either its mean value within soil samples taken at a similar depth, offsite, immediately east of the block, or else the minimum value[6] of the characteristic within the soil samples taken in the block. The latter was used only when the mean offsite value was found to be inappropriately large as an estimate of

---

6. Very low values that occurred far out in the lower tail of the distribution of values for the characteristic for samples taken from the block, and that obviously represented experimental error, were excluded from the analysis. They were not used as minimum values of the characteristic.

TABLE 85
Soil characteristics of several different use-areas around the house at Crane site

Stratum Number*	Kind of Use-Space	Corresponding Surface Collection Unit(s)	Number of Artifact Types in the Collection Unit(s)	Local High or Low in Number of Artifact Types	Average Midden Depth in the Collection Unit(s)	Mean Total Anomalous Nutrient Availability (kg/ha)	Swept Area or a Dump?	Mean Anomalous Organic Matter Content (%)	Mean Anomalous pH
4	dense cluster of pits	379	10	L	34.72	7568.	dump	1.851	.730
5	dense cluster of pits**	350, 351, 364	15,16,17	H	34.37	7428.	dump	.603	.566
7	blank area	431	13	H	32.61	7192.	dump	.720	.666
8	blank area	413	9	L	30.67	6990.	swept	.287	.455
10	blank area	397	10	H	29.17	6870.	dump	.284	.730
3	ring of pits	448,449	6, 12	H	40.16	6834.	dump	.617	.413
11	blank area	380	7	L	31.95	6259.	swept	.428	.608
9	blank area	396	5	L	30.40	4762.	swept	.266	.613
2	ring of pits	447	7	L	36.24	3995.	swept	.587	.447
6	blank area	411	11	L	30.38	3795.	swept	.684	.430
1	ring of pits	427	4	L	30.04	2380.	swept	.707	.480

*Strata are located spatially in Fig. 61.
**Indicates a portion of the Middle Woodland house

TABLE 86
Background levels of nutrient availabilities and organic matter contents within the block at Crane site

pH	6.02
P	61.92 kg/ha
Mg	234.7
Ca	3582.6
Na	28.3
K	143.5
NH_4	24.9
SO_4	2.58
organic matter	0.7 %

Proveniences of off-site soil samples used in estimating background levels:

Transect II: 2003, 2008, 2016, 2024, 2033, 2404, 2405, 2421, 2429
Transect IV: 7103, 7122, 7131, 7506, 7514, 7527

---

background noise—in particular, when it was found to be greater than the value of the characteristic for many of the soil samples within the block. Characteristics with background noise levels determined in this manner include the availabilities of Ca, NH_4, and SO_4, and organic matter content.

The particular soil cores used to characterize each of the 11 strata are listed in Table 87. Soil samples that occurred at the peripheries of the spatial strata, or within or very near to pits within the strata, were not used in the analysis. The latter were not used because the pit fill could have been derived from debris produced by a more limited range of activities than would general midden deposits. Soil samples having extremely high or low values for a number of attributes, and which obviously reflected lab errors, also were excluded from the analysis.

For each stratum, the average anomalous availabilities of N, PO_4, K, Ca, Mg, Na, and SO_4 were determined. These values were then summed to give an estimate of the mean total anomalous nutrient availability for each stratum (Table 85). The mean anomalous organic matter content and mean anomalous pH of the soil in each stratum also were determined (Table 85). As predicted, the soils within those strata used as dumping locations were found to have greater total nutrient enrichments than the soils within strata used as work areas, which were swept periodically. Strata used as dumps have total nutrient

TABLE 87
Provenience of soil samples used to characterize chemical attributes of the several different use-areas around the house at Crane site, plus background noise levels

Stratum	Provenience
1	4865, 6265
2	6964, 6966, 97R63
3	2137, 2539, 2550
4	4827, 4833, 6226, 124R535
5	2213, 2216, 2223, 2231, 2240, 2619, 2633, 6191, 6202, 137.5R54, 131R51.5, 130.5R53, 114R74
6	4841, 4846, 6240, 6248
7	2153, 2161, 111R69, 109R68.5, 109R69.5, 106.5R66, 102R72
8	112R64.5, 110.5R61, 110R64, 108R64.5
9	2180, 2183, 2185
10	121R66, 118R69
11	2192, 2207, 2595, 2606, 129.5R57, 127.5R59.5, 127R58
Off Site	2003, 2008, 2016, 2024, 2033, 2404, 2405, 2421, 2429, 7103, 7122, 7131, 7506, 7514, 7527

enrichments between 6834 and 7586 kg/ha. Swept areas have enrichments of 2380 to 6259 kg/ha. The single exception to this pattern is Stratum 8, which has all the characteristics of a swept area but a total anomalous nutrient availability of 6990 kg/ha.

Examination of the degree of correspondence between anomalous soil pH and total anomalous nutrient availability, and between anomalous organic matter content and total anomalous nutrient availability, gives some insight into the processes by which the various midden deposits within the block were formed. Anomalous soil pH correlates with the magnitude of nutrient enrichments within the several spatial strata, whereas anomalous organic matter content does not. This would suggest that the different total nutrient enrichments found within dumping locations and swept work areas is not primarily a result of differences in the numbers of exchange sites and humus micelles added to the soils within the two kinds of areas, but rather, is largely a result of the different degrees to which exchangeable H has been replaced by nutrient cations on the exchange sites of clay–humus micelles already existing there. Differences in total cation accumulation, rather than the degree of humus formation and generation of exchange sites, was responsible for the differences found in total nutrient enrichments within dumping locations and swept work areas.

Finally, it can be asked which particular nutrients are most responsible for the different total anomalous nutrient availabilities found in the two kinds of use areas. Table 88 shows the average anomalous nutrient availabilities within swept areas and within dumping locations for each nutrient used in the analysis.[7] Differences between these two kinds of areas in their average nutrient enrichments are greatest and statistically most significant for Ca and PO_4. Magnesium, Na, and K all show slightly greater enrichments on the average within dumping areas than swept work areas, but only Mg shows a statistically significant difference. Sodium, K, NH_4, and SO_4 each can be assumed to occur in similar concentrations within the two kinds of use-areas.

The responsibility of primarily Ca and P in determining the different total nutrient enrichments within swept work areas and dumping locations is reasonable. Most of the strata have high densities of hearth dumping materials (and presumably ash), burnt bone, and/or unburnt bone. Bone and ash are both extremely high in Ca and P. The determining role of Ca and P also can be attributed to the great strength with which they are fixed in insoluble, unleachable, yet slowly available forms within alkaline soils such as those at Crane. Among these fixed forms are calcium phosphate compounds. The joint occurrence of Ca and P in great concentrations within bone and ash and their mutual fixation of each other as calcium phosphates within soils analyzed here is an example of the "interaction of nutrients" described in Chapter 4 as one mechanism by which anthropic nutrient enrichments are maintained in soils.

The greater importance of Mg and K over other nutrients such as Na and SO_4 in determining differences in the total nutrient enrichments within swept areas and dumping locations is probably attributable to their fixation within clays and their lesser susceptibility to leaching. Ammonium radicals also are fixed within clays but were not found significant in this analysis probably because N is not a major constituent of the refuse materials deposited in greatest quantities within the block. Nitrogen occurs in only trace amounts within wood ash and boiled bone.

In summary, when trying to distinguish different use-areas of an archeological site using soil chemistry data, the magnitudes of nutrient anomalies can be just as important as the spectra of nutrient anomalies. At the Crane site, it has been possible to distinguish swept work areas from trash dumps within the vicinity of the house by examining the total amounts of nutrients enriched to the soils within these areas, and in particular, the amounts of Ca and P. Similar results have been obtained by Dietz (1957:405) for a prehistoric habitation site in Wisconsin, using the magnitudes of total inorganic P concentrations within the

---

7. Stratum 8, which did not show greater total nutrient enrichments within dumping locations than within swept work areas, is not included in the calculated averages.

TABLE 88
Nutrients responsible for differences in the total anomalous nutrient availabilities found between swept and dumping areas around the house at Crane site

	NH_4	PO_4	K	Nutrient Ca	Mg	Na	SO_4
Swept Areas	44.6±15.2	705±420	131±20.8	3358±903	76.9±34.1	7.81±5.94	16.73±8.49
Dumping Areas	33.3±20.7	1207±110	142±34.2	5657±271	113±44.0	9.84±2.96	12.47±3.85
Differences in Grand Means	-11.1	502	11	2299	36.1	2.03	-4.26
Sample size (Swept Areas, Dumping Areas)	5,5	5,5	5,5	5,5	5,5	5,5	5,5
T-Statistic Testing Difference in Means	---	2.239	.532	9.44	1.256	.592	---
Level of Significance for One-Sided Test	---	.04	.31	.001	.04	.28	---

soils. As at Crane, hearth cleanings were specified as a major source of differences in P concentrations. The findings presented here and in Dietz's study parallel in a more general way than those of other researchers (Arhennius 1929, 1931; Cook and Heizer 1965:58; Dauncey 1952:35; Hurley and Heidenreich 1971: 185; Provan1971:43; Weide 1966:159–161) who have documented correlations between the intensity of prehistoric activity or refuse deposition within portions of sites and the magnitude of soil anomalies, or between the length of occupation of sites and the magnitude of soil anomalies (Chapter 4).

It is important that this study of the magnitudes of nutrient anomalies, like that of the spectra of nutrient anomalies, is based on data on *exchangeable* nutrient availabilities rather than *total* nutrient concentrations, that is, fixed plus available nutrients. Only the exchangeable nutrients are in dynamic equilibrium with ions in the soil water solution and have a direct effect on soil conductivity values.

Physical Alterations of the Soils at Crane Site: Pit Fills Compared to Their Natural Soil Matrices

The physical soil attributes that have been altered by human occupation at the Crane site and that are responsible for variation in soil resistivity values there include: the degree of soil aggregation, as indicated by soil pore-size distribution and total porosity; and soil moisture retention. As discussed in Chapter 3, changes in all of these variables are induced primarily by the addition of organic matter, the extent of change depending more on the *amount* of organic matter added than on its *chemical nature*. Consequently, the following analysis of the physical variations produced in the soils at Crane will relate them to differences in the *amounts* of activity and refuse deposition that occurred in different locations rather than the kinds of activity. Inasmuch as the lengths of time that the different areas of Crane were utilized and the magnitudes of soil variation are site-specific, the numerical results of the analyses to follow will serve only an illustrative purpose.

In all, three analyses, presented in the following three sections, will be made. Two are at the level of greatest contrast, comparing clearly disturbed soils (pit fills, middens) to clearly undisturbed soils (pit matrices, B horizons underlying midden deposits). The third examines differences between soils disturbed to various degrees in locations where different amounts of refuse have been deposited and different amounts of midden have accumulated.

The first analysis compares the soil fills of pits at Crane that have been anthropically enriched in organic matter to their natural soil matrices that have not been so enriched. The two soil types are examined for differences in their structural organization and moisture retention equilibria. Paired soil samples from a number of pits and their matrices were obtained according to the sampling design outlined in Chapter 6 and analyzed for their bulk densities, particle-size distributions, pore-size distributions, and organic matter contents. The raw data for the analyses are given in Appendix 7, Tables 3 and 4. Tables 89 and 90, respectively, show the average cumulative moisture characteristic curves of the sampled pit fills and their matrices, and the average moisture capacities of several pore-size classes within the fills and matrices.

In making comparisons between each pit fill and its natural soil matrix, it will be assumed that any differences found in their structural organizations or moisture-retentive powers are attributable entirely to differences in their organic matter contents or mechanical loosening, both of which in turn are attributable to prehistoric human disturbances. Pit fills and their natural soil matrices will be assumed to be similar in their *natural* determinants of soil aggregation and/or moisture retention, including: particle-size distribution; dominant clay types; their original, naturally occurring, relative concentrations of cations promoting either dispersion or flocculation of their soil colloids; and their original, naturally occurring concentrations of organic matter. This is not an unreasonable assumption, if one considers that the primary component of each pit fill probably is natural soil derived from the immediate surroundings of the pit. In most cases, differences in the particle-size distributions of the fills and their matrices are insignificant (Appendix 7, Table 3). Provenience I28–5 F1 is an exception, where the pit fill is much more coarsely textured than its matrix; it has been eliminated from the analysis.

The particular questions pertinent to this analysis are the following:

1. Do the disturbed and undisturbed soils differ in their total porosities and moisture-holding capacities?
2. Do they differ in their pore-size distributions, and, if so, are the differences concentrated within any particular pore-size range?
3. At what environmental suctions do the disturbed soils hold more, less, or the same amounts of water as their natural counterparts; that is, how do the cumulative moisture characteristic curves of the disturbed and undisturbed soils compare?
4. Can differences in the structural organization and moisture-retention equilibria of disturbed and undisturbed soils be related to differences in their organic matter contents?
5. What are the implications of these findings on resistivity surveys?

In comparing the fills of pits to their natural soil matrices, a number of patterns may be found. In every comparison between a pit fill and its natural matrix, the fill material has a greater total porosity and moisture-holding capacity than the natural soil surrounding it. These increases in porosity can be shown in several ways to be a direct consequence of anthropic enrichments of organic matter to the fill materials and the development of greater degrees of soil granulation and aggregation within the fill materials. In each case, pit fills

TABLE 89
Average moisture contents of fills of pits and their surrounding matrices at several suctions*

Provenience	0 Bar (Total Porosity)	.05 Bar	.1 Bar	.2 Bar	.5 Bar
			Suction		
Pit TI-2F1	.4260/.6479**	.2581/.3926	.2451/.3728	.2190/.3331	.2032/.3091
Matrix TI-2F1	.4124/.6421	.2314/.3914	.2522/.3927	.2334/.3634	.2057/.3203
Pit TI-403A	.4735/.6605	.2308/.3917	.2580/.3599	.2526/.3524	.2149/.2998
Matrix TI-403B	.3939/.6326	.2405/.3862	.2355/.3782	.2182/.3504	.2062/.3312
Pit TI-4F1	.5267/.6605	.4354/.5084	.3878/.4863	.3767/.4724	.3355/.4207
Matrix TI-4F1	.4113/.6416	.2652/.4137	.2602/.4059	.2502/.3903	.2330/.3635
Pit TI-13F1	.4354/.6514	.2758/.4126	.2417/.3616	.2344/.3507	.2096/.3136
Matrix TI-13F1	.4230/.6468	.2646/.4046	.2545/.3891	.2488/.3804	.2100/.3211
Pit TIV-16F1	.4448/.6543	.2404/.3536	.2300/.3383	.2217/.3261	.2038/.2998
Matrix TIV-16F1	.4124/.6421	.2403/.3741	.2268/.3531	.2163/.3368	.1934/.3011
Pit TIV-17F1	.4513/.6562	.2648/.3850	.2541/.3695	.2417/.3514	.2200/.3199
Matrix TIV-17F1	.4460/.6547	.2788/.4093	.2658/.3902	.2532/.3717	.2436/.3576
Pit TIV-18F1	.4430/.6689	.2456/.3708	.2431/.3671	.2325/.3511	.2188/.3304
Matrix TIV-18F1	.4135/.6525	.2662/.4201	.2635/.4158	.2515/.3969	.2322/.3664
Pit I28-5F1	.4923/.6623	.2652/.3564	.2460/.3306	.2426/.3260	.2147/.2886
Matrix I28-5F1	.4392/.6526	.2468/.3667	.2544/.3780	.2469/.3669	.2369/.3520

*Moisture contents are given on both a weight basis (g H_2O/g soil; first value) and a volume basis (g H_2O/cc soil; second value). Underlined numbers indicate environmental suctions at which the fills of pits hold more water than their matrices.

**Underlined numbers indicate environmental suctions at which the fills of pits hold more water than their matrices.

TABLE 90
Average moisture capacities of several pore size classes (suction) within the fills of pits and their matrices*

Provenience	Suction						Total Porosity
	0-.05 Bar	.05-.1 Bar	.1-.2 Bar	.2-.5 Bar	>5 Bar		
Pit TI-2F1	.1679/.2553**	.0131/.0199	.0260/.0395	.0158/.0240	.2032/.3091		.4260
Matrix TI-2F1	.1610/.2507	.0137/.0213	.0188/.0293	.0278/.0433	.2051/.3203		.4124
Pit TI-403A	.1927/.2688	.0196/.0273	.0039/.0054	.0378/.0527	.2149/.2997		.4735
Matrix TI-403B	.1534/.2464	.0138/.0222	.0174/.0279	.0119/.0191	.2062/.3312		.3939
Pit TI-4F1	.1473/.1847	.0758/.0951	.0577/.0724	.0412/.0213	.3355/.4207		.5267
Matrix TI-4F1	.1462/.2281	.0049/.0076	.0101/.0158	.0172/.0268	.2330/.3635		.4113
Pit TI-13F1	.1596/.2388	.0340/.0509	.0074/.0111	.0248/.0371	.2096/.3136		.4354
Matrix TI-13F1	.1584/.2422	.0100/.0153	.0057/.0087	.0389/.0595	.2100/.3211		.4230
Pit TIV-16F1	.2044/.3007	.0172/.0254	.0120/.0177	.0179/.0263	.2038/.2998		.4448
Matrix TIV-16F1	.1721/.2680	.0135/.0210	.0105/.0163	.0229/.0356	.1934/.3011		.4124
Pit TIV-17F1	.1865/.2712	.0107/.0156	.0124/.0180	.0218/.0317	.2200/.3199		.4513
Matrix TIV-17F1	.1672/.2454	.0131/.0192	.0126/.0185	.0095/.0139	.2436/.3576		.4460
Pit TIV-18F1	.1974/.2914	.0026/.0038	.0106/.0156	.0137/.0202	.2188/.3229		.4430
Matrix TIV-18F1	.1472/.2287	.0056/.0087	.0120/.0186	.0194/.0301	.2322/.3608		.4135
Pit I28-5F1	.2276/.3059	.0191/.0257	.0230/.0309	.0280/.0376	.2147/.2886		.4928
Matrix I28-5F1	.1924/.2859	.0079/.0117	.0075/.0111	.0101/.0150	.2369/.3520		.4392

*Moisture capacities are given on both a weight basis (g H_2O/g soil; first value) and a volume basis (g H_2O/cc soil; second value).

**Underlined numbers indicate pore size range over which the fills of pits have greater volumes than their matrices.

contain more organic matter than their matrices (Appendix 7, Table 3). Also, a smooth functional relationship can be found between the *excess* of organic matter within pit fills as compared to their matrices and the *increase* in porosity of the pit fills in comparison to their matrices (Figure 62). Third, the greater total porosities of the pit fills can be shown to be attributable primarily to increases in the volume of their pores within the larger size classes (0–.5 bar suction) rather than in the volume of their pores within the smaller size classes (Table 90). This is the predictable pattern if the greater porosities of the disturbed soils are attributable to their greater degrees of aggregation and their greater organic matter contents (see Chapter 3). Finally, the more specific increases in void space within the larger pore-size classes of the disturbed soils, like their increases in total porosity, can be shown to exhibit a smooth functional relationship with the excesses of organic matter found within pit fills (Figure 62). Thus, it may be concluded that anthropic enrichments of organic matter to the fills of pits at Crane site have increased their total porosities, altered their pore-size distributions, and have increased their degree of structural aggregation in comparison to their natural soil matrices.

The particular manner in which the pore-size distributions of pit fills have been altered from those of their surrounding matrices by organic enrichments varies considerably among fill matrix pairs and follows the expected pattern of change only in the most general way. It is expected that as organic matter is added to a soil, on an absolute basis, the volume of pores of each size class will be increased in proportion to the size of the pores in those classes, that is, large pores will increase in absolute numbers more than small pores. On a relative scale, standardizing by the total volume of soil under consideration, the void space of large pores per unit volume of soil should increase while the void space of small pores per unit volume of soil should decrease, the greatest relative increases occurring in the largest pore-size classes and the greatest relative decreases occurring in the smallest pore-size classes (see Chapter 3).

If one considers the excesses of moisture held by pit fills or their matrices at various suctions at the Crane site (Table 91), it is apparent that the greatest *absolute* gains in void space within the fills have, indeed, occurred among the larger pore-size classes rather than the smaller ones; on a relative scale, the void space of larger pores per unit volume of soil have increased at the expense of the void space of the smaller pores per unit volume. Pit fills do retain *more* water per unit volume than do their matrices at low suctions, and retain *less* water per unit volume than their matrices at high suctions. The gains in void space within the large pore-size classes, however, are not strictly *proportional* to the size of the pores within the classes (Table 92, Figure 63). The maximum gains in void space are not found in all cases among pores of the largest size classes (0–.05 bar suction). Where the maximum gains do occur among the largest pores, there is not always a monotonic decline in gain as pore-size decreases. Thus, the increases in void space found within pit fills as a result of the organic matter added to them can be said to have occurred more so among larger pores than

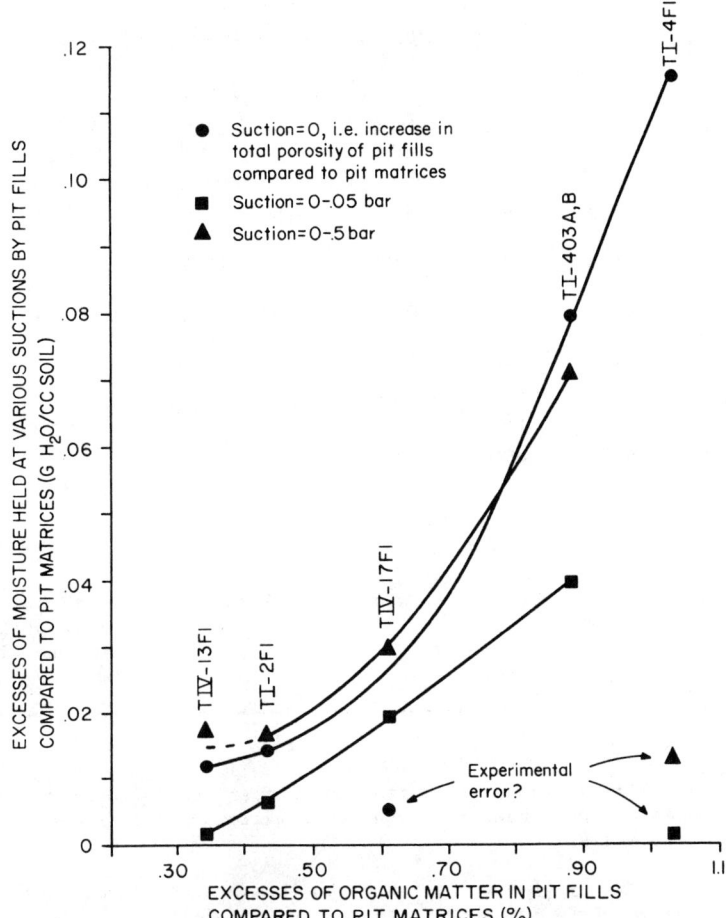

FIGURE 62. *Increases in the total porosity and in the void space of larger pore size classes within the soils filling pits compared to the natural soils surrounding the pits are a smooth function of the excesses of organic matter within the fills (Crane site).*

smaller ones, in general, but not in a manner strictly proportional to pore size.[8]

The differences in structure that characterize the pit fills and their natural soil matrices determine the manner in which and the degree to which they are

---

8 Variation in the patterns of increase in void space found within pores of the largest size classes of pit fills may in part relate to the multifold compounding of experimental errors accumulated when calculating the statistics given in Table 92. Experimental errors in the determination of the moisture held within soils at two different suctions are added when subtracting one moisture content from the other to calculate the void space of pores of a particular size class within pit fills or their matrices. Error again is compounded when subtracting the void space of that size class within a pit fill from the void space of that class within its matrix in order to calculate the excesses of void space of that class within the pit fill.

TABLE 91
Moisture by volume* held in excess by pit fills compared to their surrounding matrices at various suctions

Provenience	\	Suction				Excess Organic
	0 Bar	.05 Bar	.1 Bar	.2 Bar	.5 Bar	Matter in Pits
TIV-16F1**	.0122	-.0205	-.0148	-.0107	-.0013	.23%
TI-13F1**	.0046	.0080	-.0275	-.0297	-.0075	.34
TI-2F1**	.0058	.0012	-.0199	-.0303	-.0112	.43
TIV-17F1***	.0015	-.0243	-.0207	-.0203	-.0377	.61
TI-403***	.0279	.0055	-.0183	+.0020	-.0314	.88
TI-4F1	.0189	.0947	.0804	.0821	.0572	1.03
TIV-18F1	.0164	-.0493	-.0487	-.0458	-.0360	n.d.
I28-5F1	.0097	-.0103	-.0474	-.0409	-.0634	particle size incompatability between fill and matrix

*g H_2O/cc soil

**Distinguishability of pit fills from their matrices in moisture content decreases to zero, then increases to a maximum and then decreases as suction increases.

***Distinguishability of pit fills from their matrices in moisture content decreases to zero and then increases to a maximum as suction increases.

n.d. = no data

distinguishable from each other at various environmental suctions in their moisture contents, conductive cross-sectional areas, and the resistivities favored by these conditions. In all paired observations, pit fills hold more water than their matrices at low suctions, which would tend to give them lower resistivities, and hold less water than their matrices at moderately low through high suctions, which would tend to give them higher resistivities (differences between pit fills and their matrices in soil chemistry not considered).

At some low to moderately low suction, each fill and its matrix retain the same amount of moisture per unit volume and have the same conductive cross-sectional area. At these suctions, the contrast in resistivity between each pit fill and its natural soil matrix favored by their physical differences will be minimal,

TABLE 92
Gains in void space found for various pore size classes in pit fills relative to the void spaces of those classes in the natural soils surrounding the pits*

Provenience	Suction				Excess Organic Matter in Pits
	0-0.5 Bar	.05-.1 Bar	.1-.2 Bar	.2-.5 Bar	
TIV-16F1	.0646	.0074	.0015	-.0017	.23%
TI-13F1	.0024	.0480	.0017	-.0047	.34
TI-2F1	.0138	-.0012	.0072	-.0040	.43
TIV-17F1	.0386	-.0048	-.0002	+.0041	.61
TI-403	.0786	.0116	-0135	+.0086	.88
TI-4F1	.0072	.1418	.0476	+.0080	1.03
TIV-18F1	.1004	-.0060	-.0014	-.0019	n.d.
I28-5F1	.0704	.0224	-.0155	+.0060	particle size incompatability between fill and matrix

*Void space is measured as the volumetric moisture capacity (g H_2O/cc soil) of the pore size class (range of suctions) under consideration. Gains in void space within each pore size class are standardized as gains per .1 bar suction interval width.

n.d. = no data

depending on only differences in the tortuosity of the paths available for current conduction. At lower or higher suctions, the contrast favored between the fills and their matrices in their resistivities will be greater, depending on differences in their conductive cross-sectional areas as well as differences in the tortuosity of the paths available for current conduction.

The particular suction at which the fill and its natural soil matrix of each pit-matrix pair retain identical amounts of water and at which minimal contrast in their resistivities is favored by their physical attributes varies with the amount of organic matter that the fill contains in excess to that within the matrix. As the contrast in organic matter content between fills and their matrices increases, the suction at which they hold similar amounts of water and at which least contrast in their resistivities is favored increases (Table 93).

These differences between pit fills and their matrices in their physical characteristics and the resistivity contrast they favor at various environmental

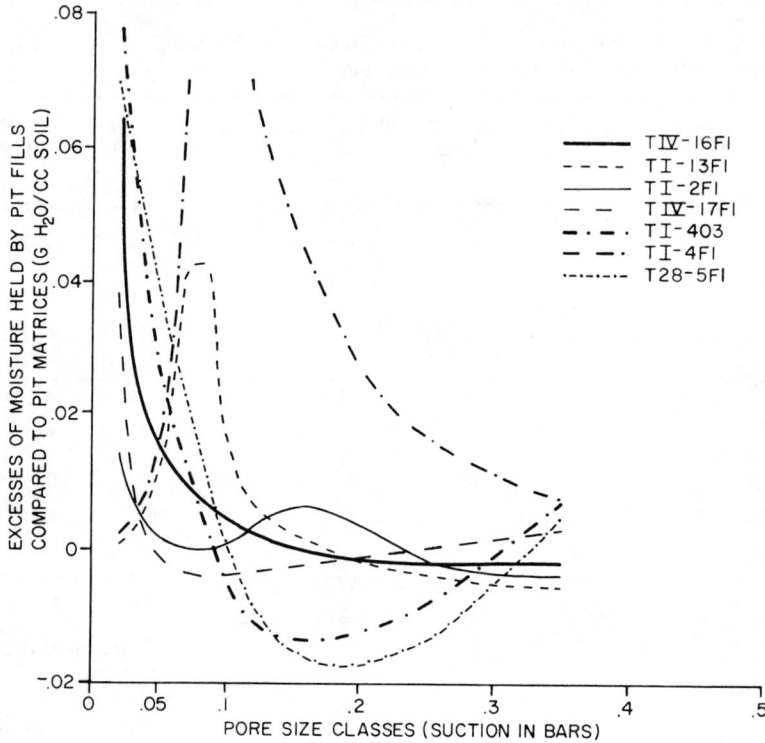

FIGURE 63. *Gains in the void space of various pore size classes within the soils filling pits compared to the natural soils surrounding pits (Crane site).*

suctions should combine with differences in their chemical properties so as to produce maximum contrast in their resistivities under the conditions specified in Chapter 3. The soil fills of nearly all pits examined have higher nutrient availabilities than do their matrices for most nutrients assayed (Appendix 7, Table 4). This chemical circumstance favors the pit fills having lower resistivities than their matrices at all environmental suctions. The effects of the physical differences and the chemical differences between the pit fills and their matrices on their resistivities will combine constructively, producing lower resistivities within the pit fills and causing the highest resistivity contrast between the pit fills and their matrices at very low suctions. As environmental suction increases, a point will be reached where the physical characteristics of the pit fills and matrices begin favoring the pit fills having higher and higher resistivities than their matrices, whereas the chemical characteristics still favor the pit fills having lower resistivities. The increasingly destructive pattern of interference between the effects of the physical and chemical properties from this point on will lead to the resistivity contrast between pit fills and their matrices decreasing until very high suctions are reached.

TABLE 93
Different suctions at which pit fills and their surrounding matrices retain the same amounts of moisture by volume* as related to differences in the excess organic matter found within the pit fills

Provenience	Excess Organic Matter Found in Pit Fill	Suction at Which Pit Fills and the Matrices Hold the Same Amounts of Water by Volume
TIV-16F1	.23%	between 0 and .05 bar
TI-13F1	.34	between .05 and .10 bar
TI-2F1	.43	between .05 and .10 bar
TIV-17F1	.61	between 0 and .05 bar
TI-403	.88	between .05 and .10 bar
TI-4F1	1.03	greater than .5 bar

*g H_2O/cc soil

Importantly, the particular low suction at which a pit fill holds the largest excesses of water compared to its matrix and at which resistivity contrast between them is maximal varies from case to case. Likewise, the particular suction at which a pit fill and its matrix holds similar amounts of water and at which their resistivity contrast becomes submaximal, as a result of destructive patterns of interference between the effects of their physical and chemical properties, varies from case to case. Decreases in contrast between fills and matrices occur at *higher suctions* for fills having *greater excesses of organic matter*. Consequently, during periods of low to moderately low environmental suctions, some archeological features will be more distinguishable from their natural soil matrices in their resistivities whereas other features will be less distinguishable. The particular features found to be more or less distinguishable in their resistivities will vary with the particular environmental suction at the time of survey.

Finally, the general range of suctions over which the fills of each pit–matrix pair retain more, the same amount, or less water than their natural soil matrices should be noted. These parameters were not specified previously, in Chapter 3, where a very general model was presented of change in the contrast between disturbed and natural soils in their moisture contents and resistivities as environmental suction increases. In most cases, the range of environmental suctions over which pit fills at the Crane site retain more or similar amounts of water and contrast most in their resistivities from their natural soil matrices is very restricted—between 0 and somewhat greater than .5 bars. These suctions correspond to times when soil waters drain from the soil under the force of gravity over a period of several days after a saturating rain (0–.1 or .2 bar environmental suction), and to times shortly thereafter when the soil waters are held very weakly by capillary forces (>.1 or .2 bar). At most environmental suctions (~.5–31 bars), those soils at the Crane site that have been anthropically disturbed by enrichment of their organic matter contents have lower moisture contents than their natural soil counterparts, favoring submaximal resistivity contrast between the disturbed and natural soils.[9]

At other sites, the environmental suctions over which organically enriched disturbed soils retain more, the same, or less water than their natural soil counterparts and contrast more or less from their natural counterparts in their resistivity will vary, depending on the average level of organic enrichment and natural soil texture. Where the level of organic enrichment within the disturbed soils is similar to that found in the features at Crane site, and where the natural soils are more coarsely textured (sandy loams, loams), the range of suctions over which disturbed soils will retain more water and over which resistivity contrast between them and their natural soil counterparts is maximal should extend higher than that found at the Crane site to yield these circumstances. The coarser texture of the soils should facilitate more aggregation and greater increases in the void space of pores within the smaller size classes (see Chapter 3). At sites having disturbed soils with levels of organic matter enrichment similar to those found at Crane and having natural soils of a finer texture than those found at Crane (clayey loams, loamy clays, clays, etc.) the range of environmental suctions over which disturbed soils will retain more water and over which resistivity contrast between them and their natural soil counterparts is maximal should be more restricted than those found at the Crane site to yield these circumstances. The finer texture of the soils should retard aggregation and lessen the increases in void space of pores within the smaller size classes. At sites having natural soils with textures similar to those at Crane and having disturbed soils more enriched with organic matter, the range of suctions over which the disturbed soils will retain more water and over which resistivity contrast between them and their natural soil counterparts is maximal should extend higher than that found at Crane.

In summary, at the Crane site, systematic differences can be found between the structural organization of pit fills and that of their surrounding natural soil matrices. These differences may cause pit fills to retain either more or less water and favor them contrasting either more or less from their matrices in their resistivities, depending on the particular environmental suction. Over most environmental suctions, the pit fills will retain less water, favoring submaximal resistivity contrast between them and their matrices.

---

9. This is not to say that pit fills at the Crane site will hold less water, favoring submaximal resistivity contrast between them and their matrices, *most of the year*. Whether or not this is true depends on the relative proportions of the year different environmental suctions prevail.

Physical Alterations of the Soils at the Crane Site: Midden Deposits Compared to Natural B Horizons

In the preceding analysis, disturbed soils that have become better aggregated and granulated as a result of anthropic additions of organic matter have been compared to natural soil horizons (A3) that are not well aggregated. The contrast in structural organization and moisture-retentive characteristics is great. Is it possible, however, that subsurface natural soil horizons, such as B horizons, that exhibit organic enrichments and strong structural development, could be mistaken for buried archeological horizons with respect to their structural and moisture characteristics, and consequently, their resistivities? If so, this circumstance could prove to be a problem in the interpretation of resistivity data from stratified archeological sites.

This question cannot be answered with respect to all earthen sites and soil conditions, but can be investigated for illustrative purposes with the soil data from Crane site. A number of midden deposits across Crane site and the B1 or B21 horizons below them were sampled in order to examine differences in their total porosities, pore-size distributions, organic matter contents, and moisture-retentive characteristics. The experimental design is summarized in Chapter 6 and the raw data are given in Appendix 7, Tables 1 and 2. Calculated average cumulative moisture characteristic curves of the soil samples are shown in Table 94. Average moisture capacities of the soils for several pore-size classes are shown in Table 95.

With respect to structural organization, the midden deposits sampled at the Crane site in most cases show greater degrees of aggregation than do the B horizons below them. This is evidenced by the greater void space found within the midden deposits than within their underlying B horizons for pores of the largest size classes. Middens exhibited greater pore space than the B horizons underlying them in four out of six locations for pores of the 0–.05 bar suction size range, in five out of six locations for pores of the .05–.1 bar suction and .1–.2 bar suction size ranges, and in five out of five locations for pores of the .2–.5 bar suction size range (Table 95). The greater degrees of aggregation of the midden deposits at Crane compared to their underlying B horizons, in turn, can be attributed in part to the larger amounts of organic matter found within the midden deposits. At all locations examined, midden deposits had greater organic matter contents than did the B horizons below them (Appendix 7, Table 2). The greater degrees of aggregation within the midden deposits, however, also probably result from their greater proximity to the surface and the greater intensity of the forces of aggregate formation to which they have been subjected (e.g., soil movements by roots, soil macrofauna, freezing, drying; see Chapter 3). Thus, with respect to structural organization, B horizons at the Crane site are significantly different from the midden deposits, there.

As a result of the different degrees of aggregation found between the midden deposits and their underlying B horizons, their moisture characteristic curves and resistivities at given suctions differ. The manner in which they differ is

TABLE 94
Average moisture content of midden deposits and the B1 or B21 horizons below them at several suctions*

Provenience	0 Bar (Total Porosity)	.05 Bar	.1 Bar	Suction .2 Bar	.5 Bar
TI-501	.4784/.6611**	.2669/.3688	.2432/.3361	.2265/.3130	.2053/.2837
TI-502	.4301/.6494	.2872/.4337	.2512/.3793	.2264/.3419	.1997/.3015
TI-5SUB	.4165/.6439	.2591/.4006	.2509/.3879	.2489/.3848	.2363/.3653
TII-801	.4916/.6622	.2755/.3711	.2522/.3397	.2366/.3187	.2035/.2741
TII-802	.4803/.6614	.2729/.3758	.2445/.3367	.2232/.3073	.1925/.2651
TII-8SUB	.4365/.6517	.2524/.3768	.2334/.3485	.2287/.3414	.2110/.3150
TIV-1601	.4547/.6570	.2804/.4052	.2556/.3693	.2290/.3309	n.d.
TIV-1602	.4592/.6580	.2472/.3542	.2321/.3326	.2218/.3178	.1926/.2760
TIV-1603	.4313/.6470	.2436/.3654	.2343/.3514	.2208/.3312	.2009/.3014
TIV-16SUB	.4218/.6462	.2626/.4023	.2426/.3717	.2290/.3508	.2082/.3190
TIV-1801	.4045/.6383	.2407/.3798	.2287/.3609	.2206/.3481	n.d.
TIV-18SUB	.4196/.6453	.2405/.3699	.2407/.3702	.2319/.3567	.2122/.3264
N100R65-01	.4430/.6539	.2645/.3904	.2414/.3563	.2159/.3187	.1909/.2818
N100R65-SUB	.4550/.6570	.2581/.3727	.2509/.3623	.2449/.3536	.2375/.3430
N104R75-01	.4554/.6571	.2832/.4087	.2493/.3597	.2375/.3427	.1941/.2801
N104R75-SUB	.4615/.6586	.2805/.4003	.2664/.3802	.2505/.3575	.2349/.3352

*Moisture contents are given on both a weight basis (g H_2O/g soil; first value) and a volume basis (g H_2O/cc soil; second value).

**Underlined numbers indicate environmental suctions at which midden deposits hold more water than the B1 or B21 horizons below them.

n.d. = no data

TABLE 95
Average moisture capacity of midden deposits and B1 horizons below them for several pore size classes (suctions)*

Provenience	0-.05 Bar	.05-.1 Bar	Suction .1-.2 Bar	.2-.5 Bar	>.5 Bar	Total Porosity
TI-501	.2115/.2923**	.0237/.0328	.0167/.0231	.0212/.0293	.2053/.2838	.4784
TI-502	.1429/.2158	.0560/.0846	.0403/.0608	.0268/.0405	.1997/.3015	.4301
TI-5SUB	.1514/.2341	.0082/.0127	.0228/.0352	.0127/.0196	.2363/.3653	.4165
TII-801	.2161/.2911	.0309/.0416	.0185/.0249	.0331/.0446	.2035/.2741	.4916
TII-802	.2074/.2856	.0284/.0391	.0124/.0171	.0306/.0421	.1925/.2651	.4803
TII-8SUB	.1841/.2749	.0190/.0284	.0069/.0103	.0177/.0264	.2110/.3150	.4365
TIV-1601	.1743/.2519	.0248/.0358	.0267/.0386	n.d.	n.d.	.4547
TIV-1602	.2120/.3038	.0151/.0216	.0103?/.0148?	.0292/.0418	.1926/.2760	.4592
TIV-1603	.1877/.2829	.0093/.0140	.0200/.0301	.019?/.0300	.2009/.3028	.4313
TIV-16SUB	.1592/.2439	.0200/.0306	.0136/.0208	.0207/.0317	.2082/.3190	.4218
TIV-1801	.1635/.2580	.0124/.0196	.0156/.0246	n.d.	n.d.	.4045
TIV-18SUB	.1791/.2754	.0034/.0052	.0088/.0135	.0178/.0274	.2122/.3264	.4196
N100R65-01	.1785/.2635	.0231/.0341	.0255/.0376	.0205/.0302	.1909/.2818	.4430
N100R65-SUB	.1969/.2843	.0072/.0104	.0142/.0205	.0107/.0154	.2375/.3430	.4550
N104R75-01	.1721/.2483	.0340/.0491	.0166/.0240	.0434/.0626	.1941/.2801	.4554
N104R75-SUB	.1707/.2436	.0128/.0183	.0209/.0298	.0155/.0221	.2349/.3352	.4615

*Moisture contents are given on both a weight basis (g H_2O/g soil; first value) and a volume basis (g H_2O/cc soil; second value).

**Underlined numbers indicate pore size ranges over which the fills of pits have greater volumes than their matrices.

analogous to ways in which the moisture characteristic curves and resistivities of the pit fills and matrices at the Crane site were found to contrast, and the ways in which disturbed soils and their natural counterparts, in general, contrast, as modeled in Chapter 3. At very low suctions (0–.05 bar), the midden deposits hold more water, which favor their having lower resistivities than the B horizons below them, whereas at higher suctions, the midden deposits hold less water, which favor them having higher resistivities than the B horizons below them (assuming the comparable chemistries of the middens and B horizons). As environmental suction increases from 0 bar through 31 bars, contrast between midden deposits and their B horizon counterparts in their moisture contents at first decreases to a minimum, and then increases (Table 96). Taking into consideration the generally greater or similar levels of nutrient availability within the midden deposits compared to the B horizons (Appendix 7, Table 2), in most cases, (a) the resistivity of the midden deposits should be less than the resistivities of the B horizons, when subjected to any same environmental suction, and (b) the contrast in resistivity between the midden deposits and B horizons should be greatest at low environmental suctions, and decrease as moderate and high suctions are reached.

Although the pattern of contrast in moisture content and resistivity found between midden deposits and B horizons at various environmental suctions is similar to that modeled for disturbed soils and their natural counterparts in general, the pattern does differ in one important respect. The suctions at which transitions are made from the midden deposits holding more water than their underlying B horizons and contrasting most from them in their resistivity to the midden deposits holding less water and contrasting less from the B horizons in their resistivity are very low. In all cases, the transition suctions fall between 0 and .05 bar (Table 94), which correspond to the times when soil waters from a saturating rainfall are draining from the soil under the force of gravity. Inasmuch as such periods are relatively infrequent over the annual cycle of moisture changes within the soils at Crane and correspond to moist periods when field conditions would tend to prohibit resistivity surveying, during most surveyable times of the year, the midden deposits at Crane hold less moisture and contrast suboptimally from their underlying B horizons in their resistivities. This pattern is more simple and temporally less variable than that holding between pit fills and their matrices at surveyable times, when sometimes the fills may hold more water than their matrices, giving the two maximal distinctness in their resistivities, whereas at other times the pit fills may hold less water than their matrices, giving the two less distinctness in their resistivities. The pattern of contrast found between midden deposits and their underlying B horizons also is less beneficial to surveying than that found between pit fills and their matrices. Maximal contrasts between midden deposits and their underlying B horizons do not occur during times when surveys can be undertaken practically. This could lead to natural B horizons being confused for buried midden deposits.

The factor responsible for the different patterns of contrast in the moisture

TABLE 96
Moisture by volume* held in excess by midden deposits compared to their underlying B horizons at various suctions

Provenience	Suction				
	0 Bar	.05 Bar	.1 Bar	.2 Bar	.5 Bar
TI-501**	.0172	-.0318	-.0518	-.0718	-.0816
TI-502**	.0055	.0331	-.0086	-.0429	-.0638
TII-801**	.0105	-.0057	-.0088	-.0227	-.0409
TII-802**	.0097	-.0010	-.0118	-.0341	-.0499
TIV-1601**	.0108	.0029	-.0024	-.0199	n.d.
TIV-1602	.0118	-.0481	-.0391	-.0330	-.0430
TIV-1603	.0008	-.0369	-.0203	-.0196	-.0176
TIV-1801	-.0070	.0099	-.0093	-.0086	n.d.
N100R65-01**	-.0031	.0177	-.0060	-.0349	-.0612
N104R75-01	-.0015	.0084	-.0205	-.0148	-.0551

*g H_2O/cc soil

**Distinguishability of midden deposits from their underlying B horizons decreases to zero and then increases as suction increases

n.d. = no data

contents and resistivities found between midden deposits and their underlying B horizons, on the one hand, and pit fills and their matrices, on the other, during surveyable conditions is the larger *excesses* of organic matter found in the pit fills (compared to their matrices) than in the midden deposits (compared to their underlying B horizons) (Table 97). Greater enrichments of organic matter to a soil increase the transitional suction at which it will hold no more or less water than its natural counterpart. This was shown to be true in the comparison of pit fills to their natural soil counterparts in the previous section.

The potential problems of distinguishable B horizons from buried midden deposits that result from the suboptimal *pattern* of resistivity contrast between them over various environmental suctions is compounded by the relatively low *magnitude* of contrast between them at surveyable times (>.2 bar environmental suction). At environmental suctions permitting survey, the excessive moisture held by B horizons compared to the midden deposits above them is greater than the excessive moisture held by the matrices of pits compared to pit fills (Table 97). These larger excesses of moisture held by the *B horizons*,

TABLE 97
Excesses of organic matter and deficits of moisture held at various suctions by pit fills compared to their natural soil matrices, and by midden deposits compared to their underlying B horizons

Feature Comparison	Excesses of Organic Matter ($\mu \pm \sigma$)	Deficits of Moisture By Volume Held at Various Suctions	
		$\geq .2$ Bar ($\mu \pm \sigma$)	$\geq .5$ Bar ($\mu \pm \sigma$)
Pit/Matrix*	.498±.255	.0178±.0137	.0178±.0158
Midden/B1 Horizon**	.200±.185	.0342±.0178	.0516±.0189

*Proveniences used include: TIV-16F1, TI-13F1, TI-2F1, TIV-17F1, TI-403.

**Proveniences used include: TI-501, TI-502, TII-801, TII-802, TIV-1602, TIV-1603, N100R65-01, N104R75-01.

which favor their having lower resistivities than the midden deposits above them, oppose the tendency of the *midden deposits* to have lower resistivities as a result of their chemical properties more so than the smaller excess of moisture held by the pit matrices oppose the chemically induced tendency of the pit fills to have lower resistivities. Consequently, the resistivity contrast between midden deposits and B horizons at Crane is less than that between the pit fills and their matrices.

The factor responsible for the greater excess of moisture held by the B horizons than the matrices of pits of Crane and the resultant lower contrast in resistivity found between midden deposits and their underlying B horizons than between pit fills and their matrices at moderately low to high environmental suctions is *not* the greater excesses of organic matter held by the pit fills than the midden deposits. This would cause the opposite ordered relation at these suctions. Rather, the conditions results from the effects of *textural* differences that occur between the midden deposits and their underlying B horizons but that do not occur between pit fills and their matrices. Clays and fine silt particles (<10 μm) have been eluviated from the upper soil horizons at Crane (including midden deposits) and have accumulated to excess within the B horizons. In all but one midden B horizon pair, the B horizons have a higher frequency of clays and fine silts than the midden deposits above them (Appendix 3, Table 1). The greater percentage of these fine particles within the B horizons have increased there the proportion of void space found within pores of the smallest size classes per unit volume of soil. These alterations favor the retention of more water within the B horizons than the midden deposits at moderate to high suctions. The greater frequency of clays within the B horizons also have increased the total hygroscopic capacities of the clays, there, relative to the total hygroscopic capacities of the clays within the midden deposits, and favors the retention of more water by the B horizons than the middens above them at moderately low to high suctions. In contrast, the pit fills and matrices at Crane have approximately the same textural distributions, which does not favor the matrices holding more water than the pits at any suction. Hence, the ordered relation of moisture contrasts and resistivity contrasts found for midden-deposit–B-horizon comparsions and pit fill–matrix comparisons.

In conclusion, as a result of differences in the organic matter content and particle-size distributions of midden deposits and their underlying B horizons at the Crane site, these soil types do not contrast in their resistivities to the extent that pit fills and their matrices do during moisture conditions when surveying is practicable. If one assumes that conditions at Crane are typical of earthen sites, in general, and considers that on earthen sites it has been difficult to discern differences in the resistivities of archeological disturbances and their natural soil matrices, then the possibility of distinguishing finer resistivity contrasts between midden deposits and B horizons is not encouraging. B horizons might be confused for buried midden deposits. This conclusion is likely to be true for other earthen archeological sites having moderate to fine textured soils, illuvial horizons more strongly developed than the weak to moderate B horizons found at Crane, and having similar to lower levels of organic matter disturbance than those found at Crane.

Physical Alterations of the Soils at the Crane Site: Comparison among Midden Deposits

In the preceding two analyses, the effects of prehistoric human activity on the physical properties of soils at the Crane site have been examined at the level of greatest contrast. Clearly disturbed soils (pit fills, middens) have been compared to natural, undisturbed soils (A3 and B horizons). In this third and final analysis, soils that have been disturbed by various amounts will be compared to each other in order to determine whether more subtle differences in their physical structures than those documented previously can be pinpointed and related to the various intensities of the activities that have modified them.

The areas of the Crane site to be compared in this analysis include the six multipurpose work areas (Strata 1, 2, 6, 8, 9, 11) and the five trash dumps (Strata 3, 4, 5, 7, 10) that surround the Middle Woodland house and were defined previously in this chapter. As described before, midden deposits have developed in both types of areas, but to greater depths in the locations of dumping. Work areas were swept to some extent, removing refuse materials generated within them. Sweepings from the work areas were deposited within adjacent dumping locations. Work areas and adjacent dumping locations thus have accumulated the same *kinds* of refuse generated from the same activities, only in differing *amounts*. Any differences, on the average, in the physical properties of the soils within work areas and dumping locations should be attributable to differences in the *amounts* of organic matter with which they have been enriched rather than the *kinds* of organic matter.

Within each areal stratum, the level examined for its soil attributes will be Barnes Layer BL2-1. As in the previous analysis this layer contains only midden accumulations within the area of study, and does not incorporate any submidden natural soil horizons for the locations examined. Consequently, any variations found among the soil samples removed from this layer should reflect the effects of variable concentrations of organic matter within the midden deposits rather than differences in the proportions of midden and natural soil occurring within the samples.

BL2-1, however, has been physically disturbed once by modern chisel plowing since its deposition in prehistoric times (see Chapter 7). This will not affect the analysis because the plowing occurred uniformly over all areal strata to be examined and because the analysis is concerned with relative differences between strata rather than the absolute characteristics of the strata, themselves. Th absolute effects that chisel plowing probably has had on the soils in BL2-1 include (*a*) an increase in the void space of *very* large size pores per unit volume of soil, (*b*) a proportional decrease in the void space of pores of all other sizes per unit volume of soil, and (*c*) an increase in total soil porosity.

With respect to the outcomes of this analysis, it is expected that the soils within the areas of dumping should have greater excesses of organic matter above the natural norm than do the soils within the areas swept free of refuse. The greater enrichments of organic matter to the soils within the areas of dumping should have produced greater degrees of aggregation, there, than within the swept areas, and consequently, higher total porosities, higher relative frequencies of large pores, and lower relative frequencies of small pores.

As a first step in the analysis, it is necessary to consider the relative amounts of organic matter found in the soils of the swept areas and dumping areas in excess to the natural concentrations found at the level of BL2-1. Any significant anthropically caused differences in the physical characteristics of the soils in the swept areas and dumping areas will be tied to differences in the amounts of organic matter augmented in them. The relevant data have been assembled previously in the study of the chemical differences between these areas, and are given in Table 85. Test statistics are shown in Table 98. As expected, the soils within the areas of trash accumulation, on the average, have significantly higher anomalous concentrations of organic matter than do the swept work areas.

Now let us consider the physical differences between the soils in these two kinds of areas. The relevant statistics are presented in Table 99. Data for the construction of moisture characteristic curves for the swept areas and dumping areas, relating soil moisture contents to *specific* environmental suctions were not available. The moisture contents of the two soil types at various times of survey representing several different but unspecified environmental suctions,

TABLE 98
Differences in the anomalous organic matter contents of soils within swept work areas and trash dumps around the house at Crane site

	Swept Areas	Dumping Areas
Strata considered	1, 2, 6, 8, 9, 11	3, 4, 5, 7, 10
Mean anomalous organic matter content	.4932±.1946	.8150±.6017
N	6	5
Student T-statistic testing difference in means	1.245	
Level of significance	10.5%	
F-statistic testing equivalence of variances	.1045	
Level of significance	.025	
Wilcoxon Rank-Sum S-statistic testing difference in medians	35	
Level of significance	21.4%	

however, were available for study. Environmental suctions are known to have increased through the time during which surveys were performed within the block and along Transects II and IV. It also would appear that during all three periods, environmental suctions fall within the low to moderately low range (.2 bar, to a suction somewhat greater than .5 bar). This judgment is based on a comparison of the moisture per unit volume held by soil samples from the two soil types to that held by the pit fills and midden deposits previously examined in this chapter.

Of the various physical characteristics described in Table 99, let us first examine total soil porosity. As expected, soils within the areas of dumping, on the average, have significantly higher total porosities than do the soils within the swept areas. This fact may be interpreted as indicating the greater degree of aggregation of the soils within the dumping areas, for two reasons. First, the total porosities of the soil samples removed from the two spatial strata can be shown to be a function of their organic matter contents (Figure 64). Second, the alternative interpretation—that the difference between the swept areas and dumping areas in soil porosity results from a systematic, natural variation in the particle-size distributions of the soils within these areas—is not appropriate. The work areas and dumping areas occur within a spatially restricted area of fairly uniform soil texture (Figure 46, Table 52).

TABLE 99
Differences between the soils in swept work areas and trash dumps around the house at Crane site in their porosites and the moisture they hold at various environmental suctions (survey times)

	Swept Areas	Dumping Areas
At the Time of the Block Survey		
H_2O by volume	31.24%±1.97	32.90%±5.28
N	6.	6.
T-statistic testing similarity of means	.7222	
Significance level	.4867	
F-statistic testing similarity of variances	7.166	
Significance level	.0249	
At the Time of Survey of Transect II		
H_2O by volume	28.49%±2.59	28.13%±2.03
N	5.	8.
Mann-Whitney U-statistic testing similarity of medians	18.00	
Significance level	.5874	
At the Time of Survey of Transect IV		
H_2O by volume	18.13%±4.57	16.15%±2.86
N	3.	2.
Mann-Whitney U-statistic testing similarity of medians	2.00	
Significance level	.70	

TABLE 99 (cont.)

	Swept Areas	Dumping Areas
Block Survey Samples		
Total porosity	14.75±5.13	18.26%±10.16
N	6.	6.
T-statistic testing similarity of means	.7622	
Significance	.4636	
F-statistic testing similarity of variances	3.917	
Significance	.0801	
Transect II Survey Samples		
Total porosity	38.17%±6.18	40.47%±2.33
N	5.	8.
T-statistic testing similarity of means	.9691	
Significance	.3533	
F-statistic testing similarity of variances	7.0453	
Significance	.0134	
Transect IV Survey Samples		
Total porosity	34.42%±8.84	29.24%±2.82
N	3.	2.
Mann-Whitney U-statistic testing similarity of medians	2.00	
Significance	.30	

Soil Samples Used in Calculating the Statistics

Block	97R63, 110R64, 129.5R57, 110.5R61	114R74, 118R69, 111R69, 109R69.5, 109R68.5, 106.5R66
Transect II	2180, 2183, 2185, 2192, 2207	2137, 2161, 2216, 2223, 2231, 2240, 2539, 2633
Transect IV	4865, 6240, 6265	4833, 6202

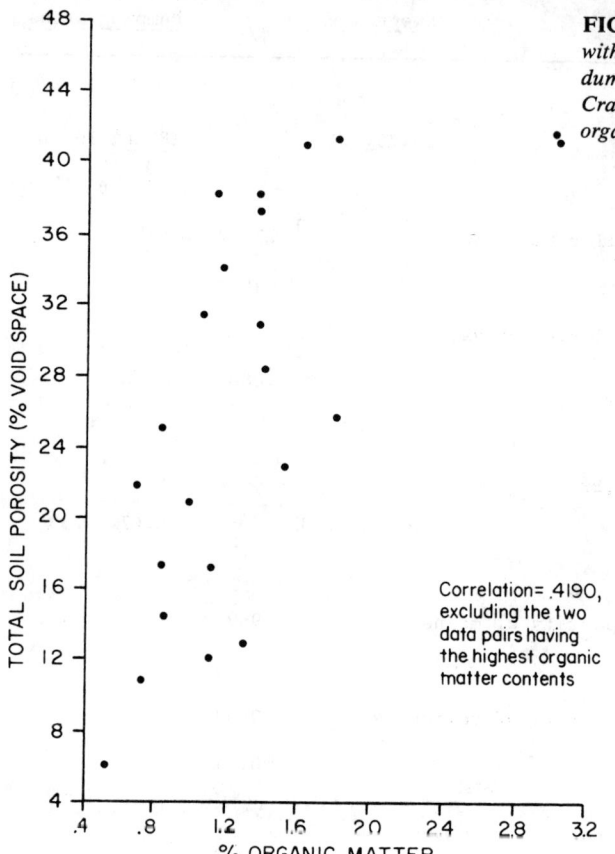

FIGURE 64. *The porosities of soils within swept work areas and refuse dumps around the house at the Crane Site are a function of their organic matter contents.*

The specific manner in which the soils from the swept and dumping areas differ in their pore-size distributions cannot be ascertained from the available data. At each environmental suction examined, soil samples from the swept areas and dumping areas held similar amounts of moisture, on the average (Table 99). This is not inconsistent with the general model presented in Chapter 3 of the effect of differences in the organic matter contents and the degrees of aggregation of otherwise similar soils on their pore-size distributions and moisture equilibria. Such soils may be expected to exhibit differences in their volumetric moisture contents at low suctions and high suctions, but not at moderately low or perhaps moderate ones, such as those at the times of the Block Survey and the survey of Transects II and IV. The equivalent moisture contents found for the soils within the swept areas and dumping areas do not, however, provide *positive* evidence that the soils within these two different spatial strata differ in their pore-size distribution in the manner described by the general model. All that can be said is that on the basis of *all* the patterns discussed previously this would not be an unreasonable interpretation.

Assuming that the soils within the swept and dumping areas do differ in their pore-size distributions according to the general model, and given the greater nutrient availabilities in the dumping areas than the swept areas, we may ask the following questions: (*a*) How would the physical differences, alone, between the soils and the two kinds of areas tend to affect their resistivities at various environmental suctions? (*b*) How would the physical and chemical differences between the soils in the two kinds of areas balance in determining their resistivities and resistivity contrast? During periods of low environmental suction (lower than those at the times of the block survey and the surveys of Transects II and IV), the soils within the dumping areas, being more aggregated, will hold more water than the soils within the swept areas and have greater conductive cross-sectional areas. This physical condition and the greater availability of nutrients in the dumping areas both will favor the dumping areas having lower resistivities than the swept areas. The physical and chemical properties will combine constructively so as to produce maximum resistivity contrast between the two areas. During periods of high environmental suction, the soils within the dumping areas should hold less water than the soils within the swept areas and have lower conductive cross-sectional areas, favoring their having higher resistivities than the soils within the swept areas. The greater availability of nutrients within the dumping areas, however, still will favor the dumping areas having lower resistivities than the swept areas. Under these conditions, it is expected that the physical and chemical properties of the soils in the two kinds of areas will combine destructively, with the chemical characteristics predominating over the physical in determining their resistivity. Soil resistivity still will be lower in the dumping areas than in the swept areas, but the contrast between them will be less than that found at low environmental suctions. At transitional times when the soils in the swept areas and dumping areas have similar moisture contents and conductive cross-sectional areas (e.g., the times of the Block Survey and the surveys of Transects II and IV), the dumping areas should again have lower resistivities than the swept areas. The resistivity contrast between them should be intermediate between that found at low suctions and that found at high suctions, depending solely on the chemical differences between the areas and less significant differences between them in the tortuosity of the paths of current flow. The expectation proposed for resistivity conditions at the transitional times are tested in Chapter 10.

In summary, different intensities of activity and amounts of refuse generation and deposition within different use-areas of earthen archeological sites can produce in the physical properties of soils within those areas alterations that vary in degree with the amount of refuse deposition and organic matter enrichment. The different degrees of soil alteration within the different use-areas may be expected to be reflected within the resistivity signatures of the soils within these use-areas.

Conclusion

Throughout this chapter, the alteration of a number of different chemical and physical properties within the soils at the Crane site have been documented and linked in a systematic way to the kinds or amounts of activity that occurred there in prehistoric times. Soils in areas of different prehistoric use have been shown to differ from each other and from their natural, undisturbed counterparts in their total nutrient availabilities; the spectra of their available nutrients; organic matter content; total porosity; pore-size distribution; and at most environmental suctions, their moisture contents. The effects of these differences upon the resistivities of the soils within the different use-areas also have been discussed, drawing on the models presented in Chapter 3. A univariate approach, in which the effects of each kind of physical or chemical alteration on soil resistivity have been considered individually, has been taken in this chapter in order to clarify the specific, numerous relationships linking prehistoric activity to soil resistivity alteration. In the following chapter, a multivariate perspective will be taken, allowing us to illustrate how the individual physical and chemical alterations combine constructively and/or destructively to determine differences in the resistivities of soils within different kinds of use-areas from each other and their natural counterparts.

References

Chemical Compositions of Foods and Raw Materials

Asch, Nancy, Richard I. Ford, and David L. Asch
    1972   Paleoethnobotany of the Koster site: The Archaic horizons. *Illinois State Museum, Reports of Investigations* 24.
Bard, P., (ed.)
    1961   *Medical physiology*. C.V. Mosby, St. Louis.
Bell, G.H., J.N. Davidson, and H. Scarborough
    1965   *Textbook of physiology and biochemistry*. Williams and Wilkins, Baltimore.
Bouillon, Jean
    1958   Quelques observations sur la nature de la coquille chez les mollusques. *Societe Royale Zoologique de Belgique, Annales* 89:229–237.
Browning, B.L.
    1963   Composition and chemical reactions of wood. In *The chemistry of wood*, B.L. Browning (ed.), pp. 57–99. John Wiley, New York.
Buchanan, E.
    1963   Extraneous components of wood. In *The chemistry of wood*, B.L. Browning (ed.), pp. 313–368. John Wiley, New York.
Chatfield, C., and L. McLaughlin
    1928   Proximate composition of fresh fruits. *U.S. Department of Agriculture, Circular* 50.

Dallemagne, Marcel J., and Leon J. Richelle
    1973   Inorganic chemistry of bone. In *Biological mineralization*, I. Zipkin (ed.), pp. 23–42. John Wiley, New York.

Duvigneaud, P., and S. Denaeyer-de Smet
    1970   Biological cycling of minerals in temperate deciduous forests. In *Analysis of temperate forest ecosystems*, D. Reichle (ed.), pp. 299–325. Springer-Verlag, New York.

Eastoe, J.E.
    1956   The organic matrix of bone. In *The biochemistry and physiology of bone*, G.H. Bourne (ed.), pp. 81–105. Academic Press, New York.

Leung, Woot-Tsuen Wu, Rivta R. Butrum, and Flora Huang Chang
    1972   Proximate composition, mineral, and vitamin contents of East Asian foods. In *Food composition table for use in East Asia*. United States Department of Agriculture and the F.A.O.U.N., Bethesda, Md.

McCance, R.A., W. Sheldon, and E.M. Widdowson
    1934   Bone and vegetable broth. *Archives of Diseases in Childhood* 9:251.

McCance, R.A., and H.L. Shipp
    1933   The chemistry of fresh foods and their losses on cooking. *Medical Research Council of London, Special Report Series* 187.

McCance, R.A., and E.M. Widdowson
    1947   *The chemical composition of foods*. Chemical Publishing, New York.

McCance, R.A., E.M. Widdowson, and L.R.B. Shackleton
    1936   The nutritive values of fruits, vegetables and nuts. *Medical Research Council of London, Special Report Series* 213.

McLean, Franklin C., and Ann M. Budy
    1964   *Radiation, isotopes and bone*. Academic Press, New York.

McLean, Franklin C., and Marshall R. Urst
    1955   *Bone*. University of Chicago Press, Chicago.

Nelson, D.J., T.C. Rains, and J.A. Norris
    1966   High purity calcium carbonate in freshwater clam shell. *Science* 152:1268–1370.

Parmalee, Paul W., Andreas A. Paloumpis, and Nancy Wilson
    1972   Animals utilized by Woodland peoples occupying the Apple Creek site. *Illinois State Museum, Research Papers* 5.

Reidhead, Van A.
    1976   Optimization and food procurement at the prehistoric Leonard Haag site, southeast Indiana: A linear programming approach. Unpublished Ph.D. dissertation, Department of Anthropology, Indiana University.

Saville, Linda D., and Sandra S. Sterrett
    1974   Metal content of Naiad shell and its relation to sex and age. *American Malacological Union, Inc., Bulletin*, 1974:44–47.

Schorger, A.W.
    1926   *The chemistry of cellulose and wood*. McGraw-Hill, New York.

Segar, D.A., J.D. Collins, and J.P. Riley
    1971   The distribution of the major and some minor elements in marine animals. *Marine Biology Association of the United Kingdom, Journal* 51:131–136.

Shapiro, Irving M.
    1973   The lepids of skeletal and dental tissues: Their role in mineralization. In *Biological mineralization*, I. Zipkin (ed.), pp. 117–137. John Wiley, New York.

Smillie, A.C.
    1973   The chemistry of the organic phase of teeth. In *Biological mineralization*, I. Zipkin (ed.), pp. 139–163. John Wiley, New York.

Watt, Bernice K., and Annabel L. Merrill
    1963   Composition of foods. *United States Department of Agriculture, Agriculture Handbook* 8. Washington, D.C.

White, R.L., E. Alvistur, C. Dias, E. Vinas, H.S. White, and C. Callazos
  1955 Nutrient content and protein quality of Quinoa and Canihua, edible seed products of the Andes mountains. *Journal of Agriculture and Food Chemists* 3:531–534.
Wilbur, Karl M.
  1960 Shell structure and mineralization in mollusks. In *Calcification in biological systems*, R. Sognnaes (ed.), American Association for the Advancement of Science, Washington, D.C.
Winters, Howard D.
  1969 *The Riverton culture*. Illinois State Museum (Springfield) and the Illinois Archaeological Survey (Urbana).
Wise, Louis E.
  1952 Miscellaneous extraneous components of wood. In *Wood chemistry*, L.E. Wise and E.C. John (eds.), pp 638–688. Reinhold, N.Y.
Zawacki, April Allison, and Glenn Hausfater
  1969 Early vegetation of the lower Illinois Valley. *Illinois State Museum, Reports of Investigations* 17.

Analyses of Soils on Archeological Sites

Arrhenius, Olaf
  1929 Die bodenanalyses im dienst der archaologie. *Zeits. fur Pflanzenernährung, Dungung und Bodenkunde*, Teil A 10:185.
  1931 In *Zeits. fur Pflanzenernährung, Dungung und Bodenkunde, Teil B* 10:427–439.
Cook, S.F., and R.F. Heizer
  1965 Chemical analysis of archeological sites. *University of California Publications in Anthropology* 2.
Dauncey, K.D.
  1952 Phosphate content of soils on archaeological sites. *Advancement of Science* 9:33–36.
Dietz, Eugene F.
  1957 Phosphorus accumulations in soil of an Indian site. *American Antiquity* 22:404–409.
Hurley, W.N., and C.E. Heidenreich
  1971 Paleoecology and Ontario prehistory. *University of Toronto Department of Anthropology, Research Report* 2.
Provan, Donald M.J.
  1971 Soil phosphate analysis as a tool in archaeology. *Norweigian Archaeological Review* 4:37–50.
Solecki, Ralph S.
  1950 Notes on soil analysis and archaeology. *American Antiquity* 16:254–256.
Weide, D.L.
  1966 Soil pH as a guide to archaeological investigation. University of California (Los Angeles) Archaeological Survey, *Annual Report* 8:155–163.

Other References

Brady, Nyle C.
  1974 *The nature and property of soils*. MacMillan, New York.
Frye, John C., H.D. Glass, and H.B. Willman

1968 Mineral zonation of Woodfordian loesses of Illinois. *Illinois State Geological Survey, Circular* 427.

Mason, Otis Tufton
1889 Aboriginal skin-dressing: a study based on material in the U.S. National Museum. U.S National Museum, *Annual Report*, pp. 553–590.

9

Soil Resistivity as a Product of Multiple Physical and Chemical Soil Properties Illustrated at the Crane Site

Human occupation can produce a number of physical and chemical alterations within soils. When considering how these alterations change the resistivity of a soil, two issues are important: (a) whether the alterations cause the disturbed soil to have a higher or lower resistivity than the natural soils; and (b) which environmental conditions (especially moisture regimes) yield the greatest contrast and which the least contrast between the disturbed soil and its natural counterpart. These issues have been discussed in detail in Chapter 3. With the background provided in Chapter 8 on the physical and chemical soil alterations produced at the Crane site, these two concerns now may also be investigated with data from Crane.

The specific form of anthropic soil disturbance of most concern in this study, for which data are available from the Crane site to document its effect on the resistivity of soils, is the addition of organic matter. It will be recalled that when organic matter is enriched to a soil, both physical and chemical changes are produced in the soil. These include: an increase in soil aggregation, total porosity, and the proportion of void space encompassed by pores of large sizes; a decrease in the proportion of void space encompassed by pores of small sizes; and an increase in the cation exchange capacity of the soil and the concentration of ions within the soil water solution. It could be deduced from current models in physics, chemistry, and soil science that as a consequence of these changes: (a) an organically enriched soil will have a lower resistivity than the natural soil from which it was derived at very low and very high environmental suctions, and (b) the contrast in resistivity between the disturbed soil and its natural counterpart will be greatest during periods of very low and very high environmental suctions, and less at times of moderate to high environmental suctions. At times of very low or very high suction, both the physical and the chemical alterations within the disturbed soil will favor its having a lower resistivity than its natural counterpart, and will combine constructively to produce the greatest contrast in resistivity between these two soil types. At moderate to high environmental suctions, the physical alteraions within the disturbed soil will favor its having a higher resistivity, whereas the chemical alterations will favor its having a lower resistivity. The opposing effects of the physical and chemical alterations within the disturbed soil will result in a smaller contrast between the disturbed soil and the natural soil than that which is found at other suctions.

It could not be predicted from extant theory whether at moderate to high environmental suction, an organically enriched soil will have a lower or higher resistivity than the natural soil from which it was derived. It is unclear whether the physical alterations to the disturbed soil or the chemical alterations to the disturbed soil, which favor changes in its resistivity in opposing directions at these suctions, take precedence in determining its resistivity. Empirical data from archeological sites, however, suggest that at these suctions, the disturbed soil will have a lower resistivity than the natural soil from which it is derived, and that chemical alterations take precedence over the physical alterations (Chapter 3).

To test these deduced and empirically derived models, an analysis was made of the correlation between soil organic matter content, volumetric water content, an estimator of the conductivity of soil water, and soil resistivity for several areas of the Crane site and for several periods of different environmental suction. The data and procedures used are as follows. Information on each of the variables to be correlated comes from a number of locations along Transects I through V, which were surveyed for their soil attributes and resistivities. The total set of locations available for study was partitioned into subsets that are internally homogeneous with respect to the environmental suction at the time of survey and with respect to natural soil texture. By structuring the data in this matter, it was possible to:

1. Examine the change in the relationships between variables as environmental suction varied—in particular, the changing balance or pattern of interference between the effects of physical soil alterations and the effects of chemical soil alterations on the resistivities of the disturbed soils
2. Examine the relationship between soil organic matter content as an agent favoring structural development of soils, and soil resistivity, free of the effects of natural soil structural variations related to textural variations
3. Examine the relationship between soil organic matter content as an agent favoring increases in the cation exchange capacities of soils, and soil resistivity, free of the effects of natural variations in cation exchange capacity related to textural variations.

Cases from different transects were separated from each other, each group representing observations at different environmental suctions. Environmental suction increased progressively from the time of survey of Transect I through the time of survey of Transects IV and V. Groups of cases from each different transect, in turn, were subdivided into sets of observations from locations similar in texture, as reconstructed in Figure 56. The subsets of observations resulting from this twofold partitioning of the data are shown in Table 100.

The stratigraphic level chosen for analysis is BL2–1. This layer encompasses midden deposits of varying organic matter contents. In none of the locations examined does it include the natural soils underlying the midden deposits. BL2–1 has been disturbed physically by chisel plowing, but this should not affect the correlation analysis. The effects of chisel plowing on soil structure and moisture-holding capacity may be assumed to be uniform over all cases within and between subsets.

The analyzed variables require definition. Soil organic matter content was corrected for laboratory variation in volume weight, as has been done in

TABLE 100
Proveniences along Transects I through V used in calculating the correlation coefficients listed in Table 102

Subportions of Transects Considered to be Homogeneous in Soil Texture and Degree of Profile Development*:	
Transect I:	116–212, 516–615
Transect II, subportion A:	2401–2516, 2001–2116
subportion B:	2517–2633, 2117–2233
Transect III, subportion A:	3401–3527, 4201–4327
subportion B:	3528–3654, 4328–4454
Transect IV, subportion A:	4501–4596, 5901–5996
subportion B:	4790–4892, 6190–6292
Transect V:	6944–7026, 7244–7426

Proveniences within these Subportions of Transects I-V Excluded from the Correlation Analysis:	
In correlating organic matter content and volumetric water content:	145, 162, 2129, 6226
In correlating organic matter content and resistivity:	550, 2129, 4292, 5977, 6985
In correlating volumetric water content and resistivity:	3500, 3507, 6985
In correlating conductive potential and resistivity:	3427, 6216, 6985, 6999

*Proveniences used in the correlation analysis in Table 102 fall within the range of proveniences considered to be homogeneous in soil texture and degree of profile development, but include only those where soil samples were removed.

TABLE 101
Lower and upper threshold values of volumetric soil moisture content (BL2–1) used in excluding observations from the correlation analyses shown in Table 102

Transect	Threshold
I	.28–.54
II	.14–.48
III	.12–.52
IV	.10–.30
V	.08–.30

previous analyses. Volumetric moisture content was calculated in the manner described in Chapter 6. Since the level of error in estimating this variable was rather high, it was necessary to exclude from analysis those observations that obviously reflected experimental error. This was done by examining histograms of the volumetric water contents of soil samples taken along each transect from BL2–1 under conditions of uniform environmental suction. Observations lying in the extreme tails of the distributions were excluded from the analysis. The upper and lower thresholds for excluding observations from each transect are listed in Table 101.

Soil resistivity observations reflecting high field measurement errors (e.g., contact resistance, metering errors) also were removed from the analysis. This was done in accord with the procedures recommended in Chapter 2, so as not to exclude observations representing subsidiary peaks created by archeological features (pits, alone). An estimate was made of the highest and lowest subsidiary peaks that might be expected as a Wenner array passes over a typical pit at Crane site, based on the following information or assumptions: (a) that pits were hemispherical in shape; (b) that the size of all pits ranged between 30 and 75 cm in radius, or between .52 and 3 times the electrode spacing used in generating BL2–1, BL3–2, and BL4–3[1]; (c) that the maximum contrast in resistivity between pits and their matrices (midden or natural soil) was between 3.4 and 9.1 times, or equivalent to the average contrast between on-site locations with deep middens and high organic contents and off-site locations lacking these characteristics, at the level of BL2–1 (that layer showing the

---

1. All three layers are considered here because: (a) pits could be expected to intrude into any of them, but not into BL5–4, based on the known depths of excavated pits, and (b) resistivity series from all three layers will be used in a number of analyses to come, and it was desirable to keep the criteria for including or excluding resistivity data points from analysis constant over all layers examined and over all analyses.

greatest average contrast in these characteristics and in resistivity between onsite and off-site locations). Using these statistics and tables and graphs provided by Van Nostrand and Cook (1966), the maximum magnitude of a subsidiary peak that might be expected to be produced by a typical archeological pit at Crane within Barnes Layer BL2–1 through BL4–3 and at any of the environmental suctions at the times of investigation was estimated as two times the average *local* resistivity of the feature matrix.

This value for the maximum expectable magnitude of subsidiary peaks then was used in the following manner to sort out those resistivity observations probably reflecting field errors from those that probably are acceptable. To each series of resistivity observations along each subtransect of Transects I–V, a running normal filter function was applied, and a smoothed resistivity series was derived. The cutoff frequency of the filter function was chosen to be 15 resistivity stations, or 8 m—the average diameter of use-areas around the house at Crane (see Chapter 5). The resultant observations along each smoothed series then were used as estimates of average local resistivities in order to calculate the maximum expectable magnitudes of subsidiary peaks above or below the local norms (two times the local norms). Next, a second running filter function of the same cutoff frequency as the first, but designed to calculate local standard deviations in soil resistivity was applied to each raw data series. Finally, the deviation of each datum point along each subtransect from the average local resistivity was compared with the maximum magnitude of subsidiary peaks expectable at the location (two times the local resistivity), and with the local standard deviation in soil resistivity. Resistivity observations were considered to reflect an acceptably low level of field measurement error and were included in the correlation analysis, unaltered, only if their deviation from the average local resistivity was less than either the maximum expectable magnitude of subsidiary peaks or the local standard deviation in soil resistivity. If an observation did not meet this standard, the average local resistivity was used in its place in the analysis.

The final variable that must be defined is a measure of soil water conductivity *as affected by only the cation exchange capacity and nutrient availability of the soil (chemical properties of the soil), and not by its physical structure or volumetric moisture content as a reflection of structure.* This variable was created by multiplying the nutrient availabilities of Ca, Mg, K, H_2PO_4, and HPO_4^2 by their equivalent ionic conductances corrected for the average soil temperature of BL2–1 during the times of survey (68.0 ± 1.6°F; Eq. (17) of Chapter 3), and then by summing these products over all nutrients. Adjustment was not made for the loss of equivalent ionic conductance resulting from the concentration of ions within the soil water above infinite dilution (Onsanger's

---

2. The availability of H_2PO_4 and HPO_4 was determined from the availability of all species of phosphate (H_3PO_4, H_2PO_4, HPO_4, PO_4) using equilibrium equations (Table 23, Chapter 4) and information on soil pH.

Eq. (19), Chapter 3, p. 90). Such an adjustment could not be made, for it would have required that variation in soil moisture content be introduced when calculating variable values and would have resulted in a variable that is not independent (by definition) of the physical variables considered. The created variable thus estimates the maximum conductive *potential* of available nutrients, not the *true* soil water solution conductivity.

The advantage of independence of physical and chemical variables gained by examining conductive potential rather than soil water conductivity itself is offset somewhat, however, by another problem. Lack of adjustment for loss of equivalent ionic conductance will affect the correlation analysis by reducing the maximum strength of correlation possible between resistivity and conductive potential as environmental potential increases and actual soil water solution concentrations rise significantly above infinite dilution. This latter circumstance, however, is predictable, and is considered less of a problem than the nonindependence of physical and chemical variables, the result of which can not be predicted.

The effect of some nutrients on soil water conductivity, including NH_4, Na, and SO_4, could not be taken into consideration because their nutrient availabilities were not known in a large number of cases. This omission, however, should not affect the correlation analysis. Variance in the availabilities of the nutrients—remaining P, Mg, Ca, and K—account for 99.98% of the variance in availability encompassed by all nutrients examined—P, Mg, Ca, K, NH_4, Na, and SO_4—within the subportions of Crane under consideration.

Having defined the variables and procedures used in the correlation analysis, let us now consider the results of that analysis (Table 102). First note the correlation between soil organic matter content and soil moisture content. As environmental suction increases from very low (Transect I) through moderately high (Transects IV, V), the correlation coefficients change from positive to negative. This agrees with what we would expect. The effect of organic matter enrichment on the pore-size distribution of a soil (sandy loam to clayey texture) is to increase the proportion of void space encompassed within pores of large size and the amount of moisture it can hold at low suctions (i.e., moisture content and organic matter content are possibly correlated at low suctions), and to decrease the proportion of void space encompassed within pores of small size and the amount of moisture it can hold at high suctions (i.e., moisture content and organic matter content are negatively correlated at high suctions). At very high suctions, the hygroscopic potential or organic matter may override the effect it has upon moisture retention through structural development, and the more organically enriched soil may again hold more water than the less organically enriched soil. Environmental suctions of this intensity do not appear to be represented in the Crane data.

Now consider the relationship between soil organic matter content and soil resistivity. At very low environmental suctions (Transect I), the correlation between these two variables is negative. At more moderate to high

TABLE 102
Variation in the strength of correlation* between soil resistivity and its determinants at various environmental suctions (times of survey)

Variables Correlated (Time Transect Surveyed)	Organic Matter Content; Volumetric Water		Organic Matter Content; Resistivity		Volumetric Water Content: Resistivity		Conductive Potential of Available Ca, Mg, K, H_3PO_4 and HPO_4;*** Resistivity	
	r	N	r	N*	r	N	r	N
TI	+.6288	7	-.0364	11.	-.2503	6	-.8444	12
TII	-.0497	15	+.2570	28.5	-.4070	21.	-.4333	30
TIII	-.1546	24	+.0711	35.5	-.4893	29.5	-.3358	34.5
TIV-V	-.3547	9	+.1379	13.7	+.2949?	9.3	-.3274	14

*Correlations listed for each transect are the average of several correlations found between the given variable-pairs in different subportions of the transects differing from each other in particle size distribution but internally homogeneous in particle size distributions (Table 100). When averaged, correlations were weighted by the sample size (reliability) of the data sets used to calculate them. The sample sizes listed in the table are the average of the sizes of the data sets used in calculating the correlations.

**g H_2O/cc soil

***Sum of the equivalent weights of Ca, Mg, K, H_3PO_4, HPO_4 found per cc of soil multiplied by their limiting ionic conductances. The limiting ionic conductance values used are those corresponding to a temperature equivalent to the average temperature of BL2-1 during the survey of Transects I-V (67.97 ± 1.56°F).

environmental suctions (Transects II–V), the correlation is positive.[3] This is in agreement with expectation and with the previous set of correlations between organic matter content and moisture content. At very low environmental suctions, sandy loam to clayey soils enriched with organic matter will hold more water and have a greater conductive cross-sectional area, favoring their lower resistivity (i.e., a negative correlation between organic matter content and resistivity). At more moderate environmental suctions, the organically enriched soils will hold less water and have a smaller conductive cross-sectional area, favoring their greater resistivity (i.e., a positive correlation between organic matter content and resistivity). At very high suctions, the organically enriched soils will once again hold more water and have a higher conductive cross-sectional area, favoring their lower resistivity (i.e., a negative correlation between organic matter content and resistivity). As before, environmental suctions of this intensity do not appear to be represented in the Crane data.

Next consider the correlations between soil moisture and soil resistivity, and between the conductive potential of available nutrients and soil resistivity. In both relationships, for most environmental suctions, negative correlations were obtained.[4] This is as expected. A soil with a greater volumetric soil moisture content will have a greater conductive cross-sectional area and offer less resistance to current flow. A soil with more available nutrients will have soil waters with a greater concentration of ions and radicals, and will be more capable of conducting an electric current.

Finally, note the decrease in the strength of correlation between soil resistivity and the conductive potential of available nutrients, as environmental suction increases. This phenomenon illustrates the change in the *pattern of interference* between the effects of physical soil properties and the effects of chemical soil properties on soil resistivity as environmental suction changes.[5] At low environmental suctions, the physical determinants of soil resistivity combine constructively with the chemical determinants to give an organically enriched soil a lower resistivity than an organically unenriched soil, and

---

3. The magnitudes of the correlation coefficients relating soil organic matter content, water content, and conductive potential to soil resistivity are not expected to be high. This derives from two circumstances: (*a*) resistivity is determined by multiple variables but was correlated here with only *single* variables that account for only a *portion* of the variation in the resistivity data, and (*b*) the relationships between resistivity and the variables with which it was correlated are nonlinear, whereas the correlation procedures used here assume a linear relationship. It is not the magnitudes of he coefficients that are important, but rather their signs or the *trends* in their values over the sequence of data sets from Transects I, II, III, and IV–V.

4. An exception is TIV–TV, volumetric moisture content versus resistivity; the positive correlation obtained is not understood.

5. The decrease in correlation affected by the changing pattern of interference between physical soil properties and chemical soil properties in determining soil resistivity may be augmented, to some degree, by my not having adjusted the equivalent conductances of Ca, Mg, K, H_2PO_4, and HPO_4 for the concentration of these ions in soil water above that of infinite dilution (see preceding).

augment the strength of correlation between soil resistivity and the conductive potential of available nutrients. At more moderate suctions, the physical determinants of soil resistivity favor the organically enriched soil having a higher resistivity than the organically unenriched soil, and partially offset the effects of the chemical determinants of soil resistivity, decreasing the strength of correlation between soil resistivity and the conductive potential of available nutrients. At very high suctions, the physical determinants of soil resistivity again would combine constructively to give the organically enriched soil a lower resistivity, and the strength of correlation between resistivity and the conductive potential of available nutrients would increase. Such high suctions do not appear to be represented in the Crane data, however, for the correlation between resistivity and conductive potential decreases steadily from Transect I through Transect IV.

The pattern of covariation found between soil resistivity and the conductive potential of available nutrients at various environmental suctions is consistent with the conclusions in Chapter 3, that the effects of chemical soils properties will always, or nearly always, override the effects of physical soil properties in determining differences in the resistivities of organically enriched and unenriched soils, and that usually, if not always, organically enriched soils will have lower resistivities than their natural counterparts. At none of the suctions examined did the correlation between soil resistivity and the conductive potential of available nutrients become positive. At the Crane site, at least, organic matter enrichments have increased the cation exchange capacities and nutrient availabilities of soils and the conductivity of soil water solutions more so than they have increased aggregation and decreased soil moisture-holding capacity and conductive cross-sectional area, with respect to their effects on soil resistivity at all environmental suctions examined.

The pattern of covariation found between soil resistivity and the conductive potential of available nutrients at various environmental suctions also is consistent with the conclusion that the degree of contrast between an organically enriched soil and its natural soil counterpart will be greatest at very high or very low suctions. At Crane, it is at times of very low environmental suction that the correlation between soil resistivity and nutrient availability is greatest and that there is evidence for physical and chemical soil properties combining *constructively* in producing a lower resistivity within organically enriched soils.

In summary, a number of relationships between soil organic matter content, soil moisture content, the conductivity of soil water, and soil resistivity were modeled in Chapter 3. These relationships and their implications on whether organically disturbed soils tend to have higher or lower resistivities than their natural counterparts and on the conditions favoring greatest and least resistivity contrast between organically disturbed soils and natural soils, have been tested at the Crane site and found accurate. Their applicability to other earthen sites occurring on soils of clayey through sandy loam texture is probable. At earthen sites located on sandy soils, however, the balance between the effects of physical soil properties and the effects of chemical soil properties in determining soil resistivity at low and moderate environmental suctions may be different from that described previously. Variation in the direction (positive or negative) and magnitude of resistivity anomalies produced by organically enriched soils compared to their natural counterparts over differing environmental suction may not follow the pattern modeled in Chapter 3 and tested at the Crane site.

Reference

Van Nostrand, Robert G., and Kenneth L. Cook
   1966 Interpretation of resistivity data. United States Geological Survey, *Professional Paper* 499. Washington.

10

The Feasibility of Using the Barnes Layer Method and Spatial Filtering Techniques to Isolate Archeologically Significant Soil Resistivity Variation Illustrated at the Crane Site

Physical and chemical variations in the soils of an archeological site and the variations they produce in soil resistivity may be the result not only of prehistoric human occupation, but also of natural geomorphological and pedological processes, topographic trends, and vegetational changes over the site; historic land-use patterns; and measurement error. In Chapter 2, it was suggested that these components of variation within resistivity data sets may be segregated over the dimensions of depth and space and that archeologically significant variation can be distinguished, using the Barnes Layer method and spatial filtering techniques of analysis. The isolated resistivity variability of an archeological nature then may be interpreted in accord with the models presented in Chapters 3 and 8, and with an understanding of the archeological phenomena expected to be found, in order to interpret intrasite structuring of prehistoric features, use-space, and activities.

To illustrate the feasibility of this analytical design, four analyses are offered, using data from the Crane site: (a) a test of the capability of the Barnes Layer method to estimate the resistivities of buried stratigraphic units, such that they are unaffected by the resistivities of more surficial layers; (b) a study of the differences in the *mean* resistivities of soils found in different kinds of use-areas surrounding the house at Crane; (c) a study of differences in the *variance* of soil resistivity values found among use-areas having different areal densities of pits and different functions; and (d) a filtering analysis of the spatial components of resistivity variation within data series collected along Transects I, II, and IV at Crane, for the purpose of defining and characterizing use-areas along these transects.

Tests of the Capability of the Barnes Layer Method

Partitioning the variability within a resistivity data set into its components of different cause and isolating the archeologically significant variation should first be done along the dimension of depth, using the Barnes Layer method. It will be recalled from Chapter 2 that with this technique, an estimate of the resistivity of an individual, buried soil horizon may be made that minimizes the effect of the resistivities of other horizons above it upon that estimate. The method stands in contrast to the standardly used Whole Volume method, in which the resistivity of an individual soil horizon is estimated by the electrical response of the horizon of interest, plus all horizons above it; the values obtained by this method may be influenced considerably by the resistivities of the more surficial horizons.

To illustrate the advantage of using the Barnes Layer method over the Whole Volume method in determining layer resistivities and its greater potential for segregating agricultural, archeological, and natural soil variation, two comparisons will be made: one between the different estimates of *mean* layer resistivities found by the two methods for various horizons at Crane site, and a second between the different estimates of the degree of *statistical independence* of several layers in their resistivity *variation*, as found by the two methods. First let us compare estimates of mean layer resistivity. For each survey station along Transect IV, resistivities of Whole Volumes WV1, WV2, WV3, WV4, and WV5 and Barnes Layers BL2–1, BL3–2, BL4–3, and BL5–4 were calculated. Extremely high and low resistivity values, which were taken to reflect measurement or field error rather than subsidiary peaks produced by archeological features, were removed from the set of points to be used in the analysis in accord with the general procedures recommended in Chapter 2 and the specific operations discussed in Chapter 9. Then, for each subtransect, mean resistivities of each of the five Whole Volumes and each of the five Barnes Layers along the entire lengths of the subtransects were calculated. These are shown in Table 103 where they are grouped in sets of threes, each set including: (*a*) a given small Whole Volume, (*b*) a second larger Whole Volume that encompasses the first plus a layer of soil below it and that, by the Whole Volume method of interpretation, would be used to estimate the resistivity of that layer, and (*c*) the Barnes Layer representing that additional layer of soil, alone.

At the time of survey of Transect IV, soil resistivity decreased progressively with depth, primarily as a function of increasing clay content, decreasing soil aggregation, and increasing soil moisture content favored at that time. Within each set of three resistivity values listed in Table 103, therefore, the resistivity of the layer represented by the small Whole Volume is greater than the resistivity of the layer represented by the Barnes Layer below it. If the Barnes Layer method is accurate in estimating the resistivity of buried soil horizons, removing the influence of the layers above them, the resistivity of the Barnes Layer in each set should be less than the resistivity of the *large* Whole Volume encompassing *both* the Barnes Layer and the small, more resistant Whole Volume above it. This is what is found in all comparisons made. It is concluded that the

TABLE 103
Testing the Barnes Layer method: Effect upon the mean resistivity of layers

Layers Compared	Mean Resistivities of the Layers		
Transect IV -- East or West			
WV1, WV2, BL2-1	17186.	13141.	10985.
WV2, WV3, BL3-2	13141.	9699.	7238.
WV3, WV4, BL4-3	9699.	6073.	4363.
WV4, WV5, BL5-4	6073.	5152.	3266.
Transect IV -- Central			
WV1, WV2, BL2-1	16758.	13141.	10060.
WV2, WV3, BL3-2	13141.	9485.	7265.
WV3, WV4, BL4-3	9485.	5892.	4223.
WV4, WV5, BL5-4	5892.	4917.	3029.
Transect IV -- East or West			
WV1, WV2, BL2-1	16804.	13109.	11020.
WV2, WV3, BL3-2	13109.	10100.	7244.
WV3, WV4, BL4-3	10100.	6203.	4133.
WV4, WV5, BL5-4	6203.	5321.	3456.

Barnes Layer method is a more accurate mean of assessing the resistivity of buried soil horizons than is the Whole Volume method.

It can be argued that the differences in mean resistivities found for the Barnes Layers and large Whole Volumes simply reflect the mathematical process of subtraction that is involved in calculating Barnes Layer resistivities, and that they do not reflect an actual improvement in the estimation of layer resistivity. A second comparison of the Barnes Layer method and whole volume method of determining layer resistivities therefore is provided. If the Barnes Layer Method does allow better estimation of layer resistivities than does the Whole Volume method, then resistivity variations within successive *segregated* Barnes Layers of a stratigraphic sequence should be statistically more independent from each other than are variations of within successive, *nested* Whole Volumes of the same sequence. This can be shown to be true. Using the same data described previously from the entire length of each of the subtransects of Transect IV, correlation coefficients were calculated between the resistivities of (a) a given, small Whole Volume and a second, larger Whole Volume encompassing the first plus an additional layer of soil below it; and (b) the given Whole Volume (which also may be envisioned as a Barnes Layer extending from the surface to its base) and the Barnes Layer representing the additional layer of soil alone (Table 104). In each comparison, the correlation coefficients obtained between the small Whole Volume and the Barnes Layer below it was less than the correlation coefficient obtained between the two nested Whole Volumes. The differences in r^2 values were significant at less than or equal to the .01 level using a one-tailed test, as outlined by Olkin (1967).

Thus, the Barnes Layer method is a more effective means of determining the resistivities of individual buried soil horizons, than is the Whole Volume method. On archeological sites under cultivation, where archeological layers and natural soil horizons may be buried below a highly resistant and spatially variable plowzone, the Barnes Layer method can be used to estimate the resistivities of the archeological and the natural horizons relatively free of noise from the plowzone. The agricultural, archeological, and natural components of soil variation can be segregated.

Use-Areas Differentiated by the Mean Resistivity of Their Soils

Stratigraphic separation of agricultural, archeological, and natural components of variation within soil resistivity data sets is the first step necessary for preparing those data for interpretation. This has been shown in the previous section to be feasible by using the Barnes Layer method. The second step required is the use of spatial filtering techniques to segregate further these several sources of resistivity variability and to distinguish different kinds of use-areas from each other within the archeological component of variability. The feasibility of the latter step depends on whether or not it is true that different kinds of use-areas as *whole geographic units* are distinguishable from each other in at least one of the following attributes: (a) their mean soil resistivities, (b) the magnitude of variation in their soil resistivities, and (c) their dimension. To illustrate that use-areas can be distinguishable in the first two manners and that spatial filtering techniques can be used to isolate and distinguish them within resistivity data sets, the following two sections of this chapter are provided. I will begin by illustrating that different kinds of use-areas may be distinguished from each other by their *mean* resistivity.

In Chapter 8, a number of use-areas around the house at the Crane site were identified. The use-areas were of two types: multipurpose work areas swept clean of the refuse materials generated within them, and trash areas where those refuse materials were dumped. The soils within the dumping areas, when compared to those within the swept work areas at the level of Barnes Layer

TABLE 104
Testing the Barnes Layer method: Effect upon the degree of independence of the resistivity of layers

Layers Compared	Correlation Coefficient	N	Significance: Correlation for First Comparison Greater Than Correlation for Second Comparison (One-Sided Test)	
			z-Score	α-Level
Transect IV -- East or West				
WV1, WV2	.7330	420	z = 35.85	α < .00001
WV1, BL2-1	.2157	420		
WV2, BL2-1	.6437	420		
WV2, WV3	.7080	420	z = 22.23	α < .00001
WV2, BL3-2	.2048	420		
WV3, BL3-2	.6043	420		
WV3, WV4	.4730	420	z = 6.125	α < .00001
WV3, BL4-3	.0269	420		
WV4, BL4-3	.3235	420		
WV4, WV5	.5629	420	z = 1.356	α < .09
WV4, BL5-4	.1008	420		
WV5, BL5-4	.4463	420		
Transect IV -- Central				
WV1, WV2	.4066	420	z = 2.394	α < .01
WV1, BL2-1	.2476	420		
WV2, BL2-1	.2913	420		
WV2, WV3	.4774	420	z = 5.941	α < .00001
WV2, BL3-2	.2176	420		
WV3, BL3-2	.6300	420		

TABLE 104 (cont.)

Layers Compared	Correlation Coefficient	N	Significance: Correlation for First Comparison Greater Than Correlation for Second Comparison (One-Sided Test)	
			z-Score	α-Level
WV3, WV4	.6444	420	z =31.78	α<.00001
WV3, BL4-3	.2549	420		
WV4, BL4-3	.7003	420		
WV4, WV5	.7250	420	z =125.79	α<.00001
WV4, BL5-4	.3606	420		
WV5, BL5-4	.6695	420		
Transect IV -- East or West				
WV1, WV2	.7091	421	z =14.12	α<.00001
WV1, BL2-1	.2353	421		
WV2, BL2-1	.6701	421		
WV2, WV3	.7452	421	z =72.50	α<.00001
WV2, BL3-2	.2994	421		
WV3, BL3-2	.6623	421		
WV3, WV4	.7541	421	z =-9.616	α<.00001
WV3, BL4-3	.4341	421		
WV4, BL4-3	.8175	421		
WV4, WV5	.6509	421	z =10.48	α<.00001
WV4, BL5-4	.1500	421		
WV5, BL5-4	.3234	421		

BL2–1, were shown to exhibit: (a) greater percentages of organic matter, (b) greater availabilities of nutrients (particularly Ca and PO_4), and (c) greater total porosities and degrees of aggregation. The soils within the dumping areas also presumably have different pore-size distributions and moisture characteristic curves than the soils within the swept areas, but this could not be verified. All soil sampling surveys performed over the several areas were made during periods of low to moderately low suctions, when the volumetric moisture contents of the soils within dumping areas and swept areas were very similar on the average.

Given these differences in the physical and chemical attributes of the soils in the swept areas and dumping areas, one would expect them to have different *mean* resistivities, in accord with the models presented in Chapter 3. In particular, the greater average availability of nutrients within the soils of dumping areas and the equivalent average volumetric moisture contents and conductive cross-sectional areas of the soils within both types of use-areas would combine to favor the soils within the dumping areas having lower resistivities on the average.

To test this expectation, the mean resistivities of the soils within several of the work areas and dumping areas around the house at Crane were calculated and compared (Table 105). Only those swept areas and dumping areas showing the greatest distinction in the nutrient availabilities of their soils—those encompassed within Transect Surveys IV and V—were analyzed (Strata 1, 2, 5, 6, 8). The layer under consideration is BL2–1, as before. The environmental suction is estimated at somewhat greater than .5 bar (see Chapter 8). Resistivity data have been filtered to exclude observations that obviously represent field measurement errors rather than subsidiary peaks, according to the methods described in Chapters 2 and 9.

In all four comparisons between swept areas and the one dumping area examined, the soils within the dumping area had lower resistivities, on the average, than the soils within the swept areas. In three out of four comparisons, the differences in the means or medians were significant at the .06 level or less. The single comparison in which a significant difference in the mean resistivities was not found was that between the soils of a swept area and the soils in the dumping area that contrasted least in their nutrient availabilities.

Thus, at the Crane site, it can be shown that different kinds of use-areas, where different amounts of refuse have been deposited and where different degrees of physical and chemical soil alterations have occurred, are distinguishable from each other as *whole geographic* zones by their mean soil resistivities. This distinctive attribute, in itself, is sufficient to allow the isolation and differentiation of use-areas with the spatial analytic techniques described in Chapter 2. If complemented by the distinction of use-areas in the variance of their soil resistivity values and in their size, however, spatial-analytic techniques can be used more effectively to isolate areas of different prehistoric function.

TABLE 105
Comparison of the mean resistivities of soils within multipurpose work areas and dumping areas surrounding the house at Crane, Barnes Layer 2-1, Transects IV, V

Strata Compared	Type of Use-Areas	Mean Resistivities	Number of Survey Stations Used to Estimate Means	Difference in Means or Medians Significant?
5,6	dump, swept	8523±4402, 10285±4661	50, 36	T= -1.786; Significance (1-sided test)=.0388 F= 1.121; Significance = .3517
5,1	dump, swept	8523±4402, 11386±5890	50, 13	Median test; Significance (1-sided test)=.0592
5,2	dump, swept	8523±4402, 15086±7779	50, 8	Median test; Significance (1-sided test)=.0632
5,8	dump, swept	8523±4402, 11370±10513	50, 30	Median test; Significance (1-sided test)=.2044

Proveniences Used in Calculating Statistics:

Stratum 5: 4791-4794, 4798, 4802, 4803, 4806-4809, 5491-5500, 5502-5503, 5505-5506, 5508-5511, 6191-6211
Stratum 6: 4837-4849, 5562-5563, 6237-6250
Stratum 1: 4860-4862, 5562-5563, 6257-6264
Stratum 2: 6960-6965, 7266, 7271
Stratum 8: 6680-6687, 6689-6691, 6694-6695, 6699, 6981-6983, 6985, 6988-6991, 7282-7284, 7287, 7289-7291

Use-Areas Differentiated by the Variance of the Resistivity of Their Soils

Mean soil resistivity is one characteristic by which different kinds of use-areas as whole geographic units may be distinguished from each other. They also may differ in the magnitude of variation in their resistivity values. Possible causes of such differences in the magnitude of variation are multiple, but in the most general terms, reduce to two. First, use-areas may differ in the *areal density of features* (e.g., pits) that occur within them, and thus, the relative volumetric proportions of disturbed and undisturbed soil encompassed within the archeological horizon having its variation assessed. Suppose soil resistivity measurements are made over several use-areas having earthen features in different densities but all of constant resistivity (idealized case) and the natural soil matrix does not vary among use-areas. Variation in soil resistivity within use-areas will increase as the relative volumetric proportion of disturbed soil within the layer being assessed increases from 0 to 50% and will decrease as the proportion of disturbed soil continues to rise from 50 to 100%. If the soils composing the archeological features are highly variable in nature and resistivity, as disturbed soils often are, however, the magnitude of resistivity variation within use-areas may continue to rise as the volumetric proportion of disturbed soil increases above 50%. The second general factor that may cause different kinds of use-areas to differ in the variance of their resistivity values is variation in the *relative proportions of features of different types* they contain (e.g., cleaned out cooking pits versus refuse-filled garbage pits), and thus variation in the relative volumetric porportions of disturbed soils of different natures and resistivities encompassed within the archeological horizon having its variability assessed.

To illustrate how soil resistivity variation within an area can be a function of archeological variation, a study was made with data from the Crane site, relating differences in the variation of soil resistivity within several areas to differences in the areal density of pits they encompassed. Five spatial strata included within the Block Resistivity Survey, and to be compared at the level of BL4–3, were chosen (Figure 65). The five strata are not natural existing use-areas, but rather, are arbitrary spatial samples with areas and shapes chosen so as to create zones that differ in pit density and *simulate* different kinds of use-areas. Barnes Layer 4–3 was selected as the layer of interest because it does not include midden deposits, only pits, and all differences between strata in their resistivity variation may be attributed to differences in pit density or type. Natural soil variation between strata probably is minimal (Figure 56).

Estimates of resistivity variation and variation in the amounts of disturbed and undisturbed soil within strata were determined as follows. Because the areas surveyed within the block survey are irregular in shape and small in size, it was not possible to apply the filtering and smoothing techniques recommended in Chapter 2 to determine which resistivity data points might represent

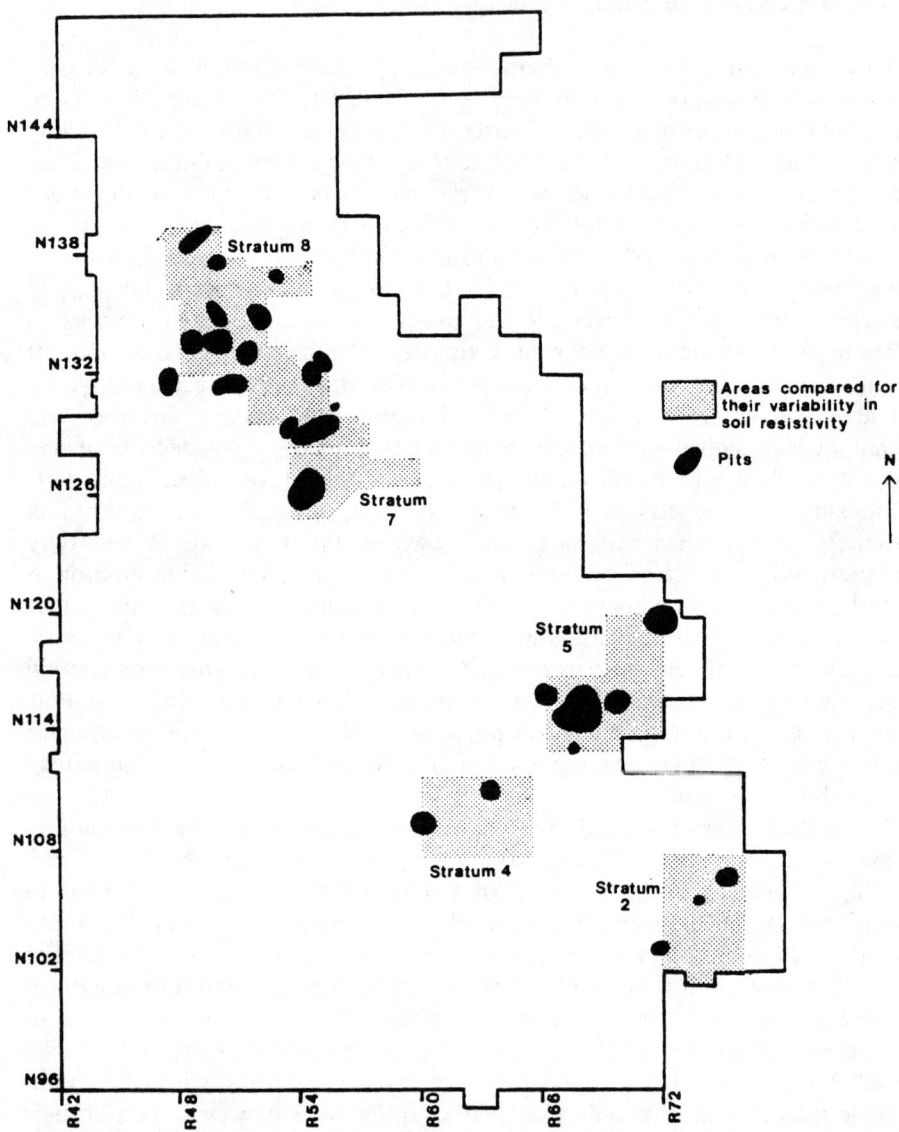

FIGURE 65. *Spatial strata having different areal densities of pits at the level of Barnes Layer BL4-3 and which were compared for the amount of variation in soil resistivity they encompassed at the time of the block survey. The irregularity of the boundaries and the different sizes of the strata reflect: (1) the necessity of avoiding areas of the block which had been disturbed by previous archeological testing, when performing the block resistivity survey, and (2) the necessity of defining strata such that they have different areal densities of pits. Note that not all pits which are shown in Figure 48 extend as deeply as BL4-3 and intrude into it.*

experimental error as opposed to subsidiary peaks and should be excluded within the analysis. Histograms of resistivity values from the survey stations in each area, therefore, were examined, and extreme outlying points that *severely* affected the estimates of variances of the distribution were excluded from analysis. Thresholds for defining unacceptable resistivity values and the number of cases excluded from analysis are shown for each station in Table 106. The number of cases excluded are few (0–3% per stratum). Standard deviations of the remaining resistivity data points then were calculated for each stratum.

Ideally, archeological variation within each stratum would have been estimated by: (*a*) determining the volumetric proportion of disturbed soil encompassed within BL4–3 at each resistivity survey station; and (*b*) calculating the standard deviations of these volumetric proportions within strata. This was not possible because the precise volumes of disturbed fill within pits at the level of BL4–3 could not be accurately assessed from field notes (which were designed for other purposes). Instead, for each survey station, the presence or absence of a pit intrusive into BL4–3 at that location was assessed. Presence was assigned only when the pit intruded more than half way into BL4–3 and was centered within it (i.e., the boundaries of the pit included the center of the electrode array used to determine the resistivity of BL4–3 at the station). Presence was assigned the value 1, absence 0. Standard deviations of these values then were calculated for each stratum.

The relationship found between archeological variability and resistivity variability as defined above is shown in Table 107 and Figure 66. Soil resistivity variation increases monotonically with variation in the presence or absence of pits for the strata examined. The correlation coefficient between these two variables is strong (.982).

Whether the observed relationship indicates one between resistivity variation and variation in the *relative proportions of disturbed and undisturbed soil* encompassed by each resistivity measurement within strata, or one between resistivity variation and the *total* amount of highly variable, *disturbed soil* within strata, is unclear. To make this distinction, it would be necessary to examine strata having pits in areal densities of greater than 50%, and to see whether resistivity variation increases or decreases with pit densitiy above to 50% threshold. These data are not available for examination. Whichever alternative interpretation is true, it nevertheless can be concluded that local resistivity variability over an area can be used to indicate the density of archeological features within it, and may be an important attribute of resistivity data sets for defining and differentiating use-areas of different functions.

In conclusion, in the previous two sections it has been shown that different kinds of use-areas as *whole geographic units* may differ in the mean and/or variance of their resistivity values. Under these conditions, and with the

TABLE 106
Resistivity observations used in calculating the statistics in Table 107

Area	Case Numbers of Observations Within the Area	Upper Threshold Defining Unacceptably High Resistivity Values; Not Used in the Analysis	Number of Cases Excluded from Analysis
2	1-5, 8-12, 15-19, 22-26, 29-33, 37-41, 44-48	3400	1
4	1-72, 76-87	4100	1
5	1-62, 68-71, 82-85, 96-99, 110-113	4000	0
7	1-90	3700	0
8	1-168, 170-251, 258-264	4100	4

TABLE 107
Variance of soil resistivity values within several areas having different densities of pits; Block Resistivity Survey, Barnes Layer BL4-3

Area	Number of Observations in Area	Proportion of Observations in Area Having Pits Present*	Standard Deviation in Presence or Absence of Pits at the Several Observations*	Standard Deviation in Soil Resistivity at the Several Observations (ohm-cm)
5	78	36.7%	.4851	539
8	253	28.0	.4544	501
7	90	15.6	.3645	451
4	83	12.5	.3330	446
2	34	8.57	.2840	390

*Presence was assigned the value 1, absence the value 0, when calculating these proportions and standard deviations.

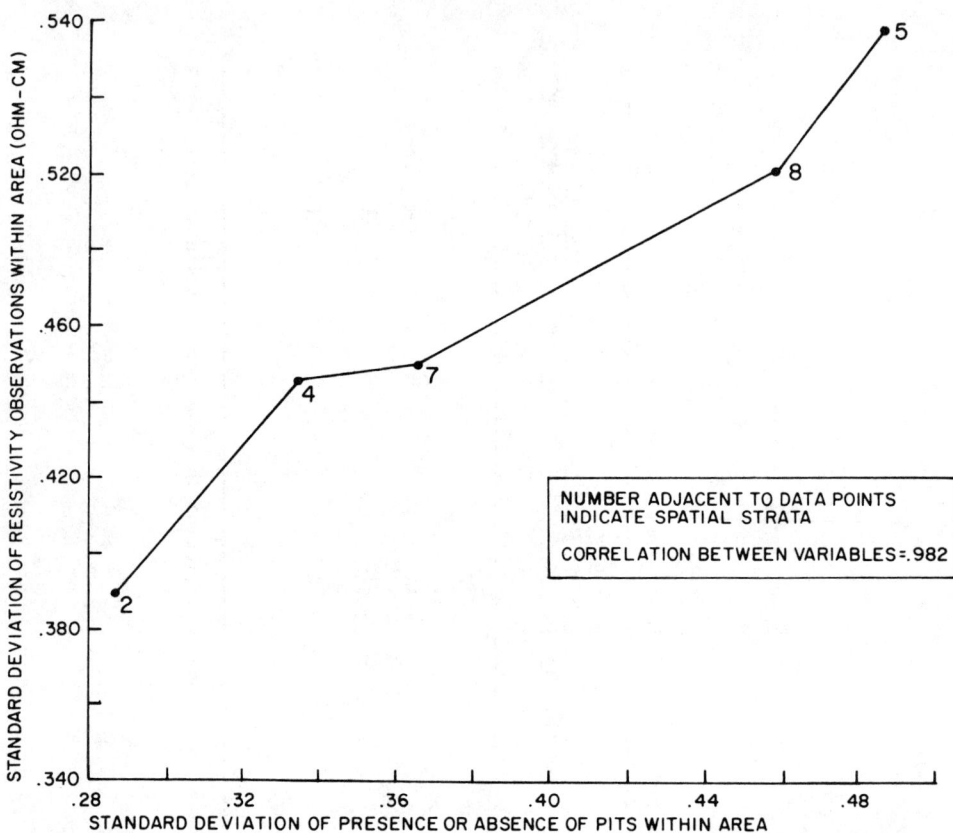

FIGURE 66. *Variation in soil resistivity values within Barnes Layer BL2-1 for several areas of the block at the Crane site, as a function of variation in the presence or absence of pits at the locations of resistivity measurement within each area.*

additional stipulation that use-areas differ in scale from natural, agricultural, or other forms of soil resistivity variation, spatial filtering techniques may be used to isolate and distinguish them. Let us now illustrate these techniques with data from the Crane site.

Use of Spatial Filtering Techniques to Remove Low-Frequency, Nonarcheological Variation from Resistivity Series

Once the variability within a resistivity data set has been partitioned into agricultural, archeological, and natural components over the dimension of depth with the Barnes Layer method, the techniques of spatial filtering and time series analysis may be applied with three objectives. First, they can be used to isolate further resistivity variation of an archeological nature from that produced by other sources according to the spatial scales of the phenomena. Second, they may be used to partition archeological variation, itself, into components of variation of different scales, the several components pertaining to use-areas of different dimensions. Finally, filtering techniques may be applied to each component series in order to distinguish areas of different means and/or variances (assessed at the scale of variation isolated within the component) that correspond to use-areas differing in function, amount of use, etc. In the following five sections, these three modes of use of spatial filtering techniques will be illustrated with data from Transects I, III, and IV.

First let us illustrate how total resistivity variation along a transect may be partitioned into its archeological and nonarcheological components, using data from Transect I and IV. Figures 67 and 68 plot total resistivity variation along Transects I and IV (central subtransects) within Barnes Layer BL2-1—that layer encompassing archeological deposits, where they exist. Note the gradual rise in resistivity from west to east along Transect I. This trend or very low frequency wave running through the data reflects several nonarcheological sources of soil resistivity variation: topography and drainage, historic land-use patterns, and vegetation at the time of survey. The zone of depressed resistivity between Survey Stations 401 and 465 reflects the poor drainage conditions, high water table, and high moisture content of the soils within the old bed of Macoupin Creek. East of Station 465, soil resistivity rises gently as surface elevation and the depth to the water table slowly increase, as drainage conditions improve, and as average soil moisture content decreases. The eastward rise in resistivity also reflects, however, the passing from historic land-use area 3 into historic land-use area 2 (Figure 58, page 409). Area 2, it will be recalled, has been deforested and used for cropping or pastoral purposes longer than has area 3. Its soils probably have lost more nutrients as a result of cropping and increased subjection to precipitation and leaching than have the soils in area 3. They also probably have lost more organic matter and consequently have lower cation exchange capacities (see Chapter 7). The soils in area 2 thus probably have lower total nutrient availabilities and soil water solutions with lower ionic concentrations and conductivities. The jump in resistivity at the boundary between areas 3 and 2 (at provenience 588) most likely reflects these changes in conditions.

A large increase and peak in resistivity occurs approximately between proveniences 640 and 660, and resistivity remains high further east to the end of the transect. The peak between proveniences 640 and 660 is attributable to a natural maximum in soil profile development, including a relatively strong structural B horizon. The maximum was observed at provenience 656 (grid locus 200R160, soil sampling station 3, test excavation unit TI–7) and is described fully in Chapter 6 and Appendix 7. As a result of this structural development, drainage conditions may be expected to be better in this vicinity. Several days after a saturating rain, soil moisture content should be less here

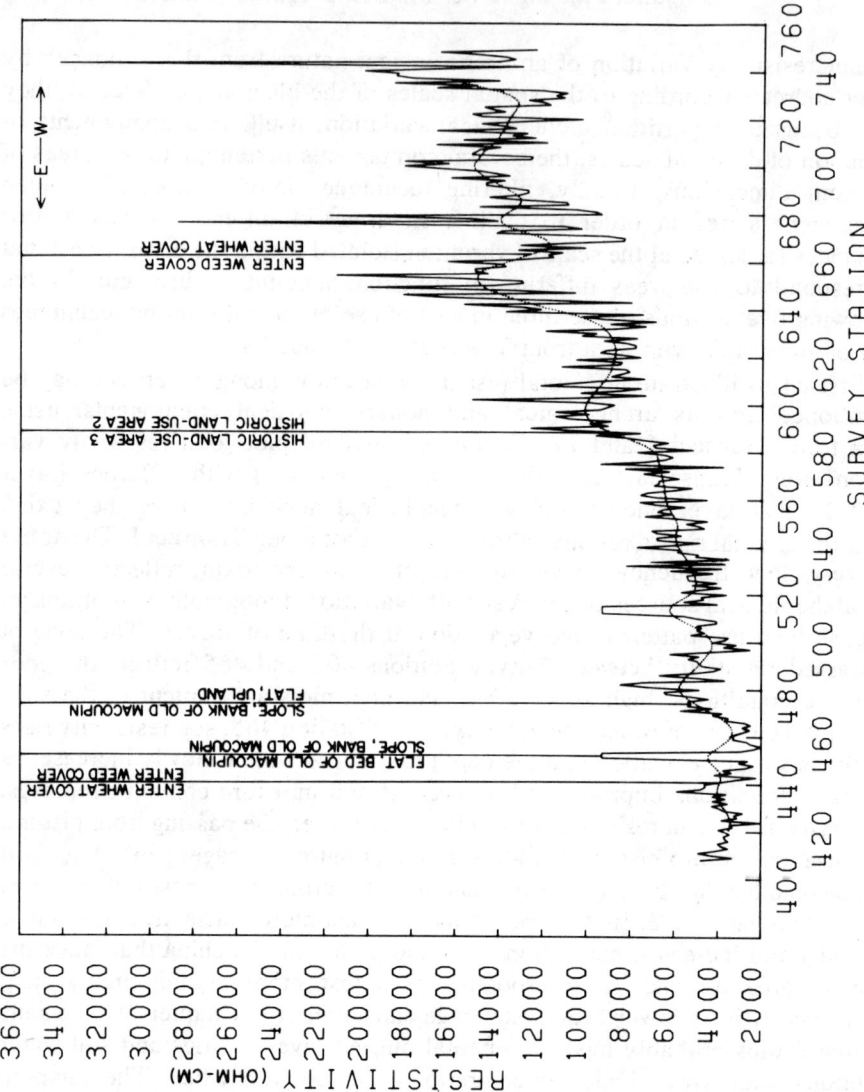

FIGURE 67. *Resistivity variation along Transect I, central sub-transect. The solid line is the original resistivity series. The smoother, dotted line is a resistivity series which was derived from the original using a running normal filter function with a cutoff frequency of 12 m. Variation in the smoothed series is largely of natural and historical origin. Intervals of the transect falling within zones of different natural topography and drainage, historic land-use, and surface vegetation are shown.*

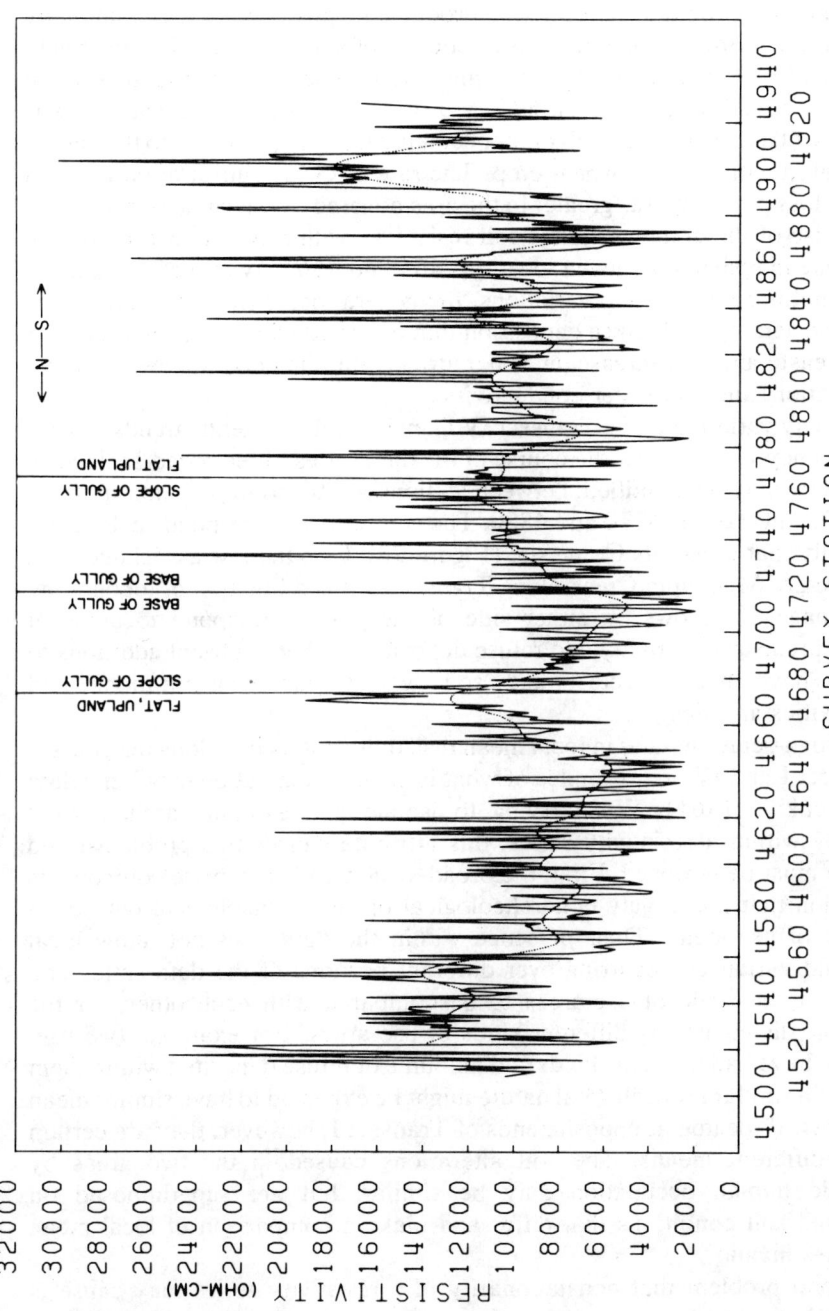

FIGURE 68. Resistivity variation along Transect IV, central subtransect. The solid line is the original series. The smoother dotted line is a resistivity series which was derived from the original using a running normal filter function with a cutoff frequency of 12m. Variation in the smoothed series reflects spatial changes in natural topography and drainage, but also broad zones where more or less refuse from prehistoric occupations were deposited and where more or less soil alteration has occurred.

than in surrounding areas, and soil resistivity should be higher. Transect I was surveyed under conditions expected to produce the positive resistivity anomaly observed in the area of maximum development of soil structure. The relatively constant, high resistivity east of the maximum reflects a change in surface vegetation. West of provenience 684, and prior to survey, the Crane site was covered with a dispersed growth of pioneer annual weeds, whereas to the east, it was covered with a dense wheat crop. The rate of evapotranspiration and soil moisture loss probably was greater in the area covered by wheat, and is the most likely cause of the greater average soil resistivity within the area. Finally, the overall rise in resistivity betwen locations 465 and 756 may reflect, in addition to these natural and historic variations, the general decrease in the amount of prehistoric activity and refuse deposition that occurred within this interval from west to east, and a decrease in soil nutrient enrichments, soil water ionic concentration, and soil water conductivities.

Resistivity variation along Transect IV likewise includes gentle trends or very low-frequency waves. Over the course of the data series, three broad-scale lows in resistivity may be identified: between stations 4560 and 4660, stations 4680 and 4730, and stations 4770 and 4880. The central low corresponds in location to the gully that crosscuts Crane site (Figure 28). Like the low associated with the old bed of Macoupin Creek along Transect I, it is a function of topography and drainage. The lows on either side of the gully correspond to areas of maximal prehistoric activitiy and refuse deposition, where nutrient additions to the soil have been greatest and soil water ionic concentrations and conductivities are high.

The broad-scale changes in local mean resistivity that occur along the courses of Transect I and IV are examples of what is called "nonstationarity" in a data series (Jenkins 1961). When trying to isolate and examine archeological variability within a resistivity series, this attribute causes two problems, and therefore must be removed. First, the broad-scale trends comprise components of variation that are largely nonarcheological or else archeological but not of the scale of use-areas. Their presence within the data does not allow local means and variances occurring over different portions of the data series and assessed at the scale of use-areas to be compared with each other, for the purpose of distinguishing different kinds of use-areas. For example, two use-areas that have had the same kinds and amounts of refuse deposited within them and are of a similar archeological nature might be expected to have similar mean resistivities. If located at opposite ends of Transect I, however, they are certain to have different means. The soil alterations caused in the two areas by prehistoric human occupation may be similar, but are superimposed on background soil conditions that differ and make a comparison of local mean resistivities invalid.

A second problem that nonstationarity in a resistivity series may cause is distortion in the results of spectral analyses of the series. It will be recalled from

Chapter 2 that spectral analysis may be used to find the truly periodic components of variation within a resistivity series. Such components could be of agricultural cause and should be removed before one attempts to isolate and distinguish use-areas. Should a trend or a portion of a very low frequency wave run through a resistivity series, the percentage of total series variance found in a spectral analysis to be contributed by low frequencies will be inflated (Jenkins 1961:153). Agriculturally caused periodic soil variations thus might be interpreted to occur over broader scales than they actually do. Then, when removing them from the resistivity series, more variation over a wider scale will be removed than is necessary, and a significant amount of variation and information of an archeological nature may be lost from the residual series.

Thus, the first step in an analysis of variation within a resistivity series should be the removal of trends or portions of very long frequencies running though the series. This may be achieved by smoothing the data series with a running average, and by subtracting the smoothed series from the original series. Smoothing isolates that variation within the series contributed by all waves having spatial frequencies less than the smoothing interval of the running operator. Subtraction effectively removes such components of variation from the original series, leaving a residual series containing information on only higher frequency variations.

In analyzing data from the Crane site, low-frequency resistivity variations were isolated with a "normal filter" smoothing function having a "cutoff frequency" or smoothing interval of 25 survey stations spaced .5 m apart, that is, a smoothing interval of 12 m. Using this filter function, each datum point of the original data series was replaced by an average of itself and the 24 points surrounding it, weighted according to the ordinates of the normal curve. Data points closest to the one under consideration were weighted most heavily in the average; those more distant were weighted less heavily. The filter weights are shown in Table 108.

The reasons for having used a normal-weighted filter as opposed to other kinds of running averages are of a general methodological nature, and have been discussed in Chapter 2. The reason for having used a cutoff frequency of 12 m is site-specific. At the Crane site, work areas around the Middle Woodland house range in size from 6.5 to 8.5 m. Away from the house, where space was less at premium, use-areas may be somewhat larger (Chapter 5). Archeologically significant variation at the scale of use-areas within the resistivity series, therefore, is most likely confined to component waves having periods of less than approximately 12 m. All longer-period, lower-frequency waves are most likely nonarcheological in origin, or only in the most general way related to prehistoric occupation (e.g., the two broad-scale lows on either side of the gully, Transect IV, Figure 68), and had to be removed from the data to be analyzed. By choosing a filter function with a cutoff frequency of 12 m, archeologically significant resistivity variations related to use-areas were segregated from

TABLE 108
Normal filter weights used in smoothing resistivity series to derive component bands of various spatial frequencies

4 m Smoothing Operator (9 Resistivity Survey Stations)	6 m Smoothing Operator (13 Resistivity Survey Stations)	8 m Smoothing Operator (17 Resistivity Survey Stations)	10 m Smoothing Operator (21 Resistivity Survey Stations)	12 m Smoothing Operator (25 Resistivity Survey Stations)
.00332	.00222	.00166	.00133	.00111
.02380	.00876	.00477	.00313	.00227
.09714	.02700	.01190	.00672	.00438
.22585	.06476	.02579	.01319	.00793
.29921	.12098	.04857	.02368	.01350
.22585	.17603	.07945	.03885	.02157
.09714	.19947	.11293	.05826	.03238
.02380	.17603	.13945	.07983	.04566
.00332	.12098	.14960	.09997	.06049
	.06476	.13945	.11442	.07528
	.02700	.11293	.11968	.08802
	.00876	.07945	.11442	.09667
	.00222	.04857	.09997	.09974
		.02579	.07983	.09667
		.01190	.05826	.08802
		.00477	.03885	.07528
		.00166	.02368	.06049
			.01319	.04566
			.00672	.03238
			.00313	.02157
			.00133	.01350
				.00793
				.00438
				.00227
				.00111

broader-scale variations not of interest, and the latter—including those component waves responsible for the trends exhibited along Transects I and IV—could be removed from the data.

Two examples of the nine low-frequency series obtained with the chosen smoothing function from the nine original resistivity series along Transects I, III, and IV are shown in Figures 67 and 68 (dotted lines). The proportions of variance within the original series that are contributed by the low frequency series are shown in Table 109. The proportions are quite significant, ranging between 26 and 74%. It should be apparent from these percentages that if an attempt had been made to isolate use-areas of different means and variances within the *original* data series, variability of nonarcheological origins would certainly have distorted, if not have completely confused, such an analysis.

Once low-frequency resistivity series were obtained from the original series along Transects I, III, and IV, they were subtracted from the original series to obtain high frequency series containing waves with periods of less than 12 m. An example of the high frequency series is shown in Figure 69. Note that the series is stationary—that is, local means are constant (zero) over the course of the series when assessed over 25 data points (12 m) or more. Thus, spectral analyses could be performed on this series to yield a *reasonable* assessment of the spatial scales at which true periodicities occur. A meaningful analysis of local means and variances for the purpose of locating and distinguishing use-areas, however, can not be performed on this series. Periodic and high-frequency, nonarcheological variation would have to be removed from it first.

Use of Spectral Analysis and Spatial Filtering Techniques to Remove Periodic and High-Frequency, Nonarcheological Variation from Resistivity Series

Trends and portions of low-frequency waves having natural or other nonarcheological origins are one form of variability that should be isolated and removed from a resistivity series before attempting to locate and differentiate use-areas within it. Periodic components of an agricultural nature resulting from plowing, banded fertilizer applications, etc., and ultra-high-frequency, random noise resulting from measurement errors or small-scale, natural soil anomalies, are other kinds of variability that likewise should be removed. Let us now illustrate these steps in analysis.

At the time of survey of Transects I, III, and IV, the Crane site was covered primarily with pioneer annual weeds. Aerial photos of the site at this time revealed lines of darker or lighter green weeds, which ran north–south or east–west, and which had periodic spacing. Three spacings between lines were found: 20 in. (.51 m), 24 in. (.61 m), and 36 in. (.91 m). On the basis of the orientation

10. Barnes Layer Method and Spatial Filtering in Archeology

TABLE 109
Percentages of total variance of resitivity series contributed by components of low spatial frequency (< 12 m period)

Transect	Total Series Variance	Variance in Low Frequency Waves	Percentage of Total Series Variance
IV West	39715280	10334321	26.02
IV Central	24554448	7180370	29.24
IV East	32874912	9602999	29.21
III South	32219520	15022337	46.62
III Central	28572640	17179984	60.13
III North	28503136	14794999	51.91
I North	17113904	12661976	73.99
I Central	27441536	17554512	63.91
I South	17815696	11877198	66.67

and spacing of the lines, it was fairly certain that these represented crop marks. It was not known, however, whether the soil moisture and/or nutrient anomalies that these vegetational patterns represented (*a*) would be reflected in the resistivity survey data, and (*b*) were restricted to the plowzone or extended into the archeological layers of interest, as well. If they were represented by periodic variations in the resistivity series from the archeological horizon, Barnes Layer BL2–1, those variations would have to be removed from the series before an attempt could be made to isolate and differentiate use-areas.

To investigate the first issue—whether the soil anomalies responsible for the vegetational patterns also produced periodic resistivity variations—spectral analyses were made of resistivity variations along the nine subtransects of Transects I, III, and IV within the plowzone (Whole Volume WV1). The resistivity series from the subtransects were smoothed with a normal filter function having a cutoff frequency of 12 m. Then the smoothed series were subtracted from the original series to derive series that: (*a*) were composed of waves with periods of only 12 m or less, and therefore had a higher proportion of their total variability attributable to agricultural causes; and (*b*) had stationary means of zero. These attributes facilitated an accurate spectral analysis. Spectal analyses were restricted to those portions of the series derived for Transects I,

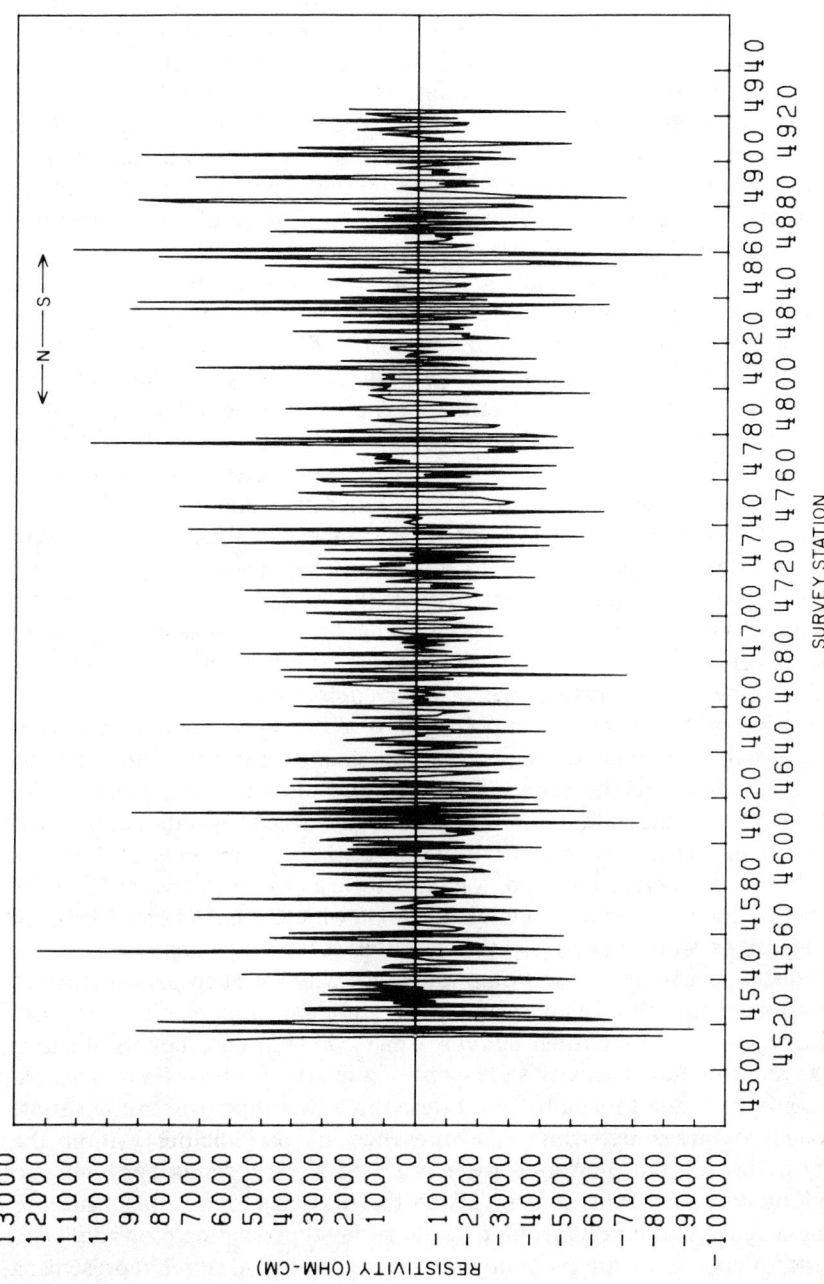

FIGURE 69. *High frequency resistivity variation along Transect IV, central subtransect. The series was derived from the two shown in Figure 68 by subtracting the smoother one from the original resistivity series. The derived series contains resistivity variations of a scale less than 12 m, including archaeological variations, periodic agriculture variations, and ultra-high frequency random variations resulting from one-station measurement errors and small scale natural soil anomalies.*

III, and IV that ran perpendicular to the crop marks (Table 110). Had sections running at various angles to the crop marks and having different sets of periodicities been included, the analyses would have been muddled. Sections of the transects that were covered by rows of wheat rather than weeds and that also were likely to exhibit different periodicities, similarily were excluded from the analysis. The resultant data sets had sample sizes of at least 68 points, which were large enough to determine accurately the presence of periodicities up to 17 survey stations, or 8.5 m in length (one-fourth the sample size). The scope of the analyses thus was great enough to determine the presence of all periodicities expected to be found, plus much larger ones.

The spectral analyses were made with a program provided by the Statistical Research Laboratory of the University of Michigan. The program uses the autocorrelation function of a series to be examined to derive a "raw sample periodogram" within the frequency range of 0 to .5 cycles per unit interval (here, the .5 m between survey stations). The sample periodogram is then smoothed with a Hanning filter (Davis 1973:269) having a "window" (cutoff frequency) of .25 times the series sample size, in order to estimate the spectral density function of the series. The spectral density function specifies the percentage of total series variance attributable to individual frequencies within the series. Those frequencies with high variance should represent the fundamental modes of periodicities within the series, compounds of the fundamental modes (i.e., the wavelength of one periodicity present in the data times the wavelength of another periodicity present in the data), or harmonics (multiples) of the fundamental modes or their compounds.

Examination of the spectral density functions of the nine resistivity series for plowzone revealed a number of frequencies that consistently contributed a large amount of variability to the series (Table 110). All of the frequencies were harmonics of the fundamental periods noted in the surface vegetational patterns at Crane (.51 m, .61 m, .91 m), or harmonics of their compounds. It thus was concluded that the periodicities in soil moisture and/or nutrient availability which the surface vegetation indicated and which were most likely of agricultural origin were reflected in the resistivity of the plowzone.

Next, the second issue had to be investigated: whether such periodicities in soil conditions and soil resistivity extended into the archeological horizon, Barnes Layer BL2–1. If so, as agriculturally caused variations, they would have to be removed from the resistivity series representing BL2–1 before a reasonable attempt could be made to isolate and differentiate use-areas within the series. The procedures used determine the presence of periodicities within the resistivity series for the plowzone were repeated for the series pertaining to BL2–1 along the same transect intervals. All frequencies found consistently to contribute a large amount of variation to the series representing plowzone also were found to contribute a large amount of variation to the series representing BL2–1. Three additional periodicities not significant in plowzone but clearly present in BL2–1 also were found (Table 111). The additional frequencies,

TABLE 110
Spatial frequencies found consistently to contribute high variability to the resistivity of plowzone (Whole-Volume 1) at Crane site, Transects I, III, IV*

Frequency (cycles/.5m)	Wavelength (.5m/cycle)	Wavelength (m/cycle)	Corresponding Fundamental Wavelength, Harmonic and/or Compound of Fundamental Wavelengths and Harmonics Found in Aerial Photo
.500	2.000	1.000	.508 m, 2nd harmonic
.240–.245	4.082–4.167	2.041–2.083	.508 m, 4th harmonic
.120–.125	8.000–8.333	4.000–4.167	.508 m, 8th harmonic
.400	2.500	1.250	.6096 m, 2nd harmonic
.260–.270	3.704–3.846	1.852–1.923	.6096 m, 3rd harmonic
.130–.135	7.407–7.692	3.704–3.846	.6096 m, 6th harmonic
.260–.270	3.704–3.846	1.852–1.923	.9144 m, 2nd harmonic
.075–.080	12.50–13.33	6.250–6.665	.9144 m, 7th harmonic
.360–.375	2.667–2.778	1.333–1.389	.508 m x .9144 m, 3rd harmonic
.180–.185	5.405–5.556	2.703–2.778	.508 m x .9144 m, 6th harmonic
.460	2.174	1.087	.6096 m x .9144 m, 2nd harmonic
.290–.300	3.333–3.448	1.666–1.724	.6096 m x .9144 m, 3rd harmonic

*Proveniences used in the spectral analyses are the following:
TI-North:916–1064; TI-Central:516–664; TI-South:116–264; TIII-North:3887–3987; TIII-Central:3487–3576; TIII-South 4287–4387; TIV-East:5233–5320; TIV-Central:4533–4620; TIV-West:5933–6020.

however, were all harmonics of compounds of the fundamental modes indentified within the vegetational patterns.

The occurrence of periodic resistivity variations of *agricultural* origin within the *archeological* horizon is an example of how agricultural, archeological, and natural soil variations and soil resistivity variations may not be segregated

TABLE 111
Spatial frequencies found consistently to contribute high variability to the resistivity of Barnes Layer BL2-1, additional to those found in plowzone (Whole Volume 1) at Crane site, Transects I, III, IV*

Frequency (cycles/.5m)	Wavelength (.5m/cycle)	Wavelength (m/cycle)	Corresponding Fundamental Wavelength, Harmonic and/or Compound of Fundamental Wavelengths and Harmonics Found in Aerial Photo
.155	6.4516	3.226	.508m x .9144m, 7th harmonic
.215–.225	4.444–4.651	2.222–2.326	.6096m x .9144m, 4th harmonic
.055–.06	16.667–18.182	8.334–9.091	.508m x .6096m, 27th harmonic or .508m x .9144m, 18th harmonic or .6096m x .9144m, 15th harmonic

*Proveniences used in the spectral analyses are the following:
TI-North:916-1064; TI-Central:516-664; TI-South:116-264;
TIII-North:3887-3987; TIII-Central:3487-3576; TIII-South:4287-4387;
TIV-East:5233-5320; TIV-Central:4533-4620; TIV-West:5933-6020.

perfectly over depth. In such cases, use of the Barnes Layer method, alone, and partitioning of resistivity variation only over the dimension of depth is not sufficient to segregate agricultural, archeological, and naturally caused resistivity variation, and to isolate the archeological component for examination. Spatial filtering also is required.

To use spatial filtering techiques for this purpose, however, the scales of the several sources of resistivity variation must differ. In the case of agriculture and archeological soil variation at the Crane site, this was true. Use-areas at Crane are expected to have varied in size between about 6.5 and 12 m. Agricultural soil variation, on the other hand, was largely restricted to smaller scales. The longest periodic resistivity variation found within BL2–1 had a wavelength of between 6.25 and 6.67 m. The most predominant periodicities having spectral densities of 2.5% or more (i.e., those whose contribution to total series variation was 2.5% or greater) had wavelengths not greater than 5.26 m.

Under these conditions, for segments of Transects I, III, and IV running perpendicular to agricultural periodicities, it was possible to make a nearly complete segregation of agricultural and archeological variability within BL2–1 using spatial filtering techniques. Using a normal filter with a cutoff frequency of 7 m and the filtering weights shown in Table 108, the original resistivity series along Transects I, III, and IV could be smoothed to obtain a series composed of frequencies only greater than 6 m and carrying information on only archeological and natural soil variation. Periodicities of agricultural origin, as well as ultra-high-frequency random noise (one-station measurement errors, small scale natural soil anomalies) were excluded from the series.

Two sets of smoothed resistivity series from BL2–1 now were available; one set derived with a 12-m-wide normal filter and carrying information on only natural soil variation, and a second set derived with a 6-m-wide normal filter and carrying information on natural and archeological resistivity variations. By subtracting the well-smoothed series from their corresponding, less well-smoothed series, data series containing variations of only 6–12 m and of the scale of use-areas were isolated. An example of one such series is shown in Figure 70 for the central subtransect of Transect IV.

Use of Spatial Filtering Techniques to Partition Archeological Variation into Components of Differing Scales

To this point, resistivity variations along Transects I, III, and IV have been envisioned as a sum of variations of agricultural, archeological, and natural origins. Spatial filters have been designed to segregate these components of variation by scale, and to isolate the intermediate-frequency archeological component. Archeological variation itself, however, also may be considered a palimpsest, reflecting soil variations of different scales corresponding to use-areas of different sizes and perhaps different functions. Spatial filters may be designed in a manner similar to that described previously to partition archeological resistivity variation into components that differ in the scale and origins of the variations they contain.

As a further step toward defining and differentiating prehistoric use areas with the Crane resistivity data, therefore, the intermediate frequency variations in soil resistivity, which were segregated previously and considered archeological rather than natural or agricultural in nature were subdivided. Three bands containing waves of periods 6–8 m, 8–10 m, and 10–12 m were derived for each of the outer subtransects of each transect. This was achieved by smoothing the *original* resistivity series along each of the subtransects four times, using running normal filter functions with cutoff frequencies of 6, 8, 10, and 12 m and with the filter weights shown in Table 108. Successively, the series derived with wider filters then were subtracted from those derived with narrower filters (6 minus 8, 8 minus 10, etc.), producing the three bands. The same segregation also could have been achieved by smoothing the composite *intermediate* frequency series derived in the last section and composed of waves with periods of 6–12 m, once with an 8-m-wide filter, and again with a 10-m-wide filter. The 6–8-m band then could have been obtained by subtracting the 8-m filtered series from the composite, intermediate frequency series. The 8–10-m band could have been obtained by subtracting the 10-m filtered series from the 8-m filtered

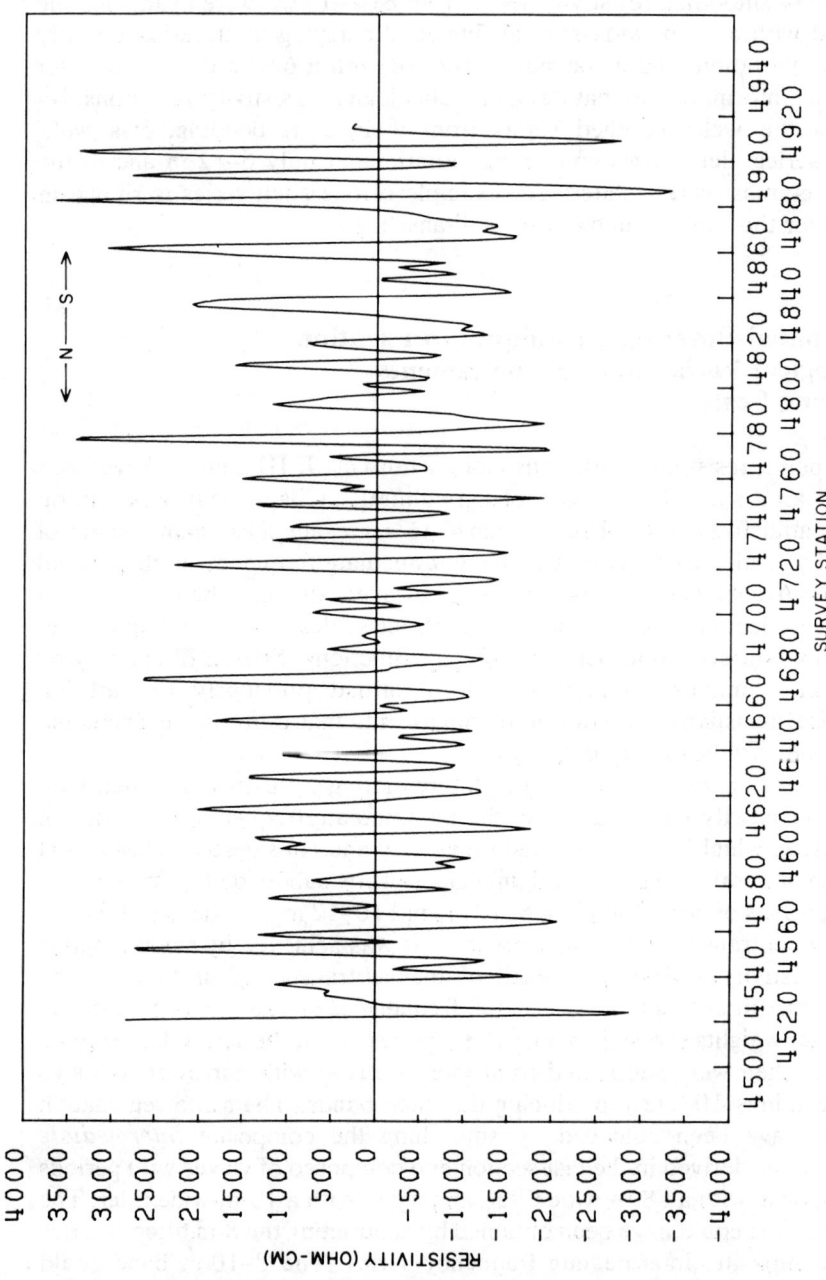

FIGURE 70. *Intermediate frequency resistivity variation along Transect IV, central subtransect. The series was derived by subtracting a well smoothed version of the original series shown in Fig 68 from a less well smoothed version. The two smoothed series had been derived with running normal filters having cutoff frequencies of 6 and 12 m, respectively. The series pictured here contains resistivity variations between 6 and 12 in scale, and largely of an archaeological nature.*

series. The 10–12-m band could have been derived by summing the 6–8-m band and 8–10-m bands and subtracting that summary series from the composite, intermediate frequency series. The first procedure using the *original* data series rather than the isolated archeological component, however, is more efficient and accurate, and therefore was used.

In addition to the 6–8-m, 8–10-m, and 10–12-m bands, a fourth band along each subtransect, composed of waves with periods of 4–6 m and containing some agricultural variability, also was derived for purposes of defining the differentiating use-areas. This extension of the range of frequencies to be investigated was undertaken *not* because it was felt that use-areas might also be of this size, but rather, because the transects might have intersected use-areas off center and included only portions of them, reducing the scale of the band in which they would manifest themselves.

The particular *cutoff frequencies* of the bands—4, 6, 8, 10, and 12 m— were chosen arbitrarily. They could have been odd meters or half meters. The *widths* of the bands (2 m), however, were designed explicitly (*a*) to be narrow enough to isolate variation of a very restricted scale, so that the dimensions of use-areas could be defined precisely, and (*b*) to be wide enough to include enough resistivity variation that inaccuracies introduced in the filtering process would not overwhelm significant archeological variation. It should be remembered that the series obtained by the smoothing and subtracting procedures only *approximate* distinct bands of frequencies and that there is some margin of error in the frequencies of variability admitted or excluded from them near their borders (see Chapter 2).

The amount of variability encompassed in each of the four bands for the six subtransects examined is shown in Table 112. Note that the variance within each band is very small—between .73 and 3.3% of the total series variances of the original series. This variation represents archeologically significant variation of interest. The remainder of the variability in the original series largely originates from agricultural or natural soil variations or from random measurement errors. These statistics should make it clear just how necessary it is to filter resistivity data from earthen sites of extraneous variations before attempting to interpret it with respect to local means and variances for archeological purposes.

Use of Spatial Filtering Techniques to Define and Differentiate Use-Areas: An Outline of Specific Procedures

Once bands of resistivity variations that are largely archeological in nature and that correspond to given spatial scales have been isolated along a transect, it is possible to transform them so as to allow the *determination of locations* of

TABLE 112
Percentages of total variance of resistivity series contributed by component bands of several intermediate spatial frequencies

Transect	Total Series Variance	Band	Variance in Band	Percentage of Total Series Variance
IV West	39715280	4–6m	741754	1.868
		6–8	312962	.788
		8–10	163481	.412
		10–12	97764	.246
	TOTALS:	(4–12)	1315961	3.314
IV East	32874912	4–6m	541739	1.648
		6–8	234106	.712
		8–10	133777	.407
		10–12	79441	.242
	TOTALS:	(4–12)	989063	3.009
III South	32219520	4–6m	156172	.485
		6–8	45600	.142
		8–10	21534	.067
		10–12	13565	.042
	TOTALS:	(4–12)	236871	.736
III North	28503136	4–6m	272459	.956
		6–8	124889	.438
		8–10	71564	.251
		10–12	43032	.151
	TOTALS:	(4–12)	511944	1.796
I North	17113904	4–6m	71098	.415
		6–8	30964	.181
		8–10	18662	.109
		10–12	12833	.075
	TOTALS:	(4–12)	133557	.780
I South	17815696	4–6m	78037	.438
		6–8	29627	.166
		8–10	14519	.081
		10–12	7816	.044
	TOTALS:	(4–12)	129999	.729

use-areas along the transect. Also the size, mean resistivity, and variation in the resistivity of the discovered use-areas may be estimated, in order to *distinguish* use-areas of different natures. The following procedures were used to locate, characterize, and distinguish use-areas along Transects I, III, and IV at the Crane site, and may be used at other sites in general.

Step 1. From each of the multiple bands of resistivity variation along the transect of interest, derive a series that monitors changes in local mean resistivity over the course of the transect. This may be accomplished with a running filter function, which replaces each datum point in the band series by the simple mean of itself and its surrounding points within an interval equivalent to the *central* frequency of the band.

Step 2. Standardize the values within each derived series of local means, a separate standardization performed for each series. Standardization of the values within derived series is necessary to make the magnitude of resistivity variations within different series of different frequencies comparable. Under most circumstances, the amount of variation encompassed within resistivity bands of different frequencies and within series of local means derived from them will decrease as frequency decreases (wavelength increases).

Step 3. For each standardized, derived series of local means, plot a number of histograms of the local means, varying the class interval widths and the location of class boundaries. Several histograms, rather than just one, should be plotted in order to determine the probable, actual shape of the distribution, unaffected by class interval width and location of class boundaries.

Step 4. Under ideal circumstances, if no use-areas are present along the transect, the distribution of standardized local mean resistivity values will approximate a smooth, symmetrical distribution centered around 0. If a few use-areas causing negative resistivity anomalies are present along the transect, ideally, negative outliers, an extended negative tail, or extra outlying modes with negative means—all reflecting the use-areas—will occur beyond (although not necessarily completely segregated from) the central, smooth, symmetrical portion of the distribution. If many use-areas causing negative anomalies are present along the transect, the central, smooth, symmetrical portion of the distribution may grade, in the negative direction, into a very jagged distribution indicating multiple, highly overlapping, outlying modes of slightly different means. A single, large, outlying mode might also occur in this circumstance.

Using the histograms of each standardized series of local means, define a negative threshold for each series that separates those standardized local mean values in the histograms which most probably reflect use-areas (negative values occurring as outliers, tails, extra negative modes, or jagged portions of the histogram) from those which most probably *do not* (values in the central, symmetrical portion of the distribution or beyond in the positive direction). Considering simultaneously all thresholds of all series, choose the threshold with the smallest absolute value (that closest to zero) and make the following categorization: (*a*) All proveniences having negative standardized local mean resistivities with absolute values greater than the threshold are locations of "significant negative resistivity anomalies" where the probability of a use-area occurring is significantly great; (*b*) all remaining proveniences are locations of insignificant resistivity variation.

The procedures outlined thus far, in Steps 1 through 4, for defining the locations where use-areas most likely occur and do not occur involve four operations and/or assumptions that should be stressed or require further explanation. First, it is assumed that for each band of standardized local mean resistivity values of a given frequency which is analyzed for the location of probable use-areas within it, at least one use-area of that scale must exist along the transect being analyzed. If this is not true, the process of standardizing the resistivity values will alter the values of some of the insignificant resistivity variations within the "empty" frequency band(s) such that they fall within the range of values that are greater (in absolute value) than the chosen threshold and that are truly significant in other "occupied" bands. The insignificant values will appear significant and use-areas will be hypothesized to exist in areas where they most likely do not exist. This situation may be avoided by dropping from the analysis those frequency bands that appear to be "empty"—frequency bands with histograms that have a smooth, symmetrical shape with no negative outliers, tails, or extra modes.

Second, it is assumed that the probability of a negative standardized local mean resistivity value correctly indicating the presence of a use-area is greater when it has a larger absolute value and occurs further out in the extremities of the histogram.

Third, not *all* negative standardized local mean resistivity values are considered to probably reflect the presence of use-areas. Only those negative values found to be significantly large relative to a threshold less than zero are considered such. The reason for judging the significance of standardized local mean resistivity values relative to a *threshold* value less than zero rather than accepting all negative values as significant is as follows. When isolating resistivity variation of a given scale within bands with the described smoothing techniques, a transformation is made from the *space* domain to the *frequency* domain. In the frequency domain, runs of positive and negative resistivity values defining the waves within the band *must* occur periodically. Some of the positive and negative runs and their corresponding maxima and minima—in particular, those of great amplitude—may actually correspond to positive or negative anomalies, respectively, within the data represented in the space domain. Others of lesser amplitude, however, may not correspond to negative or positive anomalies within the data represented in the space domain. Such low amplitude runs result from transforming the data from the spatial domain to the frequency domain. They are an artifact of the analytic procedure rather than a part of the original data structure. They must not be included within analysis of values supposedly reflecting anomalies within the space domain and representing use-areas. Consequently, the significance of values within the frequency domain must be tested against some threshold amplitude.

Finally, it should be stressed that of the several different thresholds found for different band series of standardized local means of different frequency, the threshold chosen to analyze the several series is the one with the *smallest* absolute value. Consequently, the set of locations characterized as having a high probability of containing use-areas, on the basis of their local mean resistivity values, is defined *generously* rather than conservatively.

Step 5. From the set of standardized local mean resistivity values characterized as "significant" in each band series, find the subset of those values that also are local *minima*. Call these values "significant local minima." The provenience of each significant minimum can be interpreted as the center of a probable use-area. The value of each significant minimum can be interpreted as the mean anomalous resistivity of the use-area.

Step 6. The dimensions of a probable use-area that is indicated by a significant local minimum can be defined as falling within the range of the wavelengths represented in *that band* in which the provenience of the minimum has the lowest negative value.

To this point, the analysis has allowed the determination of the *locations* of use areas along the survey transect (the proveniences of significant local minima), the *mean anomalous resistivities* of the use-areas (the values of the significant minima), and the *dimensions* of the use-areas (the bands in which the proveniences have the lowest minima). It should be stressed that no assumption has been made that the use-areas are spatially discrete. The technique outlined thus far can distinguish overlapping use-areas so long as their centers are separated by one or more survey units along the transect and their mean anomalous resistivities differ.

It next is desirable to characterize the use-areas by the *variability* of the resistivity values they encompass.

Step 7. From the *original* resistivity series, derive one high-frequency resistivity series for each intermediate-frequency band examined such that the high-frequency series contains all waves with periods less than and equal to those within the intermediate-frequency band. This may be achieved by smoothing the original data series with running normal filter function having cutoff frequencies equal to the lower cutoff frequency of the intermediate-frequency bands. Then subtract the smoothed series from the original series to obtain series containing residual, high-frequency resistivity variation.

Step 8. From each of the high-frequency resistivity series, derive a series that monitors changes in local standard deviations in resistivity over the course of the transect. This may be accomplished with a running filter function that replaces each datum point in the high-frequency series by the standard deviation of itself and its surrounding points within an interval equal to the lowest frequency contained within the high-frequency series.

Step 9. For each identified use-area, find within that particular high-frequency series containing waves with periods only less than or equal to the scale of the use area, the local standard deviation at the center of the use-area (the provenience of the "significant local minimum" defining the use area). This is the estimate of resistivity variation that may be used to characterize the use-area.

Note that the estimate of resistivity variation within a use-area contains variability of archeological origin, but also variability produced by agricultural or small-scale, natural anomalies and random measurement errors. The estimate is based on local variation within a series containing waves with periods of the scale of the use-area, but also waves with smaller periods, of the scale of small archeological features, agricultural soil anomalies, natural soil anomalies, and measurement errors. The inclusion, within the estimate, of resistivity variations produced by the nonarcheological phenomena is regrettable, but unavoidable. The purpose of the statistic is to estimate the areal density and/or variety of types of archeological features (e.g., pits) of *any* scale that occur within the use-area—from very small to the size of the use-area. Consequently, the resistivity values used in making the estimate must pertain to multiple scales, including very local scales that may encompass nonarcheological anomalies as well as archeological features.

Under some survey conditions, this circumstance may not be a problem. If nonarcheological anomalies are not numerous within the use-area of a site relative to the frequency of archeological features within use-areas, *or* if nonarcheological anomalies occur at a *uniform* density over all the use-areas (as will periodic, agricultural anomalies and random measurement errors), *differences* between use-areas in their resistivity variation may still be considered good estimates of differences in their feature densities and/or feature variety. Herein lies one of the advantages of interpreting resistivity data within a *statistical, geographic* perspective rather than a deterministic one trying to make one-to-one correspondences between anomalies and features.

Step 10. Classify each provenience along the transect where a use-area is suspected to occur (where a significant local minimum occurs) according to (*a*) the scale of the use-area, that is, the frequency band in which the "significant local minimum" defining the use-area occurs; (*b*) the mean anomalous soil resistivity of the use-area, that is, the value of the "significant local minimum"; (*c*) the variance of soil resistivity within the use-area, as calculated in Steps 7–9; and (*d*) its distance from foci of intense, multipurpose activity or reoccupation within the site, or from the center of the site if such foci are not explicitly known. Classification of use-areas by their distance from foci of intense activity or reoccupation is necessary in order to determine whether differences in mean anomalous soil resistivities should be interpreted as the result of deposition of different *kinds* of refuse materials within use-areas of different function, or the result of deposition of different *amounts* of refuse materials within areas that were used more or less time and which might or might not differ in function. It also is necessary in order to determine whether differences in the variances of resistivity values within use-areas should be interpreted as the result of different densities of features relating to the different functions of the areas, or to the different lengths of time the areas were used and reused.

The methods that might be used to classify proveniences and the use-areas that they contain are numerous. Included among them are the methods of numerical taxonomy (Sneath and Sokal 1973), multidimensional scaling, principle components analysis, etc.

Each of the four attributes by which proveniences and their use-areas are classified may vary with the archeological natures of the use-areas, and may be used to distinguish them (see Step 10). The attribute of scale, however, may also vary undesirably with a nonarcheological factor: whether or not the transect along which the resistivity series was obtained intersected the use-areas through their centers. To minimize the effect of this factor, two parallel transects several meters apart and having correspondingly spaced provenience units should be surveyed and analyzed, as was done on the Crane site. The scale of a use-area found at both corresponding proveniences of the two transects should be taken as the maximum of the two estimates of it. The precise distance between paired survey transects should be designed in relation to the expected sizes of use-areas within the site of interest, so as to maximize the probability of making correct estimates of them.

Step 11. Explain and interpret the differences between provenience units belonging to different types by the way of references to one or more of the following: (*a*) differences in the *kinds* of purposes the use-areas served, and consequently, the kinds of refuse materials deposited within them, the density and variety of features within them, and the amount of space required and used; (*b*) differences in the *amounts* of activity that occurred within them, and consequently, the amounts of refuse that were deposited within them, and the density of features within them; (*c*) differences in the *number* of purposes the areas served or for which they were used, and therefore, differences in *a* and *b*; and (*d*) differences in the *location* of the areas with respect to foci of activity or reoccupation, and thus, differences in *a*, *b*, and *c*. Note that interpretation of the differences between groups of frame-units and use-areas in terms of differences in only their *function* (kind of activity or use) is too restrictive.

When interpreting the differences in mean soil resistivity or variance in soil resistivity found between use-area and frame-units in terms of the prehistoric activities which produced them, the models presented in Chapters 3 and 8 should be used. These models, which discuss the different kinds of soil alterations that can be produced by the refuse materials deposited by different kinds of activities and the changes in soil resistivity that those soil alterations can cause, serve as the *bridging arguements* in such interpretations.

Once the use-areas along a resistivity survey transect have been located, typed, interpreted as to their general nature, then they may serve as the sampling strata of excavation designs.

Use of Spatial Filtering Techniques to Locate and Differentiate Use-Areas: Illustration

To illustrate and assess these outlined procedures, let us continue with our analysis of resistivity variation along Transects I, III, and IV. Two general modes of assessment will be used. First, while stepping through and illustrating the procedures, note will be made of whether the numerical results pattern *within themselves* in a consistent manner, and in a way one would expect them to if the procedures operate in accord with their designed purpose. This is an *internal* form of assessment. Second, a comparison will be made between the patterns found in the resistivity data from Transects I, III, and IV and the patterns one would expect to find based on the nature of the archeological remains that occur along the transects. This is an *external* form of assessment. Presentation of either one or the other of these tests, alone, would not provide a sufficient or convincing assessment of the proposed procedures. Both are required for adequate testing of the procedures.

Only Steps 1–6 and 10–11, concerned with the location of use-areas and their characterization by their anomalous resistivity and scale will be illustrated and discussed. Only the results of these steps can be assessed both internally for consistency and externally in relation to the reconstructed distribution of use-areas at Crane. Steps 7–9, concerned with the variation of resistivity within use-areas, will not be illustrated for they cannot be assessed in an external manner. The locations, types, and densities of features along Transects I, III, and IV, and the resistivity variations within use-areas that they would produce, are not known.

Note, also that external assessment of the results of Steps 1–6 and 10–11, and of the correspondences between use-areas and their resistivity responses, is subject to many of the same problems and limitations that were faced when evaluating the correspondence between use-areas and their soil attributes:

1. The *uses* that were served by given areas within particular surface collection units are not always definitely known.
2. The *amount* of activity that occurred within particular collection units has not been estimated. Only the *presence or absence* of activity groups of given types within collection units, based on the presence or absence of their diagnostic artifact types, has been reconstructed. Some use-areas within collection units for which activity groups have been coded present may not

have been used long enough to have detectably altered the physical and chemical properties of their soils, and consequently, their soil resistivity.
3. Some use-areas may have served as the location of activities that altered the soil but did not leave artifactual remains (e.g., human bodily elimination and the development of night soil).
4. Very few use-areas and collection units along Transects I, III, and IV have associated with them only one activity group. Most collection units are sites of either no activity or else overlapping activities and uses of multiple natures, and prevent the determination of one-to-one correspondences between the function of a use-area and its resistivity signature.
5. Use-areas are not necessarily centered within collection units. Single collection units may encompass portions of more than one use-area.

In all, six analyses of the Crane resistivity data were performed, one for each of the outer subtransects of Transects, I, III, and IV. Analyses were restricted to the portions of these subtransects shown in Table 113. When considered separately, these areas are uniform in (*a*) topography and drainage conditions, (*b*) historic land-use, and (*c*) the time and moisture regime at which they were surveyed, and (*d*) surface vegetation. Topography was held constant to ensure that all areas examined in a given analysis had similar rainfall infiltration/runoff ratios, held moisture at similar environmental suctions, and were comparable in the distinctness with which prehistoric anthropic soil anomalies would manifest themselves as resistivity anomalies. Areas of only similar historic land-use were examined in order to hold constant the degree to which prehistoric anthropic soil anomalies had been subdued by historic, agricultural activities and the degree to which their distinctness as resistivity anomalies had been subdued since the historic period. Time of survey and moisture regime were held constant for the same reason that topography was: in order to ensure that all areas to be compared in the analysis held moisture at similar environmental suctions. Surface vegetation and the environmental suction it can produce through evapotranspiration was essentially constant over the whole of Crane site for any one time of survey, and thus, for each subtransect examined. Uniformity of vegetationally caused suction was desirable again to keep the suction at which moisture was held within soils similar over all areas to be compared. Finally, analysis was restricted to portions of the transects which intersected surface collection units having only a few activity groups present per unit (i.e., locations of least activity overlap). The collection units included in the analyses and the activity groups present and possibly present within them are shown in Table 113.

Let us now step through the procedures outlined in the previous section. We may begin with Steps 1–6, illustrating them and assessing them internally, and then assessing them externally. In accord with Step 1, the resistivity values within each of the four band series of different frequencies for each of the six subtransects were transformed into series monitoring changes in local mean

TABLE 113
Surface collection units, resistivity survey stations falling within them, and activity groups indicated by their artifact inventories

Survey Stations		Collection Unit	Activity Groups Definitely Present within Collection Unit*	Activity Groups Possibly Present within Collection Unit**
Transect IV				
West	East			
5201-5212	5901-5912	1	9	
5213-5224	5913-5924	9	9,12,20	1,18,20
5225-5236	5925-5936	20		
5237-5248	5937-5948	33	1,12,24	
5249-5260	5949-5960	49		1
5261-5272	5961-5972	66		1,6
5273-5284	5973-5984	85		6
5285-5296	5985-5996	105	1,21	6,7
5297-5308	5997-6008	125	1,12	6,24
5309-5320	6009-6020	145		1,6
5321-5335	6021-6035	166	10	4,6,7,21,24
5336-5352	6036-6046	186+187	1,15+8,13,15,20,22	6,7,21+1,4,18,19,21
5353-5361	6047-6059	206	12,15	
5362-5368	6060-6072	225	12,15	
Transect III				
South	North			
4243-4254	3844-3855	630		
4255-4366	3856-3867	631	12,15,16	
4267-4278	3868-3879	632	1,18,20,22	
4279-4290		633	12	
	3880-3891	633+619	12	
4291-4303		634		
	3892-3903	634+620		+18,20
	3904-3915	621	4	
4304-4315		635	4,20	18,19

TABLE 113 (cont.)

Survey Stations		Collection Unit	Activity Groups Definitely Present within Collection Unit*	Activity Groups Possibly Present within Collection Unit**
	3916-3927	622	12	
4316-4327		622+636	2+	+24
4328-4339	3928-2940	624	15,17,20	19
4340-4351	3941-2952	625		1,4
4352-4363	3953-3964	626		
Transect I				
North	South			
873-885	74-83	164	2,12,16,7	
886-898	84-99	165+186	1,2,16,17+1,15	15+6,7,21
899-913	100-113	187	8,13,15,20,22	1,4,18,19,21
914-927	114-127	188	4,17,18,21,22	19,20
928-939	128-139	189	1,20,22	18,19,24
940-951	140-151	190		
952-963	152-163	191	1	
964-975	164-175	192	1,12,18,19,20	
976-987	176-187	193	13,15	

*Includes Activity Groups represented in the collection unit by artifacts: (1) diagnostic of multiple, different activities within the Activity Group (regardless of artifact density); (2) diagnostic of only a single activity within the Activity Group, but occurring in high densities (types taking multi-count states); (3) definitely diagnostic of a single activity, and present (types taking only presence-absence) or (4) possibly diagnostic of a single activity and present (types taking only presence-absence states) qualified in that the collection unit is one of a cluster of units containing artifacts definitely diagnostic of that activity.

**Includes Activity Groups represented in collection units by artifacts: (1) diagnostic of only a single activity within the Activity Group and occurring in low densities (types taking multi-count states), or (2) possibly diagnostic of a single activity within the Activity Group and present (types taking presence-absence states), qualified in that the collection unit is not one of a cluster of units containing artifacts definitely diagnostic of that activity.

resistivity. Each derived series then was standardized (Step 2), and multiple histograms of the standardized local mean resistivity values were plotted for each series (Step 3). Some example histograms are shown in Figure 71. For each of the four series of standardized local mean resistivity values of different frequencies for each of the six subtransects, threshold values were defined, separating negative outliers, tails, extra modes, or jagged sectors of the histograms from the remaining sections of the histograms (Table 114). Of the four threshold values found for each subtransect, the one with the smallest absolute value and/or the one occurring most frequently was chosen as the single threshold for segregating "significant negative resistivity anomalies" from insignificant resistivity variations for all four band series (Table 114) (Step 4).

Using the appropriate threshold, a list was made for each frequency band of each subtransect of all those negative standardized local mean resistivity values within the bands which: (*a*) have absolute values greater than the threshold, and (*b*) also are local minima. These values represent the anomalous mean resistivity of use-areas that most likely occur along the transects. The proveniences at which the anomalies occur (the locations of the use-areas) and the frequency bands within which they occur (the dimensions of the use-areas) also were assembled. These characteristics of the use-areas are summarized in Table 115.

The results to this point in the analysis, as summarized in Table 115, illustrate several features of the data reduction procedure, and allow us to assess it internally, as described before. First, at many single locations, significant local minima occur in not just one frequency band, as one might desire (one use-area at a location has one dimension), but in several frequency bands. This undesirable circumstance is predictable and correctable. It arises from two factors. First, the values of the thresholds used to define "significant negative resistivity anomalies" were chosen generously, allowing some "insignificant" resistivity anomalies to be admitted to the set of "significant" anomalies in some frequency bands. Second, the circumstance also arises from the fact that the spatial filtering techniques used here do not *perfectly* segregate bands of different wavelengths of resistivity variation. There is some overlap in the wavelengths composing adjacent bands (e.g., the 4–6-m and 6–8-m bands). That this is true can be seen in the *patterning* of the values of the significant local minima among frequency bands at single locations. At any one location, there nearly always is a *monotonic* decline in the absolute value of the minima in the frequency bands away from the band containing the lowest minimum. Over all frequency bands at one location, the values of the significant minima "focus upon" or "basin toward" one lowest minimum in some one frequency band.

To correct for this imprecision in the methodology, and to determine the single frequency band in which the one expected significant minimum should occur at a location, it is only necessary to find that frequency band containing the lowest

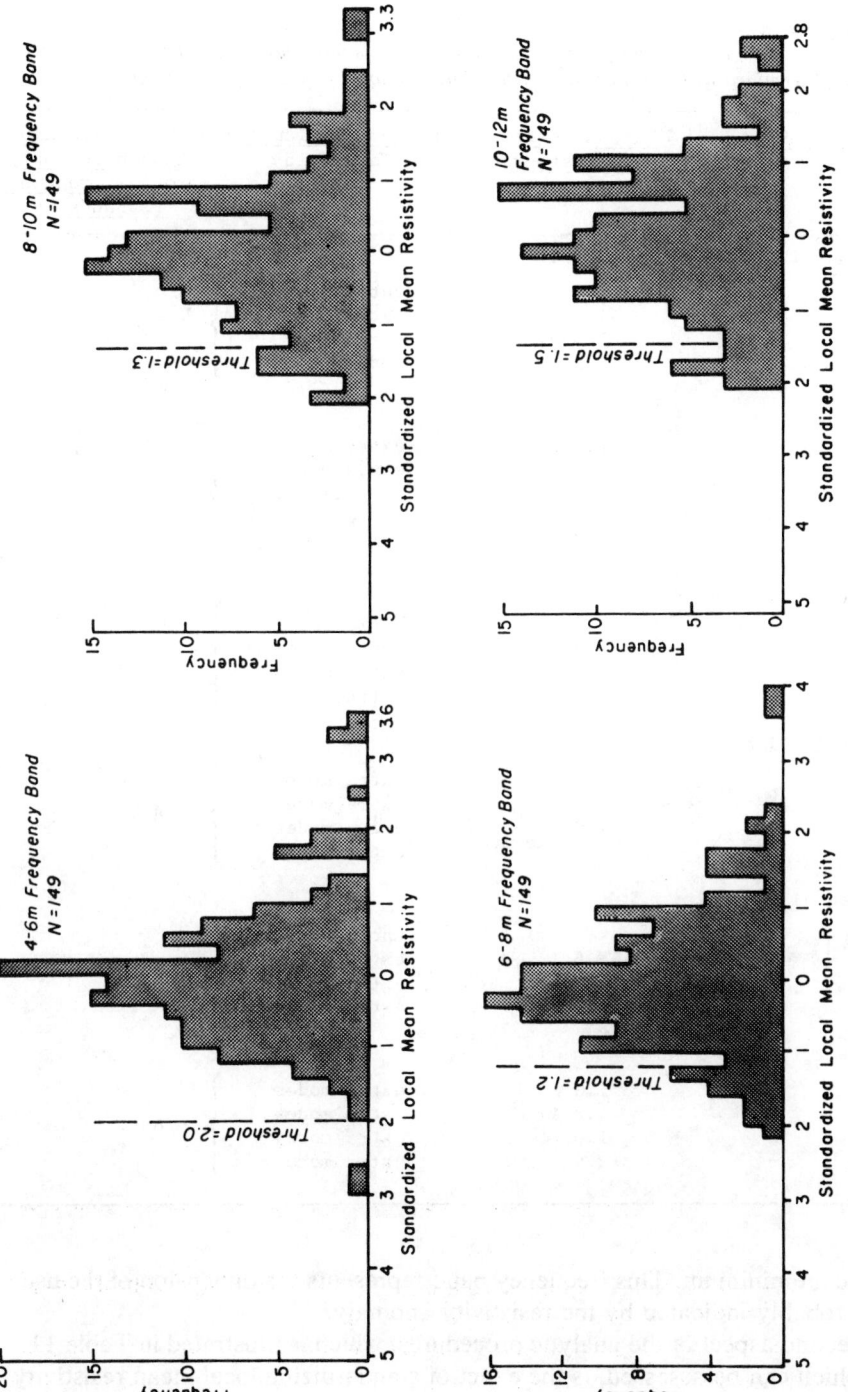

FIGURE 71. *Representative histograms of standardized local mean resistivity values within band series of various frequencies for Transect IV West, survey stations 5220 and 5368.*

TABLE 114
Thresholds used to define "significant negative resistivity anomalies" along Transects IV, III, and I, as determined within histograms of standardized local mean resistivity values

Frequency Band	Threshold Found in Histogram	Pattern Indicating Threshold	Threshold Value Chosen to Apply to All Frequency Bands
Transect IV West			
4–6 m	2.0 σ	outliers	⎫
6–8 m	1.2 σ	extra mode	⎬ 1.2 σ
8–10 m	1.3 σ	extra mode	⎪
10–12 m	1.5 σ	extra mode	⎭
Transect IV East			
4–6 m	1.5 σ	extra mode	⎫
6–8 m	1.2 σ	extra mode	⎬ 1.2 σ
8–10 m	–	no outliers or extra mode	⎪
10–12 m	1.8 σ	outliers	⎭
Transect III South			
4–6 m	1.4 σ	outliers	⎫
6–8 m	.9 σ	extra modes	⎬ .8 σ
8–10 m	.7 σ	extra modes	⎪
10–12 m	1.4 σ	extra mode	⎭
Transect III North			
4–6 m	.8 σ	extra modes	⎫
6–8 m	1.0 σ	extra modes	⎬ .8 σ
8–10 m	1.0 σ	extra modes	⎪
10–12 m	1.1 σ	extra modes	⎭
Transect I North			
4–6 m	–	unclear	⎫
6–8 m	.6 σ	extra modes	⎬ .8 σ
8–10 m	.8 σ	extra mode	⎪
10–12 m	.7 σ	extra modes	⎭
Transect I South			
4–6 m	1.0 σ	extra modes	⎫
6–8 m	.8 σ	extra modes	⎬ .8 σ
8–10 m	1.0 σ	extra modes	⎪
10–12 m	.6 σ	extra mode	⎭

significant minimum. This frequency band represents the dimension of the use-area probably indicated by the resistivity anomaly.

A second aspect of the analytic procedures, which is illustrated in Table 115 and which can be assessed, is the effect of standardizing local mean resistivity values within frequency bands (Step 2). It was said earlier that this step is

TABLE 115
Significant negative resistivity minima; their locations along transects; and the frequency band in which they occur

Surface Survey Unit	Frequency Band			
	4-6 m	6-8 m	8-10 m	10-12 m
Transect IV West				
20	5233(-1.7889)	5234(-1.5201)	5234(-1.2673)	
105			5286(-1.5535)	5287(-1.8493)
125	5304(-1.2179)	5304(-1.4359)	5304(-1.3890)	5305(-1.3156)
145	5318(-1.2420)			
166	5323(-1.9488)	5325(-1.8939)	5325(-1.6253)	
186 & 187		5336(-1.3801)	5337(-1.9381)	5337(-2.0112)
186	5350(-1.3760)	5350(-1.7025)	5350(-1.9300)	5351(-1.9841)
225	5362(-2.9729)	5362(-2.0762)		
Transect IV East				
9				5920(-1.5551)
20	5930(-1.9203)	5929(-1.3327)		
20	5936(-1.7519)	5937(-1.5271)	5938(-1.4192)	
49	5953(-1.2765)	5953(-1.4178)	5952(-1.5905)	5952(-1.5023)
66		5962(-1.2898)	5962(-1.3658)	
66	5972(-1.6596)	5972(-1.6606)	5972(-1.5658)	5971(-1.2428)
85	5983(-1.3608)	5983(-1.2999)		
105	5990(-1.6300)	5990(-1.8139)	5990(-1.4887)	
125	5998(-1.8750)	5998(-1.7005)	5999(-1.4455)	
166	6022(-2.6647)	6022(-2.3374)	6022(-1.8116)	
186 & 187	6039(-1.5326)	6039(-1.4772)	6039(-1.2992)	6038(-1.3583)
206	6053(-1.2988)	6053(-1.4901)	6052(-1.6120)	6052(-1.6913)
225	6072(-1.8068)	6071(-1.6456)	6072(-1.8122)	6072(-2.1732)
Transect III South				
630	4243(-1.3035)	4243(-1.3222)	4243(-1.1430)	4243(- .8620)
630	4248(-1.2380)			
630	4253(-2.1187)	4254(-2.1538)	4254(-1.3923)	
631	4263(-2.0970)	4263(-2.0547)	4264(-2.1069)	4265(-2.3807)
632	4269(-1.3754)	4269(-1.3216)		
632	4276(-4.2743)	4276(-2.8431)	4277(-1.432)	
634	4293(- .8771)	4293(-1.0079)	4292(-1.1286)	4292(-1.2642)
635	4307(- .8636)			
622 & 636	4316(- .9788)	4316(-1.1215)	4316(-1.2072)	4315(-1.1436)
624		4330(-1.4312)	4330(-2.1136)	4330(-2.4139)
626	4356(- .9179)			

*Numbers not in parentheses are surface survey station numbers. Numbers in parentheses are the values of "significant negative resistivity minima." Underlined proveniences and negative minima are those for which the negative minima have the greatest absolute value.

TABLE 115 (cont.)

Surface Survey Unit	Frequency Band			
	4-6 m	6-8 m	8-10 m	10-12 m
Transect III North				
630	3845(-3.2686)	3846(-3.4629)	3846(-3.3895)	3847(-3.4799)
631	3859(-2.2754)	3860(-2.4217)	3860(-2.2304)	3860(-1.8458)
632	3873(- .8996)	3873(- .8786)		
633 & 619	3886(-1.4053)			
634 & 620	3894(-1.6380)	3894(-1.8259)	3894(-1.7774)	3893(-1.7787)
621	3905(-1.0996)			
622	3925(- .8019)			
624	3934(-1.2355)			
625	3944(- .9874)	3944(- .8136)		
625	3951(- .9881)			
626	3957(- .8098)			
Transect I North				
165 & 186	894(- .9168)	894(- .9805)	894(- .9550)	895(- .9867)
187	910(-2.4328)	909(-1.8604)	909(-1.2833)	908(- .8725)
188	916(-1.0250)			
188	922(-1.8991)	923(-2.1550)	923(-2.1030)	923(-1.9233)
189	931(- .9764)			
189	937(-1.4742)	938(- .9949)		
190	944(-1.4826)	943(-1.5367)	942(-1.4371)	941(-1.5210)
191	953(1.2038)	953(-1.5189)	953(-1.5740)	954(-1.3178)
191	963(-1.5371)			
192	976(- .9257)	976(-1.0872)	974(-1.1908)	973(-1.3280)
193	983(-1.2374)			
Transect I South				
164	77(-1.3858)	77(-1.0303)		
165 & 186	84(- .9790)			
165 & 186	90(- .8153)	92(- .9873)	93(-1.2355)	93(-1.5161)
165 & 186	98(-1.1415)			
187	104(-1.0491)			
187	110(- .8308)			
188	120(-2.1081)	119(-2.3925)	119(-2.7250)	120(-2.4811)
189	128(-1.1724)	128(-1.0845)		
189	138(-1.0267)	138(-1.0821)	138(-1.1342)	138(-1.1374)
190	149(-1.9905)	149(-1.9275)	149(-1.6794)	149(-1.3790)
192	170(-1.7002)	170(-1.7577)	170(-1.8043)	170(-1.7493)
193	178(-1.3465)			
193	186(-2.2812)	185(-1.7729)	185(-1.4615)	185(-1.2026)

necessary in order to make the variability encompassed within different frequency bands comparable—to compensate for the decrease in the variance encompassed within a band series that occurs as the frequency of the band series decreases. Comparability is necessary if the "significance" of local mean resistivity values within band series of *all* frequencies examined for a transect is to be assessed in an unbiased manner with a *single* threshold value. Comparable variability within different frequency bands also is necessary if the dimensions of use-areas are to be accurately determined by the method described in the last paragraph.

If the proposed procedures did not compensate for the decrease in the variance of a band series concomitant with a decrease in the frequencies it encompasses, then the lowest significant minima found at most single locations would almost always occur in the highest frequency band examined (here, the 4–6-m band) and rarely in the lowest frequency band examined (here, the 10–12-m band). Instead, one finds in Table 115 that lowest significant minima are distributed among bands of all frequencies—an expectable distribution for an archeological site having use-areas of many dimensions and for analytical procedures that are not scale-biased. Thus, on the basis of an internal assessment involving one assumption about the nature of the data base, the procedure of standardization appears to be an adequate means for making the variability of band series of different frequencies comparable. This makes possible (*a*) the unbiased assessment of the significance of local mean resistivity minima and the probable locations of use-areas; and (*b*) the unbiased assessment of the dimensions of use-areas.

A final consequence of the data reduction procedures, which is expectable, is that they should allow the definition and discrimination of use-areas that overlap spatially somewhat. This predictable circumstance does arise in the data presented in Table 115. Some locations at which significant local minima occur are closer together than the sum of the radii of the use-areas centered at them, the radii being equal to half of the frequency of the band series in which the lowest significant minimum is found. For example, along Transect IV west, use-areas are indicated to be centered at survey stations 5337 and 5351—14 stations apart, which is equivalent to 7-m apart. Both probable use-areas have dimensions of 10–12 m and radii of 5–6 m. The two probable use-area thus overlap considerably, and yet, are distinguishable within the data.

Thus far, the technical aspects of manipulating spatially filtered resistivity data using the procedures outlined in Steps 1–6 have been illustrated and assessed internally. Numerical results pattern within themselves in a consistent manner and in ways one would expect them to if the proposed procedures operate in accord with their designed purpose. Given this positive internal assessment of Steps 1–6, it now is appropriate to test them externally, as well: to compare the patterns found in the resistivity data to those one might expect to find, based on the *nature of the archeological remains* that occur along the surveyed transects. Do the locations and characteristics of use-areas, as reconstructed from resistivity data using the procedure outlined previously, correspond with the locations and characteristics of use-areas reconstructed from the Crane surface collection?

To answer this question, for each subtransect, a tabulation was made of whether or not significant local minima occur at any of the resistivity survey stations within each controlled surface collection unit along the subtransect, and if they occur, the magnitudes of the significant local minima. These data are presented in Tables 116—118. The surface collection units are arranged in the tables according to whether significant local minima are expected to occur in them and according to the expected relative magnitudes of the minima. The expected resistivity responses within the collection units are based on the kinds and amounts of activity and refuse production, if any, which are reconstructed to "definitely have taken place" prehistorically in the units. Units having many activity groups "definitely present" in them which produce much debris (e.g., activity group 8, mollusk processing) or very nutrient-rich debris (e.g., activity groups 3 and 5 involving the deposition of wood ash and bone) were grouped together as locations where very low significant negative minima are expected to occur. Units having only a few (1 or 2) activity groups "definitely present" in them were assumed to be locales of less activity and smaller amounts of refuse deposition. These units, as well as those having more activity groups "definitely present" in them but only activity groups producing little debris (e.g., activity group 1, woodworking) or nutrient-poor debris (e.g., activity groups 16 and 17 involving the deposition of greens, roots, tubers, and fruits) were grouped together as locations where only moderately low significant minima are expected to occur. The final set of surface collection units grouped together include those for which no activity groups were coded "definitely present" in them. It is *not* expected that no significant minima will occur in these units, for the threshold used to define significant minima have generously low absolute values. Also, some activities may actually have occurred in the units but not be visible, archeologically (see preceding). However, it is expected that if resistivity minima do occur in these units (significant or not), they will not be very great in magnitude.

The nutrient richness of the refuse materials produced by the activities in an activity group was taken to be equivalent to the sum of the average percentages of N, P, K, Ca, Mg, Na, and S in the materials (Table 67). The particular activity groups classified as producing nutrient-rich or nutrient-poor refuse are listed in Table 119.

The results of the test recorded in Tables 116–118 indicate that in most cases, the expected occurred. First, significant local minima, indicating the presence of use-areas, occurred in all collection units where activity groups were coded "definitely present." In most units where no activity groups were coded "definitely present," either no significant minima occur, or else there occur significant minima with magnitudes only slightly above the threshold value defining significance. The latter are expectable, given that thresholds of

TABLE 116
Controlled surface units along subtransects IV east and IV west in which different kinds of prehistoric activity took place, and their corresponding significant local resistivity minima

Unit	Transect IV West		Transect IV East	
	Significant Minima	$\mu \pm \sigma$ of Significant Minima	Significant Minima	$\mu \pm \sigma$ of Significant Minima
Units with Many Activity Groups and/or Nutrient-Rich Debris and/or Much Debris				
186+	-2.0112,		-1.5326	
187	-1.9841			
105	-1.8493	$-1.9484 \pm .0708$	-1.8139	$-2.0037 \pm .5894$
166	-1.9488		-2.6647	
Units with Few Activity Groups and/or Nutrient-Poor Debris and/or Little Debris				
9	-1.1506*		-1.5551	
225	-2.9729		-1.8068,	
			-2.1732	$-1.7125 \pm .3356$
206	- .8113*	$-1.4742 \pm .8684$	-1.6913	
125	-1.4359		-1.8750	
33	-1.0003*		-1.1736*	
Units with No Activity Groups "Definitely Present"				
145	-1.2420		-1.1955*	
66	- .6986*		-1.3658,	
			-1.6606	
85	- .8624*	$-1.1208 \pm .4236$	-1.3608	$-1.5493 \pm .2540$
49	-1.0121*		-1.5905	
20	-1.7889		-1.9203,	
			-1.7519	

*Not a significant local minimum (< 1.2σ)

generously low absolute value were chosen. Second, for each subtransect, those units where many activities, much refuse deposition, and/or the deposition of nutrient-rich refuse occurred have, on the average, much lower[1] significant local minima than do those units where few activities, little refuse deposition, and/or deposition of nutrient-poor refuse occurred. The magnitudes of the significant local minima within the collection units are a function of the kinds and amounts of activity that occurred in them. Thus, for the set of data examined here, the locations and general characteristics of archeological use-areas do appear to be reconstructable with resistivity survey data with a fair degree of reliability.

Having illustrated and assessed Steps 1 through 6 of the analytical procedures outlined, we now may proceed to illustrate and assess Steps 10 and 11.

1. The sample of significant local minima is too small to test the significance of the difference in the magnitudes of the minima between groups.

TABLE 117
Controlled surface survey units along Subtransects III south and III north in which different kinds or amounts of prehistoric activity took place, and their corresponding significant local resistivity minima

Unit	Transect III South		Transect III North	
	Significant Minima	μ±σ of Significant Minima	Significant Minima	μ±σ of Significant Minima
Units with Many Activity Groups and/or Nutrient-Rich Debris and/or Much Debris				
631	-2.3807		-2.4217	
632	-1.3754, -4.2743	-2.2235±1.5054	- .8996	-1.4736±.8271
635	- .8636		-	
621	-		-1.0996	
Units with Few Activity Groups and/or Nutrient-Poor Debris and/or Little Debris				
633	- .5169*		-	
633 & 619	-		-1.4053	
622	-	-1.6380±.9392	- .8019	-1.1476±.3112
622 & 636	-1.2072, -2.4139		-	
624	-2.4139		-1.2355	
Units with No Activity Groups "Definitely Present"				
630	-1.3222, -1.2380, -2.1538		-3.4799	
634	-1.2642	- .9264±.3329	-	-1.1528±.4565
634 & 620	-	(excluding unit 630 which apparently misclassified)	-1.8259	(excluding unit 630 which apparently misclassified)
625	- .5985*		- .9874,- .9881 - .9881	
626	- .9179		- .8098	

*Not a significant local minimum (<.8σ)

Steps 10 and 11 request that the proveniences where use-areas are suspected to occur (where significant local minima occur) be classified according to the characteristics of the use-areas inferred from their resistivity responses—their scale, their mean anomalous soil resistivity, and the variance of their resistivity. Whether or not the proveniences occur near areas of intense, multipurpose activity or reoccupation also should be used in the classification, in order to determine whether differences between them in their mean anomalous soil resistivity are attributable to differences in the kinds of refuse deposited at them or are attributable to differences in only the amount of refuse deposited at them.

TABLE 118
Controlled surface survey units along Subtransects I north and I south in which different kinds or amounts of prehistoric activity took place, and their corresponding significant local resistivity minima

Unit	Transect I North		Transect I South	
	Significant Minima	$\mu \pm \sigma$ of Significant Minima	Significant Minima	$\mu \pm \sigma$ of Significant Minima

Units with Many Activity Groups and/or Nutrient-Rich debris and/or Much Debris

187	-2.4328		-1.0491, -.8308	
188	-1.0250, -2.1550	$-1.6169 \pm .6350$	-2.7250	$-1.5590 \pm .7483$
164	-1.1438		-1.3858	
192	-1.3280		-1.8043	

Units with Few Activity Groups and/or Nutrient-Poor Debris and/or Little Debris

165 & 186	-.9805, -.9867		-.9790, -1.5161, -1.1415	
189	-.9764, -.9949	$-1.1644 \pm .3096$	-1.1724, 1.1374	$-1.2744 \pm .4838$
193	-1.3280, -1.2374		-1.3465, -2.2842	
191	-1.5740, -1.5371		- .6207*	

Units with No Activity Groups "Definitely Present"

190	-1.5367, -1.5210	Unit 190 apparently misclassified	-1.9905	Unit 190 apparently misclassified

*Not a significant local minimum (< .8 σ)

Finally, the classification of proveniences should be interpreted in behavioral terms.

To illustrate and assess these steps, for each transect, IV, III, and I, a crossplot was made of the *magnitudes* of the lowest significant local minima found at proveniences where significant minima occur (the mean anomalous resistivities of the probable use-areas along it) against the *frequency bands* in which the lowest significant minima occur (the scales of the use-areas) (Figures 72–74). The plot for each transect represents a combination of the information held in the resistivity traces of *both* its outer subtransects. The mean anomalous resistivities and scales of use-areas indicated in the resistivity traces of one or the other or both subtransects are plotted together on one plot.

TABLE 119
Activity groups classified as producing nutrient-rich or nutrient-poor refuse

Activity Groups Producing Nutrient-Rich Refuse		Activity Groups Producing Nutrient-Poor Refuse	
3	hearth dumpings	1	wood working
4	bone working	2	pounding bark
5	crushed bone, deposition	9	dehair hides, alone
6	burnt bone, deposition	11	dehair and simmer hides
7	unburnt bone, deposition	12	grain hides alone
8	mollusk processing	13	grain and soak hides
10	dehair and soak hides	14	grain and simmer hides
21	boil bone, meat	15	uncooked animal tissue, deposition
		16	uncooked plant tissue, deposition
		17	pulverize roots, tubers, bulbs, rhizomes, fruits, dried meat
		18	uncooked seeds and seed coats, deposition
		19	uncooked nuts, deposition
		20	uncooked seeds or nuts, deposition
		22	boil seeds or nuts

Where the resistivity responses of corresponding survey stations or approximately corresponding survey stations along the subtransects (e.g., station 5323 of Subtransect IV west and station 6022 of Subtransect IV east) both indicate the presence of a use-area (most likely the *same* use-area), the mean anomalous resistivity and scale of the use-area are taken to be those indicated at the survey station suggesting the larger scale (the station along the subtransect which more likely passed closer to the center of the use-area). If the scale of the use-area indicated at the two corresponding survey stations is the same but the estimates of its mean anomalous resistivity differ, the mean anomalous resistivity of the area is taken to be the lower of the two estimates (Tables 120–122). This combining of information from both subtransects is necessary. As pointed out previously, a single resistivity transect may happen to intersect a use-area obliquely rather than through its center, in which case the scale of the use-area reconstructed from the resistivity trace will be underestimated. The use of information from two parallel subtransects to estimate the scale of use-areas decreases the probability that the estimates will be low.

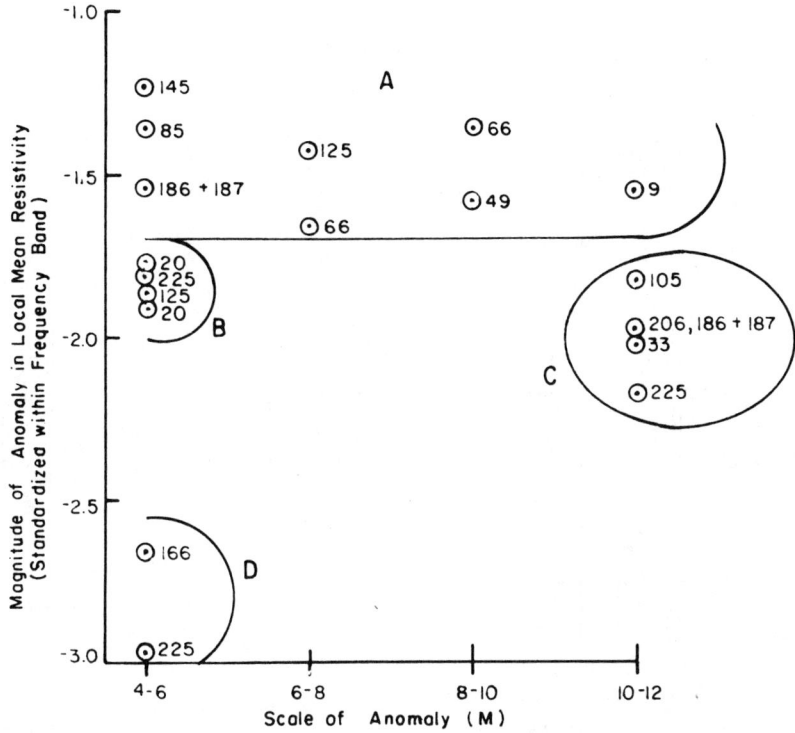

FIGURE 72. *"Probably use-areas" centered within various surface collection units along Transect IV cluster together according to the magnitudes of their mean anomalous resistivity and their dimension.*

Clusters of observations within the plots indicate those use-areas most similar to each other with regard to their mean anomalous resistivity and scale. The plots do not indicate the similarity of the use-areas with regard to the variability of their resistivity. This characteristic of the use-areas is not used here to classify them because the areal density of features within them (and responsible for their resistivity variance) is unknown and unavailable for comparison to the classification. The general setting of the use-areas within areas of intense, multipurpose activity and reoccupation or less heavily used areas of the site also is not considered in the classification. This dimension of variation is constant among all use-areas examined. The sections of Transects IV, III, and I examined here all lie within areas of Crane that are set apart from the house that were not heavily used prehistorically. Under normal, nonexperimental circumstances, the variation of resistivity values within use-areas and the placement of use-areas with regard to intensely used sectors of the site should be used to classify use-areas.

Within the plots, differences between use-areas in their mean anomalous

TABLE 120
The process of combining information from Subtransects IV west and IV east to estimate the scale and mean anomalous resistivity of use-areas along Transect IV

Uncombined Information from Both Subtransects			Combined Information from Both Subtransects		
Survey Station Having a Significant Minimum	Magnitude of the Significant Minimum	Frequency Band in Which the Significant Minimum Is Found (m)	Survey Station Where Use-Area Is Located	Mean Anomalous Resistivity of the Use-Area	Scale of the Use-Area (m)
5920	-1.5521	10-12	5920	-1.5521	10-12
5930	-1.9203	4-6	5930	-1.9203	4-6
5233	-1.7889	4-6	5233/5936	-1.7889	4-6
5936	-1.7519	4-6			
5952	-1.5905	8-10	5952	-1.5905	8-10
5962	-1.3658	8-10	5962	-1.3658	8-10
5972	-1.6606	6-8	5972	-1.6606	6-8
5983	-1.3608	4-6	5983	-1.3608	4-6
5287	-1.8493	10-12	5287/5990	-1.8493	10-12
5990	-1.8139	6-8			
5998	-1.8750	4-6	5998	-1.8750	4-6
5304	-1.4359	6-8	5304	-1.4359	6-8
5318	-1.2420	4-6	5318	-1.2420	4-6
6022	-2.6647	4-6	6022/5323	-2.6647	4-6
5323	-1.9488	4-6			
5337	-2.0112	10-12	5337/6038	-2.0112	10-12
6038	-1.3583	10-12			
6039	-1.5326	4-6	6039	-1.5326	4-6
5351	-1.9841	10-12	5351/6052	-1.9841	10-12
6052	-1.6913	10-12			
5362	-2.9729	4-6	5362	-2.9729	4-6
6072	-1.8068	4-6	6072	-1.8068	4-6
6072	-2.1732	10-12	6072	-2.1732	10-12

TABLE 121
The process of combining information from Subtransects IV west and IV east to estimate the scale and mean anomalous resistivity of use-areas along Transect IV

Uncombined Information from Both Subtransects			Combined Information from Both Subtransects		
Survey Station Having a Significant Minimum	Magnitude of the Significant Minimum	Frequency Band in Which the Significant Minimum Is Found (m)	Survey Station Having a Significant Minimum	Mean Anomalous Resistivity of the Use-Area	Scale of the Use-Area (m)
4243	-1.3222	6-8	4243/3846	-3.4629	6-8
3846	-3.4629	6-8			
4248	-1.2380	4-6	4248	-1.2380	4-6
4253	-2.1538	6-8	4253	-2.1538	6-8
3860	-2.4217	6-8	3860/4265	-2.3807	10-12
4265	-2.3807	10-12			
4269	-1.3754	4-6	4269/3873	-1.3754	4-6
3873	- .8996	4-6			
4276	-4.2743	4-6	4276	-4.2743	4-6
3886	-1.4053	4-6	3886	-1.4053	4-6
4292	-1.2642	10-12	4292/3894	-1.2642	10-12
3894	-1.8259	6-8			
3905	-1.0996	4-6	3905/4307	-1.0996	4-6
4307	- .8636	4-6			
4316	- .9788	4-6	4316	- .9788	4-6
3925	- .8019	4-6	3925	- .8019	4-6
3934	-1.2355	4-6	3934	-1.2355	4-6
3944	- .9874	4-6	3944	- .9874	4-6
3951	- .9881	4-6	3951	- .9881	4-6
4356	- .9179	4-6	4356/3957	- .9179	4-6
3957	- .8098	4-6			

TABLE 122
The process of combining information from Subtransects I north and I south to estimate the scale and mean anomalous resistivity of use-areas along Transect I

Uncombined Information from Both Subtransects			Combined Information from Both Subtransects		
Survey Station Having Significant Minimum	Magnitude of the Significant Minimum	Frequency Band in Which the Significant Minimum Is Found (m)	Survey Station Having a Significant Minimum	Mean Anomalous Resistivity of the Use-Area	Scale of the Use-Area (m)
77	-1.3858	4-6	77	-1.3858	4-6
84	-.9790	4-6	84	-.9790	4-6
93	-1.5161	10-12	93/895	-1.5161	10-12
895	-.9867	10-12			
894	-.9805	6-8	894	-.9805	6-8
98	-1.1415	4-6	98	-1.1415	4-6
104	-1.0491	4-6	104	-1.0491	4-6
110	-.8308	4-6	110/910	-2.4328	4-6
910	-2.4328	4-6			
916	-1.0250	4-6	916	-1.0250	4-6
119	-2.7250	8-10	119/923	-2.7250	8-10
923	-2.1550	6-8			
128	-1.1724	4-6	128/931	-1.1724	4-6
931	-.9764	4-6			
138	-1.1374	10-12	138/938	-1.1374	10-12
938	-.9949	6-8			
941	-1.5210	10-12	941	-1.5210	10-12
943	-1.5367	6-8	943	-1.5367	6-8
149	-1.9905	4-6	149/953	-1.5740	8-10
953	-1.5740	8-10			
963	1.5371	4-6	963	1.5371	4-6
170	-1.8043	8-10	170/973	-1.3280	10-12
973	-1.3280	10-12			
178	-1.3465	4-6	178	-1.3465	4-6
983	-1.2374	4-6	983/186	-2.2812	4-6
186	-2.2812	4-6			

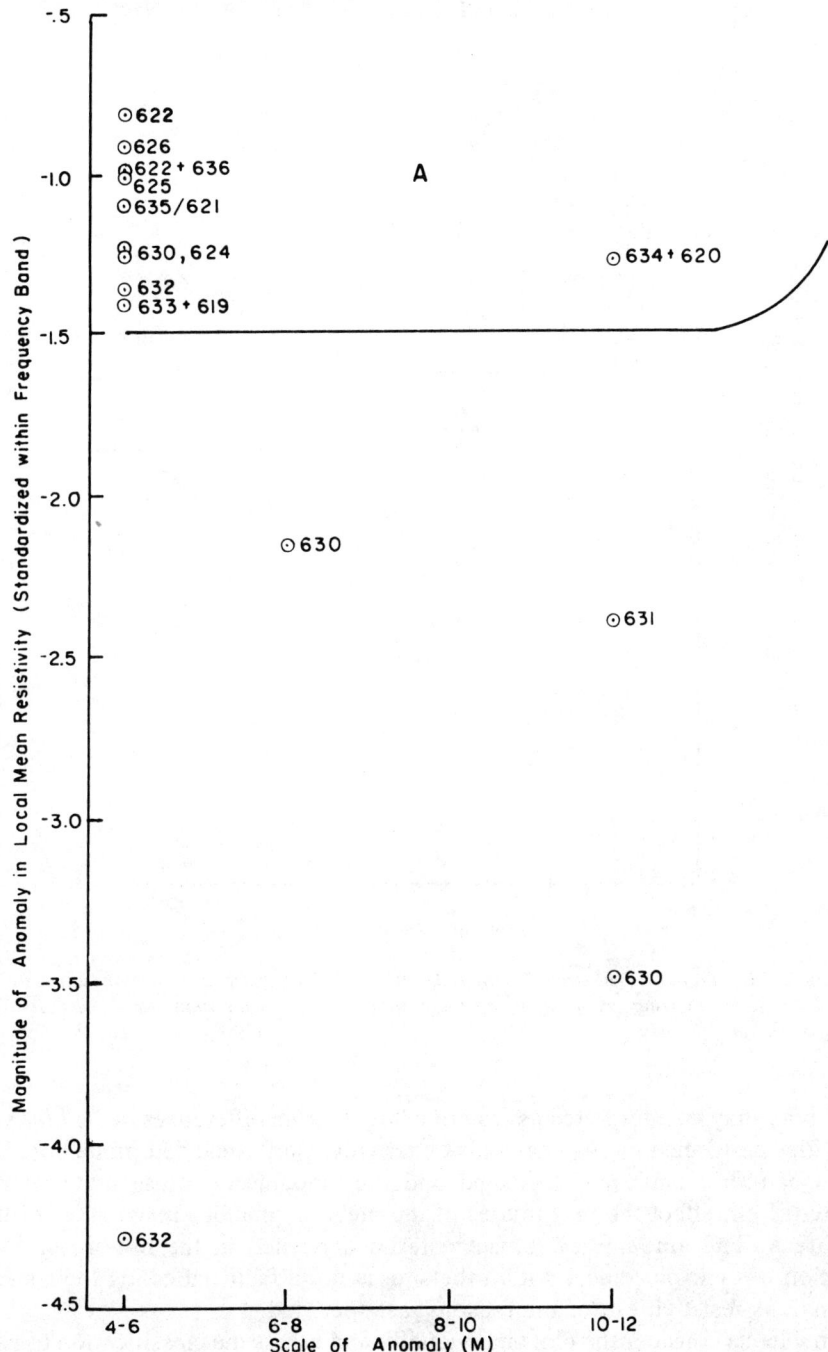

FIGURE 73. *"Probable use-areas"* centered within various surface collection units along Transect III cluster together according to the magnitudes of their mean anomalous resistivity and their dimension.

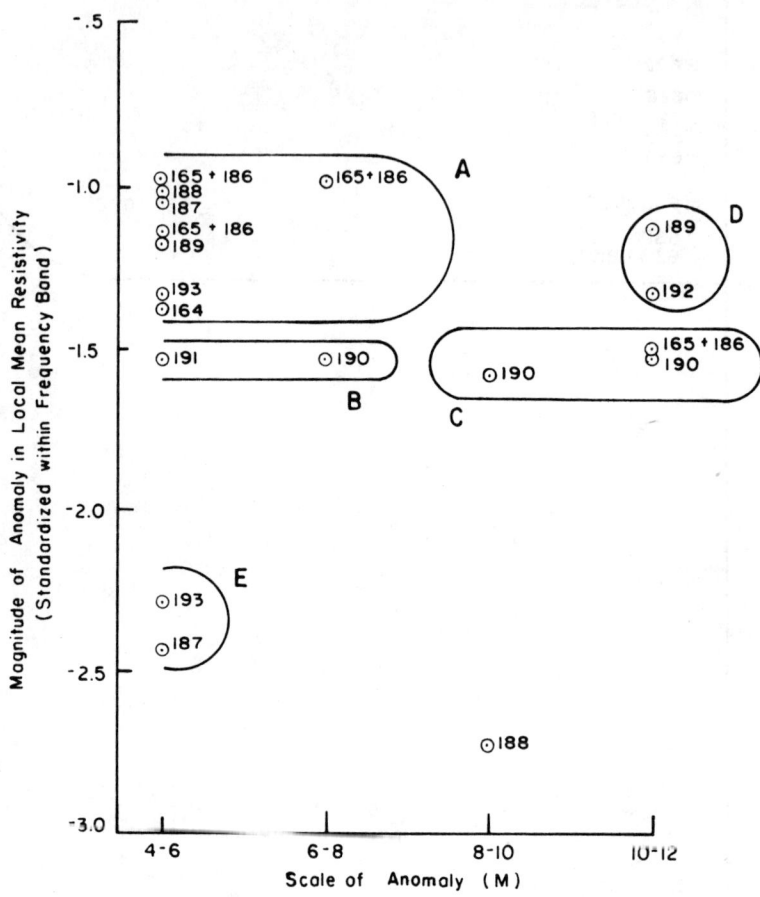

FIGURE 74. *"Probable use-areas" centered within various surface collection units along Transect I cluster together according to the magnitudes of their mean anomalous resistivity and their dimension.*

resistivity may be interpreted as resulting largely from differences in the *kinds* of activities performed there. The kinds of activities performed determine both the kinds of refuse materials deposited and the amounts of refuse material deposited. Both affect the magnitudes of the mean anomalous resistivities of the use-areas. The amounts of refuse material deposited in the use-areas, as a function of their placement within the site, is not a factor affecting their mean anomalous resistivities, for the reasons just specified.

Now let us consider the plots individually and assess the classification of use-areas that they present in light of what is known about them archeologically. Figure 72 shows the similarities and differences between probable use-areas along Transect IV. Four groups (A–D) of probable use-areas within the plot may be defined according to their distribution relative to each other, and are readily interpretable in behavior terms. Group A consists of all observations having significant minima with small absolute values—less than 1.7. Unlike the observations in the other groups, those within group A are randomly scattered; they do not tend to cluster. As might be expected from the small magnitudes of the resistivity minima of the observations in group A and their unclustered distributon, most of the observations (five of eight) fall within surface collection units having no activity groups "definitely present" in them. The observations in group A largely represent "insignificant minima" that were defined as significant when using a threshold having a generously low absolute value.

Group B is an extension of group A, both distributionally within the plot and logically. Group B contains observations that fall within surface collection units where no activity groups are definitely present, or where the activity groups produce little debris and nutrient-poor debris (woodworking, deposition of uncooked animal tissue, and graining hides alone).

Group C contains four observations, three of which fall within collection units where woodworking occurred. A variety of other activity groups also occurred in these units (Table 113). Group D contains two observations, both of which occur in collection units where hide processing and the associated activity group, deposition of uncooked animal tissue, occurred.

On the less structured side, the plot does not reveal the unique nature of the use-areas centered within collection unit 186 + 187. In this collection unit, only, the processing of mollusks and boiling of seeds or nuts is reconstructed to have occurred prehistorically (Table 113). This result is surprising, considering the plentiful and nutrient-rich debris that mollusk processing produces. The use of collection unit 105 for boiling bone and meat also is not distinguished in the plot.

In sum, the plot of probable use-areas along Transect IV distinguishes, with a fair degree of accuracy, four kinds of areas which might be used as the strata within an excavation sampling design: (*a*) areas where probably no activity occurred; (*b*) areas where there occurred no activity or activities producing little debris or nutrient-poor debris; (*c*) woodworking areas; and (*d*) hide-working areas. The end product of the analysis of the resistivity data from Transect IV is patterned and interpretable in behavioral terms, and has practical use.

The results from Transect III (Figure 73) are almost as encouraging. One cluster of observations can be defined within the plot: group A. This group contains: (*a*) *all* observations that fall in collection units where no activity groups "definitely occurred"; (*b*) *all* use-areas that are centered in collection units where there occurred activity groups that produce little debris or nutrient-poor debris (pounding bark, graining hides alone). Separated from group A are use-areas centered in three collection units, two of which one might expect to be uniqely distributed, based on the unique sets of activity groups they represent: unit 632, where seeds and possibly nuts were pounded, ground, and boiled, and where woodworking occurred; and unit 631, where uncooked plant and animal tissue were processed. The third unit, 630, is characterized as having no activity groups "definitely present" in it, and one would have expected this unit to cluster with the observations in group A. The reconstructed absence of activity groups from collection unit 630 may be a mischaracterization of it.

Three use-areas not segregated from group A that we would expect to be distinguished include those falling in collection units 635, 621, and 624. In both 635 and 621, bone working is reconstructed to have occurred. Uncooked animal tissue, pulverized roots and tubers, and uncooked seeds or nuts are thought to have been deposited in unit 624.

In summary, of the 14 observations plotted in Figure 73, 9 cluster or distribute themselves as one would expect, based on the activity groups present in the collection units they represent, and 5 observations do not distribute in an expectable manner. The results of the analysis plotted in Figure 73 could be used to define fairly homogeneous sampling strata within an excavation design.

The classification of use-areas along Transect I (Figure 74) met only limited success. Six clusters of observations and one outlier observation can be defined in the plot of the local mean resistivity of the use-areas against their scale. Group A contains use-areas that center in collection units where a large number and variety of activities are reconstructed to have occurred prehistorically. The use-areas should have scattered more widely over the plot than they do, and distributed themselves in an unclustered manner, reflecting their uniqueness. The fact that they do not, and instead are all characterized by a similar low mean anomalous resistivity and small scale, may suggest that these use-areas were intersected only peripherally by Transect I and barely monitored. This is not an unreasonable interpretation. The scale and mean anomalous resistivity of the observations in group A are *lower* than even those in groups B and C that fall within collection units where no activities are thought to have occurred or where activities yielding little debris or nutrient-poor debris (woodworking, pounding bark, deposition of uncooked plant tissue, pulverizing roots and tubers) are thought to have occurred.

The observations in groups B and C are homogeneous with regard to the activity groups that occurred in the collection units they represent, as just described. The same is true of the two use-areas in group D. Both collection units 189 and 192, within which the use-areas fall, were locations of woodworking and processing of nuts and seeds. They differ only in that collection unit 192 also was a location where hides were grained—an activity that need not have produced much refuse and caused much of a difference between the two locales. Collection units 189 and 192 are more similar to each other in the activity groups present within them than they are to any other colleciton units along Transect I.

Likewise, the two use-areas comprising group E fall in collection units that are more similar to each other in the activity groups present within them than they are to any other units along Transect I. Both collection units are locations where hides were grained and soaked, and where uncooked animal tissue was deposited. (Additional activity groups are represented in collection unit 187, however.)

Finally, the use-area centered in collection unit 188 stands alone in the plot. This is expectable. The collection unit is distinctive from all other units along Transect I in that it functioned as a generalized food-processing area, where roots and tubers were pulverized, seeds were pounded and ground, and meat and nuts were boiled.

In sum, although the classification of use-areas along Transect I presented in Figure 74 can be explained in a retrodictive manner, it would be only partially successful as the basis for an excavation sampling design. Of the 18 observations plotted in the figure, 10 cluster or distribute themselves as one would expect them to, based on the activity groups present in the collection units they represent, and 8 observations do not distribute themselves in an expectable manner.

All told, the results obtained by applying Steps 10 and 11 of the proposed analytical procedure to the Crane resistivity data suggest them to be reasonable and useful. The use-areas along Transects IV, III, and I that were classified together or separately on the basis of the magnitudes of their anomalous mean resistivity and their dimensions indicated in the resistivity data, in most cases, are those one would expect to cluster or not cluster together, based on their reconstructed prehistoric functions. Together, the assessments of Steps 1–6 and 10–11 of the proposed analytical procedure suggest that they, in conjunction with the preparatory spatial filtering observations that precede them, can be used with reliability to locate, characterize, and distinguish different kinds of use-areas as whole geographic zones within resistivity data sets.

Conclusion

It has been shown that, at the Crane site, different kinds of use-areas, as *whole geographic units*, differed in their mean soil resistivity and the variability of their soil resistivity. It was feasible and useful to view resistivity variation within the Crane site from a statistical, geographic perspective rather than a deterministic, feature-oriented one, and to apply the techniques of spatial filtering and time series analysis for the purpose of: (*a*) segregating resistivity variation at the scale of use-areas from resistivity variation of other scales and of nonarcheological origins; and (*b*) locating and distinguishing use-areas within the archeologically significant component of variability. The Barnes Layer method also was found useful in isolating archeologically significant soil variation at the Crane site. The applicability of these approaches and techniques to resistivity data collected at other archeological sites will be discussed in the next, and final, chapter.

References

Davis, John C.
   1973 *Statistics and data analysis in geology*. John Wiley, New York.
Jenkins, G.M.
   1961 General considerations in the analysis of spectra. *Technometrics* 3:133–166.
Olkin, I.
   1967 Correlations revisited. In *Improving experimental designs and statistical analysis*, J. Sterkey (ed.), pp. 111–113. Rand McNally, Chicago.
Sneath, Peter H. and Robert R. Sokal
   1973 *Numerical Taxonomy*. W.H. Freeman, San Francisco.

11
Boundary Conditions for the Proposed Methods, and Conclusions

In this study, I have argued that (*a*) different prehistoric activities on an archeological site may produce residues of different kinds and require facilities of different sizes and areal densities; (*b*) the various residues may alter in different ways and to different degrees the physical and chemical properties of soils within the areas where they were deposited; (*c*) the different soil changes within different use-areas, as well as the varying areal densities of archeological features present in them, may produce corresponding distinct zonal alterations in their soil resistivity (mean, variance); and (*d*) the Barnes Layer method and spatial filtering techniques may be used to isolate the resistivity response of archeological horizons and to identify, characterize, and differentiate use-areas as whole geographic units within that response. The techniques also may be used to define the depth and general contour of the buried site as a whole. Use-areas identified by this procedure, when combined with surface survey data, may be used to define sampling strata for excavation research designs in order to obtain representative samples of the remains (both features and artifacts) of all significant activities that occurred at the site.

This thesis has not been tested in the formal sense, nor can it be. The number of archeological, pedological, physical, chemical, and biological variables considered in this study is large. Experiments of the type conducted at the Crane site could not encompass all possible states of all the variables involved. Instead, the approach used to examine this thesis has been a combination of logical deduction, testing of select relationships, and selective illustrations. Previously tested and established models from soil science, physics, and chemistry have been logically chained in order to deduce the kinds of soil changes that might be expected to occur on archeological sites as a result of prehistoric activity and the soil resistivity responses that might be expected to

11. Boundary Conditions for the Proposed Methods

be produced by such soil changes. These deductions and the application of the Barnes Layer method and spatial filtering techniques to analyze resistivity data then were illustrated with the available site-specific data from Crane.

As a consequence of this means of examining the thesis, a complete set of boundary conditions within which the proposed methods are applicable is not known. Nevertheless, the major conditions that are known to cause problems in the interpretation of resistivity data according to the procedures proposed here can be made explicit.

Behavioral and Archeological Conditions

1. Use-areas can be located and characterized with resistivity data from an archeological site as long as they actually exist within the archeological domain. Single occupation sites having horizontally heterogeneous middens, which reflect differentiated prehistoric use of space, can be expected to exhibit areally differentiated resistivity responses. On the other hand, sites having homogeneous middens produced by multiple changes in the location of activities over time and reuse of space cannot be expected to have internally differentiated resistivity responses. This undesirable site condition is exemplified in seasonal base camps that have been reoccupied numerous times by the same social group (e.g., some deep-midden Archaic sites in eastern North America), or in encampments that have been reoccupied over the centuries by different groups.

Use-areas within a site need not be spatially segregated *completely* to locate and characterize them with the proposed geographic methods of resistivity data interpretation, provided the use-areas differ in function or amount of use. The filtering techniques discussed in Chapters 2 and 10 are designed to isolate and characterize variability from multiple, partially overlapping sources within a resistivity signature. Of course, if the overlapping use-areas are indistinguishable in their soil characteristics and resistivity, then they will be interpreted as one large use-area and can be distinguished prior to excavation only if surface collections contain characteristic artifacts.

Whether a site has a homogeneous midden or a heterogeneous midden, the proposed resistivity survey methods can be used advantageously to investigate it. The degree to which activity structure actually is present in the site can be assessed prior to excavation, and the extent and locations of subsequent excavations can be guided by the results.

2. Some use-areas cannot be expected to be identified within resistivity data sets, either as a result of the *kind* or the *amount* of the activity and refuse deposition that occurred within them, or both. (*a*) Some activities produce no refuse, or refuse which is inert. Consequently, such activities have no affect on the soil or its resistivity; examples are weaving and sewing, graining of hides without the use of auxiliary chemicals, and stone knapping. (*b*) Other

activities yield refuse materials that can alter the resistivity of the soil, but the amount of refuse deposited per unit activity is very limited. The resulting soil changes within the area of use may be too small to be detectable unless the area was used in a similar fashion over an extended period of time. Examples include bone working, meat processing, hide working, and sometimes woodworking. In contrast, cleaning hearths, shucking mollusks, and shelling nuts produce more refuse per unit activity and are more likely to produce soil anomalies. (c) Some activities which archeologists often assume to yield refuse materials capable of producing distinctive soil changes usually do not because the total amount of nutrients they contain is relatively low and their potential for increasing the concentration of ions and radicals within soil water after decomposition is minimal. Greens, fruits, meat drippings and oils, unburnt wood, and nut shells are examples. In contrast, ash, shell, and bodily waste contain large amounts of nutrients. (See Table 67 for a detailed inventory of materials and their compositions.)

3. The degree to which prehistoric activities produce soil anomalies that are maintained from the time of refuse deposition to the present and are detectable today varies with the chemical nature of their residues. On a very broad scale of generalization, activities yielding residues that are rich compared to other residues in the polyvalent cations Ca and Mg are likely to cause more permanent soil anomalies and to be represented within resistivity data sets. A number of soil mechanisms are responsible for this. As polyvalent cations, Ca and Mg tend to be held by soil colloids in exchangeable forms more strongly than are monovalent cations, and as cations, they are held in exchangeable form by clays more strongly than are anions (relative replacement power). Magnesium also is fixed within expandable clays. Calcium and magnesium maintained within the soils by these phenomena are potentially available to act as conductors of electric current within soil water solutions. Finally, activities yielding residues with high proportions of Ca and Mg relative to Na favor semipermanent physical alterations as well as chemical alterations within the soil. Calcium and magnesium can serve to link clay and humus colloids together and thereby facilitate soil aggregation and alteration of the soil water equilibira. Sodium, on the other hand, is a dispersing agent and does not facilitate the development of soil structure.

Areas where there has been deposited refuse materials high in phosphates relative to other refuse materials also are more likely to exhibit more permanent soil anomalies and to be represented within resistivity data sets. Phosphates are strongly fixed by soils as inorganic compounds of Fe, Al, Mn, and Ca, by hydrous oxides of Fe, Al, and Mg, and by silicate clays.

The likelihood that use-areas where relatively large amounts of N, K, Na, and S have been enriched to their soils, relative to other areas, will exhibit more permanent soil anomalies and will be reprensented within resistivity data sets depends on local soil and vegetational conditions. The relevant anomaly-maintaining processes have been discussed in Chapter 4, but may be highlighted, here. Nitrogen can be maintained within the soil in significant quantities in organic forms when soil pH is neutral to basic and humus colloids are not mobile and subject to leaching (e.g., in chernozems as opposed to podzols). Nitrogen also can be fixed in the soil in significant quantities in the form of NH_4 within the lattice of 2:1 clays, if they are frequent within the soil. The same is true of K. Ecosystemic cycling and "luxury consumption" of K by the surface vegetation can be an important means by which K is maintained within the soils of archeological sites, depending on the kind of vegetation. Sodium and sulfur, of all the macronutrients examined in this study, are least readily held by soils. Activities yielding refuse materials that are rich compared to other residues in only these nutrients are less likely to produce permanent soil anomalies and to be reflected in resistivity data sets.

4. Use-areas cleaned of refuse materials and residues of activity may not be detectable within resistivity data sets. This circumstance may arise in sites or subportions of sites where work space is limited, at a premium, and must be maintained. Examples include sites located in caves or on small topographic rises, sites with high population density, multipurpose use-areas centered around a focus of activity (a hearth, a hut), and the interiors of houses used all winter for many activities. Cleaning also may occur as part of the regular reuse of a facility. Hearths and earth ovens are examples of such features.

5. Some use-areas may not contain facilities or artifacts, but nevertheless may exhibit soil alterations and anomalous soil resistivity values. Such areas include (*a*) areas where liquids have been dumped, such as the solutions in which hides have been soaked or fibrous plants have been boiled as a first step in making cordage and cloth; (*b*) areas where only readily decomposable organic materials have been deposited and where any tools involved in the production of the residues have been curated elsewhere and (*c*) latrine areas.

Soil Conditions Prior to Occupation and Environmental Conditions during Occupation

1. Soil texture. All else equal, the detectability of use-areas will be greatest at sites located on sandy loam through clayey loam soils, less so at sites located on clayey soils, and least at sites located on sandy soils. Loamy texture facilitates maximum structural development within a soil and allows the effect of anthropic enrichment of organic matter on soil structure and moisture balance to reach its fullest potential. It also provides an adequate clay content for maintaining anthropic nutrient enrichments within the soil. Clayey soil texture does not facilitate anthropically caused structural development as fully as does loamy texture. This circumstance is somewhat compensated for, however, by the higher cation exchange capacities and fixation capacities of clayey soils, and their ability to maintain nutrients of anthropic origin. Sandy soils are suboptimal in both their potential for structural development and their capacity for nutrient retention. This results from the fact that organic micelles, which augment the structure and cation exchange capacities of soils, tend to be more easily leached from sandy soils than from finer textured soils.

2. Clay types. Sites located on soils having high percentages of 2:1 lattice clays (montmorillonite, vermiculite, illite) and hydrous micas will provide better opportunities for maintaining the distinctness of use-areas in which Mg, K, and/or NH_4 are the primary nutrients distinguishing them than will sites located on soils having higher percentages of 1:1 lattice clays (chlorite, kaolinite) and iron and aluminum oxides. The 2:1 lattice clays are capable of fixing Mg, K, and/or NH_4 within their structure whereas 1:1 lattice clays are not. On a broad scale of generalization, 2:1 clays will be found in soils that have developed on parent materials of moderate age in temperate climates. Chlorite is found in soils that have developed on only slightly weathered parent materials, whereas kaolinite is found in soils that have developed on more highly weathered parent materials. Iron and aluminum oxides predominate in tropical soils, where weathering is very intense (Brady 1974:88–91).

3. Soil pH. At the broadest level of generalization, soils with neutral to basic pHs favor maintenance of anthropic soil alterations and the distinctness of use-areas over time more so than do acidic soils (pH < 6). Humus colloids are more mobile and leachable within acidic soils. Anthropic enrichments of organic matter to acid soils therefore may be partially lost and not fulfill their potential for developing soil structure and for increasing the cation exchange capacity of the soils and the concentration of ions within their soil waters. Acidic soil conditions also favor the mobility and loss of cation nutrients of all kinds, which otherwise would be more tightly held on the exchange sites of silicates. Finally, Ca and Mg, which under neutral to basic conditions would be fixed by phosphates, carbonates, or sulfates, remain mobile and leachable under acidic conditions. The average pH of the natural soils surrounding a site should be determined prior to surveying it in order to assess the extent to which soil alterations produced by different kinds of activities can be expected to have been maintained within the soil, according to the models presented in Chapter 4.

4. Decompositional environment. All else being equal, sites located in climates where soils are periodically moistened and dried [e.g., where chernozems (mollisols) develop] will experience greater soil alterations and more long-term soil alterations per unit of organic residue deposited than will sites located in climates where soils are continuously moist [e.g., where podzols (spodosols) develop]. In the former climatic condition, the organic residues deposited on sites will be decomposed and resynthesized into humus micelles that tend to have aromatic structures, whereas in the latter climatic condition, humus molecules that have aliphatic structures will tend to form. Humus molecules with aromatic structures, in contrast with those with aliphatic structures, tend to have higher cation exchange capacities and facilitate the retention of anthropic nutrient enrichments to a greater degree. They also favor soil aggregation to a greater degree. The effect of decompositional environment and humus structure on the distinctness of use-areas within sites probably is most significant with respect to ephemeral sites where organic enrichments and increases in soil cation exchange capacity and aggregation are slight to begin with. Whether the decompositional environment is optimal or suboptimal for the formation of aromatic humus micelles could make the critical difference in whether or not use-areas are distinguishable in such sites. On sites with midden development, decompositional environment should not be a critical condition.

5. *Soil profile development.* Where sites are located on moderate to fine textured soils having moderately to strongly developed B horizons, the possibility of confusing the resistivity responses of the B horizons for those of buried midden deposits can be great. Under such natural conditions, the degree of contrast between the resistivity responses of the B horizons and those that would be expected for midden deposits possibly buried to a similar depth will be minimal—less than that found on the site between earthen archeological facilities and the A horizons into which they intrude. The cause of this problem is that B horizons, like midden deposits, exhibit greater structural aggregation, usually possess more hydrophilic micelles (illuviated clays), and often have higher cation exchange capacities than the other natural soil horizons that lay immediately above or below them. The quickest and most accurate solution to this problem is to verify the nature of the nondiagnostic resistivity anomaly by soil sample probe.

Weathering and Aging of Sites

The degree of distinctness of use-areas within resistivity data sets from an earthen site will depend on the *age* of the site, that is, the amount of time over which anomalous concentrations of nutrients and humus micelles within the soil have been reduced and have approached the steady-state conditions that are natural for the soil–vegetational systems of that locale. It will also depend on the *rate* at which aging has occurred, which minimally will be a function of the texture of the site soils, their pH, climate, and the nature of the local vegetation. In North America, where most sites are relatively young (Holocene-age), the degree of distinguishability of disturbed soils from their natural counterparts and of different kinds of use-areas from each other within a site will be less dependent on the absolute age of the site than on the factors governing the rates of weathering and aging. Use-areas will be found least distinct on sites located (*a*) on sandy soils, (*b*) on acidic podzols under conifer forest, or (*c*) on areas repeatedly flooded. Each of these conditions facilitate the leaching and loss of humus micelles and nutrients from soils, and the more rapid aging of a site and subduing of its anthropic soil anomalies.

Conditions of Survey

1. A critical requirement for the interpretation of resistivity survey data from archeological sites—whatever the analytical design used—is that it be collected during a single, uniform moisture regime (environmental suction) and that topography, the drainage properties of parent materials, and surface vegetation remain constant over the space where resistivity measurements are to be compared. This is necessary if the magnitudes of anomalous local means and variances in soil resistivity are to be compared with a single site.

Environmental suction must be held constant over the course of the survey because (*a*) the effects of a change in it over space (e.g., a rain during the course of the survey) on *local* resistivity means and variances cannot be factored from the resistivity landscape by spatial filtering techiques; and (*b*) simple mathematical transformations relating the mean and variance of resistivity values in a locale during one environmental suction to the mean and variance of resistivity values in the same locale at a different suction will seldom be applicable beyond the locale for which they were derived.

These circumstances occur, first, because the pore-size distribution of soils of one type, anthropically disturbed in one manner, and located in one area of an archeological site, will seldom be similar to the pore-size distribution of soils of a second type, disturbed in another manner, and located in another area of the site. The change in soil water status and resistivity that occurs at the one location with a change in environmental suction thus cannot be used to predict the change in soil water status and resistivity that occurs in the second location. Second, soils of different types in different locales will probably have different concentrations of strong electrolytes, weak electrolytes, and conductive colloidal particles within their soil waters. Because the functional relationships between soil–water solution concentration and conductivity differ for these three forms of electric carriers (square root, log–log, and linear), change in their total concentration cannot be used to predict change in their total conductive potential at a second location where their proportions differ. Consequently, resistivity data collected at different environmental suctions seldom can be made comparable.

Spatial changes in topography, in the drainage properties of parent materials, and surface vegetation over a survey area have the same negative effects as changes in moisture regimes during the course of a survey. All three affect local soil moisture content and suction. Locations having different slopes will have different precipitation runoff/infiltration ratios. Locations differing in their drainage properties will vary in the degrees to which moisture is lost from their soils under the force of the gravitational potential. Soils in areas differing in the kind and density of their surface vegetation will experience different suctions and moisture losses as a result of the varying evapotranspiration rates of their vegetation covers.

Thus, resistivity surveys for the purpose of archeological reconnaissance should be designed quite differently from the transect surveys at the Crane site, which were performed for experimental purposes. Surveys should follow the topographic contours of a site rather than crosscut them. Climatic, topographic, vegetational, and natural soil and drainage conditions should be held as constant as possible over the course of the survey, rather than allowed to vary as they did at Crane.

2. Resistivity surveys should *not* be performed immediately after a rain. Surveyors should allow at least several rain-free days to pass before surveying. This is necessary to allow the moisture to distribute itself throughout the soil and to reach an equilibrium status with respect to soil matric and gravitational potentials. If a survey is performed immediately after a rain and before soil water movement has ceased, temporal changes in the distribution of water within the soil that occur over the course of the survey will be reflected in the resistivity data along with spatial variations in soil type. This additional factor cannot be removed from the data, for the reasons given in the preceding condition, no. 1, and will hamper the interpretation of the data. The length of time required for soil water movement essentially to cease after a rain can be roughly estimated with Kozeny's and Darcy's equations for saturated and unsaturated flow (Chapter 3) and with knowledge of the approximate textural distribution of the soil.

3. In general, the optimal time for performing resistivity surveys on earthen archeological sites is during periods when environmental suctions are very low (several days after a soil-saturating rainfall), or very high (approaching the wilting point or greater). At these times, both physical and chemical soil alterations within sites will favor organically enriched, disturbed soils having lower resistivities than the natural soils from which they were derived, and the contrast between archeological disturbances and natural soils will be greatest. At moderate soil suctions, on the other hand, physical soil alterations will favor the disturbed soils having higher resistivities, whereas chemical soil alterations will favor the disturbed soils having lower resistivities than their natural counterparts, and contrast between the disturbed and natural soils will be suboptimal. Also, it is during periods of intermediate soil suction that soil resistivity variations caused by natural textural variations will be greatest and will cause the greatest problem in the interpretation of resistivity data for

archeological purposes. This latter condition reflects the fact that differences in the moisture contents of soils as a function of differences in their texture are maximized at intermediate environmental suctions.

4. With respect to soil temperature, resistivity surveys on earthen sites will be most successful when soil temperature is high. At such times, the limiting equivalent conductances of ions within soil waters will be large. Consequently, differences between disturbed and undisturbed soils, and between the soils within use-areas of different types, in their soil water solution conductivities and in their soil resistivities as a function of differences in the concentration of ions within their soil waters, will be greater.

5. Resistivity surveys are not likely to be successful on sites where the archeological layer of interest is buried under a layer of soil with a much lower resistivity than the layer of interest. Under these circumstances, a large proportion of the current applied to the ground will literally be shunted through the surficial layer of lower resistivity, and proportionately little current will flow through the layer of interest and will have an effect on resistivity measurements. Situations in which this condition may arise include sites located on coarser textured soils and capped with alluvial clays, and sites covered with forest vegetation. In the latter case, surficial root systems laden with water will serve as the surficial, highly conductive medium.

The likelihood of making a successful survey on a clay-capped site will be least during periods of intermediate environmental suction, when the moisture and resistivity contrast between the clay layer and the layer of archeological interest is greatest, as a function of soil textural differences. On a forested site, the likelihood of making a successful survey will be least during driest periods, when the moisture and resistivity contrast between root systems and the archeological layer of interest is greatest.

6. Resistivity surveys will not be successful on sites where the horizon of interest is submerged below the water table and overlain by horizons above the water table. Differences in resistivity between disturbed and undisturbed soils within the water-saturated archeological horizon will be very small compared to resistivity variations within the unsaturated, surficial horizons, and will be masked by them. The Barnes Layer method cannot be used to remove this masking effect from the data in this extreme case. The assumption of lateral homogeneity (low variability) in the resistivity of the surficial horizon, which is required by the method (see the following), will be violated.

On sites almost completely saturated with water, from the near-surface downward, the resistivity of soils within different use-areas in the buried archeological horizon theoretically may differ. Also these differences may not be masked by resistivity variation within surficial horizons. Practically, however, accurate soil resistivity measurements are very difficult to make under these wet conditions. Short circuiting of equipment and shunting of current through surficial water films that connect electrodes are the primary problems.

7. On archeological sites that have been freshly plowed and disked but not yet rained on, contact resistance between electrodes and the soil may become a significant problem. Under such conditions, there will be many pockets of air within the soil, and electrodes may contact soil particles in only a few restricted locations. As contact resistance problems increase in frequency and the amount of noise within a resistivity data set increases, use-areas will become more difficult to distinguish within the data.

Survey Efficiency and Time Limitations

As a result of the necessity of keeping soil moisture conditions and environmental suction as constant as possible over the course of a resistivity survey, a survey must be performed rapidly. Various technologies can assist in the effort. Critical are the following two.

1. Electrode arrays fixed on a board. Use of the Barnes Layer method of resistivity calculation requires that multiple, four-electrode arrays with different electrode spacings be set up at each station. Under these conditions, it is not possible to use resistivity meters with rotary switches and a leap frog system of electrode displacement (Clarke 1963:573) to speed up survey. As an alternative to this procedure, all electrodes to be used at each survey station may be mounted at their appropriate spacings on a single board of fiberglass, and inserted at once by depressing the board as a whole (after Dunk 1962:274; Palmer 1960:70). The multiple arrays on the board may be switched on and off from the resistance meter, one at a time, using a channel selector, in order to obtain different resistance readings pertaining to different depths of investigation.

It is critical that the board holding the electrodes be made of a flexible material so that microtopographic variations can be negotiated. The board must be made thick enough and of a strong enough material to hold the electrodes firmly in place against sheering pressures that the electrodes may cause to the board as it is depressed in the ground. Such pressures may develop as the electrodes attempt to follow the path of least compation within the ground and angle from the vertical as they are driven into it. Hardwood boards were found to break down rapidly as a result of such electrode pressures during the course of the Crane site resistivity surveys. Fiberglas boards are more appropriate.

2. Resistance meters, recording equipment. Use of the Barnes Layer method to calculate soil resistivity requires that multiple resistance measurements be made per survey station, and depending on the number of layers to be investigated, increases the amount of data that must be collected and recorded in the field by two or more times over what has traditionally been collected. To facilitate measurement, it is suggested that resistance meters with automatic rather than manually controlled nulling bridges be used. To expedite recording, resistance meters may be hooked up directly to mobile microcomputers such that resistance readings from the meters are transferred to the computers electrically and stored there automatically, rather than read manually and entered manually onto field forms. This recording method will save time and permit gains in accuracy in the field. Storage of data in a microcomputer rather than on field forms also will permit the surveyor to calculate Barnes Layer values in the field and to adjust survey designs where necessary on the basis of layer resistivity values instead of Whole Volume resistivity values.

Economic Considerations

1. Equipment costs. The initial investment required for the equipment needed to perform a resistivity survey is modest. A number of portable, manually nulling resistivity meters that run with low voltage batteries (<4.5V) yet are capable of surveying unconsolidated sediments to depths of 15 m (Aitken 1961; Clark 1963) presently are available for $300 to $500.[1] Another $150 should be allotted for the cost of materials and labor required for making each electrode-board set-up described previously. These are the minimum addition to an archeologist's tool list that must be made in order to perform a resistivity survey. The cost of more sophisticated, automatically nulling meters and automatic data storage–processing equipment is shown in Table 123.

2. Labor costs: preparation, survey, and analysis. The amount of time required to prepare for a resistivity survey will depend on the magnitude of the survey. The main time-consuming preparations are two: (*a*) The depth and thickness of the archeological horizon of interest and the variation of these parameters across the site should be estimated using soil probes. This step is necessary in order to choose electrode spacings that generate optimal Barnes Layers which correspond to the site stratigraphy; (*b*) the survey transect or grid must be laid out over the site with a transit and should correspond to surface collection unit.

The amount of time required to survey a given area with one resistivity meter will depend on the distance between survey stations (sampling density), the number of soundings made per station, and the sophistication of the survey equipment. At Crane, .64 ± .2 minutes were required for each survey station ($N = 2784$) in order to (*a*) move equipment from a previous survey station and set it up at the station of interest; (*b*) make five soundings using manually nulling

---

1. This figure is based on 1979 United States dollars.

TABLE 123
Costs of the equipment required for efficient resistivity surveying

Equipment	Cost*
Automatic nulling resistance meter	$700.00
Micro-computer/data storage system	
Analog-digital converter to change output of the resistance meter into a format which can be read into the system (3 significant digits)	300.00
Cassette-tape storage unit, modified from a cassette tape player	100.00
Microprocesser and portable power supply	150.00
Simple output unit supplying a hard copy similar to an adding machine's paper tape	40.00
Labor (designing and building the unit)	600.00
Programming the processer to calculate Barnes Layer resistivity values and to do time series analyses	200.00
Total cost of system	$2090.00

* Cost estimates have been obtained from a professional computer engineer and electrical engineer. They are given in 1979 United States dollars.

meters; and (c) record the data manually onto the field forms. At this rate, a 100-m-long transect with stations located .5-m apart was surveyable in approximately 2¼ hours. The use of automated equipment would have reduced this time to about 45 minutes.

No analyses of physical and chemical soil properties other than average, natural pH of the undisturbed soils surrounding the site must be made in conjunction with a resistivity survey in order to interpret survey results (see preceding). Models describing the relationships between the resistivity response of a soil, its physical and chemical properties, and the human activities that may have altered these properties, have been formulated and are available for general use in Chapters 3, 4, 8, and 9 of this study. The work necessary to determine natural soil pH may be done as part of the *routine* pH surveys that must be made over a site in order to assess the potential it offers for bone preservation. This labor does not represent an *additional* cost of survey.

Very little time is needed to analyze resistivity data by the proposed methods, if the data have been recorded in the field directly on magnetic tape and can be analyzed immediately by a preprogrammed microcomputer. One man-day should be sufficient to analyze a transect of 300–500 stations using the Barnes Layer method and all spectral analytic, spatial filtering, and clustering routines described in Chapters 2 and 10. If the survey data have been recorded manually onto field forms and are to be processed by a large computer system, more labor, of course, will be necessary to enter the data into the computer. Once entered, however, the estimated labor of one man-day per transect of 300–500 stations should be adequate for analyzing the data, given the required computer programs. The researcher should not have to develop programs to calculate Barnes Layer resistivity values nor the spectral analytic and spatial filtering programs that were used in this study and which are required to examine the data from a statistical, geographic perspective. Spectral analysis programs usually are a part of the statistical packages available at major computer centers. Programs calculating Barnes Layer resistivity values and performing spatial filtering tasks are provided in Appendix 1.

Boundary Conditions Using the Barnes Layer Method

1. For a stratified or unstratified medium that is perfectly homogeneous laterally, the Barnes Layer method can be used to estimate the resistivity of an extremely thin layer, even at great depths. Within the relatively homogeneous midden deposits at the Crane site, the resistivities of layers only 10 cm thick and at least 1 m in depth were estimated with a high degree of accuracy. As the conducting medium becomes more heterogeneous laterally and electric field lines become more warped and distorted from their theoretically elliptical shapes, however, the thickness of the layer to be evaluated must be increased to obtain an equally accurate resistivity estimate at the same depth. To determine the *minimum* layer thickness that may be used at a site to estimate accurately the resistivity of a layer of interest, several resistivity soundings at a few locations (1 or 2 per ha) should be made, allowing the calculation of resistivities of Barnes Layer having various thicknesses and centered around the mid-depth of the horizon of interest. A plot of layer resistivity against layer thickness for the horizon of interest should exhibit similar estimates of the horizon's resistivity for all layers of thickness greater than the critical layer thickness at which electric field warpage becomes a problem. Layers investigated during the survey should be kept thicker than this critical thickness.

2. In a similar manner, where lateral homogeneity of the soil cannot be assumed, there exists a *maximum* layer thickness for which layer resistivities

may be accurately estimated. This situation arises not as a result of the warpage of electric field lines, but rather, because the soil assessed for its resistivity includes a significant amount of soil to the *side* of the first whole volume used in generating the "layer" of interest in addition to soil *beneath* that first whole volume (Figure 4). The critical maximum layer thickness that may be used at a site to estimate accurately the resistivity of a buried horizon may be found in a manner similar to that used for determining the minimum layer thickness.

3. A use-area in which refuse deposition is restricted to the interiors of facilities (e.g., pits, post holes) will be more or less detectable in a resistivity series depending on not only the size and spatial density of the features within it, which may be a function of the activity which occurred in it, but also the *depth* of the layer in which the features occur. The deeper the layer is, the larger or the denser the features must be to create significant resistivity anomalies. This is a consequence of the fact that the volume of soil encompassed within a resistivity layer of a given thickness increases with the depth of the layer (Figure 46). More *total* disturbed soil is required to produce the same proportion of disturbed and undisturbed soil encompassed within a layer and the same magnitude of anomalous resistivity for larger, deeper Barnes Layers than for smaller, more shallow Barnes Layers.

Boundary Conditions Using Spatial Filtering Techniques

1. To isolate archeologically significant variability successfully within a resistivity series from agriculturally and naturally caused variation, the spatial scales of the phenomena causing these three types of resistivity variation must differ.

2. To design spatial filters that successfully isolate resistivity variation attributable primarily to use-areas within a complex resistivity series, the cutoff frequencies of the filters must be chosen corresponding to the appropriate spatial scales of the areas expected to be found. One must have knowledge of the probable dimensions of use-ares in order to search for them within a complex resistivity signature.

3. In order to assess archeologically significant differences between use-areas in the variance of their resisitivity values (e.g., as when trying to assess feature densities), agricultural variation, small-scale, natural variation, and measurement error must be constant or low in magnitude across all areas within the sampled universe. This is necessary because the variability to be assessed is that within high frequency bands containing waves with periods from 0 dimension to the size of the use-areas under investigation. Resistivity bands of such frequencies will contain agricultural, natural, and error variations as well as archeologically caused variations.

To hold agricultural variation constant over all areas to be compared, the resistivity transects crosscutting them should be oriented at the same constant angle with respect to the direction of plowing and fertilizing over the site.

Conclusion

When resistivity surveying techniques first were applied in archeology to locate individual, highly resistant or conductive features, the methods were greeted with enthusiasm and employed vigorously (particularly in Europe). This enthusiasm waned in the mid 1960s, when the methods were used on earthen sites, were extended beyond their capabilities as then known, and produced unreliable results. In this study, the problems of using resistivity techniques on earthen sites have been examined, a new approach for their use on earthen sites has been developed and shown feasible, and boundary conditions have been set for them.

In particular, it has been shown that the resistivity responses of broad areas may be examined statistically so as to allow the location and characterization of prehistoric use-areas on sites, as opposed to individual features. A number of analyses in this work suggest this approach to be feasible within the boundary conditions described in this chapter.

1. Variations in the content and concentration of ions within soils, and in their organic matter content, structure, and moisture equilibria were shown to determine systematically the resistivity of soils in accord with models developed in physics, chemistry, and soil science.

2. Prehistoric use-areas where different kinds of activities occurred (e.g, woodworking, bone working, seed grinding) have been shown to have soils distinguishable by the spectra of ions they hold in available form and in anomalous concentrations. The spectrum of anomalous, available ions characterizing particular use-areas is related to the chemical composition of the residues that were deposited within them, to natural soil processes that govern the relative strength with which given ions are retained within the soil, and to the age of the use-area.

3. Areas that served different purposes (swept, multipurpose work areas; trash dumps) were found to differ in the total concentration of ions occurring in available form within their soils. These soils also exhibited differing organic matter contents, degrees of structural development, and moisture equilibria.

4. The *mean* resistivities of the soils within use-areas were shown to be a function of the kind and magnitude of physical and chemical alterations that occurred within their soils.

5. The *variance* in resistivity of the soils within use-areas were found to be a function of the areal density of permanent facilities (pits) within them or the proportion of disturbed soil within them.

6. A number of models derived from geochemistry demonstrate how most anomalous soil conditions are maintained within sites from the time of occupation to the present by natural soil processes and general environmental processes.

To make possible the interpretation of resistivity data within a statistical and geographic framework, two mathematical techniques were introduced: the Barnes Layer method of calculating soil resistivity, and spatial filtering. These techniques allow the total variability within a resistivity data set to be partitioned along the dimensions of depth and space, permitting the isolation of archeologically significant horizons and the determination of possible use-areas within them.

It is hoped that this new approach to resistivity surveying and the advantages it provides will encourage archeologists to employ the technique frequently once again. Resistivity methods can be used to define both the nature and degree of internal structuring of a site prior to excavation, allowing the researcher to assess whether the site is worthy of excavation compared to other sites, and also to plan optimal excavations. As an evaluative survey technique, it has great potential within the areas of cultural resource management and regional archeological planning in general. As a tool for guiding the designs of intrasite excavations which representatively sample use-areas—fundamental analytical units in archeology—it is useful regardless of the excavator's ultimate purpose or theoretical orientation.

References

Aitken, M. J.
   1961 *Physics and archaeology.* Interscience Publishers, New York.
Brady, Nyle C.
   1974 *The nature and properties of soils.* Macmillan, New York.
Clark, Anthony
   1963 Resistivity surveying. In *Science in archaeology,* D. Brothwell and E. Higgs (eds.), pp. 569–581. Basic Books, New York.
Dunk, A. J.
   1962 An electrical resistance survey over a Romano–British site. *Bonner Jährbucher* 162:272–276.
Palmer, L. S.
   1960 Geoelectrical surveying of archaeological sites. *Proceedings of the Prehistoric Society* 26:64–75.

APPENDIX I

Computer Programs for Calculating Barnes Layer Resistivity Values and for Spatial Filtering Resistivity Data

```
C :::::::::::::::::::::::::::::::::::::::::::::::::::::::::::::::::::::::::::::::::::::::::
C
C                          PROGRAM RESIST
C
C :::::::::::::::::::::::::::::::::::::::::::::::::::::::::::::::::::::::::::::::::::::::::
C
C
C THIS PROGRAM:  (1) READS MULTIPLE SOIL RESISTANCE OBSERVATIONS WHICH HAVE
C BEEN COLLECTED WITH SEVERAL WENNER ARRAYS OF VARIOUS ELECTRODE SEPARATIONS
C AT EACH OF A NUMBER OF SURVEY STATIONS, (2) CALCULATES WHOLE VOLUME SOIL
C RESISTIVITY VALUES ,AND BARNES LAYER RESISTIVITY VALUES FOR EACH STATION
C (IN OHM-CM), AND (3) PRINTS (AND OPTIONALLY PUNCHES) THE WHOLE VOLUME AND
C BARNES LAYER VALUES.  THE NUMBER OF SURVEY STATIONS WHICH CAN BE PROCESSED
C IS UNLIMITED, BUT THE ELECTRODE SEPARATION DISTANCES (A) AND THE ELECTRODE
C PENETRATION DEPTHS (P) OF THE ARRAYS AT ONE STATION MUST BE THE SAME AT
C ALL OTHER STATIONS.  THE NUMBER OF WHOLE VOLUME RESISTIVITY VALUES WHICH
C CAN BE CALCULATED FOR EACH STATION (EQUIVALENT TO THE NUMBER OF WENNER ARRAY
C CONFIGURATIONS OF VARIOUS ELECTRODE SEPARATIONS USED AT THE STATION) IS
C LIMITED TO 100.  THE RESISTIVITY OF ALL POSSIBLE BARNES LAYERS OF VARIOUS
C THICKNESSES AND DEPTHS WHICH CAN BE GENERATED FROM THE RESISTANCES OBTAINED
C AT EACH STATION WITH THE UP TO 100 WENNER ARRAYS OF DIFFERENT ELECTRODE
C SEPARATIONS CAN BE DETERMINED, IF SO DESIRED.  THE USER CAN CHOOSE TO
C CALCUDLATE ONLY A SUBSET OF ALL BARNES LAYER RESISTIVITY VALUES WHICH
C COULD BE DETERMINED FOR EACH STATION.
C
C SET-UP OF CARD DECK REQUIRED BY THE PROGRAM:
C      (1) SYSTEM-REQUIRED CARDS
C      (2) PROGRAM RESIST
C      (3) SYSTEM-REQUIRED CARDS
C      (4) TWO CONTROL CARDS WITH THE FOLLOWING INFORMATION IN THE DESIGNATED
C          COLUMNS:
C              CARD 1, COL 1-3.  THE NUMBER OF WENNER ARRAYS (N) OF VARIOUS
C                 ELECTRODE SPACINGS USED TO DETERMINE SOIL RESISTANCE AT
C                 EACH SURVEY STATION (FORMAT I3).
C              CARD 1, COL 4-5.  THE MAXIMUM THICKNESS OF THE BARNES LAYERS
C                 FOR WHICH SOIL RESISTIVITY VALUES SHOULD BE CALCULATED,
C                 MEASURED IN THE NUMBER OF SMALLEST INDIVIDUAL LAYERS
C                 COMPOSING THEM (FORMATI2). FOR EXAMPLE, SUPPOSE SOIL
C                 RESISTANCES TO 5 DIFFERENT DEPTHS OF INVESTIGATION WERE
C                 MEASURED AT EACH STATION WITH 5 DIFFERENT WENNER ARRAYS. IF
C                 THE MAXIMUM THICKNESS, 1, IS SPECIFIED, THE RESISTIVITIES
C                 OF ONLY THE SMALLEST INDIVIDUAL BARNES LAYERS WILL BE
C                 CALCULATED:  BL2-1, BL3-2, BL4-3, BL5-4.  IF THE MAXIMUM
C                 THICKNESS, 2, IS SPECIFIED, THE RESISTIVITIES OF THE
C                 SMALLEST INDIVIDUAL BARNES LAYERS AND THE RESISTIVITIES
C                 OF THICKER LAYERS ENCOMPASSING TWO OF THESE WILL BE
C                 CALCULATED:  BL2-1, BL3-2, BL4-3, BL5-4 (AS BEFORE), PLUS
C                 BL3-1, BL4-2, BL5-3.  IF THE MAXIMUM THICKNESS, 4, IS
C                 SPECIFIED, THE RESISTIVITIES OF BARNES LAYERS UP TO 4
C                 OF THE SMALLEST INDIVIDUAL LAYERS IN THICKNESS WILL BE
C                 CALCULATED:  BL2-1, BL3-2, BL4-3, BL5-4, PLUS BL3-1,
C                 BL4-2, BL5-3, PLUS BL4-1, BL5-2, PLUS BL5-1.  THE
C                 VALUE OF THE MAXIMUM THICKNESS PARAMETER CAN BE NO LARGER
C                 THAN 1 LESS THAN THE NUMBER OF WENNER ARRAYS SPECIFIED
```

```
C                    IN COLUMNS 1-3 (N-1).
C               CARD 1, COL 6.  CONTAINS ANY LETTER IF WHOLE VOLUME RESISTIVITY
C                    VALUES ARE TO BE PUNCHED ON CARDS (FORMAT A1).  OTHERWISE,
C                    LEAVE BLANK.
C               CARD 1, COL 7.  CONTAINS ANY LETTER IF BARNES LAYER RESISTIVITY
C                    VALUES ARE TO BE PUNCHED ON CARDS (FORMAT A1).  OTHERWISE,
C                    LEAVE BLANK.
C               CARD 1, COL 8-72; CARD 2, COL 1-5.  TITLE (FORMAT A30).
C       (5) N CARDS, ONE FOR EACH WENNER ARRAY USED AT AN EXAMPLE SURVEY
C           STATION, EACH CARD LISTING THE ELECTRODE SEPARATION DISTANCE (A)
C           FOR THE ARRAY AND THE DEPTH (P) TO WHICH ELECTRODES WERE INSERTED
C           INTO THE GROUND FOR THAT ARRAY (FORMAT 2(10X,F7.3)).  BOTH THE
C           ELECTRODE SEPARATION AND DEPTH OF PENETRATION OF ELECTRODES SHOULD
C           BE GIVEN IN CENTIMETERS.  THE CARDS SHOULD BE ARRANGED FROM THAT
C           LISTING THE SMALLEST ELECTRODE SEPARATION TO THAT LISTING THE
C           LARGEST.
C       (6) SOIL RESISTANCE VALUES.  EACH STATION SHOULD BE REPRESENTED BY A
C           SET OF ONE OR MORE CARDS, EACH LISTING A SURVEY STATION DESIGNATOR
C           (PROVENIENCE) AND UP TO 8 SOIL RESISTANCE VALUES.  THE PROVENIENCE
C           SHOULD OCCUR IN THE FIRST 18 COLUMNS OF EACH CARD (FORMAT 9A2).
C           THE RESISTANCE VALUES SHOULD FOLLOW (FORMAT N(F7.2)) WITH THERE
C           BEING AS MANY CARDS OF 8 RESISTANCE VALUES EACH AS IS NECESSARY
C           TO LIST ALL N RESISTANCE VALUES OBTAINED AT THE STATION WITH THE
C           N WENNER ARRAYS.  FOR EXAMPLE, IF 14 RESISTANCE VALUES WERE MEASURED
C           AT EACH STATION WITH 14 WENNER ARRAYS OF DIFFERENT ELECTRODE
C           SPACINGS, THE FOLLOWING 2 CARDS WOULD BE USED FOR EACH STATION:
C
C           PROVENIENCE_____8 RESISTANCE VALUES (FORMAT 8F7.3)
C           PROVENIENCE_____6 RESISTANCE VALUES (FORMAT 6F7.3)
C
C           CARD SETS REPRESENTING AS MANY STATIONS AS WERE SURVEYED WITH THE
C           SAME SET OF ELECTRODE SEPARATIONS AND SAME SET OF DEPTHS OF
C           PENETRATION OF ELECTRODES CAN FOLLOW ONE AFTER ANOTHER.
C
C           EACH SET OF CARDS FOR EACH STATION SHOULD LIST RESISTANCE VALUES
C           BEGINNING WITH THAT RESISTANCE MEASURED WITH THE SMALLEST ELECTRODE
C           SPACING AND ENDING WITH THAT RESISTANCE MEASURED WITH THE LARGEST
C           ELECTRODE SPACING.  MISSING OBSERVATIONS SHOULD BE GIVEN THE VALUE
C           0.  THE VALUE 0 WILL BE PRINTED FOR THAT WHOLE VOLUME RESISTIVITY
C           VALUE AND THOSE BARNES LAYER RESISTIVITY VALUES CALCULATED WITH
C           THE MISSING OBSERVATION.
C       (7) SYSTEM-REQUIRED CARDS.
C
C DEFINITION OF VARIABLES, ARRAYS, AND LIMITATIONS WITHIN THE PROGRAM:
C     IPROV.  THE DESIGNATION OF A SURVEY STATION.
C.    N.  THE NUMBER OF RESISTANCE MEASUREMENTS MADE AT EACH SURVEY STATION
C           WITH WENNER ARRAYS OF DIFFERENT ELECTRODE SEPARATIONS.
C     D(100).  THE SET OF N RESISTANCE MEASUREMENTS MADE AT A GIVEN SURVEY
C           STATION WITH WENNER ARRAYS OF DIFFERENT ELECTRODE SEPARATIONS.
C     CONDW(100).  THE SET OF N CONDUCTANCE VALUES CALCULATED FROM THE
C           N RESISTANCE MEASUREMENTS MADE AT A GIVEN SURVEY STATION.
C     RSTWV(100).  THE SET OF N WHOLE VOLUME RESISTIVITY VALUES CALCULATED
C           FROM THE N RESISTANCE MEASUREMENTS MADE AT A GIVEN SURVEY STATION.
```

```
C         RSTBL(4950). THE SET OF BARNES LAYER RESISTIVITY VALUES CALCULATED
C              FROM THE N RESISTANCE MEASUREMENTS MADE AT A GIVEN SURVEY STATION.
C              THE NUMBER WILL VARY WITH THE MAXIMUM THICKNESS OF THE BARNES
C              LAYERS (IBLSEP) CHOSEN BY THE USER.
C         A(100). THE SET OF N DIFFERENT ELECTRODE SEPARATION DISTANCES DEFINING
C              THE N WENNER ARRAYS USED AT EACH SURVEY STATION TO MEASURE SOIL
C              RESISTANCE.
C         P(100). THE SET OF N DEPTHS TO WHICH ELECTRODES WERE INSERTED INTO
C              THE GROUND FOR THE N WENNER ARRAYS USED AT A GIVEN SURVEY STATION.
C         IBLSEP. THE MAXIMUM THICKNESS OF THE BARNES LAYERS FOR WHICH SOIL
C              RESISTIVITY VALUES SHOULD BE CALCULATED, MEASURED IN THE NUMBER
C              OF SMALLEST INDIVIDUAL BARNES LAYERS COMPOSING THEM.
C         IPCHWV. A LETTER INDICATING THAT WHOLE-VOLUME RESISTIVITY VALUES
C              SHOULD BE PUNCHED ON CARDS.
C         IPCHBL. A LETTER INDICATING THAT BARNES LAYER RESISTIVITY VALUES
C              SHOULD BE PUNCHED ON CARDS.
C
C DEFINITION OF SYSTEM UNITS USED BY THE PROGRAM:
C         5.   CARD DECK SOURCE
C         6.   PRINTER
C         7.   PUNCH
C
C
C**********************************************************************
      DIMENSION D(100),A(100),P(100),CONDW(100),CONDL(100),RSTWV(100),
     1RSTBL(4950)
      INTEGER*2 ITITLE(35),IPROV(9)
C CVPFMT - FORMATS FOR PRINTING COMPUTED RESISTIVITY VALUES
C DPFMT - FORMATS FOR PRINTING RESISTANCE OBSERVATIONS
C DECLARE CVPFMT AND DPFMT TO BE INTEGERS
      INTEGER CVPFMT(2),DPFMT(2)
C INITIALIZE ARRAYS
      DATA CVPFMT/'F7.0','7.0,'/,DPFMT/'F7.2','7.2,'/
      DATA IBLANK/'    '/,ITITLE,IPROV/44*'    '/
C READ THE CONTROL CARD
      READ(5,1) N,IBLSEP,IPCHWV,IPCHBL,ITITLE
    1 FORMAT (I3,I2,A1,A1,35A2)
C TEST LEGALITY OF VALUES FOR N AND IBLSEP
      IF(N.GT.100.OR.N.LE.0)GO TO 999
      IF(IBLSEP.GT.N-1.OR.IBLSEP.LT.0)GO TO 999
C READ IN ELECTRODE SEPARATION VALUES AND ELECTRODE PENETRATION DEPTHS
      DO 100 J=1,N
  100 READ(5,2,END=999) A(J),P(J)
    2 FORMAT (2(10X,F7.3))
C WRITE THE TITLE
      WRITE(6,4) ITIT
    4 FORMAT ('1',T31,35A2,//' ',T31,40A2,//A4)
C READ IN PROVENIENCE AND RESISTANCES
  500 CONTINUE
      NCARDS=N/8
      NREM=N-NCARDS*8
      IF(NCARDS.EQ.0)GOTO 510
      DO 505 JKL=1,NCARDS
      LI=1+(JKL-1)*8
```

```
          LU=LI+7
          READ(5,501,END=999)IPROV,(D(KK),KK=LI,LU)
      501 FORMAT(9A2,8F7.2)
      505 CONTINUE
          IF(NREM.EQ.0)GOTO 511
      510 CONTINUE
          LI=1+NCARDS*8
          LU=LI+NREM-1
          READ(5,501,END=999)IPROV,(D(KK),KK=LI,LU)
      511 CONTINUE
C PRINT THE RAW DATA HEADING
          WRITE(6,5) IPROV
        5 FORMAT (/'0',T31,'RAW DATA'/' ',9A2)
C CALL SUBROUTINE TO PRINT RESISTANCES
          CALL PNTOUT(N,D,DPFMT)
C CALCULATE CONDUCTANCES OF WHOLE VOLUMES
          DO 120 L=1,N
          IF (D(L)) 110,110,120
      110 D(L)=1./1000000.
      120 CONDW(L)=1./D(L)
          NN=N-1
C CALCULATE WHOLE-VOLUME RESISTIVITY VALUES
          DO 140 I=1,N
          CNWV=1.+(2.*A(I)/SQRT(A(I)*A(I)+4.*P(I)*P(I)))-(A(I)/SQRT(A(I)*A(I
         1)+P(I)*P(I)))
          RSTWV(I)=(4.*3.14169*A(I))/(CNWV*CONDW(I))
      140 IF(RSTWV(I).GT.-1.AND.RSTWV(I).LT.1.)RSTWV(I)=0.
C PRINT WHOLE-VOLUME HEADING
          WRITE(6,6)IPROV
        6 FORMAT (/'0',T31,'WHOLE-VOLUME VALUES'/' ',9A2)
C CALL SUBROUTINE TO PRINT WHOLE-VOLUME DATA
          CALL PNTOUT(N,RSTWV,CVPFMT)
C SEE IF WHOLE-VOLUME VALUES ARE TO BE PUNCHED
          IF (IPCHWV.EQ.IBLANK)GO TO 150
C CALL SUBROUTINE TO PUNCH WHOLE VOLUME RESISTIVITIES
          CALL PCHOUT(N,RSTWV,IPROV)
C CHECK TO SEE IF BARNES LAYER VALUES ARE TO BE CALCULATED
C IF NOT START ON NEXT STATION
      150 IF(IBLSEP.LE.0.OR.N.LE.1)GO TO 500
C VALUES FOR BARNES LAYER ARE TO BE CALCULATED --- SET UP COUNTER FOR
C NUMBER OF BARNES LAYERS OF VARIOUS THICKNESSES FOR WHICH CALCULATIONS
C SHOULD BE PERFORMED, AS SPECIFIED BY IBLSEP
          DO 300 M=1,IBLSEP
          NBL=N-M
C CALCULATE BARNES LAYER VALUES FOR PRESENT IBLSEP VALUE
          DO 200 MM=1,NBL
          AAV=(A(MM+M)+A(MM))/2.
          PAV=(P(MM+M)+P(MM))/2.
          CNBL=1.+(2.*AAV/SQRT(AAV*AAV+4.*PAV*PAV))-(AAV/SQRT(AAV*AAV+PAV*PA
         1V))
          AL=A(MM+M)-A(MM)
          RSTBL(MM)=(4.*3.14169*AL)/(CNBL*(CONDW(M+MM)-CONDW(MM)))
      200 IF (RSTBL(MM).GT.-1..AND.RSTBL(MM).LT.1.)RSTBL(MM)=0.
```

```
C PRINT BARNES LAYER HEADING
      WRITE(6,7) M,IPROV
    7 FORMAT (/'0',T31,'BARNES LAYER VALUES',7X,'LAYER THICKNESS=',I2/
     1' ',9A2)
C CALL SUBROUTINE TO PRINT BARNES LAYER VALUES
      CALL PNTOUT(NBL,RSTBL,CVPFMT)
C SEE IF BARNES LAYER VALUES ARE TO BE PUNCHED
      IF (IPCHBL.EQ.IBLANK) GO TO 300
C CALL SUBROUTINE TO PUNCH VALUES
      CALL PCHOUT(NBL,RSTBL,IPROV)
  300 CONTINUE
C END OF MAIN PROGRAM LOOP
      GO TO 500
C ERROR DETECTED --- SAY SO
  999 WRITE(6,8)
    8 FORMAT (////' INCORRECT CONTROL VALUE SPECIFICATION'//' JOB EXECUT
     1ION IMPOSSIBLE')
C FINISHED WITH RUN
  900 WRITE(6,9)
    9 FORMAT (////' JOB ENDS'/'1')
      STOP
      END
C
C :::::::::::::::::::::::::::::::::::::::::::::::::::::::::::::::::::::::::
C
      SUBROUTINE PCHOUT(N,PCHARR,IPROV)
C THIS SUBROUTINE PUNCHES AN ARRAY ON CARDS (UNIT 7)
C CALLING PARAMETERS :
C    N         NUMBER OF VALUES TO BE PUNCHED
C    PCHARR    ARRAY CONTAINING VALUES
C    IPROV     PROVENIENCE
C SUBROUTINE IS CALLED FROM RSTVTY MAIN PROGRAM
      DIMENSION PCHARR(100),LSTFMT(9),JFMT(4)
      INTEGER*2 IPROV(9),IBLANK/'  '/
C JFMT       FORMAT FOR LAST CARD
C LSTFMT     USED IN CONSTRUCTING LAST CARD FORMAT
C INITIALIZE FORMAT FOR LAST CARD
      DATA JFMT/'(9A2',',1(F','7.0)',',A2)'/
C COMMON FOR LSTFMT (ALSO USED BY PNTOUT SUBROUTINE)
      COMMON/FMTS/LSTFMT
C SET UP COUNTERS FOR PUNCHING
      ICOUNT=N/8
      JFIRST=1
      JLAST=8
      LSTCNT=N-8*ICOUNT
      IF(ICOUNT)200,20,10
C WRITE ALL BUT LAST LINE   RESET COUNTERS ALONG THE WAY
   10 DO 100 K=1,ICOUNT
      WRITE(7,1) IPROV,(PCHARR(L),L=JFIRST,JLAST),IBLANK
    1 FORMAT (9A2,8F7.0,A2)
      JFIRST=JFIRST+8
  100 JLAST=JLAST+8
C SEE IF LAST CARD MUST BE PUNCHED
   20 IF (LSTCNT.EQ.0)GO TO 200
```

```
C  SET UP FORMAT AND PUNCH THE LAST CARD
      JFMT(2)=LSTFMT(LSTCNT)
      WRITE(7,JFMT) IPROV,(PCHARR(M),M=JFIRST,N),IBLANK
C  LEAVE THE SUBROUTINE --- RETURN TO MAIN PROGRAM
  200 RETURN
      END
C
C  ::::::::::::::::::::::::::::::::::::::::::::::::::::::::::::::::::::::::::::::::::::::::::::::::::::
C
      SUBROUTINE PNTOUT(N,PNTARR,IFMT)
C  THIS SUBROUTINE PRINTS ARRAYS OF VALUES IN ASCENDING ORDER
C  CALLING PARAMETERS ARE :
C     N         NUMBER OF ARRAY ELEMENTS TO BE PRINTED
C     PNTARR    ARRAY CONTAINING ELEMENTS TO BE PRINTED
C     IFMT      SPECIFICATION OF FORMAT FOR INDIVIDUAL ELEMENTS
C  SUBROUTINE IS CALLED FROM RSTVTY MAIN PROGRAM
      DIMENSION PNTARR(100),IFMT(2),JFMT(6),KFMT(6),LSTFMT(9)
C  JFMT    FORMAT FOR ALL BUT LAST LINE
C  KFMT    FORMAT FOR LAST LINE
C  LSTFMT  USED IN CONSTRUCTION OF LAST LINE FORMAT
C  INITIALIZE IBLANK
      INTEGER*2 IBLANK/'  '/
C  INITIALIZE FORMATS
      DATA JFMT/'(1H+',',','T21',',','10(','F7.0',',','2X)','/A2)'/,
     1KFMT/'(1H+',',','T21',',','1(F','7.0',',','2X)/','A2)'/
C  COMMON FOR LSTFMT (USED BY PCHOUT ALSO)
      COMMON/FMTS/LSTFMT
C  SET UP COUNTERS AND FORMATS
      IFIRST=1
      ILAST=10
      ICOUNT=N/10
      LSTCNT=N-10*ICOUNT
      IF(ICOUNT)200,20,10
   10 JFMT(4)=IFMT(1)
C  WRITE ALL BUT LAST LINE
      DO 100 J=1,ICOUNT
      WRITE(6,JFMT) (PNTARR(M),M=IFIRST,ILAST),IBLANK
C  RESET COUNTERS
      IFIRST=IFIRST+10
  100 ILAST=ILAST+10
C  SEE IF LAST LINE IS NEEDED
   20 IF (LSTCNT.EQ.0)GO TO 200
C  WRITE THE LAST LINE (CONSTRUCT THE FORMAT FIRST)
      KFMT(4)=IFMT(2)
      KFMT(3)=LSTFMT(LSTCNT)
C  NOW PRINT THE LINE
      WRITE(6,KFMT) (PNTARR(L),L=IFIRST,N),IBLANK
C  FINISHED PRINTING, LEAVE SUBROUTINE
  200 RETURN
      END
C
```

```
C  ::::::::::::::::::::::::::::::::::::::::::::::::::::::::::::::::::::::::::::::::::::::::::::::
C
       BLOCK DATA
C  INITIALIZATION OF COMMON FOR OUTPUT SUBROUTINES
       COMMON/FMTS/LSTFMT
       DIMENSION LSTFMT(9)
C  CHARACTERS USED TO CONSTRUCT FORMATS
       DATA LSTFMT/',1(F',',2(F',',3(F',',4(F',',5(F',',6(F',',7(F',
      1',8(F',',9(F'/
       END
C
C  ::::::::::::::::::::::::::::::::::::::::::::::::::::::::::::::::::::::::::::::::::::::::::::::
C
C  WRITTEN BY PETER FORD (DEPARTMENT OF GEOLOGY) AND CHRISTOPHER CARR
C  (DEPARTMENT OF ANTHROPOLOGY), THE UNIVERSITY OF MICHIGAN, 1977.
C
```

```
C        13.  THE PROVENIENCE DESIGNATIONS AND OBSERVATIONS OF THE ORIGINAL
C             SPACE SERIES, STORED FOR USE BY THE PROGRAM.
C        14.  PRINTED OUTPUT DEVICE.
C        15.  CONTROL RECORDS HAVING VARIABLE, USER-SPECIFIED PARAMETERS
C             REQUIRED BY THE PROGRAM.
C
C CONTROL RECORDS TO OCCUR IN STORAGE UNIT 15:
C TEN CONTROL RECORDS ARE REQUIRED TO RUN PROGRAM FILTER.  THE INFORMATION
C REQUIRED IN EACH RECORD IS DESCRIBED BELOW, USING THE FOLLOWING SITUATION
C AS AN EXAMPLE.  SUPPOSE WE WISH TO FILTER A SPACE SERIES WITH 5 DIFFERENT
C NORMAL FILTERS.  WE WISH TO GENERATE 5 SMOOTHED SERIES HAVING FREQUENCIES
C WHICH ARE GREATER THAN OR EQUAL TO 17, 19, 21, AND 25 OBSERVATIONS IN
C WAVELENGTH; ONE HIGH FREQUENCY RESIDUAL SERIES HAVING FREQUENCIES WHICH
C ARE LESS THAN 17 OBSERVATIONS IN WAVELENGTH; AND 4 BAND SERIES HAVING
C INTERMEDIATE FREQUENCIES WITH WAVELENGTHS OF 17-19 OBSERVATIONS, 19-21
C OBSERVATIONS, 21-23 OBSERVATIONS, AND 23-25 OBSERVATIONS.  OF THE VARIOUS
C SMOOTHED SERIES WE MAY STORE ON A SYSTEM STORAGE UNIT, WE WANT TO KEEP ONLY
C THE FIRST AND LAST, GENERATED WITH FILTERS HAVING WIDTHS OF 17 AND 25
C OBSERVATIONS.  THE CONTROL RECORDS WHICH WOULD BE REQUIRED TO ACCOMPLISH
C THIS ARE AS FOLLOWS.
         RECORD 1.  THE REAL-NUMBER FORMAT OF THE PROVENIENCE DESIGNATIONS
             AND OBSERVATIONS OF THE ORIGINAL SERIES STORED ON UNIT 13.
             E.G., (1X,F5.0,1X,F10.3)
         RECORD 2.  THE NUMBER OF FILTERS TO BE APPLIED TO THE ORIGINAL SPACE
             SERIES.  THIS NUMBER SHOULD BE WRITTEN IN THE FORMAT (1X,I2).
             E.G., BB5
         RECORD 3.  THE INTEGER FORMAT TO BE USED WHEN SUPPLYING THE PROGRAM
             WITH THE WIDTHS OF THE FILTERS WHICH WILL BE APPLIED TO THE
             ORIGINAL SPACE SERIES.  E.G., (5(1X,I2))
         RECORD 4.  THE WIDTHS OF THE FILTERS WHICH WILL BE APPLIED TO THE
             ORIGINAL SPACE SERIES, GIVEN IN THE FORMAT SPECIFIED BY THE USER
             IN CONTROL RECORD 3.  FILTER WIDTHS SHOULD BE GIVEN FROM SMALLEST
             TO THE LARGEST.  E.G., B17B19B21B23B25
         RECORD 5.  THE NUMBER OF SMOOTHED SERIES WHICH ONE DESIRES TO BE SAVED
             ON A SYSTEM STORAGE UNIT.  THE NUMBER SHOULD BE WRITTEN IN THE
             FORMAT (1X,I2).  E.G., BB2
         RECORD 6.  THE INTEGER FORMAT TO BE USED WHEN SUPPLYING THE PROGRAM
             WITH THE DESIGNATIONS OF THE SMOOTHED SERIES TO BE SAVED.
             E.G., (2(1X,I1))
         RECORD 7.  THE DESIGNATIONS OF THE SMOOTHED SERIES ONE DESIRES TO SAVE.
             THE NUMBER 1 DESIGNATES THE SMOOTHED SERIES DERIVED WITH THE
             FIRST (LEAST WIDE) FILTER.  THE NUMBER 2 DESIGNATES THE
             SMOOTHED SERIES DERIVED WITH THE SECOND (NEXT LEAST WIDE) FILTER,
             ETC.  IN THE EXAMPLE, WE WISH TO SAVE THE SMOOTHED SERIES GENERATED
             WITH FILTERS OF WIDTHS 17 AND 25 OBSERVATIONS AND WOULD SUPPLY
             THE PROGRAM WITH THE RECORD B1B5.
         RECORD 8.  THE REAL-NUMBER FORMAT IN WHICH THE PROVENIENCE DESIGNATIONS
             AND OBSERVATIONS OF THE HIGH FREQUENCY SERIES SHOULD BE STORED
             ON UNIT 10.  E.G., (1X,F5.0,1X,F5.3)
         RECORD 9.  THE REAL NUMBER FORMAT IN WHICH THE PROVENIENCE DESIGNATIONS
             AND OBSERVATIONS OF EACH INTERMEDIATE FREQUENCY SERIES SHOULD BE
             STORED ON UNITS 1-9.  E.G., (1X,F5.0,1X,F5.3)
```

```
C ::::::::::::::::::::::::::::::::::::::::::::::::::::::::::::::::::::::::::::::
C
C                              PROGRAM FILTER
C
C ::::::::::::::::::::::::::::::::::::::::::::::::::::::::::::::::::::::::::::::
C
C
C THIS PROGRAM FILTERS A SPACE SERIES OF RESISTIVITY DATA WITH NORM
C FILTERS IN ORDER TO OBTAIN LOW FREQUENCY (SMOOTHED), INTERMEDIATE
C FREQUENCY, AND HIGH FREQUENCY (RESIDUAL) COMPONENTS OF ITS VARIAB
C SOME OR ALL OF THE DERIVED SERIES MAY BE STORED AS PERMANENT SYST
C FILES, DEPENDING UPON THE WISHES OF THE USER.  THE MEAN, VARIANCE
C STANDARD DEVIATION OF THE ORIGINAL SPACE SERIES; THE MEANS, VARIA
C AND STANDARD DEVIATIONS OF EACH DERIVED SERIES; AND THE PERCENTAG
C THE ORIGINAL SERIES VARIANCE CONTRIBUTED TO IT BY EACH DERIVED SE
C ARE CALCULATED AND PRINTED.
C
C THE SPACE SERIES TO BE FILTERED MAY BE UP TO 700 OBSERVATIONS LON
C UP TO 10 SMOOTHED, LOW FREQUENCY SERIES MAY BE GENERATED USING 1(
C DIFFERENT NORMAL FILTERS WITH WIDTHS OF 39 OBSERVATIONS OR LESS.
C NUMBER OF INTERMEDIATE FREQUENCY BANDS WHICH MAY BE CALCULATED F
C SMOOTHED SERIES IS LIMITED TO 9.
C
C DEFINITION OF ARRAYS AND LIMITATIONS WITHIN THE PROGRAM:
C      RESIS(700).  THE SPACE SERIES TO BE FILTERED, WITH UP TO 700
C      PROV(700).   THE PROVENIENCE DESIGNATIONS FOR EACH OBSERVATIO
C                   SPACE SERIES.
C      SMOOTH(10,700).  UP TO 10 SMOOTHED, LOW FREQUENCY SERIES DER
C                   THE ORIGINAL SPACE SERIES USING NORMAL FILTERS OF DIFF
C      W(10,40).  UP TO 10 NORMAL FILTERS USED TO DERIVE UP TO 10
C                   FREQUENCY SERIES.  THE FILTERS MAY RANGE FROM 3 TO 39
C                   IN WIDTH.  THE WIDTH OF EACH FILTER MUST BE AN UNEVEN
C                   (E.G. 3, 5, 7)
C      XHF(700).  THE HIGH FREQUENCY SERIES, CALCULATED BY SUBTRA
C                   HIGHEST FREQUENCY SMOOTHED SERFIES FROM THE ORIGINAL
C      XINTM(10,700).  UP TO 9 INTERMEDIATE FREQUENCY SERIES, EAC
C                   BY SUBTRACTING ONE OF THE WELL SMOOTHED SERIES FROM T
C                   LESS WELL SMOOTHED SERIES.
C
C SYSTEM STORAGE UNITS USED BY THE PROGRAM:
C           1-9.  THE PROVENIENCE DESIGNATIONS AND OBSERVATIONS OF INTE
C                   FREQUENCY SERIES, STORED IN A FORMAT SPECIFIED BY THE
C                   (CONTROL RECORD 9), WITH AS MANY UNITS REQUIRED AS TH
C                   INTERMEDIATE FREQUENCY SERIES REQUESTED TO BE GENERAT
C           10.  THE PROVENIENCE DESIGNATIONS AND OBSERVATIONS OF THE
C                   SERIES, STORED IN A FORMAT SPECIFIED BY THE USER (CO
C                   8).
C           11.  THE PROVENIENCE DESIGNATIONS AND OBSERVATIONS OF ALL
C                   SERIES REQUESTED TO BE SAVED (CONTROL RECORD 7), CON
C                   ONE FILE WITH A FORMAT SPECIFIED BY THE USER (CONTRO
C                   ONLY THOSE SMOOTHED SERIES REQUESTED TO BE STORED IN
C                   RECORD 7 WILL BE STORED.
C           12.  NOT ASSIGNED BY THE PROGRAM.  CAN BE USED TO STORE T
C                   FILTER.
```

```
C         RECORD 10.  THE REAL-NUMBER FORMAT IN WHICH THE PROVENIENCE DESIGNATIONS
C             AND OBSERVATIONS OF EACH SMOOTHED SERIES SHOULD BE STORED
C             ON UNIT 11.  E.G.,(1X,F5.0,1X,F5.3)
C
C
C ::::::::::::::::::::::::::::::::::::::::::::::::::::::::::::::::::::::::::::::::::::
C
C
      DIMENSION FMT1(30),FMT2(30),FMT3(30),ISMOOT(10),PROV(700),
     1RESIS(700),INT(10),SMOOTH(10,700),W(10,40),XHF(700),RLOW(700),
     2XINT(10,700),XINTM2(700),XINTM(10,700),FMT4(30),FMT5(30),FMT6(30)
C READ FORMAT OF PROVENIENCE DESIGNATORS, OBSERVATIONS IN SPACE SERIES
      READ(15,1)(FMT1(I),I=1,30)
    1 FORMAT(30A4)
C READ NUMBER OF FILTERS
      READ(15,2)ICON
    2 FORMAT(1X,I2)
C READ FORMAT OF FILTERING INTERVALS
      READ(15,3)(FMT2(I),I=1,30)
    3 FORMAT(30A4)
C READ FILTERING INTERVALS
      READ(15,FMT2)(INT(I),I=1,ICON)
C READ THE  NUMBER OF SMOOTHED SERIES TO SAVE
      READ(15,33)ISCON
   33 FORMAT(1X,I2)
C READ FORMAT OF THE DESIGNATIONS OF THE SMOOTHED SERIES TO SAVE
      READ(15,34)(FMT3(I),I=1,30)
   34 FORMAT(30A4)
C READ THE DESIGNATIONS OF THE SMOOTHED SERIES TO SAVE
      READ(15,FMT3)(ISMOOT(I),I=1,ISCON)
C READ FORMAT IN WHICH PROVENIENCE DESIGNATIONS AND OBSERVATIONS OF
C HIGH FREQUENCY SERIES SHOULD BE STORED
      READ(15,35)(FMT4(I),I=1,30)
   35 FORMAT(30A4)
C READ FORMAT IN WHICH PROVENIENCE DESIGNATIONS AND OBSERVATIONS OF
C INTERMEDIATE FREQUENCY SERIES SHOULD BE STORED
      READ(15,36)(FMT5(I),I=1,30)
   36 FORMAT(30A4)
C READ FORMAT IN WHICH PROVENIENCE DESIGNATIONS AND OBSERVATIONS OF
C SMOOTHED SERIES SHOULD BE STORED
      READ(15,37)(FMT6(I),I=1,30)
   37 FORMAT(30A4)
C READ IN DATA
      KOUNT=0
      DO 100 I=1,2000
      READ(13,FMT1,END=200) PROV(I),RESIS(I)
      KOUNT=KOUNT+1
  100 CONTINUE
  200 N=KOUNT
C CALCULATE VARIANCE IN THE WHOLE SERIES
      CALL MUSIG(S,SS,RESIS,1,N,RBAR,VAR1,SDEV)
      WRITE(14,7) RBAR,VAR1,SDEV
    7 FORMAT('1',1X,'STATISTICS FOR WHOLE SERIES',///,1X,'MEAN = ',
     1F10.2,10X,'VAR = ',F20.2,10X,'SDEV = ',F20.2,/////)
```

```
C CALCULATE NORMAL FILTER WEIGHTS
      WRITE(14,46)
  46  FORMAT(1X,'STATISTICS FOR LOW FREQUENCY SERIES')
      DO 250 I=1,ICON
      WRITE(14,4) I,INT(I)
   4  FORMAT(///,1X,'FILTERING INTERVAL ',I1,' = ',I2)
      IHALF=INT(I)/2+1
      SIGMA=(INT(I)-1.)/6.
      WRITE(14,5) I
   5  FORMAT(/,1X,'FILTER WEIGHTS FOR FILTERING INTERVAL ',I2)
      IINTRV=INT(I)
      DO 300 J=1,IINTRV
      K=J-IHALF
      W(I,J)=1./(SQRT(2*3.14159*(SIGMA**2.)))*((2.7183)**(-1.*K*K/(2.*(S
     1IGMA**2.))))
      WRITE(14,6) W(I,J)
   6  FORMAT(1X,F10.5)
 300  CONTINUE
C SMOOTH THE RESISTIVITY SERIES TO OBTAIN LOW-FREQUENCY SERIES
      IBEGIN=IHALF
      IEND=N-IBEGIN+1
      DO 500 L=IBEGIN,IEND
      SMOOTH(I,L)=0..
      DO 600 MM=1,IINTRV
      K=MM-IHALF
      SMOOTH(I,L)=SMOOTH(I,L)+W(I,MM)*RESIS(L+K)
 600  CONTINUE
      RLOW(L)=SMOOTH(I,L)
 500  CONTINUE
C CALCULATE VARIANCE IN LOW FREQUENCY SERIES
      CALL MUSIG(S,SS,RLOW,IBEGIN,IEND,RBAR,VAR2,SDEV)
C CALCULATE PERCENTAGE OF TOTAL SERIES VARIANCE CONTRIBUTED BY
C LOW FREQUENCIES
      PERCNT=PRCNT(VAR1,VAR2)
      WRITE(14,38) RBAR,VAR2,SDEV,PERCNT
  38  FORMAT(/,1X,'MEAN = ',F10.2,10X,'VAR = ',F20.2,10X,
     1'SDEV = ',F20.2,10X,'PERCENT VAR = ',F7.3)
 250  CONTINUE
C OBTAIN HIGH FREQUENCIES USING FIRST SMOOTHING INTERVAL ONLY
      IBEGIN=INT(1)/2+1
      IEND=N-IBEGIN+1
      DO 900 M=IBEGIN,IEND
      XHF(M)=RESIS(M)-SMOOTH(1,M)
      WRITE(10,FMT4) PROV(M),XHF(M)
 900  CONTINUE
C CALCULATE VARIANCE OF HIGH FREQUENCIES
      CALL MUSIG(S,SS,XHF,IBEGIN,IEND,RBAR,VAR3,SDEV)
C CALCULATE PERCENTAGE OF TOTAL SERIES VARIANCE CONTRIBUTED BY
C HIGH FREQUENCIES
      PERCNT=PRCNT(VAR1,VAR3)
      WRITE(14,49) RBAR,VAR3,SDEV,PERCNT
  49  FORMAT(/////,1X,'STATISTICS FOR HIGH FREQUENCY SERIES',///,
     11X,'MEAN = ',F10.2,10X,'VAR = ',F20.2,10X,
     2'SDEV = ',F20.2,10X,'PERCENT VAR = ',F7.3)
```

```
C OBTAIN INTERMEDIATE FREQUENCY BANDS
      ICON2=ICON-1
      DO 1000 I=1,ICON2
      IBEGIN=INT(I+1)/2+1
      IEND=N-IBEGIN+1
      DO 1100 J=IBEGIN,IEND
      XINTM(I,J)=SMOOTH(I,J)-SMOOTH(I+1,J)
 1100 CONTINUE
 1000 CONTINUE
C CALCULATE VARIANCES OF INTERMEDIATE FREQUNCIES
      WRITE(14,45)
   45 FORMAT(/////,1X,'STATISTICS FOR INTERMEDIATE FREQUENCIES')
      DO 1200 I=1,ICON2
      IBEGIN=INT(I+1)/2+1
      IEND=N-IBEGIN+1
      DO 1300 J=IBEGIN,IEND
      XINTM2(J)=XINTM(I,J)
 1300 CONTINUE
      CALL MUSIG(S,SS,XINTM2,IBEGIN,IEND,RBAR,VAR4,SDEV)
C CALCULATE PERCENTAGE OF TOTAL SERIES VARIANCE CONTRIBUTED BY
C INTERMEDIATE FREQUENCIES
      PERCNT=PRCNT(VAR1,VAR4)
      L=I+1
      WRITE(14,29) I,L
   29 FORMAT(///,1X,'FREQUENCY BAND = ',I2,'-',I2)
      WRITE(14,26) RBAR,VAR4,SDEV,PERCNT
   26 FORMAT(/1X,'MEAN = ',F20.2,10X,'VAR = ',F20.2,10X,
     1'SDEV = ',F20.2,10X,'PERCENT VAR = ',F7.3)
 1200 CONTINUE
C WRITE INTERMEDIATE FREQUENCIES TO FILES
      DO 2500 I=1,ICON2
      IBEGIN=INT(I+1)/2+1
      IEND=N-IBEGIN+1
      DO 1400 J=IBEGIN,IEND
      WRITE(I,FMT5) PROV(J),XINTM(I,J)
 1400 CONTINUE
 2500 CONTINUE
C WRITE SMOOTHED SERIES OF CHOICE TO FILES
      DO 1600 I=1,ISCON
      L=ISMOOT(ISCON)
      IBEGIN=INT(L)/2+1
      IEND=N-IBEGIN+1
      DO 1700 J=IBEGIN,IEND
      WRITE(11,FMT6) PROV(J),SMOOTH(L,J)
 1700 CONTINUE
 1600 CONTINUE
      STOP
      END
C
C::::::::::::::::::::::::::::::::::::::::::::::::::::::::::::::::::::::::::::
C
      SUBROUTINE MUSIG(S,SS,RESIS,IBEGIN,IEND,RBAR,VAR,SDEV)
      DIMENSION RESIS(700)
      S=0.
```

```
      SS=0.
      N=IEND-IBEGIN+1.
      DO 100 I=IBEGIN,IEND
      S=S+RESIS(I)
      SS=SS+RESIS(I)*RESIS(I)
  100 CONTINUE
      RBAR=S/N
      VAR=(SS-(S*S)/N)/(N-1)
      SDEV=SQRT((SS-(S*S)/N)/(N-1))
      RETURN
      END
C
C*******************************************************************************
C
      FUNCTION PRCNT(TOTVAR,PARVAR)
      PRCNT=(PARVAR/TOTVAR)*100.
      RETURN
      END
```

```
C ::::::::::::::::::::::::::::::::::::::::::::::::::::::::::::::::::::::::::::::::::::::::::::::::::::::::
C
C                              PROGRAM LOCAL
C
C ::::::::::::::::::::::::::::::::::::::::::::::::::::::::::::::::::::::::::::::::::::::::::::::::::::::::
C
C
C THIS PROGRAM FINDS THE MEAN AND STANDARD DEVIATION OF OBSERVATIONS WITHIN
C A GIVEN NEIGHBORHOOD OF EACH OBSERVATION COMPOSING A RESISTIVITY SPACE SERIES.
C THE WIDTH OF THE INTERVAL OVER WHICH LOCAL MEANS AND LOCAL STANDARD
C DEVIATIONS ARE ASSESSED CAN BE SPECIFIED BY THE USER.  THE LOCAL MEANS AND
C LOCAL STANDARD DEVIATIONS THEN ARE STORED ON A SYSTEM STORAGE UNIT.  THE
C ORIGINAL SPACE SERIES CAN BE UP TO 700 OBSERVATIONS LONG.
C
C DEFINITION OF VARIABLES, ARRAYS, AND LIMITATIONS WITHIN THE PROGRAM:
C      RESIS(700). THE SPACE SERIES TO BE FILTERED, WITH UP TO 700 OBSERVATIONS.
C      PROV(700).  THE PROVENIENCE DESIGNATIONS FOR THE OBSERVATIONS IN THE
C          SPACE SERIES.
C      MEAN(700).  THE DERIVED SERIES OF LOCAL MEANS OF THE OBSERVATIONS IN
C          THE SPACE SERIES.
C      INT.  THE INTERVAL OR NUMBER OF OBSERVATIONS OVER WHICH THE MEAN AND
C          STANDARD DEVIATION LOCAL TO A GIVEN OBSERVATION ARE EVALUATED.
C          THE GIVEN OBSERVATION LIES AT THE CENTER OF THE INTERVAL.  THE
C          INTERVAL MUST BE AN ODD INTEGER.
C
C SYSTEM STORAGE UNITS USED BY THE PROGRAM:
C       1.  THE PROVENIENCE DESIGNATIONS AND OBSERVATIONS OF THE ORIGINAL SPACE
C           SERIES, STORED IN A FORMAT SPECIFIED BY THE USER (CONTROL RECORD 1)
C      14.  THE PROVENIENCE DESIGNATIONS, LOCAL MEANS, AND LOCAL STANDARD
C           DEVIATIONS FOR THE OBSERVATIONS IN THE ORIGINAL SPACE SERIES,
C           STORED IN A FORMAT SPECIFIED BY THE USER (CONTROL RECORD 4).
C      15.  CONTROL RECORDS HAVING VARIABLE, USER-SPECIFIED PARAMETERS REQUIRED
C           BY THE PROGRAM.
C
C CONTROL RECORDS TO OCCUR IN STORAGE UNIT 15:
C      RECORD 1.  THE REAL-NUMBER FORMAT OF THE PROVENIENCE DESIGNATIONS
C          AND OBSERVATIONS WITHIN THE SPACE SERIES STORED IN UNIT 1.
C          E.G., (1X,F5.0,1X,F10.3)
C      RECORD 2.  THE INTEGER FORMAT OF THE WIDTH OF THE INTERVAL OVER WHICH
C          THE LOCAL MEAN AND LOCAL STANDARD DEVIATION ARE ASSESSED.
C          E.G., (1X,I2)
C      RECORD 3.  THE WIDTH OF THE INTERVAL OVER WHICH THE LOCAL MEAN AND
C          LOCAL STANDARD DEVIATION ARE ASSESSED.  THE WIDTH MUST BE AN
C          ODD INTEGER.  E.G., B15
C      RECORD 4.  THE REAL-NUMBER FORMAT IN WHICH THE PROVENIENCE DESIGNATIONS,
C          LOCAL MEANS, AND LOCAL STANDARD DEVIATIONS OF THE OBSERVATIONS IN
C          THE SPACE SERIES SHOULD BE WRITTEN ON SYSTEM UNIT 14.
C
C
C ::::::::::::::::::::::::::::::::::::::::::::::::::::::::::::::::::::::::::::::::::::::::::::::::::::::::
      DIMENSION RESIS(700),PROV(700),SDEV(700),FMT1(30),FMT2(30),
     1RMEAN(700),FMT3(30)
C READ FORMAT OF PROV,RESIS
      READ(15,1)(FMT1(I),I=1,30)
    1 FORMAT(30A4)
```

```
C READ FORMAT OF INTERVAL OVER WHICH LOCAL MEAN AND LOCAL STANDARD DEVIATION
C ARE TO BE EVALUATED
      READ(15,2)(FMT2(I),I=1,30)
    2 FORMAT(30A4)
C READ INTERVAL OVER WHICH LOCAL MEAN AND LOCAL STANDARD DEVIATION ARE TO
C BE EVALUATED
      READ(15,FMT2) INT
C READ FORMAT IN WHICH PROVENIENCES, LOCAL MEANS, AND LOCAL STANDARD DEVIATIONS
C SHOULD BE STORED
      READ(15,3)(FMT3(I),I=1,30)
    3 FORMAT(30A4)
C READ IN DATA
      KOUNT=0
      DO 100 I=1,2000
      READ(1,FMT1,END=200) PROV(I),RESIS(I)
      KOUNT=KOUNT+1
  100 CONTINUE
  200 N=KOUNT
C CALCULATE LOCAL STANDARD DEVIATIONS
      IBEGIN=INT/2+1
      IEND=N-IBEGIN+1
      DO 300 I=IBEGIN,IEND
      CALL SIGMA(S,SS,I,RESIS,INT,SDEV,RMEAN)
      WRITE(14,FMT3) PROV(I),RMEAN(I),SDEV(I)
  300 CONTINUE
      STOP
      END
C
C
C :::::::::::::::::::::::::::::::::::::::::::::::::::::::::::::::::::::::::::::::::::::::::::::::::::::::
C
C
      SUBROUTINE SIGMA(S,SS,I,RESIS,INT,SDEV,RMEAN)
      DIMENSION RESIS(700),SDEV(700),RMEAN(I)
      S=0.
      SS=0.
      JBEGIN=I-INT/2
      JEND=I+INT/2
      DO 100 J=JBEGIN,JEND
      S=S+RESIS(J)
      SS=SS+RESIS(J)*RESIS(J)
  100 CONTINUE
      RMEAN(I)=S/INT
      SDEV(I)=SQRT((SS-(S*S)/INT)/(INT-1))
      RETURN
      END
```

APPENDIX II

Operational Procedures for Defining Depositional Sets within the Crane Assemblage

In Chapter 5, a methodology was summarized for defining overlapping, polythetic sets of archeological deposits. In this appendix, the details of the procedures which were used in analyzing the data set from the Crane site, as well as deviations made from the proposed methodology, will be discussed.

To begin the spatial analysis, a matrix of counts of items of each artifact type per 6 × 6-m grid unit at Crane was assembled. The list of artifact types included in the matrix is similar to that which is shown in Table 31 and which was used later in tabulating the artifact types present in each surface collection unit and in defining the activities reflected by them. A few exceptions occur, however. First, no artifact types that are diagnostic of time periods other than the Bedford phase (V117–V121) were included in the matrix. The procedures used here to define depositional sets assume the synchronic nature of all items. Most of the materials from the Crane site probably relate to the Bedford phase component, and it would have confused the analysis had items from other time periods been included in the study. Second, of the several classes of hammerstones shown in Table 34, only Classes V114–V116 were included in the matrix and in the spatial analysis. Class V113 was not used. It is a summation of all the hammerstones in Classes V114–V116, plus miscellaneous hammerstones that were not analyzed and could not be sorted into morphological Classes V114–116. Class V113 is largely a redundant variable with respect to Classes V114–116, and would have distorted the cluster analysis of the spatial study had it been used along with V114–V116. Third, the matrix includes several classes of artifacts (V63, V93, V96, V109–112) that are not found in Table 31 and that were determined after the spatial analysis had been completed to be badly mixed, including artifacts of a wide range of functions. These classes, although used in the spatial analysis, were dropped from all further analyses, including the tabulating of the artifact types present in each surface collection unit and the defining of the activities reflected by them. The morphologies and functions of these mixed classes have not been described, either. Finally, the matrix includes a set of artifact classes (V79, V87, V89, V107), which likewise were found after the spatial analysis had been completed to be mixed, but not as badly as the first set. These classes were subdivided, after the spatial analysis, into functionally homogeneous groups (V79a, V79b, V87a, V87b, V89a, V89b, V89c, V107a, V107b) and were used in tabulating the artifact types present in each surface collection unit and in defining the activities reflected by them. The matrix of counts of items of each artifact type per 6×6-m collection unit used in the spatial analysis is presented in Appendix III.

As a second step, the matrix of counts was reduced to a presence–absence matrix. Counts of more ubiquitous types, which occurred in every second or third grid cell or in higher densities, and which would likely have produced spurious associations in the analysis if reduced to a *simple* presence–absence state, were given a presence–absence form by dicotomizing at density thresholds higher than 1. Different thresholds were chosen for the different ubiquitous types, depending on the spatial patterning of their counts in grid cells. Thresholds were designed to isolate locations where there clustered a number of units having high densities for the type in question. It was hoped that these locations represented places where a single activity was *repeatedly* performed or where secondary refuse from a single activity was repeatedly deposited. Such locales should contain a larger percentage of the kinds of artifacts that tended to be used together or deposited together prehistorically than locales representing single depositional episodes. The effects of factors causing polythetic depositional set organization, such as differential rates of discard, should be less apparent in such locales, allowing better assessment of whether or not types tended to be used or deposited together. Alternatively, the high density collection units might represent locations where several different activities, all involving the type in question, were performed. Inclusion of locations of this nature in an analysis with the proposed design would not help, and might hinder the analysis. Such locations do not represent places having more complete depositional sets that would help in the process of finding associations between types.

Once all values of all variables had been given a presence–absence form, between-type average nearest neighbor distances were calculated according to the proposed algorithm. At this point in the analysis, operational difficulties were encountered. The procedures outlined in Chapter 5 suggest that hierarchical clustering algorithms allowing the definition of overlapping clusters should be used in reconstructing depositional sets from Euclidean-distance dissimilarity matrices. Currently, there exists no algorithm of this kind that also is efficient. The algorithm proposed by Cole and Wishart (1970) would have required many hours of computer time to accommodate the number of variables in this study and financially was impossible to use. Thus, the analysis had to be restricted to defining nonoverlapping clusters. The Johnson rank-order complete linkage hierarchical clustering routine (School of Business Administration, University of Michigan) was used for this purpose. This routine allowed the definition of clusters of types that were most closely associated with each other, but did not allow an examination of the multiple associations of multipurpose tool types that were deposited with more than one set of types. Tools and debris that were deposited with artifacts in more than one depositional set were clustered with only those artifact types with which they were *most highly associated*.

To judge the significance of the hierarchically nested clusters of artifact types defined by the Johnson clustering program, an absolute distance threshold of 16 m was applied. This threshold was chosen in light of the expected size of use-areas at Crane (8 m) and the extent to which refuse generated within use-areas most likely was scattered beyond the confines of use-areas by the occupants of Crane and by modern plowing (4 m or more on either side of use-areas), as discussed in Chapter 5.

Using the 16-m threshold distance, 30 sets of artifact types were defined that tend to occur in closer proximity to each other than to other artifact types (Table 29). The average distances between different types within the sets are shown in Table 30. In examining Table 30, it can be noted that in several sets, a few types have AVDIST values of greater than 16 m with respect to *some* of the other members of the set within which they are included. These artifact types represent those that were not able to join *any* cluster on a complete linkage basis at distances of less than or equal to 16 m. They were appended on an average linkage basis to the multitype clusters with which they showed greatest affinity.

The use of average linkage procedures in an after-the-fact manner to cluster the residual, unlinked artifact types should not be considered a forcing of the data into patterns that do not exist. Complete linkage procedures were used instead of average linkage procedures, initially, in an attempt to isolate the tightest groupings of mutually associated artifact types that exist in the data. These groups presumably include those pairs of artifact types for which *at most one* type was polythetically distributed with respect to the other. The necessity of using average linkage procedures to append some artifact types to the initially defined cluster nuclei possibly reflects the fact that for some pairs of artifact types, *both* types were distributed polythetically with respect to each other. The statistic, AVDIST, is not always capable of resolving depositional sets having this organization, and average linkage procedures were required to compensate as best as possible for this problem.

APPENDIX III

Additional Attributes of Retouched and Unretouched Chipped Stone Tools from the Crane Site

TABLE 1
Additional Attributes of Retouched and Unretouched Chipped Stone Tools from the Crane Site

Variable Number	Kind of Artifact	Overall Morphology	Maximal Dimensions	Breakage Pattern	Bibliographic References to Previous Descriptions
KNIVES					
V83	unifacial, straight-edged knives	amorphous flakes or rectangular to subrectangular flakes	1/2 - 3 3/4 x 1/2 - 2 1/2 x 1/8 - 1	edges of broken tools with unknown total morphology	none
V84	unifacial straight-edged knives	amorphous flakes or rectangular to subrectangular flakes	1/2 - 2 1/2 x 5/8 - 2 1/2 x 1/8 - 3/4	edges of broken tools with unknown total morphology	none
V85	unifacial straight-edged knives	amorphous flakes	1/2 - 1 1/2 x 5/8 - 2 1/2 x 1/8 - 3/4	edges of broken tools with unknown total morphology	none
V86	unifacial stright-edged knives	amorphous flakes	1/2 - 1 1/2 x 5/8 - 2 1/2 x 1/8 - 3/4	edges of broken tools with unknown total morphology	none

ID	Type	Description	Size	Notes	Reference
V87a	unifacial round-edged knives	amorphous flakes	? 1/2 – 2 1/2 x 1/8 – 5/8	edges of broken tools with unknown total morphology	none
V87b	unifacial round-edged knives	amorphous flakes	? 1/2 – 2 1/2 x 1/8 – 5/8	edges of broken tools with unknown total morphology	none
V88	unifacial concave-edged knives	large and small amorphous flakes for heavy-duty and light-duty work	1/2 – 2 1/4 x 1 – 2 3/4 x 1/8 – 3/4	–	none
V108	bifacial round-edged knives	D-shaped flakes with a "back" produced by one or two fracture planes	7/16 – 1 1/2 x 7/8 – 2 x 1/8 – 1/4	–	none
V11, V58	unretouched lamellar blades and their proximal ends	true prismatic blades identified by attributes of their platforms which allowed their removal from cores by indirect percussion techniques. Parallel scars of blades previously removed from the core were not required for admittance to this class (Bordes 1961:6). Natural, unretouched edges	–	sometimes broken at tips	Montet-White (1968:28, 30) Illustration

ENDSCRAPERS

ID	Type	Description	Size	Notes	Reference
V34	unifacial endscrapers	blunted point forms (Snyders/Norton, Belknap) forming haftable endscrapers	–	–	Montet-White (1963:33 A,F) Illustration

TABLE 1 (cont.)

V35	unifacial endscrapers	blunted point forms (Snyders) forming haftable endscrapers	–	–	none
V36a	unifacial endscrapers	lamellar blades and amorphous flakes	1 – 2 x 2 5/8 – 3 3/8 x 1/4 – 3/4	–	Montet-White (1963:49) Illustration
V36b	unifacial endscrapers	amorphous flakes	1 – 2 x 2 5/8 – 3 3/8 1/4 – 3/4	–	none
V37	unifacial endscrapers	amorphous flakes	1/8 – 3/8 thick tip angle: 45 – 60°	–	none
V38	unifacial endscraper	large amorphous flakes for a heavy-duty tool	1/4 – 9/16 thick tip angles: 80°	–	none
V39	unifacial endscrapers	large amorphous flakes	1 – 2 x 2 5/8 – 3 3/8 x 1/4 – 3/4; tip angles: 90°	–	none
V90	bifacial endscrapers	blunted point forms (Snyders) forming haftable endscrapers with sides expanding at 65°	–	–	none
V103	bifacial endscrapers	similar to the contracting stems of Belknap points: sides expand at 68 – 82°	1/2 – 7/8 x 1/4 – 13/16 x 1/4 – 3/8	–	Winters (1967:27)

CONVEX-EDGED SCRAPERS, CELTS

V71, V72	turtleback scrapers	hemisphere, with or without cortex on the convex side	1 1/2 – 2 1/2 in diameter, 1/2 – 3/4 high	whole or broken	Montet-White (1968:90), Crabtee (1968; his specimens have steeper edge angles)
V81	unifacial round-edged scrapers	amorphous flakes	? x 1/2 – 2 2/2 x 1/8 – 5/8	edges of broken tools with unknown total morphology	none
V82	unifacial round-edged scrapers	amorphous flakes	?x 1/2 – 2 1/2 x 1/8 – 5/8	edges of broken tools with unknown total morphology	none
V91	choppers	bi-convex cores ("ovate preforms") with convex edges; direct percussion, alternately flaked edges are sinuous when viewed edge forward. Similar in outline to "ovate preforms" described by Montet-White but cruder in flaking technique	2 1/8 – 3 1/4 x 3 – 3 5/8 x 3/8 – 1 1/2	–	Montet-White (1968:30-47)
V92	choppers	D-shaped primary flakes with a sinuous working edge when viewed edge forward, made by direct percussion, alternate flaking; have a "back" made by a single fracture plane opposite the working edge	1 1/2 – 2 1/8 x 1 1/2 – 2 1/2 x 1/2 – 1	–	none

TABLE 1 (cont.)

V95	chisels	forms approximately the shape of a finger, bi-convex to angular-convex in cross section; straight, parallel sides or concentric, slightly curved sides, round tips in plan view; some wear on sides, most on tips	1 1/8 x 3 7/8 x 3/4	whole chisels or broken tips	none
V97	bifacial round-edged scrapers	elaborate, hafted celts, bi-lobate and tear-drop shaped in plan view	2 3/4 - 3 x 2 x 11/16 - 7/8	—	Montet-White (1968:84, #1, #2) Illustrated
V98, V99	celts	elaborate; parallel-sided with round ends in plan view	1 1/2 - 2 1/6 x 2 1/8 - 4 1/2 x 1/2 - 1 1/8	—	Montet-White (1963:36, BTI; 1968:86, #3) Illustrated
V101	celts	elaborate, expanding sides	1 1/4 - 2 1/8 x 2 3/8 - 3 x 1/2 - 5/8	—	Montet-White (1968:34, #4) Illustration
V107a	bifacial round-edged scrapers	D-shaped flakes with a "back" made by one or two fracture planes.	13/16 - 1 x 1 - 1 3/4 x 1/8 - 5/8	—	none
V40, V41	cache blades	broad, ovate to subtriangular bifaces, finely thinned and retouched	—	—	Montet-White (1968:45)
V42	cache blades	elongated blades resembling the upper half of a laurel leaf. Symmetrical sides	1 1/4 - 1 3/4 x 2 1/2 - 3 x 3/8 - 5/8	tips of 2 items in the class are broken off	none

V43a, V43b	cache blades	elongated blades with assymetrical sides, one only slightly convex, the other much more convex, giving a nearly D-shape form in plan view. Both edges show wear.	—	—	none

STRAIGHT-EDGED SCRAPERS

V73	unifacial, straight-edged scrapers	amorphous flakes	1/2 – 1 1/2 x 5/8 – 2 1/2 x 1/8 – 3/4	edges of broken tools with unknown total morphology	none
V74	unifacial straight-edged scrapers	amorphous flakes	1/2 – 1 1/2 x 5/8 – 2 1/2 x 1/8 – 3/4	edges of broken tools with unkown total morphology	none
V75	unifacial straight-edged scrapers	amorphous flakes	1/2 – 1 1/2 x 5/8 – 2 1/2 x 1/8 – 3/4	edges of broken tools with unknown total morphology	none
V76	unifacial straight-edged scrapers	amorphous flakes	1/2 – 1 1/2 x 5/8 – 2 1/2 x 1/8 – 3/4	edges of broken tools with unknown total morphology	none
V77a	unifacial straight-edged scrapers	amorphous flakes or rectangular to subrectangular flakes	1/2 – 2 x 5/8 – 2 1/2 x 1/8 – 3/4	edges of broken tools with unknown total morphology	none

TABLE 1 (cont.)

V77b	unifacial straight-edged scrapers	amorphous flakes or rectangular to subrectangular flakes	1/2 - 2 x 5/8 - 2 1/2 x 1/8 - 3/4	edges of broken tools with unknown total morphology	none
V78	unifacial straight-edged scrapers	amorphous flakes	1/2 - 1 1/2 x 5/8 - 2 1/2 x 1/8 - 3/4	edges of broken tools with unknown total morphology	none
V79a	unifacial straight-edged scapers	large, amorphous and rectangular primary flakes for heavy-duty work	–	–	none
V79b	unifacial straight-edged scrapers	large amorphous and rectangular primary flakes for heavy-duty work.	1 - 2 1/2 x 2 1/4 - 3 x 3/8 - 1	–	none
V80	unifacial straight-edged scrapers	large, amorphous, primary flakes for heavy-duty tasks	2 1/4 - 3 x 1 - 2 1/2 x 3/8 - 1	–	none
V100	rectangular celts	rectangular and subrectangular bifaces with truely parallel sides	1 - 1 1/4 x 1 - 1 7/8 x 5/16 - 12	whole and broken	none
V102	specialized planning tool	biting edge is thick and steep angled to the sides and thinner with a more acute angle in the center, where a "channel-like flake" has been removed. Outer edges are pressure-retouched, with wear on the retouched side; central portion of the edge is not retouched, has thinning wear on the side opposite from which the "channel flake" was removed		–	none

666

V104	bifacial straight-edged scraper	rectangular flakes with a "back" provided by a single fracture plane opposite the working edge	2 3/4 - 2 1/4 x 2 - 3 3/4 x 1/2 - 7/8	–	none
V105	bifacial straight-edged scrapers	rectangular flakes with a "back" provided by a single fracture plane opposite the working edge	1 1/8 - 1 3/4 x 1 5/8 - 2 7/8 x 3/16 - 3/4	–	none
V106	bifacial straight-edged scrapers	rectangular flakes with a "back" provided by a single fracture plane opposite the working edge	1 1/8 - 1 3/4 x 1 5/8 - 2 7/8 x 3/16 - 3/4	–	none
V107b	bifacial straight-edged scrapers	rectangular flakes with a "back" provided by a single fracture plane opposite the working edge	1 1/8 - 1 3/4 1 5/8 - 2 7/8 3/16 - 3/4	–	none

SLIGHTLY CONCAVE-EDGED SCRAPERS

V89a	unifacial concave-edged scrapers	large, primary flakes and small, amorphous flakes for heavy-duty and light duty work	1/2 - 2 1/4 x 1 - 2 3/4 x 1/8 - 3/4	–	none
V89b	unifacial concave-edged scrapers	amorphous flakes	1/2 - 1 1/2 x 1 - 2 x 1/8 - 1/2	edges of broken tools with unknown total morphology	none
V89c	unifacial concave-edged scrapers	Biconvex blades retouched on two opposing, concave edges	1 1/2 x 2 3/4 x 5/8 (approx)		

WEDGES

V94	wedges	tear-drop shaped in plan view and profile; biconvex, nearly circular in cross section	–	half of the items in this class have broken tips	none

References Cited

Bordes, Francois
    1961   <u>Typologie du Paleolithique Ancien et Moyen</u>.   Imprimeries Delmas, Bordeaux.

Crabtree, Don
    1968   Mesoamerican Polyhedral Cores and Prismatic Blades. <u>American Antiquity</u> 33:446-478.

Montet-White, Anta
    1963   Analytic Description of the Chipped Stone Industry from Snyders Site, Calhoun County, Illinois. In Miscellaneous Studies in Typology and Classification, edited by Anta Montet-White, Lewis R. Binford, and Mark L. Papworth. The University of Michigan Museum of Anthropology, <u>Anthropogical Papers</u> 19.

    1968   The Lithic Industries of the Illinois Valley in the Early and Middle Woodland Period. The University of Michigan Museum of Anthropology, <u>Anthropological Papers</u> 35.

Winters, Howard D.
    1967   An Archaeological Survey of the Wabash Valley in Illinois. Illinois State Museum, <u>Reports of Investigation</u> 10.

Subject Index

A

Activity areas
 definition, 195, 201
 data used to define at Crane, 191–195
Activity group
 definition, 442–449
 effects of specific activities, 491–494
 residues associated with, 443, 445–448
 spatial overlap of, 479
Activity set, 195, 199, 201
Aluminum
 linking humic materials and clays, 73
 percentage loss in decomposition, 124
Ammonium as fertilizer at Crane, 412, 414, 419–421
Anion
 availability in Crane soils, 428–430
 adsorption to clay and humus colloids, 140
Anion exchange, 136, 140
Anion fixation
 in clays, 143–145
 with surface vegetation, 145
Artifact typology
 approach used at Crane site, 212–218
 burnt bone, 242
 cache blades, 307
 celts, 306–307
  and axes, 246–249
  rectangular, 294
 chert hammers, 267
 chert nodules, 242–243
 chipped and groundstone tools, 214–218
 chipping debris, 243
 chisels, 267–268
 choppers, 305
 classification
  by edge angle, 216, 218, 273–277, 284–286
  by Knapping technique, 215, 217, 218
  by raw material, 215, 217, 218
  by shape of functional edge, 215, 216, 218
  by shaping, 215, 217, 218
  by size, 215, 217, 218
  by wear patterns, 215–218
 clay blobs, 241
 cores, 268–269
 debris, 214
 decortication material, 243
 denticulates, 263–264
 drills, 265–266
 figurines, 242
 flint ridge chert, 246
 galena, 245
 general approach, 212–218
 gouges, 267
 grinding stones, 252–258
 hammerstones, 249–252
 Havana pottery, 219–238
 hematite, 244–245
 hoes, 261–262
 hornstone, 245
 igneous rock, 243
 Jersey Bluff pottery, 239–241
 Kaolin chert, 246
 lamellar blades, 303–304
 mauls, 249

670   Subject Index

Artifact typology *(continued)*
  Missouri chert, 245–246
  notches, 262–263
  nuttingstones, 252–258
  Pike and Baehr pottery, 238
  point forms, 296–302
  pottery, 213–214
  red ochre, 243–244
  sandstone, 243
  sandstone abraders, 258–261
  saws, 266
  scrapers, 287–294, 305, 308
  sherdlets, 241
  spurs, 264–265
  tempering material, 242
  traditional approaches, 212–213
  unburnt bone, 242
  Whitehall–Weaver pottery, 238–239

B

Barnes Layer Method
  defining equations, 34–35
  design of layers at Crane, 357–359
  partitioning resistivity variability, 33, 563, 564–566
  for resistivity interpretation, 36
  tests of capability, 563–566
  versus Whole Volume method, 33, 564–565
Base nutrients
  concentration of, 136
  on clay and humus micelles p 134, 136
  exchange with hydrogen ions, 136
Block Survey
  coding data, 376
  orientation, location, 374–376
  overview, 373–376
  sampling design
    of resistivity measurements, 376
    of soil samples, 376
  time of, 374
Boundary conditions for resistivity surveying
  archeological, 624–626
  behavioral, 624–626
  chemical nature of activity residues, 625–626
  clay types in soil, 627
  cleaning of use-areas, 626
  conditions of survey, 629–632
  decompositional environment, 627–629
  economic considerations, 633–635
  equipment costs, 633
  kind and amount of activity and refuse deposition, 624–625
  labor costs, 633–635
  rainfall, 629–631
  rates of decomposition, 628–629
  soil pH, 627, 628–629
  soil profile development, 628
  soil texture, 626–627, 628
  time limitations p 632–633
  use-areas not characterized by facilities or artifacts, 626
  use area overlap, 624
  for use of the Barnes Layer Method, 635–636
  for use of spatial filtering techniques, 636–637
  vegetational cover, 628

C

Calcium
  added by fertilizing at Crane, 415–419, 421–422
  adsorption p 160
  anomalous availability, 520–523
  availability in Crane solids, 367–368
  availability in soils, 109, 415–419, 430
  chelation, 160
  concentration, 109
  cycle, 154, 159
  fixed in Humic structure, 130–131
  forms found in soils, 159
  interactions with other nutrients, 162–165
  interactions with phosphorus, 156, 159
  linking humic materials and clays, 73
  loss from soils, 159–160
  maintenance in Crane soils, 479, 483–484, 486
  percentage loss during decomposition, 124
  precipitation, 160
  ratio to magnesium, 415–419, 421–422
  ratios to other nutrients in Crane soils, 431, 434–435
  and soil aggregation, 78, 100
  sources of input to soils, 159–160
Capillary conductance, 66–68
  defining equations, 67
  effect on electrical conductivity, 67–68
  and resistivity, 98
Carbon
  as constituent of humus, 130
  ratio to nitrogen, 117
  ratio to sulfur, 152–153
Carbon-14, estimation of humus molecule stability, 128
Carbon exchange
  by clay micelles, 131, 137
  by humus micelles, 131, 141
Cation exchange capacity, 98, 138, 521–522, 553, 557
  differences in Crane soils, 424–428
Cation exchange capacity
  and organic matter content, 561

Cation fixation
  in clays, 143–145
  within surface vegetation, 145
Cations, proportions on clay colloids, 138
Chelation, 130, 131
  of calcium, 160
  of magnesium, 160
Chemical composition of raw materials, 454–466
  bark, 457, 460–466
  bone, 457, 460–466
  brains, 457, 460–465
  broths and drippings, 457, 460–466
  deer, 456, 460–466
  fruits, 456, 460–466
  greens, 456, 460–466
  meats, 456–457, 460–466
  mollusk shell, 457, 460–466
  nut meats and oil, 456, 460–466
  nut shells, 457, 460–466
  red ochre, 457
  roots and tubers, 456, 460–466
  seeds, 456, 460–466
  urine, 457, 460–466
  wood, unburnt and ash, 457, 460–466
Chemical composition of refuse materials
  cluster analysis, 495–496
  comparison to nutrient similarity in use-areas 507–514
Clay colloids/micelles
  adsorption of anions, 140–141
  adsorption of cations, 137
  absorptive capacity, 137, 140
  structure, 137, 140
Clays, fixation of hosphorus, 157
Colloidal particles affecting soil water conductivity, 92–94
Conductivity
  defining equations, 48, 79–83, 91, 94
  electrophoretic effect, 84
  function of strong electrolyte concentration and temperature, 84–91
  function of weak electrolyte concentration, 91–92
  relaxation effect, 84
  of soil water, 107, 554
    affected by kind and concentration of ions, 79, 80
    affected by nutrient availability, 557, 558
    affected by soil cation exchange capacity, 92, 557
    correlation with resistivity, 554
    function of soil colloidal particle concentration, 79, 92–94
    implications for resistivity surveying, 94–97
    and organic matter content, 561
Controlled surface survey
  at Crane, 191–195
  complementary use with resistivity survey, 19–20
Crane site
  artifact typology, 212–218
  assemblage description, 218
  block survey, 373–376
  comparison of physical soil alterations in use-areas, 541–547
  controlled surface survey, 191–195
  cropping practices, 408, 412–414
  cultivation of, 408–409, 412–414
  deforestation of, 408–411
  depositional and functional characteristics of use-areas, 308–341
  fertilizing practices, 412–414, 421
  multicomponency, 184–188
  overview, 183–191
  parent materials, overview, 387–388
  partioning into activity areas, 191–195
  physical soil alterations, 435–436
  placement in subsistence–settlement system, 188, 190–191
  point forms, 296–302
  relationships between use-areas and soil chemistry, 442–524
  soil
    anomalies within, 467–470
    overview, 387, 388–391
    profile development, 403–406
    temperature, 365
  supplementary soil surveys, 376–383
  surface collection, p 192–195
  Transect Survey, 353–373
  use-areas, comparison of soil chemistry, 441
Cropping practices at Crane, 408, 412–414
Cultivation
  of Crane site, 408–409, 412
  physical soil disturbances at Crane 435–436

D

Decomposition of Humus
  rates, 127
  stability and residence time, 127–129
Decomposition of organic matter
  amount of ions and radicals freed, 98
  amount of water-soluble organic compounds, 117–118
  and complexity of compound, 117
  documentation of rates, 119–127
  and environment, 98, 100
  and microbial action, 117, 118
  and nutrient ratios and amounts, 117
  and pH, 118

Subject Index

Decomposition of organic matter
  rates, 116–117, 119–127
  and soil aggregation, 78
  and soil megafauna, 118
  and soil moisture content, 118
  and soil temperature, 118–119
Deforestation
  of Crane site, 408–412
  physical soil disturbances, 435–436
Depositional area, definition, 196, 197
Depositional set
  definition, 196, 197, 198, 200, 655–657
  definition of by association analysis, 206–207, 655–657
  polythetic organization p 199–201, 205
Drainage conditions
  affecting soil moisture content, 62–66
  defining equations, 62–64
  saturated flow, 62–64
  unsaturated flow, 64–66

E

Electrolytes, strong, effect on soil water conductivity, 84–91
Electrolytes, weak, effect on soil water conductivity, 91–92
Energy potentials of soil water, defining equations, 52–54

F

Fertilizing practices at Crane, 412
Formation factor
  definition 48–49
  role in determining variation in resistivity, 49
  as a summary measure of soil structure, 50
  variation in, 49
Function of artifacts
  assignment for Crane specimens, 218–308
  bark-working implements, 285–286
  bone items, 242
  bone scrapers, 285–286
  description of classes, 286–295
  functionally specific chipped-stone tools, 261–269
  general methodology, 212–218
  hide grainers, 285–286
  igneous and sandstone items, 246–261
  lithic raw materials and debris, 242–246
  methods for defining knives, scrapers, celts, cache blades, and point forms, 302–308
  other chipped-stone tools, 269–270
  plant processing implements, 285–286
  pottery and clay items, 219, 236–242
  sorting by edge angle, 273–277
  sorting by retouch, 272–273
  testing of, 295–296

H

Humus
  adsorptive capacity of micelles, 69–70
  components of, 130
  decomposition of, 127, 131
  definition and adsorptive capacities, 142
  enrichments to soils, 143
  formation and anomalous nutrient availability, 421–422
  geometry of micelles, 69
  in cherozems, 69
  micelle flocculation, 78
  micelle ion exchange sites, 131, p 134, 141
  nutrients stored in, 127, 129, 130
  in podzolic soils, 69
  replacement of micelles by cycling, 168–169
  role in maintenance of soil alterations, 127–131
Hydrogen
  adsorption on humus micelles, 134
  in breakdown of parent material, 136
  constituent of humus, 130
Hydrous oxides
  definition, 141
  enrichments to soils, 143
  factors resposible for adsorptive and exchange powers, 141–142
  fixation of phosphorus, 157
  as ion exchange sites, 134, 141
  linking humic materials and clays, 73
Hysteresis
  definition, 61
  and distinguishability of soils, 62
  effect, 62
  effect of pore size distribution, 61–62

I

Interaction effects on soil alteration maintenance, 160–166
Ion concentration, 108, 553, 557
  augmenting of, 49
  change in, affecting ion replacement series 138
  and conductivity, 48, 79, 80, 84–91
  and nutrient availability, 560
  total versus exchangeable, 115
Ion exchange
  on humus micelles, 131, 134, 141
  by hydrous oxides, 131, 134, 141
Ion fixation
  contrast to absorption, 144–145
  rates, 144
Ionic conductance, defining equations, 85–90
Iron
  fixed in humic structures, 130
  linking humic materials and clays, 73
  percentage loss in decomposition, 124

Subject Index

Isomorphic substitution
  in clays, 137, 143
  fixation of nutrients, 143

K

Knives
  bifacially retouched, 282–284
  unifacially retouched, 277–279
  unretouched, 303–305

M

Magnesium
  added by fertilizing at Crane, 415–419, 421–422
  adsorption, 160
  anomalous availability at Crane, 520–523
  availability in general, 109
  availability in Crane soils, 367–368, 415–419, 430
  chelation, 160
  cycle, 154, 159
  fixation, 160
    in clays, 143
    in humic structure, 130
  forms found in soil, 159
  interactions, 162–165
  linking humic materials and clays, 73
  loss from soil, 159–160
  maintenance in Crane soils, 479, 484, 486
  percentage loss in decomposition of humus, 124
  precipitation, 160
  ratio to calcium, 415–419, 421–422
  ratios with other nutrients, 430–431, 434–435
  and soil aggregation, 78, 100
  sources of 159–160
Monothetic organization, definition, 198

N

Nitrogen
  anomalous availability, 520–523
  availability in Crane soils, 367–368
  concentration in archeological soils, 109
  constituent of humus, 130, 146, 147
  cycle, 146–150
  exchange, 147
  as fertilizer at Crane, 412, 414, 419
  fixation, 147
  forms in soil, 146
  input to soils and ecosystem, 149–150
  interaction with other nutrients, 162
  loss from soil, 149
  "luxury consumption" by plants, 145
  maintenance in Crane soils, 479, 481, 486
  nitrification, 148–149
  percentage loss in decomposition, 123–124

ratio to carbon, 117
release in humic decomposition, 127, 147
transformations in cycling, 146–147
Nutrient availability
  absolute changes in Crane soils, 414–423
  alteration by deforestation, 410–411
  altered by cultivation, 410–411
  anomalous, correlation with anomalous pH and organic matter content, 521
  calcium, 109, 415–419, 430
  correlation with resistivity, 560–561
  in Crane soils, 367–371
  differentiation of use-areas, 494–514, 515–524
  effects of red ochre on, 457, 491
  and ion concentration, 560
  magnesium, 109
  magnitudes, 423–430
    anomalous, 442, 515–516, 518–524
  maintenance of anomalous spectra in use-areas, 470–491
  nitrogen, in Crane soils, 367–368
  and organic matter content, 561
  phosporus, 157, 367–368, 430
  in pit fills and matrices, 532
  potassium, 109, 160, 430
  ratios, 423, 430–435
    anomalous, 442
  sodium, 430
  and soil conductivity, 557, 558
  sulfur, 367–368, 430
Nutrient cycling
  general model, 166–176
  overview, 131–136
  phenomena facilitating, 175
  replacement of humus micelles, 168–169
  replacement of nutrients, 168–169, 171
  role in maintenance of soil alterations, 131–136, 166–168

O

Organic matter content
  alterations in B-horizon, 535–541
  alterations in midden, 535–541
  and cation exchange capacities, 554, 561
  anomalous, 518–521
  and conductivity, 48, 561
  correlation with resistivity, 554, 556, 558, 560
  correlation with soil moisture content, 558, 560
  differences between use-areas, 541–542, 547
  effect on pore size distribution, 558
  and nutrient availabilities, 561
  of pits compared to matrices, 524–534
  and soil structure, 554, 561
  of transect samples from Crane, 367, 371

Organic matter decomposition, *see*
    Decomposition of organic matter
Oxygen, constituent of humus, 130

P

Parent materials
    lower member at Crane, 388, 396, 398
    overview, 387–388
    source of nutrients, 160
    spatial variation at Crane, 392–398
    spatial variations in upper member, 398–403
    upper member at Crane, 388, 396, 398–403
    effect on pore size distribution, 54
    effect on resistivity, 54
    effect on soil moisture content, 54
    to predict conductivity, 57
Phosphorus
    anomalous availability, 520–524
    availability in soils in general, 157
    availability in Crane soils, 367–368, 430
    concentrations, 109
    cycle, 154–159
        compared to other cycles, 154
    as fertilizer at Crane, 412, 413, 414
    fixation of, 157
        in clays, 143
    forms within soils, 154, 157
    interactions, 162–163
        with aluminum, 154, 156, 159
        with calcium, 156, 159
        with iron, 154, 156, 159
        with manganese, 154, 156
        with other nutrients, 163
    "luxury consumption" by plants, 145
    maintenance by red ochre, 491
    maintenance in Crane soils, 479, 481–482, 486
    percentage loss from humus in decomposition, 124
    relationship between form and soil pH, 154
Physical soil disturbances
    stratigraphy, 435–436
Polythetic organization
    definition, 198
    of depositional sets, 199–201, 205, 655–657
    measure reflecting spatial association of artifact types, 208
Pore-size distribution, 553, 558
    alteration by aggregation, 73–77, 98
    alterations in pit fills 524–534
    differences between use-areas at Crane, 546–547
Potassium
    abundance, 160
    adsorption, 160
    anomalous availability, 520–523
    availability, 109, 160, 430
        in Crane soils, 367–368
    cycle, 159–160
    as fertilizer at Crane, 412, 414, 419
    fixation, 160
        in clays, 143, 144
    form found in soils, 159
    interactions with other nutrients, 163
    loss of, 159–160
    "luxury consumption" by plants, 145
    maintenance in Crane soils, 479, 482–483, 486
    percentage loss in decomposition, 124
    ratio to ammonium, 419–420
    ratio to calcium, 144, 431, 434–435
    ratio to sodium, 144, 421
    ratios, 431, 434–435
    replacement, 144
    sources of, 159, 160

R

Resistivity contrast, 554
    effect on flow rates, 63, 64
    effect on hysteresis, 61–62
    granulation (aggregation) and organic matter, 58
    and porosity, 58
    relationship to soil moisture content, 54
    soil compaction, 58
    and soil moisture content, 58–59
    variation in, 58
Porosity, 553
    augmented by soil aggregation, 73–77, 98
    definition, 50
    differences between use-areas at Crane, 543
    effect on flow rates, 63
    between middens and B-horizons, 535–541, 556
    between pit fills and matrices, 3, 524–534, 556
    between use-areas at Crane, p 541, 542, 547
Resistivity survey designs, traditional, 2, 4–10
    use of Whole Volume method, 33
Running filter function
    for dissection of resistivity data, 38–43
    versus normal filter function, 40
    in place of time-series analysis, 38
    for removal of specific spatial variability, 38–43

S

Scrapers
    unifacially retouched, 279–282
    unretouched, 305–306, 308

Subject Index

Silica, percentage loss in decomposition, 124
Soil alterations on archeological sites, documentation of
  differentiation of activity areas, 112–114
  ion concentration, 109
  magnitude of soil anomalies and intensity of activity, 111–112
  organic matter content, 109, 115
  soil moisture equilibria, 115
  soil structure, 115
Sodium
  adsorption, 160
  anomalous availability, 520–523
  availability, 430
    in Crane soils, 367–368
  cycle, 154, 159
  forms found in soils, 159
  interactions with other nutrients, 162–165
  loss of, 159–160
  maintenance in Crane soils, 479, 484–485, 486
  ratio to potassium, 421
  ratios, 431, 434–435
  and soil aggregation, 78, 100
  sources of, 159–160
Soil analyses on transect soil samples, 365–371
  nutrient availability, 367–371
  organic matter content, 367, 371
  pore-size distribution, 366
  soil moisture content, 365–366
  soil pH, 367
  soil texture, 366
Soil anomalies
  within Crane site, 467–470
  definition, 467–470
  Soil moisture content, 108, p 556
  alteration in B-horizon, 535–541
  alterations in midden, 535–541
  altertions in pits, 524–534
  and conductivity, 48
  correlation with organic matter content, 558, 560
  correlation with resistivity, 554, 560, 561
  determined by infiltration effect, 77
Soil Moisture Content
  differences between use-areas at Crane, 542–547
  function of pore-size distribution, 54, 65
  hysteresis, 61–62
  influenced by humus micelles, 70
  relationship to soil structure, 51–55, 62
  of transect samples from Crane, 366
Soil organic matter
  alteration of texture, 108
  effects on rainfall infiltration, 77–78
  effects on soil aggregation, 71–77
  effects on soil chemistry, 78
  hygroscopic capactiy, 68–71, 79
  osmotic potential, 79
  soil colloid flocculation p 78, 100
Soil profile development
  at Crane p 403–406
  spatial variations, 403–406
  spatial variation compared to spatial variation in parent material, 406–407
Soil structure, 109, 553, 557
  alterations in B-horizon, 535–541
  alterations in midden, 535–541
Soil structure
  alterations in pit fills, 524–534
  and conductivity p 48
  determining potential energy relations, 52–55, 62
  effect on pore-size distribution, 55–60
  effect on rate of flow, 52, 62
  effect on resistivity, 54, 55–60
  effect on soil moisture content, 54, 55–60
  relationship to resistivity, 49
  relationship to soil moisture content, 51–55, 62
  variation in soils in general, 49
Soil surveys, supplementary
  attributes of pit fills and their matrices, 381–382
  chemical analyses, 380
  coding data, 381
  design, 377–380, 381
  natural soil variation, 382–383
  soil pore-size distribution, 377, 380
  soil texture, p 377, 380
Soil temperature
  on archeological sites, 108
  and conductivity, 48
  at Crane, 365
Soil texture
  and conductivity, 48
  effect on soil moisture content, 60
  spatial variation at Crane, 392–403
Soil–vegetational equilibrium
  nutrient concentrations, 169–171, 175
  nutrient ratios, 169–175
  parameters determining, 171–176
  rates of nutrient loss, 175
  time required to achieve, 169, 175
Spatial Analysis of the Crane assemblage, 655–657
Spatial filtering techniques
  definition and differentiation of use-areas, 563, 591–598
  Location and differentiation of use-areas at Crane, 598–621
  partitioning of archeological variation into components of differing scales, 589–591
  procedures to locate, characterize, and distinguish use-areas Crane transects, 593–598

Spatial filtering techniques *(continued)*
  removal of low-frequency nonarcheological, variation, 576–583
  removal of periodic and high-frequency, nonarcheological variation, 583, 587–589
  running filter functions, 38–43
  time series analysis, 31–38
Spectral analysis
  removal of periodic and high-frequency nonarcheological variation, 581, 583, 585–587
  use in designing filters, 41
Sulfur
  adsorption of, 152
  anomalous availability, 520–523
  availability, 430
    in Crane soils, 367–368
  cycle, 150–153
  forms found in soil, 150–151
  in humus, 130, 151
  interaction, 162
  loss of, 151, 152, 153
  maintenance by red ochre, 491
  maintenance in Crane soils, 479, 485, 486
  oxidation of, 152
  ratios, 430–431, 434–435
  reduction of, 153
  sources of, 150, 151, 153
  transformations in cycling, 150–151, 152
Survey efficiency, 632–633

T

Time series analysis to dissect resistivity data, 31, 37–38
Transect survey
  coding data, 371
  depth, 356–359
  orientation of transects, 364
  overview, 353–356
  sample design of resistivity measurements, 359–361, 364–365
  sampling design of soil samples, 361–364
  time of, 354

Use-areas
  comparison of nutrient similarity to chemical composition of refuse materials, 507–514
  comparison of physical alterations, 541–547
  definition, 3, 13–14, 32, 197
    by controlled surface surveys, 19
    by crossplots of anomalous soil resistivity, 621
    by nearest-neighbor technique, 204, 206–212
    by pit absence, 516–518
    by pit clustering, 516–518
    by spatial filtering techniques, 503, 591–598
  depositional and functional characteristics, 308–341
  differentiation
    by anomalous nutrient availability spectra, 494–514
    by magnitudes of nutrient anomalies 167, 442, 515–524
    by mean soil resistivity, 2, 566–569
    by nurient spectra versus magnitudes, 442
    by resistivity variance, 2–3, 571–576
    by spectra of nutrient anomalies, 442, 514–515
  distinguished by ion proportions, 94
  location and differentiation by spatial filtering techniques, 42
  maintenance of anomalous nutrient availability spectra within, 470–491
  methods to define, 201–212
    function of all artifact types known, 202–203
    functions of some artifact types known, 204–212
  similarity in soil nutrients, 496–507
  as unit of analysis
    operational reasons, 14–19
    theoretical reasons, 10–74

V

Vegetation as local, stabile nutrient pool, 145

ERRATUM

The following two frames reproduced poorly in the microfiche enclosed in this volume. They are reproduced here for clarity.

Frame 180

```
2671.  GBB 1.19 6.8    398.    608.    8100.    24.    298.  112.0  999.9   20.0  999.9    2.0
2682.  GCB 1.22 6.7    650.    584.    7700.    20.    282.   56.0  999.9   26.0  999.9    1.8
2686.  GDB 1.22 7.0    672.    486.    7400.    22.    278.   41.2    0.     4.0    5.2    1.4
2695.  GEB 1.23 6.8    263.    850.    5900.    29.    260.   65.2    0.     8.0    0.     1.8
2720.  GGB 1.14 7.0    114.   1142.    6500.    20.    238.   42.0  999.9    0.   999.9    2.0
3402.  3HB 1.18 6.1    102.    136.    2880.    26.    175.   74.4    9.6    4.0    0.     1.1
3416.  GKB 1.12 6.1     57.    170.    2760.    26.    128.   63.0    0.    22.0    5.2     .7
3427.  GLB 1.20 6.3     72.    156.     420.  99999.    191.   54.6    0.     8.0  999.9    1.0
3441.  GMB 1.19 5.8     81.    163.    3000.    29.    156.   46.8    0.     2.0    0.      .9
3448.  GNB 1.20 6.0    102.    194.    3600.    17.    147.   67.4    0.    22.0   15.6     .8
3449.  GOB 1.22 5.8    114.    192.    3300.  99999.    166.   56.7    0.     8.0  999.9  999.9
3457.  GPB 1.25 6.0    134.    216.    4100.    24.    178.   30.6    0.     8.0  119.7     .9
3461.  GQB 1.17 6.1    108.    231.    3300.  99999.    147.   56.7    0.     8.0  999.9     .7
3469.  GRB 1.17 6.2     84.    194.    3300.    24.    150.  999.9    0.     8.0  999.9    1.1
3474.  GSB 1.15 6.0     48.    190.    3300.    20.    104.   42.0  999.9   18.0  999.9    1.2
3483.  GUB 9.99 9.9  99999.  99999.  99999.  99999.   99999.   999.9  999.9  999.9 999.9   999.9
3485.  GVB 1.22 5.9     44.    175.    3400.    18.    128.   67.4    0.     8.0    0.      .8
3489.  GWB 1.17 5.8     60.    182.    3000.    18.    138.   50.6     .9   12.0  999.9     .9
3497.  GXB 1.16 6.0     44.    218.    3600.    20.    140.   46.0  999.9    6.0  999.9    1.1
3500.  GYB 1.14 5.9     42.      0.    3200.    18.    128.   52.0  999.9   12.0  999.9    1.2
3504.  GZB 1.16 6.1     44.    218.    3700.    20.    126.   42.0  999.9   30.0  999.9    1.3
3507.  HBB 1.14 6.2     65.    194.    3500.    17.    141.   35.8    0.    12.0  999.9     .7
3510.  HCB 1.16 6.1     63.    194.    3600.    18.    144.   54.6    0.   999.9  999.9     .7
3515.  HDB 1.22 6.2     78.    199.    3700.    18.    125.  999.9    0.    10.0    1.0    1.4
3520.  HEB 1.15 6.0     66.    268.    3800.    20.    134.   42.0  999.9   12.0  999.9    1.3
3524.  HFB 1.22 5.9    180.    238.    4800.    26.    147.   54.6    0.    12.0    0.      .9
3542.  HHB 1.18 5.9    231.    216.    3840.    29.    163.   37.5    0.     0.   999.9     .9
3551.  HIB 1.18 6.1    110.    202.    4800.    16.    144.   52.0  999.9   14.0  999.9    1.4
3556.  HLB 1.18 6.2    110.    280.    3200.    20.    148.   82.0  999.9   14.0  999.9    1.5
3564.  HMB 1.23 6.0     75.    306.    3840.    29.    147.   74.4    0.    28.0    2.1     .8
3575.  HNB 1.22 5.8     54.    400.    3700.    24.    148.   16.0  999.9   26.0  999.9    1.2
3603.  HPB 1.14 5.5     40.    522.    3400.    24.    140.   28.0  999.9  999.9  999.9    1.2
3613.  HRB 1.22 5.6     24.    292.    3200.    26.    131.   32.3    0.    12.0  999.9     .9
3616.  HSB 1.24 5.7     27.    304.    3000.    24.    119.  999.9    0.     8.0   10.8    1.3
3620.  HTB 1.26 5.6     24.    241.    3300.    29.    122.   30.6  999.9    8.0  999.9     .8
3625.  HUB 1.12 5.7     22.    202.    3000.    20.    118.   34.0  999.9   22.0  999.9    1.3
3633.  HVB 1.16 5.6     36.    182.    2800.    22.    135.   34.0    0.    12.0   15.6     .8
3636.  HWB 1.18 5.5     51.    173.    2900.    24.    150.   27.3    0.     8.0  999.9     .9
3643.  HXB 1.14 5.8     51.    173.    3000.    17.    103.   60.9    0.    20.0   13.2    1.1
3644.  HYB 1.18 5.8     69.    182.    3300.    18.    100.   35.8   19.3  999.9    7.4    1.2
3649.  IAB 1.12 5.8     47.    163.    3000.    20.     97.   34.0    2.0   12.0  999.9    1.1
4201.  IBB 1.16 6.3     96.    182.    2360.    22.    210.   63.0    0.    12.0    0.      .9
4205.  ICB 1.13 6.2     66.    187.    3400.    17.    188.   63.0    0.     4.0    0.      .8
4210.  IDB 1.17 5.9    111.    226.    4000.    22.    135.   30.6    0.    14.0    1.0     .8
4215.  IEB 1.26 6.1     48.    199.    2900.    28.    197.   74.4    0.     2.0  999.9     .7
4223.  IFB 1.15 5.9     75.    185.    3600.    15.    135.   25.6    0.     4.0  999.9     .9
4233.  IHB 9.99 9.9  99999.  99999.  99999.  99999.   99999.  999.9  999.9  999.9  999.9  999.9
4244.  IJB 1.15 6.0     74.      0.    3200.    16.    140.   44.0  999.9   18.0  999.9    1.1
4247.  IKB 1.15 6.0     90.    219.    3800.    20.    197.   28.9    0.    16.0  999.9     .9
4254.  ILB 1.17 6.4    100.    292.    4000.    18.    160.   32.0  999.9   14.0  999.9    1.2
4259.  IMB 1.18 6.3     88.    219.    4000.    20.    168.   74.0  999.9   16.0  999.9    1.3
4261.  INB 1.22 6.1     93.    241.    3000.    20.    178.   24.0    0.    18.0  115.8     .8
4263.  IOB 1.23 6.6     86.    219.    3200.    28.    175.   39.3    0.     6.0    8.6     .8
4254.  IPB 1.22 6.2     90.    219.    3840.    20.    131.   74.4    0.    20.0    0.      .8
4268.  IQB 1.18 6.4     78.    221.    4100.    18.    131.   41.2    0.     8.0    0.      .9
4274.  IRB 1.13 6.2     55.    185.    3100.    18.    153.   48.7    0.    14.0  999.9     .8
4277.  ISB 1.23 6.3     47.    170.    3400.    17.    141.   48.7    0.     8.0    0.     1.0
4282.  ITB 1.22 6.1     53.    192.    3200.    18.    153.   48.7    0.    16.0    6.3    1.3
4284.  IUB 1.14 6.1     44.    192.    3000.    18.    138.   34.0  999.9   20.0  999.9    1.1
4292.  IVB 1.17 6.0     63.    197.    3100.    26.    122.  999.9    0.     4.0  999.9    2.0
```

Frame 195

```
2059. DCE 1.22 5.6    129.    437.   4700.      35.    163. 999.9    0.      4.0 999.9     .8
2060. DDE 1.23 5.5    111.    379.   4680.      42.    147.  25.6    0.     30.0   6.3     .6
2067. DEE 1.22 4.9    118.  99999.   4000.      34.    162.  20.0  999.9   18.0 999.9     .7
2071. DFE 1.21 6.2    126.    754.   5400.      34.    172.  24.0  999.9   12.0 999.9     .8
2073. DGE 1.24 6.1    138.    498.   5300.      34.    176.   6.0  999.9   20.0 999.9     .9
2080. DHE 1.25 6.3    141.    340.   4600.      31.    172. 999.9    0.    12.0   8.6     .9
2082. DIE 1.18 5.7     96.    398.   4100.      30.    150.  24.0  999.9   30.0 999.9     .9
2083. DJE 1.19 6.1    100.    426.   3900.      34.    156.  24.0  999.9    4.0 999.9    0.
2086. DKE 1.24 6.1    190.    304.   4000.      29.    172.  34.0    0.     6.0  92.5     .7
2098. DLE 1.22 6.0    100.    486.   5200.      34.    162.  22.0  999.9   10.0 999.9     .9
2105. DME 1.23 6.3    135.    279.   4300.      37.    203.  32.3    0.     0.  999.9     .6
2109. DNE 1.20 5.4    110.    292.   2900.      32.    148.  48.0  999.9   18.0 999.9     .9
2111. DOE 1.18 5.0    100.    238.   3900.      30.    154.  32.0  999.9   20.0 999.9     .9
2113. DPE 9.99 9.9  99999.  99999.  99999.   99999.  99999. 999.9 999.9  999.9 999.9 999.9
2129. DRE 1.21 6.0    174.    400.   3900.      28.    182.  42.0  999.9   22.0 999.9     .6
2137. DSE 1.17 6.6    486.    438.   5800.      28.    190.  26.0  999.9   10.0 999.9     .7
2152. DTE 1.24 6.3    262.    400.   6300.      28.    182.  40.0  999.9   44.0 999.9     .6
2153. DVE 1.20 6.5    273.    292.   5800.      33.    213.  10.3    0.     4.0  92.5     .6
2161. DWE 1.25 6.7    378.    279.   6000.      33.    188. 999.9    0.     8.0 999.9     .7
2170. DYE 1.21 6.8    248.    462.   6500.      28.    188.  22.0  999.9   18.0 999.9     .3
2180. DZE 1.20 6.6    222.    486.   6800.      28.    176.  20.0  999.9   12.0 999.9     .5
2183. EAE 1.25 6.7    378.    304.   5500.      37.    216.  67.4    0.     4.0   0.      .6
2185. EBE 1.22 6.4    399.    224.    740.   99999.    169.  34.0    0.     8.0   0.      .6
2192. ECE 1.18 6.7    312.    241.   6500.      37.    225.  25.6    4.1   10.0 999.9     .5
2207. EDE 1.22 6.5    210.    242.   5600.      30.    188.  26.0  999.9   16.0 999.9     .6
2213. EEE 1.22 6.5    204.    426.   6300.      30.    218.  20.0  999.9    4.0 999.9     .5
2216. EFE 1.22 6.3    276.    388.   5300.      28.    206.   0.   999.9   14.0 999.9     .5
2223. EGE 1.23 6.4    216.    376.   4600.      26.    222.  22.0  999.9   12.0 999.9     .5
2231. EHE 1.28 6.3    390.    374.   4620.   99999.    191.  28.9    0.     6.0 999.9     .6
2243. EIE 1.17 6.3    338.    498.   6500.      30.    204.  18.0  999.9   18.0 999.9     .5
2270. EKE 1.14 6.8    114.    899.   8000.   99999.    338.  86.7   18.0    6.0   0.     3.3
2291. EME 1.17 6.9    123.   1283.   6360.      26.    166.  54.6   25.9   16.0 999.9    1.8
2317. EOE 1.15 6.9     86.   1070.   5200.      26.    122.  60.0  999.9   22.0 999.9    1.8
2404. EPE 1.18 5.5     36.    554.    480.   99999.    125.  32.3    0.    22.0   1.0     .6
2405. EQE 1.20 5.5     27.    595.   3400.      33.    131.  17.8    0.    16.0   0.      .6
2421. ETE 1.24 5.8     18.    584.   3300.      40.    118.  38.0  999.9   36.0 999.9     .7
2429. EVE 1.24 6.1     35.    510.   3800.      35.    147.  22.4    0.    14.0   0.      .7
2438. EXE 1.23 5.8     39.    572.   4000.      32.    128.  24.0  999.9   36.0 999.9     .9
2449. EZE 1.11 6.3     93.    335.   4320.      29.    147.  44.9    0.    16.0   0.      .7
2453. FAE 1.12 5.9    126.    481.   4560.      39.    156.  30.6    0.    24.0   0.      .6
2451. FBE 1.21 5.5    134.    450.   4000.      34.    140.  52.0  999.9   20.0 999.9     .7
2462. FCE 1.20 5.6    118.    498.   4300.      34.    166.  26.0  999.9   18.0 999.9     .6
2472. FDE 9.99 9.9  99999.  99999.  99999.   99999.  99999. 999.9 999.9  999.9 999.9 999.9
2473. FEE 1.22 5.8    142.    438.   4300.      18.     76.  26.0  999.9   16.0 999.9     .8
2480. FFE 1.24 6.1    225.    352.   4600.      33.    191.  63.0    0.    12.0   6.3     .6
2482. FGE 1.22 6.1    129.    321.   4200.      86.    156.  32.3    0.    18.0   1.0     .7
2487. FHE 1.21 6.2    118.    376.   4000.      28.    156.  22.0  999.9   16.0 999.9     .9
2497. FJE 1.22 6.1    180.    389.   4400.      31.    172.  19.3    0.    16.0  18.1     .6
2503. FLE 1.18 5.9    138.    364.   4080.   99999.    156.  41.2    0.    10.0   0.      .7
2519. FNE 1.23 5.6    130.    450.   4700.      28.    166.  22.0  999.9   12.0 999.9     .9
2525. FOE 1.19 5.9    216.    414.   4600.      28.    198.  24.0  999.9   10.0 999.9     .7
2528. FPE 1.23 6.2    231.    306.   3960.      42.    213.  20.8    0.     0.    0.      .6
2539. FQE 1.21 5.1    198.    414.   4600.      32.    194.  22.0  999.9   18.0 999.9     .5
2550. FRE 1.18 6.6    273.    328.   5900.      29.    219.  50.6    0.    12.0   6.3     .6
2569. FSE 1.21 7.5    360.    340.   7800.      34.    200.  26.0  999.9   16.0 999.9     .6
2595. FTE 1.22 6.5    390.    226.   5040.   99999.    185.  11.8    0.     8.0  78.9     .6
2605. FUE 1.26 6.3    336.    190.    760.   99999.    181.  34.0    0.     8.0   0.      .5
2619. FVE 1.21 6.6    318.    400.   5600.      28.    210.  20.0  999.9    8.0 999.9     .4
2633. OLE 1.29 6.4    390.    364.   5200.      46.    253.  19.3    0.     4.0   0.      .6
2642. FXE 1.22 6.8    375.    316.   6000.      22.    416.  28.9    0.     8.0   0.      .6
```